HANDBOOK OF SUSTAINABLE ENERGY

D1579376

Withdrawn from Stock
Dublin City Public Libraries

Handbook of Sustainable Energy

Edited by

Ibon Galarraga

Research Professor, BC3 Basque Centre for Climate Change, Spain

Mikel González-Eguino

Research Associate, BC3 Basque Centre for Climate Change, Spain

Anil Markandya

Ikerbasque Professor and Scientific Director, BC3 Basque Centre for Climate Change, Spain and Honorary Professor, University of Bath, UK

Edward Elgar
Cheltenham, UK • Northampton, MA, USA

© Ibon Galarraga, Mikel González-Eguino and Anil Markandya 2011

All rights reserved. No part of this publication may be reproduced, stored in a retrieval system or transmitted in any form or by any means, electronic, mechanical or photocopying, recording, or otherwise without the prior permission of the publisher.

Published by
Edward Elgar Publishing Limited
The Lypiatts
15 Lansdown Road
Cheltenham
Glos GL50 2JA
UK

Edward Elgar Publishing, Inc.
William Pratt House
9 Dewey Court
Northampton
Massachusetts 01060
USA

A catalogue record for this book
is available from the British Library

Library of Congress Control Number: 2011925745

ISBN 978 1 84980 115 7 (cased)

Typeset by Servis Filmsetting Ltd, Stockport, Cheshire
Printed and bound by MPG Books Group, UK

Contents

PART VI   OTHER DIMENSIONS OF ENERGY

Contributors

Luis María Abadie, BC3 Basque Centre for Climate Change, Spain.

Sabah Abdullah, University of the Basque Country, Spain and University of Bath, UK.

Alberto Ansuategi, University of the Basque Country, Spain.

Edward John Anthony, CanmetENERGY, Canada.

Rob Bailis, Yale University, USA.

Andrea Bigano, Fondazione Eni Enrico Mattei (FEEM), Italy.

Monica Bonacina, IEFE, Bocconi University, Italy.

Kenneth Button, George Mason University, USA.

Helena Cabal, Energy Systems Analysis Unit, CIEMAT, Spain.

José Manuel Chamorro, University of the Basque Country, Spain.

Julien Chevallier, Université Paris Dauphine, CGEMP-LEDa, Paris, France.

Anna Creti, Université Paris Ouest – Nanterre, La Défense – EconomiX, and École Polytechnique, Paris, France.

Susanna Dorigoni, Bocconi University, Italy.

Marta Escapa, University of the Basque Country, Spain.

Martin Evans, Agriprojects Consulting Limited, UK.

Paul S. Fennell, Imperial College of Science, Technology and Medicine, London, UK.

Roger Fouquet, BC3 Basque Centre for Climate Change, Spain.

Timothy J. Foxon, University of Leeds, UK.

Ibon Galarraga, BC3 Basque Centre for Climate Change, Spain.

M. Carmen Gallastegui, University of the Basque Country, Spain.

Tomás Gómez, Institute for Research in Technology (IIT), Universidad Pontificia Comillas de Madrid, Spain.

Antonio G. Gómez-Plana, Public University of Navarre, Spain.

Mikel González-Eguino, BC3 Basque Centre for Climate Change, Spain.

Kirsten Halsnæs, DTU Climate Centre Risø, Denmark.

Geoffrey P. Hammond, Institute for Sustainable Energy and the Environment (I-SEE), University of Bath, UK.

Marjolein J.W. Harmsen-van Hout, Institute for Future Energy Consumer Needs and Behavior (FCN), School of Business and Economics/E.ON Energy Research Center, RWTH Aachen University, Germany.

Peter B.R. Hazell, Centre for Environmental Policy, Imperial College, London, UK.

David R. Heres, BC3 Basque Centre for Climate Change, Spain.

Jose Ignacio Hormaeche, Basque Energy Board (Ente Vasco de la Energía – EVE), Spain.

Craig I. Jones, Senior Associate, Sustain, UK.

Kenneth Karlsson, DTU Climate Centre Risø, Denmark.

Xavier Labandeira, University of Vigo and Economics for Energy, Spain.

Maryse Labriet, ENERIS Environment Energy Consultants, Spain.

Alessandro Lanza, Fondazione Eni Enrico Mattei (FEEM), LUISS University (Rome), Centro Euro Mediterraneo per i Cambiamenti Climatici (CMCC) and CRENoS, Italy.

Yolanda Lechón, Energy Systems Analysis Unit, CIEMAT, Spain.

C.-Y. Cynthia Lin, University of California, Davis, USA.

Pedro Linares, Comillas Pontifical University and Economics for Energy, Spain.

Reinhard Madlener, Institute for Future Energy Consumer Needs and Behavior (FCN), School of Business and Economics/E.ON Energy Research Center, RWTH Aachen University, Germany.

Anil Markandya, BC3 Basque Centre for Climate Change, Spain and University of Bath, UK.

Emanuela Menichetti, Observatoire Méditerranéen de l'Energie, France.

Luis Olmos, Institute for Research in Technology (IIT), Universidad Pontificia Comillas de Madrid, Spain.

Ramon Arigoni Ortiz, BC3 Basque Centre for Climate Change, Spain.

Ignacio Pérez-Arriaga, Institute for Research in Technology (IIT), Universidad Pontificia Comillas de Madrid, Spain.

Roberta Pierfederici, Fondazione Eni Enrico Mattei (FEEM), Italy.

Karen Pittel, University of Munich, Germany.

Thomas Reisz, Energy Agency NRW, Germany.

Michel Rivier, Institute for Research in Technology (IIT), Universidad Pontificia Comillas de Madrid, Spain.

Renato Rodrigues, Institute for Research in Technology (IIT), Universidad Pontificia Comillas de Madrid, Spain.

Dirk Rübbelke, BC3 Basque Centre for Climate Change and Ikerbasque, Basque Foundation for Science, Spain.

Jose Luis Sáenz de Ormijana, Basque Energy Board (Ente Vasco de la Energía – EVE), Spain.

Andy Stirling, University of Sussex, UK.

Einar Strumse, Lillehammer University College, Norway.

Tatsujiro Suzuki, University of Tokyo, Japan.

Elena Verdolini, Catholic University of Milan and Fondazione Eni Enrico Mattei (FEEM), Italy.

Hege Westskog, Center for International Climate and Environmental Research (CICERO), Norway.

Tanja Winther, Centre for Development and the Environment, University of Oslo, Norway.

Go Yoshizawa, University of Tokyo, Japan.

Introduction
Ibon Galarraga and Mikel González-Eguino

Preparing an introductory chapter for a book in which the academic and professional level of the contributing authors is so high is not an easy task. A quick definition of the *Handbook of Sustainable Energy* could be something like: 'an attempt to contribute significantly to the knowledge in sustainability issues from the point of view of energy'. This book is an interesting effort to shed some light on such a complex topic as energy use and sustainability.

Part I starts by recognizing the importance of the use of energy in all past and present civilizations and socio-economic systems. In fact, the most reasonable way to forecast correctly and to drive the future use of energy in a sustainable way is possibly by looking far back to the past. This is done in Chapter 1 by Fouquet, focusing on the use of renewable energy sources and the lessons that can be learnt by looking at very long-term periods. Chapter 2 sets the scene for detailed discussion regarding what sustainability means in the field of energy systems, being aware that intergenerational and international (or interregional) justice are fundamental conceptual dimensions of the definition of sustainable use of energy resources and technologies. Hammond and Jones include very interesting concepts in the analysis, such as resource productivity (the so-called 'factor X'), the precautionary principle and the three dimensions of sustainability – economic, environmental and energy – illustrating the discussion with examples from the field of biofuels and decentralized energy resources. The historic responsibility of developed countries is stressed and magnificently clarified.

But the fact is that the use of energy is at the very core of one of the greatest challenges facing humankind: climate change and its impacts. And moreover, energy use is a fundamental part of the solutions necessary to mitigate climate change. Understanding the complexities attached to the public nature of this problem, the fact that today's efforts will generate benefits in the future and the discussion on how to delink economic growth from green house gas (GHG) emissions are the central pillars of Chapter 3 by Gallastegui, Ansuategi, Escapa and Abdullah. The authors shed some light on the questions: (1) Can economic growth be delinked from GHG emissions? (2) Can GHG emissions be cut without hurting economic growth? And (3) Is future economic growth at risk due to climate change?

Another key component for sustainable use of energy is availability, that is, the security of supply. Bigano, Ortiz, Markandya, Menichetti and Pierfederici offer in Chapter 4 a view of how energy security policies are connected to energy efficiency and saving. They show that reducing energy consumption reduces dependency on external energy sources, and thus it should be a central part of any energy security policy. Econometric analysis is used to study the effect of efficiency and saving indicators in security of supply indicators for the EU-15 and Norway. They conclude that many energy efficiency policies in the European Union (EU) are not effective by themselves, and the right policy mix is the best approach to achieve energy security goals successfully.

1

Guaranteeing the supply of energy to maintain quality of life and continue with productive activities can be managed in many different ways, but the truth is that any smart policy has to recognize the significant role that innovation has to play in this field. Climate policies need to be supported by research while designing effective carbon-pricing mechanisms; changes in energy use and the role of civil society in the promotion of a transition to a low carbon economy cannot be neglected. Planning efforts in the UK and the Netherlands are used to explain clearly what is necessary for a transition to a low carbon economy in Chapter 5 by Foxon.

Part II on energy and economics starts with Chapter 6 by Bonacina, Creti and Dorigoni. They explain how gas and electricity markets work, while describing the role that economic models have played in market design and transmission regulation. An interesting portrayal of the way in which the EU has restructured its gas and electricity markets is offered, stressing the fact that Europe still needs to define a clear common energy policy to move towards competitive, integrated and green energy markets. Chapter 7 by Pérez-Arriaga, Gómez, Olmos and Rivier analyses the role of electricity as well as the changes required to move towards a low carbon economy. They conclude that without adequate transmission and distribution networks the path towards a sustainable low carbon energy model will not be possible. The following two chapters analyse different approaches to modelling energy and economic interactions. Rodrigues, Gómez-Plana and González-Eguino, in Chapter 8, offer the reader a review of the energy–economy–environment (3E) models and the fusion of the bottom-up and top-down approaches in the so-called hybrid models. Although there exist a number of restrictions to the use of these models, they are very useful tools for informing energy and climate policy decisions. Chapter 9 by Pittel and Rübbelke illustrates how endogenous growth models allow us to understand the long-run potential of economies to overcome the scarcity of fossil energy resources, and the potential and direction of technological development. The authors differentiate between analytical solvable endogenous growth models and computable general equilibrium (CGE) models, and the role of research and development (R&D) investments. Finally, Madlener and Harmsen-van Hout analyse consumer behaviour towards energy use in Chapter 10. They study different drivers that explain human behaviour: (1) psychological drivers (cognition); (2) rational behaviour drivers (utility); (3) sociological drivers (other people); (4) ecological drivers (environment); and (5) technological drivers (innovation). The analysis enables them to identify commonalities and differences that are otherwise easily overlooked.

Parts III and IV are devoted to analysing different technology options for sustainable use of energy and transition to a low carbon economy. Chapter 11, by Yoshizawa, Stirling and Suzuki, outlines a general framework for analysing energy diversity and synergies for transitions to sustainability. It provides a multicriteria diversity analysis method as a more systematic, complete and transparent way to articulate energy portfolios. Chapters 12 and 13 are devoted to renewable energy. Cabal, Labriet and Lechón conduct a deep literature review gathering the most recent data from the most relevant studies on global and European potentials for wind power, hydropower, biomass, solar power and ocean energy. At global level, photovoltaic and thermosolar power account for 80 per cent and 90 per cent, respectively, of the total renewable power potential. At a European level, however, wind power is the technology with the biggest potential. Once the potential is estimated (with a wide range of measures) Halsnæs and Karlsson analyse the penetration of renewable energy depending on the cost of individual options and on how a portfolio of options can be integrated in energy systems in a way that energy access, energy security and climate change policy goals are met. Chapter 13 illustrates how this has been played out in international scenario studies, and in a particular study for Denmark where the goal is to cover all Danish energy consumption by renewable energy in 2050. They conclude that having 100 per cent renewable energy in Demark is not very costly given the favourable local conditions for high penetration of wind energy and large-scale electricity trade with Scandinavian countries and Germany.

Another main theme of the book relates to how to bring about energy and carbon efficiency as a central to the goal of a sustainable energy future. The book covers the two main sources contributing to CO_2 emissions – the power sector and the transport sector – and where efficiency and saving measures can be relevant. Abadie and Chamorro in Chapter 14 look at incentives to invest in enhancing energy conversion efficiency in power plants that operate under carbon constraints. Many investments to enhance energy conversion efficiency at coal plants are not undertaken, due to difficulties in determining future earning in energy savings and CO_2 emission rights. They provide some interesting results applying real option analysis. The numbers are used to provide several policy recommendations based on the idea that there is a clear role for public authorities in promoting investments in innovation and R&D in coal-based plants. Chapter 15, by Button, looks at ways of improving efficiency in the transport sector. The author proposes moving from broad perspectives (such as the 'sustainable transport' notion) to a firmer theoretical foundation that leads to policy development and implementation.

Part IV comprises three chapters devoted to nuclear energy, carbon capture and storage (CCS) technology and biofuels. Hammond in Chapter 16 deals with the rather controversial topic of nuclear energy. This is a CO_2 zero-emitting energy source that has great opposition worldwide. This chapter offers a technical and well-developed position on the topic. The author argues that the need to develop more secure and commercially viable nuclear plants will indeed be determined by the attitudes of the public sector towards this energy source. Anthony and Fennell in Chapter 17 outline the various CCS technologies that might be deployed in the next few decades to meet the requirements of a carbon-constrained world. In particular, they show that the focus has been on technologies which could reasonably be expected to be commercially available in the next ten to 20 years. According to the authors, correctly applied CCS technology will buy time for a transition to systems with increased energy efficiency, and large-scale use of renewable and nuclear power. Finally in Part IV, Chapter 18 deals with the 'promises' and 'risks' of bioenergy. After a careful analysis of all the issues, Hazell and Evans recommend that countries should be encouraged to slow down on their biofuels mandates, allowing time to reduce the existing trade-offs with food provision and to protect remaining primary forest and peatlands from conversion to agriculture.

Part V deals with energy and climate policies. Chevallier offers two very interesting chapters, on CO_2 and energy pricing, and the flexible mechanisms of the Kyoto Protocol – the EU Emissions Trading Scheme (EU ETS), the Clean Development Mechanism (CDM) and the Joint Implementation (JI). In Chapter 19 a retrospective view of the EU ETS for the 2005–07 period shows the weaknesses and strengths of this market so far. In Chapter 20 the author explains the connexion between the ETS and the CDM/ JI though the Certified Emission Reductions (CERs) of project-based instruments and the EU Emission Allowances of the EU ETS. Labandeira and Linares offer a broader perspective in Chapter 21, to tackle the complexities of first-best policy solutions in climate change issues. The authors describe a number of reasons that justify a second-best approach to climate policy. Carbon tax itself may not be the best instrument to deal with climate policy and, therefore, a combination of instruments will be necessary to address the multiple market failures and other second-best situations that arise in the real world. In Chapter 22, Westskog, Winther and Strumse deal with policies to reduce energy consumption, following a wide multidisciplinary approach by including concepts from anthropology, psychology and economics to understand behaviour in order to be successful in driving changes. They show that technology itself will not address energy saving targets, and thus behavioural aspects are essential, as changes in energy consumption are closely linked with choices and sociocultural factors that determine the choice of policy instruments to be used. Finally, to close Part V the role of R&D is deeply analysed in Chapter 23 by Lanza and Verdolini. They go through the future prospects for all the main energy technologies and relate them with patent data. They conclude that the wide energy portfolio needed to face the challenge of climate change will require significant investments in the innovation, adoption, diffusion and transfer of technologies. The role of both public and private partners is acknowledged.

Part VI, the last of this book, opens with an important issue: the impact of the transition to a low carbon future in poor countries. Chapter 24, by Bailis, looks at energy poverty in a global context and at the reforms needed to eliminate it, while also respecting the goals of improving energy and carbon efficiency. The author argues that distributional issues are critically important and that in the absence of policies to promote inclusive access to energy services and associated technologies, additional supply may simply reinforce poverty and inequality across scales.

The last theme explored in the book is the role of regions in helping to move towards a sustainable energy future. Regional governments (defined as a subnational level of governance) have an interesting advantage over national governments: the fact that they can be more innovative and can act as 'leaders' in the formation of public opinion in this field. The book offers three good examples: North Rhine-Westphalia (Germany), California (USA) and the Basque Country (Spain). Chapter 25 by Reisz focuses on the case of the North Rhine-Westphalia region, analysing the effect of a decentralization process of energy production, where electricity in future will be produced in the place where it is going to be used, offering greater scope for the regions to influence the energy markets. Chapter 26 by Heres and Lin analyses the case of California, a very interesting example of a US state with a climate policy that is much more ambitious than the federal one. Nature offers California the possibility to develop renewable energy sources, while the political will is providing the opportunity to achieve it. Chapter 27 by Hormaeche, Galarraga and Sáenz de Ormijana looks at the case of the Basque Autonomous Community to illustrate the potential of regional governments to develop their own energy policies in the broader context of the EU. According to the authors, while the expectation is that European and national regulation offer a fairly restricted playground for regional governments, the truth is that there exists plenty of room for manoeuvre for this level of governance. This represents a great opportunity that should be explored and that can surely contribute to improving energy and climate policy worldwide.

The Epilogue by Markandya offers some of the highlights and key trends in this rich collection of contributions, suggesting that this book will be of great interest for many readers and offer a lot to researchers in the field.

We cannot finish this introduction without expressing the deepest thanks of the three editors to each and every one of the contributing authors for their intense effort and excellence in presenting their analysis. We hope that you, the reader, will find it interesting and learn as much as we, the editors, have done during the journey of the preparation of this book. Enjoy it!

PART I

SUSTAINABLE USE OF ENERGY

Leabharlanna Poiblí Chathair Bhaile Átha Cliath
Dublin City Public Libraries

1 The sustainability of 'sustainable' energy use: historical evidence on the relationship between economic growth and renewable energy

Roger Fouquet

1.1 INTRODUCTION

Throughout history, energy resources have played an important role in influencing the rate of economic growth and development. It has been seen as a boost to long-term growth when new energy sources and technologies were deployed and created abundance (Rosenberg, 1998; Crafts, 2004; Ayres and Warr, 2009). They have also been responsible for slowing down economies in times of perceived scarcity (Nordhaus, 1980).

Given the interest in a transition to a low carbon economy, it is appropriate to ask about the role that energy might play in this new context. At present, one can only speculate about the relationship between economic growth and development and low carbon energy resources. A transition to low carbon energy sources may provide a boost to the economy. Alternatively, an increasing dependence on renewable energy will imply different levels of resource availability and may create new limits on economic growth. Or, meeting the economy's energy needs through renewable resources may impose substantially higher costs.

Many of the models of long-run energy use have presented a cheap, non-renewable energy source and an expensive renewable energy source as the backstop technology (Nordhaus, 1973; Dasgupta and Heal, 1974; Stiglitz, 1974; Heal, 1976; Chakravorty et al., 1997). A transition to the more expensive renewable energy source means that firms would have to charge more for their products and customers' budgets would not stretch as far. Thus, it would effectively act as a brake upon economic growth.

These models present the transition to a backstop technology as the result of a severe depletion of the non-renewable energy source, leading to high prices and a need to find substitutes. Empirical studies (Barnett and Morse, 1963; Berck and Roberts, 1996; Fouquet and Pearson, 2003; Fouquet, 2010a) and, recently, theoretical models (Tahvonen and Salo, 2001) question whether the non-renewable resource will face long-run scarcity issues and rising prices. These studies imply that if a transition to renewable energy sources were to take place, it would not be the result of dwindling fossil fuel reserves, but the result of either a preference for renewable energy or that it became cheaper than fossil fuels.

Based on historical experiences, Fouquet (2010b) argues that, although preferences are important, a transition to low carbon energy sources is unlikely without renewable energy providing energy services more cheaply than fossil fuels. This implies that a complete transition will only occur if the combined output of renewable energy and its associated technology is cheap. Thus, the transition to renewable energy sources is unlikely to impose higher prices – at least, not initially.

Yet, perhaps more in the long term, after the transition when the economy becomes dependent on renewable energy sources, it may face resource limits. Over the last 300 years, modern economies have managed to grow spectacularly and have had an almost insatiable demand for energy resources (Fouquet, 2008). So, even though supplies of renewable energy resources are potentially very large (compared with the size of the global economy), limits may indeed eventually be faced.

Traditional models of long-run energy use do not address the situation once the back-stop technology is the dominant energy source – that was not their purpose. While a transition to renewable resources is certainly decades away at the earliest, there is now a clear demand for a better understanding of the relationship between long-run economic growth and renewable energy use. In order to begin our understanding of the relationship, this chapter gathers some evidence on how past economies have managed within the confines of renewable energy systems. Given space limitations, this chapter seeks to present only snapshots of a variety of different cases, focusing on woodfuels. These cases relate to the Roman Empire, Early Modern Europe and the Far East. They investigate the 'sustainability' of the use of this 'sustainable' energy – that is, how the renewable resource was used over very long periods. This involves considering the availability of resources, the rate of use, the existence of energy crises and the various governments' attempts to manage demand and supply.

1.2   WOODFUEL CONSUMPTION DURING THE ROMAN EMPIRE

Roman daily life was highly dependent on woodfuels. In addition to consumption for cooking, hot baths, the preparation of lime for construction and cremation of bodies were major users of fuelwood. Heating may have consumed as much as 90 per cent of the timber used. Estimates suggest that at its peak, with 1.5 million inhabitants, Rome would have consumed 2.25 million m³ (equivalent to 0.7 million tonnes of oil equivalent – mtoe) and required more than 30 km² of forest per year (Williams, 2003: 93).

There is evidence that forests were coppiced or felled in rotation to be able to meet the demands of Roman energy requirements. And yet, inevitably, Rome's success and expansion imposed increasing pressures on forests, and the trade in wood spread ever further – by the third century, the largest beams were shipped to the city from the Black Sea. Although the cost of cooking, bathing, building and cremating (until the practice was abandoned with Christianity, possibly encouraged by the price of fuelwood) must have increased, no claims of a Roman energy crisis exist (Williams, 2003: 93).

The empire also required large amounts of fuelwood to meet its demands for metal smelting. For instance, Populonia (level with the isle of Elba in present Italy) produced an estimated 500000 tonnes of copper, needing 2.2 million tons of charcoal (equivalent to 1.6 mtoe) from 36.1 million tonnes of wood, over a period of 500 years. This would have needed a forest of 1875 km² if it had been stripped of its trees. However, given that the annual consumption was relatively modest by modern consumption, land requirements could have been closer to 10 to 15 km² if properly coppiced (Williams, 2003: 94).

In classical times, many industrial sites dependent on charcoal managed to produce for hundreds and even thousands of years. Examples currently in Greece, Cyprus, Italy and Spain show that very large total quantities of silver, copper or bronze (a fusion of copper and tin) were produced over long periods of time. For instance, copper smelting in Cyprus left 4 million tons of slag residues. This equates to 200000 tonnes of copper, requiring 60 million tons of charcoal (equivalent to 43 mtoe) from 960 million tonnes of wood from roughly 60000 km² of forest – a forest about five times the size of the island. While Cyprus did suffer from deforestation, this was caused more from agricultural expansion than from industrial activities. And the island managed to produce copper for 3000 years. Such large quantities of production over a very long period could only have been achieved if the use of wood for fuel had been managed in a relatively sustainable way (Williams, 2003: 94).

Thus, there is clear evidence that already in Classical times, energy requirements were often met in a sustainable manner. Modest growth could be met by managing a slightly larger area provided that the source was easily accessible by land or by water.

1.3   EARLY MODERN EUROPEAN ENERGY CONCERNS

The trend for much of human history has been encroachment on woodlands. Although in specific cases the growing use of wood for energy and timber was responsible, deforestation has been mostly due to agricultural expansion (Williams, 2003). Nevertheless, it implies a declining stock of resources for meeting woodfuel needs and an increasing distance between the source and many of the users.

Between 1700 and 1850, temperate forest cover across the world declined substantially – by 1.8 million km². In Europe, 250000 km² disappeared; in Russia, 710000 km²; in North America, 450000 km²; and in China 390000 km². Between 1850 and 1920, the rate slowed a little, with 1.29 million km² disappearing. Russia lost 800000 km², North America 270000 km² and China 170000 km². Over that period, Europe only lost 50000 km², but this reflects more than anything a lack of forests to clear (Williams, 2003: 277).

Around 1700, England and Wales was about 8 per cent woodland and the Netherlands had virtually no forests; Northern France was about 16 per cent covered, while Eastern Germany was about 40 per cent woodland (Williams, 2003: 168). By 1850, much of Europe was deforested. One-quarter of Germany was covered in forests. France was 12 per cent woodland. Most other countries, apart from Scandinavia and Russia, had very little forest left (Williams, 2003: 279).

Most European cities used woodfuels for heating. Comparing European cities in the fifteenth and eighteenth centuries, the real price of energy did depend on the proximity to forests. Austria, Germany and Poland had the cheapest energy. Interestingly, even in the fifteenth century, when still heavily dependent on woodfuel, the real price of energy was only a little higher in London. Later, when London, Antwerp and Amsterdam were dependent on coal or peat, their prices were in the middle range. This suggests that where supplies were sufficient, coal use was not necessarily cheaper than being dependent on woodfuel. Spain, which had limited forest cover and little coal or peat, had the highest energy prices (Allen, 2003: 473).

The trend in real energy prices over 400 years of major economic growth is also revealing. For a number of cities across Europe, there was no evidence of an energy crisis and only a few instances of rising real energy prices between 1400 and 1800. For this period and out of 14 cities, only Paris, Strasbourg and Florence showed signs of rising prices in the eighteenth century. Otherwise, the trends in real energy prices were stable or declining (Allen, 2003: 479).

The generally held view today is that the term 'energy crisis' is an exaggeration. There were woodfuel shortages (Sieferle, 2001; Allen, 2003), but they tended to be local problems rather than national ones affecting the whole economy. Much of the problem was associated with distribution networks. And, most likely, shortages hit different localities at different times. Overall, between the early fifteenth and nineteenth centuries, the European economy managed to grow successfully and with few constraints while being mostly dependent on woodfuel for heating (Allen, 2003).

As mentioned above, in England and Belgium, by the seventeenth century, the predominant energy source in cities was coal. This does not suggest a woodfuel shortage but only that the cost of heating using coal was cheaper than using wood in these cities (Fouquet, 2008: 75).

The main commonality amongst all economies was that once agricultural production increased or efficiency improved, the population grew, putting pressure on woodfuel resources because of both the changing land-use from forest to agriculture and the rising demand for wood products. Thus, consistently, economic growth eventually expanded to reach its resource limits. Faced with greater constraints, the reaction was either economic contraction, stagnation or even decline, better management of forest resources or a switch to another fuel. The next section considers government policies to balance the demand and supply of this renewable energy.

1.4   FOREST MANAGEMENT IN GERMANY

The multitude of local German economies benefitted from large forests close to rivers. Woodfuel provided their main source of heating for households and industries for centuries, with episodic tensions and adaptation. Evidence suggests that consumers were reluctant to switch, reflecting preferences for woodfuel and perhaps an insufficient price differential to make substitution attractive and overcome the negative aspects of coal burning. When tensions did arise, woodfuel supply adapted to rising demands, either by felling more local trees or by importing them along the river networks. In many cases, when economic growth led to pressures on resources, governments in German states did tend to intervene to assist the markets (Warde, 2003).

For example, in Northern Germany, salt production depended on large quantities of wood to evaporate seawater. The industry managed to expand substantially (more than quadrupling) from the beginning of the fourteenth to the end of the sixteenth century without suffering from higher energy costs or fuel shortages. It did depend partially on importing fuelwood, and transport networks were crucial. In fact, the promotion of road building and development of navigable rivers were promoted for the purpose of supplying the region. Yet, this industry's eventual decline was in no way a result of energy restrictions (Witthöft, 2003: 301).

The consensus about woodfuel shortages in Germany was that they were localized problems more associated with distribution of resources rather than a generalized lack of energy resources. The main problem was the lack of satisfactory transport routes to distribute. In the seventeenth and eighteenth centuries, this problem was alleviated in some regions by the expansion of rafting of timber and fuelwood to wooded uplands that had previously been undervalued (Warde, 2003: 594). Another cause of shortages was the political boundaries: many of the German states believed in energy self-sufficiency and in preserving resources for domestic use (and not exporting wood). So, in some cases, industries needed to carry wood long distances (up to 30 km) within political boundaries when nearby sources existed but were outside the state limits (Warde, 2003: 592).

German energy and woodfuel consumption until the nineteenth century was dominated by household needs for cooking and heating. Crude estimates indicate that woodfuel consumption was around 11 million m³ (equal to 5.5 mtoe) in 1500 and about 20 million m³ (9.8 mtoe) in the seventeenth and early eighteenth century. In comparison, around 1600, iron production use would have been 1.5 million m³ (about 0.8 mtoe) and, by 1700, silver, lead and salt production would have required less than 4 million m³ (about 2 mtoe) of wood. Industrial activity grew substantially in the second half of the eighteenth century, and iron production would have needed about 10 million m³ (equal to 1.t mtoe) by 1800 (Warde, 2003: 590). This indicates that general forest management policies were driven by a need to meet household needs. Nevertheless, many of the early German states responsible for introducing policies were reflecting the repercussions of concentrated demands for local industries.

Forest management in Germany began in the late medieval and early modern period. In 1368, pines were replanted in the municipal forests of Nürnberg, initiating a series of policies of managing woodlands across German states and urban centres, especially between 1470 and 1550. This trend reflected in part a growing awareness of the need for security of energy supply (either because of the welfare implications to the population or because of the legal tensions that developed over scarce resources), and for the management of stable fiscal revenue, by avoiding volatile prices associated with the changing scarcity of resources (Warde, 2003: 585).

On communal land, households had been granted rights by local authorities to extract wood. However, over time, the rate of extraction was increasingly specified, as a means of avoiding a tragedy of the commons. This was generally seen as an amount suitable to meet 'subsistence' needs, creating problems for families who sought to produce goods for markets. In noble forests, peasants were generally also allowed to collect deadwood and cut small pieces and, similarly, this practice became increasingly controlled by officials (Warde, 2003: 588–9).

Until the nineteenth century, woodland management was based on practices developed in the fifteenth century, and was not particularly innovative. It focused on felling trees by area, in relatively short succession, coppicing, and protecting (by, for instance, banning grazing around saplings) and promoting rapid regrowth. Although their control and power grew, forest officials' role was to assess the stocks and parry poaching.

When faced with demand pressures on resources, policies generally sought to minimize resource use by encouraging fuel efficiency, rather than to increase supply, such as by reafforestation and trying new tree species. This reflected the more immediate returns from improving fuel efficiency (especially to industries) than investing in programmes to increase supply (Warde, 2003: 593–5).

Institutional structure played an important role in the successful balancing of supply and demand. For instance, from the fifteenth century, regulations in Siegerland in

Germany defined the nature and the rate of smelting and forging activities (including the number of working days), and banned the exports of iron ore, raw iron or charcoal. Despite being an important mining, smelting and forging region, and needing to import some charcoal, it appears that its dependency on renewable energy supplies did not limit economic activities until the nineteenth century (Witthöft, 2003: 296).

German state governments also promoted a switch to coal, which was reluctantly adopted in the nineteenth century when industrial demands did severely outstrip supplies. By then, growing concerns of losing lumber had led to scientifically managed forests and government policies (Sieferle, 2001). Also, although the concept was vague and differed for each state, there was a general view that in addition to the need to meet ongoing demands, resources should be maintained for future generations (Warde, 2003: 595).

Before the nineteenth century, there were no doubt plenty of examples of areas where growth did lead to energy restrictions. Yet, in early modern Germany the existence and dependence on large wood reserves led to relatively successful policies of managing energy supplies and demand. This more proactive energy policy no doubt reduced the tensions, but ironically delayed the transition to fossil fuels and the potential for greater economic growth.

1.5   THE JAPANESE EXPERIENCE

The early modern Japanese economy followed a similar course. It depended heavily on woodfuel for heating purposes. Facing the risk of shortages resulting from economic growth, local policies aimed at reducing consumption, improving efficiency, increasing supply and attempting a switch to coal. Like Germany, before the nineteenth century, coal substitution was the least successful of these policies. More generally, its experience showed that through regulation governments could help boost renewable energy supplies, and balance them with demand.

During the sixteenth century, large-scale military conflict had used vast quantities of timber. From the seventeenth century onwards, the country was at peace and the population rose. Along with the encroachment of agriculture onto once wooded land, demand for timber for construction, shipbuilding and fuel led to severe deforestation. Soil erosion, floods, landslides and barren lands were common occurrences in seventeenth- and eighteenth-century Japan (Totman, 1989).

In the second half of the seventeenth century, feudal lords, who owned most of the forests, began efforts to reduce deforestation. The first policy was to ban wood removal except with direct authorization from the feudal lord. Other measures included seedling protection, selective cutting and more patrols. These measures reduced production substantially, but also feudal lord revenue. Swiftly, production increased again to make up the losses, and deforestation resumed (Totman, 1989: 246).

At the end of the seventeenth century, the rising price of woodfuel in Japan drove a few industrial activities, such as by salt and sugar producers, to start shifting towards coal use, where the fuel could be found and extracted easily. But in this densely populated country the external costs of coal production and consumption were felt, and created conflicts. Mining generated considerable pollution in nearby rivers, and coal burning emitted noxious fumes and sticky residues. Downstream rice fields suffered. Given the highly organized nature of society, protests, litigation, compensation and regulation followed. Eventually, in the 1780s, mines were closed due to their damaging effects on society and the environment, and wood burning was encouraged. But inevitably, and despite more complaints, the high price of woodfuel forced growing industries in the nineteenth century to use coal (Totman, 1995: 271–2).

Before this transition, however, in the eighteenth century, efforts to halt deforestation resumed and were more successful. Wood use was rationed, specifying the amount of wood that could be consumed according to social status. Timber for construction was used more sparingly. More efficient stoves were promoted for use in homes. And, at the end of the century, an active policy of planting new trees was introduced. Along with these new measures came a scientific approach to forest management. The story of Japanese forest management and broader energy policies has often been told as one of sustainable management, but it took nearly two centuries of deforestation and attempts to achieve a growth in wooded lands (Totman, 1989).

1.6   THE FIRST TRANSITION TO COAL

The Chinese experience was very different. It was a story of poorly managed woodfuel supplies and substitution to coal, but 'in the wrong order'. China was probably the first location where coal was used to address the problem of insufficient woodfuel. Since the Han period (25–220 CE) and perhaps as early as the fourth century BC, anthracite coal was used for a number of industrial activities. However, the potential for substitution was limited by technological developments, and the methods for using coal remained relatively crude (Wagner, 2001; Thomson, 2003: 8).

During the Song Dynasty (960–1270 CE), far more sophisticated techniques were developed. In that period, political stability and economic prosperity had generated a rapid growth in the demand for metals and iron, in particular. In 1078, Chinese iron production was about 125 000 tonnes – similar to iron production in England and Wales in 1790 (Hartwell, 1966: 34).

Despite use of coal for some industrial activities, iron had traditionally been smelted with charcoal. However, the expansion of iron production in the tenth century had led to deforestation problems, and alternative sources of fuels and technologies were sought. Much of the Chinese coal was found in the north, near the centres of iron works. Coke, derived from bituminous coal, was used for large-scale iron smelting in the north from the eleventh century, and possibly earlier. Compared with other regions, that could hardly have expanded due to a lack of solutions and access to resources, northern Chinese iron production increased to meet much of the growing demand. From the ninth century, coal also appears to have been used in domestic activities such as cooking (Hartwell, 1966: 55–6)

During the thirteenth century, however, the Chinese empire suffered from a number of adverse events, most notably the Mongol invasions, which led to economic decline. When the Chinese economy redeveloped in the seventeenth century, the economic base was in southern regions with very little access to coal, and perhaps without the methods needed to turn coal into coke and to use coke for smelting. Northern China became an 'economic backwater' with little influence over the thriving economic south. And the cost of shipping coal south was too high to make it commercially viable (Pommeranz, 2000: 62).

The modest Chinese expansion of the eighteenth century was dependent on woodfuel for iron production. As production grew and deforestation increased, the authorities failed to develop successful policies to manage and balance demand and supply, placing a break on the potential for growth. 'Medieval' China found major technological solutions to the woodfuel problem, allowing its economy to expand. Yet, because of a period of economic decline, China lost the knowledge or ability to grow in a long-term way. Upon its return to woodfuel dependence, it failed to realize that strong energy policies were necessary to grow its economy within a renewable energy system.

1.7   A FUTURE ECONOMY DRIVEN BY RENEWABLE ENERGY

The previous sections have presented histories of economies dependent on renewable energy. Being able to sustain economic growth depended on sound management of the demand, supply and trade of woodfuel. Where governments failed to develop appropriate policies, growth and development was limited. Inevitably, the vast demands of full industrialization, coupled with inefficient energy technologies and primitive transport networks, implied that a transition to fossil fuels was critical for higher levels of economic growth and development, as seen during the nineteenth century in Britain, Germany and even Japan.

At the beginning of the nineteenth century, 95 per cent of global primary energy use came from renewable resources. By the beginning of the twentieth century, this fell to 38 per cent. And at the beginning of this century, it was down to 16 per cent (Fouquet, 2009: 15). Clearly, for many years to come the proportion of renewable energy in primary energy consumed at a global level will continue to decline, as the quantity of fossil fuels used increases (especially from contributions in developing economies) more than that of renewables.

Nevertheless, the tide may be turning: in a number of industrialized countries the proportion of renewables is rising. Indeed, as Tahvonen and Salo (2001) propose in their model, it is possible that in the process of economic development an agrarian economy uses renewable energy resources, moves to fossil fuels for a phase associated with industrialization and then, reaching a higher level of technological and economic capability, returns to renewable energy sources.

The important drivers for energy transitions of the past were the opportunities to produce cheaper and better energy services (Fouquet, 2010b). They may well be the drivers for a transition to low carbon energy sources. Internalization processes, such as carbon taxes or tradeable permit schemes, can improve their competitiveness. But it is likely that, for a transition to occur, low carbon energy sources and technologies will have to provide cheaper energy services.

If renewables manage to outcompete fossil fuels, then economies (industrialized or developing ones) will in time become dependent on these low carbon sources. Fossil fuels may, in the future, be seen as the 'necessary evil' – that is, a cheap and dirty energy source

– that allowed economies of the past to reach a higher level of economic well-being. This fits with the concepts underlying the environmental Kuznets curve that environmental pollution needs to get worse before getting better.

But, will individual economies and the global economy be able to grow in the very long run within the confines of a renewable energy system? Although huge uncertainties about the future prevail, an exercise that considers the currently estimated global energy resources can help to indicate the distance between the current global economy and notional limits.

For instance, one estimate of oil reserves of all types (nearly 2 million mtoe) suggests that they are currently 12 times the amount of oil that has been consumed in the industry's 160-year history (157000 mtoe), or 486 times the 2009 global consumption of petroleum (Farrell and Brandt, 2006; BP, 2010). In 2009, the current global primary (modern) energy consumption was a little over 11000 mtoe (BP, 2010), and the global primary energy consumption, including biomass, is likely to be around 12000 mtoe (Fouquet, 2009). One estimate of global fossil fuel reserves is close to 30 million mtoe (Rogner, 2000: 168); this is nearly 2500 times the current annual global primary energy consumption. Unconventional natural gas reserves are especially large – roughly 80 per cent of the total. But, as indicated above, even for oil reserves the estimate is more than 450 times the current annual global oil consumption. Thus, even allowing for economic and population growth, fossil fuels are abundant and so the dwindling of fossil fuel reserves is unlikely for a very long time. Without full carbon capture, atmospheric limits (to assimilate greenhouse gases) will have been reached far before resource limits.

An estimate of the technical potential for global renewable energy resources is over 180000 mtoe (Rogner, 2000: 168). Two-thirds of this potential would be generated by geothermal sources; one-fifth from solar; one-twelfth from wind; one-twentieth from biomass. So, for example, the potential for wind energy is estimated to be 25 per cent greater than the current global energy consumption. And the total technical limit is 15 times the global economy's primary energy requirements.

Just as a reference, the current global primary energy consumption is 15 times its level in 1900. Thus, it took around 100 years to grow 15-fold. Although we may not expect similar growth rates or a full dependence on renewable energy sources at the beginning of the twenty-second century, these renewable energy limits could be threatened in that century.

This is not an exercise in showing that a transition to renewable energy sources is dangerous for the economy. After all, an estimate of the 'theoretical' limits of renewable energy resources was nearly 3.5 billion mtoe – almost 300000 times the current global primary energy requirement (Rogner, 2000). These are potentially meaningless numbers, given the developments in energy technology that we can expect over the next 100 years and more. Presumably, the limit is somewhere between 15 and 300000 times current consumption. However, they do help us to think about magnitudes.

Some have argued that increases in resource discoveries and improvements in energy technology were an important source of economic growth in the past (Ayres and Warr, 2009). The ability to exploit new energy reserves, such as Colonel Drake's oil discovery in Pennsylvania or the extraction of oil in the Middle East, were also boosts to economic growth. It is possible that within the limits of a renewable energy system there will be less potential for new discoveries to be made. Some might argue that the location of these resources is known with greater certainty for most renewables than for fossil fuels, so the potential for great new discoveries in the future is less likely. Effectively, the limits are known and the global economy will work its way towards them.

However, it is clear that renewable energy technologies of the future will be heavily dependent on research and development to improve their ability to harness natural forces. Technological developments will enable the economy to increase the limits of commercially viable renewable energy resources (from 15 times current global primary energy requirements towards 300 000 times). Thus, a crucial process within the renewable energy system will be the quality of signals that indicate that existing (commercially viable) limits are being reached, and technological improvements will be needed to avoid serious constraints on economic growth.

Probably more important than the limits will be the governments' energy policies. Historically, sound policies towards energy demand, supply and trade were critical to extending the ability to use renewables. This may offend certain ideologies but, based on historical evidence, a return to renewable energy sources would be more successful if properly managed, instead of a laissez-faire approach. Policies will probably need to address short- and long-term demand, supply and distribution issues. Yet it is possible, and hopeful (from an economist's perspective), that the policies will be 'light-handed' and will use incentives rather than heavy regulation.

1.8   FINAL REMARKS

This chapter considers economic growth in a renewable energy system. Previous sections in this chapter tried to show that economies of the past survived, evolved and even grew within a renewable energy system. Indeed, the first key observation is that, in particular locations, industries were operative and dependent on renewable energy sources for centuries and even millennia. Secondly, growth in demands does clearly put pressure on resources. For instance, the expansion of the European and Far East Asian population and economy between 900 and 1350 and 1400 and 1800 led to a growth in demand for energy resources.

Thirdly, in many localities woodfuel supplies were able to meet the growing demand. This was, in the cases considered, the result of government intervention and the promotion of better resource management. Often they were coupled with efforts to reduce demand and improve the efficiency of consumption. Based on the historical evidence, balancing demand and supply was crucial to achieving growth within a renewable energy system. Fourthly, in many circumstances trade was the solution, importing the necessary resources. Where local energy shortages were a problem, the main cause was due to the high cost of transporting resources, rather than an overall energy crisis.

Fifthly, another solution was a substitution towards other fuels, such as coal. However, this was often undesirable due the harmful external effects of coal mining and combustion. Finally, the Chinese experience showed that solutions can be forgotten or may no longer be appropriate. Thus, economies can first make the transition from traditional energy sources to fossil fuels, and later return to renewable energy sources.

One possible fear associated with a transition to a low carbon economy is the limits to economic growth that renewable energy sources might impose, due to the availability of resources. After all, the standard narrative about the Industrial Revolution is that woodfuel could not have met the high energy demands associated with industrialization (Cipolla, 1962; Landes, 1969; Wrigley, 1988). That version of history is, no doubt, correct; but this chapter turns this argument on its head. It gathers evidence on past economies that managed within the confines of renewable energy systems. Focusing on woodfuels, it shows how in the Roman Empire, Early Modern Europe and the Far East, renewable energy resources were the drivers of economic activity for very long periods.

These were admittedly at slow growth rates by modern standards. But they were also at times when technologies were very inefficient and transport networks poorly formed, compared with the twenty-first century. For instance, with current technologies, transport infrastructures and institutional arrangements, the energy service demands of the Industrial Revolution would probably have been met quite easily with renewable energy sources.

This chapter does not argue that resources were always managed properly, or that resource limits did not hinder economic growth. Instead, it argues simply that renewable energy systems are not necessarily doomed to stagnation and collapse. Indeed, to be successful, economies need to balance their demands with their supplies and be 'sustainable'. But, if correctly managed, it may be possible to make a transition to a low carbon economy and to grow within a renewable energy system for a very long time.

A great emphasis has been placed on a transition to a low carbon economy. This is appropriate, given the threat of climate change and the difficulties and uncertainties of a transition. However, less research has gone into investigating what happens once we reach a low carbon economy.

This chapter begins this investigation by considering how economies in the past grew within the confines of a renewable energy system. It proposes two gaps in our knowledge. First, traditional models of long-run energy use have not addressed the situation once the backstop technology, such as a renewable energy source, becomes the dominant energy source again. Although a transition to a low carbon economy is a long way off, if it ever occurs, it is now time to improve our understanding of the relationship between long-run economic growth and renewable energy use. Second, we need to identify effective new policies that would be relevant for managing 'sustainably' (that is, in the long run) 'sustainable' energy sources. This would need the development of incentives that would meet energy service demands within technically and commercially viable renewable energy supply limits that would be distributed effectively. Careful investigation of renewable energy systems may be crucial to determining whether a transition to a low carbon economy becomes a new golden age in economic history, or another dark age.

REFERENCES

Allen, R.C. (2003), 'Was there a timber crisis in Early Modern Europe?', in S. Cavaciocchi (ed.), *Economia e Energia*, Florence: Le Monnier, pp. 469–82.
Ayres, R.U. and B. Warr (2009), *The Economic Growth Engine: How Energy and Work Drive Material Prosperity*, Cheltenham, UK and Northampton, MA, USA: Edward Elgar Publishing.
Barnett, H.J. and C. Morse (1963), *Scarcity and Growth: The Economics of Natural Resource Scarcity*. Washington, DC: Resources for the Future.

Berck, P. and M. Roberts (1996), 'Natural resource prices: will they ever turn up?', *Journal of Environmental Economics and Management*, **31** (1), 65–78.

BP (2010), 'BP statistical review of world energy', http://www.bp.com/productlanding.do?categoryId=6929&contentId=7044622.

Chakravorty, U., J. Roumasset and K. Tse (1997), 'Endogenous substitution among energy resources and global warming', *Journal of Political Economy*, **195**, 1201–34.

Cipolla, C.M (1962), *The Economic History of World Population*, London: Pelican Books.

Crafts, N. (2004), 'Steam as a general purpose technology: a growth accounting perspective', *Economic Journal*, **114** (2), 338–51.

Dasgupta, P. and G. Heal (1974), 'The optimal depletion of exhaustible resources', *Review of Economic Studies*, **41** (5), 3–28.

Farrell, A.E. and A.R. Brandt (2006), 'Risks of the oil transition', *Environmental Research Letters*, **1**, 1–6.

Fouquet, R. (2008), *Heat Power and Light: Revolutions in Energy Services*, Cheltenham, UK and Northampton, MA, USA: Edward Elgar Publishing.

Fouquet, R. (2009), 'A brief history of energy', in J. Evans and L.C. Hunt (eds), *International Handbook of the Economics of Energy*, Cheltenham, UK and Northampton, MA, USA: Edward Elgar Publishing, pp. 1–19.

Fouquet, R. (2010a), 'Divergences in the long run trends in the price of energy and of energy services', BC3 Working Paper Series 2010-03, Basque Centre for Climate Change (BC3), Bilbao, Spain.

Fouquet, R. (2010b), 'The slow search for solutions: lessons from historical energy transitions by sector and service', *Energy Policy*, **38** (10), 6586–96.

Fouquet, R. and P.J.G. Pearson (2003), 'Five centuries of energy prices', *World Economics*, **4** (3), 93–119.

Hartwell, R. (1966), 'Markets, technology and the structure of enterprise in the development of the eleventh century Chinese iron and steel industry', *Journal of Economic History*, **26** (1), 29–58.

Heal, G. (1976), 'The relationship between price and extraction cost for a resource with a backstop technology', *Bell Journal of Economics*, **7**, 371–8.

Landes, D.S. (1969), *The Unbound Prometheus: Technological Change and Industrial Development in Western Europe from 1750 to the Present*, Cambridge: Cambridge University Press.

Nordhaus, W. (1973), 'The allocation of energy reserves', *Brookings Papers*, **3**, 529–70.

Nordhaus, W.D. (1980), 'Oil and economic performance in industrial countries', *Brookings Papers on Economic Activity*, **2**, 341–99.

Pommeranz, K. (2000), *The Great Divergence: China, Europe and the Making of the Modern World Economy*, Princeton, NJ: Princeton University Press.

Rogner, H.H. (2000), 'Energy resources and technology options', in J. Goldemberg and T.B. Johansson (eds), *World Energy Assessment*, New York: UNDP, pp. 135–71.

Rosenberg, N. (1998), 'The role of electricity in industrial development', *Energy Journal*, **19** (2), 7–24.

Sieferle, R.P. (2001), *The Subterranean Forest: Energy Systems and the Industrial Revolution*, Cambridge: White Horse Press.

Stiglitz, J.E. (1974), 'Growth with exhaustible natural resources: efficient and optimal growth paths', *Review of Economic Studies*, Symposium on the Economics of Exhaustible Resources, 123–38.

Tahvonen, O. and S. Salo (2001), 'Economic growth and transitions between renewable and nonrenewable energy resources', *European Economic Review*, **45**, 1379–98.

Thomson, E. (2003), *The Chinese Coal Industry: An Economic History*, London: RoutledgeCurzon.

Totman, C. (1989), *The Green Archipelago: Forestry in Preindustrial Japan*, Honolulu: University of Hawaii Press.

Totman, C. (1995), *Early Modern Japan*, Berkeley, CA: University of California Press.

Wagner, D.B. (2001), *The State and the Iron Industry in Han China*, Copenhagen: Nordic Institute of Asian Studies Publishing.

Warde, P. (2003), 'Forests, energy, and politics in the early modern German states', in S. Cavaciocchi (ed.), *Economia e Energia*, Florence: Le Monnier, pp. 585–98.

Williams, D. (2003), *Deforesting the Earth: From Prehistory to Global Crisis*, Chicago, IL: University of Chicago Press.

Witthöft, H. (2003), 'Energy and large scale Industries (1300–1800)', in S. Cavaciocchi (ed.), *Economia e Energia*, Florence: Le Monnier.

Wrigley, E.A. (1988), *Continuity, Chance and Change: The Character of the Industrial Revolution in England*, Cambridge: Cambridge University Press.

2 Sustainability criteria for energy resources and technologies

Geoffrey P. Hammond and Craig I. Jones[1]

2.1 INTRODUCTION

2.1.1 Background

The evolution of modern human society has been inextricably linked to the discovery of various energy sources for heat and power (Hammond, 2000). Early societies have become identified by their state of technological development, which are known by terms such as the 'iron age', the 'steam age' and, more recently, the 'nuclear age'. Hammond (2000) observed that it is perhaps ironic in the context of the contemporary debate over alternative low carbon energy futures that in the pre-industrial period humans relied on what are now called 'renewable' energy sources; those principally derived from solar energy. Thus, wind drove sailing ships and windmills from medieval times, to be followed by the widespread use of water wheels. Wind is induced, in part, by the diurnal solar heating of land and sea. Likewise, solar heating of the sea leads to evaporation of water vapour, which subsequently precipitates over high land resulting in the flow of water in rivers, or its storage in lakes. Such resources became reservoirs for what are now known as large-scale and small-scale hydropower schemes. The early use of these renewables powered an essentially 'low-energy society' (Buchanan, 1994), with the power output being employed in the immediate vicinity of the resource (that is to say, they have a high 'energy gradient'; Nakicenovic et al., 1998).

A number of observers have studied the way in which the world has undergone a transition over time between various energy sources. These cycles, or Kondratieff long-waves (Nakicenovic et al., 1998), are illustrated in terms of world primary energy shares in Figure 2.1, along with future projections out to 2050 according to the Shell 'Dynamics as Usual' scenario (Davis, 2001). 'Traditional' energy sources include animal mature, fuelwood, water wheels and windmills. Over the next 40 years or so there is likely to be a major growth in energy demand, resulting principally from the development of rapidly industrializing countries (such as China and India). The depletion of finite fossil fuel resources like oil and natural gas, and the need for climate change mitigation, will therefore require a portfolio of energy options (Hammond and Waldron, 2008): energy demand reduction and energy efficiency improvements, carbon capture and storage from fossil fuel power plants, and a switch to other low or zero carbon energy sources; various sorts of renewables (including bioenergy) or nuclear power (see Figure 2.1).

The development of advanced, industrialized societies in the 'North' of the planet was underpinned by the discovery of fossil fuel resources and the construction of the associated energy system infrastructure. Fire, the earliest energy source used for heating and cooking, utilized fuelwood. But the fossil fuel resource that drove the first or

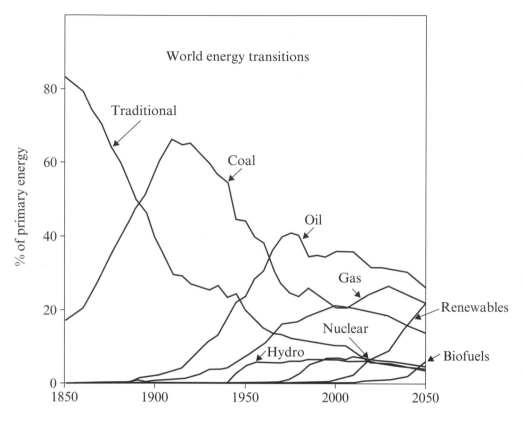

Source: Hammond and Waldron (2008); adapted from Nakicenovic et al. (1998) and Davis (2001).

Figure 2.1   World primary energy shares, 1850–2050: future projections based on the Shell 'Dynamics as Usual' Scenario

British 'industrial revolution' of the 1700s was coal; the first 'grand' energy transition (Nakicenovic et al., 1998; see also Figure 2.1). Britain was at the heart of this revolution, and was endowed with abundant coal reserves, with a nascent free market system of commerce, and a number of technological innovators. The latter agents of change (Buchanan, 1994; Hammond, 2000), people like Thomas Newcomen, James Watt and Charles Parsons, laid the foundations for the power industry. They developed 'high-energy converters' (Buchanan, 1994) such as steam engines and turbines, and subsequently various internal combustion engines. Another grand energy transition therefore resulted from the diverse range of energy end-uses that could be met by electricity (Hammond, 2000; Hammond and Waldron, 2008). Electricity is a high-grade energy carrier in the sense that it can be used to provide either power or heat. In a thermodynamic sense it has a high 'exergy' (as outlined by, for example, Hammond and Stapleton, 2001). Large energy losses occur during generation unless this takes place in conjunction with combined heat and power (CHP) systems. It is wasteful in thermodynamic terms to convert fuels to electricity, only to employ it for heating. If process or space heating were required, then it would be far more efficient to burn fossil fuels (for example) to produce heat directly. Electricity is also difficult to store on a large scale, and is mainly used instantaneously. But electricity has other benefits. There is an increasing end-use demand for high-grade and controllable energy carriers (Hammond, 2000; Hammond and Waldron, 2008).

Energy sources underpin human development via the provision of heat and power services, but they also put at risk the quality and longer-term viability of the biosphere as a result of unwanted or 'second-order' effects (Hammond, 2000). Many of these side-effects of energy production and consumption give rise to resource uncertainties and potential environmental hazards on a local, regional and global scale. Examples include the depletion of global oil and natural gas resources, the formation of acid rain via pollutant emissions from (primarily) fossil fuel power stations, the complexity of long-term safe storage of radioactive wastes from nuclear power plants, and the possibility of enhanced greenhouse effects from combustion-generated pollutants. Changes in atmospheric concentrations of 'greenhouse gases' (GHGs) affect the energy balance of the global climate system, and are arguably the key environmental burden constraining moves towards global sustainability. Human activities have led to quite dramatic increases in the 'basket' of GHGs incorporated in the Kyoto Protocol since 1950, concentrations rising from 330 ppm to about 430 ppm currently (IPCC, 2007). Prior to the first industrial revolution the atmospheric concentration of 'Kyoto gases' was only some 270 ppm. The cause of the observed rise in global average near-surface temperatures since the Second World War has been a matter of dispute and controversy. But the most recent (Fourth Assessment Report, AR4, in 2007) scientific assessment by the Intergovernmental Panel on Climate Change (IPCC) states with 'very high confidence' that humans are having a significant impact on global warming (IPCC, 2007). They argue that GHG emissions from human activities trap long-wave thermal radiation from the Earth's surface in the atmosphere (not strictly the same phenomenum as happens in a greenhouse), and that these are the main cause of rises in climatic temperatures. Approximately 30 per cent of UK carbon dioxide (CO_2) emissions, the principal GHG (with an atmospheric residence time of about 100 years; Hammond, 2000), can be attributed to electricity generation (Hammond and Waldron, 2008). In order to mitigate against anthropogenic climate change, the UK Royal Commission on Environmental Pollution (RCEP, 2000) recommended at the turn of the millennium a 60 per cent cut in UK CO_2 emissions by 2050. But eventually, on the recommendation of its independent Committee on Climate Change (CCC, 2008), the British government adopted a target of an 80 per cent reduction (against 1990 levels) by 2050. The CCC also argued that the steepest reductions in emissions must occur before 2030. But these are very challenging targets and they should be viewed against the difficulty that the UK government has been having in achieving its own short-term 'domestic' target of just a 20 per cent reduction in CO_2 emissions by 2010 (Hammond and Waldron, 2008).

Globally, humans were almost wholly dependent on finite fossil and nuclear fuels for energy resources at the turn of the millennium (see again Figure 2.1), amounting to about 77 per cent and 7 per cent of primary energy needs, respectively (Boyle et al., 2003). 'Traditional' renewable energy sources, such as burning fuelwood and dung or using water and windmills, accounted for 11 per cent of these worldwide requirements. Large-scale hydroelectric power contributed 3 per cent, and other renewables (including modern wind turbines and liquid biofuels) contributed just 2 per cent. Sustainable development in a strict sense requires a reversal of these roles, but it is unlikely that renewable energy technologies could meet a high proportion of industrial countries' energy demand before at least the middle of the twenty-first century (RCEP, 2000). This is partly due to the conflict between the needs of environmental sustainability and the downward economic pressures on energy prices arising from moves towards energy market liberalization in the industrialized world. Even in the European Union (EU), which has had a long-term policy of encouraging modern renewables, the target of 20 per cent renewables use by the year 2020 (with 10 per cent of 'green fuels', principally biofuels, for transport) is seen by many analysts as being ambitious. Although renewables are growth technologies across much of Europe, they have not played a dominant role in achieving the GHG mitigation target of 8 per cent reduction against a base year of 1990 by 2008–12 agreed under the Kyoto Protocol. The EU-15 countries (EEA, 2010) are on track to meet this target, but mainly via improvements in energy and end-use efficiency rather than the take-up of renewables (except for biofuels in the transport sector).

2.1.2   The Issues Considered

The present chapter seeks to examine the principles and practice of sustainability in the context of the energy sector, as well as the sustainability criteria that stem from them. It is shown that sustainability can be disaggregated into three elements or 'pillars': the economic, environmental and social aspects. They can be viewed both from an intergenerational and a global interregional ethical perspective. A useful distinction is drawn between sustainable development and sustainability: the journey and the destination. Several attempts have been made to devise what some have termed 'sustainability science'. This embraces, for example, the notion of resource use productivity (Factor X) as a component of the 'sustainability equation'. A number of principles are also said to underpin sustainable development, and the so-called 'precautionary principle' is arguably the most significant of these. Its implications are highlighted, along with some controversial issues surrounding its adoption within international environmental treaties, as well as in the domains of engineering design and of policy formulation. Methods of appraisal vary between the three pillars of sustainability, particularly in regard to the extent to which they can be evaluated in quantitative and qualitative terms. Different approaches are discussed, with a focus on the way that the energy sector can be evaluated in terms of its energy, environmental, and economic performance. Examples are drawn from the fields of liquid biofuels for transport, decentralized energy resources (DERs), and a simple power network in order to illustrate the use of quantitative sustainability metrics.

2.2   TOWARDS SUSTAINABILITY

2.2.1   Sustainable Development versus Sustainability

The concept of sustainability has become a key idea in national and international discussions following publication of the Brundtland Report (WCED, 1987) published under

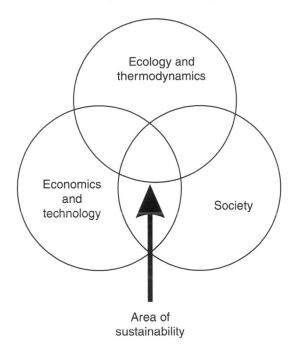

Source: Hammond (2004a); adapted from Clift (1995) and Parkin (2000).

Figure 2.2 *Venn diagram representation of 'the three pillars' of sustainability*

the title *Our Common Future*; the outcome of four years of study and debate by the World Commission on Environment and Development led by the former Prime Minister of Norway, Gro Harlem Brundtland. This Commission argued that the time had come to couple economy and ecology, so that the wider community would take responsibility for both the causes and the consequences of environmental damage. It envisaged sustainable development as a means by which the global system would satisfy 'the needs of the present without compromising the ability of future generations to meet their own needs'. The notion therefore involves a strong element of intergenerational ethics; what John Gummer, former UK Secretary of State for the Environment (1993–97), encapsulated in the popular phrase: 'Don't cheat on your children' (Parkin, 2000). More recently, sustainability has been the subject of renewed interest and debate in the context of the 2002 World Summit on Sustainable Development in Johannesburg. Here the strapline 'People, Planet, Prosperity' was adopted to reflect the requirement that sustainable development implies the balancing of economic and social development with environmental protection: the 'three pillars' model (Hammond, 2006). The interconnections between these pillars are illustrated by the sustainability Venn diagram shown in Figure 2.2 (Hammond, 2004a; adapted from a version originally developed by Clift, 1995 and extended by Parkin, 2000). Sustainability is reflected in the central portion of the diagram, where the three types of constraints are met. The originators themselves recognized that this is a simplified model (see, for example, Azapagic et al., 2004). An alternative concept still involving these three elements is the so-called 'Russian dolls' model in which the economy is viewed as being surrounded by human society, which is in turn enclosed by the natural environment (Chambers et al., 2000). Recently the UK government has added two additional principles of sustainable development to the three pillars (Defra, 2005): promoting good governance, and using sound science responsibly (that is, adopting 'evidence-based' approaches). But science and technology cannot be deployed without regard to their environmental and social implications, or 'side-effects' (Hammond, 2000). In the long term, Planet Earth will impose its own constraints on the use of its physical resources and on the absorption of contaminants, whilst the 'laws' of the natural sciences (including, for example, those of thermodynamics; Hammond, 2004a) and human creativity will limit the potential for new technological developments. Many writers and researchers have acknowledged that the concept of 'sustainable development' is not one that can readily be grasped by the wider public (see, for example, Hammond, 2000). However, no satisfactory alternative has thus far been found. Further confusion about this modern paradigm is added by the large number of formal definitions for sustainable development that can be found in the literature; Parkin (2000) refers to more than 200.

Parkin (2000) and Porritt (2000) have stressed that sustainable development is only a process or journey towards a destination, which is 'sustainability'. The endgame cannot easily be defined from a scientific perspective, although Porritt (2000) argues that the attainment of sustainability can be measured against a set of four 'system conditions'. He draws these from 'The Natural Step' (TNS); an initiative by the Swedish cancer specialist, Karl-Henrick Robèrt (see, for example, Broman et al., 2000). Its system conditions put severe constraints on economic development, and may be viewed (Hammond, 2004a) as being impractical or utopian. One of them, for example, suggests that finite materials (including fossil fuels) should not be extracted at a faster rate than they can be redeposited in the Earth's crust on geological timescales. This may be contrasted with the present rapid rate of fossil fuel depletion on the global scale, leading to estimates for resource to production ratios of 20–40 years for oil, 40–70 years for natural gas, and 80–240 years for coal (Hammond, 2000). Upham (2000) argues that TNS moves beyond (scientific and other) knowledge in signposting action for the business sector. He contends that it represents a political and ethical statement rather than any justifiable scientific consensus. TNS certainly implies that the ultimate goal of sustainability is rather a long way off when compared with the present conditions on the planet. Parkin (2000) suggests 2050–2100 or beyond.

2.2.2   Sustainability Appraisal Methods: Interdisciplinary Perspectives

The 'three pillars' of sustainability (see again Figure 2.2) imply that differing professional disciplines and insights are required in order to address each dimension:

- The environmental pillar: this can be tackled in quantitative terms via energy and environmental performance appraisal (see, for example, Hammond and Winnett, 2006); typically on a life-cycle or 'full fuel cycle' basis. This can be undertaken by using the techniques of thermodynamic (energy and exergy) analysis and environmental life-cycle assessment (LCA), outlined in more detail below. Typically the uncertainty band in the resulting estimates of energy system performance parameters are of the order of perhaps ± 20 per cent.

- The economic pillar: this is once more a pillar that can be addressed in quantitative terms via methods such as environmental cost–benefit analysis (CBA). However, Hammond and Winnett (2006) found that estimates of environmental costs and benefits associated with energy technologies exhibited a wide variation. These were found to reflect variations of several orders of magnitude, that is, factors of ten. They consequently argued that this demonstrated the frailty of the present generation of monetary valuation methods.
- The social pillar: here the approaches that can be applied are mainly qualitative. They include analytic and deliberative processes (for example stakeholder engagement), the mapping of socio-technical systems, customer surveys (in response to new technologies such as smart meters, and business models), and the ethical reflection on energy system impacts and futures. Clift (2007) observes that this pillar should encompass inter- and intragenerational equity concerns.

Attempts have been made to bring the above perspectives together using a variety of approaches, including a simple sustainability checklist, 'ecological' or environmental footprinting (see, for example, Chambers et al., 2000; Eaton et al., 2007; Hammond, 2006; Cranston and Hammond, 2010), multi-criteria decision analysis (MCDA; Elghali et al., 2007), sustainability maps or 'tortilla' diagrams, and a sustainability appraisal framework (as advocated by the UK sustainability non-governmental organization Forum for the Future, founded by Sara Parkin and Jonathan Porritt). The participatory multi-criteria mapping and decision-conferencing approach developed by Elghali et al. (2007) for the sustainability assessment of bioenergy systems is perhaps the most comprehensive thus far devised. They drew on the lessons from modern operational research methods and aim to integrate these with the use of LCA (ISO, 2006a, 2006b). Elghali et al. (2007) produced a framework for future use, but did not actually apply it to a specific bioenergy route. MCDA typically aggregates various distinct impacts arising from alternative technological options. Thus, Allen et al. (2008) argued that there are a number of reasons for discouraging such aggregate methods (including, amongst them, CBA). Decision-makers are presented with a single, aggregate decision criterion, which actually hides many disparate environmental impacts. Allen et al. (2008) suggest that it is vitally important that the implications of these impacts are faced, particularly by politicians, rather than obscured by the methodology.

2.2.3   Sustainability Science, the Sustainability Equation and Resource Productivity

A significant step forward in the development of 'sustainability science' has been taken by Graedel and Klee (2002) in trying to establish a quantifiable, long-term target for sustainability from an 'industrial ecology' perspective. They suggest a framework, or series of steps, to permit the establishment of the sustainable (or limiting) rate of natural resource use, which can then be contrasted with the current rate of consumption. The process is illustrated for the case of three common materials employed or emitted by industrial societies: zinc, germanium and greenhouse gases. Unfortunately, the Graedel and Klee procedure requires the establishment of equal planetary shares of materials or emissions on a 50-year timescale. They acknowledge that the idea of an 'Earthshare' or quota of this sort is controversial, and that the chosen timescale is somewhat arbitrary. Hammond (2006) has suggested that an alternative quantitative indicator that may be better able to track humanity's pathway towards sustainability is the ecological or environmental footprint. On a global scale, Loh and Goldfinger (2006) have utilized it for this purpose in the World Wide Fund for Nature's (WWF) biannual *Living Planet Report*. Cranston and Hammond (2010) recently used this approach to estimate the transition pathways of regional environmental footprints for the peoples of the indus-trialized North and populous South that would be needed in order to secure climate-stabilizing carbon reductions out to about 2100. Such indicators are in keeping with an interpretation of sustainable development devised by several leading international nature conservation and environmental organizations (IUCN, UNEP and WWF, 1991) as 'improving the quality of human life while living within the carrying capacity of sup-porting ecosystems'.

An alternative representation of the elements of sustainable development can be obtained using the so-called 'IPAT' equation devised by Holdren and Ehrlich (1974) for analysing environmental disruption:

$$(Environmental)\ impact = population \times affluence \times technology \qquad (2.1)$$

This expression has been termed the 'sustainability equation' by Jacobs (1996), and Tester et al. (2005) suggest that it underpins the mathematical representation of sustain-ability. Affluence, or economic consumption per person, is normally measured by gross domestic product (GDP) per capita. GDP is the traditional measure of wealth creation adopted by economists at the level of the nation state: the total output of goods and services in money terms produced within a national economy. In the period since the early 1960s this has tended to increase over time in the wealthy countries of the indus-trialized North of the planet, whilst typically falling in the poorer developing nations of the majority South. The situation with demographic growth has been quite different; for example, with almost stable populations in many affluent countries of Northern Europe. In contrast, rapid population growth has been observed in many parts of the developing world: Africa, continental Asia, and Central and South America. The 'technology' com-ponent in equation (2.1) represents the environmental damage per unit of consumption.

According to Meadows et al. (1992) the scope for reducing the various terms on the right-hand side of the IPAT equation is very large over a 50–100-year timescale. Table 2.1 is adapted from this work (see Hammond, 2004c), although they attribute the esti-mates of the potential for long-term change and the associated timescales to Amory Lovins (in a paper that Hammond, 2004b, 2004c, was unable to locate, even from the originators). Obviously the individual columns in this table reflect global aggregate figures or averages. Each socio-economic region or nation state would need to place a different emphasis on which component of the sustainability equation they tackled. The focus in the industrialized world, where the population is stable, would have to be principally on resource productivity (the 'technology' element). In developing countries with rapidly growing populations, both population and resource productivity changes would be required in order to secure sustainable development. There may also need to be a more equitable sharing of wealth in the long run (Hammond, 2004c). This implies some

Table 2.1   *The environmental impact of population, affluence and technology*

Population	Affluence		Technology		Environmental Impact
Population	$\times$ $\dfrac{\text{Capital stock}}{\text{Person}}$	$\times$ $\dfrac{\text{Material throughput}}{\text{Capital stock}}$	$\times$ $\dfrac{\text{Energy}}{\text{Material throughput}}$	$\times$ $\dfrac{\text{Environmental Impact}}{\text{Energy}}$	
		Applicable tools			
Family planning	Values	Product longevity	End-use efficiency	Benign sources	
Female literacy	Prices	Material choice	Conversion efficiency	Scale	
Social welfare	Full costing	Minimum materials design	Distribution efficiency	Siting	
Role of women	What do we want?	Recycle, reuse	System integration	Technical mitigation	
Land tenure	What is enough?	Scrap recovery	Process redesign	Offsets	
		Approximate scope for long-term change			
~2 $\times$	?	~3–10 $\times$	~5–10 $\times$	~10^2–10^{3+} $\times$	
		Time scale of major change			
~50–100 years	~0–50 years	~0–20 years	~0–30 years	~0–50 years	

Source: Hammond (2004c); adapted from Meadows et al. (1992) and based on estimates by Amory Lovins.

29

convergence in GDP per capita between developed and developing countries; a task that is obviously fraught with political difficulties. Nevertheless, the multiplier effect of the IPAT equation suggests that significant reductions in environmental impact (via falls in the rate of population growth or increases in resource productivity like those indicated in Table 2.1) are possible overall.

If the GDP of a particular country can be reliably projected into the future, then the corresponding energy demand can be estimated using a simple relation adapted from the IPAT equation (2.1), where the monetary unit is notionally the international dollar:

$$Energy\ consumption\ (PJ) = population\ (millions) \times GDP\ per\ capita\ (\$)$$

$$\times\ energy\ intensity\ (PJ/\$) \tag{2.2}$$

In order to determine pollutant emissions, equations (2.1) and (2.2) can be coupled to yield the so-called Kaya identity (see Hoffert et al., 1998) which, in the case of carbon dioxide emissions, becomes:

$$CO_2\ emissions\ (MtC) = CO_2/energy\ ratio\ (MtC/PJ) \times energy\ consumption\ (PJ) \tag{2.3}$$

Thus, the sustainability equation, as well as its energy or pollutant emission equivalents, suggest a multiplier effect between population, economic welfare, and emissions or resource intensity.

There is a widely recognized need to stimulate improvements in resource use efficiency generally, and to encourage energy conservation, as part of a sustainable (and low carbon) energy strategy. Such an approach would need to be coupled with measures to reduce the rate of consumption of fossil fuels, and stimulate an expansion in the use of renewable energy sources (Hammond, 2004c). It would involve a consumer-oriented market approach, coupled with intervention by way of a portfolio of measures to counter market deficiencies: economic instruments, environmental regulation and land use planning procedures (Hammond, 2000). Scenarios such as the 'dematerialization' or 'Factor Four' project advocated by Ernst von Weizsacker and Amory and Hunter Lovins (von Weizsacker et al., 1997) suggest that economic welfare in the industrial world might be doubled while resource use is halved; thus the Factor 4. This resource use productivity is reflected in the 'technology' component of the IPAT or sustainability equation (2.1) above, or in the energy intensity term within its energy consumption equivalent (equation 2.2). Dematerialization would involve a structural shift from energy-intensive manufacturing to energy-frugal services (Hammond, 2000). Increases in resource use efficiency at the Factor 4, 5 (an 80 per cent improvement in resource productivity, as more recently advocated by von Weizsacker et al., 2009) or 10 (as proposed by the UK Foresight Programme, Fore*sight*, 2000) levels would have an enormous benefit of reducing pollutant emissions that have an impact, actual or potential, on environmental quality over the long term (see also Table 2.1). In reality such a strategy requires a major change ('paradigm shift'; Hammond, 2000) to an energy system that is focused on maximizing the full fuel or energy cycle efficiency, and minimizing the embodied energy and carbon in materials and products (Hammond and Jones, 2008) by way of reuse and recycling. In order to make such an approach a practicable engineering option, it would be necessary to use methods of thermodynamic analysis in order to optimize the energy cascade (Hammond, 2004a).

2.2.4 Taking Precautions

Underpinning the notion of sustainable development is a set of guiding principles. Four of these were incorporated into European law in the Maastricht Treaty (Clause 130r) (see, for example, Eurotreaties, 1996), albeit in a rather ill-defined form. There it states that European Union policy on the environment should be: 'based on the precautionary principle and on the principles that preventative action should be taken, that environmental damage should as a priority be rectified at source and that the polluter should pay'. The first of these, the so-called 'precautionary principle', suggests that in the face of a significant environmental risk, lack of scientific certainty should not be used as a pretext to delay taking cost-effective action to prevent or minimize potential damage. The origins of the principle can be traced back (EEA, 2001) to the work of a London physician, Dr John Snow, on the link between cholera and polluted drinking water (*circa* 1850). He advocated the removal of the Broad Street water pump on the grounds of 'precautionary prevention'. But the concept of taking precautionary action itself really came to prominence when it formed part of the (West) German Clean Air Act of 1974. It then rose up the environmental agenda to constitute an important element of several major international treaties, including the UN World Charter on Nature (1982), the Montreal Protocol on ozone depletion (1987), the UN Rio Declaration on Environment and Development (1992), the Framework Convention on Climate Change (1992) and the Cartagena Protocol on Biosafety (2000).

The precautionary principle has caused some controversy amongst the scientific community, and between it and environmentalists generally (Hammond, 2004b). The application of the principle has often been seen (for example, by *The Economist* magazine; see Porritt, 2000) as a mechanism for restricting innovation and driving up regulatory costs. This misrepresents the precautionary approach in terms of what is sometimes viewed as its extreme, or 'strong', formulation. Lewis Wolpert (Professor of Anatomy at University College London) has also disparaged the principle (Wolpert, 1993), arguing that it is not scientifically based. This is quite true, but that is rather to miss the point. It is simply a set of guidelines of the type that engineers are well accustomed to employing in industry: 'art' or practice as opposed to pure science. Indeed environmental campaigners like Jonathon Porritt (2000) and Greenpeace see the precautionary approach: 'as the most effective way of combining science and ethics'.

A pioneering study was undertaken by the European Environment Agency (EEA, 2001) to examine lessons for precautionary action from hazards caused by human activity over the period 1900–2000. They reviewed some 14 case studies where early warnings were evident of significant environmental damage to species and ecosystems. These cases were drawn from both European and North American experience, and included acid rain, ionizing radiation, and halocarbons and ozone depletion. The EEA scientific team then used these histories to devise a set of 12 'late lessons' about how the precautionary principle should be applied in future. In the present context, the most important of these lessons were that:

- 'Blind spots' and gaps in the scientific knowledge should be identified.
- More robust, diverse and adaptable technologies should be promoted so as to minimize the costs and maximize the benefits of innovation.
- Claimed justifications and benefits should be systematically scrutinized alongside the potential risks.
- A range of options for meeting needs should be evaluated alongside the option under appraisal.
- Full account should be taken of lay and local knowledge as well as relevant specialist expertise in the appraisal process.
- Risk and uncertainty should be acknowledged and form part of the process of technology assessment and public policy-making.
- 'Paralysis by analysis' should be avoided by acting to reduce potential harm when there are reasonable grounds for concern.
- Adequate long-term environmental and health monitoring and research should be provided to ensure early warnings.

The EEA team recognized that many of these lessons are clearly interlinked.

Practising engineers working in the energy sector typically operate in an industrial setting that requires them to design products and systems on the basis of what the management thinker Igor Ansoff (1970) termed 'partial ignorance'. They are therefore unable to foresee, or take account of, the second-order side-effects of their endeavours. Systems need to be put in place that will hold out the prospect of identifying potentially harmful side-effects of particular technologies before they are introduced into the marketplace (Hammond, 2004b). This would be consistent with the 'precautionary principle', and with the late lessons identified by the EEA (2001). It is only in this way that humanity can ensure that its development is sustainable.

2.3   THERMODYNAMIC CONSTRAINTS ON ENERGY SYSTEMS

Parkin (2000) highlighted the significance of thermodynamic analysis, which she sees as underpinning the environmental pillar of sustainable development; see Figure 2.2. The two most important of its 'laws' – the First and Second Laws of Thermodynamics – lead to properties that enable process improvement potential to be identified (see, for example, Hammond, 2004a, 2004c; Hammond and Stapleton, 2001). These are 'enthalpy' (from the First Law) to represent the quantity of energy consumed, as well as 'exergy' (from the First and Second Law) to reflect its quality. They place fundamental constraints on energy systems and the energy sector more generally.

Hammond and Stapleton (2001), for example, used exergy analysis to examine the thermodynamic performance of the United Kingdom in the late 1990s. They found that final demand in the domestic and transport sectors, together with electricity genera-

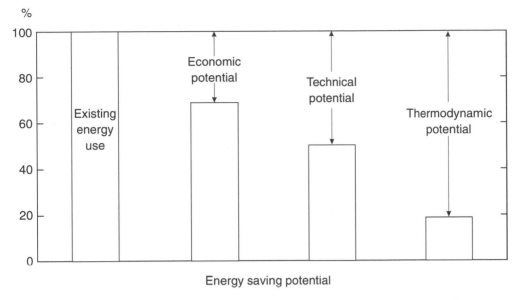

Source: Hammond (2004b); adapted from Jaffe and Stavins (1994) and Eyre (2002).

Figure 2.3 *The energy efficiency gap between theory and practice*

tion, accounted for nearly 80 per cent of 'exergetic' improvement potential. In order to achieve efficiency gains, it would be necessary to focus attention principally on making better use of space heating systems, improving the operating efficiency of power plant, and reducing thermodynamic losses in transportation systems that are presently dependent on internal combustion (IC) engines. The background studies on energy efficiency and energy productivity prepared for the 2002 UK Energy Review, conducted by the British Government's then Performance and Innovation Unit (PIU) (see Eyre, 2002; PIU, 2002), suggest that the thermodynamic findings of Hammond and Stapleton (2001) represent the maximum theoretical improvement, or energy saving, potential. However, Jaffe and Stavins (1994) rightly drew a distinction between such an optimum and what can feasibly be achieved in practice. Hammond (2004c) suggests that, although the thermodynamic (or exergetic) improvement potential is around 80 per cent (see Figure 2.3), roughly in line with the findings of Hammond and Stapleton (2001), only about 50 per cent of energy currently used could be saved by technical means. When economic barriers are also taken into account, this reduces to perhaps some 30 per cent. Notwithstanding this, the PIU team (2002) still argued that the current level of energy services could be secured using just 20 per cent of the energy used at present; something that illustrates the very great scope that there is for innovation in energy efficiency over the longer term (Hammond, 2004c). Von Weizsacker et al. (2009), in their recent book on achieving Factor Five improvement in resource and energy productivity, are rather more optimistic about the prospects of securing major gains. They examined developments and case studies from around the world (including China and India) over the 15 years since Ernst von Weizsacker published his earlier text with Lovins. They argue that the global economy could be transformed over time through 80 per cent improvements in resource productivity.

2.4 THE APPLICATION OF SUSTAINABILITY CRITERIA IN THE ENERGY SECTOR

2.4.1 Example 1: Biofuels

Transport underpins the mobility of people around the world, but presently accounts for around 20 per cent of global anthropogenic CO_2 emissions (Royal Society, 2008), an unwanted side-effect. The adoption of liquid biofuels in the transport sector (Hammond et al., 2008) has therefore been seen, particularly by the European Union, as a means for meeting climate change mitigation targets (EEA, 2010), enhancing regional energy security and contributing to rural development (through the provision of an alternative source of income in otherwise depressed agricultural communities). Biomass can be converted into premium-quality liquid biofuels and biochemicals (Tester et al., 2005). Bioethanol and biodiesel also hold out the prospect of retaining the existing transport infrastructure (for example refuelling or petrol stations), in contrast to other low carbon options such as hydrogen or electric vehicles. That has significant benefits in terms of limiting capital expenditure and the potential speed of take-up. But the deployment of biofuels may have significant impacts in terms of direct and indirect land use change, loss of biodiversity and the provision of ecosystem services (Royal Society, 2008; Tester et al., 2005), and competition with food production. Potential feedstocks and conversion routes (Hammond et al., 2008) need to be assessed against the full range of sustainability considerations and over the full life cycle of the biofuel supply chain (Elghali et al., 2007; Royal Society, 2008), 'from seed to wheel'. Only in this way will the true consequences of a given biofuel – environmental, economic and social – be determined.

Driven by the 2003 EU Directive on promoting the use of biofuels for transport (2003/30/EC), the UK government introduced a Renewable Transport Fuel Obligation (RTFO) and established a Renewable Fuels Agency (RFA) to promote modest biofuel blends in automotive fuels. This has led to a take-up of just above the RTFO target of a 2.5 per cent supply of road transport fuel for 2008/09, mainly via the adoption of biodiesel (RFA, 2010). During this period, the biofuels were supplied into the UK from some 18 different countries and from about a dozen feedstocks. However, concern over indirect land use change (iLUC) and the pressure on food prices led to an RFA review (the so-called Gallagher Review). The British government accepted its recommendation that future targets should be reduced until the full implications of these effects can be evaluated. Biofuels from conventional feedstocks offer the opportunity to reduce CO_2 emissions at the tailpipe and over their life cycle (Hammond et al., 2008), although they are not completely 'carbon neutral'. According to the RFA's own environmental LCA methodology, the biofuels delivered under the RTFO in 2008/09 amounted to a 46 per cent carbon saving compared to the equivalent fossil fuels (RFA, 2010). The LCA methodology has been encapsulated in a 'Carbon Calculator' (devised by the RFA) for emissions released across the whole production chain. However, the Gallagher Review acknowledged that the development of the biofuel market worldwide had had an adverse

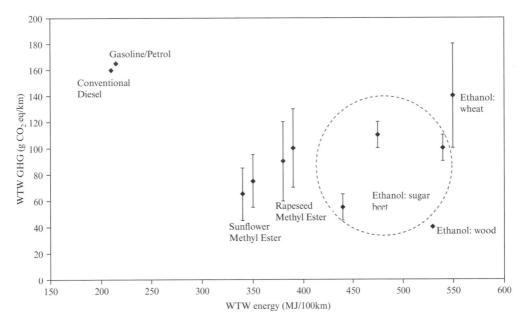

Source: Hammond et al. (2008); adapted from an earlier version of the Concawe Report (Concawe, EUCARE and JRC, 2006).

Figure 2.4   *'Well-to-wheel' (WTW) equivalent analysis of energy use and greenhouse gas (GHG) emissions for various biofuels*

impact both on the GHG emissions claimed and on food prices. Carbon savings come with a potential cost due to their detrimental effects on the rural environment (Hammond et al., 2008). These effects will depend on the type of crop grown and its subsequent management. The RFA has developed a so-called 'Meta-Standard' for environmental and social sustainability reporting by fuel suppliers under the RTFO (RFA, 2010) that sits alongside its quantitative carbon reporting system. In order to satisfy the environmental requirements, the biofuels must be grown with due regard to protecting biodiversity, carbon stocks and ecosystem (soil, air and water) quality. Workers' rights and land rights must be respected in order to meet the social obligations of the standard. The biofuel suppliers can opt to have their feedstocks and conversion pathways assessed directly, or indirectly via certification under an existing scheme that meets sufficient of the RFA sustainability criteria to be regarded as a 'Qualifying Standard' (RFA, 2010). Mandatory sustainability standards will be required under the new EU Renewable Energy Directive (RED) that also includes the 2020 target of 10 per cent of 'green fuels' specifically for the transport sector. These standards were to be implemented by December 2010.

The quantitative life-cycle analysis of biofuels provides an indication of the trade-offs that are necessary between energy requirements and carbon savings. A 'well-to-wheel' (strictly 'seed-to-wheel' in the case of biofuels) analysis of energy used by different biofuels and the corresponding GHG emissions has been produced as part of the so-called Concawe Report (Concawe, EUCARE and JRC, 2006). Hammond et al. (2008) reproduced data from an earlier version of that study to highlight the energy and

GHG requirements of the bioethanol production process (see Figure 2.4). A shift from gasoline/petroleum to bioethanol in the transport sector appears to offer significant potential GHG reductions, but requires more energy. The difference in the amount of energy required for the three bioethanols based on sugar beet results from the varying extent to which the biomass feedstock is used for energy or other purposes (Hammond et al., 2008). Whilst some options (for example using the pulp to fuel the conversion process) might be more favourable from an environmental point of view, that is unlikely to take place at present because pulp is currently a valuable animal feed. The wheat-based bioethanol shows higher GHG emissions, due to the use of fertilizers (and therefore nitrous oxide emissions) in its agricultural production. Such emissions are also associated with a large uncertainty or error band. The production of bioethanol from, for example, wheat or barley straw is likely to be more sustainable, because it utilizes what would otherwise be a waste product. However, the determination of seed-to-wheel GHG emissions in that case is more complex, due to the need to partition the inputs and outputs. The crops may be used for food, animal bedding or biofuel feedstock at the upstream boundary, while the output could be a combination of bioethanol and biochemical co-products.

2.4.2   Example 2: Domestic Microgenerators

The use of microgeneration and other decentralized or distributed energy technologies has the potential to reduce power generation and transmission and distribution (T&D) network losses. When fossil fuels are used, for example in small-scale combined heat and power (micro CHP) plants, the heat generated in the process of localized electricity production can be usefully captured and employed for space and water heating. Heat or electricity can also be produced locally via renewable energy sources, such as solar thermal water heaters, solar photovoltaic (PV) systems, and micro wind turbines. 'Distributed generation' is site-specific in relation to both energy resources and energy demand. It refers to energy supply close to the point of use by way of DERs or microgenerators. They can be in a range of generator sizes, from community or district level down to individual households. Typically they represent anything below 50–100 kW, with most household electricity supply installations being below 3 kW_e; slightly more for heat supply (Allen et al., 2008; Hammond and Waldron, 2008).

A range of 'integrated' appraisal techniques was recently utilized by Allen et al. (2008) to study the performance of various domestic micro-generators that have been proposed as possible decentralized energy resources for 'low carbon' buildings. Energy, environmental impact and cost–benefit analysis methods were employed on a whole-system basis. They effectively represented the quantitative elements of sustainability: a subset of criteria for the economic and environmental pillars (but obviously not for the social dimension). They can be viewed as being 'integrated' or interrelated in the sense that life-cycle energy analysis (EA) was one of the precursors for environmental life-cycle assessment (LCA), and is typically performed in parallel with environmental appraisal in most modern LCA software packages (see Hammond and Winnett, 2006). Both EA and LCA avoid the examination of products on a subsystem basis, whereby only one part of the life cycle is examined. They were also employed by Allen et al. (2008) to estimate impact inventories that can then be coupled with environmental cost–benefit analysis (CBA)

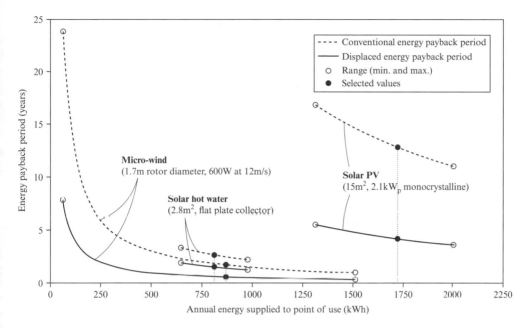

Source: Allen et al. (2008).

Figure 2.5   Energy payback periods (EPPs) for three specific micro-generators

to yield their environmental costs. The application of this 'toolkit' is illustrated via the evaluation of three specific micro-generators: a micro-wind turbine located on an open, or rural, site; a solar PV array; and a solar hot water (SHW) system.

All microgenerators considered by Allen et al. (2008) were found to pay back their energy investments well within their lifetimes (see Figure 2.5). The energy payback period (EPP) is a useful metric that can be derived from an energy analysis, and is analogous to a financial payback period (often termed the 'break-even point'). It represents the number of years that a system must operate until its cumulative energy output equals the 'whole-life' or 'life-cycle' primary energy requirement, the latter being calculated via the LCA software. When the cumulative energy output is accounted for in terms of the absolute quantity of electricity or hot water supplied, the 'conventional' energy payback period is produced. But the energy supplied by micro-generators displaces energy that is otherwise provided by conventional means. The displaced EPP illustrated in Figure 2.5 has previously been described as the 'opportunity cost convention', from its precursor in the economic literature (Allen et al., 2008; Hammond and Winnett, 2006). Here the micro-wind turbine and SHW system pay back their whole-life, primary requirements faster than the solar PV array, despite lower estimated annual energy supply. This is as a result of their lower primary energy requirements. An alternative metric to the EPP is the energy gain ratio (EGR), defined as the ratio of the energy delivered (to point of use) during a technology's lifetime to the life-cycle primary energy requirement. Data on this basis have also been reported by Allen et al. (2008) for these three specific DERs.

In the corresponding LCA study by Allen et al. (2008), the energy and materials used,

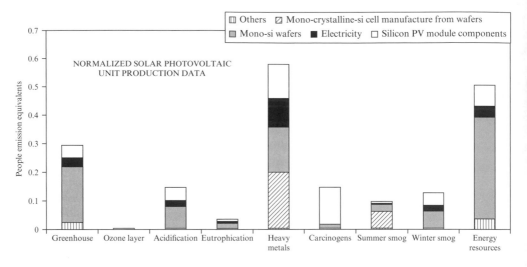

Source: Allen et al. (2008).

Figure 2.6   Normalized solar photovoltaic unit LCA production data

and pollutants or wastes released into the environment as a consequence of a product or activity, are quantified over the whole life cycle, 'from cradle to grave' (ISO, 2006a, 2006b). Production data for both the micro-wind turbine and the solar hot water system were obtained from manufacturers, whereas that for the solar photovoltaic array were taken from the literature (see Figure 2.6). LCA impact categories (ISO 2006a, 2006b) are very difficult to compare directly, and so the data were normalized with respect to European emissions. This allowed a comparison of the importance of each category to be made without attributing subjective valuation. The solar PV unit was found to require significantly more energy (and to produce more GHGs) in its production phase than either of the other two systems examined by Allen et al. (2008). It also has significantly higher impacts in terms of the other environmental categories considered. The production process of high-grade silicon, as used in PV cell manufacture, requires the consumption of a large amount of energy, and this gives rise to these relatively high production impacts. In all DER cases the greatest life-cycle environmental impact resulted from energy use, and the emission of greenhouse gases and heavy metals. The use of aluminium leads to the emission of both greenhouse gases and heavy metals (Allen et al., 2008). The utilization of other metals, such as copper, can cause the release of carcinogens, as well as leading to high energy consumption.

Allen et al. (2008) employed a risk-free discount rate of 3.5 per cent, typical of the UK government's Test Discount Rate (TDR), to undertake a CBA of the three micro-generators (see Figure 2.7). However, this discounting did not impact on the costs of the DERs, because almost all of these are in the form of capital and installation costs. Environmental externalities were quantified by coupling the LCA results with 'damage costs' taken from the ExternE project (Dones et al., 2005). There is a lack of information about some impact categories, for example, eutrophication and summer smog. The economics of the three micro-generators are highly dependent on their location, and

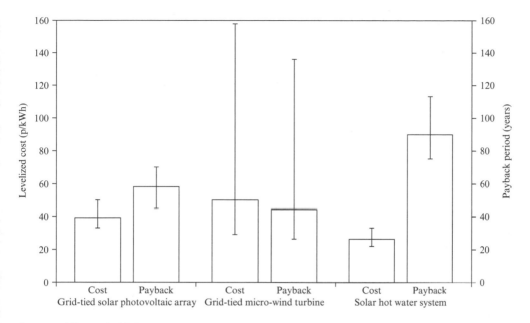

Source: Allen et al. (2008).

Figure 2.7   Levelized cost and CBA payback periods for three specific micro-generators

on the weather conditions. However, they are all currently unattractive in the liberalized UK energy market (Allen et al., 2008). Their financial payback periods are well beyond the lifetimes of the devices (see again Figure 2.7). The SHW system displaces natural gas used by the boiler, and results in an even longer payback period. However, the uncertainty ranges underlying these economic parameters are mostly a function of energy output values and therefore subject to their respective energy resource variations (that is, solar or wind). Nevertheless, the environmental externalities quantified formed a small, yet considerable, part of the total cost–benefit analysis and are therefore open to additional uncertainties. It is not unlikely that a wide range of external costs, as much as five orders of magnitude, could be estimated by adopting different valuation procedures (as discussed by Hammond and Winnett, 2006).

2.4.3   Example 3: The Lebanese Electricity System as a Simplified Power Network

In a recent study, El-Fadel et al. (2010) examined the Lebanese electricity system (LES) as an example of a simple power network in order to demonstrate the use of quantitative sustainability metrics. This investigation aimed at establishing a baseline or benchmark of several selected sustainability indicators for the LES, against which any future action could be monitored. Once again an integrated approach was adopted to assess the life-cycle technical, environmental, energy and economic attributes of the system. The risk of experiencing supply deficits can be measured by the 'loss of load probability' (LOLP), which is the probability of load not being met. Reducing this probability to near zero is prohibitively expensive (and theoretically impossible), and would require excessive capacity and back-up network routes. The generally accepted capacity margin, or amount by which capacity should exceed net peak demand, differs between nations or regions. However, the LOLP is usually set so that interruptions of supply do not exceed nine or ten winters out of 100 (9–10 per cent), which requires a capacity margin of about 24 per cent for the LES (El-Fadel et al., 2010). In reality, it is close to zero. The UK power sector, in contrast, operates with a capacity margin in the range of 16.5–22 per cent, and this could rise to 25 per cent if mothballed plants are bought back online (Hammond and Waldron, 2008).

The ability of an electricity system to respond to disturbances or perturbations is increased and more secure as the network becomes more 'diverse'. According to Grubb et al. (2006) this diversity is a combination of 'variety' (or the number of generator categories), 'balance' (or a pattern in the spread of that quantity across the relevant categories) and 'disparity' (or the nature and degree to which the categories themselves are different from each other). One of the main indices for measuring diversity is the Shannon–Weiner (S-W) index, which includes variety and balance, although not disparity. An S-W value of below 1 indicates a system that is highly concentrated and dependent upon one or at most two sources (El-Fadel et al., 2010). That would threaten security of supply, whereas a S-W value above 2 indicates a system with numerous sources, which could be considered relatively secure (Grubb et al., 2006). The current generation mix in Lebanon suggests that a real S-W index is approximately 0.83–1.13 (El-Fadel et al., 2010), depending on whether imports are included. Alternatively, the index value would be approximately 1–1.24, based on nominal capacity, depending again on the inclusion of electricity imports.

Environmental performance of the LES was evaluated through an LCA (ISO, 2006a, 2006b) and the 'CML 2001' LCA impact assessment method (Sonneman et al., 2004). The latter is a widely applied and well-respected method, and is what is known as a 'midpoint method'. The results were displayed in physical units (that is, kg, MJ), rather than an 'endpoint' method, which may employ units, such as one based on 'disability adjusted life years' (DALYs). Predefined impact categories of the CML 2001, such as abiotic depletion and human toxicity, were presented in a similar manner to the normalized LCA results for micro-generators (Figure 2.6). But Lebanon was found to exhibit higher environmental impacts in eight of the nine categories (El-Fadel et al., 2010) when compared to the European average. Significant progress is needed in Lebanon in order to lower its impacts in terms of abiotic depletion, acidification, global warming potential, and marine aquatic ecotoxicity to the European level. In contrast, Lebanon has a slightly lower impact in terms of freshwater aquatic ecotoxicity than the average for European electricity generation, although the difference was comparatively small. All other indicators displayed a comparatively large, adverse difference compared to the European average (El-Fadel et al., 2010), signifying the comparatively poor environmental performance of the LES.

Energy performance was measured via the 'energy gain ratio' (EGR), rather than the EPP illustrated in relation to the analysis of micro-generators above (see Figure 2.5). A power generator should produce more energy over its entire lifetime than is required to build, maintain and fuel this energy source. Thus, its EGR, the 'full fuel cycle' energy output divided by the corresponding energy input, should be greater than 1–1.5 (Gagnon, 2008). An EGR too close to 1 represents a poor lifetime efficiency of

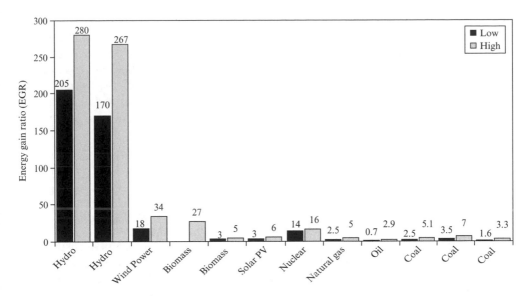

Source: El-Fadel et al. (2010); adapted from Gagnon (2008).

Figure 2.8 Typical expected energy gain ratios (EGR) for electricity generators based on life-cycle assessment

fuel conversion – this is particularly the case for those technologies consuming depleting fossil fuel resources. EGRs differ both within the same and between different technology types, depending on location or delivery distances, transportation mode of fuels and their actual accessibility and quality, as well as other parameters such as the use of end-of-pipe scrubbing technology. The EGRs for several technologies suitable for use in Lebanon, including oil and hydro, are displayed in Figure 2.8 (El-Fadel et al., 2010). This shows that there are substantial energy benefits attributable to renewable energy sources, particularly hydropower and wind farms. However, the EGR values can only be taken as indicative and comparative as each generating technology will have its own particular characteristics. Moreover, the EGR values above (Figure 2.8) were calculated without accounting for the inherent operational energy associated with the fuel consumed. Such metrics are sometimes referred to as 'external gain ratios' (Gagnon, 2005). Allen et al. (2008), for example, provide EGRs in the UK context that include the inherent energy content of fuel, which result in substantially lower EGRs. For coal, gas, oil and nuclear power these were equal to 0.29, 0.43, 0.22 and 0.28, respectively (Allen et al., 2008).

There are various indicators that can be used to measure the economic dimension of sustainability. El-Fadel et al. (2010) employed a comparative cost–benefit appraisal to similar (yet not exact) cases as expressed through the LCA, taking into account the social cost of carbon (SCC) only due to the fact that carbon emission damages are not site-specific, and to the extensive literature present on the SCC. The economic performance was measured in terms of the levelized costs, net present values (NPVs), and cost–benefit ratios of four Lebanese power sector carbon abatement cases, contrasted with those of the existing centralized electricity system. Levelized costs were presented in a similar manner to those for the micro-generators shown in Figure 2.7. It was found (El-Fadel et al., 2010) that the Lebanese electricity system exhibited large economic inefficiencies. The costs and benefits of optimizing the performance of the centralized system point to substantial net benefits from improving the T&D networks, maintaining conventional existing plants to achieve their design standards, and shifting towards the use of natural gas. Moreover, the expected levelized cost of various energy sources in Lebanon indicated that renewable energy sources are highly competitive alternatives to consider that could support to the attainment of reliability objectives.

2.5   CONCLUDING REMARKS

Sustainable development, balancing economic and social development with environmental protection, has become a modern paradigm in our technological age. There are various ways in which the energy sector interacts with the requirements of sustainable development. A strict interpretation would mean a rapid changeover to renewable energy, and the conservation of non-renewable sources (fossil fuels and uranium). This in turn could lead to a significant reduction in pollutant emissions, the unwanted side-effects of the energy sector that have a damaging impact at a local, regional and global scale. Only in this way could the biosphere be protected for future generations. The present contribution has examined the principles and practice of sustainability in the context of the energy sector, as well as the sustainability criteria that stem from them.

Sustainability is commonly disaggregated into three elements, or 'pillars', covering the economic, environmental and social aspects. They can be viewed from both an intergenerational and a global interregional ethical perspective. A useful distinction can be drawn between sustainable development and sustainability: the journey and the destination. Several attempts have been made to devise what some have termed 'sustainability science'. These embrace, for example, the notion of resource use productivity (Factor X) as a component of the 'sustainability equation'. Von Weizsacker et al. (2009) have recently argued that the global economy could be transformed through 80 per cent improvements in resource productivity: Factor 5. Some support for this view can be gleaned from the use of advanced methods of thermodynamic (energy and exergy) analysis (see, for example, Hammond, 2004a, 2004c; Hammond and Stapleton, 2001).

A number of principles are also said to underpin sustainable development, and the so-called 'precautionary principle' is arguably the most significant of these. Its implications have been highlighted, along with some controversial issues surrounding its adoption within the context of international treaties, as well as for process design and in policy-making. It is clear that the late lessons derived from the early use of the precautionary principle over the last century, identified by the EEA (2001), provide very useful guidance in this regard.

Methods of appraisal vary between the three pillars of sustainability, particularly in regard to the extent to which they can be evaluated quantitatively. This can be achieved in the environmental and economic domains, but the social pillar is typically evaluated in qualitative terms. Even the quantitative methods can exhibit large variations in uncertainty. They have been discussed with a focus on the way that the energy sector can be evaluated in terms of its energy, environmental and economic performance. Examples are drawn from the fields of liquid biofuels for transport, decentralized energy resources (DERs), and a simple power network in order to illustrate the use of quantitative sustainability metrics.

Several quantitative 'whole-systems' appraisal techniques applicable to the energy sector have been described. These include thermodynamic (energy and exergy) analysis, environmental life-cycle assessment (LCA), and environmental cost–benefit analysis (CBA). They reflect principally the ecological and economic domains of the sustainability assessment Venn diagram (Figure 2.2). Such approaches are interrelated (Hammond and Winnett, 2006) in the sense that life-cycle energy analysis was one of the precursors for LCA, and is typically performed in parallel with environmental appraisal in most modern LCA software packages. Although energy analysis enables the determination of the energy balance across an engineering system, exergy analysis is required in order to ascertain the ways in which the energy flows are qualitatively degraded (Hammond, 2004a). LCA is a very useful tool for determining global and regional environmental impacts of a product or system 'from the cradle to the grave' (ISO, 2006a, 2006b), but is currently unable to incorporate local impacts. However, it is possible that some means to achieve this will be forthcoming in the not too distant future. In any event, LCA avoids the examination of products on a 'snapshot' basis, whereby only one part of the life cycle is examined. When employed with other environmental management tools, such as environmental risk assessment, it can form a comprehensive impact assessment package. Life-cycle assessment is also sometimes employed, as illustrated here for the case of DERs, to estimate impact inventories that can then be coupled with CBA to yield their environmental costs. Sustainability assessment techniques certainly need to be used in consultation with both expert and lay opinion. Only in this way can the sort of criticisms levelled by Stirling (1998) and others at the technocratic nature of appraisal techniques be properly addressed. Wider community participation is part of a deliberative process that is a necessary prerequisite for stakeholder buy-in towards sustainability.

It is clearly important for developed or industrialized countries to play their full part in maintaining sustainability. They have accounted for the bulk of cumulative or historic carbon emissions into the atmosphere worldwide since the industrial revolution (see, for example, Cranston and Hammond, 2010). But sustainable development must also be viewed in a global context. Clearly the industrial nations of the wealthy North will need to take the lead. It will require difficult decisions for them in terms of market intervention to stimulate the development of sustainable technologies, and possibly to induce changes in lifestyles. However, the task facing the nearly 80 per cent of the world population that live in developing countries, the so-called majority South, is daunting. They have, in most cases, rapidly growing populations that will drive up energy consumption and environmental pollution. This will feed back to the whole planet, and thereby alter the climate in the wealthier nations (Hammond, 2000). Consequently they need assistance from industrial countries to promote economic growth in less developed countries, which will in time induce a 'demographic transition' (WCED, 1987), as well as improving the efficiency of their energy systems. These are matters of interregional and intergenerational ethics, rather than purely scientific debate. A more equitable sharing of world income and resources is likely to be a prerequisite for sustainable development in the long term. Environmental sustainability would certainly be aided by the transfer of best-practice, or 'leapfrog', energy technologies from the richer to the poorer regions (Goldemberg, 1996). This will ultimately be in the interests of all the citizens of 'Spaceship Earth' (Hammond, 2000, 2006).

ACKNOWLEDGEMENTS

This work is part of a programme of research at the University of Bath on the technology assessment of low carbon energy systems and transition pathways that is supported by a series of UK research grants and contracts awarded by various bodies. Professor Hammond is jointly leading a large consortium of university partners (with Professor Peter Pearson, now Director of the Low Carbon Research Institute in Wales) funded via the strategic partnership between E.ON UK (the electricity generator) and the EPSRC to study the role of electricity within the context of 'Transition Pathways to a Low Carbon Economy' (under Grant EP/F022832/1). Craig Jones's contribution to the present work has been partially funded via this grant. In addition, Hammond is also a Co-Investigator of the EPSRC SUPERGEN 'Highly Distributed Energy Futures' (HiDEF) Consortium (under Grant EP/G031681/1). This consortium has been coordinated by Professor Graeme Burt and Professor David Infield, both now with the Institute for Energy and Environment at the University of Strathclyde. Finally, Hammond is a Co-Investigator of the Biotechnology and Biological Sciences Research Council's (BBSRC) Sustainable Bioenergy Centre (BSBEC), under the 'Lignocellulosic Conversion to Ethanol' (LACE) project (Grant Ref: BB/G01616X/1). Professor Katherine Smart and Professor Greg Tucker of the University of Nottingham lead this consortium, whilst BSBEC is directed by Duncan Eggar (the BBSRC Bioenergy Champion). Both authors are grateful to external colleagues for their role in the coordination of large consortia of university and other partners. They also wish to thank their sustainable energy research colleagues at Bath for stimulating discussions on some of the issues addressed here, particularly Paul Adams, Steve Allen, Gemma Cranston, Hassan Harajli, Marcelle McManus, Will Mezzullo and Adrian Winnett. However, the views expressed in this chapter are those of the authors alone, and do not necessarily reflect the views of the collaborators or the policies of the funding bodies. The authors are grateful to Gill Green (University of Bath) for the care with which she prepared some of the figures.

NOTE

1.   The authors' names are listed alphabetically.

REFERENCES

Allen, S.R., G.P. Hammond, H.A. Harajli, C.I. Jones, M.C. McManus and A.B. Winnett (2008), 'Integrated appraisal of micro-generators: methods and applications', *Proceedings – Institution of Civil Engineers, Energy*, **161** (2), 73–86.

Ansoff, H.I. (1970), *Strategic Management*, Basingstoke: Palgrave Macmillan.

Azapagic, A., S. Perdan and R. Clift (eds) (2004), *Sustainable Development in Practice: Case Studies for Engineers and Scientists*, Chichester: John Wiley & Sons.

Boyle, G., B. Everett and R. Ramage (eds) (2003), *Energy Systems and Sustainability: Power for a Sustainable Future*, Oxford: Oxford University Press.

Broman, G., J. Holmberg and K.-H. Robert (2000), 'Simplicity without reduction: thinking upstream towards the sustainable society', *Interfaces*, **30**, 13–25.

Buchanan, R.A. (1994), *The Power of the Machine*, Harmondsworth: Penguin Books.

Chambers, N., C. Simmons and M. Wackernagel (2000), *Sharing Natures Interest: Ecological Footprints as an Indicator of Sustainability*, London: Earthscan.

Clift, R. (1995), 'The challenge for manufacturing', in J. McQuaid (ed.), *Engineering for Sustainable Development*, London: Royal Academy of Engineering, pp. 82–7.

Clift, R. (2007), 'Climate change and energy policy: the importance of sustainability arguments', *Energy*, **32** (4), 262–8.

Committee on Climate Change (CCC) (2008), *Building a Low-Carbon Economy: The UK's Contrbution to Tackling Climate Change*, London: TSO.

Concawe, EUCARE and JRC (2006), 'Well-to-wheels analysis of future automotive fuels and powertrains in the European context', Version 2b, Brussels, Belgium: European Commission – Joint Research Centre (JRC).

Cranston, G.R. and G.P. Hammond (2010), 'North and south: regional footprints on the transition pathway towards a low carbon, global economy', *Applied Energy*, **87** (9), 2945–51.

Davis, G. (2001), *Evolving Sources or Revolutionary Technology: Exploring Alternative Energy Paths to 2050*, London: Shell International.

Department for the Environment, Food and Rural Affairs (Defra) (2005), *One Future – Different Paths*, London: TSO.

Dones, R., T. Heck, C. Bauer, S. Hirschberg, P. Bickel, P. Preiss, L.I. Panis and I. de Vlieger (2005), 'ExternE-Pol – Externalities of energy: extension of accounting framework and policy applications', Final Report on Work Package 6, Villigen, Switzerland: Paul Scherer Institut.

Eaton, R.L., G.P. Hammond and J. Laurie (2007), 'Footprints on the landscape: an environmental appraisal of urban and rural living in the developed world', *Landscape and Urban Planning*, **83** (1), 13–28.

El-Fadel, R.H., G.P. Hammond, H.A. Harajli, C.I. Jones, V.K. Kabakian and A.B. Winnett (2010), 'The Lebanese electricity system in the context of sustainable development', *Energy Policy*, **38** (2), 751–61.

Elghali, L., R. Clift, P. Sinclair, C. Panoutsou and A. Bauen (2007), 'Developing a sustainability framework for the assessment of bioenergy systems', *Energy Policy*, **35** (12), 6075–83.

European Environment Agency (EEA) (2001), *Late Lessons from Early Warnings: The Precautionary Principle 1896–2000*, Environmental Issue Report No. 22, Luxembourg: Office of Official Publications of the European Commission.

European Environment Agency (EEA) (2010), *Annual European Union Greenhouse Gas Inventory 1990-2008 and Inventory Report 2010*, Technical Report No. 6, Luxembourg: Office of Official Publications of the European Commission.

Eurotreaties (1996), *The Maastricht Treaty in Perspective: Consolidated Treaty on European Union*, Stroud, UK: British Management Data Foundation.

Eyre, N. (2002), 'Energy Efficiency', Imperial College/Warwick Business School Seminar: 'The Energy Review: Drivers Behind the Report', unpublished, IMechE, London, 11 March.

Fore*sight* (2000), *Stepping Stones to Sustainability*, London: Department of Trade and Industry.

Gagnon, L. (2005), 'Electricity generation options: energy payback ratio – Factsheet', Quebec, Canada: Hydro-Québec.

Gagnon, L. (2008), 'Civilisation and energy payback', *Energy Policy*, **36** (9), 3317–22.

Goldemberg, J. (1996), *Energy, Environment and Development*, London: Earthscan.

Graedel, T.E. and R.J. Klee (2002), 'Getting serious about sustainability', *Environmental Science and Technology*, **36** (4), 523–9.

Grubb, M., L. Butler and P. Twomey (2006), 'Diversity and security in UK electricity generation: the influence of low-carbon objectives', *Energy Policy*, **34** (18), 4050–62.

Hammond, G.P. (2000), 'Energy, environment and sustainable development: a UK perspective', *Trans IChemE Part B: Process Safety and Environmental Protection*, **78** (4), 304–23.

Hammond, G.P. (2004a), 'Engineering sustainability: thermodynamics, energy systems, and the environment', *International Journal of Energy Research*, **28** (7), 613–39.

Hammond, G.P. (2004b), 'Science, sustainability and the establishment in a technological age', *Interdisciplinary Science Reviews*, **29** (2), 193–208.

Hammond, G.P. (2004c), 'Towards sustainability: energy efficiency, thermodynamic analysis, and the "two cultures"', *Energy Policy*, **32** (16), 1789–98.

Hammond, G.P. (2006), '"People, planet and prosperity": the determinants of humanity's environmental footprint', *Natural Resources Forum*, **30**, 27–36.

Hammond, G.P. and C.I. Jones (2008), 'Embodied energy and carbon in construction materials', *Proceedings – Institution of Civil Engineers: Energy*, **161** (2), 87–98.

Hammond, G.P., S. Kallu and M.C. McManus (2008), 'The development of biofuels for the UK automotive market', *Applied Energy*, **85** (6), 506–15.

Hammond, G.P. and A.J. Stapleton (2001), 'Exergy analysis of the United Kingdom energy system', *Proceedings of the Institution of Mechanical Engineers Part A: Journal of Power and Energy*, **215** (2), 141–62.

Hammond, G.P. and R. Waldron (2008), 'Risk assessment of UK electricity supply in a rapidly evolving energy sector', *Proceedings of the Institution of Mechanical Engineers Part A: Journal of Power and Energy*, **222** (7), 623–42.

Hammond, G.P. and A.B. Winnett (2006), 'Interdisciplinary perspectives on environmental appraisal and valuation techniques', *Proceedings – Institution of Civil Engineers: Waste and Resource Management*, **159** (3), 117–30.

Hoffert, M.I., K. Caldeira, A.K. Jain, E.F. Haites, L.D.D. Harvey, S.D. Potter, M.E. Schlesinger, S.H. Schneider, R.G. Watts, T.M.L. Wigley and D.J. Wuebbles (1998), 'Energy implications of future stabilization of atmospheric CO_2 content', *Nature*, **395**, 881–4.

Holdren, J.P. and P.R. Ehrlich (1974), 'Human population and the global environment', *American Scientist*, **62**, 282–92.

Intergovernmental Panel on Climate Change (IPCC) (2007), *Climate Change 2007: The Physical Science Basis*, Cambridge: Cambridge University Press.

International Standards Organization (ISO) (2006a), 'Environmental management – life cycle assessment – principles and framework', EN ISO 14040, 2nd edition, Geneva: ISO.

International Standards Organization (ISO) (2006b), 'Environmental management – life cycle assessment – requirements and guidelines', EN ISO 14044, Geneva, Switzerland: ISO.

IUCN, UNEP and WWF (1991), *Caring for the Earth: A Strategy for Sustainable Living*, Gland, Switzerland: WWF International.

Jacobs, M. (1996), *The Politics of the Real World*, London: Earthscan.

Jaffe, A.B., and R.N. Stavins (1994), 'The energy efficiency gap: what does it mean?', *Energy Policy*, **22** (10), 804–11.

Loh, J. and S. Goldfinger (eds) (2006), *Living Planet Report 2006*, Gland, Switzerland: WWF.

Meadows, D.H., D.L. Meadows and J. Randers (1992), *Beyond the Limits*, London: Earthscan.

Nakicenovic, N., A. Grubler and A. McDonald (1998), *Global Energy Perspectives*, Cambridge: Cambridge University Press.

Parkin, S. (2000), 'Sustainable development: the concept and the practical challenge', *Proceedings of the Institution of Civil Engineers: Civil Engineering*, **138**, 3–8.

Performance and Innovation Unit (PIU) (2002), *The Energy Review*, London: Cabinet Office.

Porritt, J. (2000), *Playing Safe: Science and the Environment*, London: Thames & Hudson.

Renewable Fuels Agency (RFA) (2010), *2008/09 Annual Report to Parliament on the Renewable Transport Fuel Obligation*, London: TSO.

Royal Commission on Environmental Pollution (RCEP) (2000), *Twenty-second Report: Energy – The Changing Climate*, London: TSO.

Royal Society (2008), 'Sustainable biofuels: prospects and challenges', Policy Document 01/08, No. 22, London: Royal Society.

Sonneman, G., F. Castells and M. Schuhmacher (2004), *Integrated Life-Cycle and Risk Assessment for Industrial Processes*, Boca Raton, FL: CRC Press.

Stirling, A. (1998), 'Valuing the environmental impacts of electricity production: a critical review of some "first generation" studies', *Energy Sources*, **20**, 267–300.

Tester, J.W., E.M. Drake, M.J. Driscoll, M.W. Golay and W.A. Peters (2005), *Sustainable Energy: Choosing Among Options*, Cambridge, MA: MIT Press.

Upham, P. (2000), 'Scientific consensus on sustainability: the case of The Natural Step', *Sustainable Development*, **8**, 180–90.

von Weizsacker, E., K. Hargroves, M.H. Smith, P. Stasinopoulos and C. Desha (2009), *Factor Five: Transforming the Global Economy Through 80 per cent Improvements in Resource Productivity*, London: Earthscan.

von Weizsacker, E., A.B. Lovins and L.H. Lovins (1997), *Factor Four: Doubling Wealth, Halving Resource Use*, London: Earthscan.

Wolpert, L. (1993), *Unnatural Nature of Science*, London: Faber & Faber.

World Commission on Environment and Development (WCED) (1987), *Our Common Future*, Oxford: Oxford University Press.

3 Economic growth, energy consumption and climate policy

*M. Carmen Gallastegui, Alberto Ansuategi,
Marta Escapa and Sabah Abdullah*

3.1 INTRODUCTION

Carbon-based energy is one of the major natural resources that have driven economic growth. It was coal that made the Industrial Revolution possible, and the ever-increasing use of carbon-based fuels since that time has rapidly improved the quality of life of humankind.[1] However, we have recently realized that, in pursuing these ends, we have released enough carbon dioxide into the air to affect the climate and potentially the well-being of people and nations all over the planet, for centuries to come. According to the Fourth Assessment Report of the Intergovernmental Panel on Climate Change (Parry et al., 2007), anthropogenic emissions of greenhouse gases (GHGs) will increase global average temperature, which will lead to a wide range of impacts including a rise in sea levels, increased intensity of precipitation, more frequent and more intense storms, a loss of biodiversity, continued loss of Arctic ice and glaciers, increasing salinity of freshwater aquifers, coastal erosion and flooding, and an increase in heat- and precipitation-related diseases such as malaria. Consequently, climate change is currently one of the most important issues on the international political agenda.

Global warming poses a unique mix of problems that arise from the 'public bad' nature of the problem, from the major scientific and economic uncertainties involved and from the fact that the costs of controlling GHGs must be borne in the present while the benefits will accrue decades, perhaps even centuries, down the line. There is a vast body of articles that have focused on the economics of climate change; Kolstad and Toman (2005) and Stern (2007) offer good reviews of the relevant literature. Some of the questions posed by economic approaches to climate change are the following. By how much should industries and countries reduce their GHG emissions? How fast should emissions be reduced? What instruments should be used to obtain those cuts in GHG emissions? How should the effort of emission reduction be distributed among rich and poor people and countries? How could self-enforcing international agreements on the control of GHG emissions be achieved? Less attention has been paid to the relationship between economic growth and GHG emissions, even though this is an underlying issue behind almost any climate policy decision.

The debate about the relationship between economic growth and GHG emissions revolves around three main questions. First, can economic growth be delinked from GHG emissions? Second, can GHG emissions be cut without hurting economic growth? And third, is future economic growth at risk due to climate change? The present chapter seeks to shed some light on these three questions.

After this introductory section, section 3.2 critically reviews the economic literature analysing the relationship between per capita gross domestic product (GDP) and per capita CO_2 emissions, in an attempt to answer the question of whether economic growth can be delinked from GHG emissions. Section 3.3 focuses on the impact of climate policy on economic growth and section 3.4 discusses the impact of climate change (lack of climate policy) on economic growth. Section 3.5 briefly concludes and summarizes the chapter.

3.2   THE CO_2–GDP RELATIONSHIP

The relationship between CO_2 emissions and economic growth has been the focus of a number of studies. Interest in the link between income per capita and environmental quality arose from the pioneering work of Grossman and Krueger (1991) on the North American Free Trade Agreement (NAFTA), which immediately led to a very extensive body of literature on what Panayotou (1993) termed the environmental Kuznets curve[2] (EKC). The EKC hypothesizes an inverse U-shaped relationship between a country's per capita income and some indicators of environmental degradation. The reason why policy-makers have paid so much attention to this hypothesis is because it argues that growth is not only the cause of, but can also provide the cure for environmental degradation.

Stern (2004) and Aslanidis (2009) provide comprehensive reviews of the empirical literature on the EKC and the carbon Kuznets curve (CKC), respectively. A simple plot of the relationship between CO_2 emissions and income for individual countries appears to support the CKC theory. Figure 3.1 plots the energy-related CO_2 emissions relationship to GDP (on a per capita basis) for the rapidly growing developing economies (BRICS[3]), the European Union (EU-25) and the United States of America (US). If we assume that the BRICS will follow the development path of more advanced economies such as the EU-25 or the US, the plot of data seems to suggest some sort of CKC.[4] However, most econometrically sound analyses of the relationships observed do not find conclusive support for the existence of an inverse U-shaped pattern between economic growth and carbon emissions. In fact, some authors claim that the CKC may be a 'cloudy picture emitted by bad econometrics' (Wagner, 2008).

Most empirical work in this field has little connection with explicit theory. However, if policy-makers are to find ways to lead the economy towards low carbon growth paths, both economic theory and the political economy have more important roles to play in explaining what empirical regularities suggest.

The simple reduced-form estimation of the CKC hypothesis reduces the set of possible explanatory variables to essentially just one: income. However, there is no need to assume that a single variable can capture all the underlying forces that can potentially determine the carbon intensity of growth.[5] Some of these 'forces' are closely linked to growth, and others are only indirectly linked to it.

The reasons for the inverted U-shaped relationship between economic growth and environmental degradation are thought to include income-driven changes in: (1) the preference for environmental quality; (2) the composition of production and/or consumption; (3) institutions that are needed to internalize externalities; and/or (4) increasing returns to scale associated with pollution abatement. Thus, some attempts have been

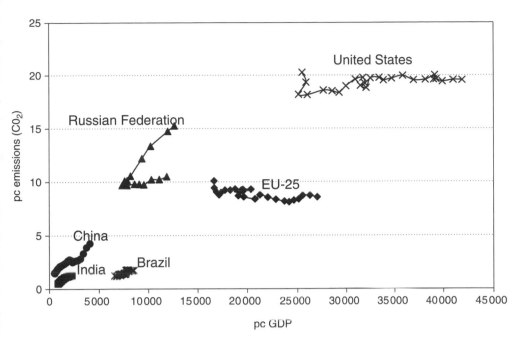

Note: * In the case of the Russian Federation, the data refer to the 1990–2005 period.

Source: World Bank (2009), GDP per capita (PPP, constant 2005 $); per capita CO_2 emissions (metric tons).

Figure 3.1 *CO_2 emission/GDP relationship for the US, EU-25 and BRICs from 1980 to 2005**

made in the relevant literature to formalize models that illustrate the role of these factors in generating inverse-U-shaped pollution–income patterns. Brock and Taylor (2005) provide an extensive review of the theoretical literature on EKCs. Here we take a very simple, comparative static approach to illustrate the underlying forces and gain insight into the general nature of EKCs.

Let us start by considering carbon emissions reduction as an ordinary market commodity whose allocation is determined in the interaction between demand and supply for it (see Figure 3.2). The demand for emission reductions (DD) is a downward sloping curve, reflecting the fact that in a highly polluted scenario (low levels of emissions reduction) society's willingness to pay for emissions reduction is higher than in a moderately polluted scenario (high levels of emissions reduction). The supply of emissions reduction (SS) is a positively sloped curve, reflecting the fact that marginal costs of emission reductions rise with the level of abatement. The intersection of supply and demand yields the level of emissions, *ceteris paribus*.

To study the CO_2 emissions–income relationship for a single country we have to relax the *ceteris paribus* condition and allow for a change in per capita GDP. Then we estimate what happens to the SS and DD curves and therefore to the equilibrium level of emission reductions. As income rises we can expect both the supply and the demand functions to

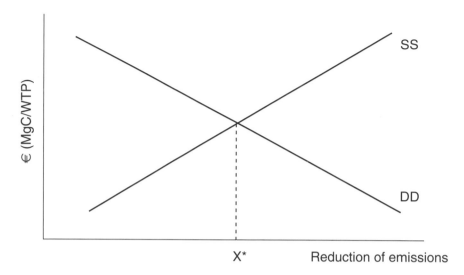

Figure 3.2   Supply (SS) and demand (DD) for the reduction of emissions

shift. Factors affecting the position of the DD curve for reduction of emissions include the preferences of the 'representative consumer', the political system (ability to reflect popular sentiment in the government's regulatory policy), institutional development to implement an effective environmental policy, and free-riding incentives reflecting the global nature of pollution. Most of these factors are highly correlated with income: as a country gets richer its citizens demand higher environmental quality, political institutions are likely to be more democratic and reflect citizens' demands better, and sufficient institutional capacity may have been built up to implement effective environmental regulations.

The position of the SS curve changes with the structure of the economy and with technology. In turn, both technology and the structure of the economy change with income. Thus, historically, most developed countries have moved from agricultural activity to heavy industry and then to light industry and services in the course of their development processes. Technological level also seems to rise with income (and over time).

These theoretical efforts to unravel the different mechanisms by which economic growth affects environmental quality have led to two types of 'second-generation' empirical approaches. One branch of the literature has 'enriched' the most basic econometric regressions, adding other variables such as measures of corruption (Cole, 2007), democratic freedoms (Barrett and Graddy, 2000), international trade openness (Atici, 2009) and even income inequality (Torras and Boyce, 1998; Heerink et al., 2001). Other researchers meanwhile have sought to use decomposition analysis to measure the effect of factors such as the fuel mix, technological change and structural change in the changes, over time in the pollution intensity of economic activity (Bruvol and Medin, 2003).

Decomposition analysis is a particularly useful approach for understanding the role of energy markets in generating CKCs. It enables changes in per capita emissions to be factored into changes in the mix of fossil fuels, the share of fossil fuels in total energy consumption, the energy intensity of production and GDP per capita. Moreover, decomposition analysis can also be employed to decompose energy intensity into changes in technology and changes in the composition of economic activity (Ang, 1994).

We can use a Kaya-type identity (Kaya, 1990) to illustrate how per capita CO_2 emissions can be broken down into the above-mentioned contributing factors. First we express per capita emissions as a function of per capita income and emission intensity:

$$\frac{CO_2}{POP} \equiv \frac{GDP}{POP} \times \frac{CO_2}{GDP} \qquad (3.1)$$

where CO_2, GDP and POP stand for emissions, income and population, respectively. Next, emissions intensity can be expressed as a function of fuel mix and energy intensity:

$$\frac{CO_2}{GDP} \equiv \frac{CO_2}{E} \times \frac{E}{GDP} \qquad (3.2)$$

where E stands for total energy use. Then, provided that access is available to data that permits disaggregation of economy-wide emissions and GDP by sectors, the change in energy intensity can be separated into two factors: structural change (changes in sector share of total GDP) and technological change (changes in sectoral energy intensity):

$$\frac{E}{GDP} \equiv \sum_i \frac{GDP_i}{GDP} \frac{E_i}{GDP_i} \qquad (3.3)$$

where E_i and GDP_i stand for energy consumption in sector i and sector i's contribution to GDP, respectively. By combining equations (3.1), (3.2) and (3.3), per capita emissions can be formulated as a function of per capita income, fuel mix, structure of the economy and technology.

Thus, decomposition analysis has recently been used to compare and understand the forces that underlie changes in carbon emissions from energy use in a wide set of countries all over the world (Kojima and Bacon, 2009). This study concludes that, at a global level, the increase in carbon emissions was greater in the 2001–06 period than in the 1996–2000 period. It is also found that absolute decoupling tended to occur more in upper-middle and high-income countries, whereas countries in the early stages of development tended to show less offsetting, and virtually none showed absolute decoupling.[6] But there is still a message of hope in the analysis: the good performance of several countries across the entire income spectrum indicates the importance of policy and government engagement with the goal of increasing energy efficiency, so that low-income countries can choose development paths that could lead them towards stronger decoupling of carbon emissions and economic growth. In the following section we deal with opportunity costs in terms of growth potential for choosing low carbon paths.

3.3   GHG CONTROL AND ECONOMIC GROWTH

In a recent article by Krugman (2010) the author answers the question that we have posed in this section, as follows:

> There is no credible research suggesting that taking strong action on climate change is beyond the economy's capacity. Even if we don't trust the models –and you shouldn't – history and logic both suggest that the models are overestimating, not underestimating, the costs of climate action. We can afford to do something about climate change.

Obviously, from what can be learnt from the previous section, any successful programme of action on climate change that seeks to stabilize atmospheric GHGs below 'safe' levels and to maintain economic growth must entail dramatic changes in the carbon intensity of production (see equation 3.1). Thus, Beinhocker et al. (2008) estimate that the twofold objective of curbing emissions and maintaining growth would require a tenfold increase in carbon productivity[7] by 2030, which means that this 'carbon revolution' would have to proceed three times faster than the rise in labour productivity that accompanied the Industrial Revolution.

In turn, changes in carbon productivity can be achieved through changes in the fuel mix and/or the energy intensity of production (see equation 3.2). Therefore, decarbonization of energy sources and exploitation of energy efficiency opportunities are crucial changes to be achieved in the near future.

Current estimations at the worldwide level of actions to prevent climate change are presented in Figure 3.3. The mitigation costs of achieving 450 ppm CO_2e (550 ppm CO_2e) are estimated at 0.3–0.9 per cent (0.2–0.7 per cent) of GDP in 2030. Figure 3.4 shows the full range of abatement actions that are either available today or are highly likely to be available by 2030 (ordered from left to right from the lowest-cost to the highest-cost).[8] Figure 3.4 predicts that under its cost curve assumptions the global economy can achieve an abatement of 27 $GtCO_2e$ by 2030 (abatement required to keep concentrations between 450 and 550 ppm) for a marginal cost of less than €50/tCO_2e. Obviously, there are many reasons to regard these estimates with caution. First, even though a significant portion of the abatement potential would be at a negative cost to society (the first 7 $GtCO_2e$ in Figure 3.4), major upfront investment is required before 2020 and this is a big challenge for developing countries, which may find it hard to direct capital to low carbon investments. Second, delaying global actions would lock the global economy into carbon-intensive technology and would significantly increase the costs of abatement. Third, if a significant number of countries decided not to participate in the mitigation strategy, the portfolio of mitigation strategies would shrink and the cost of reaching a mitigation goal would rise.

As mentioned above, any action that seeks to mitigate emissions of GHGs implies a change in the energy paradigm. If no measures affecting the energy needed per unit of output are introduced, it will be very difficult to find solutions for climate change. As equation (3.2) shows, emission intensity can be expressed as a function of fuel mix and energy intensity. This is one of the reasons why Directive 209/28/EC addresses the achievement of three aims:[9] (1) to reduce GHG emissions unilaterally by 20 per cent (on 1990 levels); (2) to reduce energy consumption by 20 per cent by promoting energy efficiency; and (3) to raise the percentage of production that comes from renewable sources to 20 per cent.

But some social scientists argue that green policies may involve some undesired consequences for the environment. It may thus be of some interest to analyse what is known as the Green Paradox, first advanced by Sinn (2008).[10] His article analyses the

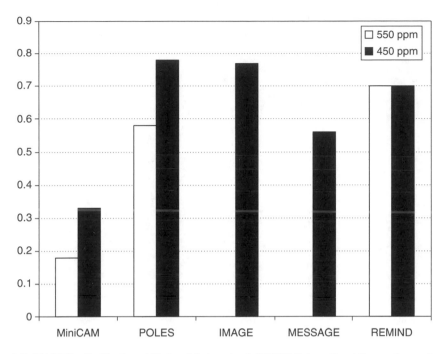

Note: MiniCAM (Pacific Northwest National Laboratory), POLES (International Energy Agency), IMAGE (Netherlands Environmental Assessment Agency), MESSAGE (IIASA) and REMIND (Postdam Institute for Climate Impact Research) are the global energy-climate models from Europe and the United States.

Source: Adapted from World Bank (2010).

Figure 3.3   Some estimates of macroeconomic costs (% GDP) of achieving 550 ppm and 450 ppm stabilization targets in 2030

limits imposed by the Kyoto Protocol on CO_2 emissions, taking into account not only the demand side (the demand for fossil fuels) but also the supply side. He maintains that suppliers have an important role to play because if demand reductions by some countries are not followed by a reaction by suppliers, then the price of carbon will decrease. If the world carbon price is cheaper, countries that do not plan to reduce their emissions will consume more, or if supplying countries believe that future prices may drop they may manage resources differently: they may well extract more today and deplete their stocks more quickly, so that the final result may be a worsening of global warming. This pioneering work was followed by other papers, such as those by Hoel (2008) and van der Ploeg and Withagen (2010). An important insight of these articles is that climate costs may increase as a consequence of improvements in renewable energy technology.

Another interesting question that arises when dealing with whether we can afford the costs of stabilizing GHG concentrations has to do with the timing and speed of climate action. For example, we can choose between the climate policy suggested by Nordhaus (2007a), known as 'the climate policy ramp' as it builds gradually over a long period of time, and the climate policy defended by Nicholas Stern, known as a 'big-bang' approach

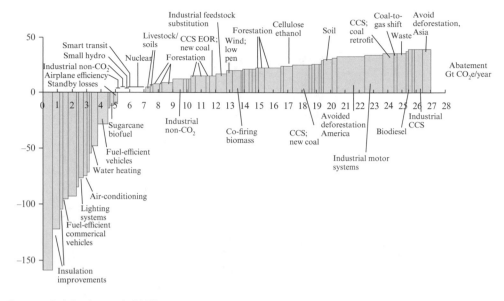

Source: Beinhocker et al. (2008).

Figure 3.4   The cost curve of GHG abatement in 2030 (€/tCO₂e)

because it consists of taking aggressive action to limit emissions in a short time. This debate also opens up an interesting line of research for the near future.

Nevertheless, the main issue is not merely whether we can do it or not, but also whether we should do it or not. That is, we need to answer the question of whether the cost of taking action against climate change is greater or lower than the cost of inaction. A comparison of the costs of action and inaction should therefore be made.

3.4   THE COST OF INACTION

In the introduction we mention some of the consequences of failing to act against climate change: more droughts, floods and severe storms, and the risk of losing ecosystems and biodiversity. These climate and environmental changes are expected to have wide-ranging impacts and economic effects. To answer to the question of what the cost of inaction would be, some important models have been developed and some difficulties have had to be overcome. Calculating the total or marginal costs of inaction is a complex task. Detailed integrated assessment models (IAMs) linking emissions, climate impacts, economic costs and adaptation need to be used.

The key studies centre on four IAMs which are well known in the relevant literature: FUND (Anthoff and Tol, 2009), PAGE (Hope, 2004), RICE/DICE (Nordhaus, 2010) and MERGE (Manne et al., 1995). The results obtained using these IAM models are summarized in Figure 3.5. The coverage of these models is however partial, and they may tend to underestimate the true cost of climate change.

In the Stern Review (Stern, 2007), a reference point in the relevant literature, the

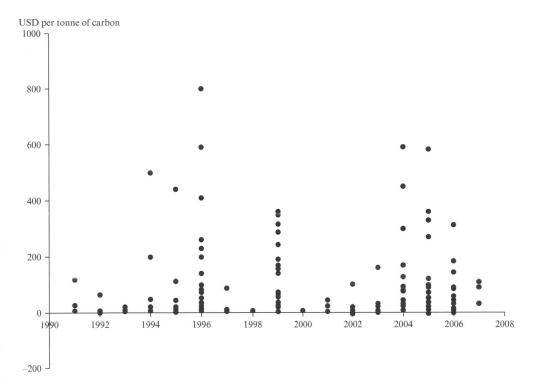

USD per tonne of carbon

Source: Tol (2005), updated by EEA (2007).

Figure 3.5   Estimates of the marginal costs of inaction

author maintains that 'climate change threatens the basic elements of life for people around the world – access to water, food production, health, and use of land and the environment' and urges that immediate action be taken. The Stern Review uses the results of the PAGE 2002 model and estimates that in a business-as-usual (BAU) scenario, the overall costs and risks of climate change will be equivalent to losing at least 5 per cent of global GDP each year, now and forever. The Review argues that if a wider range of risks and impacts is taken into account, the estimates of damage could rise to 20 per cent of GDP or more.

On the other hand, Nordhaus (2007b) defends a different view and believes that the Review's worrying conclusions are severely overstated because it assumes a near-zero discount rate and that the marginal utility does not decline rapidly enough as technical progress causes consumption to rise: 'An examination of the Review's radical revision of the economics of climate change finds, however, that it depends decisively on the assumption of a near-zero time discount rate combined with a specific utility function.' Thus, Nordhaus (2010), using the BAU scenario of the DICE model, estimates that average temperatures will be about 5°C higher in 2100 than they were in 2000 and that this will translate into a reduction in GDP of almost 5 per cent. If the temperature rises by 2.2°C – the consensus projection for 2100 – the losses estimated by the model are around 2 per cent of GDP.

This debate brings us to a discussion of what discount rates should be used when discussing climate policy and, hence, when estimating the cost of action and inaction regarding the climate problem. This has been and indeed still is being widely discussed. The debate concerns ethics, as it has to do with the way in which present generations are valued with respect to future ones. The range of positions that can be found in economic theory about what the discount rate should be is closely connected with cost–benefit analysis (CBA). This has also been taken into account in the economic literature related to environmental problems in general, and to climate change and emissions of GHGs in particular.

In the economics of climate change, CBA is usually described as a trade-off between generations. It is possible to choose to live in a world that does not have to suffer the damage of climate change in the future because costly action can be taken in the present to prevent warming. The trade-off is across generations because climate change is taken as a long-term problem. Hence the choice of the discount rate used when assessing the trade-offs between losses today and tomorrow is crucial; so crucial that Arrow (2007) shows how the differences in discount rates used in the analysis of different alternatives override any possible differences in the estimation of future losses generated by global warming.

Some interesting papers that take into account environmental aspects of discounting have appeared recently. Hoel and Sterner (2007) include changes in relative prices in a model that distinguishes between environmental goods and consumption goods. There are also papers (Gollier, 2010) that pose the question of what rates should be used to discount costs and benefits of different natures at different time horizons. In a model with a representative agent who consumes two goods, Gollier concludes that the use of an environmental discount rate lower than the economic discount rate may be justified. Both articles provide useful results and insights on the topic of what discount rate should be used in models in which environmental considerations are included.

One final argument that should be taken into account is the one made by Weitzman (2007), who states that if there is a positive probability (although it may be low) of a catastrophe, the details of CBA are not very relevant. If a catastrophe is a real possibility, then climate policy should be implemented. Any other action would be irresponsible: the world should not be exposed to this risk, however small the probability of its occurrence is. In Weitzman's words: 'There is little doubt that the worst-case scenarios of global-warming catastrophes are genuinely frightening.'

3.5   CONCLUDING COMMENTS

We draw on theory and empirical evidence to show that there is no 'single' relationship between economic growth and GHG emissions. It has been argued that the optimal use of energy and environmental policy is crucial in finding ways to curb emissions without compromising economic growth. Moreover, the numbers presented in sections 3.3 and 3.4 imply that from an economic point of view, making policy decisions to prevent climate change is a good option if the policies implemented are well designed.

However, knowing that we should and could act is just the first step. We also need to consider how to proceed in many different aspects to find the solution to the problem. Obviously, the change in the energy paradigm and the consideration of fossil fuels as toxic assets implies mitigation costs that not all countries are willing, or indeed able, to bear in the short term. Hence, not surprisingly, the alternative of exploring geo-engineering solutions is being taken seriously. Environmental scientists (Crutzen, 2006; Keith, 2000; Shepherd et al., 2009) are increasingly paying attention to these proposals.

Geo-engineering solutions open up the debate on acting or not acting to a third possibility, of acting differently. Geo-engineering is defined as the deliberate large-scale manipulation of the environment to reduce undesired anthropogenic climate change (Keith, 2000). Technically simple, reversible measures to reduce mean global temperatures directly seem to be available.[11] Until recently most economic analyses on climate change ignored geo-engineering. This situation is currently changing and some studies that focus the attention on the economic aspect of geo-engineering and climate change can now be found (Barrett, 2008; Moreno-Cruz and Smulders, 2007; Moreno-Cruz, 2010). There have also been some recent attempts to analyse the issue from a multi-disciplinary perspective including environmental, economic, social and ethical aspects (Shepherd et al., 2009).

As recognized by Barrett (2008), we have to deal with a problem of governance as long as one of the fundamental questions is who should decide whether and how geo-engineering should be attempted. Meanwhile, other important issues are being analysed by economists. These include discussions on whether geo-engineering options are substitute measures (Barrett, 2008) or complementary measures (Moreno-Cruz, 2010) with regard to mitigation and adaptation options. Moreno-Cruz (2010) analyses the strategic interactions that arise when geo-engineering and mitigation are considered jointly. He shows that geo-engineering does not necessarily increase the well-known free-riding effect on mitigation. He concludes that it is difficult to determine whether or not geo-engineering technologies are essential tools for managing climate change, and further research is needed. However, we cannot revisit the act-then-learn versus learn-then-act debate that marked the progress of climate policy in the late 1990s. Looking at the figures provided in sections 3.3 and 3.4, and taking into account Weitzman's argument that the non-negligible probability of utter disaster should dominate our policy analysis, it seems clear that it pays to design policies that address the climate change problem and aim at curbing GHG emissions.

NOTES

1. As reported by the US Energy Information Administration (2010), in 2008 approximately 80 per cent of the world's primary energy demand was met by fossil fuels.
2. The environmental Kuznets curve (EKC) echoes the inverse U-shaped relationship found by Simon Kuznets (1955) between economic growth and income inequality.
3. The BRICS are Brazil, Russia, India and China.
4. In fact, it may be argued that it depicts two possible CKCs, depending on whether the BRICS follow the development path of the EU-25 or that of the USA. The CKC would be flatter if the development path of the EU-25 were chosen.
5. For instance, Ansuategi and Escapa (2002) show how the international and intergenerational dimensions of climate change are important factors to be taken into account in understanding the relationship between economic growth and GHG emissions.
6. Offsetting measures the extent to which net decreases in the fuel mix and the energy intensity of the economy offset net increases in per capita income and population. Absolute decoupling exists if total emissions fall while GDP increases (offsetting is more than 100 per cent).

7. 'Carbon productivity' (level of GDP per unit of CO_2e – carbon dioxide equivalent) is the inverse of 'carbon intensity' (level of CO_2e per unit of GDP).
8. This cost curve would be the estimation of the global supply curve of abatement effort (SS in Figure 3.2).
9. A general description of the measures and instruments implemented by the EU to achieve these objectives can be found in Gallastegui and Galarraga (2010).
10. Previous contributions from which the Green Paradox was developed include Sinclair (1994), Ulph and Ulph (1994) and Withagen (1994).
11. In a recent study by Shepherd et. al. (2009) geo-engineering methods are shown as divided into two categories: carbon dioxide removal (CDR) and solar radiation management (SRM). CDR techniques include biomass and carbon capture storage and direct capture of CO_2 from the air, which means that they allow for stabilization of GHG concentrations. CDR techniques are slow to act and expensive. On the other hand, SRM techniques seek to reflect sunlight to reduce global warming, for example by pumping sulphur aerosols into the sky. SRM methods seem to be cheap and fast-acting but they do not control GHG concentrations. Moreover, further research is needed to analyse not only direct but also indirect consequences of these methods.

REFERENCES

Ang, B.W. (1994), 'Decomposition of industrial energy consumption: the energy intensity approach', *Energy Economics*, **16** (3), 163–74.

Ansuategi A. and M. Escapa (2002), 'Economic growth and greenhouse gas emissions', *Ecological Economics*, **40** (1), 23–37.

Anthoff, D. and R.S.J. Tol (2009), 'The impact of climate change on the balanced growth equivalent: an application of FUND', *Environmental and Resource Economics*, **43** (3), 351–67.

Arrow, K. (2007), 'Global climate change: a challenge to policy', *Economists' Voice*, Berkeley Electronic Press, **4** (3), 1–5.

Aslanidis, N. (2009), 'Environmental Kuznets curves for emissions: a critical survey', Nota di Laboro 75.2009, Fondazione ENI Enrico Mattei.

Atici, C. (2009), 'Carbon emissions in Central and Eastern Europe: environmental Kuznets curve and implications for sustainable development', *Sustainable Development*, **17**, 155–60.

Barrett, S. (2008), 'The incredible economics of geo-engineering', *Environmental and Resource Economics*, **39** (1), 45–54.

Barrett, S. and K. Graddy (2000), 'Freedom, growth and the environment', *Environment and Development Economics*, **5** (4), 433–56.

Beinhocker, E., J. Oppenheim, B. Irons, M. Lahti, D. Farrell, S. Nyquist, J. Remes, T. Nanclér and P.A. Enkvist (2008), 'The carbon productivity challenge: curbing climate change and sustaining economic growth', McKinsey Global Institute, McKinsey & Company.

Brock, W. and S. Taylor (2005), 'Economic growth and the environment: a review of theory and empirics', in S. Durlauf and P. Aghion (eds), *Handbook of Economic Growth*, Amsterdam: North Holland, pp. 1750–1821.

Bruvoll, A. and H. Medin (2003), 'Factors behind the environmental Kuznets curve: a decomposition analysis of the changes in air pollution', *Environmental and Resource Economics*, **24**, 27–48.

Cole, M.A. (2007), 'Corruption, income and the environment: an empirical analysis', *Ecological Economics*, **62** (2–3), 637–47.

Crutzen, P.J. (2006), 'Albedo enhancement by stratospheric sulfur injections: a contribution to resolve a policy dilemma?', *Climatic Change*, **77**, 211–19.

EEA (2007), 'Climate change: the cost of inaction and the cost of adaptation', European Environmental Agency Technical Report No. 13/2007.

Gallastegui, M.C. and I. Galarraga (2010), 'La Unión Europea frente al Cambio Climático: El Paquete de Medidas Sobre Cambio Climático y Energía (20-20-20)', in I. Sánches-Galán, F. Becker, J. Martines-Simancas and C.M. Cazorla-Prieto (eds), *Tratado de Energías renovables*, Vol. I. Cizur Menor, Spain: Aranzadi-Thomson Reuter, pp. 525–54.

Gollier, C. (2010), 'Ecological discounting', *Journal of Economic Theory*, **145**, 812–29.

Grossman, G.M. and A.B. Krueger (1991), 'Environmental impacts of a North American free trade agreement', NBER Working Papers 3914, National Bureau of Economic Research.

Heerink, N., A. Mulatu and E. Bulte (2001), 'Income inequality and the environment: aggregation bias in environmental Kuznets curves', *Ecological Economics*, **38** (3), 359–67.

Hoel, M. (2008), 'Bush meets Hotelling: effects of improved renewable energy technology on greenhouse gas emissions', Memorandum No 29/2008, Department of Economics, University of Oslo.

Hoel, M. and T. Sterner (2007), 'Discounting and relative prices', *Climatic Change*, **84**, 265–80.

Hope, C. (2004), 'The marginal impact of CO_2 from PAGE2002: an integrated assessment model incorporating the IPCC's five reasons for concern', mimeo, University of Cambridge.

Kaya, O. (1990), 'Impact of carbon dioxide emission control on GNP growth: interpretation of proosed scenarios', paper presented at the IPCC Energy and Industry Subgroup, Response Strategies Working Group, Paris.

Keith, D.W. (2000), 'Geo-engineering the climate: history and prospect', *Annual Review of Energy and the Environment*, **25**, 245–84.

Kojima, M. and R. Bacon (2009), 'Changes in CO_2 emissions from energy use: a multicountry decomposition analysis', Extractive Industries for Development Series #11, World Bank.

Kolstad, C.D. and M. Toman (2005), 'The economics of climate policy', in K.G. Mäler and J. Vincent (eds), *Handbook of Environmental Economics*, Vol. 3, Amsterdam: Elsevier, pp. 1561–1618.

Krugman, P. (2010), 'Building a green economy', *New York Times*, 5 April.

Kuznets, S. (1955), 'Economic growth and income inequality', *American Economic Review*, **45** (1), 1–28.

Manne, A., R. Mendelsohn and R. Richels (1995), 'MERGE: a model for evaluating regional and global effects of GHG reduction policies', *Energy Policy*, **23** (1), 17–34.

Moreno-Cruz, J.B. (2010), 'Mitigation and the geo-engineering threat', mimeo.

Moreno-Cruz, Juan B. and S. Smulders (2007), 'Geo-engineering and economic growth: making climate change irrelevant or buying time?', mimeo.

Nordhaus, W.D. (2007a), 'A question of balance: economic models of climate change', New Haven, CT: Yale University Press.

Nordhaus, W.D. (2007b), 'A review of the Stern Review on the economics of climate change', *Journal of Economic Literature*, **45** (3), 686–702.

Nordhaus, W.D. (2010), 'Economic aspects of global warming in a post-Copenhagen environment', *PNAS*, **107** (26), 11721–6.

Panayotou, T. (1993), 'Empirical tests and policy analysis of environmental degradation at different stages of economic development', Working Paper WP238, Technology and Employment Programme, International Labour Office, Geneva.

Parry, M.L., O.F. Canziani, J.P. Palutikof, P.J. van der Linden and C.E. Hanson (eds) (2007), *Contribution of Working Group II to the Fourth Assessment Report of the Intergovernmental Panel on Climate Change*, Cambridge, UK and New York, USA: Cambridge University Press.

van der Ploeg, F. and C.A. Withagen (2010), 'Is there really a green paradox?', CESifo Working Paper Series 2963, CESifo Group Munich.

Shepherd, J., K. Caldeira, J. Haigh, D.W. Keith, B. Launder, G. Mace, G. MacKerron, J. Pyle, S. Rayner, C. Redgwell, P. Cox and A. Watson (2009), *Geo-engineering the Climate: Science, Governance and Uncertainty*, London: Royal Society.

Sinclair, P.J.N. (1994), 'On the optimum trend of fossil fuel taxation', *Oxford Economic Papers*, **46**, 869–77.

Sinn, H.W. (2008), 'Public policies against global warming: a supply side approach', *International Tax Public Finance*, **15**, 360–94.

Stern, D. (2004), 'The rise and fall of the environmental Kuznets curve', *World Development*, **32** (8), 1419–39.

Stern, N. (2007), 'The economics of climate change: the Stern Review', Cambridge: Cambridge University Press.

Tol, R.S.J. (2005), 'The marginal damage costs of carbon dioxide emissions: an assessment of the uncertainties', *Energy Policy*, **33** (16), 2064–74.

Torras, M. and J.K. Boyce (1998), 'Income inequality and pollution: a reassessment of the environmental Kuznets curve', *Ecological Economics*, **25** (2), 147–60.

Ulph, A. and D. Ulph (1994), 'The optimal time path of a carbon tax', *Oxford Economics Papers*, **46**, 857–68.

US Energy Information Administration (2010), 'Annual Energy Outlook 2010 early release overview', http://www.eia.doe.gov/oiaf/aeo/.

Wagner, Martin (2008), 'The carbon Kuznets curve: a cloudy picture emitted by bad econometrics?', *Resource and Energy Economics*, **30** (3), 388–408.

Weitzman, M.L. (2007), 'A review of the Stern Review on the economics of climate change', *Journal of Economic Literature*, **45** (3), 703–24.

Withagen, C.A. (1994), 'Pollution and exhaustibility of fossil fuels', *Resource and Energy Economics*, **16**, 235–42.

World Bank (2009), *World Development Indicators 2009*, Washington, DC: World Bank.

World Bank (2010), *World Development Report 2010: Development and Climate Change*, Washington, DC: World Bank.

4   The linkages between energy efficiency and security of energy supply in Europe

Andrea Bigano, Ramon Arigoni Ortiz, Anil Markandya, Emanuela Menichetti and Roberta Pierfederici

4.1   INTRODUCTION

In recent decades, increasing demand for energy, fluctuating oil prices, uncertain energy supplies and global warming made European Union (EU) citizens and governments realize that secure and safe supplies of energy can no longer be taken for granted. Security of energy supply has been widely debated, mostly in relation to upstream (security of supply for a specific geographical region or single country). However, it can be argued that one way to reduce the dependence on external energy sources, or the exposure to energy prices volatility and increase, is simply to reduce the demand for energy. Energy savings may thus be considered a policy priority when concerns for energy security are particularly strong. In addition, improved energy efficiency can play a critical role in addressing not only energy security, but also environmental and economic objectives.

In order to understand fully how energy security affects the European society and how demand-side policies can be geared, it is important to know the energy intensity in different economic sectors of European countries, and to investigate their potential for efficiency improvement.

This chapter collects the main results of the analyses of energy efficiency in an energy security perspective, looking in detail into energy use in Europe. To this purpose an original econometric approach is applied to the EU-15 countries and Norway. Drawing on Ortiz et al. (2009), which focused solely on energy and carbon efficiency indicators, we check whether policies and measures (P&M) that affect indicators of energy efficiency performance have an analogous effect on security of supply indicators, both at the whole economy level and within the main sectors of energy use.[1]

The analyses have shown that the indicators studied are affected by a number of policies and measures. However, very few P&Ms seem to be able to tackle energy efficiency, carbon efficiency and energy security effectively and simultaneously. The main lesson to be drawn from this analysis is that there are a number of effective energy-efficiency policies in the EU, but there is no one single policy or measure able successfully to address different policy objectives. Taking a more general perspective, what seems to work is the policy mix rather than specific policies separately.

The rest of the chapter is organized as follows. Section 4.2 gives a general overview of energy consumption in Europe in the last three decades and describes in more detail the indicators studied. Section 4.3 looks at the energy reduction potential and at the European policy framework for the promotion of energy efficiency, and at national policies in the various sectors of energy use. Section 4.4 explains the methodology applied in our panel analysis and the dataset used, while section 4.5 discusses the results. Section 4.6 concludes. Appendix Table 4A.1 lists the variables used in the econometric analyses.

4.2 MAIN ENERGY EFFICIENCY INDICATORS FOR THE EU

Energy efficiency is evaluated by macro and specific indicators defined at the level of the economy as a whole, of a sector, of an end use. Three indicators are considered to compare energy efficiency performances and to monitor energy efficiency trends: energy intensity index (EI); energy efficiency index (EE) and carbon intensity (CI). These indicators can also be used to help monitor the success of key policies that attempt to influence energy consumption and energy efficiency. Before discussing the energy efficiency indicators, let us look briefly at the general situation of energy consumption in Europe in order to frame our discussion in its appropriate context.

4.2.1 Energy Consumption in the EU-27

Despite being the largest economy worldwide in terms of gross domestic product (GDP), the growth in energy consumption of the EU-27 has been rather limited, contributing to 15.9 per cent of total world energy consumption (Eurostat), which is as much as China (15.2 per cent), and less than the amount consumed by the USA (20.5 per cent). The primary and final energy consumption increased at approximately the same rate between 1990 and 2004 (1 per cent per year on average) in the EU-15 and amounted to around 1000 Mtoe and 1500 Mtoe, respectively (Odyssee, 2007). However, the period 1993–2000 was characterized by faster growth in energy consumption (1.5 per cent per year), driven by a steady and rapid expansion of the economy (2.7 per cent per year for GDP and 2.3 per cent per year for industry). Since 2000, there has been a slowdown in economic activity, which has resulted in a lower progression of energy use. Electricity demand underwent a more rapid progression of around 2 per cent per year on average.

In 2007, the final energy consumption of the EU-27 reached 1196 Mtoe (Eurostat). The industrial sector accounted for 25 per cent of final energy consumption and the residential sector 25 per cent; the remainder was shared among services, transport and agriculture. The share of renewable energies in the total final energy consumption was 9 per cent (Enerdata).

Indexing the level of energy consumption in 1990, the European consumption decreased, then from 1996 it increased smoothly at a rate of 10 per cent in 15 years, which is sensibly lower than the rates shown by the other world economies (Figure 4.1).

Disaggregating demand by energy fuels, European (EU-27) consumption is mainly composed of oil, gas and electricity, and their shares are equal, respectively, to 42, 25 and 20 per cent. Solid fuels, in spite of being historically an important source of energy, at present contribute only marginally to the total energy mix. Renewable energy sources and industrial waste have a limited share of total consumption and their contribution remained invariant between 1990 and 2005. In terms of categories of final users, the services, agricultural and household sectors taken together contribute the largest share of total final energy consumption, followed by industry and, finally, by transport. Over

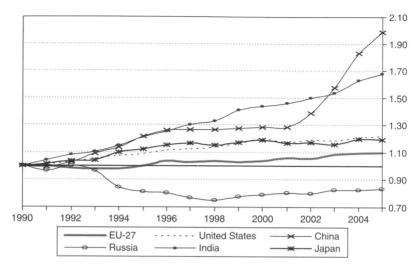

Note: 1990=1.

Source: Eurostat data.

Figure 4.1 *Gross inland energy consumption: EU-27 and selected regions*

the 15-year period, the demand in the industry sector has slightly decreased, while an opposite trend characterizes the transport sector.

As to electricity generation, solid fuels remain a significant energy source, contributing to 28 per cent of total generation, although their use has diminished a little over time. The largest source is represented by nuclear, making more than 30 per cent of total production. A sustained upward thrust is displayed by gas, which at present guarantees 21 per cent of total production, while renewables have a relevant share (14 per cent in 2005).

4.2.2 Energy Intensity

Energy intensity is an economic indicator of energy used in the production activity of a country. The index is defined as the ratio between energy consumption and an indicator of activity measured in monetary units (for example gross value added, GVA). This indicator can be used whenever energy efficiency is assessed at a high level of aggregation (that is, at the level of the whole economy or at a sector level), since in this case it is not possible to characterize economic activity with technical or physical indicators. High (low) EI indicates a high (low) price or cost of converting energy into GVA. The classical EI index is calculated by dividing energy consumption by GVA, on a sector basis.

In this study the final and sectoral energy consumption have been obtained from the International Energy Agency (IEA) balance sheet (ktoe). The sectoral values added result from a combination of data from Eurostat national accounts and the Organisation for Economic Co-operation and Development (OECD) database.

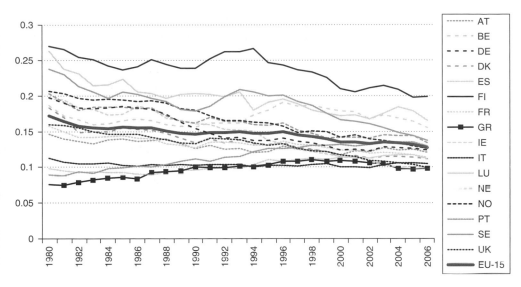

Source: Author's computation on data from IEA, Eurostat, OECD.

Figure 4.2 *Final energy intensity in European countries + Norway, 1980–2006,*
ktoe/00$ppp

Figure 4.2 shows the pattern of final energy intensity of the overall economy in the EU-15 plus Norway from 1980 to 2006. The average of European countries exhibits a smooth decrease over the entire period under scrutiny. The largest improvements are displayed by Luxembourg and Finland, the latter registering a sharp decrease in the EI index. By contrast, the Portuguese EI index shows a stable upward trend, interrupted by a drop starting from 2005. In Spain, after a period of decrease, the index starts to increase from the 1990s. On the other hand, Italy exhibits a four-phase pattern. In the first phase can be noticed a stable decrease in the EI index until the mid-1980s. From this period the index remains nearly constant up to 2002, when it starts to rise. In the latest phase, starting in 2005, the index drops again.

4.2.3 Energy Efficiency

Residential sector
The energy intensity index cannot capture the efficiency of the residential sector, since household activities do not generate value-added directly. For this sector, one needs to resort to indexes unrelated to economic values, such as the energy efficiency index. In contrast with energy intensity indicators, in fact, the energy efficiency index is based on measures of unit consumption, that is, on physical or technological measures.

Hence, it follows that the influence of economic structural changes, as well as the impact of other factors which are not directly associated to a strict definition of energy efficiency, are not considered in the construction of the indicators. The classical energy efficiency (EE) index ranges between 0 and 100. A decrease in the index is to be interpreted as an improvement in energy efficiency.

The EE index can be calculated by weighting the changes in unit consumptions (UC), according to the consumption's share of the sector they refer to. UC are defined at a more disaggregated level by relating energy consumption to an indicator of activity measured in physical terms. UC are expressed in different units, depending on the subsector or end use, in order to provide the best proxy of energy efficiency. The final EE index is a pure number (that is, it is not expressed in terms of any unit of measure).

UC for the households sector are not of course pure numbers, but are expressed in physical units: toe per dwelling or per m^2 for heating; toe per dwelling or per capita for water heating and cooking; and kWh per dwelling or per appliance for electrical appliances as televisions, fridge, freezers, washing machines or dishwashers.

The EE index is calculated as a weighted average of unit consumption indices by subsectors. Its interpretation is easier, as the value obtained is directly linked to the variation of EE within each subsector. The idea is to calculate the variation of the weighted index of UC between a base year and year t, as follows:

$$I_t/I_0 = \left(\sum_i EC_{i,t}*(UC_{i,t}/UC_{i,0}) \right), \tag{4.1}$$

where UC_i indicates the unit consumption index of a sub-sector i and EC_i is the share of subsector i on total consumption. The EE index is then calculated by taking the data starting point as the base year.

Table 4.1 shows the percentage change in the energy efficiency index in the EU-15 and Norway between 1980 and 2004 by considering separately the subsamples 1980–92 and 1993–2004. That is, it shows whether significant changes have occurred in the residential sector. The resulting ranking of these countries does not necessarily single out the most or least 'virtuous' countries in terms of energy efficiency: the table displays the countries that have been able to benefit from their potential of energy efficiency improvement, irrespective of their original level of energy efficiency in the base year. For example, the most significant improvements in the energy efficiency of the household sector have been achieved in Portugal and Norway. Although in Norway energy efficiency has decreased by 15.8 per cent during the period between 1980 and 1992, this country was able to raise energy efficiency standards. Consequently, during the period between 1992 and 2004, energy efficiency has increased by approximately 11.7 per cent.

Transport sector
Table 4.2 shows the percentage change of energy efficiency for the transport sector. Over the whole sample (1980–2004), the countries that reported the best performances have been Ireland and Greece. Across subsamples the most significant improvements have been achieved by the Belgian transport sector. On a smaller scale, France, Sweden and Norway have reported similar changes. By contrast, performances in the energy transport sector have worsened in Spain.

Disaggregating the EE index by transport modes, it can be noticed that a regular improvement of the energy efficiency of transport (12 per cent) took place in the EU-27 over the period 1990–2006. The lowest progress can be blamed on the road transport of goods, while the best performance in the index took place in air transport (Figure 4.3).

Table 4.1   *Percentage change of energy efficiency in the EU-15 countries and Norway, 1980–2004: household sector*

Household (% change in EE index over period)					
1980–2004		1980–1992		1992–2004	
PT	−49.8%	−31.7% DK	PT	−42.4%	
DK	−43.4%	−18.0% SE	DK	−17.2%	
SE	−28.5%	−12.9% PT	AT	−16.3%	
AT	−24.9%	−10.3% AT	SE	−v12.8%	
FR	−17.1%	−10.0% FR	NO	−11.7%	
FI	−16.1% Median	−7.9% FI	FI	−8.9%	
DE	−10.5%	−6.9% UK	DE	−8.5%	
UK	−8.7%	−2.2% DE	FR	−7.9%	
IT	−4.2%	0.5% IT	IT	−4.7%	
NO	2.2%	15.8% NO	UK	−1.9%	
ES	142.7%	40.5% ES	ES	72.7%	
BE	n/a	n/a BE	BE	n/a	
EL	n/a	n/a EL	EL	n/a	
IE	n/a	n/a IE	IE	n/a	
LU	n/a	n/a LU	LU	n/a	
NL	n/a	n/a NL	NL	n/a	
Average =	−5.3%	−3.9%		−5.4%	
Median =	−16.1%	−7.9%		−8.9%	
St. Dev. =	0.516	0.188		0.280	
Minimum =	−49.8%	−31.7%		−42.4%	
Maximum =	142.7%	40.5%		72.7%	

Notes: Countries are ordered according to their energy efficiency performance in descending order. Arrows show significant movements between quartiles over time.

Source: Authors' calculations on Odyssee (Enerdata) data.

4.2.4   Carbon Intensity

Carbon intensity (CI) is an indicator akin to energy intensity, and measures the degree of carbonization of an economy or of a given productive sector. At the aggregated level, carbon intensity is computed as the ratio of CO_2 emission equivalents generated (in terms of Mtonne of CO_2) to the indicator of economic activity, GVA. The same sectoral disaggregation as in the case of energy intensity can be performed. The carbon content of consumed energy measures the quantity of CO_2 (or, in its more general format, CO_2 equivalents[2]), per unit of energy consumed. It can happen that energy intensity increases while carbon intensity decreases, for instance in the presence of a massive switch from oil to natural gas, the latter being 'cleaner' and allowing a decrease in CO_2 equivalents emitted while leaving unchanged the quantity of energy consumed. The carbon content can thus be regarded as a technological parameter which takes into account changes in the fuel mix of a country or a sector.

Table 4.2   *Percentage change of energy efficiency in the EU-15 countries and Norway, 1980–2004: transport sector*

Transport (% change in EE index over period)					
1980–2004		**1980–1992**		**1992–2004**	
IE	−45.0%	−35.4% ES	BE		−49.4%
EL	−43.7%	−25.8% IE	IE		−26.0%
AT	−33.2%	−24.1% EL	EL		−25.9%
ES	−31.1%	−21.7% AT	NO		−21.4%
NO	−27.4%	−13.8% IT	PT		−14.8%
PT	−24.4%	−12.2% DE	AT		−14.7%
DE	−23.1%	−11.3% PT	DE		−12.4%
DK	−16.9% Median	−10.9% DK	FR		−12.0%
IT	−13.4%	−7.6% NO	SE		−11.2%
SE	−12.8%	−3.4% NL	FI		−7.4%
FR	−12.0%	−2.8% LU	DK		−6.8%
BE	−11.2%	−1.9% UK	NL		−4.7%
NL	−7.9%	−1.8% SE	UK		−3.4%
FI	−7.5%	−0.2% FI	IT		0.5%
UK	−5.2%	0.0% FR	ES		6.7%
LU	123.5%	75.4% BE	LU		129.9%
Average =	−12.0%	−6.1%			−4.6%
Median =	−15.2%	−9.2%			−11.6%
St. Dev. =	0.382	0.241			0.381
Minimum =	−45.0%	−35.4%			−49.4%
Maximum =	123.5%	75.4%			129.9%

Notes: Countries are ordered according to their energy intensity. Arrows show significant movements between quartiles over time.

Source: Authors' calculations on Odyssee (Enerdata) data.

Available information on CO_2 emissions starts from 1990, hence carbon intensity indexes cover a period shorter than energy intensity and energy security indexes. In Europe, total CO_2 emissions registered a slight increase from 1990, with a growth rate of 5.8 between 1990 and 2006. In 2006, Germany contributed the most to total CO_2 emissions in Europe, followed by the United Kingdom, Italy and France. The shares of CO_2 emissions by country remain rather stable during the period considered. Germany and the United Kingdom are the only EU countries which show a decrease of emissions during the period under scrutiny, by 14 per cent and 1 per cent respectively, while the largest increase is registered in Spain.

Figure 4.4 shows the trend of carbon intensity in European countries between 1990 and 2006. Looking at the average of EU-15 countries, carbon intensity decreased from 1990 to 2006 by about 20, although in Spain and Portugal the index increased. The best performances are attained by Ireland and Germany, which show a variation of about −45 and −33 per cent respectively between 1990 and 2006.

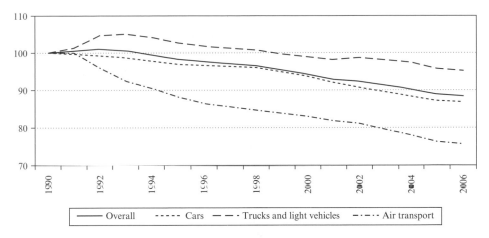

Note: Technical ODEX index calculated on seven modes: cars (litres/km), trucks and light vehicles (toe per tkm), air (toe per passenger); rail and water (toe/tkm or pkm); motorcycles and buses (toe/vehicle).

Source: ODYSSEE.

Figure 4.3 Energy efficiency index for transport EU-27 (ODEX)

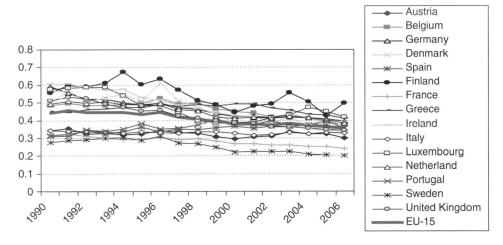

Source: Authors' computation on data from Enerdata, Eurostat, OECD.

Figure 4.4 Total carbon intensity in European countries, 1990–2006, kt CO_2/00$ppp

4.2.5 Energy Security

In the scientific literature, different approaches for studying energy security can be identified. Some studies focus on a country's current diversification of energy sources or import sources as a measure of energy security, for instance Neff (1997) and Jansen et al. (2004). Other studies look at the future development of oil supply and imports using

Table 4.3   Energy security indicators

	Vulnerability	Dependence
Physical dimension	• Imported oil used in transportation (Mtoe)/Total energy used in transportation (Mtoe) • Imported oil and gas-fired electricity generation (gWh)/Total electricity consumed (gWh) • Per capita oil consumption (Ktoe) • Degree of supply concentration for oil and gas • Shannon–Weiner Index for supply	• Imports of energy/Total primary energy supply • Country's oil gross and net imports/Total oil consumption • Country's gas gross and net imports/Total gas consumption
Economic dimension	Value of oil (or gas) imports/Value of total exports	Oil or gas consumption (toe) per $ of real GDP

bottom-up energy systems models, for example Constantini et al. (2007) and Turton and Barreto (2006). A number of researchers have tried to develop a set of security indicators (IEA, 2001; Kendell, 1998; von Hirschhausen and Neumann, 2003). These measures can be further grouped into two categories: dependence and vulnerability represented in both physical and economic terms.

Dependence is a measure of how much the domestic economy relies on sources of energy that are not under its control. Physical measures of dependence include: (1) imports of energy as a percentage of total imports; (2) oil imports as a percentage of total oil consumption; (3) gas imports as a percentage of total gas consumption. Economic measures of dependence are oil and gas consumption in physical units per US dollar of real GDP.

Vulnerability is a measure of the likelihood of domestic disruption in case some external energy source is reduced or cut off. Physical measures of vulnerability include: (1) the amount of imported oil used in transportation relative to total energy used in transportation; (2) amounts of imported oil- and gas-fired electricity generation relative to total electricity generation; (3) degree of supply concentration; and (4) the Shannon–Weiner diversity index.[3]

A non-exhaustive but fairly extensive list of indicators can be found in Table 4.3. Subject to data availability these indicators were tested in our panel analyses. Those that yielded the best results in terms of responsiveness to energy policies were oil intensity, gas intensity, the ratio of gas imports to gas consumption, and the ratio of net imports of energy to total primary energy supply.

Oil intensity is given by consumption (Ktoe) per dollar of real GDPs, which we choose to measure as purchasing power parity (PPP), in constant 2000 international millions of US dollars.[4] The bulk of oil products are used in transportation (light and middle distillates); currently the most important alternative fuels – LPG and natural gas – hold minuscule shares.

All EU countries have improved their energy ratio since 1975 with growth in GDP outstripping that of oil consumption. Most likely this is due to energy switching toward

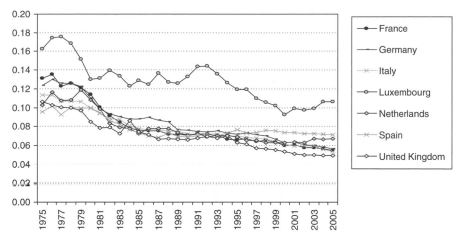

Figure 4.5   Oil consumption (Ktoe) per US$M of real GDP

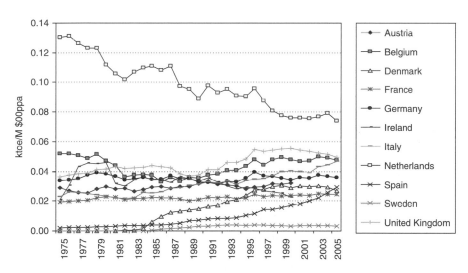

Figure 4.6   Gas primary consumption (Ktoe) per US$M of real GDP

other fuels (mainly gas), and to an increase in the efficiency in the transport sector. Figure 4.5 shows a progressive convergence of the index among the European countries.

All countries have seen an increase in gas intensity since 1975 to 2005, with the exception of the Netherlands. Ireland and Denmark registered a remarkable upward trend, while Austria and Belgium have seen the smallest increase in percentage terms. In Italy the value of the indicator almost tripled over the period considered. Figure 4.6 illustrates the performance of this indicator for gas over the period 1975–2005. Differences between countries reflect many factors including climatic and industrial structure characteristics. The residential sector is the largest-consuming sector of natural gas, followed by the industrial, electricity and commercial sectors. The use of gas in power generation is growing rapidly and for this reason in the early 1990s, before the use of gas for electricity

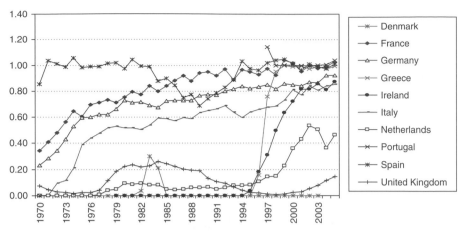

Figure 4.7   Gross imports of gas over total primary gas consumption, by selected countries

generation, gas demand was more seasonal and the daily average demand was only around half the winter maximum.

Figure 4.7 shows the ratio of gross natural gas imports to natural gas consumption. Greece registered the most noticeable upward trend over the period considered. In Italy, the indicator exhibited a steady increase during the period under consideration. United Kingdom (0.14)[5] was the country with the lowest ratio in 2004, while Portugal registered the highest index. Notice that Ireland, Greece and Portugal are rather new to the gas market, introduced only recently.

The last indicator is not the most appropriate index to measure the dependence on imported energy. A more appropriate indicator can be calculated using net imports of energy. In fact according to Skinner (1995): 'with total [gross] imports in the numerator rather than net imports, not only is the computed dependence higher due to the quantity of exports, but also comparisons in dependence over a number of years can be substantially distorted due to changes in export patterns'. In order to have an indicator with an upper bound equal to 1 (that indicates the maximum level of dependence) we include in the denominator the TPES,[6] stock variations and marine bunkers. As can be seen in Figure 4.8, Luxembourg,[7] Ireland, Portugal and Italy registered the highest dependence ratio in energy imports in 2004. By contrast Norway, a net exporter of energy, registered the lowest ratio followed by Denmark and the UK. In the period 1980–90 all the EU-15 countries and Norway registered a downward trend in the energy dependence indicators (bar Luxembourg and the Netherlands). In the period 1970–2004, Germany registered the largest increase in this indicator (+41 per cent).

4.3   ENERGY SAVING POTENTIAL AND ENERGY POLICIES IN THE EU

Several European Directives to improve energy efficiency have been implemented during the recent past years. Milestone policies are listed in Table 4.4. Until 2006, most initia-

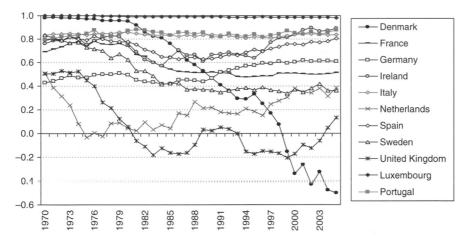

Figure 4.8   Ratio of net imports to TPES in selected EU countries

Table 4.4   Key energy saving policies in the EU

1992	2000	2002	2005	2006	2008
EU Directive on labelling of the energy consumption of household appliances	Action Plan for Energy Efficiency 2000–2006	EU Directive on buildings' efficiency	Eco-Design Directive concerning all new products outside of the transport sector	EU Action Plan for Energy Efficiency (2007–13)	Climate Action and Renewable Energy Package

Source: ADEME (2008).

tives target specific modes or sectors of energy use in Europe, setting the general framework in which national policies of member states should then develop in accordance with the subsidiarity principle.

The Green Paper, 'Energy', adopted by the EC in March 2006, lays the basis for a European Energy Policy; this document highlights that the development of a common policy is a long-run project whose ultimate purpose is to balance three core objectives: sustainable development, competitiveness and security of supply. As a foundation for this process the European Commission (EC) proposes establishing a Strategic EU Energy Review to be presented to the Council and Parliament on a regular basis, covering all the energy policy issues. Through the Strategic EU Energy Review, the EC aims at covering all aspects of energy policy, analysing all the advantages and drawbacks of different energy mixes. Although a country's energy mix is and will remain a question of subsidiarity, related decisions have consequences for other countries and for the EU as a whole, in terms of both pollution and energy security. All in all, this should eventually lead to the definition of a EU's overall energy mix to ensure security of supply and sustainability, whilst respecting the right of member states to make their own energy choices.

What emerges from the Green Paper is that the three policy objectives – competitiveness, security of supply and sustainability – are closely interlinked and complementary. In January 2007 the European Commission presented an Energy and Climate Change Package including a Strategic Energy Review. This package was finally agreed upon in December 2008. In March 2007 the EU Summit of Heads of States agreed upon an action plan, including among others:

- To save 20 per cent of the EU's total primary energy consumption by 2020.
- A binding target to raise the EU's share of renewables to 20 per cent by 2020.
- An obligation of 10 per cent biofuels in the transport fuel mix by 2020 for each EU member.
- A European Strategic Energy Technology Plan for low carbon technologies.

The Green Paper on 'Energy efficiency' (European Commission, 2005) points out that the EU could effectively save at least 20 per cent of its present energy consumption. In order to support a better integration of energy efficiency measures into national legislation, the European Commission has proposed several Directives which have been adopted and are now in force. These concern broad areas where there is significant potential for energy savings, such as:

- End-use Efficiency and Energy Services;
- Energy Efficiency in Buildings;
- Eco-design of Energy-Using Products;
- Energy Labelling of Domestic Appliances;
- Combined Heat and Power (Cogeneration).

Among the main EU legislation for buildings are the Boiler Directive (92/42/EEC), the Construction Products Directive (89/106/EEC) and the buildings provisions in the SAVE Directive (93/76/EEC). The Directive on the energy performance of buildings (EBPD 2002/91/EC), enforced since January 2003, builds on those measures with the aim to improve further the energy performance of public, commercial and private buildings in all member states.[8]

The Commission has published an impact assessment report for the Action Plan for Energy Efficiency, which allows quantification of the effects of the action proposed (Tipping et al., 2006). The estimates, however, contain a certain degree of uncertainty as a wide range of topics, at all levels of policy-makers and decision-makers, is involved. After evaluating a large set of possible instruments, some priority actions have been selected on the ground of their impact on energy savings. By far the most promising measure seems to be the extension of white certificate schemes,[9] after evaluation of present national schemes, to all EU countries, coupled with energy efficiency obligations on energy suppliers (80 Mtoe of potential savings); followed by maximum CO_2 emission standards for different type of cars coupled with more stringent agreements with car and truck producers after 2008–09 (28 Mtoe of potential savings) and end-user price increase to discourage fuel use (20 Mtoe of potential savings). Taken together the 18 policy options identify up to 353 Mtoe of potential primary energy savings over and above the current 'business-as-usual' projection without taking into account antagonistic or synergetic interactions (overlap) between the different policy options. Taking into account the separate policy options overlap, the gross estimated aggregate energy savings potential estimate reduces by 26 per cent to 262 Mtoe in 2020.

We have also investigated key policies in specific economic sectors but do not discuss them here. The reader can refer to Bigano et al. (2010) for a discussion on sectoral policies.

4.4 PANEL ANALYSIS: METHODOLOGY

This section describes the techniques applied in this study to identify and characterize the energy intensity, carbon efficiency, carbon intensity and energy security determinants by means of panel data econometric analyses, focusing on the following factors suggested by the literature:

- Structural changes in the economy: GDP, sectoral GDP shares changes, R&D expenditure.
- Policies: national and supranational energy policies (for example EU Directives, presence of national carbon and energy taxes).
- Measures: fiscal, education and information initiatives, legislation (mandatory standards or labelling), cooperative measures, cross-cutting measures.
- Energy: energy prices, energy balance sheet.

The goal is hence to assess the economic variables which could have a significant effect in improving the energy intensity, energy efficiency, energy security and carbon intensity and to identify the policies and measures implemented in European countries which have been effective for the same purpose. A further goal is to compare the significant drivers resulting from regressions, in order to understand whether there are some factors which affect both energy intensity and energy security, and if improvements in carbon intensity match with lower energy intensity.

In order to achieve these goals, we have chosen to apply econometric models which exploit the panel data format. The estimates are obtained by regressing the energy intensity index (EI), the energy efficiency index (EE), the energy security index (Total Imports/TPES – ES) and the carbon intensity index (CI) on a set of explicative variables X (such as energy prices, GDP, research and development expenditure) and policy measure variables (PM). Our analysis included 18 different panel models, with alternative specifications for energy security,[10] focusing on the EU-15 countries and Norway in the period between 1980 and 2006.[11] We only present here our best-fit models. The econometric models have the following functional form:

$$EI_{it} = \alpha_i + \lambda X_{it} + \beta_1 PM1_{it} + \ldots + \beta_K PMk_{it} + u_{it} \tag{4.2}$$

$$EE_{it} = \alpha_i + \lambda X_{it} + \beta_1 PM1_{it} + \ldots + \beta_K PMk_{it} + u_{it} \tag{4.3}$$

$$ES_{it} = \alpha_i + \lambda X_{it} + \beta_1 PM1_{it} + \ldots + \beta_K PMk_{it} + u_{it} \tag{4.4}$$

$$CI_{it} = \alpha_i + v X_{it} + \beta_1 PM1_{it} + \ldots + \beta_K PMk_{it} + u_{it} \tag{4.5}$$

$$CC_{it} = \alpha_i + \lambda X_{it} + \beta_1 PM1_{it} + \ldots + \beta_K PMk_{it} + u_{it} \tag{4.6}$$

where *EI* is the Energy Intensity index, *EE* is the Energy Efficiency index, *ES* is the Energy Security index, *CI* is the Carbon Intensity index, and CC are the carbon emissions per capita. The matrix X_{it} includes the explanatory variables related to economic structural changes, society and energy market. The variables PM_j, $j = 1, \ldots, K$, represent the policies included in the regression, which are dummy variables equal to 1 if the policy is in force in the *i-th* country and *t-th* year.

The double pointer (i, t) shows the panel structure of the dataset. In particular the index $i = 1, \ldots, N$ represents the country (16 in total), while the index $t = 1, \ldots, T$ refers to time (27 years). The parameters λ and β_j, $j = 1,\ldots,K$, are constant across countries and over time, while the parameters α_i change only with the country, are known as fixed effects and capture the individual heterogeneity that characterizes panel data models.

The individual heterogeneity is unknown, systematic and correlated with regressors. To solve this issue we have chosen a fixed-effect model, where the individual heterogeneity is modelled by means of country-specific constants. Such models differ from random-effects models, where instead the individual heterogeneity is a random variable μ_i, included in the disturbance term, $\alpha_i = \alpha$ and $u_{it} = \mu_i + e_{it}$.

The random-effect model implies the use of a random sample of individuals. We used instead a dataset where the selection of countries under scrutiny is not random; this makes the fixed-effects models more useful for our purpose than the random-effects models.

We have tested also one-year and two-year lags for all the P&M variables, and one-year lags for the main economic variables. The approach consisted in testing models which cover all macro-variables and policies, as well as their lags, cutting out variables with non-statistically significant coefficients. This process has been iterated until a set of significant explicative variables has been obtained.

For the estimates of the energy indexes and the economic variables we have combined a set of different data sources. The Energy Intensity index has been calculated by using the IEA[12] database for energy final consumption, and Eurostat[13] and OECD[14] databases for the estimates of sectoral value-added. Energy security indexes have been obtained employing data extracted from Enerdata[15] and the IEA. Data for the carbon intensity index have been extracted from the Enerdata[16] and Eurostat/OECD databases, while per capita CO_2 emissions for the residential sector have been computed by combining data from World Development Indicators (WDI)[17] and Enerdata.

In this study we consider final energy consumption to calculate the EI index. Regarding the indicator of economic activity, used both in the energy intensity and in the carbon intensity indexes, we have chosen GVA in US dollars at constant prices, calculated at PPP using 2000 as the base year. The indexes therefore have GVA, rather than GDP, as the denominator since taxes and subsidies, included in GDP, are not relevant for our purposes. The energy efficiency indexes have been computed by combining data extracted from IEA and MURE-Odyssee databases. IEA energy balances provide data on final and sectoral energy consumptions (Mtoe), while Odyssee (MURE) database includes the data on unit consumption (physical and technological data).

The economic time series are obtained from different sources, mainly the WDI, Eurostat[18] and the IEA.[19] Energy prices data have been extracted from IEA databases, R&D expenditures have been obtained from Eurostat, while WDI has provided information on the remaining macro-variables.

Policies and measures data are taken from the MURE[20] database. MURE (Mesures d'Utilisation Rationnelle de l'Energie) provides information on energy efficiency policies and measures that have been carried out in the member states of the European Union. The database collects the energy efficiency measures relevant to the four main energy demand sectors – namely household, transport, industry and tertiary – and on general energy efficiency programmes and general cross-cutting measures. Dummies variables have been created by subcategory of policy, that is, the dummy variable is equal to 1 if any kind of policy included in the same subcategory is implemented in the country under scrutiny during the period considered. Appendix Table 4A.1 provides a glossary of data with a description of economic variables and policy dummies.

4.5   RESULTS

Our aim is to check whether the implementation of energy efficiency policies has had an effect on indicators of energy efficiency, carbon efficiency and security of supply in the EU-15 and Norway. In particular we are interested in checking whether some policies had a sort of 'double dividend' by having a positive effect on more than one of these indicators. Besides policy dummies, we also look at the effect of the macro drivers: GDP, energy prices, research and development (R&D), and so on. All econometric results are presented in Table 4.5.

Energy intensity at the aggregated level is affected by a number of policies. It is interesting to note that besides general cross-cutting policies on energy efficiency, promotion of renewable energy sources or climate change mitigation (particularly if using market-based instruments), sector-specific policies also have a beneficial effect on overall energy intensity. In the residential sector, mandatory standards for electrical appliances and the deployment of grants, subsidies or soft loans have proven particularly effective. Measures supporting information, education and training in the industrial sector and tax exemptions in the tertiary sector also seem to improve overall energy intensity.

As expected, increasing the residential electricity price induces a small but significant reduction in overall energy intensity. As shown in the first column of Table 4.5, the effect of the macro driver 'share of industry on value-added' on energy intensity is very similar to that of energy prices, both in terms of arithmetical sign and in terms of magnitude. The sign of the variable GDP shows that GDP reduces energy intensity, suggesting that richer economies, at least in Europe, tend to use their energy more efficiently, while an increase in R&D expenditures tends to increase energy intensity, a somewhat puzzling result. Note however that the R&D variable does not capture R&D in the energy sector, but overall R&D. It is thus not implausible that these expenditures steer the overall economy towards a slightly more energy-intensive configuration.

A similar picture characterizes carbon intensity. Household electricity prices and GDP have roughly the same effect as on energy intensity, both in terms of sign and in terms of order of magnitude. R&D expenditures and industry's share in value-added have no significant effect, while energy production slightly worsens this indicator (although the significance of this variable is weak). A number of sector-specific policies improve this indicator: legislative or informative measures for the industry sector, mandatory stand-

Table 4.5   Econometric results: the whole economy

Coefficients			Dependent variables			
		Unit	Energy intensity	Energy security		Carbon intensity
			eifin	esfin*	esfin2	Cifin
Macro drivers	Energy price	US$/unit	−0.001	0.0047	–	−0.002
	GDPppp	US$	−0.020	0.333	−22.41	−0.067
	R&Dppp	US$	0.0166	−0.1	6.843	–
	Share industry	%	−0.002	–	−0.473	–
	Energy production	ktoe	–	−0.178	−11.2	0.0341
Industry policy	In03		–	–	–	−0.060
variables	In08		−0.012	–	–	–
Household	Hh04		−0.02	–	–	−0.0431
policy	Hh06		−0.011	–	–	–
variables	Hh07		−0.01	–	–	–
	Hh11		–	–	–	−0.030
	Hh12		–	–	–	−0.019
Transport policy variables	Tr11		–	–	−12.59	–
Tertiary policy	Te05		–	–	−3.29	–
variables	Te06		–	–	−9.126	–
	Te07		−0.012	–	–	–
	Te08		–	−0.041	−3.878	–
	Te09		–	–	–	−0.0175
Cross-cutting	Cc01		−0.006	–	–	–
policy	Cc02		–	−0.042	–	–
variables	Cc05		–	−0.0754	–	–
	Cc06		−0.007	–	–	–
	Cc07		−0.009	–	−5.379	−0.0196
R^2			0.72	0.64	0.71	0.67

Notes: All reported coefficients are statistically significant. Negative numbers indicate an improvement in energy security or reduction in energy intensity and carbon intensity and vice-versa. *esfin1* = Total import/TPES; *esfin2* = Total oil consumption/GDP.

ards for household electrical appliances, cooperative measures in the household and tertiary sectors, and cross-cutting policies.

As to energy security, after testing various candidates we have chosen to focus on two indicators for aggregate energy security (total energy imports/TPES and oil consumption/GDP). The first indicator displays a relatively low sensitivity to energy efficiency policies. In fact, only cross-cutting measures (legislative and cooperative) and, curiously, information initiatives in the tertiary sector have a significant beneficial effect, reducing the imports of energy as expected; energy production reduces import dependence, while it is less clear why a similar effect is produced by increasing R&D

expenditures. Higher GDP and higher household energy prices stimulate imports, not unexpectedly.

If, instead, vulnerability is assessed by looking at how important oil is in the economy, EU-15 countries have some more tools at their disposal to reduce it: general cross-cutting measures, soft loans for the adoption of renewable energy sources and efficiency improvements in the transport and tertiary sectors, grant subsidies and, again, informative measures in the tertiary sector. Increases in electricity and industrial production, which are not very oil-intensive in Western Europe, tend to reduce the weight oil has on the economy and hence the vulnerability of the latter. Also, there is a significant positive relationship between a higher level of GDP per capita and higher energy security of the overall economy, as oil becomes increasingly substituted by other energy sources.

The impact of GDP on energy system vulnerability therefore seems to be twofold, depending on the indicator we use to measure the aggregate energy security. On the one hand, indeed, an increase in GDP reduces the dependence on oil, improving the security of energy supply; while on the other hand it increases imports, strengthening the dependence on foreign energy suppliers. Looking at the regression coefficient values, however, the effect of decreasing the consumption of oil in favour of a less vulnerable energy mix seems to be more significant.

4.5.1 Discussion

In general, the fit of the econometric models analysed in this study is reasonable (R-square ranging from 0.64 to 0.76). A number of policies have a beneficial influence across EU countries on specific policy target indicators. There is however very little overlap among policies in terms of their effectiveness on both energy efficiency indicators and energy security indicators. This seems to confirm the traditional economic policy wisdom dating back to Jan Tinbergen (1952, 1956) that multiple policy objectives require multiple instruments. However, there is an exception to this general rule in our case: general cross-cutting policies appear to have beneficial effects on aggregate energy intensity, carbon intensity and energy security.

Between energy intensity and carbon intensity the overlaps are more widespread, and also some sector-specific policies improve the performance of both indicators. This is hardly surprising, given the high correlation between the two indicators, and holds in particular for the household sector. In addition, cooperative measures in the industry sector also affect both carbon and energy intensity at the aggregated level. It is quite striking that energy efficiency policies aimed at the residential, tertiary and agricultural sector have very little effectiveness in improving energy security. Cross-cutting policies, which are very relevant in terms of multidimensional effectiveness in the aggregate case, play a less relevant role in the residential, tertiary and agricultural sectors: only general programmes related to energy efficiency, climate change mitigation and renewable energy have this double beneficial effect, and only in terms of the ratio of gas consumption and GDP, and household energy efficiency.

For the transport sector, although not shown in this chapter (see Bigano et al., 2010), our analysis has shown that while there are quite a number of cross-cutting policies and policies aimed at the transport sector that improve energy efficiency, energy intensity and carbon efficiency, only cross-cutting policies (both with and without sector-specific characteristics) have a significant impact on oil security, the only facet of energy security that, according to our descriptive analysis, is relevant for this sector. The indication here seems to be that while energy efficiency can be significantly improved in this sector by well-designed policies, the sector is still too tightly bound to oil products for any of these policies to result in significant change in its oil security. This result is also underpinned by the fact that our analysis did not find any significant overlapping between security and other indicators. One significant overlap among energy efficiency, carbon intensity and energy intensity was singled out, as carbon intensity and energy intensity overlap twice.

4.6   CONCLUSIONS

In this study we have explored the relationships between energy efficiency and energy security for the economy of the EU-15 and Norway. To this purpose we have provided a descriptive analysis of a few energy efficiency indicators and of the energy efficiency potentials. The most original contribution of this study, however, is the development and application of an econometric approach to a dataset of policies and measures in the EU that applies panel analysis methods to test the effect of such policies on energy efficiency, carbon efficiency and energy security.

The descriptive analyses of sections 4.2 and 4.3 have highlighted a fairly convergent trend in the EU-15 towards a more efficient configuration of energy use, both at the aggregate level and in the industry sector, albeit with varying results in terms of performance and speed across countries and sectors. Our survey of energy efficiency policies in the EU has shown that there is indeed a significant commitment, both at the EU level and at the national level, to devise and implement policies and measures to promote energy efficiency.

For the residential sector, varying results in terms of performance and speed across countries are noticeable, but they are difficult to assess in terms of pure energy efficiency due to the intrinsic cross-country incomparability of the index, that by construction mainly allows us to track energy efficiency progress of a given country across time, but cannot tell us for any given pair of countries, which one has ever been more efficient than the other.

In the transport sector there is more homogeneity across Europe due to the overwhelming preponderance of road transport, both for passenger and freight traffic, and the fact that road transport is the mode that has improved the least over the period considered in this study.

Certainly since the 1990s there has been growing policy activity in this area in the EU. While it has surely led to a number of success stories in terms of unit efficiency (take for instance the energy efficiency labelling for electrical appliances or the mandatory standards for lighting), their ultimate effectiveness has been limited by a significant presence of the rebound effect in the residential sector. The Green Paper 'Energy' explicitly recognizes the great potential for energy efficiency gains in the transport sector, and indeed it appears clear that there is still a lot to do, in particular in terms of rethinking the pecking order of the transport mode in Europe, still severely unbalanced towards road transport.

The current situation is thus the result of a complex evolution towards not fully achieved but increasing coordination between energy efficiency policies among member states, in which EU Directives have played a major role as catalysts and harmonizing devices, but in which some significant heterogeneity is still present. It is thus interesting to draw on this diversity across countries to look at the effectiveness of energy efficiency policies in different national contexts and in terms of different indicators. A panel analysis is the ideal tool to explore this issue as it exploits a large amount of heterogeneous information by combining cross-sectional data and time series data, to obtain a gain in the efficiency of estimates.

Our panel analyses cover energy efficiency indicators, carbon efficiency indicators and energy security indicators. It turns out that quite a number of policies have had a beneficial impact on energy efficiency and carbon efficiency, measured respectively as energy intensity and carbon intensity, at the aggregated level. However, only one category of these policies (general cross-cutting policies) has also proven useful to improve the performance of aggregated energy security indicators.

The main lesson to be drawn from this analysis is that energy efficiency policies in the EU do work, but there is no one single policy able to address different policy objectives successfully, unless it is a policy so general that it naturally encompasses different sectors and modes of energy use. Thus, only broadly defined cross-cutting policies seem to have this double effect. The other seemingly surprising lesson is that there are policies, designed to improve energy efficiency, that are more effective in terms of improving energy security than in terms of their original goal. This may have to do with our choice of energy security indicators: we may have focused on the consumption of fuels that are more sensitive to certain policies, but may not have enough weight to improve the efficiency of the overall or sectoral energy mix. This is the case for instance with cross-cutting policies focused on the transport sector that have a significant effect on discouraging the consumption of oil products and therefore improve the performance of the energy security indicator that measures the dependence of the economy on oil.

Taking a more general perspective, what seems to work is the policy mix rather than this or that policy in isolation; the good news is that currently in Western Europe a policy menu is in place that has produced significant improvements in energy efficiency, has reduced the amount of carbon emissions generated by the economic system, and has contributed to a more secure energy supply for Europe.

The main limitation of this study has been data availability. In particular, policy indicators and energy efficiency indicators for new accession countries were not available, or only available for a decade or less of observations. For policy variables, the MURE database is mostly qualitative, and reports the presence and the category of the policies and measures implemented in a given country, but it does not provide systematically quantitative information about these policies (such as the funds earmarked for a given policy or the financial impact of a given tax). Future analyses can be pursued by investigating the country-specific P&Ms that contributed to energy efficiency improvements. We have looked at such P&Ms at the regional level (EU-15 plus Norway), but analyses of single countries can help us to understand whether selected policies are more effective in different countries than others.

Another limitation is that the policy database covers only efficiency- and carbon emissions-related policies, while the policy areas related to competitiveness and market liberalization are not captured. This is potentially a problem given that a more competitive market can in principle spur efficiency through more correct price signals. An indirect hint that the market reforms of the EU energy markets may also have had a role from the energy efficiency point of view is the significant impact of prices on energy efficiency.

Finally, given the unavoidable lag in data collection, the effects of the global economic crisis that started in 2008 could not be incorporated into this analysis. The crisis has resulted in a noticeable decrease in energy consumption, thus temporarily reducing the case for policy support to energy efficiency and carbon emission reduction. On the one hand, it has also temporarily reduced the momentum of the investment process in new technologies, thus slowing down the penetration of efficiency-improving technologies, particularly in the industrial sector and in new infrastructures. On the other hand the strong commitment of the EU to climate change mitigation, confirmed at the 15th Conference of the Parties (COP) in Copenhagen, suggests that the positive consequences of the crisis will not result in a relaxation of these policies in the EU.

ACKNOWLEDGEMENT

This work is part of the SECURE Project (No 213744), 7th Framework Programme. The financial support of the DG Research of the European Commission is gratefully acknowledged. The authors are grateful to Claudia Checchi and Stefan Hirschberg for their comments and suggestions. All remaining errors are ours.

NOTES

1. This chapter shows the results of analyses for the economy as a whole. Sectoral-specific analyses are available in Bigano et al. (2010).
2. CO_2 emission equivalents are computed on the basis of the global warming potential of each greenhouse gas, that is, the contribution to global warming of each gas relative to CO_2 ($CO_2 = 1$, $CH_4 = 21$, $N_2O = 310$).
3. The Shannon–Weiner index can be used to evaluate how the diversity of a given market is changing over time. The minimum value the Shannon–Weiner index can take is zero, which occurs when imports come from a single country. In this case, there would be no diversity of supply. The index places weight on the contributions of the smallest participants in various fuel markets as they provide the options for future fuel switching. Unfortunately this indicator did not yield significant results in our panel regressions.
4. Gas and oil consumption in Ktoe (thousand tonnes of oil equivalent) provided by Enerdata. GDP data provided by World Bank (2008).
5. 2004 data not available for Denmark, which registered in 2003 an indicator equal to zero.
6. TPES is defined by IEA as the sum of: *Indigenous production + imports – exports – international marine bunkers +/– stock changes.*
7. Obviously for Luxembourg the TPES has been calculated not considering marine bunkers, as this land-locked country has none.
8. The existing implemented Directives for eco-design of energy-using products are related to ballasts for fluorescent lighting (2000/55/EC), household electric refrigerators and freezers (96/57/EC), and hot-water boilers fired with liquid or gaseous fuels (92/42/EEC). These Directives were amended in July 2005 by Article 21 of Directive 2005/32/EC. The latter defines conditions and criteria for setting requirements regarding environmentally relevant product characteristics (such as energy consumption). In principle, the Directive applies to all energy-using products (except vehicles for transport) and covers all energy sources. For energy demand in households, relevant Directives are the energy labelling for electric refrigerators (2003/66/EC), electric ovens (2002/40/EC), air-conditioners (2002/31/EC), dishwashers (1999/9/EC) and household lamps (98/11/EC). Others Directives are related to household dishwashers (97/17/EC), washing machines (96/89/EC), household combined washer-driers (96/60/EC), household electric tumble driers (95/13/EC), household washing machines (95/12/EC), household electric refrigerators, freezers and their combinations (94/2/EC) and household appliances (92/75/EEC).

9. White certificates are issued by national energy authorities and certify energy efficiency improvements by eligible economic agents. They are tradeable in order to minimize the overall costs of reaching a given overall national energy efficiency target.
10. Given the vast range of possible energy security indicators, we have tested a few alternative options.
11. For the EE indexes the analysis focuses on the period 1980–2004.
12. IEA World Energy Statistics and Balances – Extended Balances, Vol. 2008, release 01.
13. Eurostat – National Accounts by 6 and 31 branches – aggregates at current prices.
14. OECD.Stat – Gross domestic product (output approach) US dollars, constant prices, constant PPPs, OECD base year (2000), millions.
15. Enerdata – World Energy database, 2007.
16. Enerdata – EmissionStat, 2007.
17. 'World Development Indicators' (WDI), World Bank (2008).
18. Eurostat – Statistics on research and development – R&D expenditure at national and regional level.
19. IEA – Energy Prices and Taxes, Vol. 2009, release 02.
20. http://www.isisrome.com/mure/.

REFERENCES

ADEME (2008), 'L'efficacité énergétique dans l'Union Européenne: panorama des politiques et des bonnes pratiques', Paris.
Bigano, A., R.A. Ortiz, A. Markandya, E. Menichetti and R. Pierfederici (2010), 'The linkages between energy efficiency and security of energy supply in Europe', Fondazione Eni Enrico Mattei FEEM Nota di Lavoro, 64-2010.
Constantini, V., F. Gracceva, A. Markandya and G. Vicini (2007), 'Security of energy supply: comparing scenarios from a European perspective', *Energy Policy*, **35**, 210–26.
European Commission (2005), Commission Green Paper, 'Energy efficiency – or doing more with less', COM(2005) 265 final, 22 June, Brussels.
Eurostat (2009), *Panorama of Transport*, 1990–2006, 6th edition, Luxembourg: Eurostat Statistical Books.
von Hirschhausen, C. and A. Neumann (2003), 'Security of "gas" supply: conceptual issues, contractual arrangements, and the current EU situation', paper presented at the INDES Workshop on Insuring against Disruptions of Energy Supply, Amsterdam, The Netherlands.
Jansen, J.C., W.G. van Arkel and M.G. Boots (2004), 'Designing indicators of long-term energy supply security', ECN-C–04007, ECN, Petten, The Netherlands.
Kendell, J.M. (1998), 'Measures of oil import dependence', EIA-DOE.
IEA (2001), *Toward a Sustainable Energy Future*, Paris: OECD/IEA.
Neff, T.L. (1997), 'Improving energy security in pacific Asia: diversification and risk reduction for fossil and nuclear fuels', PARES (Pacific Asia Regional Energy Security Project)
Odyssee (2007), 'Evaluation of energy efficiency in the EU-15, indicators and policies', ADEME/IEEA, Paris.
Ortiz, R.A., A. Bastianin, A. Bigano, C. Cattaneo, A. Lanza, M. Manera, A. Markandya, M. Plotegher and F. Sferra (2009), 'Energy efficiency in Europe: trends, convergence and policy effectiveness', MPRA Paper, 15763.
Skinner, W.C. (1995), 'Measuring dependence on imported oil', Energy Information Administration, *Monthly Energy Review*, August.
Tinbergen, J. (1952), *On the Theory of Economic Policy*, Amsterdam: North-Holland.
Tinbergen, J. (1956), *Economic Policy: Principles and Design*, Amsterdam: North-Holland.
Tipping, P., R. Antonelli, P. Boonekamp, M. Donkelaar Kroon, T. Longstaff, K. McLeod, G. Srinivasan and C. Tigchelaar (2006), 'Impact assessment on the future action plan for energy efficiency', CLWP: 2006/TREN/032, ECN, The Netherlands, http://www.ecn.nl/docs/library/report/2006/e06041.pdf.
Turton, H. and L. Barreto (2006), 'Long-term security of energy supply and climate change', *Energy Policy*, **34** (15), 2232–50.
World Bank (2008), 'World Development Indicators', World Bank Databank, available at http://data.worldbank.org/indicator.

APPENDIX

Table 4A.1   Data dictionary

Variable	Description
Country	EU-15 countries + NO
Year	1980–2006
EIfin	Energy intensity index; Final (all sectors)
EIind	Energy intensity index; Industry sector
EIoth	Energy intensity index; Other sectors
EItra	Energy intensity index; Transport sectors
EEhouOdy	Energy efficiency index; Residential sector; 1980–2004, Odyssee data
EEtraOdy	Energy efficiency index; Transport sector; 1980–2004, Odyssee data.
ESfin1	Energy security index (Total imports/TPES); Final (all sectors)
ESfin2	Energy security index (Total oil consumption/GDP); Final (all sectors)
ESind1	Energy security index (Total oil consumption/GDP); Industry sector
ESind2	Energy security index (Total gas consumption/GDP); Industry sector
ESoth	Energy security index (Gas import/Gas consumption); Other sectors
ESagter	Energy security index (Gas import/Gas consumption); Agriculture & Tertiary sectors
EShou	Energy security index (Total gas consumption/GDP); Residential sector
EStra	Energy security index; Transport sectors;
CIfin	Carbon intensity index; Final (all sectors)
CIind	Carbon intensity index; Industry sector
CIoth	Carbon intensity index; Other sectors
CIagter	Carbon intensity index; Agriculture & Tertiary sectors
Citra	Carbon intensity index; Transport sectors
CO_2hou	Per capita CO_2 emissions; Residential sector
PReleHH	Price in US$ of electricity residential (incl. taxes), Total price (US$/unit)
PReleIND	Price in US$ of electricity industry (incl. taxes); Total price (US$/unit)
PRdiesel	Price in US$ of diesel (incl. taxes); Total Price (US$/unit), Household
ShINDwdi	Industry, value added (% of GDP) (NV.IND.TOTL.ZS) WDI
R&Dpps	Total intramural R&D expenditure (GERD). Millions of PPS (purchasing power standard). All sectors. EUROSTAT
GDPppsCur	GDP per capita, PPP (current international $) (NY.GDP.PCAP.PP.CD), WDI
EnProdWdi	Energy production (kt of oil equivalent) (EG.EGY.PROD.KT.OE), WDI
PMhhT1	P&Ms Household sector – Mandatory standards for buildings
PMhhT2	P&Ms Household sector – Regulation for heating systems and hot water systems
PMhhT3	P&Ms Household sector – Other regulation in the field of buildings
PMhhT4	P&Ms Household sector – Mandatory standards for electrical appliances
PMhhT5	P&Ms Household sector – Legislative/Informative
PMhhT6	P&Ms Household sector – Grants/Subsidies
PMhhT7	P&Ms Household sector – Loans/Others
PMhhT8	P&Ms Household sector – Tax exemption/Reduction
PMhhT9	P&Ms Household sector – Tariffs
PMhhT10	P&Ms Household sector – Information/Education
PMhhT11	P&Ms Household sector – Co-operative measures
PMhhT12	P&Ms Household sector – Cross-cutting with sector-specific characteristics
PMtrT1	P&Ms Transport sector – Mandatory standards for vehicles

Table 4A.1 (continued)

Variable	Description
PMtrT2	P&Ms Transport sector – Legislative/Informative
PMtrT3	P&Ms Transport sector – Grants/Subsidies
PMtrT4	P&Ms Transport sector – Tolls
PMtrT5	P&Ms Transport sector – Taxation (other than eco-tax)
PMtrT6	P&Ms Transport sector – Tax exemption/Reduction/Accelerated Depreciation
PMtrT7	P&Ms Transport sector – Information/Education/Training
PMtrT8	P&Ms Transport sector – Co-operative measures
PMtrT9	P&Ms Transport sector – Infrastructure
PMtrT10	P&Ms Transport sector – Social planning/Organizational
PMtrT11	P&Ms Transport sector – Cross-cutting with sector-specific characteristics
PMinT1	P&Ms Industry sector – Mandatory demand side management
PMinT2	P&Ms Industry sector – Other mandatory standards
PMinT3	P&Ms Industry sector – Legislative/Informative
PMinT4	P&Ms Industry sector – Grants/Subsidies
PMinT5	P&Ms Industry sector – Soft loans for energy efficiency, Renewable and CHP
PMinT6	P&Ms Industry sector – Fiscal/Tariffs
PMinT7	P&Ms Industry sector – New market-based instruments
PMinT8	P&Ms Industry sector – Information/Education/Training
PMinT9	P&Ms Industry sector – Co-operative measures
PMinT10	P&Ms Industry sector – Cross-cutting with sector-specific characteristics
PMteT1	P&Ms Tertiary sector – Mandatory standards for buildings
PMteT2	P&Ms Tertiary sector – Regulation for building equipment
PMteT3	P&Ms Tertiary sector – Other regulation in the field of buildings
PMteT4	P&Ms Tertiary sector – Legislative/Informative
PMteT5	P&Ms Tertiary sector – Grants/Subsidies
PMteT6	P&Ms Tertiary sector – Soft loans for energy efficiency, Renewable and CHP
PMteT7	P&Ms Tertiary sector – Tax exemption/Reduction
PMteT8	P&Ms Tertiary sector – Information/Education/Training
PMteT9	P&Ms Tertiary sector – Co-operative measures
PMteT10	P&Ms Tertiary sector – Cross-cutting with sector-specific characteristics
PMccT1	P&Ms Cross-cutting – General energy efficiency/Climate change/Renewable programmes
PMccT2	P&Ms Cross-cutting – Legislative/Normative measures
PMccT3	P&Ms Cross-cutting – Fiscal measures/Tariffs
PMccT4	P&Ms Cross-cutting – Financial measures
PMccT5	P&Ms Cross-cutting – Co-operative measures
PMccT6	P&Ms Cross-cutting – Market-based instruments
PMccT7	P&Ms Cross-cutting – Non-classified measure types

5 Governing a low carbon energy transition: lessons from UK and Dutch approaches
Timothy J. Foxon

5.1 INTRODUCTION

At the Copenhagen Climate Change Conference in December 2010, the major industrialized nations, including the USA and China, set a goal of limiting the increase in global temperature due to man-made greenhouse gas (GHG) emissions to below 2°C. The scientific evidence presented in the Fourth Assessment Report of the Intergovernmental Panel on Climate Change (IPCC, 2007) and the Stern Review on the Economics of Climate Change (Stern, 2007) strongly indicates that this will require reductions in global GHG emissions of the order of 50 per cent from 1990 levels by 2050, with much higher reductions needed in industrialized countries. As energy use gives rise to the largest share of GHG emissions, this would imply a transition in the systems for the provision and consumption of energy services, particularly in industrialized countries that currently rely heavily on fossil fuels of coal, oil and gas for their primary energy sources. As Chapter 1 by Fouquet (this volume) illustrates, past energy system transitions have generally taken a long period and involved a great deal of upheaval.

This chapter examines the scale of the challenge in redesigning energy technologies, institutions and markets to meet ambitious carbon emissions reduction targets, whilst at the same time ensuring secure supply of energy services at affordable costs to consumers. It argues that the development of policies and measures should be based on a whole-systems analysis, in order to provide incentives for investment, innovation and changes to practices of energy use. As the Stern Review (2007) and others have argued, this will require a combination of measures:

- to put a price on carbon emissions, through taxes or trading schemes;
- to support research and development (R&D), demonstration and early-stage commercialization of low carbon technologies; and
- to remove institutional and other non-market barriers to the deployment of new technologies and practices.

The chapter examines and compares recent policy measures to simulate low carbon innovation in the UK, under the Low Carbon Transition Plan, and in the Netherlands, under the Energy Transition Approach. It argues that, though these approaches represent an important step forward, they still largely focus on technological changes and on the roles of government and market actors, and neglect changes in practices of energy use and the role of civil society in promoting a low carbon transition. Finally, it argues that ongoing work to develop and analyse transition pathways to a low carbon energy system in the UK could usefully aid the further development of these approaches.

5.2   THE UK LOW CARBON TRANSITION PLAN

Since the early 1990s, successive UK governments have aspired to play a world-leading role in efforts to raise awareness of the challenge of climate change and to put in place international and domestic efforts to promote climate change mitigation. As we argue below, this has coincided with a period when the political consensus has largely favoured market-based solutions to social and environmental challenges. By the mid-2000s, an all-party consensus had also been reached for the need to set out a long-term, legally binding target for carbon emissions reductions, following successful lobbying by environmental groups. This led to the passing of the Climate Change Act by the UK Parliament in 2008.

The Climate Change Act requires the UK government to set five-yearly carbon budgets, starting with the period 2008–12, to put the UK on a path to reaching an 80 per cent reduction in carbon (greenhouse gas) emissions by 2050 (compared to 1990 levels), with significant progress by 2020. The Act also established the expert advisory Committee on Climate Change to provide independent advice to the government on setting and meeting these carbon budgets and targets. The committee's First Report, published on 1 December 2008, gave recommendations for the levels of the first three carbon budget periods 2008–12, 2013–17 and 2018–22, and set out a path towards the 80 per cent reduction target for 2050 (CCC, 2008). The committee was guided by the agreement between European Union (EU) countries to set a unilateral target of an average 20 per cent reduction in GHG emissions by 2020, relative to 1990 levels, which would be increased to a 30 per cent reduction if other industrialized countries proposed similar reduction targets under a global agreement at the Copenhagen Climate Conference in December 2009 or thereafter. The committee therefore recommended interim carbon budgets that would result in a 34 per cent reduction in UK GHG emissions by 2020, relative to 1990 levels, to be increased to meet an intended target of a 42 per cent reduction by 2020, if there was a global agreement at Copenhagen. In the financial Budget in April 2009, the UK government accepted the interim carbon budgets, equivalent to a 34 per cent reduction by 2020 on 1990 levels (or an 18 per cent reduction on 2008 levels by 2020).

The UK Low Carbon Transition Plan, published by the government in July 2009, sets out a national strategy for energy and climate, in order to deliver the carbon budgets and meet the reduction target for 2020 (HM Government, 2009). As the Secretary of State of Energy and Climate Change describes in the Foreword to the Plan: 'this Plan sets out a route-map for the UK's transition from here to 2020', as part of the transition to a low carbon economy which 'will be one of the defining issues of the 21st Century' (HM Government, 2009, Foreword).

The UK Low Carbon Transition Plan sets out a range of measures for putting the country on track to meet its 2020 target, building on existing UK and EU policies, including the European Emissions Trading System (ETS). The largest new contributions to meeting the 2020 target are planned to come from a significant decarbonization of electricity supply. The Plan sets a target of 40 per cent of UK electricity from low carbon sources by 2020, including around 30 per cent of electricity from renewables, up to four demonstration carbon capture and storage (CCS) plants, and facilitating the building of new nuclear power stations. Policy measures to promote this decarbonization of electricity include: an expansion of the Renewables Obligation, which requires electricity supply companies to source an annually increasing proportion of renewable generation; financial support for CCS demonstration plants; and eased planning and licensing rules for new nuclear power stations. The Plan provides support for improving the energy efficiency of existing households and businesses, the roll-out of 'smart meters' to every home by 2020, and the introduction of 'clean energy cash-back schemes', through the application of new feed-in tariffs so that people and businesses will be paid for the use of low carbon sources for small-scale electricity and heat generation. The Plan also outlines ambitions to make the UK a centre of green industry, promising £120 million investment in offshore wind and £60 million in marine energy. Measures for reducing emissions from transport include: the EU agreement on reducing CO_2 emissions from new cars by 40 per cent from 2007 levels by 2020; a demonstration project for electric vehicles; and sourcing 10 per cent of UK transport energy from sustainable renewable sources by 2020.

In its first annual progress report, published in October 2009, the Committee on Climate Change welcomed the government's Low Carbon Transition Plan as a very comprehensive account of the opportunities for reducing emissions by 2020 and for providing an overview of the policy framework for realizing these opportunities (CCC, 2009). However, it argued that the pace of emissions reductions in the period 2003–07 was slower than that needed to meet the budget commitments, and that reductions in 2008 due to the effects of the economic recession could produce an over-rosy impression of progress against budgets and undermine steps to drive long-term reductions, in particular by reducing the carbon price within the EU ETS. Hence, the committee argued that a step-change in the pace of emissions reductions will be needed to achieve the deep emission cuts required through the first three carbon budget periods and beyond. In particular, it recommended that changes would be needed in electricity market arrangements in order to strengthen incentives for investment in low carbon generation; that current incentives for household energy efficiency improvements should be replaced by a new government-led policy for delivering improvements on a 'whole-house' and neighbourhood basis; and that support should be given for a roll-out of 1.7 million electric and plug-in hybrid cars by 2020.

In its official response to the committee's progress report, published in January 2010, the government accepted the need for a step change and renewed its commitment to implementing the Low Carbon Transition Plan. It set out the additional measures that it has subsequently taken, including the publication for consultation of six draft National Energy Policy Statements, which will guide planning decisions for large energy projects by the new Infrastructure Planning Committee (IPC), and work to assess the energy market framework to ensure that it can effectively deliver the low carbon investment needed to meet the long-term goals. Ongoing work by the energy regulator Ofgem under its Project Discovery has recently argued that far-reaching energy market reforms will be needed to ensure the security of UK energy supplies whilst achieving the climate change targets and keeping costs as low as possible for consumers and business. In particular, Ofgem noted that: 'the outcome of Copenhagen, in terms of lower carbon prices, reinforces the climate of significant uncertainty just when an unprecedented level of investment is required' (Ofgem, 2010).

In July 2010, the Department for Energy and Climate Change published work on '2050 pathways', analysing different pathways to reaching the 80 per cent emissions reduction target by 2050 (DECC, 2010a). These pathways explored different mixes of technology deployment and behavioural changes across energy supply, energy demand and non-energy sectors. These highlighted the huge challenges involved, with ambitious per capita energy demand reduction needed alongside a rapid growth of low-carbon electricity generation to enable replacement of coal and gas generation for both current uses and increasing electrification of heating and transport. They also highlighted the key uncertainties and trade-offs, including the shape of future energy infrastructure, the precise 2050 electricity generation mix and the availability of sustainable bioenergy. This work will inform the further development of energy policies by the new UK coalition government, which took office in May 2010. These policies will focus on four key areas (DECC, 2010b):

1. Saving energy through the Green Deal and supporting vulnerable consumers.
2. Delivering secure energy on the way to a low carbon future.
3. Managing the UK's energy legacy responsibly and cost-effectively.
4. Driving ambitious action on climate change at home and abroad.

Further details are to be set out in future policy documents.

This brief review of the UK Low Carbon Transition Plan and related activities highlights that the idea of a low carbon transition has been institutionalized through a series of political actions. These include the setting up of the Department of Energy and Climate Change and the passing of the Climate Change Act, which set legally binding steps for the UK government to begin the transition and created institutions and mechanisms by which it can be informed and held to account on its actions. As I shall discuss further below, this is beginning to involve a larger role for government, beyond just setting the overarching framework and leaving the implementation to 'the market', but it is still largely a top-down, expert-driven process, with the interactions between the government and the Committee on Climate Change playing a central role.

5.3   THE DUTCH ENERGY TRANSITION APPROACH

The Energy Transition Approach was enacted in the Netherlands, following the 4th Netherlands Environmental Policy Plan (NEPP) in 2000. The Plan argued that a set of persistent environmental problems – climate change, biodiversity issues, depletion of resources, threats to human health – remain to be addressed; and they require a systems approach to policy-making, in order to stimulate transitions towards sustainable energy, transport, resource use and agriculture.

This drew on reports produced by Dutch academics, working closely with policy-makers in the 'co-production' of a new strategic framework for energy innovation policy (Rotmans et al., 2001; Kemp and Rotmans, 2005, 2009; Grin et al., 2010). The approach is designed to be shaping or modulating, rather than controlling, to be oriented towards long-term sustainability goals and visions, and to be iterative and flexible, with a steering philosophy of 'goal-oriented incrementalism' (Kemp and Loorbach, 2006). This builds on a long strand of political science thinking that is sceptical about the merits of centralized planning and advocates the value of governments pursuing incremental, trial-and-error learning approaches as they attempt to reconcile different priorities and interests, notably referred to by Charles Lindblom as 'muddling through' (Lindblom, 1959, 1979). The extent to which such an incrementalist approach can deliver the radical systems transition in energy systems in the short timescales that climate change scientists argue is necessary to prevent catastrophic climate change is, arguably, a key question for the governance of energy systems.

Following the publication of the NEPP, the ministries responsible have initiated transition programmes for these four areas, with the 'Energy Transition' programme being led by the Ministry of Economic Affairs (2004). The key characteristics of this approach are systems thinking and a long-term orientation, combined with specific projects or 'transition experiments', which are typically public–private partnerships between government and stakeholders. This is based on a 'learning-by-doing' approach: undertake experiments; design learning goals into experiments; and feed back lessons into subsequent measures.

The Dutch government sees the transition approach as a way of dealing with uncertainties and avoiding apparent certainties. In its view, the government is not 'choosing' specific options, but organizing its policy around a cluster of options: the 'transition paths' (main roads). These should then enable the government to give direction to the market, whilst giving market players the opportunity to develop their own products based on their own market analysis, ambitions and entrepreneurship.

The Ministry of Economic Affairs (2004) argues that this requires a new form of concerted action between market and government ('policy renewal'), based on:

- Relationships built on mutual trust: stakeholders want to be able to rely on a policy line not being changed unexpectedly once adopted, through commitment to the direction taken, the approach and the main roads formulated. The government places trust in market players by offering them 'experimentation space'.
- Partnership: government, market and society are partners in the process of setting policy aims, creating opportunities and undertaking transition experiments, for example through ministries setting up 'one-stop shops' for advice and problem-solving.
- Brokerage: the government facilitates the building of networks and coalitions between actors in transition paths.
- Leadership: stakeholders require the government to declare itself clearly in favour of a long-term agenda of sustainability and innovation that is set for a long time, and to tailor current policy to it.

The transition paths are organized under seven themes, each led by a public–private partnership platform (Dietz et al., 2008):

1. Green raw material.
2. New gas.
3. Sustainable electricity supply.
4. Transport (sustainable mobility).
5. Chain efficiency.
6. Build environment.
7. The greenhouse as an energy source.

There are over 80 transition experiments now under way under these paths.

The Dutch government argues that these innovation processes can contribute to achieving a 30 per cent reduction in CO_2 emissions by 2020, compared to 1990 levels (Creative Energy – Energy Transition, 2008) through a combination of:

- research and development of sustainable techniques and systems;
- applying new sustainable energy systems, and learning from this experience, thus reducing the complexity and reducing costs;
- integrating sustainable systems by removing costs.

5.4   GOVERNING ENERGY TRANSITIONS

The plans of both the UK and the Netherlands represent ambitious approaches to promoting a sustainable low carbon energy transition. They have successfully developed institutional structures for promoting action and holding governments to account. The Dutch transition approach, in particular, has some desirable features (Foxon and Pearson, 2008):

1. The formulation of clear, long-term sustainability goals. This reduces uncertainties and creates a positive climate for investment in more sustainable technologies and processes. The UK's approach of creating legally binding carbon reduction targets should be even more desirable in this respect, but this also depends on creating a shared credible belief in the targets.
2. Promoting a diversity of options through experimentation. There is a value in supporting the creation of options, which may later be further pursued or discontinued.
3. Mixes of both policy instruments and technological options are needed. Measures to internalize environmental externalities through taxes or tradeable permit schemes need to be complemented by measures that promote early-stage technologies and improve the flow of information about potential solutions between actors.
4. The role of 'learning-by-doing'. The transition approach is seen as a learning process by all the actors involved. Governments are able to learn about what works and what does not, whilst private actors are given the space to try out alternatives, in the context of knowing that there are clear overall strategic goals.

However, they both also exemplify the severe challenges facing the promotion of an energy transition. Current fossil fuel-based energy systems have benefited from a long period of increasing returns to the adoption of technologies and associated institutions, due to learning, scale, adaptation and network effects, so that these systems are said to be in a state of 'carbon lock-in' (Unruh, 2000, 2002, 2006). This implies that low carbon energy technologies cannot easily substitute for high carbon alternatives, without related changes in institutions, business strategies and user practices (Foxon, 2010). Experience from past system transitions suggests that this is likely to be a slow and lengthy process, because of the time needed to build new enthusiasm, infrastructures and institutions, turn over capital stock and overcome the lock-in and alignment of the physical and human skills of the current actors to the existing dominant regime (Pearson, 2010).

The UK Low Carbon Transition Plan clearly marks a break with past UK energy and climate policies in that it seeks to provide a clear overarching framework for meeting a socially defined end-goal – an 80 per cent reduction in GHG emissions by 2050, relative to 1990 levels – whilst also incorporating other policy goals, such as ensuring security of supply. It outlines emissions reduction potentials across the main sectors of the economy, and assigns responsibility to each government department for meeting its share of the emissions reductions. There is a recognition that industry and local communities will also need to be engaged to deliver the necessary reductions. The Low Carbon Transition Plan gives a greater role for central government, but still with a strong emphasis on solutions driven by market actors.

Similarly, the Dutch energy transition plan has ambitious long-term goals for promoting a sustainable energy transition but, in practice, has emphasized technological solutions and the role of business in delivering solutions. It has been criticized for the key role that leading actors within the current regime, such as the chief executive officer (CEO) of Shell, are playing within the public–private partnership platforms (Kern and Smith, 2008).

This leads to a number of key issues for a low carbon transition that are not fully addressed within the current UK and Dutch approaches. Firstly, developing credible long-term targets and pathways. Both the UK Low Carbon Transition Plan (HM Government, 2009) and the most recent Dutch Energy Innovation Agenda (Creative Energy – Energy Transition, 2008) only contain detailed policy measures up to 2020, whereas a full transition to a low carbon economy is likely to take until at least 2050. Given that governments are often criticized for failing to look beyond the next general election, a timescale of 11–12 years into the future from publication represents an extended forward view. The scale of the 2020 target has been criticized both by environmentalists for not being sufficiently stringent to put the UK on a path to a low-carbon future (Stop Climate Chaos Coalition, 2009), and by industrial energy users for leading to unrealistic expectations for the rate of installation of renewable energy technologies and for affecting the competitiveness of UK industry through the likely resulting rises in energy prices (Nicholson, 2009). The recent publication of pathways to 2050 in the UK (DECC, 2010a) attempts to address some of these concerns, but the way in which these long-term views of different energy futures will inform current energy policy decision-making is still not clear. The transitions management literature (Loorbach, 2007) argues for the importance of developing shared goals and visions of a desired future amongst different actors, and for adopting a learning-based approach to achieving these. This suggests a greater need to discuss and come to some provisional consensus about what a longer-term future low carbon energy system might look like and how this could be reached, in order to achieve greater buy-in to current changes. Of course, there are high levels of uncertainty about future progress in different energy technologies; whether changes in other technologies such as information and communication technologies (ICTs) could help to facilitate change through some form of 'smart grid'; and the extent to which actors' practices and behaviours will change, driven by individual incentives such as price changes, and changes in wider social values and attitudes. This suggests a value in the use of scenarios to analyse the implications of these issues (Hofman et al., 2004; Hughes, 2009).

Secondly, too much emphasis on 'front-of-pipe' technological changes. Though the UK Low Carbon Transition Plan does contain incentives for energy efficiency improvements, particularly for households, the greatest emissions reductions to 2020 are proposed to come from technological changes at the 'front-of-pipe', mainly through a massive expansion of renewable electricity generation and facilitating the building of new nuclear power stations, as well as at the 'end-of-pipe' through the demonstration of capture and storage of carbon from coal-fired power stations. Similarly, the criteria for choosing niche experiments in the Dutch transition approach, such as cost-effectiveness and creating business opportunities, have been criticized for accentuating marketable technological fixes, rather than social or institutional innovation (Kern and Smith, 2008). This implicitly assumes that the future practices of individuals, communities and industries will be like those of the past, only with lower carbon inputs and outputs. The historical evidence from past system transitions shows that large-scale technological changes both shape and are shaped by wider social and cultural changes. Of course, this is not to argue that governments should seek to set out the type of future society as well as technological mix, but it does suggest the importance of facilitating debates about the type of future that people want. Again, this suggests a valuable role for scenario analysis.

Thirdly, insufficient focus on the role of civil society. Following on from the above two points is the recognition that a low carbon transition would go beyond merely a change in a few key technologies and would be driven by the need to achieve a social goal of reducing carbon emissions that would not emerge naturally out of behaviour driven by purely private goals within a market framework. In economic terms, the latter arises since market incentives to introduce low carbon technologies face two interacting market failures: the environmental market failure relating to unpriced carbon emissions and the innovation market failure relating to the fact that social returns to innovation are greater than private returns because of spillover effects (Jaffe et al., 2005). The low level of carbon pricing created by current emissions trading schemes is likely to be insufficient to drive a rapid rate of low carbon innovation and investment, as Ofgem and others have argued. From a political science perspective, the ability to set stringent enough financial and regulatory incentives for low carbon innovation is limited by the intensity of social, business and political interests lobbying against their introduction or strengthening. This suggests that a strong coalition of public, business and political interests advocating in favour of incentives for low-carbon innovation would be needed to overcome this opposition.

Of course, governments can not create social movements but they can recognize and help to stimulate wider social change. Interestingly, this recognition is apparent in statements by UK government ministers, but is difficult to put into practice in the face of widespread public distrust of governments' actions and motives. For example, the then UK Energy and Climate Change Secretary, Ed Miliband, argued in a lecture shortly before the Copenhagen conference:

> But it is people demanding change that has, throughout history, changed the world. Nowhere is this more true than in relation to climate change, where the green movement has already moved opinion in so many countries. That movement will face big challenges in the years ahead as it reaches out to a wider constituency but it is a vital part of winning the battle to create a wider consensus on climate change. (Miliband, 2009)

However, government information programmes have largely focussed on trying to persuade people to make small behavioural changes, which are often seen by people as not matching the rhetoric of the level of change needed.

The UK and the Netherlands are amongst the world leaders in establishing frameworks designed to promote a transition to a sustainable low carbon economy by stimulating innovation in energy systems. The programmes under way in the two countries have many appealing features, but their deficiencies described above highlight the scale of the challenge facing these and other countries. High levels of innovation and deployment of low carbon technologies will be necessary to achieve a low carbon transition. However, equally dramatic social and institutional changes are also likely to be necessary. These will require the application of political will and the creation of wide social movements to ensure that the incentives for low carbon innovation are not watered down or captured by current vested interests.

As we noted, the Stern Review (Stern, 2007) argued that promoting low carbon technological innovation will require:

- to put a price on carbon emissions, through taxes or trading schemes;
- to support R&D, demonstration and early-stage commercialization of low carbon technologies; and
- to remove institutional and other non-market barriers to the deployment of new technologies and practices.

The UK and Dutch approaches aim to set out a clear framework within which the support of low carbon technological innovation and early-stage commercialization and the removal of institutional barriers to their deployment can be achieved. However such frameworks are vulnerable to manipulation by corporate and other interests, as has been argued to be the case in relation to carbon trading (Spash, 2010), unless there is strong and continuing social and political support for them.

5.5 UNDERSTANDING ENERGY TRANSITIONS

The above concerns suggest the need to develop energy scenarios that address the challenges of governance of energy systems (Mitchell, 2008; Smith, 2009), as well as the technological challenges. In a project being undertaken with both engineers and social scientists from a number of UK universities, the author and colleagues are developing and analysing a set of scenario or transition pathways to a low carbon energy system in the UK (Foxon et al., 2008, 2009, 2010). This is applying the multilevel framework, developed by Dutch transitions researchers (Geels, 2002, 2005; Verbong and Geels, 2007, 2010), to formulate and examine the plausibility and acceptability of different transition pathways for a low carbon electricity-based energy system in the UK by 2050. These pathways explore different governance patterns, depending on the relative power and influence of the different categories of actors, and the mix and balance of centralized and decentralized decision-making within energy systems. Three key categories of actors are identified:

1. Government actors: this covers government departments, advisory and regulatory bodies, and the legislation they create.
2. Market actors: this covers the major vertically integrated supply companies, but also smaller market-based actors, for example emerging energy service companies (ESCOs).

3. Civil society actors: this includes not only 'end users', but also other civil society actors such as trade unions, the media and organized environmental protest movements.

These three different kinds of actors create a broadly defined 'action space' in which the current energy regime sits. Different kinds of relationships between actors exist and different forms of transition may develop, depending on the evolving balances of 'power' between these actors. Viewing these relationships and the interplay between them through this interpretive lens should provide insights into how the initial phases of transition pathways might play out within the current energy regime, and how different actors might be likely to react to transition processes. This will then inform the further development of the pathways. The specification of these pathways draws on the multifaceted experience of the project team, the insights provided by stakeholders at workshops and through 'gatekeeper' interviews, and insights from other modelling and foresight exercises. Our three initial outline pathways are (Foxon et al., 2010):

- Market rules: this envisions the broad continuation of the current market-led governance pattern, in which the government specifies the high-level goals of the system and sets up the broad institutional structures, in an approach based on minimal possible interference in market arrangements.
- Central coordination: this envisions greater direct governmental involvement in the governance of energy systems, applying some of the principles of transition management.
- 'Thousand flowers': this envisions a sharper focus on more local, bottom-up diverse solutions ('let a thousand flowers bloom'), driven by innovative local authorities and citizens groups, such as the Transition Towns movement (Hopkins, 2008), to develop local micro-grids and energy service companies.

Initial analysis of the 'gatekeeper interviews' with 32 stakeholders covering the range of energy actors has identified how different representations of UK 'public' and 'government' by different types of actors could influence which pathway will emerge (Hargreaves and Burgess, 2009). Thus, for instance, market actors tend to view the public as more or less rational consumers, sometimes in need of education to help them make more rational energy management decisions. Government actors, however, see the public as both consumers and citizens, concerned with the price of energy services as well as with their local community and environment, but facing real limits to their power to influence change purely as consumers. Civil society actors see the public as a complex and varied group, with multiple roles and identities, but as being marginalized in wider debates about energy futures.

Market actors tend to see government as best placed to set policy, but as suffering from incompetence, and so call for government to set a strong policy framework and then to disappear to let the market deliver. Civil society actors tend to see government as suffering from bias and lack of transparency rather than incompetence, and so they call for strong government and leadership rather than 'dancing to the tune of industry'. The self-representations of government actors changed over the period of the interviews from September 2007 to October 2009. In the early interviews, the emphasis was on the government withdrawing and handing over decision-making to expert bodies, such as the Committee on Climate Change, and to the market. In later interviews, the emphasis had changed to a view that markets alone are unlikely to deliver the radical changes needed to meet the targets, and that stronger government action was beginning to be put in place, stimulated by the strengthening climate science and the economic analysis of the Stern Review. Which of these representations of 'public' and 'government' gains wider credence could strongly influence which pathway is followed (Hargreaves and Burgess, 2009).

The plausible mix of generation technologies and demand reduction measures under each of these pathways is now being investigated by the project team, and a whole-systems sustainability appraisal is being conducted using the techniques set out in Chapter 2 by Hammond and Jones (this volume).

5.6   CONCLUSIONS

This chapter has reviewed policy frameworks being applied in the UK and the Netherlands to promote innovation in low carbon energy technologies for a transition to a sustainable energy future. It has argued these have begun to create new institutional structures that form a credible incentive structure for long-term low carbon innovation, recognizing the need for a diversity of options, for learning what does and does not work, and for promoting public–private partnerships. However, it has argued that most of the focus so far has been on technological innovation, and a complementary focus on wider social innovation is needed, with greater involvement from civil society to create credible and sustainable pathways to a low carbon future. The work of the author and colleagues on transition pathways to a sustainable low carbon energy future for the UK aims to contribute to this process.

REFERENCES

Committee on Climate Change (CCC) (2008), 'Building a low-carbon economy – the UK's contribution to tackling climate change', First Report, 1 December, http://www.theccc.org.uk/reports.
Committee on Climate Change (CCC) (2009), 'Meeting carbon budgets – the need for a step change', First Annual Report to Parliament, http://www.theccc.org.uk/reports.
Creative Energy – Energy Transition (2008), 'Energy innovation agenda', http://www.senternovem.nl/energytransition/.
Department of Energy and Climate Change (DECC) (2010a), '2050 pathways analysis', http://www.decc.gov.uk/en/content/cms/what_we_do/lc_uk/2050/2050.aspx.
Department of Energy and Climate Change (DECC) (2010b), 'Annual energy statement: DECC departmental memorandum', http://www.decc.gov.uk/en/content/cms/what_we_do/uk_supply/aes/aes.aspx.
Dietz, F., H. Brouwer and R. Weterings (2008), 'Energy transition experiments in the Netherlands', in J. Van den Bergh and F. Bruinsma (eds), *Managing the Transition to Renewable Energy: Theory and Practice from Local, Regional and Macro Perpsectives*, Cheltenham, UK and Northampton, MA, USA: Edward Elgar, pp. 217–44.
Foxon, T.J. (2010), 'A co-evolutionary framework for analysis a transition to a sustainable low carbon economy', SRI Working Paper, University of Leeds, http://www.see.leeds.ac.uk/research/sri/working_papers/.
Foxon, T.J., G.P. Hammond and P.J. Pearson (2008), 'Transition pathways for a low carbon energy system in the UK: assessing the compatibility of large-scale and small-scale options', 7th BIEE Academic Conference, St Johns College, Oxford, 24–25 September.

Foxon, T.J., G.P. Hammond and P.J. Pearson (2010), 'Developing transition pathways for a low carbon elec-
tricity system in the UK', *Technological Forecasting and Social Change*, **77**, 1203–13.
Foxon, T.J., G.P. Hammond, P.J. Pearson, J. Burgess and T. Hargreaves (2009), 'Transition pathways for a
UK low carbon energy system: exploring different governance patterns', paper for 1st European Conference
on Sustainability Transition: Dynamics and Governance of Transitions to Sustainability, Amsterdam, 4–5
June, www.lowcarbonpathways.org.uk.
Foxon, T.J. and P.J. Pearson (2008), 'Overcoming barriers to innovation and diffusion of cleaner technologies:
some features of a sustainable innovation policy regime', *Journal of Cleaner Production*, **16** (1), Supplement
1, S148–S161.
Geels, F.W. (2002), 'Technological transitions as evolutionary reconfiguration processes: a multi-level perspec-
tive and a case-study', *Research Policy*, **31**, 1257–74.
Geels, F.W. (2005), *Technological Transitions and System Innovations: A Co-evolutionary and Socio-Technical
Analysis*, Cheltenham, UK and Northampton, MA, USA: Edward Elgar.
Grin, J., J. Rotmans, J. Schot, F. Geels and D. Loorbach (2010), *Transitions to Sustainable Development: New
Directions in the Study of Long Term Transformative Change*, New York, USA and Abingdon, UK: Routledge.
Hargreaves, T. and J. Burgess (2009), 'Constructing energy publics', presentation at Transition Pathways
workshop, Loughborough, 9–10 November.
HM Government (2009), *The UK Low Carbon Transition Plan: National Strategy for Climate and Energy*,
London: Stationery Office, http://www.decc.gov.uk/en/content/cms/publications/lc_trans_plan/lc_trans_plan.
aspx.
Hofman, P., B. Elzen and F. Geels (2004), 'Sociotechnical scenarios as a new tool to explore system innova-
tions: co-evolution of technology and society in the Netherlands' energy system', *Innovation: Management,
Policy and Practice*, **6** (2), 344–60.
Hopkins, R. (2008), *The Transition Handbook: From Oil Dependency to Local Resilience*, Totnes: Green Books,
http://www.transitiontowns.org/.
Hughes, N. (2009), 'Using scenarios to bring about low carbon energy transitions: lessons from transitions
theory and the scenario building tradition', Transition Pathways Working paper, http://www.lowcarbon-
pathways.org.uk/lowcarbon/publications/.
Intergovernmental Panel on Climate Change (IPCC) (2007), *Third Assessment Report*, Cambridge: Cambridge
University Press.
Jaffe, A., R. Newell and R. Stavins (2005), 'A tale of two market failures: technology and environmental
policy', *Ecological Economics*, **54** (2–3), 164–74.
Kemp, R. and D. Loorbach (2006), 'Transition management: a reflexive governance approach', in J.-P. Voss,
D. Bauknecht and R. Kemp (eds), *Reflexive Governance for Sustainable Development*, Cheltenham, UK and
Northampton, MA, USA: Edward Elgar, pp. 103–30.
Kemp, R. and J. Rotmans (2005), 'The management of the co-evolution of technical, environmental and social
systems', in M. Weber and J. Hemmelskamp (eds), *Towards Environmental Innovation Systems*, Berlin:
Springer Verlag, pp. 33–55.
Kemp, R. and J. Rotmans (2009), 'Transitioning policy: co-production of a new strategic framework for
energy innovation policy in the Netherlands', *Policy Sciences*, **42**, 303–22.
Kern, F. and A. Smith (2008), 'Restructuring energy systems for sustainability? Energy transition policy in the
Netherlands', *Energy Policy*, **36**, 4093–103.
Lindblom, C. (1959), 'The science of 'muddling through'', *Public Administration Review*, **19**, 79–88.
Lindblom, C. (1979), 'Still muddling, not yet through', *Public Administration Review*, **39**, 517–26.
Loorbach, D. (2007), *Transition Management: New mode of governance for sustainable development*, Utrecht:
International Books.
Miliband, E. (2009), 'The politics of climate change', the Ralph Miliband Lecture, London School of
Economics, 19 November.
Ministry of Economic Affairs (The Netherlands) (2004), 'Innovation in energy policy – energy transition: state
of affairs and way ahead', http://www.senternovem.nl/energytransition/.
Mitchell, C. (2008), *The Political Economy of Sustainable Energy*, Basingstoke: Palgrave Macmillan.
Nicholson, J. (2009), speech by Director, Energy Intensive Users Group at CBI Energy Conference, 22
October.
Ofgem (2010), 'Action needed to ensure Britain's energy supplies remain secure', Ofgem press release, reference
number R/5.
Pearson, P. (2010), 'Past and prospective energy transitions: insights from historical experience', presentation
at Transition Pathways workshop, 24 February.
Rotmans, J., R. Kemp and M. van Asselt (2001), 'More evolution than revolution: transition management in
public policy', *Foresight*, **3** (1), 15–31.
Smith, A. (2009), 'Energy governance: the challenges of sustainability', in I. Scrase and G. MacKerron (eds),
Energy for the Future: A New Agenda, Basingstoke: Palgrave Macmillan.

Spash, C. (2010), 'The brave new world of carbon trading', *New Political Economy*, **15** (2), 169–95.
Stern, N. (2007), *The Economics of Climate Change – the Stern Review*, Cambridge: Cambridge University Press.
Stop Climate Chaos Coalition (2009), 'Act fair and fast', policy statement, http://www.stopclimatechaos.org/policy/act-fair-and-fast.
Unruh, G.C. (2000), 'Understanding carbon lock-in', *Energy Policy*, **28**, 817–30.
Unruh, G.C. (2002), 'Escaping carbon lock-in', *Energy Policy*, **30**, 317–25.
Unruh, G.C. (2006), 'Globalizing carbon lock-in', *Energy Policy*, **34**, 1185–97.
Verbong, G. and F. Geels (2007), 'The ongoing energy transition: lessons from a socio-technical, multi-level analysis of the Dutch electricity system (1960–2004)', *Energy Policy*, **35**, 1025–37.
Verbong, G. and F. Geels (2010), 'Exploring sustainability transitions in the electricity sector: with socio-technical pathways', *Technological Forecasting and Social Change*, **77**, 1214–21.

PART II

ENERGY AND ECONOMICS

6 How energy works: gas and electricity markets in Europe

Monica Bonacina, Anna Creti and Susanna Dorigoni

6.1 INTRODUCTION

Electricity is produced and delivered in a four-stage, vertically interdependent process involving generation (the production of electrical energy), transmission (transportation of this energy along high-voltage cables), distribution (transportation at lower voltages to final customers) and retailing (advertising, branding, contract bundling and billing for final customers). The same structure applies to gas, except that Europe mainly imports this commodity from foreign countries, as domestic production is very low.

Around the turn of this century, electricity and gas market reforms opened up closed markets to competition within the European Union (EU). Electricity and gas transmission and local distribution, involving large sunk capital costs, remain natural monopolies and there is little scope for actual competition; typically, each European country has one company operating its national transmission network and a number of regional local monopolies operating its distribution networks.

The electricity and gas liberalization programme can be divided into three separate stages. The first stage was the adoption of the Directives on price transparency and on the transit of electricity and gas. The second stage consisted of Directive 96/92/EC laying down rules for the internal market in electricity, and Directive 98/30/EC concerning the internal market in natural gas. Each Directive reflected the peculiarities of the sector concerned, but both followed similar approaches including introducing phased minimum opening levels of liberalization of demand, non-discriminatory third-party access to networks and essential facilities such as gas storage, and unbundling. Although the original gas and electricity Directives made significant contributions towards the creation of an internal market, material shortcomings were identified. Consequently, new electricity and gas Directives (respectively 2003/54/EC[1] and 2003/55/EC[2]) were adopted in June 2003. This is the third stage, for which the following was agreed: in 2004, full opening of the gas and electricity markets for professional users; in 2007, full opening of the gas and electricity markets for domestic consumers.

A competition enquiry into the electricity and gas sector, published in January 2007, revealed some 'serious malfunctions' that prevented competition. Corrective action was promised by the EU executive, which tabled a further package of proposals in September 2007. After long negotiations, Directive 2009/72/EC for electricity and Directive 2009/73/EC for gas actually shape the new development of EU energy markets and networks. The 'third energy package' provided companies in the member states with three options for unbundling gas and electricity production from supply provision.[3] Moreover, the powers and duties of national regulators are reinforced, as well as the cooperation between national transmission system operators for gas and electricity.[4]

This chapter illustrates how energy works, in the scene set by the European Directives since 2000. We focus on those issues that illustrate the parallel evolution of gas and electricity markets. Although an assessment of the implementation of the institutional changes under way in the European Union is beyond the scope of this chapter,[5] we explain the lesson learned from the liberalization experience of electricity and gas markets in Europe (sections 6.2 and 6.3). We turn then to the discussion of some topics pertaining to gas and electricity networks in the European context (sections 6.4 and 6.5). We briefly conclude with some remarks on the future of energy markets in the context of new environmental challenges (section 6.6).

6.2 ELECTRICITY MARKETS: HOW TO EXCHANGE A NON-STORABLE COMMODITY

The institutional framework set by the European Directives lays out the need for competitive electricity markets. However, identifying a benchmark design for competitive power markets is a never-ending concern in the debate on electricity sector restructuring (Joskow and Schmalensee, 1983; Schweppe et al., 1988; Wilson, 1998; von der Fehr and Harbord, 2002; Stoft, 2002; Rious, 2009). Uncertainty about how best to support competition, political barriers and, last but not least, physical and economic attributes of the carrier,[6] have all contributed to this enduring trend. Putting an end to this dispute is of the utmost importance and this section is a first step towards this direction. One may think that ill-conceived electricity markets would have little chance of surviving. Unfortunately, this may not be the case. Experience provides numerous examples of basic market design flaws which endure for extended periods, even after mistakes are identified. Why? The motive is in the distribution of costs and benefits (if any) among market participants. Flaws may not be such for every interested party. Errors may advantage some market participants over the others and, if the former group is sufficiently organized, it can successfully lock-in the system along unsuitable trends (Newbery and Pollitt, 1997; Cramton, 2003). After setting the guidelines for effective and efficient markets for power, as they have emerged from the 'reform of the reforms' (Joskow, 2006), we discuss the consistency of actual market designs in the EU with theoretical benchmarks.

Academics have come to an agreement on the reference frame (Hunt, 2002; Joskow, 2005, 2008a, 2008b; Sioshansi, 2008). The creation of wholesale energy market institutions is among the prerequisites[7] for the overall – textbook – architecture for the development of competitive markets for power. And the manifold scopes of these institutions should take in the day-ahead and real-time balancing of power requests, cost-reflective allocation of scarce network transmission capacity, timely and consistent response to accidental outages of both generation and transmission facilities and, more extensively, any facet linked to efficient trading of power. Doubts remain on implementation details which, nevertheless, are responsible for actual performance: the market can be centralized or decentralized; it can include ancillary services or not; it can be based on physical or financial obligations; these contractual obligations may be of a financial or a physical nature; they could be customized or standard; participation in wholesale markets can be mandatory or voluntary; secondary markets can be favoured or discouraged; and so on. The solution to what could have been a theoretic stalemate has come from national experiences. Since the mid-2000s, empirical evidence from early market restructuring has provided the criteria to disentangle the most promising alternatives among the contending options, as we detail in the following.

6.2.1 Centralized Dispatch and Decentralized Bilateral Trading: From Substitutes to Complements

Decentralized bilateral exchanges for financial contracts should constitute a core component of a centralized dispatch system for physical trading as these institutions meet complementary but substitutive needs. Let us review the core features of these challenging tools to stress how they add on to each other.[8]

Bilateral trading involves only two agents – a buyer and a seller – which enter into contracts without involvement, interference or facilitation from a third party. If there is one word to describe this institution, it would be 'discretion'. Buyers and sellers may choose among customized long-term contracts, over-the-counter trading and electronic trading, depending on their actual constraints.[9] Parties are free to set the price of their transaction; no official price exists. Summary statistics are gathered and published by independent reporting services but no obligation exists in reporting the details of negotiated contracts, which may be kept private.[10] The flexibility of bilateral trading is all the while its main strength and weakness. Participation is enhanced by tailoring and long-term hedging opportunities. As the contract expires, cash flows take place among contracting parties. If the committed price exceeds the actual spot market price, the buyer reimburses the seller for the difference; otherwise the reverse occurs. Transaction costs and strategic attitudes have not proved to be significant (European Commission, 2007). However bilateral trading encounters fundamental problems with physical flaws and real-time imbalances which cannot be managed properly by decentralized tools. We can therefore conclude that, albeit a large share of power can be traded through unmanaged open markets, these are unsuitable to keep power systems reliable. Here enters the role of central dispatches.

The organization and functioning of a central dispatch system is more challenging than that of bilateral trading. A centralized dispatch system provides standardized, physical – both day-ahead (spot) and real-time – obligations for handling power shortages (and excesses), thus complementing the advantages of decentralized bilateral markets with the effective and efficient, centralized, delivery of the underlying commodity. Positions must be physical and binding at central dispatch. Moreover, as imbalances must be corrected faster than in a conventional market, bid-based auction markets are used.

Even though variants are possible, the operation is essentially as follows. Generating companies submit bids to supply a certain amount of power at a certain price for a certain period. At gates closure, the system operator – which is the body in charge of the clearing of positions – collects and ranks bid schedules (in order of increasing price) to get the supply curve of the market. As demand is poorly elastic, it is assumed to reduce to a vertical line at the value of the load forecast. The equilibrium price in such a system is not the conventional intersection of demand and supply requests, but it is obtained as the 'shadow value' of a constrained maximization programme which includes among its inputs operating details, requests and purported costs. Therefore the schedule of real-time daily equilibriums is obtained by the system operator running security-constrained dispatch algorithms which incorporates both requests and the physical topology of the power network.[11] The resulting market clearing price is the official price for transactions. Generators having submitted offers at a price lower than or equal to this equilibrium price are instructed to produce, and consumers are informed of the amount of energy that they are allowed to withdraw from the system at the clearing price. As in the case of flexible decentralized mechanisms, inflexible tools have their own limitations and strengths. Centralized dispatches are timely, transparent and integrate every aspect of power system operations, thus minimizing transaction costs, capturing productive efficiency goals and ruling out imbalance concerns.[12] But the cost is high: mandatory participations, exposure to widespread market risks, no hedging opportunities, limited time span (transactions may refer, at least, to day-ahead operations), uniform contract forms, binding physical commitments, computer-driven equilibria,[13] hard procedures, and so on. All these in turn may unevenly affect liquidity and public acceptance.

Concerns about the long-term performance of power systems reinforce the idea that centralized and decentralized institutions complement each other. Questions on whether wholesale markets would produce adequate and accurate generation and transmission investment incentives to balance supply and demand (so as to match consumer valuations of reliability) have been raised since the transition to competitive power systems began. Uniform energy pricing and operating reserves, if any of the latter exist,[14] offer a first (physical and short-time oriented) answer to the issue; long-term contracts are the decentralized financial counterpart to the same problems. In this respect, uniform pricing outperforms pay-as-bid rules[15] in that it leaves inframarginal generators what is usually referred to as a 'scarcity rent' which helps to recover not only generators' operating costs, but also fixed capital expenditures. Operating reserve services may add further rewards to generators[16] and work as out-of-market means either to accelerate restructuring or to correct spot market signals.[17] The flaw of uniform pricing (and operating reserves) is that it provides short-term signals which – as for the properties of energy carriers – are highly unstable, while long-term, sunk investment decisions dictate for stable pricing frameworks. Bilateral systems may compensate this fallacy by securing the environment and thus providing the right incentives to investors. Therefore a competitive power market consisting of a centralized (fully integrated) energy market for short-run, standardized, physical needs and a decentralized system of bilateral trading for long-term, tailored, financial requests may be a sound guide for successful reform.

6.2.2   Electricity Markets Design in the EU

Keeping in mind the benchmark design above, we discuss the extent to which theoretical guidelines have evolved into practice for the European Union, and the likely consequences of actual flaws. Rademaekers et al. (2008) report that in 2007 wholesale markets accounted for 1.92 TWh, while bilateral trading contributed with almost 6.30 TWh. The European electricity market is not equally divided among pools and bilateral trading. Years have passed since Directives 96/92/EC and 98/30/EC entered into force, but we still have no single power market in the EU but a conglomerate of seven heterogeneous regional markets (that is, Baltic, Central East, Central South, Central West, Northern, South West and France–UK–Ireland) that are more or less physically interconnected with each other. Moreover, buyers' and sellers' needs are poorly consistent with current jeopardized bilateral paths,[18] and unsatisfactory investment levels can obtain (Glachant, 2005). Bilateral tradings are essential drivers for sunk and not recoverable decisions; therefore, actual scepticism about long-term contracts is not only inconsistent with empirics (European Commission, 2007) but may also cost even more than the fallacies it would hamper.[19] For instance, current interest in operating reserves may be explained by the attempt to circumvent the policy mistrust towards bilateral trading.

A further issue is the heterogeneity among wholesale designs (Rious, 2009). However, waiting for an unconstrained market to recover from design flaws could be infeasible and extremely expensive.[20] Quoting Joskow (2008a): 'electricity sector reforms have significant potential benefits but also carry the risk of significant potential costs if the reforms are implemented incompletely'.

6.3   GAS MARKETS: FROM LONG-TERM CONTRACTS TO HUBS

Due to the high costs of transportation, be it via pipeline or liquefied natural gas (LNG), the natural gas market is segmented on the basis of distance. The main world markets are in the US, Japan and Europe, but all of them remain regional in scope. This structure is reflected in the way in which natural gas is priced.

The 'market value' or replacement value principle is the basis for gas marketing. According to this principle the price of gas is linked to the price of alternative fuels that differ according to different gas consumers (gas uses): for instance, gas oil for small-scale users and fuel oil for large-scale users. The market value principle means that consumers do not have to pay more for gas than for competing alternative fuels and, on the other hand, that they do not pay much less. In the framework of the market value principle the negotiation between importers and exporters occurs with reference to the 'netback value' of gas which is calculated as the price of the cheapest alternative fuel diminished by the cost of transporting gas from the border to the customer, and the cost of storing gas to meet fluctuations of demand.

The netback value is in other words calculated on the basis of the value of competing energies backed to the border of the buyer's country by deducting the costs of transportation and distribution of the buyer. In this way the netback value can be defined as the maximum selling price of gas: should the latter be higher, consumers would switch to the backstop fuel. The minimum selling price of natural gas consists in the price that allows the producer to cover extraction and transportation costs (the so-called 'cost plus value'). The difference between the netback value and the cost plus value constitutes a rent that is shared among exporters and importers according to their bargaining power.

This concept would ensure a reliable sales volume for the seller at prices as close as possible to what can be sold in competition with other energies in the market. This way the netback calculated back to the wellhead provides for the maximum specific rent which can be obtained from the market, supplied without losing competitiveness. On the other hand, it allows marketing of the gas while offering a reasonable margin to the buyer. Risks related to price movements of the competing energies are mainly carried by the producing country, while the buyer takes the volume risk linked to marketing.

In fact, apart from price the major elements incorporated in gas export contracts consist in a long-term supply obligation balanced by a long-term off-take obligation ensured by the minimum-pay concept (the so-called take-or-pay clause) and the usually long duration of the commitment (up to 30 years) in order to assure the payback of the investment (in infrastructures and production).

Also, the destination clause is an instrument often included in the contract. These clauses exclude the reselling of the gas to a third country, thereby protecting the exporter's position by preventing arbitrage operations to the detriment of the seller on the basis of any price differentials in different downstream markets.

Up to a few years ago the competitive situation of gas was dominated to an extent by heavy fuel oil used by large customers; more recently, with the increasing penetration of gas in the residential and commercial market, the mix shifted to light fuel oil. Today a gas import price formula would typically have a share of 60–65 per cent pegged to light fuel, with the rest pegged to indices reflecting the competitive position in the industrial and power sector, mainly against heavy fuel oil.

While gas oil and heavy fuel oil are the most common competing fuels, the concept also works with reference to other competing energies, like coal or electricity, but also gas itself in competitive markets. In particular it was with the Interconnector becoming operational in 1998 that the issue of gas-to-gas competition was tackled in the price reviews by introducing a limited share in the formula reflecting gas-to-gas competition that in the UK led to the development of the sole European gas hub, the National Balancing Point (NBP) where gas is traded on a spot basis with a price that is decoupled from the price of oil. Nevertheless this market shows a limited liquidity and a low churn rate, suggesting that competition in Europe has not taken off yet.

It should however be considered that new trends are developing on the European gas market, as confirmed by the decreasing duration of import contracts and the decreasing extent of the take-or-pay clause, both being explained by the increase in the regulatory risk brought about by the liberalization process (Creti and Villeneuve, 2004). Such changes are taking place mostly in the LNG market.

Almost all European LNG markets are organized as a bilateral monopoly, given that a single gas importer buys gas from a single gas exporter. With the exception of Great Britain, the number of companies operating in import and production of natural gas is very limited. The level of concentration seems to be particularly high in the Netherlands and France, followed by Italy, Spain and Germany as confirmed by the Benchmarking Report of the European Commission (2009).

Nevertheless considering planned and under-construction investments, both on the liquefaction side and on the regasification side, it is possible to argue that the LNG market will be characterized by the presence of new competitors in the very near future. Moreover, most of the new plants are owned by fairly new operators on the gas market. Moreover, very often just part of the regasification capacity is covered by a long-term take-or-pay contract and it is very likely that a considerable part of it will be used for spot transactions, adding in this way to hub liquidity.

Since 2005, the share of LNG on global trade movements has increased from 23 to 32 per cent. In Europe the growth has been faster with the weight of LNG trade on total movements moving from 8 to 14 per cent. In Europe spot transactions have been growing even more rapidly: their weight on total LNG trade has moved from 10 to 25 per cent (Brito and Hartley, 2007).

Despite these new trends on the market, the DG Competition of the European Commission, in its Energy Sector Inquiry, laments a persisting lack of competition on the European gas market. According to its analysis the concentrated structure of the sector can be explained on the basis of still existing barriers to entry on the market mainly represented by long-term import contracts and by vertical foreclosures, calling for a revision in regulation towards a more aggressive approach, especially with reference to networks and the allocation of their capacity.

6.4   ELECTRICITY TRANSMISSION NETWORKS: RELEVANT ISSUES IN THE EUROPEAN CONTEXT

An electric transmission network consists of high-voltage power lines that connect different locations, referred to as nodes.[21] Electricity transmission, which defines the activity of transporting electricity over a high-voltage network (typically over long distances), is an unusual product, because the marginal costs at one location depend in what is happening elsewhere on the transmission system. This specific aspect makes the economic evaluation of transmission infrastructure quite difficult.

One approach to transmission pricing, developed by Schweppe et al. (1988) and known as 'spot pricing' (or 'nodal pricing'), attempts to base prices on these real-time marginal costs. Nodal prices are the prices that allow the decentralization of the optimal dispatch of power through a network.[22]

According to the nodal price theory, when the network is optimally dispatched, at each node the marginal utility of power is equal to its marginal cost. From one node to another this marginal valuation of electricity can vary, depending on the capacity of the connecting lines as compared with the flows of energy. If no line is congested and there are no power losses, electricity is valued the same throughout the network, so that there is a unique energy price at all nodes. When some lines are congested, instead, differences in marginal nodal valuations reflect what one might think of as the 'shadow value' of the lines. In any event, if the market is perfectly competitive at each end and the grid operator is neutral, the resulting allocation is one of first-best. Therefore, nodal prices are entry–exit tariffs, computed on the basis of optimized marginal valuations.

In more complex networks, because flows of energy cannot be controlled in real time, and because electricity follows the path of least resistance (by Kirchoff's laws), congestion on one line not only affects nodal prices at its own two ends, but also at every other node somehow affected by the flows originating from or finishing at these two buses. Thus, in a so-called 'meshed network' all nodal prices vary continuously with the load, regardless of where injections and withdrawals of electricity actually take place: this is the result of the phenomenon of loop flows (Crampes and Laffont, 2001).

The marginal transmission pricing, although theoretically optimal, does not provide enough revenues to compensate the transmission system owner; it is very complicated and non-transparent and, finally, it may not be politically implementable. The two simple alternatives to the marginal pricing are postage stamp and contract path pricing. Postage stamp methodology is the simplest as it allocates a uniform pro-rata transmission price to all the transactions without regard to the location of the buyer and the seller. As such, the methodology completely neglects any transmission effects. The contract paths methodology works as follows. If there is a contract from country A to country B, the two countries determine arbitrarily the physical paths on which the electricity flows. In fact, however, electricity flows on a number of physical paths which usually are not properly reflected in the contract path agreement.

6.4.1   Postage Stamp Tariffs and Contract Paths: The Choices of the European Union

The regulated postage stamp pricing system is now imposed in Europe for pricing electricity at the national level. This system involves fixing a toll independent of the distance separating the supplier and the consumer. The toll depends on both the power capacity reserved and the rate of utilization of this capacity and it is generally paid at the exit of the network. In order to avoid foreclosure behaviour, the rule 'use it or lose it' is now imposed by the European Commission for reserved capacity (Percebois, 2008).

The postage stamp tariff creates some inefficiencies. By developing a 13-node model of the transmission system in England and Wales, incorporating losses and transmission constraints, Green (2007) shows the inefficiencies of postage stamp prices. Green compares the scenario with optimal prices to that with uniform prices for demand and for generation, redispatching when needed to take account of transmission constraints. Moving from uniform prices to optimal nodal prices could raise welfare by 1.3 per cent of the generators' revenues, and would be less vulnerable to market power. It would also send better investment signals, but create politically sensitive regional gains and losses.

Cross-border transmission capacity allocation in Europe relies on a contract path model and a physical transmission rights (PTR) framework.[23] According to EC regulation 1228/2003 and subsequent decision 2006/770, explicit or implicit auction mechanisms are an appropriate market-based measure to allocate available cross-border capacities to market participants.

Explicit auctions commonly describe the concept that a transmission system operator (TSO) auctions off available cross-border transmission capacity to market participants. According to this model the two TSOs of the systems between which congestion exists sell their interconnector capacity to the party with the highest bid. The auction can be designed in different ways with regard to bidding mechanisms and time periods for auctioning (days, weeks, months, years). The explicit auction can be carried out on a load-flow and non-load-flow basis. The flow-based auctions are superior and not only include the typical commercial part but also account for the simultaneous physical constraints on the different transmission borders resulting from possible schedules of cross-border exchanges. The allocation of revenue resulting from the auction among the TSOs concerned remains the most essential issue and it is immediately associated with the regulatory incentives to extend the interconnection and relief congestions.

Based on the impression that the sequential operation of capacity and energy markets may lead to suboptimal results as market parties would need to anticipate future energy market outcomes (for example one year ahead) when buying PTRs, the concept of 'implicit auctions' was brought forward. The underlying idea of implicit auctions is that capacity and energy are auctioned simultaneously. Market parties would buy and sell energy on a market platform, and the market operator together with TSOs would implicitly ensure that grid capacity is sufficient to guarantee the feasibility of the trades.

These cross-border implicit auctions are usually referred to as either market coupling (if two or more power exchanges of national electricity markets couple their price zones), or market splitting (if one power exchange splits an area into several price zones in the case of congestion between them).

A look at the currently running implicit auctions confirms that all of them have been established in radial parts of the European electricity grid, that is, over cables such as between Germany and Denmark, or between radially aligned countries such as Spain and Portugal, or France, Belgium and the Netherlands. However, in a meshed grid, as the European one, the contract-path model becomes increasingly unwieldy. At this point, the typical reaction is to try to track and trace somehow the flows associated with electricity exchanges and include them in the transmission rights. This leads straight to the flow-based approach.

Actually, there are two ongoing projects in Europe that aim at introducing a flow-based capacity allocation based on a zonal grid model, namely the flow-based explicit capacity auctions of the Central–East Europe regional initiative and the flow-based market coupling of the Central–West Europe regional initiative, started in 2010 (ERGEG, 2008).

Finally, let us mention that in 2008 a Project Coordination Group of experts was given the task to develop an EU-wide 'target model' for the integration of the regional electricity markets. In this target model the exchange of energy and the access to interconnection capacity are bundled, that is, a single price-coupling mechanism is to be implemented across all European countries.

The economic literature indicates that both implicit and explicit auctions lead to a welfare-maximizing outcome in a competitive market, with full information, no uncertainty and perfect foresight – same prices and quantities, consumer and producer surplus as well as congestion revenues (Ehrenmann and Neuhoff, 2009).[24] Although commonly applied in theoretical work to derive rigorous results, these assumptions are not realistic and any deviation from them can create inefficiencies. For instance, when generators' strategic behaviour is taken into account, the welfare properties of market integration become extremely complex to predict: market structure has a crucial role in determining the level of competition. Several authors have used simple (two-node) and meshed (three-node) networks to study the behaviour of generators in monopoly and oligopoly markets, and have examined the effects of different capacity allocation mechanisms on competition, generally assuming perfect foresight. The way transmission capacity is allocated to market participants is relevant in defining the efficiency properties of the integrated market and it is related, in no simple manner, to the question of market power.[25]

Computational models are extremely useful to evaluate different market designs under realistic assumptions. To this end, large-scale market models are built, calibrated on real markets and solved with mathematical programming techniques.[26]

An interesting branch of the literature has specifically focused on the inefficiencies of explicit auctions. In general the empirical literature (Newbery and McDaniel, 2002; Tornquist, 2006; Ehrenmann and Neuhoff, 2009; Kristiansen, 2007a, 2007b) shows that with implicit auctions: netting of flows in opposite directions becomes feasible, which significantly increases the cross-border capacity; cross-border capacity is allocated as a function of the price differential in the two market areas; and correct signals prevail regarding the value of interconnector capacity.

Another matter that influences the efficiency of an integrated market is the choice of the transmission model. Although suboptimal with respect to a nodal representation, a zonal model is an acceptable simplification when: (1) certain network areas can be identified as internally well meshed and can thus be considered as 'single' nodes (or single price areas) for the calculation of the day-ahead prices; and (2) a balanced transaction within a single zone does not significantly affect interzonal flows (Perez-Arriaga and Olmos, 2005). Nonetheless, the literature has indicated that there might be gains from increased coordination between TSOs – an indication well received by the European Commission with the Third Energy Package (Bjorndal and Jornsten, 2007; Glachant et al., 2006). In this perspective, ISO is a better choice if coordinating regional interconnected power systems generates benefits through the increase in cross-border competition, and the internalization of cross-border externalities is the most important criterion (Leveque et al., 2009).

6.5   GAS NETWORKS: RELEVANT ISSUES IN THE EUROPEAN CONTEXT

Problems arise in designing the regulation of essential facilities regarding both the natural gas sector and the electricity sector. Nevertheless, networks play a more important role in the natural gas industry. In fact, while it is true in both cases that competition can develop in contestable segments of the chain if access on a non-discriminatory basis is granted on networks to third parties, it is also true that electricity can be produced everywhere and, for this reason, liberalizing the electricity sector could actually mean making different producers (technologies) compete, while natural gas is located in a few countries outside Europe and far from final markets. On this issue it could be argued that in the gas market transport plays a twofold role: as in the power sector it carries out a technical function consisting in the delivery of the service to the market, but it is also essential in providing the raw material availability which represents the basic condition for the existence of the market: 'steel is molecule' (Percebois, 2003). In other words, due to the fact that gas has to be imported to Europe, competition in the downstream market could occur if opportunistic behaviours of the transport operator of the vertically integrated gas company can be prevented. The debate on this point is mainly related to the degree of network unbundling and particularly to the need for ownership separation.

6.5.1   Network Unbundling

When talking about vertical relations, this refers to both vertical integration and vertical foreclosure. Vertical integration is: 'the organization of successive production processes within a single firm, a firm being an entity that produces goods and services' (Riordan, 2005). For better understanding, a firm can be seen as a unified ownership of assets used in production (Grossman and Hart, 1986), or as a nexus of contracts linking its owners to production factors, managers and creditors (Jensen and Meckling, 1976).[27] Foreclosure, however, refers to a dominant firm's denial of proper access to an essential good that it produces, with the intent of extending monopoly power from the segment of the market to an adjacent segment (Rey and Tirole, 2007). A foreclosure can be considered a vertical one when the essential facility is upstream (or downstream) with respect to the competitive segment.

Vertical integration and vertical foreclosure are unequivocally correlated. Every time a firm decides to integrate either upstream or downstream, competition authorities investigate the possibility that this operation raises barriers to the market. Vertical integration, though, could also bring efficiency gains, which would have beneficial effects for consumers. There are many arguments made in favour but also in disfavour of ownership separation.

The gas industry can be divided into three segments: production, transportation and sales.[28] In this simple division, the network can be seen as the essential facility needed by both producers and sellers. Theoretical analysis, though, generally considers the gas market as composed by just two segments: the network and the competitive downstream market (Vickers, 1993; Buehler et al., 2004; Cremer et al., 2006); only in Baranes et al. (2003) is a three-segment structure presented. The exclusion of the upstream segment seems to be crucial for competition in the market and reveals that the focus on the other two segments is due to the fact that economists consider all network industries to be the same. The exclusion of production from theoretical analysis concerning vertical integration and vertical foreclosure could lead to misleading results as it neglects one crucial aspect: that is, who gets the scarcity rents. Due to the international gas pricing mechanism, namely the netback value, producers beyond the European border are the residual claimants of the rents. This calls for the introduction of countervailing power theory.[29] Moreover, a producer who owns the network can clearly discriminate access, as can any other operator owning an essential facility. This can occur especially if a producer is integrated in the downstream market.[30]

6.5.2 Ownership Separation

From an empirical point of view, it seems that it is worth focusing on the main concerns about vertically integrated undertakings in the natural gas market. They are represented by the possible creation of an (under)investment problem on the grid, and by access discrimination. One of the most important topics to investigate is security of supply. Security of supply can be achieved only with investments in new infrastructures that could bring a sufficient amount of gas to final consumers. The EU is worried that vertically integrated firms have less incentives to invest in infrastructures. This underinvestment problem would raise barriers to entry and, at the same time, would reduce security of supply. Two interesting papers (Buehler et al., 2004; Cremer et al., 2006) analyse this issue and demonstrate that, contrary to common thinking, only integrated operators have more incentive to invest. They unequivocally suggest not continuing on the road that leads to ownership separation. Their findings are based on the fact that, if the system operator is excluded from the profits gained in the final market, it will have no incentive to make the optimal network investment (in both size and quality). This is a typical vertical externality argument which states that a non-integrated upstream monopolist ignores the positive effects on downstream profits. Nevertheless these two papers make some unrealistic assumptions about the gas industry. In fact, their findings are correct only in a deregulated environment (Buehler et al., 2004) or where authorities just regulate the access tariff to the essential facility. Actually the gas network is fully regulated, also with respect to investment remuneration and timelines. Their conclusions, though, can still be considered as a caveat by regulatory authorities, which have to find the right incentives for the investments needed. Moreover, it is possible to say that these papers remind us of the need for regulation whenever there is an essential facility (be it integrated or not) that can become a bottleneck to the market.

But do we still need to fear foreclosure (access discrimination) if the operator is fully regulated? In such a case the foreclosure cannot be put into practice unless a consistent information asymmetry exists (Vickers, 1993). This states the uselessness of ownership separation (OS) in the case of efficient price and non-price regulation of the network, unless the information asymmetry would persist or if it would be too costly to reduce it. It is then worth noting that, even in the case of OS, the market would be left with a new operator benefiting from this asymmetry. Besides, as argued by Polo and Scarpa (2003), it is normal that information asymmetry will reduce (and perhaps disappear), given that authorities move quickly on the learning curve. The reduction (elimination) of the information asymmetry would reduce the chance of discrimination, shrinking (or even eliminating) the benefits of ownership separation. So if the Sector Inquiry laments a lack of competition due to vertical foreclosure, it should be argued that the problem with gas market liberalization has to do with the upstream segment.

More particularly, competition along the European border is not sufficient to guarantee a decrease in the final price paid by consumers. In fact, in a situation where many European importers face a sole exporter, who is likely to charge the same price to every purchaser, the competition for the scarcity rent would turn in favour the purchaser. This situation is likely to be emphasised by the liberalization process which will weaken importers turning into a context in which the monopolist (producer) deals with several players, further increasing its countervailing power compared to a situation in which it had to face a monopsonist for each member state.[31]

6.6   CONCLUSIONS

As Monti (2010) argues: 'Europe needs a functioning single market for energy to ensure secure and affordable supplies for its consumers and business. It has to harness its potential to turn its political leadership on climate change in a concrete chance for its innovative industries.' For these reasons, electricity and gas markets in Europe stand at something of a crossroads. Many countries have made some progress with market-based reforms, but serious bottlenecks remain, as our analysis has shown. During the reform period, governments have reduced their direct involvement in the energy sector. Now however rising environmental concern about global warming requires reduction of carbon dioxide emissions from the same sector and calls for effective environmental regulation. Therefore, addressing climate change seriously has the potential to introduce significant challenges. The question that remains to be debated is how to finalize an 'all-in-one' solution in order to evolve toward fully competitive, integrated and green energy markets.

NOTES

1. Directive 2003/54 was implemented to deal with those areas which were still causing difficulties: differential rates of market opening; disparities in access tariffs between network operators; a high level of market power amongst existing generating companies, thereby contributing to the impeding new entrants to the market; and insufficient interconnection in the infrastructure between the different member states. This electricity Directive mandates the legal separation (and the independent operation) of transmission and distribution grids from production and sales activities. However, small-scale distribution companies are exempt from this. Network access tariffs are to be set, published and approved by national regulators before entering into force. Each member state is to have an energy regulator.
2. Directive 2003/55 establishes common rules for the transmission, distribution, supply and storage of natural gas. This gas Directive contains further measures such as unbundling (it requires the legal unbundling of network activities from supply), establishing national regulatory authorities (one in each member state) with well-defined functions, publishing of network tariffs, reinforcing public service obligations (especially for vulnerable customers) and introducing monitoring of security of supply.
3. Under the Commission's preferred option, companies that control both energy generation and transmission would be obliged to sell part of their assets. Investors would be able to keep their participation in the dismantled groups via a system of 'share-splitting', whereby two new shares are offered for each existing share. The Independent System Operator (ISO) option was a Commission compromise proposal whereby companies involved in energy production and supply would be allowed to retain their network assets, but would lose control over how they are managed. During the negotiations, a 'third option' was introduced in response to the successful efforts of France and Germany. Like the ISO option, the Independent Transmission Operator (ITO) model allows integrated companies to retain ownership of their gas and electricity grids. However, they would have to give up daily management of the grids to an independent transmission operator.
4. The 'third package' promotes the creation of an Agency of the Cooperation of the Energy Regulators and a coordination body among the European Networks and Transmission System Operators.
5. Readers interested in this issue can refer to Jamasb and Pollitt (2005) and Pollitt (2009), as well as to the official documents by yearly 'benchmarking report' of the European Commission that describe progress achieved in the development of the internal market of electricity and of gas as from 2004.
6. For an extensive review of the problems associated with the trading of power see, among others, Joskow (2005).
7. The extensive list of the basic tools for good (short- and long-term) performance includes: privatization of state-owned monopolies; vertical separation and unbundling of potentially competitive segments (generation, marketing and retail) from regulated segments (that is, distribution, transmission and system operation); horizontal restructuring of the generation segment; horizontal integration of transmission facilities and network operation; design of wholesale spot energy and operating reserve market institutions; creation of active 'demand-side' institutions; introduction of transmission access; unbundling of retail tariffs and bills; establishment of market oversight and monitoring; and development and implementation of a transition strategy (see Joskow 2005, 2008a, 2008b).
8. More extensive analysis is in Schweppe et al. (1998), Stoft (2002) and Kirschen and Strbac (2004).
9. Electronic trading is the less time-consuming option; customized forward (long-term) contracts are extremely flexible but entail large transaction costs; over-the-counter trading is the standard small-size package for clients willing to refine their position as delivery time approach.
10. Further details on the working of this institution may be found in Schweppe et al. (1998), Stoft (2002), Littlechild (2006) and Wilson (1998).
11. Network constraints and transmission congestion pricing are extensively debated in Hogan (1992, 1993).
12. Power market designs disconnecting energy from ancillary services have performed poorly and proved to be exposed to unilateral strategic behaviours.
13. According to Wilson (1998), generators may be reluctant to participate in a trading game whose outcome depends on complex computer programs more than on submitted bids.
14. Further details on operating reserves may be found in Joskow (2005) and Stoft (2002).
15. Under pay-as-bid rules, each generator that is instructed to produce by the system operator receives a price equal to the price in its own bid, which may well fall short of the market clearing price.
16. If the expected spot price is low, suppliers may decide to reserve capacity to the system operator for shortages occurring after gate closures; bids are submitted specifying the individual willingness to accept repayments for withholding a certain amount of power. If adverse events occur (real demand exceeds programmed supply, or dispatched power units are worn out) and the system operator has to dispatch reserved units, operating reserve suppliers are instructed to produce and receive compensation for having withdrawn capacity.

17. This may occur as a consequence of regulated caps on energy prices which limit upward pressures, discretionary behaviours by system operators during true scarcity conditions, and so on (Tirole, 1988; Joskow, 1987).
18. Only EEX (Germany and France wholesale market) and Endex (the Dutch power exchange) provide both peak and base-load financial products for up to six and four years, respectively. OMIP in Portugal covers a shorter time span (one year) and a more limited portfolio in that peak-hour hedging is excluded. The remaining markets offer financial products (if any) for day-ahead transactions.
19. Opponents to long-term contracts rely upon one or more of the following arguments: contracts may contribute to foreclosing energy markets with clauses such as exclusivity when they cover a large share of actual demand or have long lifespans; contracts may endanger the liquidity of wholesale markets; contracts may curb the transparency of power systems by keeping a large share of information private; and so on (Bonacina and Cretì, 2010).
20. Academics notice that any market design works reasonably well in the short run at base-load demand levels when supply conditions are smooth. The challenge arises in the tiny bundle of peak-load hours when inelastic (demand and supply) scheduling combines with dispersed transmission congestions. It is during these hours – when market power concerns are most serious, system operator discretion is of utmost importance, non-price rationing is compulsory to keep the network's physical parameters within acceptable levels, and so on – that markets have to work hard to get their targets, facilitate the effective and efficient allocation of scarce resources, and provide truthful price signals to investors (Joskow, 2005).
21. Where several lines meet or where a line terminates at a generator or load, there is a 'bus'. This is a piece of electrical equipment (a bus bar) that is used to make connections. A 'node' is a more general mathematical term applied to the intersection of connecting paths in any type of network. The two terms are often used interchangeably in power system economics.
22. The optimal dispatch of a system is the net quantity that should be injected and withdrawn at each node in order to maximize social welfare, given a certain demand and the technical, economic and locational features of the generation and transportation infrastructures of a system. By definition, therefore, the optimal dispatch represents the allocation that would be decided by a benevolent and perfectly informed grid operator exclusively concerned with the efficiency of the system, but untouched by any distributional concern.
23. An owner of a physical transmission right would be guaranteed free usage of a congested path between zones A and B (up to a level equal to the number of rights the owner has). The owner would have the option of using the rights or of putting them up for sale in a secondary market that would (possibly temporarily) transfer this right-of-way to another agent.
24. Bohn et al. (1983) showed this for implicit auctions. Chao and Peck (1996) showed a similar results with explicit auctions and continuous trading of energy and transmission contracts.
25. Borenstein et al. (2000) use a simple two-node network to show how limited transmission capacity induces withholding strategies on the part of generators with market power; Harvey and Hogan (2000), and also Nehuoff (2003), explore the comparative effects on competition of nodal pricing with financial transmission contracts versus bilateral trading with physical transmission contracts; Joskow and Tirole (2000) provide a comprehensive treatment of the effects of transmission contracts in two- and three-node networks; Willems (2002) studies the welfare effects of rules to allocate demand for scarce transmission capacity in the presence of market power; Gilbert et al. (2004) study welfare effects when transmission rights are obtained in an auction or inherited as legacy rights.
26. Hobbs et al. (2005) measure the welfare effects of interconnection between Belgium and the Netherlands. Their results show that an increase in social surplus is driven by two elements: flows in opposite directions are allowed to net each other out and an explicit spot market is set up in Belgium, initially the high-price area; however, the size and distribution of the gains depend crucially on companies' pricing behaviour. Kube and Wadhwa (2007) find that market integration leads to a decrease in prices due to efficiency gains. Also Lundgren et al. (2008) conclude that a larger electricity market seems to reduce the probability of sudden price jumps. Malaguzzi Valeri (2009) studies the welfare effects of additional interconnection between Northern Ireland and Great Britain and finds that Northern Ireland (which starts off with higher wholesale electricity prices) enjoys larger net benefits than Great Britain.
27. A clear example is given by Riordan (2005): 'Consider a supply chain in which raw materials and other inputs are used to produce an intermediate good, which in turn is a component input into the production of a final good, which in turn is distributed to consumers through a retail channel. Forward vertical integration occurs when a firm expands the scope of its activities to both produce and distribute the final good.'
28. Actually, there are five segments: production and import, transportation, storage, distribution, and sales. For modelling reasons, though, we can unify transportation, storage and distribution into just one segment.
29. 'Countervailing power' was a term coined by Galbraith (1952) to describe the ability of large buyers in

concentrated downstream markets to extract price concessions from suppliers. Galbraith saw countervailing power as an important force offsetting suppliers' market power. The concept of countervailing power was controversial in Galbraith's day (see Stigler's 1954 criticism), and continues to be so today. There are a number of theories explaining why large buyers obtain price discounts from sellers. A simple theory is that the cost of serving large buyers is lower per unit. For example, if the supplier's production function exhibits increasing returns to scale (as that of a gas producer does) and the supplier serves one buyer at a time in each production period, per-unit production costs will be lower when serving a large buyer.

30. This is the case of Gazprom, which is free to sell directly on the Italian final market. The Russian gas giant offered Eni the abolition of destination clauses as compensation. Destination clauses formerly prevented the Italian incumbent from selling the gas purchased from Russia outside its national borders.

31. At the other end of the value chain (with respect to production), along the European border, liberalization has brought many operators into the gas market. This has given producers the possibility to choose their partners, since in any country there is more than one operator in the competitive final market.

REFERENCES

Baranes, E., F. Mirabel and J.C. Poudou (2003), 'Analysis of the vertical structure in the European gas industry', *Energy Studies Review*, **12** (1), 27–52.

Bjorndal, M. and K. Jornsten (2007), 'Benefits from coordinating congestion management: the Nordic power market', *Energy Policy*, **35** (3), 1978–91.

Bohn, R.E., M.C. Caramanis and F.C. Schweppe (1983), 'Optimal pricing in electrical networks over space and time', *RAND Journal of Economics*, **15** (3), 360–76.

Bonacina, M. and A. Creti (2010), 'A note on forward contracts in leader–follower games', *Economics Bulletin*, **30** (2), 1539–47.

Borenstein, S., J. Bushnell and S. Stoft (2000), 'Competitive effects of transmission capacity in a deregulated electricity industry', *RAND Journal of Economics*, **31** (2), 294–325.

Brito, D.L. and P.R. Hartley (2007), 'Expectations and the evolving world gas market', *Energy Journal*, **28** (1), 1–24.

Buehler, S., A. Schmutzler and M.A. Benz (2004), 'Infrastructure quality in deregulated industries: is there an underinvestment problem?', *International Journal of Industrial Organization*, **22**, 253–267.

Chao, H.P. and S. Peck (1996), 'A market mechanism for electric power transmission', *Journal of Regulatory Economics*, **10** (1), 25–60.

Crampes, C. and J.J. Laffont (2001), 'Transport pricing in the electricity industry', *Oxford Review of Economic Policy*, **17** (3), 313–28.

Cramton, P. (2003), 'Electricity market design: the good, the bad, and the ugly', Proceedings of the Hawaii International Conference on System Sciences.

Cremer, H., J. Crémer and P. De Donder (2006), 'Legal vs. ownership unbundling in network industries', Working Paper 5767, Centre for Economic Policy Research.

Creti, A. and B. Villeneuve (2004), 'Long-term contracts and take-or-pay clauses in natural gas markets', *Energy Studies Review*, **13**, 75–94.

Ehrenmann, A. and K. Neuhoff (2009), 'A comparison of electricity market designs in networks', *Operations research*, **57** (2), 274–86.

ERGEG (2008), 'ERI coherence and convergence report', www.energy-regulators.eu.

European Commission (2009), 'Benchmarking report: correct implementation of EU energy law and infrastructure investment top priority'.

European Commission (2007), 'DG Competition report on energy sector inquiry', http://ec.europa.eu.

von der Fehr, N-H. and D. Harbord (2002), 'Competition in electricity spot markets: economic theory and international experience', EconWPA Industrial Organization, 0203006.

Galbraith, J.K. (1952), *American Capitalism: The Concept of Countervailing Power*, New York: Houghton Mifflin.

Gilbert, R., K. Neuhoff and D. Newbery (2004), 'Allocating transmission to mitigate market power in electricity networks', *RAND Journal of Economics*, **35** (4), 691–711.

Glachant, J.-M. (2005), 'Implementing the European internal energy market in 2005–2009: a proposal from Academia', Working Paper ADIS, 7.

Glachant, J.-M., R. Belmans and L. Meeus (2006), 'Implementing the European internal energy market in 2005–2009', *European Review of Energy Markets*, **1** (3), 51–85.

Green, R.J. (2007), 'Nodal pricing of electricity: how much does it cost to get it wrong?', *Journal of Regulatory Economics*, **31** (2), 125–49.

Grossman, S. and O. Hart (1986), 'The costs and benefits of ownership: a theory of vertical integration', *Journal of Political Economy*, **94**, 691–719.

Harvey, S.M. and W.W. Hogan (2000), 'Nodal and zonal congestion management and the exercise of market power', Harvard University Working Paper.

Hobbs, B., F. Rijkers and M. Boots (2005), 'The more cooperation, the more competition? A Cournot analysis of the benefits of electric market coupling', *Energy Journal*, **26** (4), 69–97.

Hogan, W. (1992), 'Contract networks for electric power transmission', *Journal of Regulatory Economics*, **4**, 211–42.

Hogan, W. (1993), 'Markets in real networks require reactive prices', *Energy Journal*, **14** (3), 171–200.

Hunt, S. (2002), *Making Competition Work in Electricity*, New York: Wiley.

Jamasb, T. and M. Pollitt (2005), 'Electricity market reform in the European Union: review of progress toward liberalization and integration', *Energy Journal*, **26** (Special Issue), 11–41.

Jensen, M.C. and W.H. Meckling (1976), 'Theory of the firm: managerial behavior, agency costs and ownership structure', *Journal of Financial Economics*, **3**, 305–60.

Joskow, P.L. (1987), 'Contract duration and relationship specific investment', *American Economic Review*, **77** (1), 168–85.

Joskow, P. (2005), 'The difficult transition to competitive electricity markets in the US', in J. Griffin and S. Puller (eds), *Electricity Restructuring: Choices and Challenges*, Chicago, IL: University of Chicago Press, pp. 31–91.

Joskow, P. (2006), 'Introduction to Electricity sector liberalization: lessons learned from cross-country studies', in F.P. Sioshansi and W. Pfaffenberger (eds), *Electricity Market Reform: An International Perspective*, Oxford: Elsevier, pp. 1–32.

Joskow, P. (2008a), 'Lessons learned from electricity markets liberalization', *Energy Journal*, Special Issue, 9–42.

Joskow, P. (2008b), 'Capacity payments in imperfect electricity markets: need and design', *Utilities Policy*, **16**, 159–70.

Joskow, P. and R. Schmalensee (1983), *Markets for Power*, Cambridge, MA: MIT Press.

Joskow, P. and J. Tirole (2000), 'Transmission rights and market power on electricity power networks', *RAND Journal of Economics*, **31** (3), 450–87.

Kirschen, D. and G. Strbac (2004), *Power Systems Economics*, Chichester: John Wiley & Sons.

Kristiansen, T. (2007a), 'A preliminary assessment of the market coupling arrangement on the Kontek cable', *Energy Policy*, **35** (6), 3247–55.

Kristiansen, T. (2007b), 'An assessment of the Danish–German cross-border auction', *Energy Policy*, **35** (6), 3369–82.

Kube, M. and P. Wadhwa (2007), 'Does size matter? The effect of market integration on wholesale prices in the Nordic electricity market', Master's thesis, Stockholm School of Economics.

Leveque, F., J.-M. Glachant, M. Saguan and G. de Muizon (2009), 'How to rationalize the debate about "EU Energy Third Package"? Revisiting criteria to compare electricity transmission organizations', Loyola De Palacio Working Paper, 9.

Littlechild, S. (2006), 'Foreword: The market versus regulation', in F. Sioshansi and W. Pfaffenberger (eds), *Electricity Market Reform, An International Perspective*, Oxford: Elsevier, pp. xvii–xxix.

Lundgren, J., J. Hellstrom and N. Rudholm (2008), 'Multinational electricity market integration and electricity price dynamics', 5th International Conference on European Electricity Market, Lisbon, 28–30 May.

Malaguzzi Valeri, L. (2009), 'Welfare and competition effects of electricity interconnection between Ireland and Great Britain', *Energy Policy*, **37** (11), 4679–88.

Monti, M. (2010), 'A new strategy for the single market', report to the President of the European Commission, J. Barroso.

Neuhoff, K. (2003), 'Combining transmission and energy markets mitigates market power', DAE Working Paper 310, University of Cambridge.

Newbery, D. and T. McDaniel (2002), 'Auctions and trading in energy markets: an economic analysis', DAE Working Paper 233, University of Cambridge.

Newbery, D. and M. Pollitt (1997), 'The restructuring and privatization of Britain's CEGB – was it worth it?', *Journal of Industrial Economics*, **45** (3), 269–303.

Percebois, J. (2003), 'Ouverture à la Concurrence et Régulation des Industries de Réseaux: Le Cas du Gaz et de l'Electricité', Cahiers de recherche, CREDEN, 03.11.40.

Percebois, J. (2008), 'Electricity Liberalization in the European Union: balancing benefits and risks', *Energy Journal*, **29**, 1–20.

Pérez-Arriaga, I.J. and L. Olmos (2005), 'A plausible congestion management scheme for the internal electricity market of the European Union', *Utilities Policy*, **13** (2), 117–34.

Pollitt, M.G. (2009), 'Evaluating the evidence on electricity reform: lessons for the South East Europe (SEE) market', *Utilities Policy*, **17** (1), 13–23.

Polo, M. and C. Scarpa (2003), 'Entry without competition', prepared for the Workshop on Antitrust and Regulation, Naples.

Rademaekers, K., A. Slingenberg and S. Morsy (2008), 'Review and analysis of EU wholesale energy markets: historical and current data analysis of EU wholesale electricity, gas and CO_2 markets', Final Report, DG Commission.

Rey, P. and J. Tirole (2007), 'A primer on foreclosure', in M. Armstrong and R.H. Porter (eds), *Handbook of Industrial Organization III*, Amsterdam: North-Holland, pp. 2145–2221.

Riordan, M.H. (2005), 'Competitive effects of vertical integration', prepared for LEAR Conference on Advances in the Economics of Competition Law, Rome, 23–25 June.

Rious, V. (2009), 'The design of the internal market in relation to energy supply security and climate change', EUI Working Paper RSCAS.

Schweppe, F.C., M.C. Caramanis, R.D. Tabors and R.E. Bohn (1988), *Spot Pricing of Electricity*, Boston, MA: Kluwer Academic Publishers.

Sioshansi, F.P. (2008), 'Competitive electricity markets: questions remain about design, implementation, performance', *Electricity Journal*, **21** (2), 74–87.

Stigler, G.J. (1954), 'The economist plays with blocks', *American Economic Review Papers and Proceedings*, **44**, 7–14.

Stoft, S. (2002), *Power Systems Economics: Designing Markets for Electricity*, Piscataway, NJ: IEE Press, Wiley Interscience.

Tirole, J. (1988), *The Theory of Industrial Organization*, Cambridge, MA: MIT Press.

Törnquist, J. (2006), 'More transmission capacity for European cross border electricity transactions', 5th Conference on Applied Infrastructure Research, Berlin, 6–7 October.

Vickers, J. (1993), 'Competition and regulation in vertically related markets', *Review of Economic Studies*, **62** (1), 1–17.

Willems, B. (2002), 'Cournot competition in the electricity market with transmission constraints', *Energy Journal*, **23** (3), 95–125.

Wilson, R. (1998), 'Design principles', in H. Chao and H.G. Huntington (eds), *Designing Competitive Electricity Markets*, Norwell, MA, USA and Dordrecht, The Netherlands: Kluwer Academic Publishers, pp. 159–83.

7 Transmission and distribution networks for a sustainable electricity supply

Ignacio Pérez-Arriaga, Tomás Gómez, Luis Olmos and Michel Rivier

7.1 THE ENABLING ROLE OF ELECTRIC NETWORKS

A sustainable economy has to be based on a sustainable energy model, where the power sector is a key component. This adds a new perspective, and a shift in priorities, to future energy policy and the regulation of the electric power sector in particular. The current regulatory paradigm has to be reconsidered in this new context, where intense political oversight is anticipated. Security and sustainability will have at least the same priority as efficiency in the regulatory design.

In addition, some inescapable trends will likely change the landscape of power systems in the medium term: a strong presence of intermittent and typically distributed generation, mostly from renewable sources; widespread availability of affordable communication and control technologies which will facilitate the active participation of demand in the functioning of the power system; and the integration of existing power systems and markets into larger organizations, because of economic rationality, technical feasibility and political convenience.

The main message of this chapter is simple. In the foreseeable future's low carbon economy, electricity that is almost carbon free will have to play a major role. This will require drastic changes in how electricity is produced, transported, distributed, commercialized and used by the end consumers. And the distribution and transmission networks will have to adapt to the new situation so that this revolution towards a sustainable low carbon energy model can take place. New enhanced electric networks – the so-called 'smart grids' – will enable the required technologies and activities to take place.

7.1.1 The Evolving Challenges

Transmission and distribution networks have to be considered separately, since their functions and the challenges they will have to face are so different. Transmission networks are the meeting place of all the agents in a wholesale electricity market. Generally they have been developed to improve the efficiency of the process by which generation meets demand and to ensure an adequate quality of supply. Their increasingly new role will also include reaching to those places where the best large renewable resources are located, enlarging the footprint of intermittent generation – therefore increasing their economic value and their contribution to the reliability of the power system – and permitting the integration of otherwise quasi-independent electricity markets.

Transmission policy might seem easy at first sight. Transmission typically makes a comparatively small contribution – 5 to 10 per cent – to the total electricity cost. Its technical and economic characteristics indicate a regulated monopoly approach. The number of new major investments per year in a system of the size of Spain or California can be handled easily on an individual basis. It seems then appropriate that a highly qualified independent entity – the system operator – proposes an annual expansion plan, to be approved by the regulatory authorities, implemented, and the costs passed to end consumers as a component of regulated electricity tariffs.

However, there are some unrelenting transmission policy issues, which will become more acute under the new conditions. If very large amounts of power have to be trans- ported from distant places – offshore wind production from the North Sea, solar power from Northern Africa to Europe, large wind resources from the sparsely populated Midwest in the USA – and very broad market integration is an objective, then just reinforcements of the existing high-voltage grid (400 and 220 kV in Europe) may not be sufficient and some sort of overlay or supergrid will have to be built, perhaps using higher voltage levels and direct current (DC) technology. How will these decisions be made, by whom (some planning authority with such a wide reach) and how (a method that can cope with a problem of such a huge dimensionality and uncertainty)? How are wide interconnection interests – the European Union (EU) or USA in scope, for instance – to be reconciled with national or local interests? Another open issue is the best use of existing or novel technologies to minimize the environmental impacts and to make maximum use of the existing or future transmission capacity: gas insulated cables, superconductors, low-sag conductors, phase measurements, wide-area monitoring, flexible alternating current transmission systems (FACTS), and so on. Transmission networks and flows will criss-cross interconnected power systems, where some agents, companies, states or entire countries will benefit clearly from these flows while others will not obtain much benefit from the lines sited in their territories. Should the cost of these lines be socialized, or should these costs rather be allocated to the beneficiaries of the transmission facilities? How can the benefits and beneficiaries be identified in an objective way? How can the hostility of those who do not benefit from the installation of transmission facilities in their vicinity be minimized? Given the large uncertainty and the diversity of interests that exist in the expansion of the transmission network, should all decisions be left to a central planner under a regulated monopoly scheme, or are other business models with more participation of stakeholders also possible? And, finally, how far is it meaningful to extend the reach of wholesale electricity markets on the basis of efficiency, reliability and better utilization of renewable resources? How is the coordination of the operation of large interconnected power systems to be addressed?

We turn now to distribution. Integration of renewable generators, higher efficiency of energy consumption in homes and commercial buildings, deployment of future plug-in electric vehicles, and higher reliability of supply are some of the drivers that are demand- ing a profound transformation in the way that electricity distribution grids are designed and operated. This transformation would result in an enhanced distribution grid, more sophisticated and complex than the actual one but providing new or improved services to end consumers, and new opportunities to energy stakeholders moving forward in the direction of a low carbon economy (Pérez-Arriaga, 2009).

A critical difference between distribution and transmission networks is the number of physical facilities to be considered, which is at least two orders of magnitude larger in distribution. Typically, a few thousand connection points are monitored and controlled in transmission networks. In distribution grids the number of active connection points, customers and distributed generators could in some areas easily reach several hundred thousands or even millions, therefore requiring new advanced decentralized architectures for real-time supervision, data acquisition and security analysis.

Present distribution electricity grids have been designed to carry electricity from the meshed transmission grid, where most of the generation resources were connected, to final electricity consumers. Distribution grids are characterized by one-direction flows from sources to loads, radial structure, simple operation rules and acceptable reliability of supply. Planning and operation of distribution grids are based on 'fit and forget' practices. Distribution grids are currently planned to supply the future peak demand with ample design margins. And they are operated in a passive mode, meaning that once the distribution grid facilities have been installed, for the most part, medium-(several kilovolts) and low-voltage grids are not monitored or controlled in real time. Customer meters are used for energy settlement and commercial services, but not for network operation. Automatic control systems almost do not exist, and grid operators are mainly focused on ensuring continuity of supply and reconnecting affected customers in the case of grid failures or maintenance works.

However, much of the expected volume of electricity production with renewables will be connected in distribution networks, either in low-voltage (small wind, rooftop solar panels), or in medium- and high-voltage distribution. This will force distribution utilities to change the customary procedures for design and operation and, in most cases, to incur additional costs. Distribution is treated as a regulated monopoly worldwide, although it has been always difficult to determine the adequate level of remuneration and the proper incentives to promote reduction in losses and an optimal level of quality of service. This will become much more challenging with significant amounts of distributed generation, signalling the need for advanced network models and an in-depth revision of the remuneration procedures, as Ofgem is currently doing in the UK (Ofgem, 2010). There is the need, again, for an in-depth revision of the regulation of the distribution activity, in this case to assign roles to distributors, retailers and energy service companies on who is doing what in metering, aggregation of consumers, relationship with the distribution system operator, interaction with the end consumers and control of their appliances, as well as improving the models of remuneration while taking into account quality of supply.

7.1.2 Smart Grids

Both in the technical literature and in the non-specialized media the term 'smart grid' is frequently used, suggesting a radical departure from the present transmission and distribution networks. Smart grid is a loosely defined concept, which includes a diversity of technologies and innovations. The European Union Smart Grids Platform[1] defines smart grids as: 'electricity networks that can intelligently integrate the actions of all the users connected to them – generators, consumers, and those that do both – in order to efficiently deliver sustainable, economic and secure electricity supplies'. The USA Energy Independence Security Act of 2007 provides a very comprehensive definition.

The expected benefits for consumers and society in general of smart grids deployment are multiple, see Stuntz et al. (2010) for instance: reduction of the environmental impact and carbon emissions of electricity and transportation, integration of high penetration levels of renewable and electric vehicles, higher reliability and quality of supply, reduced network energy losses and the active participation – by means of aggregators – of millions of end consumers whose demand, supply and storage capabilities could be managed in a coordinated fashion to provide useful system services in multiple time ranges.

The transition from the present electricity grids to transmission and distribution networks with enhanced capabilities requires very significant volumes of investment in new facilities as well as in innovation efforts. Most of them are mainly related to the implementation of much more complex and sophisticated information, communication and control systems. In addition, investment in grid infrastructure will be also needed to replace old assets, to increase network redundancy and to connect new generation sites and demand users. Finally, operational and maintenance costs should be re-evaluated, taking into account the new structure and functionalities provided by smart grids.

Existing electricity grids are already smart, but they need to become much smarter to cope with the new realities of a much more complex, decentralized and interactive power sector, in facilitating an efficient, reliable and carbon-free electric supply. It will be a long, evolutionary process that will use and expand existing network capabilities and add new ones. The design and implementation of adequate regulation at both distribution and transmission levels will be essential in guiding the financial resources and technical capabilities of the private firms towards this common objective.

7.2   TRANSMISSION NETWORK POLICY

In general terms the major policy issues in the regulation of the transmission activity are: the criteria for transmission expansion; who is responsible for network planning and what methodology is employed; the decision about line siting and its implications; the adoption of a cost allocation scheme; the actual implementation of the plan and the business model for the transmission investors; and, finally, the rules and supervision of the coordinated operation of the interconnected power system.

7.2.1   Criteria

Transmission expansion may respond to several criteria that typically are mutually reinforcing. Trying to achieve some prescribed standards of reliability of electricity supply is the issue of major concern. Efficiency – that is, reducing losses and economic distortions due to the network in the continuous process of matching supply and demand – is another major concern. The lack of reliability results in economic costs for the consumers, and therefore both criteria have much in common. The current reliability metrics will have to be reconsidered in face of the anticipated active demand response, since they just assume a passive demand that has to be supplied with any available means (Rodilla and Batlle, 2010).

Other network expansion criteria, less frequently used and more difficult to reflect in monetary terms, are support to the functioning and geographical extension of markets, mitigation of market power and implementation of energy policies, such as making possible the deployment of large amounts of renewable generation in faraway places or the creation of a broad transmission overlay or Supergrid, as a strategic decision by the corresponding policy-makers. Recent documents that examine transmission planning criteria in depth in the USA and the EU – although so far they have not been implemented – are RealiseGrid (2010), ENTSO-E (2010) and FERC (2010).

Logically, both planning and cost allocation – to be discussed next – critically depend on the criteria that have been explicitly adopted for transmission network expansion.

7.2.2   Planning

By transmission planning is understood a recursive process of generation and evaluation of potential transmission expansion plans in the search for a preferred solution that best meets a prescribed set of criteria. The high dimensionality of the search space, its high uncertainty, the lumpiness and longevity of the decision variables and the multiplicity of criteria, typically characterize this process.

In simple terms, the objective of transmission planning is to determine when and where new transmission facilities should be built so that any prescribed criteria are met. Assuming that only reliability and economic criteria are considered, a network investment by itself, or as part of a suite of investments, will be justified if it is necessary to meet any prescribed reliability targets or if it results in more benefits for the network users than the associated transmission costs (investment plus operation and maintenance of the facility).

More precisely, rather than having to define a complete optimal plan, the objective of transmission planners is to define the transmission facilities that should be built now to create a robust system going forward, in the face of the strong prevailing uncertainty. This is why planning has to combine two complementary approaches or timescales (see de Dios et al. 2009): 'strategic', that is, the exploration of what the future grid will look like in the long run, twenty years from now, for instance; and 'tactical', where the interest is to identify the reinforcements that are consistent with the strategic plan and whose implementation process – environmental permits, acquisition of rights-of-way, and so on – must start immediately.

A realistic representation of the problem imposes exacting modelling requirements in several dimensions. First, a correct representation of the facilities in the interconnected system with a significant transport function. Depending on the acceptable and feasible level of detail, AC, DC or even transportation models are used. Second, the larger the geographical footprint of the plan, the more opportunities can be captured for efficient operation and resource utilization, combining somehow bottom-up (incorporation of proposals made by local planning entities or stakeholders) and top-down (fully integrated view) perspectives. Third, consideration of non-transmission solutions, such as storage or demand side management. Fourth, adequate representation of uncertainty, in generation expansion, demand growth, fuel prices or policy measures. Scenarios are customarily used, although a probabilistic characterization is better. Fifth and finally, the model should allow the evaluation of a 'figure of merit' – either a scalar or multidimensional – that captures the desired set of criteria for the plan.

The dominating issue in transmission planning is dimensionality, due to the multiplicity of expansion options with high uncertainty and a long time horizon. The search for the preferred plan can be formally posed as a mathematical optimization problem (see Latorre et al., 2003). However, the most frequent industry practice (see RealiseGrid, 2010) is trial and error, with evaluation of individual reinforcements or suites of lines for a prescribed ensemble of scenarios. Scenario analysis allows the consideration of uncertainty in a bounded manner, but relies on assumed correlations between variables that may be incorrect or that change dramatically over the analysis period. When using multiple period models, robust lines will be those that appear across optimal plans and different scenarios.

The present state of the art of transmission planning has been able to address medium-sized power systems with moderate uncertainty, such as an USA regional transmission organization (RTO) or a large European country, but is currently unable to cope with the entire EU system or the Eastern Interconnection in the USA and the large uncertainty involved. There are preliminary efforts under way, both in the EU and the US (see ENTSO-E, 2010; RealiseGrid, 2010; EIPC, 2010), but so far they have only been able to gather bottom-up plans or to hypothesize and evaluate a few suites of lines.

Besides the methodology, the other major issue is who has the responsibility for transmission planning in a wide region like the US Eastern Interconnection region or the EU, encompassing multiple systems with their own planning authorities. The Federal Energy Regulatory Commission (FERC) in the USA is proposing regulation that basically amounts to establishing guidelines for soft coordination of the multiple planning authorities (see FERC, 2010), while simultaneously these authorities have already associated in a collaborative planning effort (see EIPC, 2010). The EU has moved further ahead by establishing two institutions at European level: the European Network of Transmission System Operators for Electricity, ENTSO-E, and the Agency for the Coordination of Energy Regulators, ACER, with responsibilities for European-level transmission planning. While ENTSO-E has to prepare non-binding EU-wide plans with a ten-year horizon every two years, ACER has to supervise, but without true executive powers, to ensure that the national plans are consistent with the EU-wide plan (see European Union, 2009). This is a pragmatic solution to the thorny problem of coordination of EU interests and national sovereignty.

7.2.3 Siting

Siting of transmission facilities has become a thorny problem in most developed countries, since transmission facilities in general cause inconvenience and do not provide any direct benefits for those who live in the environs. Siting is a less technical and more institutional issue, but it is interdependent with planning and cost allocation and adds another challenge to network expansion. Siting requires the proper consideration of the local concerns of those who will be affected by the presence of transmission lines, together with the objective of implementing a project that has been found beneficial for society. When several jurisdictions exist (local, province or state or autonomous region, supranational or federal) it is necessary to delimit responsibilities and to make sure that there is a clear decision-making procedure where all stakeholders are somehow represented.

It is expected that siting will become easier to address once satisfactory solutions are found for the previous topics of criteria, planning, cost allocation, investment and cost recovery. At least siting will be reduced to what really is, and no more.

7.2.4   Cost Allocation

The allocation of the cost of a transmission network among its users must obey some basic principles that result from the combination of microeconomic theory and power systems engineering (Pérez-Arriaga and Smeers, 2003). First, cost causality – which is equivalent to 'beneficiary pays', since transmission is built when it results in more aggregated benefits than the incurred costs – should be the conceptual basis of any cost allocation methodology, although in general it is difficult to implement. This implies that, in principle, both generators and consumers should pay. And, when it is not feasible to apply strict allocation to beneficiaries, some proxy to benefits could be used instead, such as some measure of 'network utilization'. Second, transmission charges should depend on the location of the users in the network and on the temporal patterns of injection (for generators) and withdrawal (for loads), but not on the commercial transactions – that is, who trades with whom. Therefore transmission charges should be levied on those who benefit from the existence of any given transmission facility, regardless of any trading relationships. Third, transmission network charges for new network users should be determined *ex ante* and not updated, or at least not for a reasonably long time. This is the only way to send the stable economic locational signals that investors need in order to choose with a low financial risk the most convenient sites; this is of particular interest for wind and solar generators, which usually have many potential installation sites. Fourth, unless the margin of benefits of a specific transmission investment over its investment costs is very large, an incorrect allocation of the cost – for instance, charging only consumers when generators also benefit – will create opposition from those who are told to pay.

The international practice of transmission cost allocation at national or system operator level is very diverse. The most common scheme is the plain 'postage stamp' method, whereby every load pays a flat charge per kWh of consumed energy at any time, or per contracted kW of capacity. In some instances generators also pay, on a per kW or per kWh basis – the latter is not recommended, since wholesale market bids would be distorted. A few systems have introduced some sort of locational transmission charges, but more are now considering doing it, because of the anticipated large penetration of wind and solar plants that could unnecessarily stress the transmission grid in the absence of any locational signal. The principle of 'beneficiary pays' is commonly accepted in official documents in the US (see FERC, 2010) for instance, although its practical implementation is so far very rudimentary, to say the least. In the EU the term 'locational signals' is commonly used in regulatory documents as a desideratum, but no progress has been made in this regard at European level[2] and only the UK and, up to a certain point, Sweden have implemented it at country level.

In the USA no serious attempts have been made so far to extend intraregional (RTO) cost allocation methods to inter-regional level. On the contrary, in the EU an Inter-Transmission system operator (TSO) Compensation (ITC) mechanism has been in place since 2002 with the following characteristics (Olmos and Pérez-Arriaga, 2007). Countries – represented typically by one TSO, sometimes more than one – compensate one another for the utilization of their networks, using some metric that is based on network usage. The net balance of compensations and charges for each country, either positive or negative, is added to its total network cost from which the transmission tariffs are computed. Every country is free to design its internal network tariffs. Payment of the national transmission tariff gives every agent the right to access the entire EU transmission network, without any additional charge. Although some computational aspects of this method could be much improved, this overall hierarchical approach has been a major contributor to facilitating electricity trade in the EU and, despite its simplicity, has a solid conceptual basis. Note that this method implicitly and automatically allocates the cost of any new transmission investment in the EU territory.

7.2.5   Business Model

A sound transmission policy should ensure that all beneficial lines are built.[3] This requires that some company or institution decides to build these transmission facilities, with the expectation of receiving an attractive remuneration. This section takes the perspective of the investor and examines different business models. Coxe and Mccus (2010) shows that several investment models may coexist within any given power system or in a wider interconnected domain. Most frequently, once a line is part of an expansion plan that is approved by the regulatory authority, the investment cost can be placed by the regulator in the rate base of the investor – which would be a regulated monopoly – and paid with transmission tariffs. This is typically the case of investments made by the 'incumbent' company, be it a vertically integrated utility, a transmission system operator (TSO) in Europe or an established transmission developer. Most investments in most parts of the world belong to this type. New entrants in transmission development could also build lines that regulators may accept to include in the regulated rate base. 'Policy lines', which are built to satisfy some high-level energy policy, despite their economic justification, always fall into this category.

In other cases either the incumbent firm or, more likely, a new entrant may agree with a group of prospective beneficiaries that they will finance the new line. This is 'merchant model type A'. Since the beneficiaries may be many and very dispersed, and they would love to free-ride, in general it will be difficult for a promoter to build a transmission facility that is financed by long-term contracts. And the charges to those who finance the line will likely be higher than under regulated transmission rates. In a few instances – Argentina since the early 1990s and very recently New York Independent System Operation (ISO) – transmission expansion relies on coalitions of beneficiaries of the prospective new line, who propose to the regulator and actually pay for the costs of the facility, and may receive some compensation for the use of the line by third parties.

Still, some transmission developers may decide to build a line so that the income will only come from arbitraging the energy prices at both ends of the line, that is, from buying cheaper energy from one side and selling it to the other. Given that differences in locational marginal prices – also called energy nodal prices (see Schweppe et al., 1988) – or any contracts that are based on these price differences, in general systematically underrecover (barely 20 per cent of the total) the total transmission costs once the investment is in place (Pérez-Arriaga et al., 1995), the conclusion is that only rarely would a 'merchant line of this type B' be financially viable. Exceptions to this situation are lines joining quasi-independent systems with price differences that will only be barely affected by the new interconnection.

In conclusion, although all these business models should be allowed to coexist and each one of them may contribute to a comprehensive transmission expansion, it must be clear in any sound transmission policy that most lines should be built under regulated conditions, with the costs being allocated to the network users by regulated rates. Therefore, once it has been decided – based on a sound planning procedure – to build a new line, the main role of the regulator is to make sure that the line is built, therefore trying to reduce the risk of cost recovery of the investor as much as possible. If transmission planning has followed a well-designed and transparent process, the risk of building non-beneficial lines is minimized. And the negative consequences of underinvesting in transmission are far greater.

7.2.6   Coordinated Operation

An efficient dispatch of generation and demand that also takes into account any network constraints can be only achieved under a system of nodal energy prices (Schweppe et al., 1988), as it is currently used in several Latin American countries and the US RTOs. European countries, typically with more meshed networks covering not very large distances, have opted for single energy prices at national level, applying not sound enough ad hoc fixes to cope with losses and network constraints *ex post*. In this way the market clearing is more transparent, but there remains the problem of checking *ex post* the compatibility of the market results with the reality of the network limitations.

The challenge for any of these two models is to get as close as possible to a seamless coordination of several power systems (RTOs in the US, countries in Europe) who want to preserve their individual pricing schemes, market institutions and system operators. Obviously, the ideal solution would be based on a single nodal pricing scheme for the entire interconnection, as in the case of the Central American Electricity Market, encompassing six countries (MER, 2010). In the US several RTOs of the Eastern Interconnection have started a project to integrate the operation of their systems by sharing information about the network conditions in neighbouring systems. In the EU the initial approach has consisted of encouraging the coordination of the TSOs in order to maximize the volume of interconnection capacity among the individual systems that is available for trade (see ETSO, 2001). The next step is to ensure an efficient utilization of this network capacity by suitable market mechanisms and coordination of the access to scarce capacity.

Under any of the preceding approaches, the utilization of the grid can only be maximized if access to scarce capacity is managed in several time horizons. In the medium and long term, rights to use scarce transmission capacity should be allocated through a single auction platform, with harmonized rules, information technology (IT) interfaces and products. The capacity products that are traded in long- to medium-term auctions can refer to the physical use of the corresponding transmission capacity (physical rights) or they can be financial (entitling the owner to the corresponding congestion rents). They can also be defined between any two points (point-to-point rights) or referred to specific lines or corridors (flow-gate-based). Finally, the access to the capacity by the rights owners can be firm or non-firm. The different types of capacity products, as well as their advantages and drawbacks, are discussed in Hogan (1992, 2002) and Chao and Peck (2000).

In the day-ahead time frame, the scarce capacity that has not been previously committed physically should be allocated together with energy in a system-wide implicit auction, which jointly considers energy and network capacity, if this is institutionally acceptable. Gilbert et al. (2004) show that implicit auctions maximize the use of transmission capacity. If system-wide implicit auctions are out of question, either decentralized implicit auctions (Belpex, 2010), or a coordinated explicit capacity auction followed by decentralized local energy-only auctions, could be applied (Pérez-Arriaga and Olmos, 2005).

System-wide intra-day auctions should also be made available to allow agents to balance their positions in the presence of undesired deviations from their daily programme. A group of European power exchanges is working on the implementation of a regional intra-day energy and capacity allocation mechanism based on continuous trading. And at an even shorter time range, the integration of balancing or regulation markets would allow sharing operation reserves, reducing the cost and facilitating the integration of intermittent generation sources (see EuroPEX, 2009).

7.3   DISTRIBUTION NETWORK POLICY

The major drivers behind the anticipated changes in distribution networks are: the strong presence of distributed generation, the active consumer response and the search for energy efficiency, the advent of electric vehicles and energy storage, and the request for 'digital quality' reliability of supply.

The integration of distributed generation – wind, solar, micro-cogeneration or combined heat and power (CHP) plants for industrial, agricultural or residential applications – into the current distribution grids, which were not designed to accept internal generation sources, poses new challenges in terms of new infrastructure investments and operational problems.

Demand response shifting consumption from peak to off-peak hours, and energy savings in final uses, are the most efficient ways of reducing the need for new power installations and carbon emissions. However achievements up to now in this direction have been modest. Smart meters open the door for a massive response from homes and businesses in modifying their energy patterns according to cost-reflective time-varying electricity prices. In addition, customers would value replacing home appliances by more efficient ones or locating on-site generation to decrease their electricity bills and improve their carbon footprint.

For efficient integration of electric vehicles into distribution grids the role of smart grids would also be relevant. Charging electric vehicle batteries would require grid reinforcements and extensions, but investment would be much less if smart charging strategies, driven by time-varying electricity prices, were in place with the aim of minimizing charging at peak hours. Moreover, storage capability of electric vehicles connected to the grid could provide valuable vehicle-to-grid (V2G) services, such as peak power in emergency situations or regulation reserves. Smart grids would allow the pooling of a large number of those distributed resources to procure these services.

Finally, smart grid configurations with higher redundancy and automation, including distributed generation and storage resources, would deliver the reliability of supply levels required by a digital society. The micro-grid concept, within smart grids, has proposed new decentralized grid control structures to keep the lights on should a part of the distribution grid be isolated due to system failures. Within the electrical island loads and generators would be locally balanced until the interconnection to the rest of the grid could be restored.

These four major drivers will bring threats and opportunities to the distribution system operators, energy retailers or suppliers, aggregators and new energy service providers. With distribution being treated as a regulated monopoly, the regulatory authorities must establish the allowed regulated revenues that compensate distributors for operating costs and give a fair return on investments, set any appropriate performance-based incentives, and try to make sure that in the long term the benefits of the adopted network enhancements appear to exceed costs. A sound regulation should procure that both suppliers and consumers share the costs and expected profits of this transformation (ERGEG, 2009), with profits exceeding costs.

There follows a concise discussion of the major new policy issues that are anticipated in the regulation of the distribution activity in the medium and long term.

7.3.1   New Agents and Roles

The introduction of smart grids poses new challenges for regulators defining the market rules and roles of the agents under the new context. After the electricity sector reforms during the 1990s and 2000s, in many countries, and in the European Union in particular, the distribution activity has been unbundled from supply or retail, which is considered a competitive activity (European Union, 2009). The discussion that follows assumes this scenario.

Distributors or distribution system operators (DSOs) are the regulated agents in charge of planning, building, operating and maintaining distribution grids. Therefore it is expected that they will play a pivotal role in the introduction of smart grids. In recent years, traditional cost-of-service regulation based on audited costs has been replaced by incentive regulation or performance-based regulation. Incentive regulation puts pressure on DSOs to reduce costs and obtain higher profits, but this could be at the expense of investment cuts that in the medium term would lead to quality-of-service degradation. As we will see, the type of DSO regulation and how regulators acknowledge the investment made by DSOs will play a key role in the introduction of smart grid technologies.

Retailers and energy service providers will face new opportunities and challenges too. These deregulated businesses can be directly affected by regulatory decisions. For instance, regulators should enforce the adoption of smart meter architectures and open communication standards to facilitate retail competition. Even if the legislation requires distributors to install smart meters, they have to provide reliable meter information access to retailers without discrimination or preferences.

New energy service providers are the obvious candidates to offer products and services directed to residential and small businesses, in order to reduce their energy bills and improve carbon footprints through energy efficiency and demand price response programmes. Regulation should be adapted to facilitate the entrance of these new agents and to ensure fair competition among them. For instance, recent legislation in Spain introduced a new type of market agent who would be responsible for deploying the charging infrastructure for electric vehicles and also for selling them electricity for charging and/or storage (Spanish Royal Decree-Law 6, 2010).

7.3.2   Revenue Decoupling

Under the new paradigm of penetration of distributed resources and energy efficiency it is absolutely critical that regulated revenues for distribution and transmission companies should be correctly decoupled from actual energy sales. Under cost-of-service regulation higher sales meant higher incomes and likely higher profits. Incentive regulation implemented through revenue caps is a first step in the right direction, since it acknowledges that distribution revenues and volume of electricity sales should be decoupled. Revenue caps acknowledge the intrinsic nature of network costs, where fixed costs are predominant over variable costs.

In Europe, revenue caps for regulating DSOs is a common practice. However in the US a situation report indicated that only 13 out of 50 states have approved or are pending approval of revenue decoupling mechanisms for electric utilities (Edison Foundation, 2009).

7.3.3   DSO Incremental Costs Due to Integration of Distributed Generation (DG)

As a consequence of the success of renewable energy and DG support schemes, in Europe the DG impact on distribution grids is becoming more and more important. The installed capacity of DG in the EU-25 is expected to grow from 201 GW to about 317 GW in 2020 (Nieuwenhout et al., 2010).

High DG penetration levels result in the need for incremental network investments and incremental energy network losses. Both involve higher costs for DSOs. Only in the case of low DG penetration – that is, less than 20 per cent of the load supplied by local generation – can DSOs obtain benefits from energy losses reductions.

Regulators should be aware of those economic implications of DG. Traditionally the main drivers for distribution costs have been the number of customers and their demand. However, DG connections are becoming relevant as a new distribution cost driver and they should be adequately considered.

New tools can help regulators to assess the economic impacts of DG. For instance 'reference network models' have been developed in Spain to include distributed generation or demand response actions as new cost drivers that affect grid design and investment (Mateo et al., 2010). It is not the same in terms of new grid infrastructure to connect a 1 MW photovoltaic concentrated power plant as to connect 200 5 kW photovoltaic installations distributed in houses in a residential area.

In Great Britain, the regulated revenues that each DSO is allowed to collect are increased proportionally to the DG capacity connected to its grid (Ofgem, 2009). This is a move in the right direction, although the rule is still too basic.

7.3.4   Cost-reflective Network Charges

Progressive integration of DG, efficient response of demand according to system needs, and future network users such as electric vehicles, would require rethinking the design of network charges. Network charges should be cost-reflective. This will incentivize the efficient short-term operation of loads and generators, recognizing that injecting energy into the grid at peak hours when the network is congested has more value than at off-peak hours with no congestion. This will also provide signals for the location of new generation sources. The grid is more valuable if generation is integrated with already existing loads than if it is installed in remote areas where the grid is weak and needs to be reinforced. In addition, total symmetry of network charges allocated to network users, no matter whether they are consumers, generators or both (prosumers), is recommended.

Connection charges paid by new network users, especially DG, should be averaged, regulated and shallow: that is, only include the direct connection installations to the grid. If additional grid reinforcement were required its costs would be socialized and recovered via use-of-system (UoS) charges. This practice is transparent and would avoid grid and market access conflicts between DG promoters and DSOs.

On the other hand, UoS charges would be paid by all network users depending on where they are connected (voltage level, rural or urban area) and with time-of-use differentiation. The same charge that would apply to 1 kWh withdrawn from the network would apply as a payback to 1 kWh injected into the network at the same time and at the same location. The cost causality criterion implies that UoS charges could be either positive or negative for injected energy into the network, since DG may achieve cost savings through losses reduction, investments deferral, voltage control, and so on; but also it may increase costs when it results in the opposite effects (Cossent et al., 2009).

7.3.5   Feed-in Tariffs and Priority Access for DG

Feed-in tariffs and priority rules for access and dispatch have been successful policies to promote the initial deployment of renewable and CHP DG in European countries, like Germany or Spain. However, flat feed-in tariffs that remunerate generation at a constant rate no matter when and where it takes place, interfere with market energy prices with hourly changes and well-designed network locational charges. Therefore, other support schemes such as feed-in premiums with time discrimination on top of the market price, or incentives partially covering DG investment costs, are recommended instead.

In case of network congestion, priority dispatch rules should be combined with DG redispatch and demand response actions with adequate economic compensation. In the long term both generation and demand, no matter their size and location, must be fully integrated in the electricity market subject to market prices and cost-reflective network charges. Smart grid technologies would play a key role in this achievement.

7.3.6   Smart Meter Benefits and Cost Allocation

Smart meters would bring multiple environmental and economic benefits. They would allow DSOs to improve grid maintenance and operation, reduce metering costs, and monitor DG production and network flows. Moreover, DSO asset planning and management would benefit from knowing disaggregated consumption patterns across the whole year in different parts of the grid. Detection of supply interruptions and reduction of supply restoration times would be facilitated by real-time information provided by smart meters. Frequent network reconfigurations would help to minimize energy losses, since distribution losses account for 90 per cent of total losses, and also to reduce carbon emissions.

Smart meters will benefit consumers too by identifying poor energy performance appliances, or by facilitating users to manage domestic appliances and integrating distributed generation in order to reduce their total energy bill and carbon footprint. Finally, smart meters would benefit retailers and aggregators who could have valuable information regarding load patterns of end users to negotiate energy contracts, to design and manage energy curtailment offers in emergency situations, or to provide energy services to residential customers with potential high energy savings.

In many countries smart meters, as any other smart grid technology deployment, are mandated to be under the responsibility of DSOs. The regulatory issue here is how to allocate smart meter costs among the different beneficiaries. It is recommended that in the short term DSOs as regulated entities should be entitled to include smart grid investment in their revenue allowances, so that regulated network charges would pay for those expenses. On the other hand, in the medium and long term DSO allowances should be progressively decreased because of the expected DSO operational cost reductions. In addition, DSOs should be also allowed to charge for metering services to other market agents, retailers and aggregators who benefit from them.

7.3.7   Incentives for Innovation

The experience with DSO incentive regulation accumulated in some countries from the beginning of the 1990s has demonstrated that DSOs have made important achievements reducing their operational costs. However infrastructure investment and technology innovation have not been a DSO priority. Current DSO regulatory practices should be reviewed in this regard. The challenge is to design a new framework promoting the required changes for a sustainable and low carbon energy sector.

For instance, Ofgem, after more than 20 years of successful incentive regulation (from the beginning of the 1990s) of the transmission and distribution activities, has opened a process of rethinking and consultation with stakeholders about network regulation for the next decade (2010–20). Some of the preliminary ideas are: (1) continue to use a revenue cap using an *ex ante* approach to estimate efficient grid costs per company; (2) put greater focus on the delivery of outcomes related to safe, secure, high-quality and sustainable network services; (3) strengthen incentives for cutting costs in a longer term than the five-year customary price control period, by applying specific regulatory instruments to a longer time horizon; and (4) provide a separate time-limited innovation stimulus and specific incentives for delivering a low carbon energy sector (Ofgem, 2010). In the last distribution price control review in 2009, Ofgem introduced some new features in this direction (Ofgem, 2009).

As another example, in California, the Southern California Edison (SCE) performance-based rate-making plan includes a provision for accounts devoted to specific developments for a low carbon energy sector: advanced metering infrastructure, a demand response programme, procurement energy efficiency and the California solar initiative programme (SCE, 2010).

7.4   CONCLUSIONS

Transmission and distribution grids will play a key and enabling role in the path toward a sustainable, almost carbon-free electricity sector. The paradigm of the so-called 'smart grids' integrates many of the technology challenges and opportunities that this needed transformation of the actual networks has to focus on.

This chapter has identified and examined the policy and regulatory issues that regulators and governments have to address regarding the transmission and distribution grids. Policy recommendations regarding planning, cost allocation and business models for developing new transmission facilities have been provided. Coordinated operation and maximization of the use of interconnection capacities among regional interconnected systems has been highlighted as another key regulatory transmission issue. Regarding distribution grids, it has been emphasized how incentive regulation for setting revenues to distributors should be revisited in order to promote technology innovation. Finally, recommendations for designing new support schemes for renewable generation, and for the allocation of costs and benefits associated with the deployment of smart grid technologies, have been provided.

NOTES

1. See http://www.smartgrids.eu/.
2. The Inter-transmission System Operator Compensation (ITC) mechanism that is described below cannot truly be considered to provide locational signals.
3. Note that much uncertainty exists in classifying a line as beneficial; justification will be unclear in borderline cases.

REFERENCES

Belpex (2010), 'Market coupling', http://www.belpex.be/index.php?id=4.
Chao, H.-P and S. Peck (2000), 'Flow-based transmission rights and congestion management', *Electricity Journal*, **13** (8), 38–59.
Cossent, R., T. Gómez and P. Frías (2009), 'Towards a future with large penetration of distributed generation: is the current regulation of electricity distribution ready? Regulatory recommendations under a European perspective', *Energy Policy*, **37**, 1145–55.
Coxe, R. and L. Meeus (2010), 'Survey of non-traditional transmission development', 2010 IEEE Power and Energy General Meeting in Minneapolis, Minnesota, July.
de Dios, R., S. Sanz, J.F. Alonso and F. Soto (2009), 'Long term grid expansion: Spanish Plan 2030', CIGRE Conference, http ://www.cigre.org.
Edison Foundation (2009), 'Status of revenue decoupling for electric utilities by state', report of the Institute for Electric Efficiency, Washington, DC, March.
EIPC (2010), 'Eastern interconnection planning collaborative', http://www.eipconline.com/.
ENTSO-E (2010), 'Ten Year Network Development Plan, TYNDP', https://www.entsoe.eu/index.php?id=282.
ERGEG (2009), 'Position paper on smart grids', an ERGEG public consultation paper', European Regulators Group for Electricity and Gas, December, http://www.energy-regulators.eu/.
ETSO (2001), 'Key concepts and definitions for transmission access products', report by the Association of

European Transmission System Operators, http://www.entsoe.eu/fileadmin/user_upload/_library/publications/etso/Congestion_Management/Key%20Concepts%20for%20Transmission%20Access%20Products.pdf.

European Union (2009), 'Directive 2009/72/EC concerning common rules for the internal market in electricity and repealing Directive 2003/54/EC'.

EuroPEX (2009), 'Development and implementation of a coordinated model for regional and inter-regional congestion management', http://www.europex. org/default.asp?kaj=news&id=277.

FERC (2010), 'Transmission planning and cost allocation Notice of Proposed Rulemaking, NOPR', http://www.ferc.gov/whats-new/comm-meet/2010/061710/E-9.pdf.

Gilbert R., K. Neuhoff and D. Newbery (2004), 'Allocating transmission to mitigate market power in electricity markets', *RAND Journal of Economics*, **35** (4), 691–709.

Hogan, W.W. (1992), 'Contract networks for electric power transmission', *Journal of Regulatory Economics*, **4** (3), 211–42.

Hogan, W.W. (2002), 'Financial transmission right formulations', http://www.ksg.harvard.edu/hepg, Center for Business and Government, John F. Kennedy School of Government, Harvard University.

Latorre, G., R.D. Cruz, J.M. Areiza and A. Villegas (2003), 'Classification of publications and models on transmission expansion planning', *IEEE Transactions on Power Systems*, **18** (2), 938–46.

Mateo, C., T. Gómez, A. Sánchez, J. Peco and A. Candela (2010), 'A reference network model for large-scale distribution planning with automatic street map generation', *IEEE Transactions on Power Systems*, **26** (1), 190–97.

MER (2010), 'Mercado Eléctrico Regional Centroamericano' (Central American Regional Electricity Market), http://www.crie.org.gt/, http://www.enteoperador.org/.

Nicuwenhout, F. et al. (2010), 'Market and regulatory incentives for cost efficient integration of DG in the electricity system', final report of European project IMPROGRES, May, http://www.improgres.org/.

Ofgem (2009), 'Electricity distribution price control review. DPCR5 final proposals', Office of Gas and Electricity Markets, http://www.ofgem.gov.uk.

Ofgem (2010), 'Regulating energy networks for the future: RPI-X@20. Emerging thinking', Office of the Gas and Electricity Markets, http://www.ofgem.gov.uk.

Olmos, L. and I.J. Pérez-Arriaga (2007), 'Evaluation of three methods proposed for the computation of inter-TSO payments in the internal electricity market of the European Union', *IEEE Transactions on Power Systems*, **22** (4), 1507–22.

Pérez-Arriaga, I.J. (2009), 'Regulatory instruments for deployment of clean energy technologies', MIT-CEEPR Working Paper 2009-009, http://web.mit.edu/ceepr/www/publications/workingpapers.html.

Pérez-Arriaga, I.J. and L. Olmos (2005), 'A plausible congestion management scheme for the internal electricity market of the European Union', *Utilities Policy*, **13** (2), 117–34.

Pérez-Arriaga, I.J., F. Rubio-Odériz, J.F. Puerta Gutiérrez, J. Arcéluz Ogando and J. Marín (1995), 'Marginal pricing of transmission services: an analysis of cost recovery', *IEEE Transactions on Power Systems*, **10** (1), 65–72.

Pérez-Arriaga, I.J. and Y. Smeers (2003), 'Pricing of electrical transmission and distribution networks', in Francois Leveque (ed.), *Transport Pricing of Electricity Networks*, Dordrecht, The Netherlands: Kluwer Academic Publishers, pp. 175–204.

RealiseGrid (2010), 'Research project in the EU 7th Framework Program', http://realisegrid.erse-web.it/.

Rodilla, P. and C. Batlle (2010), 'Redesigning probabilistic production costing models and reliability measures in the presence of market demand elasticity', IIT Working Paper IIT-10-028A, February, www.iit.upcomillas.es/batlle/Publications.html.

Schweppe, F.C., M. Caramanis, R. Tabors and R. Bohn (1988), *Spot Pricing of Electricity*, Boston, Dordrecht, London: Kluwer Academic Publishers.

Southern California Edison (SCE) (2010), 'Regulatory information – SCE Tariff Books. Preliminary statements', http://www.sce.com/AboutSCE/Regulatory/tariffbooks/.

Spanish Royal Decree-Law 6 (2010), 'On measures for the economic and employment recovery', *Boletin Oficial del Estado*, April. (In Spanish)

Stuntz, L., S. Tomasky and L. Hermann (2010), 'An electricity grid for the 21st century', ASPEN Institute, Washington, DC.

8 Energy–economic–environmental models: a survey
Renato Rodrigues, Antonio G. Gómez-Plana and
Mikel González-Eguino

8.1 INTRODUCTION

The use of economic models for purposes of general policy analysis has changed considerably in the past few decades. Increasing concern about the scarcity of some natural resources and about environmental problems have led to the development of a discipline that it is now an important part of mainstream economics (Perman et al., 2003). Models have begun to incorporate energy as a relevant production input, alongside capital and labour. In the same way, many pollutants have been incorporated as undesired output from production, such as acidifying substances and, more recently, greenhouse gases (Galarraga and Markandya, 2010). The incorporation of these elements has led to the development of so-called energy–economic–environmental or E3 models (Faucheux and Levarlet, 1999; Kemfert and Truong, 2009). E3 models are useful tools for analysing policies whose purpose is to shift economic activities onto a more sustainable path.

The 1973 energy crisis motivated the first energy–economic models, which focused on the macroeconomic consequences of energy shortages and the optimal allocation of energy resources (Manne et al., 1979; Nordhaus, 1980) The increasing demand for energy and the soaring prices of fossil fuel in 2007–08 led to a revival in the literature of studies on the macroeconomic consequences of an increase in energy prices and on energy security issues (Markandya and Pemberton, 2010; Tang et al., 2010).

Anthropogenic climate change and its links with energy consumption have also increased the interest in modelling the interactions between energy, economic variables and greenhouse gas emissions. Various types of model began to be developed in the 1990s (see the surveys by Weyant, 1993 and Springer, 2003). Many models focus on the optimal emissions abatement path, following a cost–benefit analysis, stemming from the pioneer DICE model (Dynamic Integrated Model of Climate and the Economy) by Nordhaus (1993). Integrated assessment models for climate change have also been developed which incorporate feedback effects from changes in natural systems into the economy (Alcamo, 1994; Manne et al., 1995). Finally, E3 models are also being applied to the power sector to provide insights into trade-offs between competitiveness, security of supply and environmental effects when selecting appropriate technologies (Soloveitchik et al., 2002).

E3 models are highly relevant in energy and climate policy-making. Governments are interested in future energy prices and demand, technology prospects and CO_2 emissions when setting their main policies. Normally, this information comes from reports from specialist agencies such as the International Energy Agency (IEA, 2009) or the Intergovernmental Panel for Climate Change (IPCC, 2007). The results of these

reports are usually based on different types of E3 models. Reliability and proper interpretation of their results are essential if the correct signals are to be sent to decision-
makers.

There are many different E3 models, but three main groups can be distinguished:
(1) bottom-up (BU) or engineering models, which represent in detail the energy sector
or a specific part of the economy; (2) top-down (TD) or economic models, which
represent all sectors of the economy, and are usually general equilibrium models; and
(3) hybrid models, which seek to reach a compromise between the other two types.
There are three trade-off areas in any E3 model (see Hourcade et al., 2006a): technical
explicitness, macroeconomic completeness and microeconomic realism. The challenge
of E3 modelling is to advance to the ideal model where these characteristics are fully
incorporated.

This chapter presents a survey of the evolution of E3 models and their applications.
The aim is to present the main characteristics behind these models. The chapter is structured as follows. Section 8.2 presents BU model approaches, and section 8.3 the TD
approach with special attention to general equilibrium models. Section 8.4 shows the
integration of the two approaches in hybrid models. Section 8.5 concludes.

8.2 BOTTOM-UP MODELS

8.2.1 An Outline of some Characteristics of Bottom-up Models

BU or engineering E3 models are usually partial equilibrium models which strive to
produce a detailed characterization of the energy sector. Recently, environmental pressures have started to be incorporated into these models, linking economic activity with
various pollutants, such as SO_2 from coal or CO_2 from fossil fuel combustion.

Most of these models were originally optimization or linear programming models,
with a high level of technological detail. In fact, technology disaggregation is the main
characteristic that agglutinates this type of models. As computer capabilities have
increased, these models have started to incorporate non-linear functions that allow for a
better representation of microeconomic behaviour. Non-linear functions make it possible to capture substitutability between factors and inputs so as to represent more realistic
energy demand functions.

Many BU E3 models are derived from traditional energy system planning models.
These models focused on providing a detailed characterization of the energy sector
(Rath-Nagel and Voss, 1981). The building blocks in Figure 8.1 represent a simplification of this type of complex system. Energy system models cover different technologies
that convert specific inputs into final useful outputs or energy services. Primary or raw
energy commodities such as crude oil, coal, uranium or solar radiation are typically
converted through different processes and conversion technologies into marketable
products that can be consumed by end users. These products or energy carriers may be
storable (such as petrol, diesel or biofuels), or non-storable (such as electricity and heat).
One example is the process of raw crude oil, which is generally converted into petrol that
can be used in the transport sector. Similarly, in the case of nuclear power the process
includes uranium mining, conversion, enrichment and final use in a reactor to produce

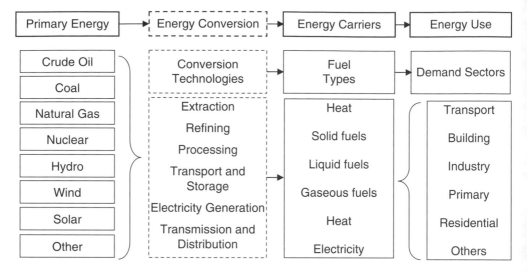

Figure 8.1   Typical structure of an energy system model

electricity. Finally, the demand for energy services may be disaggregated by sector (for example transport, industry, residential, commercial) and also by specific functions within a sector (for example lighting, residential air conditioning, heating, hot water). It is also usual for energy flows to be conserved in an energy system model following thermodynamic laws.

Technologies in an energy system model cover the transformation of all inputs into outputs. These technologies include mining, refineries, pipelines, power plants and end-use appliances. They are connected with the different energy carriers and can be classi-fied by different characteristics, although in some cases the final output is the same. For example, gas turbines in gas-fired power plants and photovoltaic cells in solar plants both produce electricity, although the former uses chemical transformation and the latter electromagnetic transformation.

Energy system models may represent hundreds of competing technologies, so their realistic representation is therefore critical for the modelling framework. The basic infor-mation required to characterize a technology is the following:

- initial investment;
- operation and maintenance (O&M) costs;
- fuel costs;
- lifetime;
- technical efficiency;
- emission coefficients.

The general features of a technology description include the initial investment necessary to put the technology into service, O&M costs (fixed and variable) and fuel costs (in the case of renewable energy this cost may be zero, for example with wind and solar radia-tion). Other variables are also relevant, such as the lifetime (that is, depreciation rate) of the technology and its efficiency. Inclusion in the model of emission coefficients associated with each energy carrier allows environmental pressures to be tracked. Moreover, if for example CO_2 emissions have a price (as in the European Trading Scheme – ETS) this may represent another relevant cost. There are many other variables that should be accounted for and that depend on each technology, such as the capacity factor or the plant type (that is, peak or base load plant) in the case of power plants. In addition, technologies can change over time in their use, efficiency, costs and energy needs.

Some variables in energy system models are often defined as exogenous. These generally include population, gross domestic product (GDP) and primary energy prices. GDP growth determines differences in energy demand and, therefore, levels of activity in the disaggregated sectors (industry, transport, and so on). For these levels of activity, the model calculates the best technology mix option for meeting demand, and the energy service prices. The demand corresponds to different forms of secondary energy (electricity, petrol, diesel, and so on) and the production of primary energy (fossil fuels, renewables, and so on). Technological change is occasionally considered, especially in dynamic models, but in the form of exogenous factors such as changes in technological efficiency.

Finally, there are also some BU models in the relevant literature that extend technological details to other subsectors. This is the case for the many models for the power sector (Uri, 1976; Hillsman et al., 1988; Hoster, 1998; Soloveitchik et al., 2002) and the transport sector (Ortuzar and Willumsen, 2001). Although they are less common, there are also specific models for capturing processes in energy-intensive industries such as the steel industry (Hidalgo et al., 2005), the cement industry (Szabo et al., 2006), the petrochemical industry (Calloway and Thompson, 1976) and the paper industry (Bloemehof-Ruwaard et al., 1996).

8.2.2 Types of Model

BU models can be classified as optimization models or as simulation models. Optimization models use an objective function which seeks to minimize energy costs or maximize consumer utility, subject to technological possibilities and a wide range of other restrictions (capacity, emissions, and so on). Their solution is the best of all possible alternatives. In simulation models the variables are related statistically and try to represent in detail how the real system evolves under given conditions. These models are used to evaluate effects for a scenario or policy. Both types of model are surveyed below.

Optimization models
Optimization techniques are common in energy system planning. These models find the optimal solution based on cost and constraints defined by technology characteristics. An example of this type of model, representing a partial equilibrium model, is illustrated in Figure 8.2. The figure refers to a single energy service in a single time period. Consumers' willingness to pay is displayed as a decreasing, continuously differentiable function of the amount of energy available to them, and producers' incremental costs are shown as an increasing step-function of the amount to be supplied. The energy cost function is estimated through a technological process analysis with linear or non-linear programming. Consumers' demands are typically estimated statistically through econometric

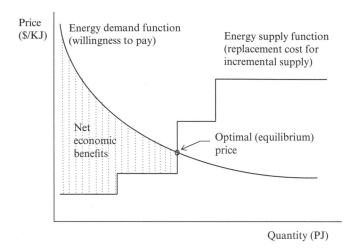

Figure 8.2   Energy market mechanism and maximization

techniques. Supplies and demands are then balanced through an equilibrium price using an algorithm to maximize the net economic benefits (hatched area). This process resembles market mechanisms for optimisation models.

Relevant models using optimization frameworks are for example ETA, MARKAL and MESSAGE. The ETA (Energy Technology Assessment) model (Manne, 1976) was one of the earliest energy system models, and was originally developed to evaluate the US nuclear energy programme. ETA is a non-linear model with an objective function focused on maximizing the sum of consumer and producer surpluses over a time horizon (75 years). Supply is represented through a set of technologies with upper bounds (imposed to control the rates of market penetration for new supply technologies) and lower bounds (to ensure that older technologies are not phased out too rapidly). Energy demands are divided into two final composites – electricity and non-electric energy with imperfect substitution – which are specified as an econometric function of the (US) economy. The MARKAL (Market Allocation) model (Fishbone and Abilock, 1981) is probably the most widely used energy system model. MARKAL is a linear programming model with a very high technology disaggregation covering the life-cycle cost of each technology. The model has been extended into many different areas and now incorporates aspects such as an elastic energy demand, externalities and a climate module. The MARKAL family also covers different geographical scales such as national, regional and global (TIMES model). Finally, MESSAGE (Messner and Strubegger, 2001) is another engineering optimization model focused on long-term energy planning that has been used by the World Energy Council. The model is global, is disaggregated into 11 regions and incorporates the international trade in energy commodities. The current version has recently been expanded to include endogenous learning for various technologies and to cover all six Kyoto green house gases (GHGs). It was used in the International Institute for Applied Systems Analysis (IIASA) study of Global Energy Perspectives to define long-term energy scenarios (Grübler et al., 1996).

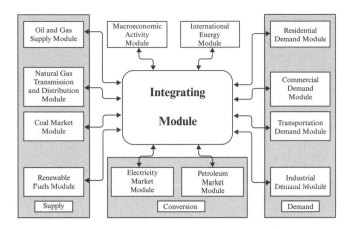

Source: NEMS.

Figure 8.3   A bottom-up simulation model

Simulation models

BU simulation models represent the dynamics of energy, economic and environmental variables. The representation seeks to be very exhaustive and the relations between variables are carefully determined by statistical methods. However, these models are not suited to providing 'least-cost' solutions, but rather highly detailed responses of probable outcomes related to changes in variables. They are useful for analysing policy implications.

Relevant models using simulation frameworks are, for example, NEMS and POLES. NEMS (National Energy Modeling System) is a computer-based, energy-economy modelling system designed and implemented by the Energy Information Administration (EIA) of the US Department of Energy. It is used to prepare the projections for 20–25 years for the Annual Energy Outlook (EIA, 2009) and to evaluate alternative policies in new energy programmes. The POLES (Prospective Outlook for the Long-term Energy System) model is supported by the European Commission and is designed to develop long-term scenarios (to 2050) that describe the supply and demand of energy in different regions of the world. The structure is similar to NEMS, but the scope is global. The POLES model has recently been used in the World Energy Technology Outlook (Lapillonne et al., 2003) and serves to support the development of long-term European policies on issues such as security of energy supply, energy research and development (R&D) programmes, and Kyoto and post-Kyoto target implementation.

BU simulation models can be illustrated as in Figure 8.3, which represents the behaviour of energy markets and their interactions with the economy. The system reflects markets, industry structure, existing energy policies and regulations that influence market behaviour. In this case (the NEMS model) it consists of four supply modules (oil and gas, natural gas, coal market and renewable fuels), two conversion modules (electricity and petroleum markets), four end-use demand modules (residential, commercial, transportation and industrial demands), one module to simulate energy–economy interactions (macroeconomic activity), and one module to simulate international

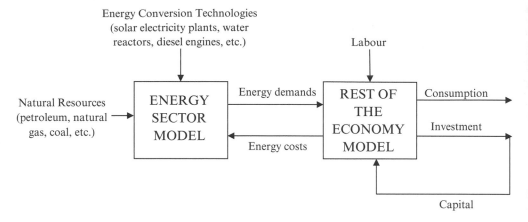

Figure 8.4   Basic connections in an energy–economic model

energy markets (international energy). Finally, there is one module that provides the mechanism for achieving market equilibrium between all the other modules (Integrating Module).

8.2.3   Bottom-up Models with Macroeconomic Linkage

BU models are designed to represent relations within the energy sector in the most extensive way possible. Energy demand is generally considered to be exogenous and independent from prices, so leakage with the rest of the economy is often not fully captured. To overcome this shortcoming some models have incorporated an economic module into the energy model.

Figure 8.4 provides an overview of the basic connections in energy–economic models. There are physical flows of energy services from the energy sector model to the rest of the economy model, and the corresponding energy cost payments. Energy demand is exogenous to the energy sector model, but endogenous to the rest of the economy model. The costs of energy supply appear in the objective function of the energy sector model, but enter the rest of the economy model through period-by-period constraints governing the allocation of the aggregate output of the economy between consumption, investment and energy cost payments. In these models the rest of the economy is generally aggregated into a single non-energy sector.

ETA-MACRO (Manne 1978) and MARKAL-MACRO (Manne and Wenne, 1992) are examples of these merged models. For example, in MARKAL-MACRO a representative consumer maximizes the discounted value of consumption over time. Basically, production is to be used for consumption (C_t), investment for building up the stock of capital (I_t), and interindustry payment for energy cost (EC_t), as equation (8.1) shows. This equation implies that an increase in energy costs (EC_t) will reduce the net amount of output available for meeting current consumption and investment demands. Production is assumed to depend upon three inputs: capital (K_t), labour (L_t) and energy (D_{jt}), which are combined through a nested constant elasticity of substitution CES production function, as displayed in equation (8.2):

$$Y_t = C_t + I_t + EC_t \tag{8.1}$$

$$Y_t = \left[aK_t^{\rho\alpha}L_t^{\rho(1-\alpha)} + \sum_j b_t(D_{j,t})^\rho \right]^{\frac{1}{\rho}} \tag{8.2}$$

where a is a scale factor, b_t is energy efficiency adjustment parameter, α is the share parameter for capital, and ρ is a parameter associated to the elasticity of substitution between energy and value added aggregates.

The principal advantage of these models is that they enable a direct link to be established between the analysis of a physical process and a standard long-term macro-economic growth model. However, models implicitly assume that expenditures on energy affect the marginal utility related to the consumption of non-energy products. Moreover, these models cannot make a direct connection with sectoral leakages or therefore calculate economy-wide effects when energy policies affect the whole economy. In the following sections more details are offered regarding TD or general equilibrium models and their hybridization with BU models to solve these problems.

8.3 TOP-DOWN MODELS

8.3.1 An Outline of some Characteristics of Top-down Models

TD models are able to capture economy-wide market interactions endogenously. In this sense, they overcome this weakness of BU models, although TD models lack the technological details that may be relevant for the analysis and assessment of energy strategies. The most widely used TD models for integrated E3 assessment are computational general equilibrium (CGE) models or applied general equilibrium (AGE) models. It is on these models that this section focuses. They are Walrasian models based on the perfectly competitive Arrow–Debreu general equilibrium framework. They also often include some extensions departing from the Arrow–Debreu assumptions, such as imperfect competition in goods markets, unemployment, different expectations, financial assets, and so on.

The number of books and articles surveying and introducing CGE modelling is vast, and to repeat that work would take us outside the scope of this chapter. Perhaps the most cited introductory paper on the subject is Shoven and Whalley (1984). Among the many publications that have also contributed to developing the theoretical basis and the rationale of CGE for policy simulations, a non-exhaustive list would include Adelman and Robinson (1978) on economic development, Shoven and Whalley (1992) and Ginsburgh and Keyzer (1997) on applied and theoretical topics, Francois and Reinert (1997) and Hertel (1997) on trade policy, and Kehoe et al. (2005) on recent developments.

The main thrust of CGE models is policy-analytical, as simulation models, rather than oriented towards forecasting. These applied simulations include policies on taxation, international trade, development, migration, and energy and environmental issues. In order to develop a simulation with this framework it is first necessary to construct a benchmark scenario with a set of structural assumptions and a dataset that represents the equilibrium reference. Next, a particular policy scenario is constructed. CGE models are then capable of endogenously evaluating the impact of such policies through changes in prices or by setting quantity constraints. The simulation results reflect adjustments in resource allocations, movements in supplies and demands, changes in relative prices, welfare effects, and so on, with respect to the reference scenario.

CGE models consider the circular flow of factors, goods and incomes and interactions between economic agents. They seek to approach the main features of a real economy (at regional, national or global level) with real data. Nevertheless, Walrasian models only determine relative prices, and are homogeneous at degree zero in prices: thus changing all prices by the same amount does not change real results (that is, money neutrality). The absolute price level is indeterminate although some models set out it through assumptions on its exogenous specification.

8.3.2   CGE Modelling Framework

A simplified standard CGE model is presented in this subsection. Two classes of agent are represented in this economy: households and firms. More agents can be incorporated, such as the public sector or some foreign countries. The Walrasian paradigm assumes optimizing agents (that is, households maximize welfare, firms maximize profits) but most extended models specify the public sector with non-maximizing rules of behaviour. For example, a common specification is that public expenditure is fixed in real terms, public revenue comes from fixed tax rates, so public savings are determined residually.

The core model
Three core conditions constitute the equilibrium result of a CGE model: welfare maximization, profit maximization and market clearing. An allocation of demanded and supplied quantities and price levels constitutes an equilibrium when those conditions are satisfied. This definition of equilibrium is often widened to include extensions of the model: price and quantity constraints in goods markets, imperfect competition in commodity markets, labour market imperfections, macroeconomic closure for the foreign and public sectors, investment in dynamic models, and so on. Next, we focus on the three core conditions and then we describe some extensions related to E3 models.

The first core conditions affect the demand side of the model. Households behave rationally and optimally with a welfare objective. Their decision problem is to choose consumption levels of the various goods that are available on the market according to their preferences. Some constraints limit consumption choices. The most common are the positive price of goods and the wealth bound because of factor endowments. A typical welfare or utility maximization problem where a representative household fully spends its income is:

$$Max\ W(x_1, \ldots, x_n)$$

$$subject\ to\ \sum_{f=1}^{m} w_f l_f \geq \sum_{i=1}^{n} p_i x_i$$

(8.3)

where W represents welfare for the representative household, $x_1, \ldots, x_n$ are the n goods, w_f are factor f unitary rents, l_f is the endowment of factor f, and p_i is the price of good i. The result of the maximization problem is a set of demand functions responding to changes in prices and incomes. From them, the household chooses a consumption bundle.

The second set of core conditions lies on the supply side of the model. A number of production units or firms produce the goods demanded. They are able to transform inputs into outputs. The firms also behave rationally and optimally, and their objective is profit maximization. CGE models assume this objective rather than other objectives such as the maximization of sales revenues, managers' bonuses, the minimizing of emissions or the maximizing of the size of the firms' labour forces. This assumption is standard: individuals who are also consumers own firms in this type of economy. If those individuals have different objectives, the problem can become intractable. It seems, under reasonable assumptions (Mas-Collel et al., 1995), that this is the goal that all owners would agree upon.

Each firm has a technology commonly described by means of a production function that gives the maximum output that can be produced using input amounts. If there is only one firm producing each good, a standard profit maximization problem is:

$$Max \; \pi_i = p_i q_i - c_i(q_i)$$
$$subject \; to \; q_i = \varphi_i(w_1, \ldots, w_m) \tag{8.4}$$

where π_i are profits for firm i, q_i is the output of good i, c_i is the total cost of good i, f_i is the production function. The result of the problem is a production plan that maximizes firms' profits, taking prices of inputs and outputs as given.

The final set of conditions harmonizes the demand and supply sides of the model. A matching between the desired consumption (x_i) and the production level (q_i) is required. Hence, aggregate demand for each commodity equals aggregate supply of it (see equation 8.5). If there is excess supply, some producers will find it worthwhile to change prices and offer a discount on the current price. And if there is excess demand, some consumers who are not getting the desired commodity may be better off offering a higher price for it. Price becomes the mechanism for adjustments in this core model:

$$q_i = x_i \tag{8.5}$$

Next, we focus on a set of specific extensions of CGE models that can commonly be found in E3 models. The grouping mainly stems from Sue Wing (2009) where a more detailed presentation is provided. The extensions are mathematical mechanisms price- and quantity-related to reproduce real or possible policies, for example taxes and subsidies related to E3 policies that affect prices, and constraints on demand and supply due to E3 policies influencing commodity markets. These mechanisms are responsible for the transmission of price and quantity adjustments between markets.

Extensions related to price mechanisms

Taxes and subsidies affect prices because they introduce a wedge between consumer and producer prices. They can burden the output of energy sectors, the final consumption of energy commodities or even energy as an intermediate input. Their welfare effect depends on the interactions in the model. A tax or a subsidy changes relative prices, and some substitution and income effects take place. The substitution effects reflect that the more expensive outputs or inputs will be replaced by cheaper ones. The income effect reflects a lower real income when a tax is levied, and a higher real income when a subsidy is set. These effects are accounted for in this general equilibrium framework.

The effects of tax revenue or subsidy expenses can also depend on the way the public sector agent is inserted. A simple way is to treat the government as a passive entity that collects (spends) the revenue (subsidy) and recycles it to (from) the households as a non-distorting transfer. Other ways involve more interactions. For example, the distorting effect of the introduction of a green tax can be offset with the reduction or elimination of other distorting fiscal instruments. This has resulted in all the E3 models related to the theory of the second (or third) best and the double dividend hypothesis (that is, some gains in addition to environmental benefits, see, for example Schöb, 2005 for an empirical survey). Another example is the use of tax revenue to compensate the subset of households and/or firms negatively burdened with the tax. In the case of subsidies, in general, the effects can also be distorting, with a subset of households and/or firms favoured at the expense of other agents. Again, a general equilibrium framework is appropriate to analyse transmission adjustments.

Extensions related to quantity mechanisms

The variety of quantity mechanisms is wide-ranging. As Sue Wing (2009) points out, in comparison with taxes or subsidies, quantity instruments vary widely in their characteristics and methods of application. It is useful to draw a distinction between the instrument itself and its effect on supply or demand in a particular market or set of markets. Quantity distortions generate a stream of rents that must be allocated somewhere in the economy, as taxes do.

The setting of upper or lower bounds on the supply and/or use of energy commodities is a common quantity constraint. Such constraints may be direct or indirect, and relative or absolute. They are direct when the energy commodity is limited in some way (for example a Renewable Portfolio Standard that imposes a lower bound on the production of renewable energy). They are indirect when the control affects some attributes of the commodity (for example a mitigation policy limiting the emissions from a CO_2-intensive fossil fuel, which ends up curtailing demand for it). They are absolute when the target in energy or its attributes is economy-wide (for example a GHG emissions cap), and relative when the target is in relation to other variables in the economy (for example the carbon content in imported manufactured goods with respect to domestic goods). Three examples of quantity mechanisms follow.

A quota on an energy commodity is a pure rationing instrument. It restricts its output to a level below the competitive equilibrium. A common way to model it is through a virtual tax. The tax changes the energy commodity price to the point where the quota is obtained. The virtual revenue comes back to the household as a lump-sum or non-distorting transfer. Hence, the effect of the quota involves an endogenous change in relative prices.

Another rationing instrument is a Renewable Portfolio Standard. This mechanism redistributes revenue from conventional to renewable energy producers, with indirect impact on aggregate income, which operates through the prices of energy commodities. It acts as a tax on conventional energy commodities recycled to finance renewable energy commodities. Setting a lower bound in renewable commodities involves reducing the share of conventional energy, lowering its price. It can be modelled as a tax on all energy commodities. The revenue would go to renewable energy producers to subsidize their output.

A third rationing instrument is a cap on emissions. A simple way to model this is through a commodity-specific emission coefficient. The demand for and consumption of a fuel determine the level of emissions, given those coefficients. There are several methods for constraining emission levels: one standard method is through emission permits. Permits can be allocated by auctioning and grandfathering firms. A public auction generates revenues for the government that can be recycled to households and/or firms as distorting or non-distorting transfers. A grandfathering allowance is equivalent to defining a new factor of production that increases the profitability of firms but at the same time is also owned by the tenant, who would receive the income from permits. In both cases redistributive effects take place, and relative prices are thus able to change.

In short, the core model and its extensions with price and quantity mechanisms reflect a relevant virtue of TD models with respect to BU models: economic interactions are consistently represented.

8.3.3   Data Requirements for Top-down CGE Models

The data requirements for TD models are, as for CGE models, dependent on the mathematical functions chosen to simulate the policies. Several sets of data are commonly used to run an E3 CGE model: a social accounting matrix (SAM), environmental data and behavioural parameters (as well as some calibrated parameters). CGE models are calibrated to a benchmark equilibrium dataset (that is a SAM, completed with environmental data and the behavioural parameters). The calibration process computes some parameters for the model's functions, to reproduce the SAM as an equilibrium solution of the model (Mansur and Whalley, 1984; Dawkins et al., 2001).

The SAM
A SAM (Reinert and Roland-Holst, 1997; LEG SAM, 2003) is a 'snapshot' of the economy that embodies information normally included in national accounts and other sources. It interrelates the main national accounts macro-statistics with micro-statistics on suppliers and households often extracted from the input–output framework and household budget survey. The SAM delineates the circular flow of income in the economy. The data are presented in a matrix format, which elaborates on the linkages between supply and use: for every income or receipt there is a corresponding expenditure or outlay. These are the interrelations characterizing TD models.

A SAM account ensures that the corresponding row and column totals, the income and expenditure for each account, must be equal. As a result, SAMs satisfy a variant of Walras's Law: if all accounts but one balance, then the last account must also balance. This property hints at the relationship between SAMs and CGE models. The representative SAM presented in Figure 8.5 divides economic activity into five main areas: production, consumption, public sector activity, investment and a link with the rest of the world.

Suppliers receive their revenue from selling consumption goods to households (C) and to the public sector (G), investment goods (I) to the capital account, and exports (X) to the rest of the world. The revenue from these sales passes to the consumption account as income paid to the factors of production (Y) and imports (M) from the rest of the

		Expenditures					
		Suppliers	Consumers	Public Sector	Capital	Foreign Sector	Total
Receipts	Suppliers	–	C	G	I	X	Demand
	Consumers	Y	–	–	–	–	Income
	Public Sector	–	T	–	–	–	Receipts
	Capital	–	Sh	Sg	–	Sf	Savings
	Foreign Sector	M	–	–	–	–	Imports
	Total	Supply	Expenditure	Expenditure	Investment	Foreign Exchange	

Figure 8.5   A social accounting matrix of an open economy with a public sector

world. Equation (8.6) reflects the balance. Household outlays take the form of consumption expenditures (*C*), tax payments (*T*) and private domestic savings (*Sh*), as stated in equation (8.7). Government outlays take the form of consumption goods (*G*) and government savings (*Sg*) (equation 8.8). Inflows from the rest of the world take the form of export demand (*X*) and foreign savings (*Sf*). Finally, foreign savings (*Sf*) are the negative from the trade balance (equation 8.9). A macroeconomic balance identity on savings and investment completes the specifications, as presented in equation (8.10). Hence, the accounting identities are:

$$C + G + I + E = Y + M \tag{8.6}$$

$$Y = C + T + Sh \tag{8.7}$$

$$T = G + Sp \tag{8.8}$$

$$X + Sf = M \tag{8.9}$$

$$Sh + Sp + Sf = I \tag{8.10}$$

This macroeconomic SAM can be extended to a multi-household SAM using mainly a household budget survey. It can be also extended to a multi-supplier SAM using mainly the input–output framework. Other accounts can be split, and SAMs with a variety of primary factors, countries and even public sector levels abound.

E3 data

The environmental data must be reconciled with the SAM to calibrate model equations. There is a huge variety of environmental data that can be linked to a CGE model. Each policy simulation requires specific data, so the possibilities are broad. Next, we describe an example of this data matching: the Global Trade Analysis Project (GTAP; see Hertel, 1997). This is the most widely used global economic database for CGE modelling. GTAP includes a SAM for the world economy which is highly disaggregated at commodity and country levels. Every 3–4 years a new version is developed. The latest is GTAP version 7, referenced to year 2004, described by Narayanan and Walmsley (2008). Several sets of E3 data have been harmonized with the GTAP framework, such as CO_2 emissions, non-CO_2 gas emissions and land use, described below.

Lee (2008) explains the link between GTAP data and CO_2 emissions. He describes the compilation of carbon emissions from fossil fuel combustion by users (sectors) in 113 regions. Combustion-based CO_2 emissions are calculated from the energy volume data. GTAP adopts the Tier 1 method of the revised 1996 IPCC Guideline (IPCC, OECD and IEA, 1997).

Rose and Lee (2008) expound the link of GTAP with non-CO_2 greenhouse gas emissions (NCGG). The GTAP NCGG emissions dataset highlights NCGG emissions associated with land-based activities, and the heterogeneity of sectoral and regional NCGG emissions. The NCGG dataset complements the previous CO_2 database and a forest carbon stock dataset. Together, these datasets provide a fairly comprehensive GHG emissions and carbon sink profile for each sector within each region. Unlike other NCGG databases, the data were specifically developed for direct integration with economic activity datasets.

Another GTAP link with E3 data is with land use statistics. Hertel et al. (2009) include a collection of pioneering papers on the applied economics of land use in CGE models. They open up the chance to determine analytically the potential role for agriculture and forestry in climate change mitigation. The scarcity of data in this worldwide database limits the approach to the use of existing data on land rent for crops and livestock.

Behavioural parameters

Behavioural parameters can be divided into two types: those related to model structure, and those related to functional forms. The level of aggregation also determines the number of parameters. Shoven and Whalley (1992) extensively develop the following comments on the relevance of those parameters.

Structural parameters The appropriate general equilibrium model for any particular application depends largely on the policy issues being addressed. E3 models often confront dynamic issues, so CGE dynamic models are common in the relevant literature. This involves the assumption of values of parameters related to dynamics: factor growth rates, discount rates, intertemporal preference parameters, and so on; see van der Mensbrugghe (2008), for example, for the set of assumptions and related parameters of the World Bank dynamic CGE model ENVISAGE (Environmental Impact and Sustainability Applied General Equilibrium), and Walmsley and Strutt (2009) for the GTAP framework.

If some kind of imperfect competition in good markets is modelled, the ordinary parameters are linked to margins, market shares, fixed costs, profit levels, and so on. Examples of work on the role and relevance of parameters of imperfect competition in CGE models are Hoffmann (2002), Willenbockel (2004) and, in the E3 framework, Böhringer et al. (2008). More parameters must be specified when imperfections in factor markets are present in the model, as in labour markets: unemployment rates, wage rigidities, matching processes, and so on. A well-known example of these imperfections in labour markets applied to CGE models is the Multiple Indicator, Multiple Cause (MIMIC) model, developed by the CPB Netherlands Bureau for Economic Policy Analysis (Graafland et al., 2001).

Functional parameters   The specific forms chosen for utility and production functions depend upon how elasticities are to be used in the model. This point is best illustrated by considering the demand side of the model. Demands derived from Cobb–Douglas utility functions are easy to work, but have the restrictions of unitary-income and uncompensated own-price elasticities and zero uncompensated cross-price elasticities. These restrictions are typically implausible, given empirical estimates of elasticities applicable to any particular model, but can only be relaxed by using more general functional forms. The general approach adopted by most modellers is to select the functional form that best allows key parameter values to be incorporated, while retaining tractability.

Hierarchical or nested functions are another device widely employed in applied models. The nested structure in E3 models requires a large number of elasticities, often not available in the relevant literature. Figure 8.6 shows a nesting structure in the MIT Emissions Predictions and Policy Analysis (EPPA) model, as in Paltsev et al. (2005). Panel (a) shows the structure of production technology for services (SERV), industrial transportation (TRAN), energy intensive (EINT) and other industries (OTHR). Panel (b) displays the technology for agriculture (AGRIC). Each nest is associated with an elasticity of substitution among or between inputs. The design of the nesting can bias the results because there may be other ways of representing technologies, as Jacoby et al. (2006) demonstrate.

Kemfert and Truong (2009) pinpoint the energy elasticities in those nestings as crucial parameters that determine the answer to the policy simulated. The relevance of this assertion can be explained with an example on the elasticity of substitution between energy inputs and capital in production functions. Empirical evidence on its value has been rather mixed. Estimated values of this parameter have tended to depend not only on the level of aggregation, but also on the type of data used and the specification of the empirical production function. If the policy issue is dynamic in nature there is a maximization of an intertemporal welfare function. The focus of attention is on the division of output between consumption and investment, and the main issue here may be the optimal rate of (energy) resource depletion to sustain economic growth and consumption in the long term. If the elasticity of substitution between capital and energy inputs is greater than or equal to one, then sustainable economic growth and consumption is achievable even if the energy resource is in fixed supply. When substitution elasticity is less than one, this implies that there are diminishing returns in the process of substitution of human-made capital for natural resources. In this case, sustainable economic growth may still be achievable if technological progress can be made to offset the effect of diminishing returns.

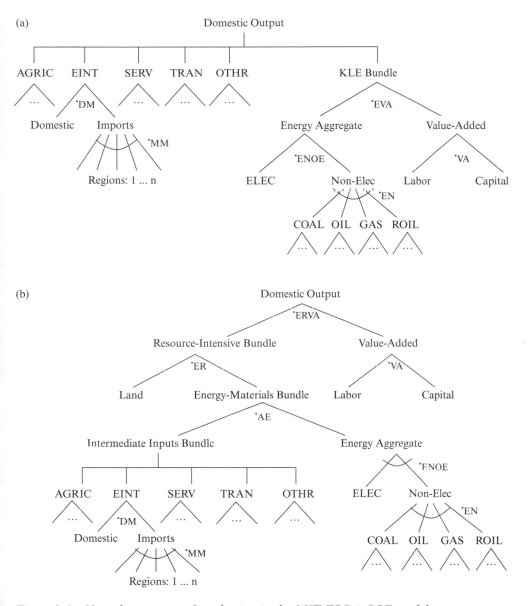

Figure 8.6   Nested structures of production in the MIT EPPA CGE model

In general, modellers borrow elasticity estimations from CGE literature or directly from econometric studies. This encourages the use of sensitivity analysis on the simulation results, given that some of the assumed elasticities can bias the results. Beckman and Hertel (2009) examine key CGE model parameters, and thus the validity, of one of the widely utilized CGE models for E3 policies: the GTAP-E. Welsch (2008) also examines the importance of CGE parameters and their subsequent impact on energy policy analysis. There are some attempts to correct this shortcoming. For example, WIOD (2010) develops an input–output database intended to provide not only the SAM for CGE models but also a time series of input–output tables to ameliorate the robustness of the parameters for TD models.

8.3.4   Critiques and Caveats

There are some critiques regarding the methodology of CGE models, as well as the lack of detail in the representation of the energy sector. Furthermore, as with any other economic methodology, some caveats should be taken into account when building a TD model and deriving policy conclusions.

CGE models are deterministic non-linear systems of equations. Non-linear equations can often involve problems such as indetermination, non-existence of equilibrium, multiple equilibria, instability or even corner solutions. Most CGE literature disregards these problems and fails to demonstrate the existence, stability and uniqueness of the equilibrium. Model equations are often too complex, making demonstrations implausible. The complication inherent in the possibility of multiple equilibria, added to their potential instability, points to a need to analyse whether the equilibrium is at least locally unique. That is why it is crucial to draw up a sensitivity analysis of the equilibrium point obtained in order to be able to apply the model in comparative statics analysis, as long as the shocks to the system are not large enough to involve substantial changes. This is the most frequent way of checking such problems.

CGE models are capable of representing different economic closure model options (for the foreign sector, public sector, investments, and so on). This provides the ability to describe not only pure neoclassical models, but also variations that incorporate alternative neo-Keynesian postulations, and even a partial representation of the assumptions of the structuralist school. The closure rules are chosen by the researcher, and so can always be subject to criticism. Hence, the theoretical and empirical grounds of the closure rules chosen should be in accordance with the simulation performed, as in Kehoe et al. (1995).

Deterministic models do not model random behaviour, and as such are incapable of directly modelling even measurable risk and uncertainty. This is directly related to the critique of the use of apparently arbitrary values for behavioural parameters and a lack of model validation (Jorgenson, 1984; McKitrick, 1998; Kehoe, 2005). A common way of getting around this problem is to apply the CGE deterministic approach under a group of scenario evaluations or Monte Carlo simulations (Webster et al., 2002). Other papers have tried to ensure robust parameter estimations (Liu et al., 2003; Hertel et al., 2007). An example of an attempt to design a stochastic model validation is Valenzuela et al. (2007). At the same time, an alternative for handling this uncertainty endogenously is to adopt a different modelling approach named dynamic stochastic general equilibrium modelling (DSGE). In DSGE models, some important economic parameters – such as GDP, consumption, investment, prices, wages, employment and interest rates – are estimated using Bayesian statistical techniques in order to approximate their levels to observed behaviour while still making use of micro-foundations in the determination of agents' behaviour.

The assumption that the economy remains in equilibrium at benchmark and after the simulated shocks is a vital assumption for CGE models. Although the SAM database departs from balanced national accounts, the assumption of equilibrium in the economy at this initial time does not mean being in a stationary equilibrium or one with full market clearing (for example unemployment is characteristic in most countries, and inventories vary continually). Attempts to design intertemporal national accounts (WIOD, 2010) and their derived SAMs drive the answer to this limitation.

Another relevant issue is that every CGE model is a relative prices model, as stated above, which assumes neutrality of money in the economy. As such, adjustments in money stocks and monetary shocks cause effects in the economy but are impossible to represent in CGE models, directly invalidating their application to more complex issues related to money such as inflation patterns and estimation of future price levels. One branch of CGE models includes some extensions regarding monetary markets, as Robinson (2006) shows.

Finally, it is very important to interpret the results of the model properly. The complexity of CGE models can turn them into 'black boxes', so a careful description of the mechanisms behind the general equilibrium effects, grounded on theoretical assumptions, is essential. In fact, the 'black box' critique is unexceptional, so interpretative exigencies as delimited by Adams (2005) seem a good research strategy.

8.4   HYBRID MODELS

8.4.1   Hybrid Models in the Context of Bottom-up and Top-down Models

As stated in the previous sections of this chapter, the integrated assessment of environmental–economic–energy issues led to the development of two main disjoint modelling approaches: BU and TD models. BU models are capable of addressing detailed information about production technologies and the decision structure specificities inside a specific sector of the economy or market. Macroeconomic direct and indirect effects of policies can be evaluated under a TD approach.

The choice of the framework to be adopted clearly depends on the issue in question. Regarding the BU alternative, no more than a partial equilibrium approach would be necessary if the interactions between the studied sector and the remaining economy were negligible. However, when feedbacks to other agents and rebound effects are considerable, TD models are more suitable for the job. The question to be asked is: what happens in the climate assessment problem dealt with by E3-like models?

One can argue that choosing between addressing the technological richness of BU or the indirect effects evaluation of TD models can represent a significant commitment when dealing with environmental issues. Undoubtedly, the detailed description provided by BU models of the set of technologies available is crucial in an analysis of pollutants, specifically in the case of energy sectors. At the same time, energy sectors can cause substantial indirect spillovers to other markets and, simultaneously, many climate issues can be presented as problems of a global nature, highlighting the importance of a comprehensive macroeconomic approach such as the ones provided by TD models.

The ambiguity on the modelling choice paradigm for E3 model assessment emphasizes the failures of both isolated BU and TD models to represent the linkage between the economic forces ultimately driving demand and production choices and their environmental consequences. Therefore, potential benefits could be achieved if a hybrid structure is adopted.

8.4.2   An Outline of some Characteristics of Hybrid Models

In order to overcome the inherent limitations of BU and TD frameworks when dealt with individually, a new integrated modelling approach was proposed in the relevant literature: hybrid modelling. Hybrid models attempt to integrate the individual strengths of the detailed treatment of specificities in crucial markets into the problem (in the BU case), and at the same time deal with indirect effects (in the TD case), under a unified modelling approach.

The BU modelling lack of indirect effects appraisal usually leads to overly optimistic simulated results. In the meantime, the typical TD modelling failure on disregarding complex technological alternatives could lead to pessimistic estimated results of real-world behaviour. Hybrid models include the macroeconomic representation contained in TD models and at the same time aim to portray the detail in production and exchange decision structure contained in BU models for the most sensitive sectors for the problem in question. The objective is to achieve a more realistic simulation of the interactions between human actions, production decisions and environmental inter-relations.

Typically, economic-based TD models represent the most important activities as aggregated commodities, ocurring only once in each period (usually once a year), at the efficient frontier of a specific production function (that usually adopts a CES functional form, as in Figure 8.6), by the combination of diverse production factors and supplementary commodities. The choice of representing one of these activities in a BU format, where the production function form is substituted by a detailed description of a set of specific technologies and the aggregate production time period can be divided into smaller time blocks more representative of the actual production process and of the specific demand conditions, entails a number of limitations and consequent modelling alternatives. These difficulties can be divided into two main fields: the set of theoretical choices for formulating the linkage between BU and TD models, and the problem of data incompatibilities.

An integrated hybrid approach requires a certain compatibility in communication between values provided by each model. The aggregate information used in TD models should in principle reflect, or at least be compatible with, the results obtained in BU models to allow the convergence of the model. Furthermore, the actual link between the different models can be represented through different degrees of interaction. Depending on the situation analysed, the computational requirements and the data availability of the problem can vary substantially, as detailed in the next two sections.

Whether by adapting the models employed or by making changes in the databases that they use, much has been achieved in the relevant literature in successive attempts to reconcile BU electricity operational detail and TD indirect effects evaluation. The search for integration at the modelling level can be summed up in number of recent research papers on hybrid modelling (Wene, 1996; Böhringer, 1998; Böhringer and Loschel, 2006; Hourcade et al., 2006b; Schumacher and Sands, 2007; Böhringer and Rutherford, 2008; Strachan and Kannan, 2008; Turton, 2008; Böhringer and Rutherford, 2009; Labandeira et al., 2009; Tuladhar et al., 2009).

On the other hand, the database requirements in such hybrid integrations have also resulted in a number of papers that pursue the desirable representation of technological compatibility (Koopmans and Velde, 2001; McFarland, 2004; Ghersi and Hourcade, 2006; McFarland and Herzog, 2006; Sue Wing, 2006, 2008; Rodrigues and Linares, 2011).

8.4.3   Hybrid Modelling Framework Choices

In an exclusive TD economy-wide model such as a CGE model,[1] all sectors are represented as production functions with a nested structure. To incorporate specificities of certain activities, it is possible to shape the representation of a specific sector utilizing more descriptive ways (for example utilizing BU partial equilibrium models).

The main objective of this approach is to represent better the sectors for which the process descriptions and the data available are more plentiful and more detailed, with the intention of allowing a more refined theoretical structure for the process of determining the production of prices and quantities. Moreover, a more detailed representation of the interrelations within the sector enables specific sectoral policies and their consequences in the entire economy to be studied, which may extend the possible uses of TD and E3 models.

The first alternative for providing a more detailed representation of the production structure of the energy sector cannot strictly be considered as a hybrid model. It consists of formulating a TD CGE model with detailed energy demand decisions represented directly by economic production functions with n-nested levels and specific technology substitution elasticities (see Figure 8.6).

The construction of a reduced-form sector model according to the CGE tradition adds a few complications to the original data requirements. For example, economic production functions and especially the elasticities involved do not consider physical limitations of fuel thermodynamic transformation limits. Therefore, relative fuel price changes in relation to factor prices (for example capital, labour) could potentially lead to more than thermodynamically efficient production of energy products (such as electricity) because of the elasticities of substitution assumed, underlining the need to avoid such cases.

It is also possible to use an independent BU model to calibrate its own TD elasticities. The model from Drouet et al. (2008) follows this approach, using a nested CES reduced-form model involving capital, labour, energy and materials, with elasticities estimated by detailed partial equilibrium models for electricity, transportation and industrial sectors. Again, the same approach is taken by Pizer et al. (2006) when addressing an economic analysis of climate change policies.

Another hybrid alternative is to assemble a soft-linking approach (see Figure 8.7). A soft-linking approach employs sequential models to obtain a solution (that is, soft-linking involves generating outputs from one model to serve as inputs to another model without physically connecting the two). As Mitra-Kahn (2008) points out, the 'idea of having a "chain of models" where a set of exogenous variables would be endogenous further down the chain was formulated in Robinson (1976) and described in Adelman and Robinson (1978) . . . and this idea has become very influential since.'

A sequential soft-link approach allows for a more detailed exploration of the parameters determined in the first model applied, with few additional data requirements.

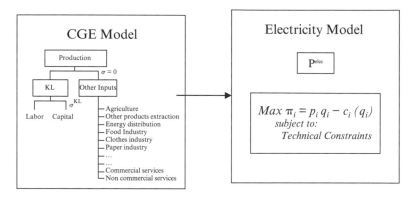

Figure 8.7   Soft-link, sequential, hybrid formulation

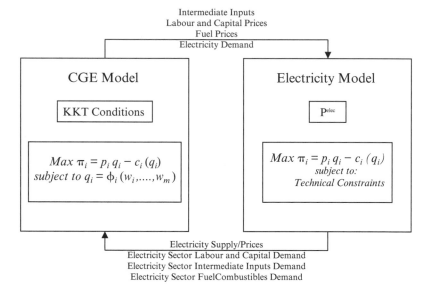

Figure 8.8   Soft-link, with feedback, hybrid formulation

Nonetheless, this formulation does not fully consider the cross-influences of demand and income effects contained in CGE formulation. The works of Wene (1996), Labandeira et al. (2006) and Rodrigues et al. (2011) are examples of models with a soft-link formulation. Wene (1996) specifically analyses a soft-link approach to the BU engineering model MESSAGE and the TD macroeconomic model ETA-MACRO.

Nevertheless, to take advantage of cross and indirect effects between economic models it is necessary to implement a feedback instrument connecting BU and TD models, as in the case of an improved soft-link approach such as the one shown in Figure 8.8.

The idea consists of iteratively linking the two models until convergence is reached. Turton (2008) makes use of this approach in linking the ECLIPSE and MESSAGE-MACRO models, while Böhringer and Rutherford (2006) present a similar iterative

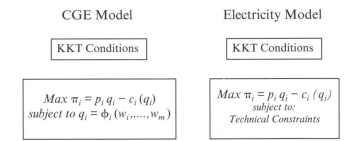

CGE Model   Electricity Model

KKT Conditions   KKT Conditions

$$Max \ \pi_i = p_i \, q_i - c_i \, (q_i)$$
subject to $q_i = \phi_i \, (w_i, \dots, w_m)$

$$Max \ \pi_i = p_i \, q_i - c_i \, (q_i)$$
subject to:
Technical Constraints

Figure 8.9   Hard-link, mixed complementarity problem, hybrid formulation

decomposition and Labandeira et al. (2009) evaluate the effects on the Spanish economy of carbon policies based on the European Trading Scheme, applying a CGE model linked with a detailed BU electricity model.

Turton (2008) presents a basic economic model that considers the output of the energy system (production of energy and transport) separately from the output of the rest of economy. That way, information about energy and transport produced in the BU part of the model (Energy Research and Investment Strategy, ERIS) is needed to obtain the results for the macroeconomic model. However, in contrast to a simple soft-link approach, the linkage is made interactively. A simulated cost function determined by BU dependent's parameters is applied to the macroeconomic model. The solution to these parameters is obtained by iterating energy demands into ERIS, which determines the energy shadow prices that are then fed into the macroeconomic model, which determines new demands. This process is repeated until convergence criteria are satisfied.

In the case of a BU model formulated as in Turton (2008), full integration between models would require the construction of the energy cost functions implied in the BU model for each possible point along the supply curve. The huge number of possibilities derived from the combination of every possible macroeconomic variable level and the corresponding efficient energy cost functions would entail impractical computation requirements for the complete integration of BU and TD models in Turton's case. As an additional trade-off, such a soft-link with feedback models has particular obstacles in the achievement and assurance of a convergence level between model results.

The best of both worlds – integration of TD and BU – can only be achieved through a hard-linking formulation (that is, physically connecting two or more models). The linkage adopted by Böhringer and Rutherford (2008) is the one that most closely resembles a hard-link approach, where the solutions to the models are obtained simultaneously through a mixed complementary problem (MCP) (see Figure 8.9). The Karush–Kuhn–Tucker conditions of the CGE equilibrium model and of the BU engineering optimization are incorporated into a unique non-linear equilibrium problem.

A real integration can only be obtained by formulating the BU structure in a similar input procedure to that used in CGE modelling, especially in terms of the production factors utilized. By doing this it is possible to replace the economic–technological description of elasticity by a more realistic, richer BU formulation. Therefore, rather than describing production technologies in the form of many levels of CES production functions, production possibilities could be described as in detailed engineering BU models and inserted into the CGE formulation through the use of Leontief technologies, used depending on their profitability, or even inserted directly if the output data of each model are compatible. Furthermore, the MCP formulation may entail additional computational requirements and may thus represent a limitation of the BU model size.

Even so, linkage with the TD model would be made directly only if the cost functions in the BU representation were compatible with the primary economic factors used in the TD structure. However, there are several differences between the formalized language describing BU (partial equilibrium models) and TD (CGE models) that make it difficult to link them. Compatibility between the inputs and results of each BU technology and TD costs and destinations is not a trivial issue, as will be described in detail in the next section.

8.4.4   Data Compatibility Issues

As described above in section 8.4.3, TD models make use of aggregate information about economic sectors and their use of production factors and intermediate inputs, which is not necessarily compatible with the description of the cost of BU models based on technologies available, variable and fixed operation, maintenance and investment costs, and fuel use.

In the real world, neither the availability nor the compatibility of data can be guaranteed because of the use of different sources for the aggregate data of TD models and for the technological information on BU, deriving from physical properties of production processes and transformation of combustibles, losses in the production process and competition between companies within the sector.

Consequently, data compatibility can be considered a major issue when dealing with hybrid models, especially in the cases of approaches using a hard link or a soft link with feedback. The necessity of linkage for expenditure and income sources and destinations in TD CGE models is a further complication. BU models have no need to describe in full the cycle of economic activities, as they seek only to determine the most efficient cost techniques and technologies to be used according to the alternatives available.

Take as an example a BU model for the electricity sector. In this case, marginal operation models aim to choose the most inexpensive technologies to produce enough electricity to meet demand. This is usually represented by stacking the sequence of each power plant's production capacity in the order of their respective variable production costs (fixed construction costs are also considered in the case of investment decisions), as shown in Figure 8.10. The intersection between the stepwise marginal cost and the demand curve provides the most efficient production mix.

As may be expected, an electricity sector-only analysis does not necessarily represent all destinations of the income acquired by the electricity sector, as the choice is based only on the decision of the most efficient way to produce a product, focusing on a variable-only costs analysis. However, a general equilibrium approach of the type usually employed in TD models takes into account all the destinations.

Assume an increase in the variable production costs of carbon power plants due to an increase in fuel prices, to emissions right prices, to operating costs or to any other feasible cause (see Figure 8.11). If the effects are not sufficient to push carbon technology costs to above marginal technology costs, the final price settled by the market will not change

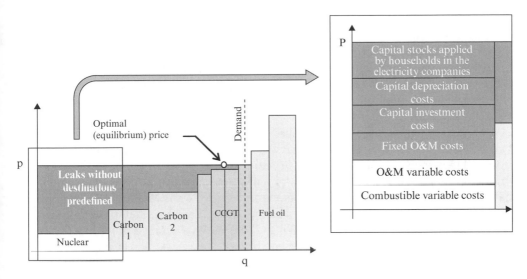

Figure 8.10   Differences in BU and TD data in a detailed model of electricity operation

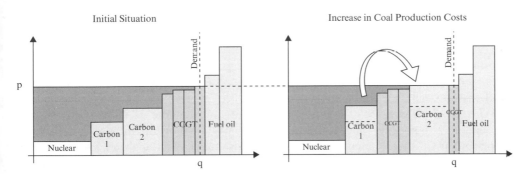

Figure 8.11   Effects of an increase on carbon production costs in a simplified competitive electricity market

in the BU framework. The BU analysis will still produce the same amount of revenue, under a new situation of higher total production costs.

In a partial equilibrium analysis the effects would end here, but a more general analysis would have additional repercussions. As can be seen, the difference between total revenue and the variable operating costs (the darkest area in Figure 8.11) is smaller in the second situation. In a perfectly competitive market, the sum of this area throughout the lifetime of the power plant should correspond exactly to the capital requirements for constructing the corresponding power plant capacity. Therefore, the change in the operation, and consequently in the dark area, should have a well-defined counterpart effect on the capital payments of the installed capacity in order to allow good communication between the BU operation model and the TD production factor numbers.

Even if the costs of each production technology are reliably known, the problem of defining the revenue destinations is crucial in TD models and substantially increases in difficulty if more realistic assumptions are made which are common to energy markets, such as a non-perfect competitive market with market power and excess profits. The important point is that compatible measures need to be taken into account to allow quality communication between models in the hybrid alternative. Variable and fixed costs need to have a compatible production factor and intermediate input representation. Furthermore, all revenue and expenditure interrelations should be well defined from the beginning.

A number of papers have been published which seek to match the representation of the two frameworks and allow their behaviour to converge. More specifically, the adaptation of TD models to deal with BU information and the calibration of future high-cost 'backstop' technologies are fields of major importance for data compatibility issues and have been extensively discussed (Koopmans and Velde, 2001; McFarland, 2004; Ghersi and Hourcade, 2006; McFarland and Herzog, 2006; Kypreos, 2007). Seeking to deal with the data compatibility issue, Sue Wing (2008) shows a detailed procedure for disaggregating the TD macroeconomic representation of the electricity sector into the subactivities that integrate the sector – generation, transmission and distribution – in a manner consistent with the characteristics of BU engineering technologies.

Even when it is possible to make the necessary adaptations, a further complication could arise if the time aggregation of BU and TD models differs. This is a common issue because TD models usually entail annual estimations, while BU models may provide weekly, hourly or even minute-by-minute details. An annual approach makes use of aggregate average quantities and prices to avoid problems of lack of information and computational size. However, the aggregate average representation is not capable of reflecting the same behaviour in more disaggregated schemes in competitive markets.

Take the electricity sector again as an example and assume an increase in electricity demand at a specific time. If this demand level change occurs in base periods (that is, periods with lower demand levels, which consequently make use of less expensive, cost-efficient units to supply electricity), the increase in demand will cause an increase in costs lower than the initial aggregate average price. Consequently, assuming a competitive market where prices reflect marginal costs, the increase in demand could actually decrease the aggregate result of the electricity average price. A completely opposite effect will occur if the same demand increase happens in peak demand periods.

Because of the different circumstances faced by base and peak electricity production, the results obtained in a BU approach would clearly contradict the results of an aggregate analysis where typically increasing demand increases prices. Rodrigues and Linares (2011) analyse the consequences of this data compatibility issue under a CGE approach and propose a solution based on a more disaggregated TD representation.

Besides its clear advantages in the integrated assessment of technological detail and indirect effects, hybrid modelling is not an issue to be dealt with trivially. The increasing computational requirements of dealing with more and more information in the same modelling framework, and the compatibility issues between the different frameworks used, are evidence of the need for a specific case analysis in order to develop instruments capable of providing satisfactory measurement procedures for policy evaluation scenarios.

8.5 CONCLUSIONS

This chapter gives an overview of the types of model used in the analysis of energy–economic–environmental iterations, the so-called E3 models. We review the two most widespread approaches used in E3 modelling – top-down and bottom-up – along with an integrated approach that offers a hybrid of the two. The effort to bridge the gap by integrating engineering and economic approaches into a more multidisciplinary approach is one of the best alternatives for assessing energy and climate policies. What technologies may serve this policy purposes, how to promote their development and rapid dissemination and how the economy can or may adapt to them are questions on which E3 modelling tools can give important insights and quantitative assessment. We hope that the E3 modelling will help policy-makers in taking decisions when many different trade-offs and objectives need to be assessed in order to shift energy systems to a more sustainable path.

NOTE

1. Most of the concepts developed in this section can be easily extended to other kinds of hybrid modelling that make use of different TD modelling approaches, such as the Cambridge econometrics hybrid model MDM-E3 (Junankar et al., 2007).

REFERENCES

Adams, P.D. (2005), 'Interpretation of results from CGE models such as GTAP', *Journal of Policy Modeling*, 27, 941–59.

Adelman, I. and S. Robinson (1978), *Income Distribution Policy in Developing Countries: A Case Study of Korea*, Oxford: Oxford University Press for the World Bank.

Alcamo, J. (1994), *IMAGE 2.0: Integrated Modeling of Global Climate Change*, Dordrecht: Kluwer Academic Press.

Beckman, J. and T. Hertel (2009), 'Why previous estimates of the cost of climate mitigation are likely too low', GTAP Working Paper 2954, Center for Global Trade Analysis, Purdue University, USA.

Bloemhof-Ruwaard, J.M., L.N. Van Wassenhove, H.L. Gabel and P.M. Weaver (1996), 'An environmental life cycle optimization model for the European pulp and paper industry', *Omega*, 24, 615–29.

Böhringer, C. (1998), 'The synthesis of bottom-up and top-down in energy policy modelling', *Energy Economics*, 20, 233–48.

Böhringer, C. and A. Loschel (2006), 'Promoting renewable energy in Europe: a hybrid computable general equilibrium approach', Special Issue 2, *Energy Journal*, 135–50.

Böhringer, C., A. Löschel and H. Welsch (2008), 'Environmental taxation and induced structural change in an open economy: the role of market structure', *German Economic Review*, 9, 17–40.

Böhringer, C. and T.F. Rutherford (2006), 'Combining top-down and bottom-up in energy policy analysis: a decomposition approach', Discussion Paper 06-007, ZEW.

Böhringer, C. and T.F. Rutherford (2008), 'Combining bottom-up and top-down', *Energy Economics*, 30, 574–96.

Böhringer, C. and T.F. Rutherford (2009), 'Integrated assessment of energy policies: decomposing top-down and bottom-up', *Journal of Economic Dynamics and Control*, 33, 1648–61.

Calloway, J.A. and R.G. Thompson (1976), 'An integrated industry model of petroleum refining, electric power, and chemicals industries for costing pollution control and estimating energy prices', *Engineering and Process Economics*, 1, 199–216.

Dawkins, C., T.N. Srinivasan and J. Whalley (2001), 'Calibration', in J.J. Heckmann and E. Leamer (eds), *Handbook of Econometrics*, 5, 3653–3703.

Drouet, L., M. Labriet, R. Loulou and M. Vielle (2008), 'A master program that will drive the coupling of

GEMINI-E3 and MARKAL TIMES models', *REME Working Paper-2009-005*, Ecole Polytechnique Federale de Lausanne, Switzerland.

EIA (2009), *The National Energy Modeling System: An Overview*, Washington, DC: Energy Information Administration.

Faucheux, S. and F. Levarlet (1999) 'Energy–economy–environment models', in C.J.M. van den Berg (ed.), *Handbook of Environmental and Resource Economics*, Cheltenham, UK and Northampton, MA, USA: Edward Elgar Publishing, pp. 1060–89.

Fishbone, L.G. and Abilock, H. (1981), 'Markal a linear-programming model for energy system analysis: technical description of the BNL version', *International Journal of Energy Research*, **5**, 353–75.

Francois, J.F. and K.A. Reinert (eds) (1997), *Applied Methods for Trade Policy Analysis: A Handbook*, Cambridge: Cambridge University Press.

Galarraga, I. and A. Markandya (2010), 'Climate change and its socioeconomic importance', *Ekonomiaz*, **71**, 14–39.

Ghersi, F. and J. Hourcade (2006), 'Macroeconomic consistency issues in E3 modeling: the continued fable of the elephant and the rabbit', Special Issue 2, *Energy Journal*, 39–61.

Ginsburgh, V. and M. Keyzer (1997), *The Structure of Applied General Equilibrium Models*, Cambridge, MA: MIT Press.

Graafland, J.J., R.A. de Mooij, A.G.H. Nibbelink and A. Nieuwenhuis (2001), *MIMICing Tax Policies and the Labour Market*, Contributions to Economic Analysis, Amsterdam: North-Holland.

Grübler, A., A. Jefferson and N. Nakicenovic (1996), 'Global energy perspectives: a summary of the joint study by the International Institute for Applied Systems Analysis and World Energy Council', *Technological Forecasting and Social Change: An International Journal*, **51** (3), 237–64.

Hertel, T.W. (ed.) (1997), *Global Trade Analysis. Modelling and applications*, Cambridge: Cambridge University Press.

Hertel, T.W., D. Hummels, M. Ivanic and R. Keeney (2007), 'How confident can we be in CGE-based assessments of free trade agreements?', *Economic Modelling*, **24**, 611–35.

Hertel, T.W., S. Rose and R. Tol (eds) (2009), *Economic Analysis of Land Use in Global Climate Change Policy*, London: Routledge.

Hidalgo, I., L. Szabo, J.C. Ciscar and A. Soria (2005), 'Technological prospects and CO_2 emission trading analyses in the iron and steel industry: a global model', *Energy*, **30**, 583–610.

Hillsman, E.L., D.R. Alvic and R.L. Church (1988), 'A disaggregate model of the US electric utility industry', *European Journal of Operational Research*, **35**, 30–44.

Hoffmann, A.N. (2002), 'Imperfect competition in computable general equilibrium models: a primer', *Economic Modelling*, **20**, 119–39.

Hoster, F. (1998), 'Impact of a nuclear phase-out in Germany: results from a simulation model of the European power systems', *Energy Policy*, **26**, 507–18.

Hourcade, J.-C., M. Jaccard, C. Bataille and F. Gershi (2006a), 'Hybrid modeling: new answers to old challenges', *Energy Journal*, **2**, 1–12.

Hourcade, J.-C., M. Jaccard, C. Bataille and F. Gershi (eds) (2006b), 'Hybrid modeling of energy – environment policies: reconciling bottom-up and top-down', *Energy Journal*, Special Issue 2, 1–178.

IEA (2009), *World Energy Outlook 2009*, Paris: International Energy Agency.

IPCC (2007), *Climate Change 2007: Fourth Assessment Report*, Cambridge: Cambridge University Press.

IPCC, OECD and IEA (1997), *Revised 1996 IPCC Guidelines for National Greenhouse Gas Inventories*, Paris: Intergovernmental Panel on Climate Change (IPCC), Organisation for Economic Co-operation and Development (OECD), International Energy Agency (IEA).

Jacoby, H.D., J.M. Reilly, J.R. McFarland and S. Paltsev (2006), 'Technology and technical change in the MIT EPPA model', *Energy Economics*, **28**, 610–31.

Jorgenson, D. (1984), 'Econometric methods for applied general equilibrium analysis', in H.E. Scarf and J.B. Shoven (eds), *Applied General Equilibrium Analysis*, Cambridge: Cambridge University Press, 139–203.

Junankar, S., O. Lofsnaes and P. Summerton (2007), *MDM-E3: A Short Technical Description*, Cambridge: Cambridge Econometrics.

Kehoe, T.J. (2005), 'An evaluation of the performance of applied general equilibrium models of the impact of NAFTA', in T.J. Kehoe, T.N. Srinivasan and J. Whalley (eds), *Frontiers in Applied General Equilibrium Modeling*, Cambridge: Cambridge University Press, pp. 341–77.

Kehoe, T.J., C. Polo and F. Sancho (1995), 'An evaluation of the performance of an applied general equilibrium model of the Spanish economy', *Economic Theory*, **6**, 115–41.

Kehoe, T.J., T.N. Srinivasan and J. Whalley (eds) (2005), *Frontiers in Applied General Equilibrium Modeling*, Cambridge: Cambridge University Press.

Kemfert, C. and T. Truong (2009), 'Energy–economy–environment modelling: a survey', in J. Evans and L.C Hunt (eds), *International Handbook on the Economics of Energy*, Cheltenham, UK and Northampton, MA, USA: Edward Elgar Publishing, pp. 367–82.

Koopmans, C.C. and D.W. Velde (2001), 'Bridging the energy efficiency gap: using bottom-up information in a top-down energy demand model', *Energy Economics*, **23**, 57–75.

Kypreos, S. (2007), 'A MERGE model with endogenous technological change and the cost of carbon stabilization', *Energy Policy*, **35**, 5327–36.

Labandeira, X., J.M. Labeaga and M. Rodríguez (2006), 'A macro and microeconomic integrated approach to assessing the effects of public policies', Documento de Trabajo 02-2006, FEDEA, Madrid.

Labandeira, X., P. Linares and M. Rodríguez (2009), 'An integrated approach to simulate the economic impacts of carbon emissions trading', *Energy Journal*, **30**, 217–37.

Lapillonne, B., D. Gusbin, P. Tulkens, W.V. Ierland, P. Criqui and P. Russ (2003), *World Energy, Technology, and Climate Policy Outlook: WETO 2030*, Brussels: Directorate General for Research, European Commission.

LEG SAM (2003), *Handbook on Social Accounting Matrices and Labour Accounts*, Luxemburg: Eurostat Secretariat.

Liu, J., C. Arndt and T.W. Hertel (2003), 'Parameter estimation and measures of fit in a global, general equilibrium model', *Journal of Economic Integration*, **19**, 626–49.

Manne, A.S. (1976), 'ETA: a model for energy technology assessment', *Bell Journal of Economics*, **7**, 379–406.

Manne, A.S., R. Mendelsohn and R. Richels (1995), 'A model for evaluating regional and global effects of GHG reduction policies', *Energy Policy*, **23**, 1–105.

Manne, A.S., R.G. Richels and J.P. Weyant (1979), 'Energy policy modeling: a survey', *Operations Research*, **27**, 1–36.

Manne A.S. and C.O. Wenne (1992), *MARKAL-MACRO: A Linked Model for Energy–Economic Analysis*, New York: Brookhaven National Laboratory

Mansur, A. and J. Whalley (1984), 'Numerical specification of applied general equilibrium models: estimation, calibration, and data', in H.E. Scarf and J.B. Shoven (eds) *Applied General Equilibrium Analysis*, Cambridge: Cambridge University Press, pp. 69–127.

Markandya, A. and M. Pemberton (2010), 'Energy security, energy modelling and uncertainty', *Energy Policy*, **38**, 1609–13.

Mas-Colell, A., M.D. Whinston and J.R. Green (1995), *Microeconomic Theory*, Oxford: Oxford University Press.

McFarland, J.R. (2004), 'Representing energy technologies in top-down economic models using bottom-up information', *Energy Economics*, **26**, 685–707.

McFarland, J.R. and H.J. Herzog (2006), 'Incorporating carbon capture and storage technologies in integrated assessment models', *Energy Economics*, **28**, 632–52.

McKitrick, R.R. (1998), 'The econometric critique of computable general equilibrium modeling: the role of parameter estimation', *Economic Modelling*, **15**, 543–73.

van der Mensbrugghe, D. (2008), 'The ENVironmental Impact and Sustainability Applied General Equilibrium (ENVISAGE) model', mimeo, World Bank.

Messner, S. and M. Strubegger (2001), 'MESSAGE IV model', International Institute for Applied Systems Analysis, Laxenburg, Austria.

Mitra-Kahn, B.H. (2008), 'Debunking the myths of computable general equilibrium models', SCEPA Working Paper 01-2008, Schwartz Center for Economic Policy Analysis (SCEPA), The New School, New York.

Narayanan, B. and T.L. Walmsley (eds) (2008), 'Global Trade, Assistance, and Production: The GTAP 7 Data Base', Center for Global Trade Analysis, Purdue University, USA.

Nordhaus, W.D. (1980), 'The energy crisis and macroeconomic policy', *Energy Journal*, **1**, 11–20.

Nordhaus, W.D. (1993), 'Rolling the "DICE": an optimal transition path for controlling greenhouse gases', *Resource and Energy Economics*, **15**, 27–50.

Ortuzar, J. de and L.G. Willumsen (2001), *Modelling Transport*, Chichester: Wiley.

Paltsev, S., J.M. Reilly, H.D. Jacoby, R.S. Eckaus, J. McFarland, M. Sarofim, M. Asadoorian and M. Babiker (2005), 'The MIT Emissions Prediction and Policy Analysis (EPPA) model: Version 4', Report 125, MIT Joint Program on the Science and Policy of Global Change, Cambridge, MA.

Perman, R., M. Common, J. Mcgilvray and Y. Ma (2003), *Natural Resource and Environmental Economics*, Boston, MA: Pearson-Addison Wesley.

Pizer, W., D. Burtraw, W. Harrington, R. Newell and J. Sanchirico (2006), 'Modeling economy-wide vs sectoral climate policies using combined aggregate-sectoral models', *Energy Journal*, **27**, 135–68.

Rath-Nagel, S. and A. Voss (1981), 'Energy models for planning and policy assessment', *European Journal of Operational Research*, **8**, 99–114.

Reinert, K.A. and D.W. Roland-Holst (1997), 'Social accounting matrices', in J.F. Francois and K.A. Reinert (eds) *Applied Methods for Trade Policy Analysis*, Cambridge: Cambridge University Press, pp. 94–121.

Robinson, S. (1976), 'Income distribution in developing countries, toward an adequate long-run model of income distribution and economic development', *American Economic Review*, **66**, 122–7.

Robinson, S. (2006), 'Macro models and multipliers: Leontief, Stone, Keynes, and CGE models', in A. de

Janvry and R. Kanpur (eds), *Poverty, Inequality and Development: Essays in Honor of Eric Thorbecke*, Heidelberg: Springer, pp. 205–32.

Rodrigues, R. and P. Linares (2011), 'Improving the representation of the electricity sector in computable general equilibrium models', IIT Working Paper, Instituto de Investigación Tecnológica, Madrid.

Rodrigues, R., P. Linares and A.G. Gómez-Plana (2011), 'A CGE assesment of the impacts on electricity production and CO$_2$ emissions of a residential demand response program in Spain', *Estudios de Economía Aplicada*, **29**.

Schumacher, K. and R.D. Sands (2007), 'Where are the industrial technologies in energy–economy models? An innovative CGE approach for steel production in Germany', *Energy Economics*, **29**, 799–825.

Shoven, J.B. and J. Whalley (1984), 'Applied general-equilibrium models of taxation and international trade: an introduction and survey', *Journal of Economic Literature*, **22**, 1007–51.

Shoven, J.B. and J. Whalley (1992), *Applying General Equilibrium*, Cambridge: Cambridge University Press.

Soloveitchik, D., N. Ben-Aderet, M. Grinman and A. Lotov (2002), 'Multiobjective optimization and marginal pollution abatement cost in the electricity sector – an Israeli case study', *European Journal of Operational Research*, **140**, 571–83.

Springer, U. (2003), 'The market for tradable GHG permits under the Kyoto Protocol: a survey of model studies', *Energy Economics*, **25**, 527–51.

Strachan, N. and R. Kannan (2008), 'Hybrid modelling of long-term carbon reduction scenarios for the UK', *Energy Economics*, **30**, 2947–63.

Sue Wing, I. (2006), 'The synthesis of bottom-up and top-down approaches to climate policy modeling: electric power technologies and the cost of limiting US CO$_2$ emissions', *Energy Policy*, **34**, 3847–3869.

Sue Wing, I. (2008), 'The synthesis of bottom-up and top-down approaches to climate policy modeling: electric power technology detail in a social accounting framework', *Energy Economics*, **30**, 547–73.

Sue Wing, I. (2009), 'Computable general equilibrium models for the analysis of energy and climate policies', in J. Evans and L.C. Hunt (eds), *International Handbook on the Economics of Energy*, Cheltenham, UK and Northampton, MA, USA: Edward Elgar Publishing, pp. 332–66.

Szabo, L., I. Hidalgo, J.C. Ciscar and A. Soria (2006), 'CO$_2$ emission trading within the European Union and Annex B countries: the cement industry case', *Energy Policy*, **34**, 72–87.

Tang, W., L. Wu and Z.S. Zhang (2010), 'Oil price shocks and their short- and long-term effects on the Chinese economy', *Energy Economics*, **32** (1), 3–14.

Tuladhar, S.D., M. Yuan, P. Bernstein, W.D. Montgomery and A. Smith (2009), 'A top-down bottom-up modeling approach to climate change policy analysis', *Energy Economics*, **31**, S223–S234.

Turton, H. (2008), 'ECLIPSE: an integrated energy–economy model for climate policy and scenario analysis', *Energy*, **33**, 1754–69.

Uri, N.D. (1976), 'Optimal investment, pricing and allocation of electrical energy in the USA', *Applied Mathematical Modelling*, **1**, 114–8.

Valenzuela, E., T.W. Hertel, R. Keeney and J.J. Reimer (2007), 'Assessing global computable general equilibrium model validity using agricultural price volatility', *American Journal Agricultural Economics*, **89**, 383–97.

Walmsley, T.L. and A. Strutt (2009), 'A baseline for the GDyn model', https://www.gtap.agecon.purdue.edu/resources/download/4497.pdf.

Webster, M.D., M. Babiker, M. Mayer, J.M. Reilly, J. Harnisch and M.C. Saromin (2002), 'Uncertainty in emissions projections for climate models', *Atmospheric Environment*, **36**, 3659–70.

Welsch, H. (2008), 'Armington elasticities for energy policy modeling: evidence from four European countries', *Energy Economics*, **30**, 2252–64.

Wene, C. (1996), 'Energy–economy analysis: linking the macroeconomic and systems engineering approaches', *Energy*, **21**, 809–24.

Weyant, J.P. (1993), 'Cost of reducing global carbon emissions', *Journal of Economic Perspectives*, **7**, 27–46.

Willenbockel, D. (2004), 'Specification choice and robustness in CGE trade policy analysis with imperfect competition', *Economic Modelling*, **21**, 1065–99.

WIOD (2010), *World Input–Output Database*, http://www.wiod.org/.

9 Energy supply and the sustainability of endogenous growth
Karen Pittel and Dirk Rübbelke

9.1 INTRODUCTION

Regarding the use of energy as an input, two threats to sustainability are regularly highlighted: the source problem, that is, the supply of energy; and the sink problem, that is, pollution generated by the consumption of energy resources. Ever since the Club of Rome's publication of the *Limits to Growth* and the first oil crisis these problems have been discussed extensively in the economics literature. Empirical and theoretical analyses, static as well as dynamic approaches, can be found in abundance, some of them featuring hundreds of equations and restrictions while others are highly stylized, analytically solvable models with only a handful of mathematical relations. All of these approaches have merits and shortcomings which are dealt with extensively in Edenhofer et al. (2006). The number of approaches used in the discussion mirror clearly the complexity of the issues at hand. In this chapter we will focus especially on one strand of the related literature: endogenous growth models that deal with the source and sink problems of energy – or more general – resource use. Since energy supply as well as sustainable development – that is, non-decreasing welfare over time – give rise to intertemporal problems, employing dynamic approaches whose focus is on the very long run seems straightforward. A look at the literature indeed shows that ever since *Limits to Growth* (Meadows et al., 1972), growth models have been used to identify conditions under which sustainable development is technically feasible and optimal over a long time horizon.

In the beginning of the 1970s the focus was primarily on the input and optimal timing of resource use. The aim was to derive dynamic allocation rules as a prerequisite for sustainable development, that is, non-decreasing welfare. However, the consequences of resource use and economic activities on environmental systems in the form of pollution and waste were initially of secondary interest. It was only in the course of increasing environmental degradation that the focus changed. Nevertheless, the interest in the field dwindled over time as the methodological tools applied, specifically those of neoclassical growth theory, were quite unsatisfactory. Two aspects in particular finally induced a revival of this line of research. On the one hand, new research fields were identified due to a formerly unknown scale of repercussions of anthropogenic activities on the environment (for example climate change). On the other hand, more sophisticated methodological approaches like the endogenous growth theory were developed that enabled researchers to reconsider and re-evaluate some of the rather strict results formerly derived.

In contrast to the neoclassical growth literature, endogenous growth approaches allow for a feedback effect of energy shortages and pollution-induced productivity and welfare losses on the long-run growth performance of an economy. In neoclassical growth models, long-run growth is essentially driven by exogenous technological development. Thus feedback effects of resource scarcity and environmental externalities on the growth engine – the ultimate source of the dynamic development of an economy – do not arise. Overcoming this shortcoming by determining the rate of long-run growth within the model was the main contribution of the so-called 'new growth revolution'. By introducing non-rivalry of knowledge, learning and imperfect competition, endogenous growth models resolved the problem of decreasing returns to capital which is at the core of the failure to sustain long-run endogenous growth. The induced increase in the explanatory power of the new generation of growth models not only revived the dwindling interest in the overall growth literature, but also allowed a more satisfying analysis of the effects triggered by resource scarcity and pollution on the long-run development of economies.

In the first decade after the new growth revolution, endogenous engines of growth were mainly considered in highly stylized and analytically solvable growth models. The merits of these approaches lie mainly in their ability to delineate clearly the dynamic channels through which energy and resource scarcity impact long-run development and growth. Effects of decreasing energy inputs and rising energy prices on, for example, the speed and direction of technological development can be understood as well as the repercussions of pollution externalities on the incentives to accumulate capital and conduct research.

The drawback to this traceability is the restricted modeling scope. The derivation of closed-form solutions limits the functional forms of technologies and preferences that can be considered as well as the degree of heterogeneity between agents. This especially holds if the aim is to derive a balanced growth path along which the economy grows at a constant rate in the long run. Numerical forecasts about the impact of specific policies on the long-run growth performance of an economy require a more disintegrated approach. Different economic sectors react very differently to energy shortages, policies and pollution. In order to reproduce the diverse reactions within a model, different production technologies – especially with respect to the importance and substitutability of energy as an input – have to be considered. Yet, this sectoral heterogeneity more often than not prevents the derivation of closed-form solutions and requires to resort to simulations. While the fast-growing capacity of computers allows for running more and more sophisticated simulations, one crucial problem remains; due to the complex structure of the models, it is difficult to trace the effects of policies and scarcities on economic performance through the model. Consequently, economic processes sometimes seem to take place in a black box.

In this chapter we aim to give an introduction to both types of modeling approaches in the context of endogenous growth. Section 9.2 deals with highly stylized frameworks in the tradition of, especially, Romer (1986) and Acemoglu (2002). We give an overview of the topics treated which is separated according to the input and output side – that is, pollution – of energy use. We present the most important insights obtained from the analyses. Regarding disintegrated models employing an endogenous growth engine, we give an overview of the literature in section 9.3. Section 9.4 provides an outlook on future research, before a short summary in section 9.5 closes the chapter.

9.2   THEORETICAL FRAMEWORK

In the context of energy use and climate change, the endogenous growth literature's aim is to identify policies and incentives that lead to sustainable development, that is: '"development that lasts" and that is supported by an economically profitable, socially responsive and environmentally responsible energy sector with a global, long-term vision' (IEA, 2001: 4). More specifically, sustainable development is associated with non-decreasing utility in the long run.

Regarding the use of fossil energy sources the absolute limit on exhaustible energy reserves seems to give rise to a fundamental dichotomy. On the one hand, the limited availability of fossil energy will eventually induce energy prices to rise. It is often postulated that this price increase will lead to a downturn of economic activities, as was observed, for example, after the first oil crisis. In this respect, the scarcity of non-renewable energy sources is seen as negative for welfare and sustainability. On the other hand, the unavoidable decrease of fossil energy use that follows from its limited stocks will reduce CO_2 emissions and thereby the threat to the environment. In this sense, the limited availability of exhaustible resources is positive for welfare and sustainability. So, it might seem as if a fundamental tension exists between economic and environmental prosperity.

This view of the problem does not, however, take into account that the scarcity-induced rise of energy prices fosters incentives to develop alternative energy sources and to reduce the resource intensity of production (Bretschger, 2010). The endogenous growth literature identifies the mechanisms at work and shows that rising resource scarcity might even lead to an increase in growth if the resulting efficiency gains are sufficiently strong. It aims at showing ways to reduce the dependency on fossil energy, promote alternative investment in carbon-free energy-sources, more resource-extensive production processes and possibly sectoral change towards less energy intensive goods.

In the following we will discuss whether the tension between resource use and scarcities of resources on the one hand, and pollution on the other hand, necessarily exists and which mechanisms could overcome it. We start by shedding some light on the input side of the energy sustainability debate in the endogenous growth literature. Specifically, we focus on approaches that incorporate energy from fossil sources whose supply is absolutely limited in the long run. Subsequently, we take a look at the output problem of the energy debate, that is, the pollution generated by the input of energy. The section is closed by a short look at policies aiming at an optimal extraction and pollution path.

9.2.1   The Input Side

A large variety of approaches exist in the endogenous growth literature that analyze the dependency on scarce natural inputs like energy stemming from fossil sources. Most of these models, however, do not focus exclusively on energy, but rather more generally on natural resources that can be of a renewable or a non-renewable nature. Energy in this sense is just one possible type of these resources. Due to the high degree of abstraction, sectoral differences in energy intensity and substitutability play a role, yet not to the same extent as in the disintegrated approaches.

The models introduced in this subsection can be distinguished along different lines. First, we can differentiate according to the engines that drive growth. These may include the accumulation of physical and/or human capital, learning-by-doing and technological progress. Second, models differ with respect to the number of sectors they consider. In the simplest case, the economy features only one production sector – as, for example, in AK-type models (see for example Gradus and Smulders, 1993; Baranzini and Bourguignon, 1995; Withagen, 1995; Smulders and Gradus, 1996). Yet, especially more recent models often encompass a number of sectors that produce goods, conduct research and development (R&D) and/or extract resources.

In the one-sector economies of many early approaches, the only way to reduce fossil energy use is to invest in some other type of capital, for example by investing in physical capital) (see Groth and Schou, 2007). Yet, with respect to physical capital this substitution is necessarily limited by the second law of thermodynamics. (See also Pittel et al., 2010 for a discussion of material balances and their relation to the accumulation of capital.) As a consequence, some other source of accumulable asset is required. However, without considering explicitly human capital or R&D, accumulation usually results from either learning-by-doing in the tradition of Romer (1986) or from public infrastructure following Barro (1990). Either explanation has its merits and empirical evidence can be found to support that both factors attribute to growth. Regarding public infrastructure see, for example, Aschauer (1989), Baxter and King (1993), and Easterley and Rebelo (1993); with respect to learning-by-doing see, for example, Arrow (1962) and Sheshinski (1967). With respect to the energy sector, learning-by-doing has especially been extensively analyzed empirically and, for example, McDonald and Schrattenholzer (2001) support that cumulative experience influences production costs favorably. Nevertheless, these approaches remain unsatisfying as they seem to suggest that there is no room – or rather no need – for private activities to promote a change in the energy regime. Yet, the evidence on learning curves as well as on public investment also shows that induced productivity increases might be limited (see for example Barro and Sala-i-Martin, 1991; Thompson, 2010). Consequently, the interesting task lies in the exploration of incentives to develop new technologies and, specifically, to promote R&D in less (fossil) energy-intensive technologies.

Much of the literature in this area builds upon the papers of Romer (1990), Grossman and Helpman (1991) and Aghion and Howitt (1992), who explicitly model research activities. Incentives to engage in research stem from profits arising from monopolistic competition in combination with patents on the blueprints developed. In their basic versions, these models consider either horizontal differentiated goods (Romer, 1990), that could be interpreted as new product varieties, or vertically differentiated goods (Grossman and Helpman, 1991; Aghion and Howitt, 1992), that can be thought of as process innovations. As the mechanisms driving growth in these two models are quite similar, we focus in the following on only one of the two – the Romer-type approach.

Growth in this model is driven by the expanding variety of goods available as intermediates in the production of final output or, alternatively, for consumption purposes. Research leads to the development of new product varieties. As R&D is considered to be labor-intensive, labor L_R is often considered to be the only rival input to research (for example Scholz and Ziemes, 1999; Pittel, 2002; Schou, 2002). Alternatively it can be assumed that natural resources and/or capital are additional inputs to research (Groth, 2007; Bretschger, 2008). Furthermore, research productivity is assumed to depend positively on the amount of past research. In the simplest version of this 'standing on the shoulders of giants' approach, research is linear in spillovers from past research, such that the production function for new intermediates reads:

$$\dot{N} = \frac{dN}{dt} = \zeta L_R N \tag{9.1}$$

where ζ is a productivity parameter and N is the 'number' of intermediates that equals the stock of knowledge from past research.

The assumption of linearity of research in spillovers from past research has often been heavily criticized – not only in the context of resource economics. It is often argued that past knowledge only fertilizes new research subject to decreasing returns (the 'fishing out' phenomenon). If, however, research is less than linear in knowledge spillovers, productivity growth peters out in the long run. In this case, long-run growth requires population growth, such that the increase in the size of the labor force compensates the decreasing returns from research. Often this population growth is assumed to be exogenous (leading to so-called semi-endogenous growth; see Jones, 1999). For a model with endogenous population growth that depends on economic conditions, see Bretschger (2008). A similar compensating force is required if research is modeled to depend on the input of exhaustible resources. As the input of non-renewable resources has to decline in the long run, linear spillovers from past knowledge are in this case not sufficient to generate sustainable productivity growth.

In the Romer (1990) model, incentives to develop new product varieties arise from profits earned by selling these varieties on a monopolistic market. Competition in the production of new products is prevented by patent protection of new blueprints. Combination of the expanding variety approach with Ethier (1982) production functions (or Dixit–Stiglitz preferences, Dixit and Stiglitz, 1977) forms the basis for the sustainability of long-run growth in these models. Based on the work of Spence (1976), the tendency of diminishing returns with respect to individual products is overcome due to gains from specialization, that is, the larger the variety of goods, the more productive the aggregate. The aggregate output of intermediates in efficiency units (that is, the physical output of intermediates weighted by their productivity) can be written as:

$$\tilde{X} = \int_0^N x_i^{\alpha_1} di, \qquad 0 < \alpha_1 < 1 \tag{9.2}$$

with x_i denoting the output of the individual intermediates' varieties. Given that all varieties are produced with the same production technology, $x_i = x$ holds in equilibrium and (9.2) reduces to $\tilde{X} = N^{1-\alpha_1} X$ with $X = \int_0^N x_i di$. So, even if the amount of intermediates X is constant over time (that is, $g_X = \dot{X}/X = 0$, where g_X denotes the growth rate of X) increasing specialization due to the development of new varieties gives rise to growth of the aggregate in efficiency units.[1] In the context of sustainable energy use, this implies that long-run growth might be feasible even if the input of energy is constant or decreases over time. This is shown by, for example, Scholz and Ziemes (1999) in which R&D leads to an increasing variety of capital intermediates ($x(i) = K(i)$). The positive productivity effect of this increasing variety can overcome the scarcity of the essential input of natural resources (for example fossil energy) in the production of final output Y:

$$Y = A \int_0^N K_i^{\alpha_1} di L_Y^{\alpha_2} R^{\alpha_3} = N^{1-\alpha_1} f(K, L_y, R), \quad \alpha_2, \alpha_3 > 0, \sum_{k=1}^{3} \alpha_k = 1 \qquad (9.3)$$

where L_Y, K and R are the inputs of labor, aggregate capital and exhaustible resources in final output production.

Although R&D is in this case not directly aimed at reducing the non-renewable input, research decreases the energy intensity of output as it increases the productivity of all factors. Scholz and Ziemes show that long-run growth under the increasing scarcity of non-renewable inputs is feasible in these types of models given that research is sufficiently productive and the implementation of new ideas increases marginal productivity enough. The drag on growth which is due to the decreasing input of an exhaustible factor is over-compensated by the rising number of differentiated outputs and the induced increase in productivity. This type of model shows one basic mechanism by which resource scarcity can be overcome in the long run. It is, however, not entirely satisfying as the forces at work are not resource- or energy-specific.

Pittel (2002) models research to be directly aimed at increasing the variety of scarce material intermediates. Material intermediates in this model are a composite of virgin renewable resources R and recycled materials W_R. Final output production is thus given by:

$$Y = A \int_0^N (W_{R_i}^\beta R_i^{1-\beta})^{\alpha_1} di L_Y^{\alpha_2} = N^{1-\alpha_1} g(W_R, R, L_Y). \qquad (9.4)$$

The last expression on the right-hand side in (9.4) shows clearly that, although research is directly aimed at increasing the efficiency of scarce material inputs, it affects all inputs symmetrically due to the assumed Cobb–Douglas production technology; that is, technological development is Hicks-neutral. As the elasticity of substitution between different inputs is unity, natural resource-enhancing technological progress has the same implications as technological progress in the Scholz and Ziemes (1999) model in (9.3). Research in this case does not induce substitution processes between natural and man-made inputs and therefore leaves the optimal input mix unchanged. The same holds for the model of van Zon and Yetkiner (2003) who consider an economy in which intermediates are produced from capital services and energy. In their model, research leads to an enhancement of the quantity as well as the quality of intermediates. Assuming exogenously increasing energy prices, the authors show that the rise in energy prices has a negative effect on growth. Due to the increase in the costs of intermediate production, the profitability of research declines along with the profitability of intermediates production.

In reality, different economic sectors display very different resource intensities and the interesting question is not only whether technological development can overcome the non-increasing input of natural resources, but also how the rising scarcity of natural resources might affect sectoral production and the sectoral composition of an economy as well as the direction of research. To answer these types of questions models are needed that not only comprise different sectors but also allow for endogenous sector shares and directed technological change. One option by which to attain these goals is a more flexible production function of the CES type. Acemoglu (2002) has shown that combining the Romer (1990) approach with a CES technology induces technological change that is directed at the relatively scarcer input. In the case where neither factor of production is non-essential, that is, if the elasticity of substitution between factors is below unity, the resulting long-run growth path is stable.

Smulders and de Nooij (2003) were among the first to employ Acemoglu's approach to an energy economics model in which, however, the supply of energy is exogenously given. Di Maria and Valente (2008) extend the analysis to the case of an endogenous supply of non-renewable resources. They show that Acemoglu's result remains valid in a model with capital and non-renewable resources as inputs to production. In the long-run equilibrium, research is purely resource-augmenting. Pittel and Bretschger (2010) generalize the analysis to the realistic case of heterogeneous resource intensities across production sectors. The production function for final output in their model reads:

$$Y = \left(\tilde{X}^{\frac{v-1}{v}} + \tilde{Z}^{\frac{v-1}{v}} \right)^{\frac{v}{v-1}}, \qquad v < 1 \qquad (9.5)$$

where X and Z are two types of intermediates that are produced in two sectors that differ with respect to the resource intensity of production. In contrast to (9.3) and (9.4), the elasticity of substitution between the two inputs to final production is less than unity. Without increases in resource productivity, output growth would peter out due to the limited availability of natural resources. In the case of exhaustible resources like fossil energy (as in Pittel and Bretschger), output would even go to zero in the long run.

As in the previous models, this drag on growth and the level of output can be overcome by research-induced productivity increases. In contrast to the previous models, however, the direction of technological change matters. Due to the CES production technology, technological progress is not Hicks-neutral as in the Cobb–Douglas case but rather sector-specific. Pittel and Bretschger show that resource-intensive sectors need not vanish in the long run. Due to increasing resource scarcity, the profitability of conducting research in these sectors increases. As a result, productivity is enhanced which overcompensates the drag of declining resource inputs. The shares of resource-intensive sectors remain unchanged in the long run, solely productivity develops differently across sectors with resource-intensive sectors conducting more research. Anecdotal empirical evidence seems to support this result, as investment in energy-related R&D has been observed to increase faster than research activities in general (see for example OECD, 2008 for Hungary). Also, the International Energy Agency (IEA) emphasizes the large potential for improving energy efficiency in the energy-intensive sectors (see IEA, 2008: 112).

The result of a long-run bias of technological change towards non-renewable resources is confirmed by André and Smulders (2008). In contrast to the previous papers they specifically consider dynamics of extraction costs. Models that assume either no or constant extraction costs typically show that energy prices (extraction) increase (decreases) continuously over time. For the long run this prediction seems straightforward due to the rising scarcity of resources. In the short run, however, empirics have shown that energy prices might decrease. Andre and Smulders show that this phenomenon may well be in line with the endogenous growth literature. Calibrating their model such that improvements in mining efficiency are sufficiently large and factor-augmenting technological change is initially neutral, the energy share in factor income can decrease temporarily.

Yet, over time the decrease in extraction costs is overcompensated by the increasing scarcity of energy, which induces a bias towards energy-saving technological change. In the long run, the energy share in factor income is again constant, thus confirming the results of the previous literature.

Most of the literature on endogenous growth and resources focuses on the input of one type of resource, for example fossil energy, without considering substitution processes between non-renewable and renewable or backstop resources. One of the few exceptions is Grimaud et al. (2007). Grimaud and co-authors consider the input of energy as a mix of fossil energy and energy stemming from a 'backstop' resource. This backstop resource is produced from final output and knowledge by a concave production technology. Research is dedicated to the overall efficiency of energy use as well as the efficiency of the backstop resource. It is shown that, as to be expected, the growth path of the economy is characterized by substitution out of fossil energy towards the backstop resource. As fossil and backstop fuels are assumed to be imperfect substitutes, both resources are, however, employed in the long run. Due to the assumption that research cannot be aimed directly at fossil fuels, nothing is said with respect to the optimality of investing in the efficiency of the non-renewable resource.

9.2.2 The Output Side

Burning fossil fuels is the main cause for the emission of the most important greenhouse gas, CO_2. According to the US Energy Information Administration (EIA, 2009: 111), the energy-related global carbon dioxide emissions will rise by 1.4 per cent annually between 2006 and 2030. Regarding the recent Intergovernmental Panel on Climate Change (IPCC, 2007) projections of future global warming, the so-called best estimates for six emissions scenarios are in the range 1.8–4° C at 2090–99 relative to 1980–99. This warming may cause a substantial sea level rise; the snow cover is forecasted to contract and it is likely that the frequency of weather extremes will continue to rise.

Despite these forecasts of severe consequences of climate change, most of the models introduced in the previous section solely concentrate on the input side of non-renewable resource (for example fossil energy) use. Pollution is often neglected in this strand of literature (see for example Scholz and Ziemes, 1999; Grimaud and Rougé, 2003; Pittel et al., 2010). The most straightforward explanation for this neglect is probably that the focus of the models is on the very long run. As the input of fossil resources declines over time, so does the generated flow of pollution, thus making pollution generation from fossil sources a temporary problem.

Among those papers considering environmental externalities that explicitly consider pollution from non-renewable inputs are Schou (2000, 2002) and Grimaud and Rougé (2005). The pollution flow P that is modeled as a function of the extracted exhaustible resources:

$$P = P(R), \quad P_R > 0 \qquad (9.6)$$

affects households' utility and/or production negatively. As shown by Schou (2002), whether or not pollution is modeled as a flow (as in 9.6) or stock S:

$$\dot{S} = P(R) - n(S) \tag{9.7}$$

does not matter in the long run as long as the pollution stock has sufficient degenerative capacity, $n(S)$, and ecological thresholds, $\bar{S}$, after which the environmental degradation becomes irreversible, play no role.

Pollution can harm either production and/or the utility of households. In the case where production is affected negatively, the positive contribution of resources to output is diminished, thus lowering the social return to resource extraction. In the case where households are affected, their intertemporal utility is lowered. Assuming that households live forever (which could also be interpreted as an infinite succession of generations), and derive utility from consumption C and disutility from pollution P, their utility would thus be given by:

$$\int_0^\infty U(C_t, P_t)e^{-\rho t}dt \tag{9.8}$$

with $\rho > 0$ being the rate at which households discount future utility. The utility function satisfies the usual properties ($U_C > 0$, $U_{CC} < 0$, $U_P < 0$, $U_{PP} > 0$).

Sustainability in the sense of non-decreasing welfare usually requires pollution to be non-increasing, at least in the long run. For pollution stemming from non-renewable sources this is automatically fulfilled as the extraction of non-renewable resources necessarily decreases over time (that is, $g_P = P_R g_R < 0$), although a temporary increase in extraction and thereby an increase in pollution is conceivable. Nevertheless, pollution gives rise to externalities that lead to suboptimal growth and can therefore call for environmental policies (see next subsection).

Pollution however, can not only be generated directly from the input of fossil energy but also from other economic variables that increase in a growing economy, that is, output or the input of capital. The flow of pollution in these cases would be given by:

$$P = P(Y), \quad P_Y > 0, \quad resp. \quad P = P(K), \quad P_K > 0. \tag{9.9}$$

In a growing economy, this pollution would increase over time in the absence of environmental policies or abatement, and therefore threaten sustainability. Equivalently, pollution that accumulates over time and only degenerates at a very low (or zero) rate ($n(S) \to 0$) or exceeds ecological thresholds is not compatible with sustainable development. The endogenous growth literature on pollution deals extensively with these types of pollution and derives a number of policy rules that aim at internalizing pollution externalities and assuring for sustainability. As this discussion is, however, only indirectly related to the energy sustainability debate, we do not discuss these cases here at length. The interested reader is referred to Pittel (2002) who gives an introduction to the different types of pollution and policies and also provides a review of the relevant literature.

9.2.3   Energy and Resource Policies

Most of the literature in the field of endogenous growth and non-renewable resources assumes that resource extraction is conducted by perfectly competitive firms such that resource prices follow the Hotelling rule. Thus, in the absence of other resource-related market failures such as pollution, the resource sector itself does not warrant governmental intervention. Typical non-resource-related market failures that arise in the models of subsection 9.2.1 are related to research spillovers and monopolistic competition in the intermediate goods sectors. As a result, optimal policies rules comprise the standard policies for the Romer (1990) model.

Regarding policies aiming at influencing the level and time path of resource extraction, a number of policies are discussed in the literature. Some papers derive the need for such policies from pollution caused by the use of fossil resources (for example, Schou, 2000, 2002; Groth and Schou 2007), others abstract from pollution, yet justify the policy analysis from targets currently discussed in the field of energy and climate policies (Pittel and Bretschger, 2010). Besides resource taxes these policies include taxes on capital gains and interest income (Groth and Schou, 2007). Assuming that the politically defined aims are to save on resource use and to support resource-extensive sectors, Pittel and Bretschger (2010) furthermore check whether or not these goals can be attained by the provision of productive public goods and labor or research subsidies.

Let us take a closer look at resource taxation as this policy is not only among the most frequently analyzed (for example Groth and Schou, 2007; Pittel and Bretschger, 2010) but also among the most commonly adopted instruments. In the context of climate change, the usual aim of resource taxation is to reduce fossil energy use at present and in the near future in order to lengthen the extraction phase and move emissions of CO_2 at each point in time closer to the absorptive capacity of the environment. It is shown that resource taxation only affects long-run growth if the rate of taxation changes over time. Given a constant (*ad valorem*) tax rate, resource taxation leaves the intertemporal arbitrage of resource owners unaffected. Taxation in this case only leads to a rent transfer from the producer to the taxing institution. A rising rate of resource taxation on the other hand induces the speed of resource extraction to rise as resource owners foresee the future decrease in the non-taxed share of resource revenues. Consequently, postponing resource extraction requires taxing resource use at a decreasing rate – thus the endogenous growth literature confirms the result of the 'green paradox' (Sinn, 1982, 2008).

It should be noted that while a decreasing tax rate slows down resource extraction and therefore has the potential to increase growth (see for example Groth and Schou, 2007), this might not always be optimal from a welfare perspective. Grimaud (2004) and Grimaud and Rougé (2005) show that the optimality of this result depends on the preferences of households.

Consider a model in which pollution has a negative amenity effect on utility (as in 9.8) and fossil resources are an essential input to final goods production. The optimal environmental policy rule derived by Grimaud and Rougé for this case reads:

$$g_\tau = -\frac{U_P P_R}{U_c F_R}(g_{(U_P P_R)} - \rho) \qquad (9.10)$$

where F_R is the marginal product of the resource in production. Clearly, the rate at which the tax rate should grow over time depends crucially on the relative disutility of resource extraction, $U_P P_R / U_c F_R$. The higher the marginal disutility from pollution generated by resource extraction is, compared to the marginal utility of consumption that can be produced from the extracted resource, the faster the tax rate should grow. Whether the optimal tax should, however, increase or decrease over time depends on the term in brackets (as the relative disutility of resource extraction is strictly negative). Given that the marginal disutility of pollution decreases over time or increases only moderately, the growth rate of the tax is optimally negative, implying the previously described shift of resource extraction from the present into the future. Yet, if the marginal disutility from pollution increases over time (such that $g_{(U_P P_R)} > \rho$), a positive growth rate of the tax can be optimal.

Whether or not environmental policy is called for at all in the case of pollution from fossil energy use depends crucially on the shape of the production and utility functions. Employing a Cobb–Douglas production function for final output and a constant relative risk aversion (CRRA) utility function, Schou (2000, 2002) shows that environmental policy is not required to attain the optimal growth path. It should be noted though that this result is due to the specific choice of technology and preferences in Schou's paper and does not carry over to more general specifications (Grimaud, 2004).

9.3   CGE MODELS

In contrast to the endogenous growth models of the previous section, computable general equilibrium (CGE) models that integrate energy markets usually strive to provide scientists as well as politicians with numerical estimations of policy impacts for specific economies or world regions. While the models of section 9.2 aim primarily at deriving general policy impacts by identifying the relevant transmission channels, the models of this section focus on giving concrete policy advice. As already stated in the introduction, this sometimes comes at the expense of a seemingly black-box approach where policy enters on the one side and economic implications on firms, sectors and households emerge at the other side. Tracing policy effects through the model becomes difficult if not impossible due to the multitude of interrelations.

Integrating endogenous growth into these models has proven to be rather challenging due not only to the more complex economic structure but also due to implementation problems when simulating these models. As a consequence a large part of the CGE literature has relied on exogenous growth processes (for example Burniaux et al., 1992; Nordhaus, 1992; Peck and Teisberg, 1992). The drawback to this set-up is the same as in the analytically solvable models: the engine of growth is independent of energy policies such that no feedback effects of policy on the growth engine arise. As CGE models are usually constructed in order to allow estimations of policy effects, leaving an important transmission channel of environmental policy out of the picture can lead to wrong conclusions and policy advice. Empirical evidence that energy price changes – as, for example, induced by policy measures – affect innovation (for example Newell et al., 1999, Popp, 2001, 2002; Bretschger, 2010) and thus the economies' growth engine, support this view.

Models in which growth is driven endogenously comprise approaches that incorporate learning-by-doing as well as R&D and gains from specialization. The drawback to learning-by-doing is again that it does not disclose the decision-making processes behind technological development and investment in research but rather takes technological progress as an automatism. CGE models that include learning-by-doing are, for example, Messner (1997), van der Zwaan et al. (2002) and Gerlagh et al. (2004).

Beyond learning-by-doing a number of papers also include investment in R&D (for example Goulder and Schneider, 1999; Nordhaus, 2002; Popp, 2004). For a more extensive review of the literature see Bretschger et al. (2010). The only paper known to us that incorporates investment in R&D and gains from specialization as in Romer (1990) is, however, Bretschger et al. (2010). Their paper predicts the effects of Swiss carbon policies on consumption, welfare and sectoral development where growth is driven endogenously by a sector-specific increasing specialization in capital varieties. Due to the incorporation of gains of specialization in an endogenous growth model with research, their economy reacts differently to energy and carbon taxation than an economy with exogenous growth. The growing capital stock not only provides a substitute for energy but also raises productivity. While substitution helps to decrease fossil energy use, that is, helps to achieve the environmental goals, the simultaneous productivity increase attenuates detrimental effects on growth and welfare. As Bretschger et al. employ a model which is closest to the Romer (1990) model on which we focused in section 9.2, we present their approach in a little more detail.

The model comprises ten 'regular' economic sectors plus an oil sector and an energy sector. Production technologies are nested and firms in each sector conduct sector-specific R&D. In their research activities firms employ labor, L, and a share of sectoral output, $Y_{R\&D}$, as inputs and generate investment in non-physical capital as I_{NP}:

$$I_{NP_t} = \left[\gamma L_t^{\frac{\sigma_N-1}{\sigma_N}} + (1-\gamma) Y_{R\&D_t}^{\frac{\sigma_N-1}{\sigma_N}} \right]^{\frac{\sigma_N}{\sigma_N-1}}. \tag{9.11}$$

σ_N is the sectoral elasticity of substitution in the production of non-physical capital. I_{NP} is then combined with investment in physical capital I_P and previously accumulated capital to a composite capital stock per firm K:

$$K_{t+1} = \left[\gamma I_{P_t}^{\frac{\sigma_I-1}{\sigma_I}} + (1-\gamma) I_{NP_t}^{\frac{\sigma_I-1}{\sigma_I}} \right]^{\frac{\sigma_I}{\sigma_I-1}} + (1-\delta) K_t \tag{9.12}$$

where δ is the depreciation rate of capital. The CES specification in (9.11) and (9.12) allows for a high degree of flexibility and sector-specific modeling. Elasticities of substitution vary for sectoral R&D as well as with respect to sectoral intermediates, final goods and energy production. Also, due to the CES specification, the optimal input mix in R&D as well as production can react to policy-induced price changes. Thus not only the engine of growth but also the direction of technological change, the sectoral structure and the optimal resource allocation are completely endogenized.

Typically for a CGE model, the paper's focus is on numerical policy scenarios for which the impact on a specific economy, in this case Switzerland, is to be estimated. Two taxation scenarios that are modeled to be compatible with actual policy goals are compared with respect to their effects on growth and sectoral structure. In the first scenario, a CO_2 tax is levied that is inspired by the reduction scenarios discussed at the UN Climate Change Conference in Copenhagen in 2009. The second scenario builds upon the goal of transforming the Swiss economy into a '2000 watt society' by 2090.

Bretschger et al. derive results on, among others, the development of welfare, individual sectors and energy use. Due to the model's complexity, the exact channels through which policies are transmitted are hard to follow. Yet, it is exactly the disaggregated nature of the model that allows a precise analysis of sector-specific reactions to climate policies that is usually not feasible within the highly stylized models of section 9.2. Specifically, sectoral reactions are driven by the differences in investment intensities, in resource intensities and sectoral linkages.

The paper finds that starting from a benchmark scenario with balanced growth and no damages from climate change, a CO_2 policy following the Copenhagen Accord entails moderate yet not negligible welfare losses. Welfare losses in case of the 2000 watt society scenario are predictably lower as this policy is less stringent. This comparison focuses on the costs of climate policy (that is, the 'costs of action'), while an analysis of its benefits, in the form of avoided climate change damages (that is, the 'costs of inaction'), is left for future research. The integration of the benefits might of course affect the welfare ranking of the two policies due to the different time paths of CO_2 emissions.

When comparing the costs of climate policies obtained from different models and also across policies, some caution is advised. The respective degree of disaggregation as well as assumptions regarding production technologies and preferences of course crucially affect the results obtained. Regarding limitations of (and potential biases in) the cost estimations in climate policy models, see also Tavoni and Tol (2010).

9.4   FUTURE RESEARCH FIELDS

Although many topics have been addressed extensively by the endogenous growth literature on sustainability and energy use, there are also aspects that have so far often been neglected. The security of energy supply and ancillary benefits arising from climate policies are two of these topics which, due to their importance, will be addressed in the following.

9.4.1   Security of Energy Supply

In its Green Paper 'A European strategy for sustainable, competitive and secure energy' (EC, 2006: 17–18), the European Commission recognizes that two of the main objectives of Europe's energy policy should be environmental sustainability (already addressed in this chapter) and the security of supply.

Among the most discussed questions regarding the security of supply are the problem of long-run (non-)availability due to decreasing fossil energy stocks as well as problems arising from market failures like market power or the risk of supply disruptions for geopolitical reasons. While the first aspect, the exhaustibility of fossil energy, has been at the core of the analysis of subsection 9.2.1, the second aspect is regularly disregarded by the endogenous growth literature.

A lack of energy security in the second interpretation might give rise to welfare losses. A shortage of energy suppliers, for example, influences the functioning of markets negatively by constituting an oligopoly. Energy supply disruptions, of course, also involve components of other sustainability dimensions. An energy shortage-induced price increase, for example, tends to have negative social implications. In the long-run perspective, energy security influences especially the incentives to invest in the techno-logical development of alternative energy sources as well as energy-saving technologies. Therefore an analysis within the endogenous growth framework could yield interesting results.

Outside the endogenous growth literature, recently several studies have investigated the linkages between pollution and energy security. The IEA (2007) assesses interactions between energy security and climate change. Turton and Barreto (2006) provide a study which examines the interrelations and synergies between climate change mitigation and supply security risk management policies. Furthermore, they investigate the role of technology in achieving these two policy goals. They observe that there are some syner-gies between policies pursuing the combat of global warming, and policies intending to mitigate insecurity of energy supply, but point out that the interaction is complex. Markandya et al. (2003) provide an analysis of energy policy in Russia, taking into account energy security and climate change aspects.

9.4.2   Ancillary Benefits of Climate Policies

In the past, the evaluation of climate policies has mainly focused on the costs and ben-efits of the mitigation of climate change. Yet, climate policies inducing a decline in the burning of fossil fuels also have additional effects that are often ignored. Among those so-called ancillary effects – that is, effects which do arise from climate policies, but not from the mitigation of climate change itself – are air quality improvements. The Appendix displays some of the pollutants emitted in conjunction with CO_2 which are, of course, also mitigated if climate policies reduce the burning of fossil fuels. An extensive discussion of the divergences between the characteristics of primary and ancillary can be found in Rübbelke (2002) as well as Markandya and Rübbelke (2004).

A comprehensive analysis of climate change policies should include benefits from the reduction of all types of externalities; that is, climate change mitigation benefits (primary benefits) as well as ancillary benefits should be taken into account. For the design of optimal policies this is especially important as ancillary effects often exhibit characteristics which are very different from those of climate change mitigation. While the mitigation of CO_2 exerts the global effect of climate change mitigation, the abatement of other pollutants like SO_2 or particles has more limited effects geographically. Also the delay between emission of the pollutants and the point of time when they effectively start to harm the environment diverges between the individual pollutants. There is a delay in the reaction of climate to greenhouse gas (GHG) emission changes: 'because of the thermal inertia of the oceans, the climate appears to lag perhaps a half century behind the changes in GHG concentrations' (Nordhaus, 1994: 4–5). In contrast, ancillary ben-efits of local and regional air pollution reductions can largely be enjoyed shortly after the climate policy implementation.

The endogenous growth literature has so far mostly ignored these additional ben-efits, although their inclusion could substantially affect the optimal time path as well as the optimal level of pollution policies. An approach should be chosen that allows differentiation between long-run and short-run effects of pollution reduction and also takes an international and regional perspective in order to differentiate between different geographical scopes. To our knowledge, the only paper addressing both questions in an endogenous growth framework is by Pittel and Rübbelke (2010), who derive implications for optimal pollution taxation. Another growth paper that considers primary and secondary benefits is Bahn and Leach (2008) in which technological development is, however, exogenous. They consider secondary effects of climate policy due to the reduction of SO_2 emissions in an overlapping generation model. Their model is not analytically solvable, such that transmission channels of secondary benefits and costs are not clearly identifiable.

Alternatively to the reduction of the burning of fossil fuels, climate can also be protected by the substitution of less carbon-intensive fuels for carbon-intensive ones, for example by substituting natural gas for coal. Also with respect to the optimal design of such substitution processes, ancillary effects play an important role since trade-offs may arise. For example, a switch in electricity generation from fossil fuels like oil, gas or coal to nuclear technologies reduces greenhouse gas emissions in the shape of CO_2 emissions but raises other negative externalities. External costs of nuclear electricity generation accrue, for instance, from the higher risk of catastrophic accidents in power plants (Ewers and Rennings, 1996). Furthermore, the switch from gasoline as a fuel for cars to diesel reduces the emissions of CO_2 but raises PM2.5 emissions (Mayeres and Proost, 2001). When considering the benefits of a change in the energy mix, these types of ancillary benefits – or in this case ancillary costs – should also be considered.

9.5   SUMMARY

In the past decades the strand of literature employing endogenous growth models to the energy and sustainability debate has made some important contributions to understanding the long-run potential of economies to overcome the scarcity of fossil energy resources and the potential and direction of technological development.

Following the UN's (2001: 19) classifications of the four primary dimensions of sustainable development (economic, environmental, institutional and social), the focus of this chapter has been especially on the environmental dimension. More specifically we have dealt with challenges arising from the use of non-renewable resources, as the burning of fossil energy resources is the main driver of climate change. As these challenges arise from the input side as well as from the output side of energy use, both source and sink issues have been addressed.

In this chapter we have further differentiated between analytical solvable endogenous growth models and CGE models that can only be solved by simulations. We have shown that for economic development and growth to be sustainable, both types of models identify technological progress and efficiency improvements as the main drivers. We have focused largely on models employing the disintegrated approach of Romer (1990) as this approach models R&D investments as the result of decision-making processes (in comparison to the quasi-automatic efficiency improvements in learning-by-doing frameworks). Although endogenous growth in many CGE models still relies largely on learning-by-doing, we have also presented an approach in which intentional research investments in combination with gains from specialization drive growth.

Beyond questions regarding the use of fossil resources, on which the focus has been in this chapter, the endogenous growth literature has also extensively addressed problems regarding the use of renewable resources (see for example Bovenberg and Smulders, 1995, 1996; Aghion and Howitt, 1998; Grimaud, 1999) of which renewable energy sources are one possible type. The use of renewable resources can diminish both source and sink problems simultaneously and therefore the integration of such resources constitutes an important aspects of the analysis of sustainable energy. As the scope of this chapter is limited, however, we have concentrated mainly on the worst-case scenario, that is, a regeneration rate of zero.

NOTE

1. To see this, consider that the growth rate of $\widetilde{X}$ is given by $g_{\widetilde{X}} = (1 - \alpha_1)g_N + g_X$. For a constant X (that is, $g_X = 0$) and increasing specialization ($g_N > 0$), $\widetilde{X}$ grows at a positive rate: $g_{\widetilde{X}} = (1 - \alpha_1)g_N > 0$.

REFERENCES

Acemoglu, D. (2002), 'Directed technical change', *Review of Economic Studies*, **69**, 781–809.
Aghion, P. and P. Howitt (1992), 'A model of growth through creative destruction', *Econometrica*, **60**, 323–51.
Aghion, P. and P. Howitt (1998), *Endogenous Growth Theory*, Cambridge, MA: The MIT Press.
André, F. and S. Smulders (2008), '*Fueling growth when oil peaks: growth, energy supply and directed technological change*', mimeo.
Arrow, K.J. (1962), 'The economic implications of learning by doing', *Review of Economic Studies*, **29**, 154–74.
Aschauer, D.A. (1989), 'Is public expenditure productive?', *Journal of Monetary Economics*, **23**, 177–200.
Bahn, O. and A. Leach (2008), 'The secondary benefits of climate change mitigation: an overlapping generations approach', *Computational Management Science*, **5**, 233–57.
Baranzini, A. and F. Bourguignon (1995), 'Is sustainable growth optimal?', *International Tax and Public Finance*, **2**, 341–56.
Barro, R.J. (1990), 'Government spending in a simple endogenous growth model', *Journal of Political Economy*, **98**, S103–S125.
Barro, R.J. and X. Sala-i-Martin (1991), 'Public finance in models of economic growth', *Review of Economic Studies*, **59**, 645–61.
Baxter, M. and R. King (1993), 'Fiscal policy in general equilibrium', *American Economic Review*, **83**, 315–34.
Bovenberg, A.L. and S. Smulders (1995), 'Environmental quality and pollution-augmenting technical change in a two-sector endogenous growth model', *Journal of Public Economics*, **57**, 369–91.
Bovenberg, A.L. and S. Smulders (1996), 'Transitional impacts of environmental policy in an endogenous growth model', *International Economic Review*, **37**, 861–93.
Bretschger, L. (2008), 'Population growth and natural resource scarcity: long-run development under seemingly unfavourable conditions', CER-ETH Working Paper 08/87, ETH Zürich.
Bretschger, L. (2010), 'Energy prices, growth, and the channels in between: theory and evidence', CER-ETH Working Paper 06/47, ETH Zürich, revised version March 2010.
Bretschger, L., R. Ramer and F. Schwark (2010), 'Long-run effects of post-Kyoto policies: applying a fully dynamic CGE model with heterogeneous capital', CER-ETH Working Paper 10/129, ETH Zürich.
Burniaux, J.-M., J.P. Martin, G. Nicoletti and J.O. Martins (1992), 'GREEN a multi-sector, multi-region general equilibrium model for quantifying the costs of curbing CO_2 emissions: a technical manual', OECD Economics Department Working Papers, no. 116, OECD Publishing.
Di Maria, C. and S. Valente (2008), 'Hicks meets Hotelling: the direction of technical change in capital-resource economies', *Environment and Development Economics*, **13**, 691–717.
Dixit, A.K. and J.E. Stiglitz (1977), 'Monopolistic competition and optimum product diversity', *American Economic Review*, **67**, 297–308.
Easterley, W. and S. Rebelo (1993), 'Fiscal policy and economic growth: an empirical investigation', *Journal of Monetary Economics*, **32**, 389–405.
EC (2006), 'A European strategy for sustainable, competitive and secure energy', Green Paper, Brussels.

Edenhofer, O., C. Carraro, J. Koehler and M. Grubb (2006), 'Endogenous technological change and the economics of atmospheric stabilisation', Special Issue, *Energy Journal*.

EIA (2009), '*International energy outlook 2009*', Washington, DC.

Ethier, W.J. (1982), 'National and international returns to scale in the modern theory of international trade', *American Economic Review*, **72**, 389–405.

Ewers, H.J. and K. Rennings (1996), 'Quantitative Ansätze einer rationalen umweltpolitischen Zielbestimmung', *Zeitschrift für Umweltpolitik und Umweltrecht (ZfU)*, **4**, 413–39.

Gerlagh, R., B. van der Zwaan, M.W. Hofkes and G. Klaassen (2004), 'Impacts of CO_2 taxes in an economy with niche markets and learning-by-doing', *Environmental and Resource Economics*, **28**, 367–94.

Goulder, L.H. and S.H. Schneider (1999), 'Induced technological change and the attractiveness of CO_2 abatement policies', *Resource and Energy Economics*, **21**, 211–53.

Gradus, R. and S. Smulders (1993), 'The trade-off between environmental care and long-term growth pollution in three prototype growth models', *Journal of Economics*, **58**, 25–51.

Grimaud, A. (1999), 'Pollution permits and sustainable growth in a Schumpeterian growth model', *Journal of Environmental Economics and Management*, **38**, 249–66.

Grimaud, A. (2004), 'Note on the Schou's paper: "When environmental policy is superfluous: growth and polluting resources, *Scandinavian Journal of Economics*", IDEI Working Paper no. 261, University of Toulouse.

Grimaud, A., G. Lafforge and B. Magné (2007), 'Innovation markets in the policy appraisal of climate change mitigation', IDEI working paper no. 481, University of Toulouse.

Grimaud, A. and L. Rougé (2003), 'Non-renewable resources and growth with vertical innovations: optimum, equilibrium and economic policies', *Journal of Environmental Economics and Management*, **45**, 433–53.

Grimaud, A. and L. Rougé (2005), 'Polluting non-renewable resources, innovation and growth: welfare and environmental policy', *Resource and Energy Economics*, **27**, 109–9.

Grossman, G. M. and E. Helpman (1991), 'Quality ladders in the theory of growth', *Review of Economic Studies*, **58**, 43–61.

Groth, C. (2007), 'A new-growth perspective on non-renewable resources', in L. Bretschger and S. Smulders (eds), *Sustainable Resource Use and Economic Dynamics*, Dordrecht: Springer, pp. 127–63.

Groth, C. and P. Schou (2007), 'Growth and non-renewable resources: the different roles of capital and resource taxes, *Journal of Environmental Economics and Management*, **53**, 80–98.

IEA (2007), 'Energy security and climate policy – assessing interactions', Paris.

IEA (2008), *Energy Technology Perspectives: Scenarios and Strategies to 2050*, Paris: OECD.

IEA (2001), 'Toward a sustainable energy future', www.iea.org/textbase/nppdf/free/2000/future2001.pdf.

IPCC (2007), *Climate Change 2007: The Physical Science Basis*, New York: Cambridge University Press.

Jones, C.I. (1999), 'Growth: with or without scale effects?', *American Economic Review*, **89**, 139–44.

Markandya, A., A. Golub and E. Strukova (2003), 'The influence of climate change considerations on energy policy: the case of Russia, *International Journal of Global Environmental Issues*, **3**, 324–38.

Markandya, A. and D.T.G. Rübbelke (2004), 'Ancillary benefits of climate policy', *Jahrbücher für Nationalökonomie und Statistik*, **224**, 488–503.

Mayeres, I. and S. Proost (2001), 'Should diesel cars in Europe be discouraged?', *Regional Science and Urban Economics*, **31**, 453–70.

McDonald, A. and L. Schrattenholzer (2001), 'Learning rates for energy technologies', *Energy Policy*, **29**, 255–61.

Meadows, D.H, D.L. Meadows and J. Randers (1972), *The Limits to Growth*, New York: Universe Books.

Messner, S. (1997), 'Endogenized technological learning in an energy systems model', *Journal of Evolutionary Economics*, **7**, 291–313.

Newell, R.G., A.B. Jaffe and R.N. Stavins (1999), 'The induced innovation hypothesis and energy-saving technological change', *Quarterly Journal of Economics*, **114**, 941–75.

Nordhaus, W.D. (1992), 'An optimal transition path for controlling greenhouse gases', *Science*, **258**, 1315–19.

Nordhaus, W.D. (1994), *Managing the Global Commons: The Economics of Climate Change*, Cambridge, MA: MIT Press.

Nordhaus, W.D. (2002), 'Modeling induced innovation in climate-change policy' in A. Grübler and N. Nakicenovic (eds), *Technological Change and the Environment*, Washington, DC: Resources for the Future, pp. 182–209.

OECD (2008), 'GERD Data', titania.sourceoecd.org.

Peck, S.C. and T.J. Teisberg (1992), 'CETA: a model for carbon emissions trajectory assessment', *Energy Journal*, **13**, 55–77.

Pittel, K. (2002), *Sustainability and Endogenous Growth*, Cheltenham, UK and Northampton, MA, USA: Edward Elgar.

Pittel, K., P. Amigues and T. Kuhn (2010), 'Recycling under a material balance constraint', *Resource and Energy Economics*, **32**, 379–94.

Pittel, K. and L. Bretschger (2010), 'The implications of heterogeneous resource intensities on technical change and growth', *Canadian Journal of Economics*, **43**, 1173–97.

Pittel, K. and D.T.G. Rübbelke (2010), '*Local and Global Externalities, Environmental Policies and Growth*', BC3 Working Paper Series 2010-15, Basque Centre for Climate Change (BC3), Bilbao.

Popp, D. (2001), 'The effect of new technology on energy consumption', *Resource and Energy Economics*, **23**, 215–39.

Popp, D. (2002), 'Induced innovation and energy prices', *American Economic Review*, **92**, 160–80.

Popp, D. (2004), 'ENTICE: Endogenous technological change in the DICE model of global warming', *Journal of Environmental Economics and Management*, **48**, 742–68.

Romer, P.M. (1986), 'Increasing returns and long-run growth', *Journal of Political Economy*, **94**, 1002–37.

Romer, P.M. (1990), 'Endogenous technological change', *Journal of Political Economy*, **98**, 71–102.

Rübbelke, D.T.G. (2002), *International Climate Policy to Combat Global Warming: An Analysis of the Ancillary Benefits of Reducing Carbon Emissions*, Cheltenham, UK and Northampton, MA, USA: Edward Elgar.

Scholz, C.M. and G. Ziemes (1999), 'Exhaustible resources, monopolistic competition and endogenous growth', *Environmental and Resource Economics*, **13**, 169–85Schou, P. (2000), 'Polluting non-renewable resources and growth', *Environmental and Resource Economics*, **16**, 211–27.

Schou, P. (2002), 'When environmental policy is superfluous: growth and polluting resources', *Scandinavian Journal of Economics*, **104**, 605–20.

Sheshinski, E. (1967), 'Tests of the "Learning by doing" hypothesis', *Review of Economics and Statistics*, **49**, 568–78.

Sinn, H.-W. (1982), 'Taxation, growth and resource extraction: a general equilibrium approach', *European Economic Review*, **19**, 357–86.

Sinn, H.-W. (2008), 'Public policies against global warming', *International Tax and Public Finance*, **15**, 360–94.

Smulders, S. and R. Gradus (1996), 'Pollution abatement and long-term growth', *European Journal of Political Economy*, **12**, 505–32.

Smulders, S. and M. de Nooij (2003), 'The impact of energy conversation on technology and economic growth', *Resource and Energy Economics*, **25**, 59–79.

Spence, M. (1976), 'Product selection, fixed costs, and monopolistic competition', *Review of Economic Studies*, **43**, 217–35.

Tavoni, M. and R.S.J. Tol (2010), 'Counting only the hits? The risk of underestimating the costs of stringent climate policies', *Climatic Change*, **100**, 769–78.

Thompson, P. (2010), 'Learning by doing', in Bronwyn H. Hall and Nathan Rosenberg (eds), *Handbook of the Economics of Innovation*, Vol. 1, Amsterdam, The Netherlands and Oxford, UK: Elsevier/North-Holland, pp. 429–76.

Turton, H. and L. Barreto (2006), 'Long-term security of energy supply and climate change', *Energy Policy*, **34**, 2232–50.

UN (2001), 'Indicators of sustainable development: guidelines and methodologies', www.un.org/esa/sustdev/natlinfo/indicators/indisd/indisd-mg2001.pdf.

Withagen, C. (1995), 'Pollution, abatement and balanced growth', *Environmental and Resource Economics*, **5**, 1–8.

van Zon, A. and I.H. Yetkiner (2003), 'An endogenous growth model with embodied energy-saving technical change', *Resource and Energy Economics*, **25**, 81–103.

van der Zwaan, B.C.C., R. Gerlagh, G. Klaassen and L. Schrattenholzer (2002), 'Endogenous technological change in climate change modelling', *Energy Economics*, **24**, 1–19.

APPENDIX

Table 9A.1   Selection of pollutants emitted in conjunction with CO_2 and examples of impacts

Pollutant	Sources	Health effects	Visibility and other effects
Carbon monoxide (CO)	Fuel combustion; industrial processes; natural sources like wildfires	Reduction of oxygen delivery to the body's organs and tissues; visual impairment; reduced work capacity; reduced manual dexterity; poor learning ability; difficulty in performing complex tasks	Acceleration of the greenhouse effect indirectly by reactions with other substances
Lead (Pb)	Fuel combustion; metals processing	Adverse affection of the kidneys, liver, nervous system, and other organs; neurological impairments such as seizures; mental retardation, and/ or behavioral disorders; changes in fundamental enzymatic, energy transfer, and homeostatic mechanism; high blood pressure and subsequent heart disease	Deposition on the leaves of plants, and with it, representing a hazard to grazing animals
Methane (CH$_4$)	Burning of natural gas; coal mining; oil production; decomposition of waste; cultivation of rice; cattle breeding		Acceleration of the greenhouse effect; contributes to increased level of tropospheric ozone
Nitrogen oxide (NO$_x$)	Combustion processes in automobiles and power plants; home heaters and gas stoves also produce substantial amounts	Irrigation of lungs and causing lower resistance to respiratory infections; increased incidence of acute respiratory diseases in children	Gaseous absorb light, reduce the visual range; important precursors to ozone and acidic precipitation; impact on particulate matter (PM) concentration; causing severe injury to plants; acceleration of the greenhouse effect by contributing to ozone generation

Table 9A.1   (continued)

Pollutant	Sources	Health effects	Visibility and other effects
Nitrous oxide (N_2O)	Burning of fossil fuels; agricultural soil management		Acceleration of the greenhouse effect; reduces the stratospheric ozone layer
Ozone (O_3)	No direct emission but formation by the reaction of violatile organic compounds (VOCs) and NO_x; therefore, ozone is indirectly caused by combustion processes	Increased hospital admissions and emergency room visits for respiratory causes; higher susceptibility to respiratory infection and lung inflammation; aggravation of pre-existing respiratory diseases; significant decreases in lung function; increase in respiratory symptoms; irreversible changes in the lungs	Reduction in agricultural and commercial forest yields; reduced growth and decreased survivability of tree seedlings; plants' higher susceptibility to diseases, insect attack, harsh weather and other environmental stresses; acceleration of the greenhouse effect
Particulate matter (PM)	Emission directly by a source or formation by the transformation of gaseous emissions; combustion processes cause direct emissions	Premature death; increased hospital admissions and emergency room visits; increased respiratory symptoms and disease; decreased lung function; alterations in lung tissue and structure and in respiratory tract defence mechanisms; lung cancer	Important cause of reduced visibility; airborne particles cause soiling and damage to materials
Sulfur dioxide (SO_2)	Burning of coal and oil; metal smelting and other industrial processes	Effects on breathing; respiratory illness; alterations in the lungs' defences, and aggravation of existing cardiovascular disease	A major precursor to PM, which is a main pollutant impairing visibility together with NO_x, a main precursor to acidic deposition

Source: Rübbelke (2002).

10 Consumer behavior and the use of sustainable energy

Reinhard Madlener and Marjolein J.W. Harmsen-van Hout

10.1 INTRODUCTION

> Those of us who call ourselves energy analysts have made a mistake . . . we have analyzed energy. We should have analyzed human behavior. (Lee Schipper, in Cherfas, 1991)

A better understanding of consumer behavior has long been a central research goal in a number of social sciences, including social, economic and environmental psychology, neoclassical and behavioral economics, diffusion of (technological) innovation research, and sociology. Numerous methodologies have been used and research strands and traditions developed in various (sub)disciplines. A main goal of research aimed at a better understanding of energy consumer behavior (ECB), at least from a social sciences perspective, is to come up with a comprehensive behavioral account of energy consumption that allows for designing policies that ensure sustainable energy use[1] and to better understanding the needs, attitudes, motives and behavior of energy consumers.

In actual fact, much of the analysis done was based on what sociologist Loren Lutzenhiser refers to as the 'physical–technical–economic model' (Lutzenhiser, 1993). In this tradition, which often served policy planning purposes, the actual human behavior in all its variety (and its understanding) is seen as being only of secondary importance (that is, merely as a 'disturbance' of the technical system that could otherwise be optimized much easier, faster and better). However, because of the great impact of actual consumer needs and behavior in specific situations, and the often very slow diffusion of innovations, potentials for energy efficiency and energy conservation have often been grossly over- or underestimated, casting serious doubts on the eventual practical usefulness of such technology- and innovation-biased policy advice and calling for *ex post* policy evaluation.

Historically, the topic of energy use and consumer behavior has been particularly intensively researched in the aftermath of the two oil shocks of the 1970s, especially for the residential sector, and in reaction to major policy efforts to save energy (see for example Claxton et al., 1981; Stern and Aronson, 1984; Ester, 1985; Kempton and Neiman, 1987, for compilations of and reflections on that early literature). Much of this early research has dealt with energy efficiency (technology-oriented policies) and energy conservation (policies targeting lifestyle and behavioral change). In particular the latter was seen as 'low-hanging fruits', that is, providing options to save energy at no or low additional cost. In these early days, as Claxton et al. (1981: 1) note pessimistically: 'neither energy conservation programs, nor the initial consumer research appears to have had much impact on energy consumption'. Already it was recognized that researchers' understanding of policy-makers' perspectives is important, and that 'links which will bridge the worlds of both policy maker and researcher' should be fostered (Ester et al.,

1984; Lutzenhiser, 1993). It was claimed that the economic, engineering and legal disciplines have had the dominant inputs into energy conservation policies, with often disappointing outcomes (Ester et al., 1984: 15) and that: 'The behavioral scientist, thus, has a range of concepts to use separately and in combination to approach problems in energy conservation, with an overall marketing and consumer behavior perspective serving an integrative function' (Ester et al., 1984: 15).

In recent years, there has been a strongly revived interest in ECB research, for various reasons. First of all, there is much technological innovation, such as smart metering, programmable thermostats and intelligent white goods, and the increasing automation of homes and the energy grids by means of novel information and communications technology (ICT) solutions and new energy-related services (for example Madlener et al., 2009; Sun et al., 2010; Wade et al., 2010). Second, the liberalization of the energy markets is gradually reaching the final bastion, the residential household, where transaction costs are particularly high and the still untapped economic energy efficiency potentials are (seemingly) large. Third, at a more general level, the call for sustainable energy use (for example in light of climate change and other environmental strains, peak oil, overpopulation, and further severe societal and policy challenges ahead) addresses mainly energy efficiency and renewable energy issues, and also the transition from a mainly fossil- and nuclear-based energy supply system towards a more sustainable one. Implicit in these requirements for change towards sustainable development are its three dimensions to be tackled – economy, society, environment – upon which any truly sustainability-oriented energy policy has to be designed *ex ante* and evaluated *ex post*. Fourth, different research communities have evolved over time and made considerable progress, developing new models and identifying new and untapped research fields. Today, a number of social and behavioral sciences – even relatively new ones such as ecological economics and behavioral economics – can contribute to a more comprehensive analysis and understanding of ECB for more sustainable development, and have not inspired and influenced energy research much until today.

In fact, a number of useful and fairly comprehensive (within their self-defined boundaries) literature reviews on ECB already exist, such as the particularly thorough ones by Stern (1992) and Lutzenhiser (1993), which have been widely acknowledged. A review that also provides some explicit recommendations for research priorities is Brewer and Stern (2005), although it omits economics, which they claim has received much more attention so far from decision-making bodies anyway than any others (such an argument, however, we think should not be used as an excuse for not dealing with this literature). Further comprehensive reviews of interventions to change behavior have recently been provided by Guerin et al. (2000), DiClemente and Hantula (2003), Gillingham et al. (2006), Wilson and Dowlatabadi (2007), Owens and Driffill (2008) and Maréchal (2010), among others.

This review distinguishes itself from earlier reviews in at least five important ways. First, it seems desirable to structure a review along the disciplinary lines of all main disciplines that have traditionally been engaged in ECB research, plus those with a great potential for ECB research but no research tradition in this direction so far. Second, we discuss the literature surveyed along five different types of drivers underlying human behavior: (1) psychological drivers (cognition); (2) rational behavior drivers (utility); (3) sociological drivers (other people); (4) ecological drivers (environment); and (5) technological drivers (innovation), enabling us to identify commonalities and differences otherwise easily overlooked. Third, we aim at finding potentials for more integrated and interdisciplinary research and how this could better serve policy-makers' needs. Fourth, we do not deliberately restrict our analysis to private households, as many others have done, although of course we have to admit that most research so far has been undertaken for this sector. Finally, our review also includes selected new references that more dated reviews, for obvious reasons, could not yet cover.

More multidisciplinary research appears to be necessary and useful to understand better the drivers and impacts of energy consumer behavior, and how (single or a combination of) drivers can effectively and cost-efficiently be tackled by policy-makers aiming at sustainable energy use. This is why we aim at reviewing the ECB literature in a broader sense; that is, we do not restrict ourselves to residential energy consumers, as the distinction between producers and consumers becomes increasingly blurred anyway ('prosumers'), and we do not confine ourselves to energy saving and energy efficiency research only. With some exceptions (for example relevant articles still unpublished in journals, seminal papers, dedicated survey articles), we screened articles published mainly in high-quality scientific journals, and literature since 1990. To avoid duplication, we often refer to discussions of important literature done in other reviews. Many research traditions base their analysis on theoretically grounded decision models that are founded on informed rationality, psychological factors, physical factors or contextual factors, and focus on many different scales (individuals versus groups of individuals or society, local versus global or national impacts, and so on). Therefore, our motivation is to address, for each of the research disciplines discussed, typical research areas, methodologies and characteristics, and to identify which of the above-mentioned five types of drivers are considered particularly relevant.

Figure 10.1 shows the structural essence of our literature review. It depicts the different research traditions regarding ECB, which are then discussed in turn and subdivided into particular research themes or approaches. Note that some research areas, such as diffusion of innovation or energy rebound research, are not listed in separate sections because they were undertaken by several research disciplines. In such cases, we have tried to discuss the cross-disciplinary themes in one section and to establish cross-references to other sections. Also, some topics are discussed in much more detail in dedicated chapters of this *Handbook*. In such cases we have tried to include cross-references as a guidance for further readings.

The remainder of this chapter is organized as follows. Section 10.2 covers social and environmental psychological research, section 10.3 sociological and socio-technical research, section 10.4 engineering–economic analysis, section 10.5 neoclassical economics models, section 10.6 economic psychology and behavioral economics, and section 10.7 ecological economics. Section 10.8 discusses the policy implications found in our review synthesis, and section 10.9 concludes.

10.2 SOCIAL AND ENVIRONMENTAL PSYCHOLOGY STUDIES

Research by social and environmental psychologists dates back to the 1970s and has a long tradition of analyzing residential energy behavior and efficiency. While in the 1970s

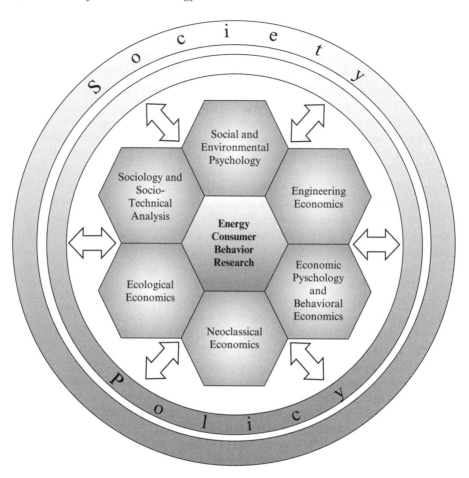

Figure 10.1   Multidisciplinary research on energy consumer behaviour that (hopefully) serves policy and societal needs and learning

interest was mainly on deficient information, it shifted later to the role of psychological constructs (values, attitudes, norms) framed by environmental concerns (see Wilson and Dowlatabadi, 2007: 181). Much of the socio-psychological literature is rooted in either value–belief–norm (VBN) theory or the theory of planned behavior (TPB), briefly discussed in the following.

10.2.1   Value–Belief–Norm Theory

In value–belief–norm (VBN) theory, a causal chain is proposed from the stable essentials of personality (values, views of the world) to specific beliefs about the consequences and responsibilities of particular actions, and on to attitudes and norms (see Dietz et al., 1998; Stern, 2000). VBN is rooted in activated norm theory, which itself stems from earlier work on the elicitation and characterization of values (Wilson and Dowlatabadi, 2007). In VBN, activated norms directly influence behavior, while other psychological constructs act indirectly through activated norms. Activated norms are personal obligations to act in a way that reduces adverse consequences to things assigned a value, creating a predisposition for behavioral change and having a link to self-expectations. In VBN theory, the norm activation pathway was modified to include also altruistic values, both towards humans and the biosphere. Both types of altruism, in addition to self-enhancement (for example status) values and egotism (for example financial return), have been shown empirically to predict different types of pro-environmental behavior, including residential energy conservation (see Wilson and Dowlatabadi, 2007). Altruistic values, however, may not be relevant in contexts where individuals lack perceived self-efficacy or where action is associated with self-sacrifice or a sense of helplessness.

Scherbaum et al. (2008) explore individual factors related to employee energy conservation behavior at work. The study employs VBN theory and examines the individual factors related to energy conservation behaviors among employees of a large state university. By using path analysis, the authors find that environmental personal norms predicted self-reported energy conservation behavior and behavioral intentions. Environmental personal norms also mediated the relationship of environmental worldviews with self-reported energy conservation behaviors, as well as behavioral intentions. Finally, implications for theory and organizational energy conservation interventions are discussed.

10.2.2 Theory of Planned Behavior

The theory of planned behavior (TPB) is an extension of the earlier theory of reasoned action in which attitudes and perceived social norms explain behavior (see Fishbein and Ajzen, 1975; Ajzen, 1991). In the TPB, attitudes are formed from an individual's beliefs about a behavior as well as an evaluation of its outcomes. Together with normative beliefs about what valued peers might think of a particular behavior, these attitudes lead to an intention to act, which in turn predicts behavior. To address decision contexts in which action is constrained or individuals do not otherwise have full control over volition, perceived behavioral control was introduced as a third precursor of intention to act and as a direct precursor of behavior (for further discussion see Wilson and Dowlatabadi, 2007). The TPB has been applied to a great many consumer issues, including transport choices, green consumers, recycling and public health. However, studies relating directly to energy use are still relatively rare (for example Michelsen and Madlener, 2010).

Empirically, TPB studies rely on the elicitation of psychological or psycho-economic constructs, typically by (questionnaire) surveys. Perceived behavioral control in the TPB is a subjective assessment of how contextual factors influence behavior (see Wilson and Dowlatabadi, 2007; Michelsen and Madlener, 2010). A number of TPB studies have investigated either 'green' or conservation (for example Kaiser and Gutscher, 2003; Kaiser et al., 2005) or energy (Abrahamse and Steg, 2009) consumer behavior. Specifically, Abrahamse and Steg (2009) acknowledge that private households constitute an important target group for energy conservation, and that they use energy both directly (gas, electricity, fuel) and indirectly (embedded in the production, consumption and disposal of goods). Direct and indirect energy use and savings of 189 Dutch households were monitored over five months, with the aim to scrutinize the relative importance of socio-demographic and psychological factors related to household energy consumption and changes thereof. Variables from the TPB and the norm-activation model were used. The results indicate that energy use is determined by socio-demographic variables, whereas changes in energy use (which may require some cognitive effort) appear to be more related to psychological variables. The authors show that adding variables from the norm-activation model, the explanatory power of the model to explain energy savings could be significantly enhanced over the TPB model, and that different types of energy use and energy savings seemed to be related to different sets of determinants.

Apart from the VBN theory and the TPB, other theoretical models have been introduced in the socio-psychological literature, including the following behavioral perspective model and the attitude–behavior–external conditions model.

10.2.3 Behavioral Perspective Model

Foxall (1994), in a generic context (that is, not related to ECB in particular), investigates the epistemological status of a comprehensive model of purchase and consumption derived from a critique of behavior analysis. The model, dubbed the 'behavioral perspective model', comprises consumers' learning history, the consumer behavior setting, purchase and consumption responses, and their reinforcing and punishing consequences. Consumer behavior is divided into four operant classes, defined by the environmental contingencies controlling them: maintenance, accumulation, pleasure and accomplishment. Consumer behavior is then described as a hierarchy of these operants over the consumer life cycle, exemplified by reference to household saving and financial asset management. Further, the operant classification is used to interpret consumer behavior as an evolutionary process, exemplified by the adoption and diffusion of innovations. Finally, the model is evaluated according to the criteria of description, delimitation, generation and integration.

10.2.4 Attitude–Behavior–External Conditions Model

The attitude–behavior–external conditions (ABC) model articulates the importance of external conditions. Attitudes lead to behavioral change only if contextual variables provide either weak incentives or disincentives (see Guagnano et al., 1995). Put differently, there are boundary conditions, determined by the context, that limit the ability of attitudes to predict behavior; the ABC model explicitly draws together attitude-based decision models and findings on the influence of external factors, such as incentives and information, on behavior (see Wilson and Dowlatabadi, 2007).

10.2.5 Selected Empirical Studies

In what follows, we summarize some empirical studies on ECB from the social and environmental psychology literature. Stern et al. (1986) is an early study of the effectiveness of incentives for residential energy conservation. It shows that larger incentives can indeed help to raise participation, that marketing (for example, by word-of-mouth communication) and implementation (for example, by a trusted organization) may be more important than the size of the incentive, and that the preferences for grants versus loans vary with income. Black et al. (1985) investigate how external (contextual) conditions influence residential energy behavior, both directly by defining available choices and their relative attractiveness, and indirectly through attitude formation.

Gatersleben et al. (2002) present a measure of environmentally significant household consumer behavior that is based on direct and indirect energy use. From two large-scale field studies conducted in the Netherlands, the authors find that respondents who indicate that they behave more pro-environmentally do not necessarily use less energy. They conclude that pro-environmental behavior is more strongly related to attitudinal variables, while household energy use is primarily related to variables such as income and household size.

McMakin et al. (2002) use a broad socio-psychological model to investigate the impact of customized approaches designed to motivate residents to conserve energy without financial incentives. The case is for two US military installations where residents do not pay their own utility bills. End-use behavior of the residents was surveyed before and after energy use was measured. According to the residents, who declared themselves to be motivated by 'the desire to do the right thing', set good examples for their children, and have comfortable homes, they recommended continued awareness and education, disincentives and incentives to sustain change. Behavioral change is apparently motivated by both altruistic and egoistic motives. The insights gained from this study could also be useful for other types of residents not billed for individual energy use.

Poortinga et al. (2004) explore the role of values in household energy use by using the concept of quality of life (QOL). The authors distinguish seven value dimensions (determined from importance judgments on 22 QOL indicators). These, together with general and specific environmental concerns, contributed significantly to the explanation of policy support, and for market strategies aimed at managing environmental problems, as well as to the explanation of the acceptability of specific home and transport energy-saving measures. Home and transport energy use are particularly related to socio-demographic variables (for example income, size of household). The findings show that it is relevant to distinguish between different measures of environmental impact and different types of environmental intent, and that using attitudinal variables only may be way too limited to explain all types of environmental behavior.

Jager (2006) takes a behavioral perspective and discusses consumer motives for adopting photovoltaic (PV) systems, both theoretically and empirically (via a survey after a promotional and support campaign in a Dutch city). Financial support and general problem awareness are found to be critical motives, but the strong positive effect of information meetings, technical support meetings and social networks are also identified. Suggestions for improved policy measures to stimulate the diffusion of PV systems are proposed.

Dietz et al. (2009) investigate how household action can provide a behavioral wedge to reduce CO_2 emissions rapidly. The behavioral approach adopted contrasts the many debates on long-term options arising from the diffusion of innovative low-carbon energy technologies and creating a cap-and-trade regime for greenhouse gas emissions. Their behavioral approach is for examining near-term reductions in carbon emissions by altered adoption and use of available technologies in US homes and non-business travel. Seventeen household action types and five behaviorally distinct categories are investigated by use of data on the most effective documented interventions that do not involve new regulatory measures. The interventions vary by type of action and often combine several policy tools and strong social marketing. The authors estimate that some 20 per cent of household direct emissions (or 7.4 per cent of national emissions) could be saved with little or no reduction in household well-being. Finally, the authors conclude that future analysis of the potential should incorporate behavioral as well as economic and engineering elements.

Overall, socio-psychological research (as relevant here) mostly addresses individual decisions with consequences on (mostly) residential energy consumption or, more broadly, the environment. Independent variables that explain those decisions are specific to both context and the behavior in question. The influence of psychological variables is constrained by external conditions. As with the expected utility and attitude-based models described above, systemic influences are treated as exogenous, so the time scale over which decisions are considered is short (for example weeks) (see Wilson and Dowlatabadi, 2007). A distinct class of models also frequently employed in socio-psychological research is diffusion of innovation models. These are rooted and used in a number of disciplines aiming to explain innovation behavior, and are discussed briefly in the following section on sociological and socio-technical studies, where one of their main roots lies. Revisiting the five categories of factors driving human behavior that we listed in the introduction, clearly the psychological drivers are important for social and environmental psychology. Furthermore, in many studies sociological and ecological drivers are incorporated as well.

10.3   SOCIOLOGICAL AND SOCIO-TECHNICAL STUDIES

Technological (or innovation) adoption and diffusion theories in the social sciences are typically rooted either in sociology (Rogers, 2003) or in economics (Stoneman, 2001). Innovations can be ideas, certain practices or technologies; that is, they are broadly defined. Diffusion processes, according to Rogers, describe social communication processes where both word-of-mouth communication and media channels play a role. The decision-making process follows distinct stages, from a change in knowledge to a change in behavior (see Wilson and Dowlatabadi, 2007). According to Rogers, five perceived attributes determine the rate of adoption: relative advantage, compatibility, complexity, trialability and observability. A number of studies have shown the relevance of the attributes for residential energy use (for example Darley and Beniger, 1981; Brown, 1984; Dennis et al., 1990; Madlener and Artho, 2005; Mahapatra et al., 2007).

Generally speaking, sociological studies emphasize the influence of social context on decision-making. Shove (1998), for instance, draws on the sociology of science and technology literature, and unpacks conventional beliefs about the diffusion of energy-efficient technologies, proposing an alternative approach which acknowledges the social structuring of technical innovation. The focus of the study is on theories of technology transfer, non-technical barriers to adopt energy-efficient technologies, and energy use in buildings. In another study, Shove has argued that the real wedge between individual decision models and the social dimension (of energy use) is embeddedness (Shove, 2003). For example, the needs and expectations of private households for thermal comfort have evolved over time. So have housing design (for example room size, window areas), energy appliances (for example heating systems, thermostats), institutions and supporting infrastructure (electricity and heat grids, utility tariffs, services) and social norms (room occupancy, indoor temperatures). These changes in norms and technologies are interdependent and drive further change (Wilhite et al., 2000). Note that the embeddedness of energy use in household routines is reinforced by the counter-marketing of newly available and desirable energy devices and the services they provide. The marketing strategies used to sell these services indicate the myriad social roles played by energy technologies: display, status, self-expression, conventionality, convenience, security, independence and flexibility (see Wilson and Dowlatabadi, 2007).

Science, technology and society (STS) studies emphasize the role of societal actors in technological development, and belong to the innovation rather than the consumer behavior literature. Several approaches can be distinguished, such as the constructivist approach (for example Bijker, 1995), where innovations are described through the eyes of social group members; actor–network theory (for example Rip and Kemp, 1998), where the linkages between actors and artefacts are emphasized; or large technical systems theory (for example Hughes, 1983). Mostly, STS studies analyze how new technology evolves and is introduced, but less so how established technologies are used or abandoned. In this respect, STS studies have a close relation to the diffusion of innovation literature. Overall, this type of studies can be categorized as to focus on sociological drivers of ECB with often an important role for technological factors, too.

10.4   ENGINEERING–ECONOMIC ANALYSIS

Traditionally, engineering–economic analysis has dealt mostly with the role of energy efficiency improvements to save, or 'conserve', energy. Commonly, bottom-up process analysis is used to determine energy demand, which is considered a derived demand that is strongly influenced by the rate of utilization of the energy-consuming capital stock (for example appliances, devices, cars). Aggregate energy demand is differentiated by type of fuel, type of user and type of end-use energy provided. If costs are attached to the calculations, engineering–economic assessments are possible. For example, in studies on energy efficiency (for example concerning the retrofit of buildings), payback periods for capital investments are calculated from expected energy savings (see Sutherland, 1996). Engineering–economic analysis has been widely used in the literature and is politically very influential. Decision-making is typically guided by some cost–benefit analysis, in the simplest case with a constant discount rate, and the heterogeneity of decision-makers, and the particular situational context is often ignored. Much of the literature has focused on market barriers that hinder the exploitation of the technical or economic potential of energy-efficient technologies.

Since the engineering–economic energy analysis literature is vast, in this section we restrict ourselves to six thematic areas with a close link to ECB, and provide an arbitrary selection of studies spanning a large spectrum. Section 10.4.1 focuses on the role of information and labeling to overcome the energy efficiency 'gap'; section 10.4.2 on lifestyle, ownership and sociodemographic considerations (from an engineering–economic modeler's perspective); section 10.4.3 addresses the literature on energy decomposition and the use of energy indicators; section 10.4.4 deals with literature on energy rebound or take-back effects; and section 10.4.5 with that on smart grids and smart meters.

10.4.1   Information and Labeling to Overcome the Energy Efficiency Gap

Engineering–economic studies often point to the energy efficiency 'gap' or 'energy paradox', quantifying existing and seemingly economically attractive potentials to raise energy efficiency and thus to reduce energy consumption. As a matter of fact, there has been a long-lasting discussion between engineers and economists about the true potentials of energy efficiency, and the relevance of economic mechanisms and the existence and role of market failure, justifying governmental intervention. In recent years, likely driven by national and international policy efforts, the role of information and labeling in overcoming barriers to energy efficiency has received increasing attention (for example Sorrell et al., 2004; Krarup and Russell, 2005; Owen, 1999).

Economists have provided a plethora of explanations for the efficiency gap or paradox, including the lack of information, limited capital and organizational barriers. Policy action to narrow the gap needs to address the many pervasive factors of human behavior, and it is widely recognized amongst economists that only market failures can justify governmental intervention. Numerous economists have attempted to investigate the paradox identified by engineering scholars that many energy consumers apparently act in a very irrational manner, ignoring profitable energy efficiency investment opportunities. For example, Jaffe and Stavins (1994a, 1994b) examine the factors that affect the diffusion of sustainable energy technologies, focusing on potential market failures such as information problems, principal–agent slippage and unobserved costs, and non-market failures such as information cost, high discount rates and heterogeneity among potential adopters. The authors simulate how effectively alternative policy instruments, both economic incentives and direct regulation, could speed up the diffusion of energy-conserving technologies.

Howarth and Andersson (1993) study the energy efficiency gap, and point out that imperfect information and transaction costs may bias rational consumers to purchase devices that use more energy than those that would be adopted by a well-informed social planner guided by the criterion of economic efficiency. However, consumers will base their purchase decisions on observed prices and expectations of post-purchase equipment performance. The authors conclude that such 'market barriers' suggest a role for regulatory intervention to improve market performance at prevailing energy prices.

Schleich and Gruber (2008), by using econometric analysis for 19 subsectors of the commercial and services sectors in Germany, assess the empirical relevance of various barriers to the diffusion of energy-efficient measures. The results show that the most important barriers are the investor–user dilemma and the lack of information about energy consumption patterns. Multiple types of barriers are found to be statistically significant but to vary considerably across subsectors. Policy implications are discussed for the most relevant barriers.

Recently, Ansar and Sparks (2009) introduced a new model to explain the 'energy paradox' that focuses on investment irreversibility, the uncertainty of the future pay-off streams, and the investor's anticipation of future technological progress.

In another recent study, Munns (2008) investigates four different incentive methods to foster electric energy efficiency: shared savings, bonus return on equity, energy service companies (ESCOs) and virtual power plants. According to the author: 'the time has come to find out how to make energy efficiency a sustainable, profitable business for the electric company.' (p.20)

10.4.2 Lifestyle, Ownership and Sociodemographic Effects

Weber and Perrels (2000) investigate the impact of lifestyle factors on current and future energy demand by input–output modeling, addressing both directly environmentally relevant consumer activities (car use, heating) but also induced environmental damage through the production of the consumed goods. From a household survey dataset, a wide range of activities is covered, and a variety of different behavioral patterns derived from the available socio-economic household characteristics. Ownership effects, for example, as occur when residences are rented rather than owned (landlord–tenant dilemma), have been found to be relevant in many studies on the adoption of energy-saving technology (for example Curtis et al., 1984; Walsh, 1989; Scott, 1997).

Regarding the impact of age of energy consumers (or energy technology adopters), education, minority and other sociodemographic variables, most studies again have been conducted on residential energy consumption. Lindén et al. (2006) find that younger people have better knowledge about energy-efficient measures, and Carlsson-Kanyama et al. (2005) find that younger households tend to prefer up-to-date technology, and that lower uptake of energy-efficient technology by older people may correlate with older people's fewer years of formal education.

10.4.3 Energy Indicators and Decomposition Analysis

A substantial body of the engineering–economic literature has dealt with indicators of energy use, aimed at describing the links between energy use and human activity in a disaggregated manner (for a useful review see, for example, Schipper et al., 2001). There are many basic concepts of various indicators and methodologies to derive them (a recent review of concepts, indicators and terminology is provided in Ang et al., 2010), including the use of decomposition methods (for example Sun, 1998; Bor, 2008), and more recently also covering simplified carbon indicators for the amount of greenhouse gases released to the atmosphere.

Munksgaard et al. (2000), for instance, use decomposition analysis to study the direct and induced impact of household consumption on CO_2 emissions. The authors find that overall growth in private consumption, and not changes in the composition of consumption, explain increase in CO_2 emissions, and that the effect of consumption growth has been partly offset by substantial energy conservation in the energy supply and manufacturing sectors.

10.4.4 Energy Rebound

The rebound effect (Khazzoom, 1980; Brookes, 1990; Saunders, 1992) denotes the (counter-intuitive) increased consumption of energy services as a reaction to increased energy efficiency in providing those services, so that part of the expected energy savings cannot be realized due to changed ECB (the Khazzoom–Brookes postulate). The explanation is that increases in energy efficiency can render energy services cheaper, thus encouraging the consumption of those services (direct rebound effect), and maybe increase the demand for other services or products as well (indirect rebound, macroeconomic rebound). Research interest in the rebound effect in recent years has risen, partly due to a thorough study on its magnitude and methodological issues that was carried out in the UK (Sorrell and Dimitropoulos, 2008; Sorrell et al., 2009). An earlier review of the magnitude of the rebound effect estimated in different studies was provided by Greening et al. (2000). Both investigations find that the direct rebound effect alone in many situations is not negligible (possibly in the range of 10–30 per cent), and is especially high in developing countries and among the poorer population. An explanation for the rebound effect is that individuals, households and firms may act only boundedly rationally due to the high complexity of assessing all the consequences of behavioral changes arising from a change in energy efficiency. A better understanding of ECB will thus also help to understand the rebound effect better, to moderate the ongoing debate between economists and engineers, and to guide policy-makers (see section 10.8; Chapter 19 in this volume).

10.4.5   Smart Grids and Smart Metering

Changes in the ways that consumers use electricity and other energy sources, and the increasing share of often intermittent electricity generation from renewables, spur the need for a more intelligent way to distribute grid-based energy, and also to provide incentives for consumers to adjust their demand behavior and possibly to become more actively engaged in system optimization. With increasing significance of the 'smart grids' (electricity, gas, heating), distributed generation and electric cars, the need grows to find ways to design and manage these grids and consumer behavior efficiently, in order to avoid demand patterns shifting at a large scale in unpredictable and undesirable ways, thus destabilizing the system and calling for the stabilizing forces of grid automation of the increasingly complex power systems.

A smart meter is a device that regularly and interactively provides feedback about energy consumption, and as a result might influence it. The underlying perspective of research in connection with this enabling technology is therefore focused on information as a driver of ECB. However, it can also be combined with economic incentives like dynamic pricing (for example, Alexander, 2010) or tradable green certificates (for example Bertoldi and Huld, 2006), structural changes like the smart grid (for example Brown et al., 2010; Chao, 2010; Faruqui et al., 2009), attitude (for example Darby, 2006), societal benefits and complex technology–behavior interaction (for example Neenan and Hemphill, 2008), habits and other heuristics (for example Darby, 2006) or environmental impact assessment (Hledik, 2009).

Economic incentives and structural changes in principle fit in the neoclassical economic approach, as does information provision in a world without perfect information. Moreover, if any of the psychological, ecological and technological drivers are indeed relevant, the analysis could gain from other approaches as well. If suppliers would go so far as also to provide information on other consumers' behavior (in aggregated form for privacy reasons), smart metering could even offer a promising application field for sociological research or behavioral economics research on social preferences, which is not even mentioned in the systematic survey of Wilson and Dowlatabadi (2007).

We conclude that of the five types of drivers underlying human behavior identified in the introduction, engineering–economic models largely address rational and technological drivers.

10.5 NEOCLASSICAL ECONOMICS MODELS

In neoclassical economics, the microeconomic decision model based on utility maximization and fixed preferences is the central pillar for the analysis. The microeconomic theories of consumer (firm) choice are based on the assumption that consumers (firms) are perfectly rational, and that they maximize their utility (profits) given a certain budget constraint. Outcomes with higher utility (profit) are preferred to those with lower utility (profit). In this world, consumers are assumed to behave rationally in a normative sense, according to preferences that are ordered, known, invariant and consistent (see Wilson and Dowlatabadi, 2007: 172). In this respect, utility is a construct that serves as a proxy for 'well-being' that, apart from self-interest, can also include, say, perceived fairness and altruism.

Utility theory and the rational choice model, or theory, that is based on preference orderings, are the foundation for a broad range of economic theories, and many different applications of rational choice models exist. These include stated and revealed preference models (the former studies responses to hypothetical questions, while the latter is based on actual purchasing behavior), discrete choice models and (computable) general equilibrium models. In the following, we will provide a review of some of the literature that follows these three strands.

10.5.1 Conjoint or Discrete-choice Analysis and Other Experiments

Experiments constitute a common method in behavioral economics as well as in psychology. So far only rarely, and relatively recently, have researchers started to apply this method to ECB. Conjoint or discrete-choice analysis uses hypothetical choice situations, whereas other experiments focus on revealed choice in a controlled laboratory environment. Experiments could investigate any drivers of energy behavior, but so far have mainly been incorporated as a single manipulation factor within a broader field setting in order to research the effect of some intervention, measuring other drivers of energy behavior (for example Abrahamse and Steg, 2009; Alexander, 2010; Benders et al., 2006; Bertoldi and Huld, 2006; Faruqui et al., 2009) and leading to so-called 'field experiments'. In the following, we review a range of studies in the discrete-choice analysis tradition.

Vaage (2000) investigates household energy demand in Norway by means of a combined discrete and continuous choice approach. The discrete choice for appliance (heating technology) adoption is specified as a multinomial logit model that contains appliance attributes and individual characteristics as explanatory variables. Conditional on the appliance choice, a continuous choice model of energy use is applied. The estimated price elasticity for energy from the use of detailed micro-data is substantially higher (exceeding −1) than found in studies that ignore the explicit appliance dependence.

Scarpa and Willis (2010) investigate the willingness-to-pay (WTP) for renewable energy by a choice experiment approach that focuses on various micro-generation technologies (solar PV, micro-wind, solar thermal, heat pumps, biomass boilers and pellet stoves). The results from conditional and mixed logit models are compared, the latter of which estimate the distribution of utility coefficients. The study derives WTP values as a ratio of the attribute coefficient to the price coefficient, with a model in which the WTP distribution is estimated directly from utility in the money space. The results indicate that while the adoption of renewable energy is significantly valued by households, this value is insufficient for the majority of households to cover the higher capital costs of micro-generation energy technologies.

Poortinga et al. (2003), Sammer and Wüstenhagen (2006) and Banfi et al. (2008) all consider the importance of structural changes as drivers of energy behavior (that is, attributes), which would fit in the neoclassical framework, whereas Moxnes (2004) allows for heuristic decision-making as in behavioral economics. More specifically, Banfi et al. (2008) study energy efficiency measures for residential buildings in Switzerland; Poortinga et al. (2003) preferences of Dutch households regarding technical and behavioral energy-saving measures; Sammer and Wüstenhagen (2006) eco-labels for washing machines in Switzerland; and Moxnes (2004) energy efficiency standards for refrigerators in Norway. Two recent non-residential discrete choice studies on ECB are, for instance, Newell and Pizer (2008) (commercial buildings), and Achtnicht et al. (2008) (automobile transport).

Several researchers have developed aggregate ('top-down') economic and detailed ('bottom-up') engineering models in order to link the energy with the economic system, and in particular to improve the analysis of energy policy impacts by endogenizing the investment decisions of energy consumers (for example Frei et al., 2003). Rivers and Jaccard (2005) and Horne et al. (2005), for instance, use a hybrid energy–economy model that applies discrete-choice modeling to the empirical estimation of key behavioral parameters representing technological choice (in one case on steam generation, in the other case on vehicle and commuting decisions).

Menges et al. (2005) is an exceptional study in the sense that it attempts to replicate a controlled and artefactual laboratory environment in shopping malls (to recruit participants from the field), in order to investigate motivations to pay a premium for electricity generated from renewables.

Gleerup et al. (2010) study the effect of providing feedback on electricity consumption to residential households by SMS text messages and email, that is, low-cost instant feedback options. In the experiment, 1452 households were divided into three experimental groups and two control groups. Results indicate that email and SMS messages providing timely information about a household's 'exceptional' consumption periods, such as the week with the highest electricity consumption in the past quarter year, produced average reductions in total annual electricity consumption by about 3 per cent.

10.5.2   Econometric Studies of Energy Demand

Empirical analysis of energy demand (and thus indirectly also ECB) still seems to be dominated by econometric techniques, whereas the type of model and data used varies a lot. Reviews of energy demand elasticity studies and modeling approaches used include

Bohi and Zimmermann (1984), Dahl and Sterner (1991), Madlener (1996) and Pindyck (1979, 1980).

Aggregate-level studies make use of the benefit that individual preferences of a population are relatively stable, so that preferences on the aggregate level can often be estimated successfully. Long (1993), for instance, reports on an econometric analysis of residential expenditures on energy conservation and renewable energy sources. Ibrahim and Hurst (1990) construct aggregate oil and energy demand functions for a number of oil-importing and oil-exporting developing countries, in order to identify (by econometric estimation) the factors that have determined the level and patterns of energy demand in the 1970s and 1980s. They find the price elasticity of energy demand to be low in the short and the long term, and a strong link between income and aggregate energy demand. Oil demand behavior they find to be more complex, since oil products can be rationed and domestic energy production can affect demand for oil.

Assimakopoulos (1992) introduces an approach for modeling residential energy demand in developing countries, in which energy demand equations apply to endogenously defined (structural analysis of energy demand) 'homogeneous' groups of consumers. Principal components analysis and discriminant analysis are the main methods used. The set of groups obtained is linked to a set of equations through a qualitative response model simulating the households' decisions. Equations for energy consumption, the choice of energy-using equipment, and the repartition of energy products are estimated for each consumer group.

Dumagan and Mount (1992) investigate consumer welfare effects of carbon penalties. The authors use a generalized logit demand system that conforms better to the theory of consumer behavior than standard flexible functional forms used frequently by others (for example translog, 'almost ideal demand system' – AIDS, generalized Leontief). The model is applied to New York state-level and company-level data on residential consumption of electricity, natural gas and fuel oil.

Poyer and Williams (1993) investigate residential energy demand by minority household type. In particular, the authors estimate electricity and total energy demand elasticities by minority and majority household type for both the short and the long run. The demand for electricity is found to be relatively price-inelastic irrespective of the household group.

Puller and Greening (1999) examine the dynamics and composition of the household adjustment to gasoline price changes using a panel of US households. Demand for gasoline is decomposed into demand for vehicle miles traveled and the demand for household composite miles per gallon. Total price elasticity estimates are within the range found in previous studies, but the authors found that consumers initially respond to a price rise with a much larger decrease in consumption than would be indicated by the total elasticity. Moreover, households are found to respond to price changes by adjusting vehicle miles traveled more than composite miles per gallon in the year after the price change.

Nesbakken (1999) studies the relationship between the choice of heating equipment and residential energy consumption using micro-data for Norway. The energy–price sensitivity is found to be higher for high-income households than for low-income households.

De Groot et al. (2001) investigate the promotion of investments in energy-saving technologies by firms, thus adding to the still scarce empirical evidence on success conditions of associated policy measures. The authors conducted a survey among firms in the Netherlands, in order to identify factors that determine the investment behavior of firms, their attitude towards various types of energy policy and their responsiveness to changes in environmental policy. By using discrete-choice modeling, the authors aim at investigating whether and how such strategic features vary over firm characteristics and economic sectors.

Nicol (2003) estimate elasticities of demand for gasoline (in Canada and the US) for different household groups, based on household level data from the Canadian family expenditure survey and the US consumer expenditure survey. As in earlier studies, demand is found to be inelastic both for own-price and income, and elasticities are found to vary across regions of Canada and the US, but these variations are smaller than those with respect to household size and housing tenure.

Nyborg et al. (2006) explore the responsibility of green consumers for the provision of public goods, which is assumed to depend on beliefs about the behavior of others, even for consumers motivated by internalized moral norms, rather than social sanctions. Permanent increases in green consumption are shown to be achievable by imposing temporary taxes or subsidies, or through advertising that affects beliefs about either others' behavior or external effects. The influence of moral motivation is shown to diminish with taxation, if a tax is interpreted as taking responsibility away from the individual.

Brännlund et al. (2007) examine how exogenous technological progress (increase in energy efficiency) affects household energy consumer choice and hence emissions in Sweden. The necessary change in the CO_2 tax is estimated to avoid the rebound effect (see section 10.4.4; in this context, that CO_2 emissions remain constant), and its impact on SO_2 and NO_x emissions. The results indicate that a 20 per cent increase in energy efficiency would increase CO_2 emissions by about 5 per cent, requiring an increase in the CO_2 tax by 130 per cent. Such a tax increase would reduce the emissions of SO_2 below the initial level, but would raise NO_x emissions. The authors conclude that if marginal damage from SO_2 and NO_x are non-constant, additional policy instruments are required.

Serletis and Shahmoradi (2008) provide semi-parametric elasticity estimates by semi-non-parametric estimation techniques. The focus is on interfuel substitution between crude oil, gas and coal in the US.

Gundimeda and Köhlin (2008) investigate how fuel demand elasticities for India can be used for energy and environmental policy-making. By using micro-data for more than 100 000 households and applying an AIDS model, they estimate price and expenditure elasticities of demand for four main fuels for both urban and rural areas and different income groups. The authors argue that the results can be used for projecting energy demand and CO_2 emissions for different rates of growth and population segments, but also to evaluate recent and ongoing energy policies.

A great many studies have investigated the reaction of energy consumers to induced price signals, for example provided by energy taxes or subsidies. For instance, Berkhout et al. (2004) is an econometric study on household energy demand in the Netherlands. An estimation of the actual impact of an energy tax introduced in 1996 is presented first, using panel data and estimating a demand function that controls for a large set of variables (for example outside temperature, type of house and insulation, household cooking behavior and a large number of durable goods). The results show that in the short term the energy tax had a small but significant impact on energy consumption. The influence of variables other than price and income on household energy demand is also discussed. This information is used to compare the impact of the energy tax on household energy use with the energy reduction that could have been achieved when using other policy instruments.

Boonekamp (2007) starts off with the expectation that the large number of new policy measures introduced in the Netherlands since the late 1990s has influenced the response of households to energy price changes. To verify this, the author uses a bottom-up model and alternative price scenarios, for which elasticity values found are explained by using the bottom-up changes in energy trends. The specific set of savings options explains much of the price responses. The price effect is also analyzed in combination with standards, subsidies and energy taxes. The simulation results show that the elasticity value could possibly be 30–40 per cent higher than without these measures.

Ghalwash (2007) studies the differences in consumer reaction to the introduction of (or the change in) environmental taxes as a signaling device. His hypothesis is that an environmental tax conveys new information about the properties of the directly taxed goods, which in turn may affect consumer preferences for the goods, hence altering the choice of consumption. From his econometric analysis for household demand in Sweden, he finds that all goods have negative own-price elasticities and positive income elasticities. The signaling effect is found to be ambiguous: whereas the tax elasticity for energy goods and for heating seems to be markedly higher than the traditional price elasticity, the opposite appears to be the case for energy goods used for transportation.

Dynamic pricing of energy (for example by means of real-time, time-of-use and critical peak tariffs) is a topic that attracted considerable attention in the 1980s and early 1990s. In more recent years, electricity market liberalization and technological progress (for example smart metering) has renewed both business and research interest in the way that energy consumers respond to price changes and alternative tariff designs (for time-of-use-pricing studies of the 1990s, see for example Aigner and Ghali, 1989; Train and Mehrez, 1994; Aubin et al., 1995). While most of the attention has been dedicated to electricity so far, smart metering (and submetering in multi-family houses) coupled with dynamic pricing is also increasingly being tested and introduced for heat energy and water consumption.

Along with the revived research interest, the awareness of methodological problems related to non-linear energy pricing, heterogeneity of price elasticities amongst consumers, and aggregation of consumption over appliances and time is also growsing. Reiss and White (2005), for example, find a highly skewed distribution of household price elasticities among a representative sample of 1300 Californian households, and that a very small fraction of the sampled households accounts for most of the aggregate demand response. Borenstein (2005) investigates the long-run efficiency implications of real-time electricity pricing, and shows that time-of-use pricing (that is, a simple peak and off-peak pricing system) captures much lower efficiency gains than a retail real-time pricing scheme with hourly price changes. Lijesen (2007) investigates real-time price elasticities of electricity demand, while Woo et al. (2008) review the options for advanced metering infrastructure (AMI) in California and different types of electricity pricing schemes.

10.5.3   Computable General Equilibrium (CGE) Modeling

Computable general equilibrium models for energy studies first became popular among energy modelers in the 1980s. Energy policy issues are related to a number of economic aspects, including price formation, output determination, income generation and distribution, consumer behavior and governmental operation. For policy design and impact studies a systematic and coherent framework of analysis is needed. Bhattacharyya (1996) surveys the literature of applied CGE models applied to energy issues, and reports their special features, their evolution through time and their limitations.

Edenhofer and Jaeger (1998) study the role of induced technical change for tackling the problem of timing in environmental policy. The authors conceptualize power in a non-linear model with social conflict and induced technical change. The model shows how economic growth, business cycles and innovation waves interact in the dynamics of energy efficiency. Three different ways of government control are investigated: energy taxes, energy and labor subsidies, and energy caps. Energy taxes help to select more energy-efficient technologies. As for energy subsidies, energy-efficient technologies helps to explain why in contemporary economies labor productivity grows faster than energy efficiency. With an energy cap, the social network of the relevant agents may be stabilized via social norms. It seems plausible to the authors that innovation waves comprise several business cycles, and that such a wave was in the making in the 1990s. However, proposals to postpone policies for improving energy efficiency increase the risk of energy-inefficient lock-in effects.

Bjertäes and Fäehn (2008) explore the welfare effects of energy taxes in a small open economy by means of a CGE model for Norway. In particular they examine the social costs of compensating the energy-intensive export industries for profit losses incurred because of the imposition of the same electricity tax on all industries (uniform tax rates usually perform better than differentiated schemes, especially when revenues can be recycled by cutting other taxes that are more distortionary). The authors find that the costs are surprisingly modest, which they explain by the role of the Nordic electricity market, which is still limited enough to respond to national energy tax reforms. Hence an electricity price reduction in part neutralizes the direct impact of the tax on profits. The authors also investigate the effects of different compensation schemes, finding markedly lower compensation costs when the scheme is designed to release productivity gains.

Obviously, neoclassical economics by definition assumes only rational drivers of human behavior.

10.6   ECONOMIC PSYCHOLOGY AND BEHAVIORAL ECONOMICS

In the previous sections we dealt with several disciplines that consider ECB assuming different underlying drivers. Clearly, neoclassical economics constitutes the most parsimonious approach in this respect, whereas the other disciplines can often describe real-world behavior more accurately. Economic psychology is a discipline trying to combine these two benefits by using rational as well as psychological drivers as an explanation for consumer energy behavior. Some recent (energy) consumer behavior research published in the leading journal dedicated explicitly to economic psychology (the *Journal of Economic Psychology*) include the following.

McCalley and Midden (2002) investigate product-integrated feedback as a means to increase energy conservation behavior. The authors explore the roles of goals and goal-setting, and social orientation. The former was studied via a simulated, technologically advanced, washing machine control panel (100 subjects with 20 simulated washing trials each), and the effectiveness with regard to conservation behavior of self-set and assigned goals were compared with each other when they are combined with energy feedback. Social orientation was found to interact with goal-setting mode, with pro-self individuals saving more energy when allowed to self-set a goal, and pro-social individuals saving more energy when assigned a goal.

Poortinga et al. (2003) study preferences for different types of energy-saving measures by using additive part-worth function conjoint analysis. Energy-saving measures are found to differ in the domain of energy savings, the energy-saving strategy and the amount of energy savings. The energy-saving strategy is found to be the most important characteristic influencing the acceptability of energy-saving measures, and especially shifts in consumption. Furthermore, home energy-saving measures were more acceptable than energy-saving measures in transport. The amount of energy savings is the least important characteristic. Except for respondents differing in their environmental concern, no differences are found in average acceptability of the energy-saving measures among the respondent groups. Finally, some interesting differences in relative preferences for different types of energy-saving measures are found between the respondent groups.

DiClemente and Hantula (2003) review the applied behavioral literature on consumer choice. They start from early work by Watson (1908), Watson and Rayner (1920) and Lindsley (1962) on the role of behavior analytic theory and application in consumer behavior. The applied behavior analysis movement, in their view, brought operant-based applications into the 'consumer' research area, focusing largely on pro-social and social marketing applications. Increased interest in behavioral theory then sparked a continued interest in classical conditioning of consumer attitudes and behavior. Recent theoretical work on the behavioral perspective model (see section 10.2.3) and in the field of behavioral ecology of consumption both draw from Watson's early work but also from new developments towards a comprehensive behavioral account of consumption. DiClemente and Hantula (2003) point out that: 'Research showed that behavioral techniques such as prompting, a rebate system, video modeling, and feedback can effectively increase residential energy conservation . . . as well as [that] in office buildings.'

However, researchers in economic psychology are facing a trade-off between the benefits of the neoclassical approach and those of the psychological approach, rather than complementarity: when psychological variables are added to fit the data better on a specific application, the model becomes more ad hoc. Moreover, our set-up with drivers underlying ECB obviously suggests that better results could be achieved by adding even more drivers to the rational–psychological combination.

Behavioral economics is a rapidly developing new research discipline that tries to improve on both these drawbacks of economic psychology. It systematically attempts to model how people think, which is subsequently (also) expressed in consumer behavior. A central concept in this approach is bounded rationality (Simon, 1955; Conlisk, 1996), which deviates from full rationality due to cognitive restrictions but keeps striving for it as a benchmark steering mechanism of behavior. Seminal contributions of behavioral economics can be found in prospect theory (Kahneman and Tversky, 1979), which models behavior under uncertainty; mental accounting (Thaler, 1985), which models how economic outcomes are cognitively framed; intertemporal choice (Frederick et al., 2002), which models valuation of economic outcomes over time; and social preferences (Fehr and Schmidt, 1999), which models how other people's preferences are taken into account (for useful introductions to behavioral economics see for example Camerer et al., 2008; Wilkinson, 2008).

To the best of our knowledge, up to now hardly any research has been published that empirically applies this approach to the domain of energy behavior. Wilson and Dowlatabadi (2007) illustrate that it would nevertheless be feasible to do so, although they do not refer to social preferences, whereas that could be the unique path to systematically adding sociological drivers to the economic analysis apart from psychological ones. Therefore, behavioral economics offers highly promising short-term research potential for energy behavior researchers.

Menges et al. (2005) can again be considered an exception, since it refers to behavioral economics literature on social preferences (specifically, several articles by J. Andreoni) when investigating the role of 'warm glow' motivations rather than pure altruism in paying a premium for renewable electricity and the resulting effects on crowding-out. Furthermore, the authors adopt an experimental approach, as is common in behavioral economics.

10.7   ECOLOGICAL ECONOMICS

Ecological economics goes one step further than psychological and behavioral economics by integrating physics and biology into economics, thereby moving beyond environmental economics in the sense that it considers nature as an independent entity rather than just part of the human environment, from where resources are taken and where the wastes from economic activity are being deposited. However, the ecosystems are highly interconnected. Therefore, ecological economics does not introduce a new driver of human behavior, but rather stresses the normative instead of empirical importance of a specific driver: a positive attitude towards the environment (ecological driver). Moreover, this emerging discipline could involve the most holistic approach of ECB research if it incorporates all five driver types as identified in the introduction.

Up to now, the ecological economics literature does not seem to have developed a strong research tradition on ECB (casual observation: a keyword search for 'energy consumer' and 'behavior' only yielded 16 hits of papers addressing very diverse topics in the community's flagship journal *Ecological Economics*; an example study is Longo et al., 2008, which is essentially restricted to ecological and rational drivers). Therefore, this approach would be a promising longer-term research path.

10.8 INTERVENTIONS TO AFFECT ENERGY CONSUMER BEHAVIOR: POLICY INSIGHTS AND LEARNING

The fact that the energy demand of individuals, apart from the endowment with technical equipment, income, age and so on, is deeply embedded in social and individual norms of comfort, cleanliness and convenience, has important implications for the ability of policy interventions to impact energy use by behavioral change. Hence it is quite clear that deliberate targeting of psychological or contextual variables can only achieve limited success in terms of affecting behavioral change in the short run. Moreover, contextual variables are seen as important drivers in social psychology, and as malleable and legitimate targets for interventions. In contrast, sociologists see contextual variables as elements of highly structured systems that shape, stabilize and constrain ECB, and that have often evolved over a long time alongside technologies, creating path-dependent socio-technical regimes (see Wilson and Dowlatabadi, 2007). For policy-makers who take such socio-psychological aspects seriously, it is thus important to find out where on the individual–social, instinctive–deliberative, psychological–contextual and short-term–long-term decision continuums their policy interventions are targeted, and which determinants of decision-making they aim to influence.

In what follows, we discuss some literature that has explicitly dealt with policy issues related to ECB. Again, rather than being comprehensive, the aim is to shed light on the spectrum of issues dealt with in the various studies and the methodologies used.

Pohekar and Ramachandran (2004) review the literature dealing with applications of multicriteria decision-making as a tool for sustainable energy planning. The merits and challenges of combining participatory multicriteria analysis (MCA) with scenario-building for analyzing and aiding decision-making in a public (energy) policy context, have been shown in Kowalski et al. (2009) (see especially Table 1 therein for a comprehensive review of the literature on MCA applied to energy issues).

Potoski and Prakash (2005) investigate the role of 'voluntary program' (conceptualized as 'club goods' providing non-rival but potentially excludable benefits to members) as an instrument for governments and non-governmental organizations (NGOs) aiming to improve industry's environmental and regulatory performance. The analysis of about 3700 US facilities shows that joining ISO 14001, an important non-governmental voluntary program, improves facilities' compliance with governmental regulation, due to the reputational benefit reaped that helps induce facilities to take costly progressive environmental action not undertaken otherwise.

Literature on the success of a variety of different energy efficiency policies (in a US context) has been reviewed in Gillingham et al. (2006), including appliance standards, financial incentive programs, information and voluntary programs, and management of government energy use. Stern (1992) provides an interesting discussion on policy analysis failures. Darby (2006) recommends strengthening energy policy by improving the 'continuous learning' at all levels by means of investment in feedback, training and public education, strengthened by product policy. Such learning, in her view, is essential for sustainability.

Laitner et al. (2003) discuss the problem of frequent and large energy demand forecasting errors derived from energy–economy models that do not properly model ECB and the complex interactions between a critical number of energy-related issues (for example energy security, air pollution and climate change, electricity market liberalization). They criticize such model-based studies on the basis that they may inadequately inform policy-makers, and such biases in appropriately capturing behavioral responses may substantially alter the conclusions of policy evaluation.

Oikonomou et al. (2009), starting from microeconomic theory and the concept of the rational use of energy, study the relationships between economic variables that determine energy efficiency behavior, and energy savings that are either the result of frugality (change in ECB, energy conservation) or efficiency measures (no change in ECB, technological substitution). The paper outlines the role of the parameters that determine energy-saving behavior for the outcome of energy efficiency policies, and maintains that policy design should address these properly to be effective.

In recent years, a substantial body of literature has evolved surrounding the learning or experience curve concept applied to energy technologies (for a recent review of the state of the art, see Junginger et al., 2010). Learning curves measure cost decreases due to technological learning in the entire system, that is, learning-by-doing, learning-by-using and so on. In order to accelerate the process of technological learning related to sustainable energy use, it is also important to relate learning to ECB, and in particular the interaction between consumers of energy and investors in new energy technologies, and suppliers thereof. Currently, research is largely lacking that focuses on the impact of ECB on the slope of the learning curve. Also, there is a close relationship between economies of scale and learning effects that is difficult to untangle when learning curves are observed. Isoard and Soria (2001) investigate the effects of learning and returns to scale in the capital costs reduction patterns experienced by renewables. The findings on the role of learning effects and economies of scale are found to be essential to the dynamics of innovation and market structure, and hence also for policy design. The effect of non-constant and flexible returns to scale on the diffusion dynamics is shown to be considerable, although returns to scale are suggested to be constant in the long run.

A number of studies have focused on the relevance of providing information to energy consumers that goes beyond the directly measurable energy consumption, also including indirect or 'embodied' energy, in order to raise the awareness and concern of the consumers and thus make energy policies more effective. Reinders et al. (2003), for instance, study the direct and indirect energy requirements of private households in selected European Union (EU) countries. The analysis is based on data of expenditures of households and the associated energy intensities of consumer goods. The share of direct energy to total energy requirements in different countries is found to vary greatly (between 34 per cent and 64 per cent), which cannot be explained by differences in climate alone. Bin and Dowlatabadi (2005) propose an alternative to sectoral energy demand studies, the so-called consumer life-cycle approach (CLA), which is based on embodied energy and aimed at revealing the total impacts of consumer activities on energy use and related environmental impacts. In their study for the US, they find that more than 80 per cent of the energy used and CO_2 emitted are a consequence of consumer demands, and the economic activities to support these demands. Direct influences of consumer activities (for example home energy use and personal transport) account for only 4 per cent of gross domestic product (GDP) but account for 28 per cent (41 per cent) of energy use (CO_2 emissions). Indirect influences (for example housing operation, transport operation, food and apparel) involve more than twice the direct energy use and CO_2 emissions. The authors claim that the characterization of both direct and indirect energy use, and emissions, is critical to the design of more effective energy and climate policies, and to mitigate the dichotomy between 'them versus us' (that is, households versus industry).

Several researchers have called for the development and application of more integrated theories of ECB. Keirstead (2006), for example, studies integrated analytical frameworks for domestic energy consumption in the UK, criticizing traditional studies taking disciplinary perspectives only as missing important contextual factors, such as cultural values and behavioral interactions with technologies. The author calls for a common language as a stimulus for a renewed interest in the integrated perspective on domestic energy consumption, and presents a flexible agent-based framework to stimulate debate and clarify the role of an integrated approach to domestic energy policy. In similar vein, Faiers et al. (2007), in the context of residential energy use in the UK and energy efficiency investments, draw together some theories relevant to ECB studies in order to aid policy-making in a broader context and to foster the discussion around integrated theories of ECB. In the 1990s Lutzenhiser had argued that sociologists should play a more prominent role in interdisciplinary energy research, given growing interest by natural scientists in the human dimensions of global environmental change (Lutzenhiser, 1994). She also noted that the initiative would have to come from within the discipline, and that there are both external limits on sociological analysis and also sociology's unease with technology and the physical and natural world (p. 58). It is probably safe to say that the same line of reasoning could be applied to other social scientists dealing with ECB research.

Finally, in section 10.4.4 we have argued that policy-makers should not underestimate the rebound effect, although a thorough assessment might be at best difficult and expensive, and at worst infeasible. The same holds true for policy evaluations. For a useful discussion on policy responses to energy rebound see van den Bergh (2010), especially section 10.7.

10.9   CONCLUSIONS

In this survey of the literature on energy consumer behavior and the quest for sustainable energy use we have first divided the ECB literature along the main research disciplines involved. After that, we have identified the main characteristics, and in particular the strength and weaknesses and particular views of the various approaches applied. This helped us to assess the contribution of the various approaches in guiding policy-makers to design appropriate policies that help to steer a course towards sustainable energy development.

Both research and public interest in ECB issues waned in the 1980s, mainly due to a longer period of low energy prices coupled with the relief from significant, policy-induced energy efficiency gains. Still, a number of signs indicate that in recent years interest has been regaining momentum. Today's interest in ECB is fueled by rising energy prices in a still largely fossil energy world, the challenge of climate change (and the related urgent need to decarbonize the energy system), world population growth (and the related growth in energy consumption), the neoliberal quest for more economically efficient energy markets, technological innovation, and the increased recognition of the perceived need to steer the economy and society towards a sustainable energy pathway.

We have identified some strengths and weaknesses of models that are centered on individual choice and action, models that focus (primarily) on social aspects or the aggregate behavior and impacts of energy consumers, and economic ones that rely excessively on the assumption of perfectly rational actors. These are used especially by neoclassical economists and energy analysts with an 'engineering out of problems' mindset.

Our findings show that some new and important technological and other developments are being picked up by scientists, such as smart grids and smart metering, electric mobility and teleworking, many of which will have critical behavioral and environmental consequences of an a priori unknown sign. Smart metering seems particularly interesting, as it enables two-way communications of many kinds between energy companies and households, and as it potentially offers considerable efficiency gains through real-time pricing guided much more by marginal rather than average cost considerations and thus scarcity of resources. A wider diffusion of smart meters, and also of electric cars with smart charging stations, can be expected to bring forth a plethora of new opportunities to incentivize energy consumers to change their behavior, and obviously also many new and challenging research and policy questions.

In contrast to other reviews on ECB literature, such as Lutzenhiser (1993) or Wilson and Dowlatabadi (2007), we did not restrict ourselves to literature focusing on residential consumers only. Also, unlike for example Brewer and Stern (2005), we deliberately included the energy economics literature. Our aim was to organize both seminal and recent literature in the major research disciplines that have dealt with ECB, trying to identify differences and similarities regarding the interest in certain drivers of human behavior (psychological, rationality-based, sociological, ecological and technological). We showed that most disciplines at best focus on one or two of these drivers, while for a holistic assessment of ECB more integrated and advanced models and approaches are needed. Apart from the likely increased complexity of such analysis, however, it also requires the willingness of researchers from the different disciplines to engage in such multidisciplinary studies. In this respect, we pointed out that behavioral economics combines normative analysis based on utility theory with the psychological and sociological insights to support, or better understand, decision-making. So far, it has hardly been applied to energy topics, leaving ample room for investigations and new insights. Hence, ECB seems to be a promising new research area for both field and laboratory experiments based on behavioral economics theory.

Overall, while integrated and interdisciplinary models for the study of energy consumer needs and behavior seem desirable, we conclude that for the time being it is probably more important that the limitations of the existing models are clearly identified, and taken into account, when providing policy guidance or, conversely, seeking policy advice.

Finally, every literature review must be limited due to space constraints and can never be exhaustive. This one is no exception in this respect. Still, we have tried to add another perspective and synthesis to the literature, hoping to foster both multi- and interdisciplinary research on the needs and the behavior of energy consumers in the twenty-first century; in light of world population growth and climate change probably the most critical century so far regarding the sustainable use of energy.

NOTE

1. For a discussion of sustainability criteria and indicators for energy resources and technologies see Chapter 3 of this *Handbook*.

REFERENCES

Abrahamse, W. and L. Steg (2009), 'How do socio-demographic and psychological factors relate to households' direct and indirect energy use and savings?', *Journal of Economic Psychology*, **30** (5), 711–20.

Achtnicht, M., G. Bühler and C. Hermeling (2008), 'Impact of service station networks on purchase decisions of alternative-fuel vehicles', ZEW Discussion Paper No. 08-088, Centre for European Economic Research, Mannheim, Germany.

Aigner, D.J. and K. Ghali (1989), 'Self-selection in the residential electricity time-of-use pricing experiments', *Journal of Applied Econometrics*, **4**, S131–S144.

Ajzen, I. (1991), 'The theory of planned behavior', *Organisational Behavior and Human Decision Processes*, **50**, 179–211.

Alexander, B.R. (2010), 'Dynamic pricing? Not so fast! A residential consumer perspective', *Electricity Journal*, **23** (6), 39–49.

Ang, B.W., A.R. Mu and P. Zhou (2010), 'Accounting frameworks for tracking energy efficiency trends', *Energy Economics*, **32** (5), 1209–19.

Ansar, J. and R. Sparks (2009), 'The experience curve, option value, and the energy paradox', *Energy Policy*, **37** (3), 1012–20.

Assimakopoulos, V. (1992), 'Residential energy demand modelling in developing regions: the use of multivariate statistical techniques', *Energy Economics*, **14** (1), 57–63.

Aubin, C., D. Fougere, E. Husson and M. Ivaldi (1995), 'Real-time pricing of electricity for residential customers: econometric analysis of an experiment', *Journal of Applied Econometrics*, **10**, S171–S191.

Banfi, S., M. Farsi, M. Filippini and M. Jakob (2008), 'Willingness to pay for energy-saving measures in residential buildings', *Energy Economics*, **30** (2), 503–16.

Benders, R.M., R. Kok, H.C. Moll, G. Wiersma and K.J. Noorman (2006), 'New approaches for household energy conservation: in search of personal household energy budgets and energy reduction options', *Energy Policy*, **34** (18), 3612–22.

van den Bergh, J.C.J.M. (2010), 'Energy conservation more effective with rebound policy', *Environmental and Resource Economics*, http://www.springerlink.com/content/n2n58h744078&2gp/fulltext.pdf.

Berkhout, P.H.G., A. Ferrer-i-Carbonell and J.C. Muskens (2004), 'The ex post impact of an energy tax on household energy demand', *Energy Economics*, **26** (3), 297–317.

Bertoldi, P. and T. Huld (2006), 'Tradable certificates for renewable electricity and energy savings', *Energy Policy*, **34** (2), 212–22.

Bhattacharyya, S.C. (1996), 'Applied general equilibrium models for energy studies: a survey', *Energy Economics*, **18** (3), 145–64.

Bijker, W. (1995), *Of Bicycles, Bakelites and Bulbs: Toward a Theory of Sociotechnical Change*, Cambridge, MA: MIT Press.

Bin, S. and H. Dowlatabadi (2005), 'Consumer lifestyle approach to US energy use and the related CO_2 emissions', *Energy Policy*, **33** (2), 197–208.

Bjertnäs, G.H. and T. Fæhn (2008), 'Energy taxation in a small, open economy: social efficiency gains versus industrial concerns', *Energy Economics*, **30** (4), 2050–71.

Black, J., P. Stern and J. Elsworth (1985), 'Personal and contextual influences on household energy adaptations', *Journal of Applied Psychology*, **70**, 3–21.

Bohi, D.R. and M.B. Zimmermann (1984), 'An update on econometric studies on energy demand behavior', *Annual Review of Energy*, **9**, 105–54.

Boonekamp, P.G.M. (2007), 'Price elasticities, policy measures and actual developments in household energy consumption: a bottom up analysis for the Netherlands', *Energy Economics*, **29** (2), 133–57.

Bor, Y.J. (2008), 'Consistent multi-level energy efficiency indicators and their policy implications', *Energy Economics*, **30**, 2401–19.

Borenstein, S. (2005), 'The long-run efficiency of real-time electricity pricing', *Energy Journal*, **26** (3), 93–116.

Brännlund, R., T. Ghalwash and J. Nordström (2007), 'Increased energy efficiency and the rebound effect: effects on consumption and emissions', *Energy Economics*, **29** (1), 1–17.

Brewer, G.D. and P.C. Stern (2005), *Decision Making for the Environment, Social Behavioral Science Research Priorities*, Washington, DC: National Academies Press.

Brookes, L.G. (1990), 'The greenhouse effect: the fallacies in the energy efficiency solution', *Energy Policy*, **18** (2), 199–201.

Brown, H.E., S. Suryanarayanan and G.T. Heydt (2010), 'Some characteristics of emerging distribution systems considering the smart grid initiative', *Electricity Journal*, **23** (5), 64–75.

Brown, M. (1984), 'Change mechanisms in the diffusion of residential energy conservation', *Technological Forecasting and Social Change*, **25**, 123–38.

Camerer, C.F., G. Loewenstein and M. Rabin (eds) (2008), *Advances in Behavioral Economics*, Princeton, MA, USA and Oxford, UK: Princeton University Press.

Carlsson-Kanyama, A., A.-L. Lindén and B. Ericcson (2005), 'Residential energy behavior: does generation matter?', *International Journal of Consumer Studies*, **29**, 239–52.

Chao, H. (2010), 'Price-responsive demand management for a smart grid world', *Electricity Journal*, **23** (1), 7–20.

Cherfas, J. (1991), 'Skeptics and visionaries examine energy saving', *Science*, **251**, 154–61.

Claxton, J.D., C.D. Anderson, J.R.B. Ritchie and G.H.G. McDougall (eds) (1981), *Consumers and Energy Conservation. International Perspectives on Research and Policy Options*, New York: Praeger.

Conlisk, J. (1996), 'Why bounded rationality?', *Journal of Economic Literature*, **34**, 669–700.

Curtis, F., P. Simpson-Housley and S. Drever (1984), 'Household energy conservation', *Energy Policy*, **12**, 452–56.

Dahl, C. and T. Sterner (1991), 'Analysing gasoline demand elasticities: a survey', *Energy Economics*, **13** (3), 203–10.

Darby, S. (2006), 'Social learning and public policy: lessons from an energy-conscious village', *Energy Policy*, **34** (17), 2929–40.

Darley, J. and J. Beniger (1981), 'Diffusion of energy-conserving innovations', *Journal of Social Issues*, **37**, 150–71.

De Groot, H.L., E.T. Verhoef and P. Nijkamp (2001), 'Energy saving by firms: decision-making, barriers and policies', *Energy Economics*, **23**, 717–40.

Dennis, M., E. Soderstron, W. Koncinski and B. Cavanaugh (1990), 'Effective dissemination of energy-related information: applying social psychology and evaluation research', *American Psychologist*, **45**, 1109–17.

DiClemente, D.F. and D.A. Hantula (2003), 'Applied behavioral economics and consumer choice', *Journal of Economic Psychology*, **24** (5), 589–602.

Dietz, T., G.T. Gardner, J. Gilligan, P.C. Stern and M.P. Vandenbergh (2009), 'Household actions can provide a behavioral wedge to rapidly reduce US carbon emissions', *Proceedings of the National Academy of Sciences of the United States of America*, **106** (44), 18452–6.

Dietz, T., P.C. Stern and G.A. Guagnano (1998), 'Social structural and social psychological bases of environmental concern', *Environment and Behavior*, **30** (4), 450–71.

Dumagan, J.C. and T.D. Mount (1992), 'Measuring the consumer welfare effects of carbon penalties: theory and applications to household energy demand', *Energy Economics*, **14** (2), 82–93.

Edenhofer, O. and C.C. Jaeger (1998), 'Power shifts: the dynamics of energy efficiency', *Energy Economics*, **20** (5–6), 513–37.

Ester, P. (ed.) (1985), *Consumer Behavior and Energy Conservation*, Dordrecht, Netherlands, Boston, MA, USA and Lancaster, UK: Martinus Nijhoff.

Ester, P., G. Gaskell, B. Joerges, C.J. Midden, W.F. van Raaij and T. de Vries (eds) (1984), *Consumer Behavior and Energy Policy*, Amsterdam, The Netherlands and New York, USA: North-Holland.

Faiers, A., M. Cook and C. Neame (2007), 'Towards a contemporary approach for understanding consumer behaviour in the context of domestic energy use', *Energy Policy*, **35** (8), 4381–90.

Faruqui, A., R. Hledik and S. Sergici (2009), 'Piloting the smart grid', *Electricity Journal*, **22** (7), 55–69.

Fehr, E. and K. Schmidt (1999), 'A theory of fairness, competition, and cooperation', *Quarterly Journal of Economics*, **114** (3), 817–68.

Fishbein, M. and I. Ajzen (1975), *Belief, Attitude, Intention, and Behavior: An Introduction to Theory and Research*, Reading, MA: Addison-Wesley.

Foxall, G.R. (1994), 'Behavior analysis and consumer psychology', *Energy Economics*, **15** (1), 5–91.

Frederick, S., G. Lowenstein and T. O'Donoghue (2002), 'Time discounting and time preference: a critical review', *Journal of Economic Literature*, **40** (2), 351–401.

Frei, C., P.-A. Haldi and G. Sarlos (2003), 'Dynamic formulation of a top-down and bottom up merging energy policy model', *Energy Policy*, **31**, 1017–31.

Gatersleben, B., L. Steg and C. Vlek (2002), 'Measurement and determinants of environmentally significant consumer behavior', *Environment and Behavior*, **34** (3), 335–62.

Ghalwash, T. (2007), 'Energy taxes as a signaling device: an empirical analysis of consumer preferences', *Energy Policy*, **35** (1), 29–38.

Gillingham, K., R. Newell and K. Palmer (2006), 'Energy efficiency policies: a retrospective examination', *Annual Review of Environment and Resources*, **31** (1), 161–92.

Gleerup, M., A. Larsen, S. Leth-Petersen and M. Togeby (2010), 'The effect of feedback by sms-text messages and email on household electricity consumption: experimental evidence', *Energy Journal*, **31** (3), 113–32.

Greening, L.A., D.L. Greene and C. Difiglio (2000), 'Energy efficiency and consumption – the rebound effect – a survey', *Energy Policy*, **28** (6–7), 389–401.

Guagnano, G., P. Stern and T. Dietz (1995), 'Influences on attitude–behavior relationships: a natural experiment with curbside recycling', *Environment and Behavior*, **27**, 699–718.

Guerin, D.A., B.L. Yust and J.G. Coopet (2000), 'Occupant predictors of household energy behavior and consumption change as found in energy studies since 1975', *Family and Consumer Sciences Research Journal*, **29** (1), 48–80.

Gundimeda, H. and G. Köhlin (2008), 'Fuel demand elasticities for energy and environmental policies: Indian sample survey evidence', *Energy Economics*, **30** (2), 517–46.

Hledik, R. (2009), 'How green is the smart grid?', *Electricity Journal*, **22** (3), 29–41.

Horne, M., M. Jaccard and K. Tiedemann (2005), 'Improving behavioral realism in hybrid energy–economy models using discrete choice studies of personal transportation decisions', *Energy Economics*, **27** (1), 59–77.

Howarth, R.B. and B. Andersson (1993), 'Market barriers to energy efficiency', *Energy Economics*, **15** (4), 262–72.

Hughes, T. (1983), *Networks of Power: Electrification in Western Society*, Baltimore, PA: Johns Hopkins University Press.

Ibrahim, I.B. and C. Hurst (1990), 'Estimating energy and oil demand functions: a study of thirteen developing countries', *Energy Economics*, **12** (2), 93–102.

Isoard, S. and A. Soria (2001), 'Technical change dynamics: evidence from the emerging renewable energy technologies', *Energy Economics*, **23** (6), 619–36.

Jaffe, A.B. and R.N. Stavins (1994a), 'Energy-efficiency investments and public policy', *Energy Journal*, **15** (2), 43–65.

Jaffe, A.B. and R.N. Stavins (1994b), 'The energy paradox and the diffusion of conservation technology', *Resource and Energy Economics*, **16** (2), 91–122.

Jager, W. (2006), 'Stimulating the diffusion of photovoltaic systems: a behavioural perspective', *Energy Policy*, **34** (14), 1935–43.

Junginger, M., W. van Sark and A. Faaij (eds) (2010), *Technological Learning in the Energy Sector. Lessons for Policy, Industry and Science*, Cheltenham, UK and Northampton, MA, USA: Edward Elgar.

Kahneman, D. and A. Tversky (1979), 'Prospect theory: an analysis of decision under risk', *Econometrica*, **47** (2), 263–91.

Kaiser, F.G. and H. Gutscher (2003), 'The proposition of a general version of the theory of planned behavior: predicting ecological behavior', *Journal of Applied Social Psychology*, **33** (3), 586–603.

Kaiser, F.G., G. Hübner and F.X. Bogner (2005), 'Contrasting the theory of planned behavior with the value–belief–norm model in explaining conservation behavior', *Journal of Applied Social Psychology*, **35** (10), 2150–70.

Keirstead, J. (2006), 'Evaluating the applicability of integrated domestic energy consumption frameworks in the UK', *Energy Policy*, **34** (17), 3065–77.

Kempton, W. and M. Neiman (eds) (1987), *Energy Efficiency: Perspectives on Individual Behavior*, Washington, DC and Berkeley, CA: American Council for an Energy-Efficient Economy.

Khazzoom, J.D. (1980), 'Economic implications of mandated efficiency in standards for household appliances', *Energy Journal*, **1** (4), 21–40.

Kowalski, K., S. Stagl, R. Madlener and I. Omann (2009), 'Sustainable energy futures: methodological challenges in combining scenarios and participatory multi-criteria analysis', *European Journal of Operational Research*, **197**, 1063–74.

Krarup, S. and C.S. Russell (eds) (2005), *Environment, Information and Consumer Behavior. New Horizons in Environmental Economics*, Cheltenham, UK and Northampton, MA, USA: Edward Elgar.

Laitner, J., S. DeCanio, J. Koomey and A. Sanstad (2003), 'Room for improvement: increasing the value of energy modeling for policy analysis', *Utilities Policy*, **11**, 87–94.

Lijesen, M.G. (2007), 'The real-time price elasticity of electricity', *Energy Economics*, **29** (2), 249–58.

Lindén, A.-L., A. Carlsson-Kanyama and B. Eriksson (2006), 'Efficient and inefficient aspects of residential energy behaviour: what are the policy instruments for change?', *Energy Policy*, **34** (14), 1918–27.

Lindsley, O.R. (1962), 'A behavioral measure of television viewing', *Journal of Advertising Research*, **2**, 2–12.

Long, J.E. (1993), 'An econometric analysis of residential expenditures on energy conservation and renewable energy sources', *Energy Economics*, **15** (4), 232–8.

Longo, A., A. Markandya and M. Petrucci (2008), 'The internalization of externalities in the production of electricity: willingness to pay for the attributes of a policy for renewable energy', *Ecological Economics*, **67** (1), 140–52.

Lutzenhiser, L. (1993), 'Social and behavioral aspects of energy use', *Annual Review of Energy and the Environment*, **18**, 247–89.

Lutzenhiser, L. (1994), 'Sociology, energy and interdisciplinary environmental science', *American Sociologist*, **25** (1), 58–79.

Madlener, R. (1996), 'Econometric analysis of residential energy demand: a survey', *Journal of Energy Literature*, **2** (2), 3–32.

Madlener, R. and J. Artho (2005), 'Technology adoption as a multi-stage process: beliefs and perceived merits of heating systems among cooperative building societies', in *Proceedings of the 14th European Biomass Conference, Biomass for Energy, Industry, and Climate Protection*, 17–21 October, Paris.

Madlener, R., J. Liu, A. Monti, C. Muskas and C. Rosen (2009), 'Metering and measurement facilities as enabling technologies for smart electricity grids in Europe', Sectoral e-Business Watch Special Report 1/2009, E.ON Energy Research Center, RWTH Aachen University, Germany, on behalf of the European Commission Directorate General Enterprise and Industry, November.

Mahapatra, K., L. Gustavsson and R. Madlener (2007), 'Bioenergy innovations: the case of wood pellet systems in Sweden', *Technology Analysis and Strategic Management*, **19** (1), 99–125.

Maréchal, K. (2010), 'Not irrational but habitual: the importance of "behavioural lock-in" in energy consumption', *Ecological Economics*, **69** (5), 1104–14.

McCalley, L.T. and C.J.H. Midden (2002), 'Energy conservation through product-integrated feedback: the roles of goal-setting and social orientation', *Journal of Economic Psychology*, **23** (5), 589–603.

McMakin, A.H., E.L. Malone and R.E. Lundgren (2002), 'Motivating residents to conserve energy without financial incentives', *Environment and Behavior*, **34** (6), 848–64.

Menges, R., C. Schroeder and S. Traub (2005), 'Altruism, warm glow and the willingness-to-donate for green electricity: an artefactual field experiment', *Environmental and Resource Economics*, **31**, 431–58.

Michelsen, C. and R. Madlener (2010), 'Integrated theoretical framework for a homeowner's decision in favor of an innovative residential heating system', FCN Working Paper No. 2/2010, Institute for Future Energy Consumer Needs and Behavior (FCN), RWTH Aachen University, Germany, February.

Moxnes, E. (2004), 'Estimating customer utility of energy efficiency standards for refrigerators', *Journal of Economic Psychology*, **25** (6), 707–24.

Munksgaard, J., K.A. Pedersen and M. Wien (2000), 'Impact of household consumption on CO_2 emissions', *Energy Economics*, **22** (4), 423–40.

Munns, D. (2008), 'Modeling new approaches for electric energy efficiency', *Electricity Journal*, **21** (2), 20–26.

Neenan, B. and R.C. Hemphill (2008), 'Societal benefits of smart metering investments', *Electricity Journal*, **21** (8), 32–45.

Nesbakken, R. (1999), 'Price sensitivity of residential energy consumption in Norway', *Energy Economics*, **21** (6), 493–515.

Newell, R.G. and W.A. Pizer (2008), 'Carbon mitigation costs for the commercial building sector: discrete– continuous choice analysis of multifuel energy demand', *Resource and Energy Economics*, **30** (4), 527–39.

Nicol, C.J. (2003), 'Elasticities of demand for gasoline in Canada and the United States', *Energy Economics*, **25** (2), 201–14.

Nyborg, K., R.B. Howarth and K.A. Brekke (2006), 'Green consumers and public policy: on socially contingent moral motivation', *Resource and Energy Economics*, **28** (4), 351–66.

Oikonomou, V., F. Becchis, L. Steg and D. Russolillo (2009), 'Energy saving and energy efficiency concepts for policy making', *Energy Policy*, **37**, 4787–96.

Owen, G. (1999), *Public Purpose or Private Benefit? The Politics of Energy Conservation, Issues in Environmental Politics*, Manchester: Manchester University Press.

Owens, S. and L. Driffill (2008), 'How to change attitudes and behaviours in the context of energy', *Energy Policy*, **36** (12), 4411–18.

Pindyck, R.S. (1979), 'International comparison of the residential demand for energy', *European Economic Review*, **13**, 1–24.

Pindyck, R.S. (1980), *The Structure of World Energy Demand*, Cambridge, MA and London, UK: MIT Press.

Pohekar, S. and M. Ramachandran (2004), 'Application of multi-criteria decision making to sustainable energy planning: a review', *Renewable and Sustainable Energy Reviews*, **8**, 365–81.

Poortinga, W., L. Steg and C. Vlek (2004), 'Values, environmental concern, and environmental behavior: a study into household energy use', *Environment and Behavior*, **36** (1), 70–93.

Poortinga, W., L. Steg, C. Vlek and G. Wiersma (2003), 'Household preferences for energy-saving measures: a conjoint analysis', *Journal of Economic Psychology*, **24** (1), 49–64.

Potoski, M. and A. Prakash (2005), 'Green clubs and voluntary governance: ISO 14001 and firms' regulatory compliance', *American Journal of Political Science*, **49** (2), 235–48.

Poyer, D.A. and M. Williams (1993), 'Residential energy demand: additional empirical evidence by minority household type', *Energy Economics*, **15** (2), 93–100.

Puller, S.L. and L.A. Greening (1999), 'Household adjustment to gasoline price change: an analysis using 9 years of US survey data', *Energy Economics*, **21** (1), 37–52.

Reinders, A.H.M.E., K. Vringer and K. Blok (2003), 'The direct and indirect energy requirement of households in the European Union', *Energy Policy*, **31** (2), 139–53.

Reiss, P.C. and M.W. White (2005), 'Household electricity demand revisited', *Review of Economic Studies*, **27**, 853–83.

Rip, A. and R. Kemp (1998), 'Technological change', in S. Rayner and E. Malone (eds), *Human Choice and Climate Change*, Vol. 2, Columbus, OH: Battelle Press, pp. 327 99.

Rivers, N. and M. Jaccard (2005), 'Combining top-down and bottom-up approaches to energy–economy modeling using discrete choice methods', *Energy Journal*, **26** (1), 83–106.

Rogers, E.M. (2003), *Diffusion of Innovations*, 5th edn, New York: Free Press.

Sammer, K. and R. Wüstenhagen (2006), 'The influence of eco-labelling on consumer behavior results of a discrete choice analysis for washing machines', *Business Strategy and the Environment*, **15** (3), 185–99.

Saunders, H.D. (1992), 'The Khazzoom–Brookes postulate and neoclassical growth', *Energy Journal*, **13** (4), 131–48.

Scarpa, R. and K. Willis (2010), 'Willingness-to-pay for renewable energy: primary and discretionary choice of British households for micro-generation technologies', *Energy Economics*, **32** (1), 129–36.

Scherbaum, C.A., P.M. Popovich and S. Finlinson (2008), 'Exploring individual-level factors related to employee energy-conservation behaviors at work', *Journal of Applied Social Psychology*, **38** (3), 818–35.

Schipper, L., F. Unander, S. Murtishaw and M. Ting (2001), 'Indicators of energy use and carbon emission: explaining the energy economy link', *Annual Review of Energy and Environment*, **26**, 49–81.

Schleich, J. and E. Gruber (2008), 'Beyond case studies: barriers to energy efficiency in commerce and the services sector', *Energy Economics*, **30** (2), 449–64.

Scott (1997), 'Household energy efficiency in Ireland: a replication study of owners of energy saving items', *Energy Economics*, **19**, 187–208.

Serletis, A. and A. Shahmoradi (2008), 'Semi-nonparametric estimates of interfuel substitution in US energy demand', *Energy Economics*, **30** (5), 2123–33.

Shove, E. (1998), 'Gaps, barriers and conceptual chasms: theories of technology transfer and energy in buildings', *Energy Policy*, **26** (15), 1105–12.

Shove, E. (2003), *Comfort, Cleanliness, and Convenience: The Social Organisation of Normality*, Oxford: Berg Publishers.

Simon, H.A. (1955), 'A behavioral model of rational choice', *Quarterly Journal of Economics*, **69**, 99–118.

Sorrell, S. and J. Dimitropoulos (2008), 'The rebound effect: microeconomic definitions, limitations and extensions', *Ecological Economics*, **65**, 636–49.

Sorrell, S., J. Dimitropoulos and M. Sommerville (2009), 'Empirical estimates of the direct rebound effect: a review', *Energy Policy*, **37** (4), 1356–71.

Sorrell, S., E. O'Malley, J. Schleich and S. Scott (eds) (2004), *The Economics of Energy Efficiency: Barriers to Cost-Effective Investment*, Cheltenham, UK and Northampton, MA, USA: Edward Elgar.

Stern, P.C. (1992), 'What psychology knows about energy conservation', *American Psychologist*, **47** (10), 1224–32.

Stern, C.P. (2000), 'Toward a coherent theory of environmentally significant behavior', *Journal of Social Issues*, **56**, 407–24.

Stern, P.C. and E. Aronson (eds) (1984), *Energy Use: The Human Dimension*, Washington, DC: National Academic Press.

Stern, P., E. Aronson, J. Darley, D. Hill, E. Hirst, W. Kempton and T.J. Wilbanks (1986), 'The effectiveness of incentives for residential energy conservation', *Evaluation Review*, **10**, 147–76.

Stoneman, P. (2001), *The Economics of Technological Diffusion*, Oxford: Blackwell Publishers.

Sun, D.-Q., J. Zheng, T. Zhang, Z.-J. Zhang, H.-T. Liu, F. Zhao and Z.-J. Qiu (2010), 'The utilization and development strategies of smart grid and new energy', *Proceedings of Asia-Pacific Power and Energy Engineering Conference (APPEEC) 2010*. pp. 1–4.

Sun, J.W. (1998), 'Changes in energy consumption and energy intensity: a complete decomposition model', *Energy Economics*, **20** (1), 85–100.

Sutherland, R.J. (1996), 'The economics of energy conservation policy', *Energy Policy*, **24**, 361–70.

Thaler, R.H. (1985), 'Mental accounting and consumer choice', *Marketing Science*, **4** (3), 199–214.

Train, K. and G. Mehrez (1994), 'Optional time-of-use prices for electricity: econometric analysis of surplus and pareto impacts', *RAND Journal of Economics*, **25** (2), 263–83.

Vaage, K. (2000), 'Heating technology and energy use: a discrete/continuous choice approach to Norwegian household energy demand', *Energy Economics*, **22** (6), 649–66.

Wade, N., P. Taylor, P. Lang and P. Jones (2010), 'Evaluating the benefits of an electrical energy storage system in a future smart grid', *Energy Policy*, **38** (11), 718–88.

Walsh, M. (1989), 'Energy tax credits and housing improvement', *Energy Economics*, **11**, 275–84.

Watson, J.B. (1908), 'The behavior of noddy and sooty terns', Carnegie Institute Publication 103, 197–225.

Watson, J.B. and R. Rayner (1920), 'Conditioned emotional reactions', *Journal of Experimental Psychology*, **3**, 1–14.

Weber, C. and A. Perrels (2000), 'Modelling lifestyle effects on energy demand and related emissions', *Energy Policy*, **28** (8), 549–66.

Wilhite, H., E. Shove, L. Lutzenhiser and W. Kempton (2000), 'The legacy of twenty years of energy demand management: we know more about individual behavior but next to nothing about demand', in E. Jochem, J. Sathaye and D. Bouille (eds), *Society, Behaviour, and Climate Change Mitigation*, Dordrecht: Kluwer Academic Publishers, pp. 109–26.

Wilkinson, N. (2008), *An Introduction to Behavioral Economics*, New York: Palgrave- Macmillan.

Wilson, C. and H. Dowlatabadi (2007), 'Models of decision making and residential energy use', *Annual Review of Environment and Resources*, **32**, 169–203.

Woo, C.K., E. Kollman, R. Orans, S. Price and B. Horii (2008), 'Now that California has AMI, what can the state do with it?', *Energy Policy*, **36** (4), 1366–74.

PART III

RENEWABLE ENERGY AND ENERGY EFFICIENCY

11 Multicriteria diversity analysis: theory, method and an illustrative application
Go Yoshizawa, Andy Stirling and Tatsujiro Suzuki

11.1 CONTEXTS FOR ANALYSING ENERGY DIVERSITY

Few themes have been more consistently prominent over successive cycles in the energy policies of so many different countries than 'diversity' (IEA, 1985; CEC, 1990, 2007; Verrastro and Ladislaw, 2007; Bazilian and Roques, 2008). The concept is characterized differently for contrasting purposes under contending perspectives in varying circumstances. Surprisingly often, it remains entirely undefined in high-level policy-making. Being curiously underexamined in analysis, it is particularly vulnerable to special pleading. Yet the substantive rationales for interest in diversity remain remarkably deep and broad. The aim of this chapter is to explore these challenges and identify a systematic, comprehensive and transparent framework through which to address them.

To this end, the chapter will begin by examining the variety of contexts and approaches for the analysis of energy diversity. Attention will then turn to the definition of some underlying common elements. Criteria will be developed for the rigorous aggregation, accommodation and articulation of these multiple dimensions. It is on this basis that a novel heuristic framework will be proposed for exploring different perspectives. A multicriteria diversity analysis method will be outlined, and illustrated using a schematic empirical example of direct relevance to current practical policy-making on energy strategies.

As a starting point, we should begin by clarifying the focus. Despite the complexities, ambiguities and expediencies, international policy discussions of energy diversity are all in various ways about the pursuit of an evenly balanced reliance on a variety of mutually disparate options. It follows from this that there lies a crucial difference between diversity and other key themes in energy policy. Unlike many aspects of financial, economic, environmental or security-of-supply performance, diversity is an inherent and irreducible feature of an energy system taken as a whole (Stirling, 1994a). For reasons that will be discussed later, evenness of 'balance', the scale of mutual 'disparities' and even the partitioning of 'variety' are all holistic system-level properties. In other words, diversity cannot be reduced to simple aggregates of the attributes of individual technological, resource or institutional 'options' within a given energy system.

The principal reasons for an interest in energy diversity have traditionally lain in concerns over security of energy supplies (for example CEC 1990, 2007; IEA, 2007; NERA, 2002; PMSU, 2002; DTI, 2003a). Indeed, this is often treated as if it were the exclusive rationale for deliberate diversification (Spicer, 1987; Parkinson, 1989; PIU, 2001; DTI, 2006). Here, diversity is seen as a means simultaneously to help prevent disruptions to energy supply or mitigate their effects should they occur (IEA, 1985). Attention tends to

Leabharlanna Poiblí Chathair Bhaile Átha Cliath

Dublin City Public Libraries

focus on what are held to be relatively well-known sources of disruption, like fuel price fluctuations, constraints on the availability of specific primary resources or a restricted number of clearly identified threats (Lucas et al., 1995). However, to focus exclusively on these relatively readily characterized parameters in some ways circumscribes the real value of diversification (Stirling, 1995). As distinct from a range of more specific and targeted preventive and mitigative strategies, diversity remains effective (at least in part) even if the sources or modalities of the prospective disruptions are effectively unknown. By maintaining an evenly balanced variety of mutually disparate options, we may hope to resist impacts on any subset of these, even if we do not know in advance what these impacts might be (Stirling, 1996).

The essential quality of diversity thus lies simply in 'putting eggs in different baskets'. Although the value of this strategy rests on a general variety of 'baskets', it applies irrespective of the particularities of any individual basket. This has profound implications for analytical methodology. There exist a host of specific, structured, targeted techniques for addressing well-determined threats – and corresponding attributes of individual options. These include sophisticated probabilistic tools like risk assessment, Monte Carlo analysis, Bayesian reasoning and portfolio theory, which offer powerful responses under conditions where both specific outcomes and their respective probabilities may each be determined with confidence (Stirling, 2003). Yet, for this same fundamental reason, such techniques offer a poor general basis for thinking about diversity. This is because the particular value of diversification lies in providing a robust response to the most intractable forms of uncertainty, ambiguity and ignorance where these probabilistic tools are, by definition, not applicable (Stirling, 1999). To paraphrase a notorious recent remark in a field not unrelated to energy policy (Rumsfeld, 2002), a key rationale for diversification is that it is what we can do 'when we don't know what we don't know'.

Depending on judgements over the relative priorities to ascribe to different sources or modalities of uncertainty, there thus arise many possible dimensions of energy diversity (Stirling, 1994a). These represent multiple parameterizations of the salient 'mutual disparities', which distinguish different options. Permutations in these disparities may variously involve: conversion and end-use technologies; types of primary resources and energy carriers; regions of origin and transport routes; facility operators and infrastructure dispositions; resource suppliers and traders; equipment vendors and component manufacturers; labour unions and professional associations; and various kinds of environmental, health or social effects (and associated regulatory exposures) (Stirling, 1996). Demand-side contexts present many further neglected aspects of energy diversity. All these factors – and others – are of relevance to the reasons for strategic interest in diversity (Farrell et al., 2004). To analyse energy diversity in more constrained ways – for instance as a simple function of conventional categories of primary fuels or fuel price variability – is to risk missing a crucial part of its strategic value (Awerbuch et al., 2006). This will be returned to in more detail later.

This said, it is important to acknowledge that despite the importance of diversity to debates over energy security, there exist many other dimensions of supply security that extend beyond the issue of diversity alone. Internationally ubiquitous aims around achieving 'availability of energy at all times in various forms in sufficient quantities and at affordable prices' (Umbach, 2004) can be pursued by many different strategies.

A number of these are entirely distinct from – and sometimes even in tension with – diversification. For instance, reliance on indigenous resources has been advocated as an energy security strategy, even if this reduces diversity (IEA, 1980). By contrast, efficient functioning of energy markets is often highlighted as a means to achieve 'optimal' levels of energy security, without the need for extraneous policy interventions to foster diversity (Helm, 2002, 2007).

Despite this, it is well established that (even under agendas of liberalization) states do continue to seek to structure more 'secure' energy markets – for instance through promotion of economic interdependence on the part of supplier interests (European Energy Charter Secretariat, 2004) or more effective international planning of responses to disruption (Adelman, 1995). In this way, strategic stockpiles are often crucial (Greene et al., 1998), as are efforts to exercise greater control over energy supply chains (Lawson, 1992). Irrespective of their diversity, a security premium is often paid for more flexible supply options (Costello, 2004); redundant infrastructures (Farrell et al., 2004) or demand-side efficiency programmes (Lovins and Lovins, 1982). Finally, although less openly acknowledged, there is the potential for violent action – both as resisted in 'resilient' domestic energy infrastructures (JESS, 2004) and as perpetrated in offensive military interventions against perceived energy security threats (Plummer, 1983).

Diversity is thus only one – albeit prominent – factor among a wide range of different dimensions of energy security. However, it is important to note that references to diversity also feature prominently in a number of parallel policy debates (Stirling, 1998). Some of these are highly relevant to the energy sector, and should therefore also be set alongside historic preoccupations with energy security as reasons for interest in energy diversity. As shown in Figure 11.1, the relationship between energy diversity and security is therefore somewhat more complex than is often assumed.

In considering claims over the multiple benefits of diversity, it must be remembered that, in any single context, diversity rarely offers a 'free lunch' (Weitzman, 1992). Those options that are marginal in any given energy system are often in this position for reasons of poor performance. Contingent differences in resource endowments or institutional environments thus condition different patterns of emergent diversity in different geographical, socio-economic or cultural contexts. To enlarge the contributions of marginal options in any given context will thus, under any given perspective, often incur some performance penalty (David and Rothwell, 1996). In addition, there are typically further trade-offs between diversity and transaction costs (Williamson, 1993) and with foregone benefits like coherence (Cohendet et al., 1992), accountability (Grabher and Stark, 1997), standardization (Cowan, 1991) and economies of scale (Matthews and McGowan, 1992). Diversification may retard enhanced learning about incumbent technologies in favour of learning about more marginal options (Jacobsson and Johnson, 2000).

The crucial challenge thus lies in striking a balance between the benefits of diversity and these countervailing aspects of portfolio performance (Geroski, 1989). The value of the 'diversity premium' (Ulph, 1988, 1989) that is warranted in any particular energy mix may be appraised under a variety of strategic criteria, including financial, economic, environmental, health or broader social impacts (Stirling, 1994b). Indeed, the situation may even arise where trade-offs are made between the general security benefits of enhanced energy diversity and more particular security penalties associated with specific

security of supply is more than diversity ...

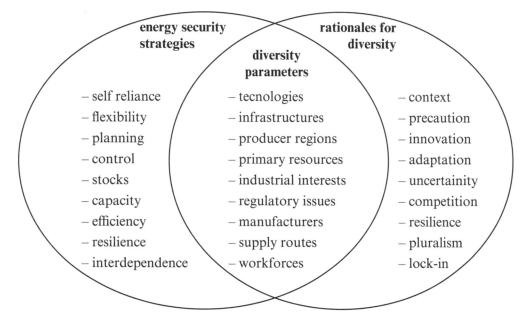

energy security strategies

- self reliance
- flexibility
- planning
- control
- stocks
- capacity
- efficiency
- resilience
- interdependence

diversity parameters

- tecnologies
- infrastructures
- producer regions
- primary resources
- industrial interests
- regulatory issues
- manufacturers
- supply routes
- workforces

rationales for diversity

- context
- precaution
- innovation
- adaptation
- uncertainity
- competition
- resilience
- pluralism
- lock-in

... diversity is more than security of supply

Figure 11.1   Schematic summary of relationships between energy diversity and security

options through which diversification is achieved (IEA, 1980). In the end, all such issues are judgemental, offering ample scope for legitimate disagreement. A crucial challenge for policy analysis of energy diversity is therefore to examine these multidimensional interactions and trade-offs.

Thus qualified, it remains to substantiate the potential benefits of energy diversity identified in Figure 11.1, beyond the traditional preoccupations with preventing or mitigating service disruptions. The first kinds of benefit arise quite naturally from the rationale conventionally discussed with respect to supply security. Putting 'eggs in different baskets' hedges against 'surprises' (Brooks, 1986), including operational, environmental, economic or wider strategic uncertainties that are not directly relevant to supply security (Rosenberg, 1996). This underscores increasing associations between diversity, sustainability and precaution in energy strategies (Stirling, 1999, 2008b; Bird, 2007; Helm, 2002, 2007; IEA, 2007). Second, it is true in the energy sector, as elsewhere, that reducing concentration in technology, service or commodity markets is an important means to promote competition (Aoki, 1996). Indeed, competitive diversity in energy markets is often presented as significant to the competitiveness of the wider economy (DTI, 2003a, 2006). Third, a growing literature shows how general technical, institutional and functional heterogeneity can help foster more effective innovation (Rosenberg, 1982; Landau et al., 1996; Grabher and Stark, 1997; Kauffman, 1995). This is of crucial interest equally where policy attention focuses on moves towards 'the knowledge society' (CEC, 2005), or in driving specific radical transformations like those involved in environmental innovation (Kaijser et al., 1991) and transitions to sustainable energy (Bird, 2007; Patterson, 1999; Mitchell, 2007; Grubb et al., 2006; Stirling, 2008b). Indeed, with moves towards sustainability intrinsically requiring enhanced context-sensitivity (Brundtland et al., 1987), there arises a further, fourth, distinct rationale for diversity in better tailoring energy systems to diverse local cultural, ecological, and geopolitical and geophysical conditions (Landau et al., 1996).

This leads to a fifth and final important role for energy diversity. By sustaining a plurality of disparate techno-institutional strategies, we may better accommodate otherwise irreconcilable socio-economic interests (Stirling, 1997). Energy policy has long served as a site for deeply entrenched and protracted general cultural, institutional and political conflicts around technology choice (Scrase and MacKerron, 2009). Issues around nuclear power provide an iconic, but not exclusive, example (Elliott, 2007). Far from being attenuated by growing consensus over the general imperatives for transitions to 'sustainable energy', the role of the energy sector as an arena for public contention may actually be amplified by the many questions that follow from this (Smith et al., 2005). What really counts as 'sustainable'? Which strategies are most viable? Far from diminishing the resulting dilemmas of social choice, the scale and urgency of the proposed transformations and associated public policy interventions actually compounds and renders more acute the essential political challenges. What roles are to be played, for instance, by alternative fluid fuels, carbon capture and storage, 'new generation' nuclear power, centralized renewable energy, distributed intelligent networks or novel 'energy service' institutions? Even within the disparate category of 'renewable energy' there lie a host of strategic choices. All the above options are variously seen as feasible or desirable routes to 'sustainable energy' under at least some influential perspective. Yet we cannot equally pursue all to their full potential (Stirling, 2009).

This crucial role for deliberate social choice is seriously downplayed by current official and incumbent discourses in energy policy. Senior figures routinely understate the scope for agency, for instance promoting nuclear strategies for the paradoxically contradictory reasons that there is 'no alternative' (King, 2006) or that policy-making should 'do everything' (King, 2007). In fact, not only is contemporary energy policy dominated by dilemmas of choice, but the stakes are rendered even higher and more urgent by the fact that each possible sustainable energy pathway displays dynamics of 'increasing returns' (Arthur, 1989, 1994). Early patterns in economic investment (Cowan, 1991), learning-by-doing (Jacobsson and Johnson, 2000), institutional momentum (Hughes, 1983), political commitment (Walker, 2000) or cultural expectations (Brown and Michael, 2003) may strongly condition the prospects of success in any one of a number of equally viable paths. With the implications of contending values and interests thus accentuated, it becomes more imperative to make these social choices in a deliberate and accountable fashion – without detracting from the efficacy of moves towards sustainability.

The way to achieve this is to pursue a judicious diversity of pathways. These will prioritize only some of the array of possible trajectories (Stirling, 2009). Just as no single option is unique in offering diversity, so none is individually imperative. Diverse strategies towards sustainability do not therefore mean 'doing everything' in a blanket fashion, but involve pursuit of a 'requisite variety' (Ashby, 1956) among possible pathways. Arguments that any single option is rendered inevitable by the general desirability of diversity are thus just as spuriously expedient as the claim that there is 'no alternative'.

Given this remarkable conjunction of reasons for interest, it should be no surprise that the concept of diversity is as prominent as it is in energy policy. There clearly exists a challenging imperative to understand the various synergies, complexities, constraints and trade-offs. Yet – despite notable exceptions (DTI, 1995; Jansen et al., 2004; IEA, 2007) – it is curious that systematic appraisal of diversity occupies only a relatively minor niche in official energy policy analysis. When compared, for instance, with entire subdisciplines and literatures formed around analysis of energy-specific financial evaluation, external costs assessment, risk analysis and environmental appraisal, rigorous attention to diversity itself remains relatively neglected. This may be understood, perhaps, partly as a reflection of political pressures for expedient exploitation of the platitudinous and tautologous connotations of diversity (Matthews and McGowan, 1990). For instance, past high-profile UK government diversity rhetoric has later been acknowledged by the ministers involved to have been 'code' for the less widely supported aims of promoting nuclear power and neutralizing organized labour (Lawson, 1992). Under this kind of political dynamic, systematic policy analysis is actually inconvenient to potential sponsors. Despite the many substantive reasons for interest in diversity, then, such overbearing instrumentalism may serve to discourage serious academic attention.

None of these political factors diminish the pressing underlying need for more comprehensive, rigorous and transparent policy analysis of energy diversity. Consequently, despite any inhibitions, many important and honourable such initiatives do, of course, exist (Ulph, 1988, 1989; Stirling, 1994a, 1994b, 1996; NERA, 1995; ERM, 1995; Feldman, 1998; Awerbuch and Berger, 2003; Jansen et al., 2004, 2007; Markandya et al., 2005; Awerbuch and Yang, 2007; Grubb et al., 2006; Hubberke, 2007; IEA, 2007; Bazilian and Roques, 2008; Yoshizawa et al., 2008). These will be discussed in the next section. As with any analysis in such a complex, dynamic, uncertain and contested field, however, different studies typically yield highly variable outcomes, under entirely reasonable divergences in input assumptions. This is often downplayed as a pathology, with individual studies and approaches remaining relatively silent on the possible sources of volatility. Yet when seen from the general view of policy appraisal, such plurality can in some ways actually be a positive general quality. Collectively, the concurrence of diverse analytical frameworks provides a useful means both to enrich and to qualify what might otherwise tend to be myopic, blinkered or manipulative institutional, political or economic interests.

Approached in a mature, transparent and plural fashion, then, openness in policy-making to a multiplicity of valid appraisals of energy diversity can help calibrate apparent risks, identify boundary conditions, reveal sensitivities to particular assumptions, values or prejudices and so triangulate prescriptive conclusions (Stirling, 2008a). Handled in the right way, this kind of more pluralistic policy discourse can enhance the robustness and accountability of high-level decision-making over contending possible energy strategies. What is needed is a framework under which the contending approaches may be articulated, such as more transparently to reveal their specific idiosyncrasies, conditionalities and possible sources of bias. This will be the topic of the following sections (based on Stirling, 2008b).

11.2   GENERAL PROPERTIES OF ENERGY DIVERSITY: VARIETY, BALANCE AND DISPARITY

Energy diversity was defined at the outset of this paper as an evenly balanced reliance on a variety of mutually disparate options. As such, diversity is a property of any system whose elements may be apportioned into categories. Energy systems are simply one example of this. Disciplines like ecology, economics, taxonomy, palaeontology, complexity and information theory have all developed sophisticated frameworks for analysing various aspects in contrasting contexts (Stirling, 1998). The analysis of energy diversity may therefore gain much through building on the approaches developed in other fields. This is all the more the case because the parameters of interest are – as we have seen – so wide-ranging within energy policy itself. This breadth in the policy salience of energy diversity presents an inherent advantage for the most generally applicable frameworks.

To take the electricity sector as an example, discussions of diversity span an array of disparate supply- and demand-side technologies and primary resources. The scope of diversity analysis is, as we have seen, further extended by a variety of other relevant factors, including: the regional sourcing of fuel and associated supply routes; concentration among trading, supplier or service companies; reliance on generic equipment or component vendors; dependencies on monopoly utilities, shareowners or labour unions; and the configurations and spatial distribution of infrastructures (PIU, 2001; Farrell et al., 2004; Verrastro and Ladislaw, 2007; Helm, 2007; CEC, 2007). Each of these parameters is potentially relevant to diversity as a means to hedge ignorance, foster innovation, mitigate lock-in or accommodate plural values and interests, in the broader senses also discussed earlier. It is therefore desirable that any framework for the analysis of energy diversity be equally applicable in principle across all these aspects.

Fortunately, it is precisely when approached in this most general fashion (as a fundamental property of any system apportioned among elements), that experience in other disciplines holds the clearest lessons for analysing energy diversity. In short, diversity concepts employed across the full range of sciences mentioned above display some combination of the three basic properties included in our present definition: 'variety', 'balance' and 'disparity' (Stirling, 1994a; Grubb et al., 2006). Each is a necessary but individually insufficient property of diversity (Stirling, 1998). Though addressed in different vocabularies, each is applicable across a range of contexts. Each is aggregated in various permutations and degrees in quantitative indices (Hill, 1973). Despite the multiple disciplines and divergent empirical details, there seems no other obvious candidate for a fourth important general property of diversity beyond these three (Stirling, 1998: 47). They are summarized schematically in Figure 11.2.

In terms of electricity supply portfolios, variety is the number of diverse categories of 'option' into which an energy system may be apportioned. It is the answer to the question: 'How many options do we have?' This aspect of diversity is highlighted (for instance) in conventional approaches based on the simple counting of named categories like coal, gas, nuclear and renewable energy. All else being equal, the greater the variety of distinct types of energy option, the greater the system diversity. For instance, in 1990, standard Organisation for Economic Co-operation and Development (OECD) statistics partitioned national member state electricity supply systems among six options: coal, oil,

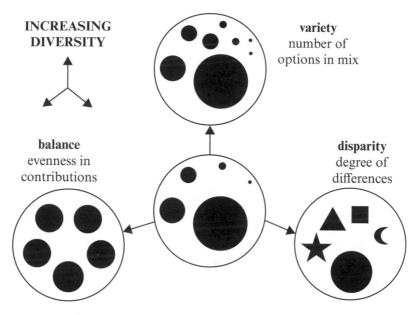

INCREASING
DIVERSITY

variety
number of
options in mix

balance
evenness in
contributions

disparity
degree of
differences

Figure 11.2   A schematic picture of three co-constituting properties of energy diversity

gas, nuclear, hydro/geothermal and 'other' (IEA, 1991). For more specific purposes in 2001, the resolution of reporting increased to 11 options: coal, oil, gas, nuclear, hydro, geothermal, solar, tide/wave/ocean, wind, combustion renewables and waste and 'other' (for example fuel cells) (IEA, 2002). Each scheme provides a different basis for counting variety.

Balance is a function of the apportionment of the energy system across the identified options. It is the answer to the question: 'How much do we rely on each option?' The denominator here may (depending on the context) be expressed as energy inputs or outputs, power capacity, economic value or services delivered. Either way, balance (like statistical variance; Pielou, 1977), is simply represented by a set of positive fractions, which sum to one (Laxton, 1978). This dimension appears most frequently in energy debates, in discussions around a possible role for designating the contributions of different supply options (Helm, 2007). It is captured in more detail in a variety of 'concentration' or 'entropy' measures that are nowadays quite widely applied in energy policy, like the Shannon–Wiener (Stirling, 1994a, 1994b; NERA, 1995; DTI, 1995, 2003b; Markandya et al., 2005; Scheepers et al., 2007) and/or (Grubb et al., 2006) Herfindahl-Hirschman (IEA, 2007; Hubberke, 2007) indices. All else being equal, the more even is the balance across energy options, the greater the system diversity. An example of the importance of considering balance lies in the contrasting stories of Japanese and French electricity supply systems following the 1973 'oil shock'. Over the 27-year period up to 2000, both Japan and France moved away from oil-dominated systems with nuclear at the margin (2 per cent of delivered electricity in Japan, 8 per cent in France). In the Japanese case, the diversification strategy led to a roughly even balance across nuclear, coal and gas as modal options (Suzuki, 2001). In France, however, the 'diversification' strategy simply substituted an initial 40 per cent dependence on oil for an even greater 77 per cent dependence on nuclear at the end (IEA, 2002), involving an eventual decrease in diversity over this period (Stirling, 1994a).

Disparity refers to the manner and degree in which energy options may be distinguished (Runnegar, 1987). It is the answer to the question: 'How different are our options from each other?' This is the most fundamental – and yet most frequently neglected – aspect of energy diversity. After all, it is judgements over disparity that necessarily govern the resolving of the categories of energy option, which underlie characterizations of variety and balance. It is this aspect of diversity that is addressed by applications of portfolio theory, which (deliberately or inadvertently) take fossil fuel price covariance as a stochastic proxy for wider disparities, such as those mentioned above (Ulph, 1988, 1989; NERA, 1995; ERM, 1995; Awerbuch and Berger, 2003; Awerbuch et al., 2006; Awerbuch and Yang, 2007). This method will be returned to below. Alternative approaches to this property in other disciplines are usually based on some more general form of scalar distance measure. Either way, all else being equal, the more disparate are the energy options, the greater the system diversity. In other words, an electricity supply system divided equally among gas, nuclear, wind and biomass power is more disparate than one divided equally among coal, oil and Norwegian and Russian gas.

The consequence of this threefold understanding of diversity is recognition that, though disparity is fundamental, each property helps constitute the other two (Stirling, 1998). This in turn highlights difficulties with diversity concepts and associated indices that focus exclusively on subsets of these properties (Eldredge, 1992), an illustrative selection of which are displayed in Table 11.1. The resulting ambiguities or hidden assumptions can exacerbate the tendency already noted for insufficiently rigorous treatments of diversity to serve as a vehicle for special pleading.

Variety and balance, for instance, cannot be characterized without first partitioning the system on the basis of disparity (May, 1990). An electricity system may be assigned a nominal variety of four, if it is divided into categories labelled coal, gas, nuclear and renewable energy (Stirling, 1994a). Yet 'renewables' might readily be further divided into numerous other nested categories (like hydro, wind, biomass and tide). The mutual disparities between many of these ostensibly subordinate taxa might reasonably be thought greater than those between large centralized thermal nuclear and fossil fuel plant.

For similar reasons, considerations of disparity also hold crucial importance for the resolving of balance. These hinge on the simple fact that the structures of proportional contributions are – like counting the categories themselves – determined by the ways and degrees in which options are divided up. Though the more complex quantitative form may confer an apparent authority, an index of balance (like Shannon or Simpson/Herfindahl–Hirschman) is no less arbitrary than the simple counting of variety. It will yield radically different results depending on the partitioning of options. The implications may be addressed by systematic sensitivity analysis and by adopting explicitly conservative assumptions on disparity with respect to the argument propounded or the hypothesis under test (Stirling, 1994a). However, the fact remains that taking measures of variety and/or balance as proxies for diversity thus remains highly sensitive to tacitly subjective taxonomies and arbitrary linguistic conventions concerning the implicit bounding of categories.

Conversely, the importance of disparity to energy diversity is itself typically qualified by the pattern of apportionment across options. For instance, an electricity supply portfolio comprising a 95 per cent contribution from one of four highly disparate resources (like Russian gas, with the residual 5 per cent made up of nuclear, wind and hydro) might reasonably under some perspectives be judged less diverse than a portfolio comprising even contributions from four much less disparate options (like piped Russian, Norwegian and UK gas with internationally traded and transported liquified natural gas (LNG) (PIU, 2001). Likewise, the balance of a portfolio is neglected where attention is restricted to variety alone, as is the case in much of the literature. At what scale of contribution is an option considered to add to system diversity? Does the installation of the first household rooftop photovoltaic panel increase the diversity of a national energy portfolio by the same degree as the construction of the first 1.5 GWe new nuclear power station? If not, at what scale of contribution does any given option begin to be counted? How do we avoid perverse threshold effects? Different indices of balance treat this crucial threshold issue in quite radically different ways (Skea, 2007). Taking disparity (or variety) as proxies for diversity ignores the balance with which a system is apportioned. It therefore seems that the only robust approach to thinking about energy diversity is to think about variety, balance and disparity together.

11.3   AGGREGATING, ACCOMMODATING AND ARTICULATING DIFFERENT ASPECTS OF ENERGY DIVERSITY

Thus far, we have established a definition of energy diversity in terms of three necessary but individually insufficient properties of disparity, variety and balance. A series of methodological questions follows from this. How can these quite distinct aspects of diversity be aggregated into a single coherent framework? How might such an analytical framework be applied such as to accommodate the range of relevant perspectives typically engaged in real debates over energy strategy? And how can the results of any diversity analysis on these lines be articulated with wider policy considerations – such as the performance of individual generating options under criteria of economic efficiency, environmental quality, social impact and security of supply, raised earlier in relation to broad sustainability goals? The present section will consider these challenges of aggregation, accommodation and articulation.

With regard to aggregation, most contemporary approaches to analysing energy diversity focus on variety and/or balance alone (for example Stirling, 1994a, 1994b, 1996; DTI, 2003b; Markandya et al., 2005; Grubb et al., 2006; Scheepers et al., 2007; Hubberke, 2007; IEA, 2007). Even where disparity is thus neglected, however, it is far from straightforward what relative emphasis to place on variety as compared with balance. How much weight should be assigned to small contributions from additional options, compared with enhanced balance among dominant options? It is a little-recognized property of widely used sum-of-the-squares concentration indices – like Simpson (1949), the Herfindahl–Hirschman index ('HHI' – Herfindahl, 1959; Hirschman, 1964) and the Gini diversity index (Gini, 1912; Sen, 2005); see Table 11.1 – that different exponent powers yield divergent rank orderings for portfolios displaying different patterns of composition across marginal and dominant options (Stirling, 1998: 56). That such divergent rankings may not arise in practice for certain particular portfolios (Grubb et al., 2006), does not

Table 11.1   Selected indices of contrasting subordinate properties of diversity

Property	Name and/or Reference	Form
Variety	Category count (MacArthur, 1965)	N
Balance	Shannon Evenness (Pielou, 1969)	$\dfrac{-\sum_i p_i \ln p_i}{\ln N}$
Variety/ Balance	Shannon Wiener (Shannon and Weaver, 1962)	$-\sum_i p_i \ln p_i$
	Simpson–Herfindahl– Hirschman (Simpson, 1949; Herfindahl, 1959; Hirschman, 1964)	$\sum_i p_i^2$
	Gini (1912)	$1 - \sum_i p_i^2$
	Hill (1973)	$\sum_I (p_i^a)^{1/(1-a)}$
Disparity	Weitzman (1992)	$\max_{i \in S}\{\mathbf{D}_w(S \setminus i) + \mathbf{d}_w(i, S \setminus i)\}$
	Solow and Polasky (1994)	$f(\mathbf{d}_{ij})$
Variety/ Balance/	Junge (1994)	$\left(\dfrac{\sigma}{\mu.\sqrt{n-1}}\right).\left(\dfrac{1}{\sqrt{N}}\right).\left(\sqrt{N-1} - \sqrt{N\sum_i p_i^2 - 1}\right)$
Disparity	Stirling (2007a)	$\sum_{Ij} \mathbf{d}_{ij}{}^a.(p_i.\,p_j)^b$

Notation	Interpretation in terms of energy portfolios
N	number of categories of energy options
$\ln$	logarithm (usually natural)
p_I	proportion of energy system comprised of option category i
a	a parameter governing relative weighting on variety and balance
n	number of attributes displayed by options
s	standard deviation of attributes within option categories
m	mean of attributes within option categories
$f(\mathbf{d}_{ij})$	function of disparity distance between option categories i and j ($\mathbf{d}_{ij}$)
$\mathbf{D}_w(S)$	aggregate disparity of energy system S
$\mathbf{d}_w(i, S \setminus i)$	disparity distance between option i and nearest option in S if i is excluded
a, b	parameters governing relative weightings on variety, balance and disparity

resolve this concern. Yet there exists no firmly grounded theoretical or empirical reason for taking the commonly used exponent value of two rather than, say, three, four or so on.

Logarithmic entropy functions (like Shannon–Wiener) avoid similar ranking sensitivities across different logarithm bases (Stirling, 1998: 56). But there still arises the question as to why the particular implicit weighting embodied in such indices should necessarily reflect the appropriate weighting for real energy systems or stakeholder perspectives. Theoretical work in mathematical ecology derives generalizations of these kinds of index, in which this crucial issue is dealt with by explicit weighting parameters (for example Hill, 1973 in Table 11.1; Kempton, 1979). Just because such parameters are not recognized in the conventional indices used in the energy sector does not remove this

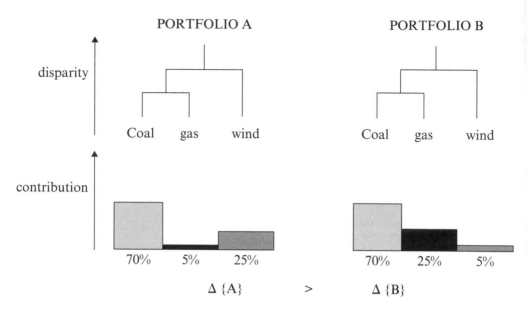

Figure 11.3   A case where varying diversity is not discriminated by conventional indices

problem. To ignore this risks straying into mathematical mysticism, where contingently privileged algorithms are ascribed transcendent authority concerning appropriate interests and priorities in the real world.

Beyond this, however, the most serious difficulty with conventional variety–balance indices concerns the neglect of the crucial property of disparity. Where attention is restricted to the enumeration or concentration of options, the assumption is effectively made that all options are equally disparate. This can yield manifestly perverse results, such as that represented in Figure 11.3. Here, any index restricted to variety and/or balance alone will entirely fail to address the fact that the enhanced representation of the more disparate option makes Portfolio A more diverse than Portfolio B. This point is returned to below.

Of course, the salience of the example illustrated in Figure 11.3 rests on the understanding (illustrated with the stylized dendrograms) that wind is more disparate from coal and gas than either of these fossil fuels are from the other. This may seem reasonable at an immediate intuitive level. But how can we be sure? And any practically useful diversity analysis must also be clear about the degree to which this is the case. In other words, in contemplating an increase in diversity in moving from Portfolio B to Portfolio A where gas is rated as the best-performing option, exactly what value of performance on the part of gas might be sacrificed as a premium for the additional diversity conferred by wind? This raises the second challenge identified above: that of accommodating the divergent perspectives on disparity (as well as performance) that typically characterize even the most specialist discussions of energy strategy.

This is essentially the problem that many seek to address through the exploratory application of probabilistic risk modelling techniques (Feldman, 1998; NERA, 2002) – and especially portfolio theory – to the analysis of contending possible electricity supply mixes (Ulph, 1988, 1989; NERA, 1995; ERM, 1995; Awerbuch and Berger, 2003; Awerbuch and Yang, 2007). In brief, what this does is take the single parameter of covariance in fuel price risk as a stochastic proxy for a range of multidimensional economic, physical, environmental, technical, institutional and geographical disparities between electricity options (Awerbuch et al., 2006). Such elegant shorthand may suffice for short-term decisions by private firms, dominated simply by financial risks due to fuel price shocks (Awerbuch, 2000). It has the benefit, at least in principle, of requiring only relatively objectively attested data (Lucas et al., 1995; Brower, 1995). But as strategic interests grow wider and time frames longer, serious questions arise over the sufficiency (and even validity) of this approach. To what extent can the historic behaviour of fuel prices be taken as a reliable guide even to the future trajectories of this one parameter, let alone the host of other factors invoked by wider sustainability agendas (Brower, 1995)? How do we address options whose attributes are simply not reflected in fossil fuel markets (Stirling, 1996)? Where diversity is undertaken as a response to strict uncertainty and ignorance, the use of probabilistic concepts like covariance coefficients is by definition invalid (Stirling, 2003). Despite its applicability where attention is confined to near-term fluctuations in fuel prices, portfolio theory is a highly circumscribed basis for addressing the full strategic scope of energy diversity (Awerbuch et al., 2006).

Though rarely integrated with consideration of variety and balance, the challenge of accommodating divergent possible understandings of disparity is quite well addressed in other disciplines. Approaches in fields like taxonomy, palaeontology, archaeology and conservation biology all routinely adopt a framework based on the notion of distance. Often, there exist in such fields some well-established or even objectively determinable criterion of disparity. This is the case, for instance, with genetic distance measures in evolutionary ecology, assumed (often incorrectly) to display a strict branching form (Weitzman, 1992; Solow and Polasky, 1994). Under such expedient circumstances, the analysis of diversity can be quite strictly codified in terms of 'disparity distances' between elements and relatively unambiguous answers derived.

Elsewhere, however, the picture is much more similar to that in the energy field, where options are not differentiated by strictly branching genetic processes and where there typically exist a variety of different views over the relevant aspects of disparity and their respective degrees of importance (Stirling, 1998). Even here, however, it is possible to use the simple concept of distance. Options are characterized in terms of whatever are held to be the salient attributes of difference, such that each can be represented as a coordinate in a multidimensional 'disparity space' (Stirling, 2007). With the different disparity attributes weighted to reflect their relative importance, the simple Euclidean distances separating options in this space can be taken as a reflection of their mutual disparity. By appropriate normalization and weighting procedures, a 'disparity space' of this kind can be constructed such as accurately to reflect any conceivable perspective on the distinguishing features of different energy options (Kruskal, 1964). Indeed, fuel price covariance might be seen just as one such possible dimension of disparity.

As a starting point for implementing this broader disparity–distance approach in the analysis of energy diversity, it is useful to consider the nature of existing datasets concerning the economic and/or wider sustainability performance of different energy options (for example Externe, 2004). Encompassing different aspects of financial, operational, environmental, health and broader social impacts, many such datasets have been

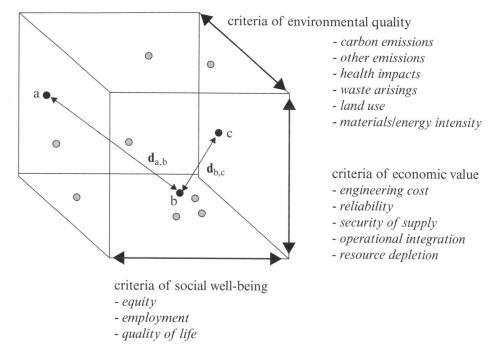

criteria of environmental quality
- *carbon emissions*
- *other emissions*
- *health impacts*
- *waste arisings*
- *land use*
- *materials/energy intensity*

criteria of economic value
- *engineering cost*
- *reliability*
- *security of supply*
- *operational integration*
- *resource depletion*

criteria of social well-being
- *equity*
- *employment*
- *quality of life*

Figure 11.4   A disparity space generated by normalized sustainability performance data

generated over past years by a range of different disciplines – including cost–benefit, environmental impact and technology assessment and comparative risk, life-cycle and multicriteria analysis. To the extent that the performance of each energy option is structured differently under the various performance criteria, each of these datasets contains potentially useful information over their disparities. If the different criteria are normalized such that all options are reassigned nominally equal performance, then the distances in the resulting multidimensional space will provide a robust indication of broader disparities according to the perspective in question on salient strategic attributes. Figure 11.4 provides a schematic illustration of this approach. Of course, none of this precludes use of non-normalized performance data, as a basis for determining trade-offs between performance and diversity (Stirling, 2007).

The kind of disparity structures that pilot work indicates may be yielded by such performance data are illustrated in Figure 11.5. Distances between successively more remote pairs of options in disparity space can be represented using standard cluster analysis techniques and represented as a dendrogram (with disparity distance indicated on the horizontal axis labelled 'd'). The actual underlying distances used in diversity analysis will not be affected by the sometimes slightly contrasting representations yielded by different clustering metrics, algorithms and procedures (Sneath and Sokal, 1973).

More importantly, however, different performance datasets can be expected to generate different disparity structures. Similarly, given the broad strategic scope required in considering diversity, it may also be expected that the different dimensions of such analy-

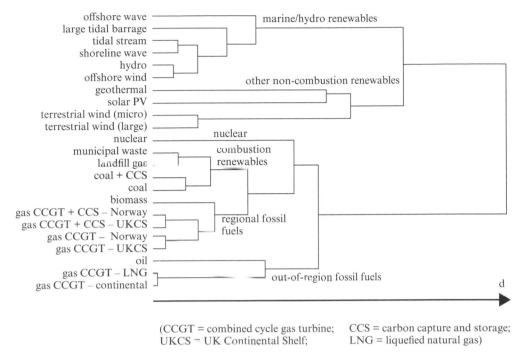

(CCGT = combined cycle gas turbine;   CCS = carbon capture and storage;
UKCS = UK Continental Shelf;   LNG = liquefied natural gas)

Figure 11.5   Indicative disparities in multicriteria performance of electricity options

sis will be weighted differently under different perspectives. This is a matter for empirical investigation, a summary of which is reported later (Yoshizawa et al., 2008). For the moment, early indications are that direct elicitation from interviewees, by reference to typical detailed sustainability performance datasets, consistently yields intuitively meaningful structures, such as those indicated in Figure 11.5. Existing performance datasets provide a useful starting point to the construction of meaningful patterns of disparity, which may readily be followed up by direct elicitation of further more specific disparity attributes in an intensive interview or deliberative group setting.

In considering the many possible queries that might be raised over detailed disparity structures such as that illustrated in Figure 11.5, two things must be remembered. The first is that the point here is not to assert that there exists any single well-defined 'objective' disparity structure that applies irrespective of context or perspective across real-world energy options. Instead, the value of this general approach lies in the possibility of more systematic and transparent ways of accommodating inevitably divergent viewpoints on disparity. It is interesting and potentially significant that existing performance datasets – backed up by direct elicitation and deliberation involving specialists – may quite readily yield intuitively robust disparity structures.

For those to whom this seems like an uncomfortably subjective or open-ended approach, the second point is that these challenges are unavoidably intrinsic to the complexities of energy disparity itself. Simply to ignore this challenge – for instance through the conventional restriction of attention to variety and balance alone or a single circumscribed parameter like fuel price covariance – does not avoid assumptions over option

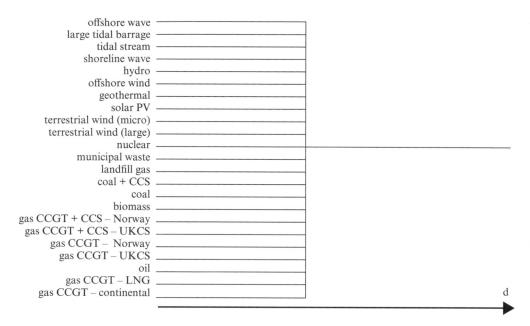

Figure 11.6   Indicative disparities implicit in use of Shannon or Herfindahl indices

disparity. It simply conceals them. If indices like Shannon or Simpson/Herfindahl–Hirschman are used to analyse diversity among a set of electricity supply options like that resolved in Figure 11.5, for instance, this simply amounts to treating all identified options as equally disparate – as represented in Figure 11.6. The choice is therefore not whether to respond to the challenge of accommodating divergent perspectives on disparity, but how – and with what degree of rigour and openness. The resulting questions concern not the absolute precision of any given directly elicited disparity structure, but its relative plausibility in relation to a default picture like that in Figure 11.6.

The third and final challenge raised at the beginning of this section, concerns the articulation of diversity with other properties of strategic interest in the appraisal of energy portfolios. As already noted, the key dimensions of portfolio performance will often be quite distinct from diversity, concerning criteria such as operational efficacy, financial performance, environmental impacts and other aspects of supply security. The challenge is particularly acute as attention extends to the full range of pressing issues invoked by sustainability agendas. Aggregate option performance under such criteria (under any perspective) will typically be a function of the performance of the individual options. This will therefore be subject to important trade-offs as deliberate diversification draws larger contributions from what appear under the perspective in question to be lower-performing options. In order to be useful, therefore, any practical framework for the analysis of energy diversity should not be applied in isolation, but should articulate directly with these broader appraisal criteria, such as to allow systematic exploration of relevant trade-offs, interactions and operational effects. It is to one possible framework for addressing these challenges of aggregation, accommodation and articulation that our attention will now turn.

11.4 A NOVEL DIVERSITY HEURISTIC FOR STRATEGIC APPRAISAL OF ENERGY PORTFOLIOS

To take seriously these problems of aggregation, accommodation and articulation does not necessarily lead to a counsel of despair over the potential for systematic general characterizations – or even quantifications – of energy diversity. A more positive starting point is the observation that the futility of deriving a single definitive diversity index need not preclude the possibility of a flexible general heuristic. Like an index, a heuristic may be quantitative. But rather than aiming to measure diversity in some unconditional objective fashion, it offers an explicit, systematic basis for exploring sensitivities to the assumptions conditioning aggregation, accommodation and articulation (Stirling, 2007).

For any particular perspective on the appropriate weightings for variety and balance and the salient dimensions of disparity, such a heuristic would behave as an index. It would accommodate different views on the relevant attributes of disparity, aggregate these with consideration of variety and balance, and allow systematic articulation with important system-level properties other than diversity (like portfolio interactions). For applications involving a range of perspectives, this heuristic would allow clear comparisons to be made between the implications of contending judgements. In other words, a heuristic characterization of energy diversity aims to combine the rigour, transparency and specificity of quantification with the scope, applicability, flexibility and symmetry of qualitative approaches. The real challenge lies in achieving this, whilst minimizing the introduction of further complexity and contingency.

No existing diversity index addresses all three properties of variety, balance and disparity in an unproblematic way. However, based on well-established criteria applied to the treatment of these individual diversity properties by researchers such as Hill (1973), Laxton (1978), Pielou (1977), Weitzman (1992) and Solow and Polasky (1994), a series of non-trivial requirements are quite readily developed. For instance, some significant desirable features of a general diversity heuristic Δ, that help explicitly to address challenges of aggregation, accommodation and articulation as defined here, are outlined in Table 11.2.

No established diversity index satisfies all the criteria summarised in Table 11.2. Yet there is one relatively straightforward quantitative heuristic, independently derived in different disciplines (Rao, 1982; Stirling, 1998), which does offer a starting point. This is the sum of pairwise option disparities, weighted in proportion to option contributions (D):

$$D = \Sigma_{ij(i_{*}j)} d_{ij}.p_i.p_j \qquad (11.1)$$

where p_i and p_j are proportional representations of options i and j in the energy system (balance), and d_{ij} is the distance separating options i and j in a particular disparity space (Figure 11.4) The summation is across the half-matrix of $((n-1)^2/2)$ non-identical pairs of n options ($i \neq j$). In the special case where all d_{ij} are equal (scaleable to unity), D reduces to half Gini (Table 11.1).

It is readily demonstrated that this heuristic, D, complies with criteria (1) to (7). Compliance with criterion (8) remains a matter of judgement, but it is difficult to imagine a solution to these criteria that is simpler or more parsimonious. As to criterion (9), this

Table 11.2   Formal conditions for a robust general heuristic of energy diversity, Δ

Condition	Description*
1: Scaling of variety	Where option variety is equal to one, Δ takes a value of zero.
2: Monotonicity of variety	Where energy options are evenly balanced and equally disparate, Δ increases monotonically with variety.
3: Monotonicity of balance	For given option variety and disparity, Δ increases monotonically with balance (i.e. Δ is maximal for equal reliance on all options).
4: Monotonicity of disparity	For given variety and balance, Δ increases monotonically with the aggregate disparity between energy options.
5: Scaling of disparity	Where aggregate disparity is zero (i.e. where all energy options are effectively identical), Δ takes a value of zero.
6: Open accommodation	Δ accommodates any perspective on salient dimensions of disparity under which energy options can be differentiated.
7: Insensitivity to partitioning	For any given perspective on taxonomy, Δ is insensitive to alternative partitionings of options into categories.
8: Parsimony of form	Δ is as uncomplicated in structure and parsimonious in form as necessary to fulfil the above conditions.
9: Explicit aggregation	Δ permits explicit aggregation of variety, balance and disparity, by reflecting divergent contexts or perspectives using weightings.
10: Ready articulation	Δ allows unconstrained articulations of diversity with other salient properties of individual options or the energy system as a whole.

Note: * See references in Stirling (2007).

raises a final notable feature of *D*, that can be illustrated by introducing just two further terms (Stirling, 2007a):

$$\Delta = \Sigma_{ij(i \neq j)} (d_{ij})^{\alpha} . (p_i . p_j)^{\beta} \qquad (11.2)$$

If exponents α and β are allowed to take all possible permutations of the values 0 and 1, this yields four variants of the heuristic Δ. Each of these usefully captures one of the four properties of interest: variety, balance, disparity and diversity (Table 11.3).

Shifting values of exponents α and β between 0 and 1 yields further variants of Δ, collectively addressing all possible relative weightings on variety, balance and disparity. Of these, the reference case, Δ ($\alpha = \beta = 1$) does the same job as other widely used non-parametric measures like Gini, Shannon and Simpson, but with the major additional feature that it also captures disparity. Unlike the disparity measures proposed by Weitzman or Solow and Polasky (Table 11.1), Δ also addresses variety and balance. Unlike the 'triple concept' proposed by Junge (1994; see Table 11.1), Δ accommodates radically divergent perspectives on disparity itself, yet is relatively parsimonious in form. An entirely novel feature of Δ, is that it systematically addresses alternative possible aggregations of these subordinate properties, according to perspective and context.

Table 11.3 Four variants of Δ and their relationships with diversity properties

Property	α	β	Δ =	Equivalents*	Interpretation
variety	0	0	$\Sigma_{ij}\, d_{ij}^0$	([category count] − 1)²/2	scaled variety
balance	0	1	$\Sigma_{ij}\, p_i \cdot p_j$	[Gini]/2	variety-weighted balance
disparity	1	0	$\Sigma_{ij}\, d_{ij}$	[Solow and Polasky]	variety-weighted disparity
diversity	1	1	$\Sigma_{ij}\, d_{ij} \cdot p_i \cdot p_j$	D	balance-weighted disparity

Note: * See Table 11.1.

11.5 ARTICULATING ENERGY DIVERSITY WITH OTHER ASPECTS OF STRATEGIC PERFORMANCE

Of the formal criteria identified for a general heuristic of energy diversity in Table 11.2, the discussion in the last section leaves only criterion (10) unaddressed – concerning the articulation of diversity with other relevant system-level properties. As already mentioned, energy diversity is rarely a free lunch. The overall strategic performance of a portfolio as a whole will be a function not only of diversity but also of other system properties and the performance of individual options. For instance, there will typically be portfolio effects resulting from interactions between subsets of options, such as: potentially negative feedbacks between high penetrations from intermittent renewables and the inflexibility of large, predominantly base-load plant like nuclear power (Gross et al., 2006); or competition between contending technology strategies or more specific institutional and industrial tensions through which, for instance, nuclear power can 'crowd out' large-scale commitments to renewable energy (Mitchell and Woodman, 2006). Of course, there can also be positive synergies between disparate energy options, such as those often argued to benefit joint pursuit of distributed generation and demand-side energy efficiency measures (Prindle et al., 2007). Diversity itself may provide a strategic response to supply security challenges, as well as hedging more general sources of ignorance, fostering innovation, mitigating lock-in and accommodating pluralism. But it will typically require some compromise on other aspects of performance like financial costs, operational efficacy, environmental impacts or wider economic factors.

In addition, many energy options will be constrained in their possible contributions. A number of renewable energy sources, for instance, are in this position. Rather than being static in nature, such constraints will take the form of a dynamic resource curve, under which successive increments are available at varying levels of performance. The shape of this curve will reflect the performance attributed to successive incremental 'tranches' for these options. This will be a function of two contending factors. On one hand, there are the combined negative effects of using successively less favourable sites (OXERA, 2004; DTI, 2005). On the other hand, there are learning effects and other increasing returns processes, which will yield countervailing positive improvements as experience accumulates with increasing use (Jacobsson and Johnson, 2000). To take account of these factors, then, the value assigned under any given perspective to any particular energy system under specific conditions ($V\{S\}$) can be expressed as the sum of the value due to the aggregate performance of individual options ($V\{E\}$) and an incremental value attached to irreducible portfolio-level properties including diversity ($V\{P\}$). If the net implications of diversity are adverse, then $V\{P\}$ can be negative:

$$V\{S\} = V\{E\} + V\{P\} \qquad (11.3)$$

It has already been mentioned in discussing disparity distances, that there exist numerous well-tried methods, and extensive bodies of data, addressing the multicriteria performance of energy options. Long experience in the field of decision analysis (Vincke et al., 1992) shows that, just as divergent notions of difference can be represented as coordinates in an n-dimensional Euclidean disparity space (Figure 11.3), so can divergent valuations of individual system elements be represented as coordinates in an m-dimensional Euclidean performance space (Stirling, 2006). The dimensions of this space represent any set of m relevant performance criteria, each weighted to reflect their respective importance (Stirling, 1998: 81). As with disparity, the selection, characterization and scaling of these criteria will vary across context and perspective (Stirling, 1997). Although it is difficult to justify any single approach to aggregating across perspectives, decision analysis has shown that any single perspective can be uniquely captured by means of the following expression for the overall value attached to the performance of individual system elements $V\{E\}$:

$$V\{E\} = \Sigma_i \Sigma_c (w_c . s_{ic}) . p_i \qquad (11.4)$$

where s_{ic} is the value attached to the performance of option i under criterion c; w_c is a scalar weighting reflecting the effective relative importance of criterion c (under the perspective and context in question) and p_i is (as in equations 11.1 and 11.2), the proportional representation of option i in the energy system in question. It follows from equation 11.2 that the corresponding value attached to irreducible portfolio-level properties including diversity ($V\{P\}$), can then be expressed follows:

$$V\{P\} = \delta . \Delta'$$

$$= \delta . \Sigma_{ij(i_*j)} (d_{ij})^\alpha . (p_i . p_j)^\beta . \iota_{ij} \qquad (11.5)$$

where Δ' represents an augmented form of the diversity heuristic Δ given in equation (11.2), which includes an additional term to reflect portfolio interactions (ι_{ij}). ι_{ij} is an array of scalar multipliers exploiting the pairwise structure of Δ' to express the effect on system value of synergies or tensions between options i and j, respectively, as marginal positive or negative departures from a default of unity ($\iota_{ij} = 1 \pm \delta\iota$: for most systems $\delta\iota << 1$). This serves as a means to capture a variety of system-level properties that, like diversity, are irreducible to individual options. The coefficient δ scales expressions of portfolio value to render them commensurable with aggregate values of individual options in equation (11.4). For positive assessments of portfolio value, $0 < \delta < \infty$. From equations (11.3), (11.4) and (11.5), we therefore obtain the following heuristic system-level articulation ($V\{S\}$) of the value attached to diversity together with that assigned to other portfolio properties ($V\{P\}$) and the value attached to the performance of individual energy options ($V\{E\}$):

$$V\{S\} = \Sigma_i \Sigma_c (w_c.s_{ic}).p_i + \delta.\Sigma_{ij(i \neq j)}(d_{ij})^\alpha.(p_i.p_j)^\beta.\iota_{ij} \qquad (11.6)$$

It is in $V\{S\}$ that we have a means to address the final criterion (ready articulation) developed above for a heuristic of energy diversity, in that it should allow systematic unconstrained articulation of system diversity with alternative characterizations of other salient properties of the energy system as a whole (interactions) or its component options (individual performance). Under such an approach, the 'systems' and 'options' in question may equally be defined to address contexts like primary energy mixes, electricity supply portfolios, energy service provision and transport systems modalities. The approach can as readily focus on specific economic performance or broader criteria of energy sustainability.

The interest of the heuristic $V\{S\}$ does not lie in any attempt to derive some unconditional 'optimal' balance between the performance of individual options, system interactions and portfolio diversity. Instead, $V\{S\}$ can be used systematically to explore different possible perspectives and assumptions concerning the contributions of these components to the overall value of an energy system. For each perspective on the available options, their individual performance, dynamic resource curves, joint interactions, mutual disparities, aggregations of diversity properties and the performance–diversity trade-off, there will exist a particular apportionment of options that yields some maximum overall value. By varying δ between zero and infinity, $V\{S\}$ yields a set of all possible conditionally optimal energy systems ranging (respectively) from those that maximize value due to aggregate performance of individual options, to those that maximize positive value due to portfolio interactions and system diversity.

Drawing on work further summarized later (Yoshizawa et al., 2008), Figure 11.7 provides a schematic illustration of the kind of picture that arises from this analysis. For purposes of exposition, it focuses on the mix of generating technologies at the level of the UK electricity supply system taken as a whole. It is constructed on the basis of economic and resource data developed for the UK government's recent energy reviews (PIU, 2001; DTI, 2005). Broader sustainability criteria are also included under one particular perspective on the weighting of different aspects of performance (Stirling, 2007). Such an approach might as easily be addressed with respect to regional or international systems, to primary energy mixes in a broad sense, or to include information on demand-side options for the provision of energy services. Either way, disparities will be conceived in a fashion similar to that represented in Figure 11.3: concerning a wide range of attributes of the resources and technologies involved, together with their geographical, commercial, institutional and socio-political contexts (Stirling, 1996). Positive and negative economic, organizational and operational synergies between different technologies inform the modelling of interactions (using term ι_{ij} in equations 11.5 and 11.6). Certain options are tightly constrained in terms of the available resource, or display reductions (from learning or scale) or increases (from depletion) in costs or impacts as the contributions rise. For now, the point is not to assert the empirical particularities, but simply to illustrate the method. Figure 11.7 shows – for a particular hypothetical perspective – how the resulting conditionally optimal electricity portfolios vary as greater or lesser priority is placed on diversity.

Low values of δ in Figure 11.7 may express high confidence in performance appraisals of individual technologies, with little concern over deep uncertainties and ignorance to

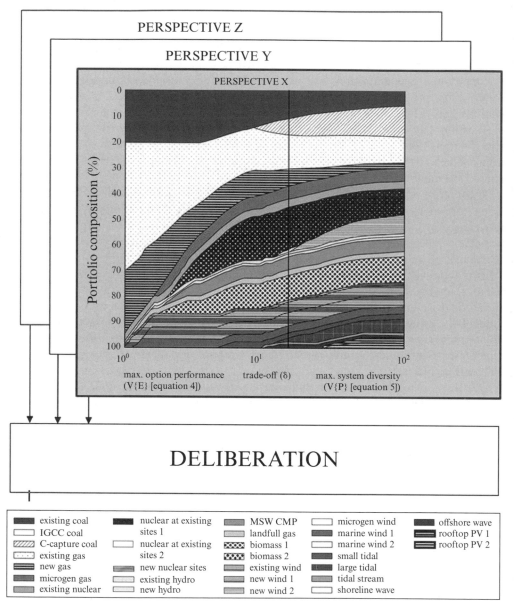

Figure 11.7   Illustrative performance–diversity trade-offs for Yoshizawa et al. (2008)

which diversity is a reasonable response (Awerbuch et al., 2006). Likewise, low values of δ may imply that priority is attached to maximizing this performance, rather than the other benefits of diversity (in fostering innovation, mitigating lock-in or accommodating pluralism). High values of δ, on the other hand, reflect a dominant interest in these benefits of diversity, with little concern over the resulting compromises on performance. Again, the value of this kind of heuristic framework is as a means more explicitly and systematically to inform analysis under individual perspectives, and to provide a basis for more effective and transparent deliberation between contending positions.

11.6 EMPIRICAL ILLUSTRATION

A pilot empirical test of this method has been undertaken and reported on in full elsewhere by the authors (Yoshizawa et al., 2008). By summarizing some of the key findings from this exercise, it is possible to give a better feel for the kind of contribution to policy deliberation that might be made by this analysis. These findings have been tested by comprehensive sensitivity analysis that it is not possible to detail in the present chapter, but which can be read in the original source.

One kind of contribution to policy discussion centres on the systematic highlighting of similarities and differences between alternative expert disparity structures. Here, Figure 11.8 compares two indicative structures from among the nine elicited through in-depth computer-assisted appraisals conducted by leading electricity system specialists in the UK and Japan. Each chart is based on a detailed assessment of the multidimensional performance of the key generating options, as reflected in the best available technical data and expert judgements as to the relative importance of contrasting evaluative criteria. The resulting disparity structures are thus not arbitrary or uninformed subjective constructs, but embody legitimate divergences of specialist interpretation, concerning the objective parameters that already inform official energy policy discussions over the relative merits of alternative electricity generating options.

The first general point illustrated here (and shared across the majority of participants in this empirical study) is the fundamental discrimination between 'conventional' energy options (as a group including coal, oil and nuclear) and the 'renewables' group. This is shown by the repeated appearance of a two-cluster solution at the right-hand (high-disparity) side of the diagrams (as illustrated in Figure 11.8). This is not remarkable in itself, but it does present an unusually detailed and systematic empirical grounding for what is often little more than casual assertion. This picture is also interesting for the light that it casts on the specific reasons and implications of certain variations on this fundamental pattern. For instance, UK appraisals tend to place hydro in the 'renewable' category, whilst there is a tendency for Japanese assessments to group this with conventional options. Conversely, there is a particular ambiguity in the UK over the 'renewable' status of biomass, with some appraisals grouping it alongside conventional options and others (like all Japanese appraisals) including it with renewables. This pattern is even more pronounced in the case of waste. These distinctions evidently arise from the perceived environmental status of hydro and combustion technology in each case, but they provide an interesting indication of the ways in which cultural views (even among experts) can interact with technical data to yield divergent representations of disparity.

A second general point that arises concerns the internal diversity of the renewable options. It is a feature of every disparity structure elicited in this exercise (without exception), that the broad category of 'renewable energy' is disaggregated at a level that is at least equal to the conventional partitioning of coal, oil and gas. This is an important finding for the analysis of energy diversity, because it substantiates how conventional analysis (based on accepted linguistic and official statistical categories

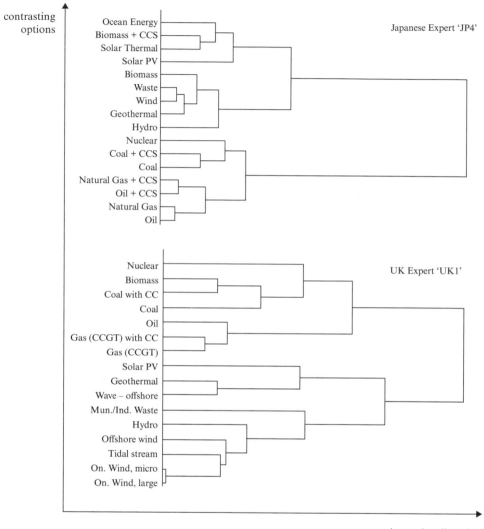

Source: Yoshizawa et al. (2008).

Figure 11.8   Illustrative disparity structures elicited from UK and Japanese experts

like coal, oil, gas, nuclear and renewables), can seriously understate the diversity of renewable options. Although details vary between different specialist perspectives, it is a common feature that renewable options (including wind, wave, tidal, hydro, biomass, waste, geothermal and solar) are effectively as disparate from each other as they are from fossil fuels – and indeed even as the fossil fuels are collectively disparate from nuclear. Whatever the differences in other respects, this general result has potentially significant implications for considerations over how to realize diversity benefits in electricity supply portfolios.

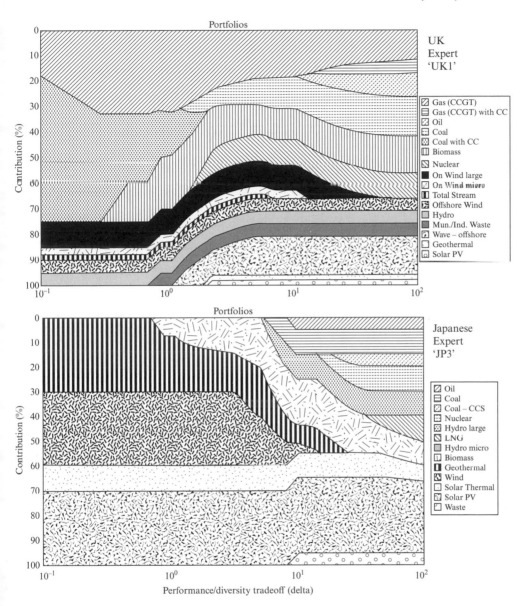

Source: Yoshizawa et al. (2008).

Figure 11.9 Illustrative contrasts in diversity–performance trade-offs elicited from UK and Japanese experts

A further kind of contribution to policy discussion rests not on the disparity structures but on the associated performance–diversity trade-offs (like that provided in large scale in Figure 11.7). The two further smaller scale diagrams in Figure 11.9 (one each from among Japanese and UK experts) illustrate contrasting pictures of the ways in which

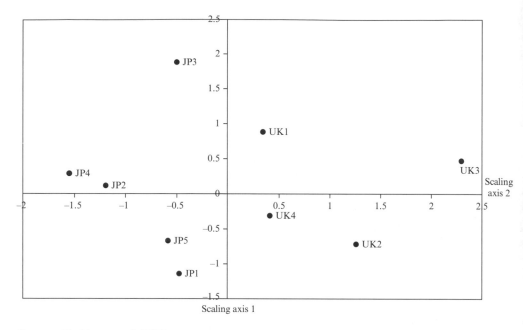

Source: Yoshizawa et al. (2008).

Figure 11.10   Distribution of UK and Japanese experts in scaling analysis of performance–diversity trade-offs

different expert appraisals determine 'optimal' portfolio compositions to vary with increasing diversity. Despite the small sample size in this pilot study and the differences displayed between experts within each national setting, it is clear from scaling analysis of the portfolios yielded across the nine UK and Japanese experts taken as a whole that there is a prominent distinction between UK and Japanese perspectives. This is displayed in Figure 11.10, which plots the distribution of UK and Japanese experts in a space defined by the two principal axes embodied in the different portfolio compositions. This striking consistency is a reflection of both the technical and resource characteristics in these different settings, as well as divergent cultural perspectives on the part of the experts involved. It is a further indication that any serious analysis of diversity should be equally rigorous with regard to both aspects.

11.7   CONCLUSION

The present chapter has outlined a novel general framework for the analysis of energy diversity. The discussion began by noting many different reasons for an interest in diversity. These include well-rehearsed issues in supply security debates, as well as some less widely recognized and variously framed general diversity benefits such as: hedging ignorance, promoting competition, accommodating plural interests, fostering innovation, nurturing context-sensitivity and mitigating lock-in. In particular, diversity is argued to feature as a prominent element in social choices over contending pathways to sustainable energy transitions.

Based on the recognition of three necessary but individually insufficient properties of diversity (variety, balance and disparity) the analytical framework proposed here is applicable in a wide range of different energy policy contexts. It can be focused equally at the level of primary energy mixes, electricity supply portfolios or energy service delivery. Unlike other approaches, it is not constrained to address only certain specific perform-ance criteria (like the fuel prices highlighted in portfolio theory). Whilst it may be applied in purely economic terms if this is thought appropriate, the method can also be readily extended to encompass any range of issues addressed in well-established multicriteria appraisal techniques. Nor is this approach dependent on assumptions that past experi-ence necessarily provides a reliable guide to future performance. The pictures of disparity that underlie this analysis can be informed by extensive bodies of existing multivariate performance data, building only on the general underlying structures in this perform-ance, rather than the specific values. The framework allows attention to be given to more detailed properties such as dynamic resource curves and system-level interactions between options.

In order to illustrate one potential mode of application, the chapter concludes with an illustrative summary of an empirical study conducted in the UK and Japan. This shows that the present framework for multicriteria diversity analysis can be applied flexibly to an unconstrained array of different specialist, institutional or stakeholder perspectives. This amenability to more open and plural processes of engagement is also central to agendas around democratization and sustainability. As a framework for quantification, the approach is compliant with a series of rigorous formal quality criteria (Table 11.2). However, as a heuristic approach, it does not purport to provide a basis for deriving a single objectively 'optimal' energy portfolio, irrespective of perspective. In any case, such aims (and claims) might be regarded as misleading and spurious in fields as complex, uncertain, dynamic and contested as contemporary energy policy.

In the end, however, the present methodology is not so much an alternative to other approaches to the appraisal of energy diversity, as it is a complement, and a potential integrative framework. It was reported at the beginning of this chapter that – even within particular approaches – the profusion of different methods and framings for assessment of diversity and option performance often yield radically different policy implications. Where policy-making is based (either contingently or strategically) on a circumscribed subset of such technical perspectives, then it risks delivering results that are at best lacking in robustness. At worst, they present a vulnerability to manipulation by special interests of the kind (as we have seen) that is not unknown in the analysis of diversity. Here, as in other areas of policy appraisal, we need to move away from simplistic and hubristic prescriptive methods, which neglect uncertainties and presume consensus around a particular asserted set of priorities and value judgements. Instead, the present general heuristic framework provides a means to articulate an unlimited variety of differ-ent approaches to the assessment of performance and appraisal of disparity. By revealing the legitimate scope for disagreement and identifying areas of common ground, multic-riteria diversity analysis may help build a framework for more robust and accountable policy-making on energy diversity.

ACKNOWLEDGEMENTS

Part of this chapter is republished with gratitude for permission from the editors and publishers, Elsevier, from: A. Stirling (2010), 'Multicriteria diversity analysis: a novel heuristic framework for appraising energy portfolios', *Energy Policy*, **38**, 1622–34. The research reported here was funded by the UK Economic and Social Research Council and by the University of Tokyo.

REFERENCES

Adelman, M. (1995), *The Genie Out of the Bottle: World Oil Since 1970*, Cambridge, MA: MIT Press.
Aoki, M. (1996), 'An evolutionary parable of the gains from international organisational diversity', in R. Landau, T. Taylor and G. Wright (eds), *The Mosaic of Economic Growth*, Stanford, CA: Stanford University Press, pp. 247–63.
Arthur, W. (1989), 'Competing technologies, increasing returns, and lock-in by historical events', *Economic Journal*, **99**, 116–31.
Arthur, W. (1994), *Increasing Returns and Path Dependence in the Economy*, Ann Arbor, MI: University of Michigan Press.
Ashby, W. (1956), *An Introduction to Cybernetics*, London: Chapman & Hall.
Awerbuch, S. (2000), 'Investing in distributed generation', in E. Bietry, J. Donaldson, J. Gururaja, J. Hurt and V. Mubayi (eds), *Decentralized Energy Alternatives: Proceedings of the Decentralized Energy Alternative Symposium*, New York: Columbia School of Business, Sustainable Development Initiative.
Awerbuch, S. and M. Berger (2003), 'Applying portfolio theory to EU electricity planning and policy making', Report No. EET/2003/03, International Energy Agency, Paris.
Awerbuch, S., A. Stirling, J. Jansen and L. Beurskens (2006), 'Portfolio and diversity analysis of energy technologies using full-spectrum uncertainty measures', in David Bodde and Karyl Leggio (eds), *Understanding and Managing Business Risk in the Electric Sector*, Amsterdam: Elsevier, pp. 202–22.
Awerbuch, S. and S. Yang (2007), 'Efficient electricity generating portfolios for Europe: maximising energy security and climate change mitigation', *EIB Papers*, **12** (2), 8–37.
Bazilian, M. and F. Roques (eds) (2008), *Analytical Methods for Energy Diversity and Security*, Oxford: Elsevier.
Bird, J. (2007), *Energy Security in the UK*, London: Institute for Public Policy Research.
Brooks, H. (1986), 'The typology of surprises in technology, institutions and development', in W. Clark and R. Munn (eds), *Sustainable Development of the Biosphere*, Cambridge: Cambridge University Press, pp. 325–48.
Brower, M. (1995), 'Comments on Stirling's "Diversity and ignorance in electricity supply investment"', *Energy Policy*, **23** (2), 115–16.
Brown, N. and M. Michael (2003), 'A sociology of expectations: retrospecting prospects and prospecting retrospects', *Technology Analysis and Strategic Management*, **15** (1), 3–18.
Brundtland, G. et al. (1987), *Our Common Future: Report of the United Nations Commission on Environment and Development*, Oxford: Oxford University Press.
Cohendet, P., P. Llerena and A. Sorge (1992), 'Technological diversity and coherence in Europe: an analytical overview', *Revue d'Economie Industrielle*, **59**, 94–128.
Commission of the European Communities (CEC) (1990), *Security of Supply*, SEC(90)548 final, Brussels.
Commission of the European Communities (CEC) (2005), *Working Together for Growth and Jobs A New Start for the Lisbon Strategy: Communication from the President*, COM (2005) 24, February.
Commission of the European Communities (CEC) (2007), *An Energy Policy for Europe*, Communication from the Commission to the European Council and the European Parliament, SEC (2007)12.
Costello, K. (2004), 'Increased dependence on natural gas for electric generation: meeting the challenge', National Regulatory Research Institute, Ohio State University, USA.
Cowan, R. (1991), 'Tortoises and hares: choice among technologies of unknown merit', *Economic Journal*, **101**, 601–814.
David, P. and G. Rothwell (1996), 'Standardisation, diversity and learning: strategies for the coevolution of technology and industrial capacity', *International Journal of Industrial Organization*, **14**, 181–201.
Department of Trade and Industry (DTI) (1995), *The Prospects for Nuclear Power in the UK: Conclusions of the Government's Nuclear Review*, Cmnd 2860, London: HMSO.

Department of Trade and Industry (DTI) (2003a), *Energy White Paper: Our Energy Future – Creating a Low Carbon Economy*, London: HMSO, March.

Department of Trade and Industry (DTI) (2003b), *Digest of UK Energy Statistics*, London: HMSO.

Department of Trade and Industry (DTI) (2005), *The Role of Fossil Fuel Carbon Abatement Technologies (CATs) in a Low Carbon Energy System – A Report on the Analysis Undertaken to Advise the DTI's CAT Strategy*, London, December.

Department of Trade and Industry (DTI) (2006), *Our Energy Future – Creating a Low Carbon Economy*, Energy White Paper.

Eldredge, N. (1992), *Systematics, Ecology and the Biodiversity Crisis*, New York: Columbia University Press.

Elliott, D. (ed) (2007), *Nuclear or Not? Does Nuclear Power Have a Place in a Sustainable Energy Future?*, London: Palgrave.

Environmental Resources Management (ERM) (1995), 'Diversity in UK electricity generation: a portfolio analysis', report commissioned by Scottish Nuclear Ltd, Environmental Resources Management, December.

European Energy Charter Secretariat (2004), *The European Energy Charter Treaty and Related Documents: A Legal Framework for International Co-operation*, Brussels: European Commission, available at http://www. encharter.org/fileadmin/user_upload/document/EN.pdf#page=211.

Externe (2004), Externe Project, 'New elements for the assessment of external costs from energy technologies', report to EC DG Research, Technological Development and Demonstration (RTD), available at http://www.ier.uni-stuttgart.de/forschung/projektwebsites/newext/.

Farrell, A., H. Zerriffi and H. Dowlatabadi (2004), 'Energy infrastructure and security', *Annual Review of Environment and Resources*, **29**, 421–69.

Feldman, M. (1998), *Final Report: Diversity and Risk Analysis in a Restructured California Electricity Market*, Resource Decisions for California Energy Commission (CEC), Contract #500-96-019, San Francisco, CA, October.

Geroski, P. (1989), 'The choice between diversity and scale', in E. Davis (ed), *1992: Myths and Realities*, London: Centre for Business Strategy, London Business School, pp. 29–45.

Gini, C. (1912), 'Variabilità e mutabilità', *Studi Economica-Giuridici della R. Università di Cagliari*, **3**, 3–159.

Grabher, G. and D. Stark (1997), 'Organizing diversity: evolutionary theory, network analysis and postsocialism', *Regional Studies*, **31** (5), 533–44.

Greene, D., D. Jones and P. Leiby (1998), 'The outlook for US oil dependence', *Energy Policy*, **26**, 55–69.

Gross, R., P. Heptonstall, D. Anderson, T. Green, M. Leach, J. Skea (2006), 'The costs and impacts of intermittency: an assessment of the evidence on the costs and impacts of intermittent generation on the British electricity network', report for the UK Energy Research Centre (UKERC), London, March.

Grubb, M., L. Butler and P. Twomey (2006), 'Diversity and security in UK electricity generation: the influence of low-carbon objectives', *Energy Policy*, **34** (18), 4050–62.

Helm, D. (2002), 'Energy policy: security of supply, sustainability and competition', *Energy Policy*, **30** (3), 173–84.

Helm, D. (2007), 'European energy policy: meeting the security of supply and climate change challenges', *EIB Papers*, **12** (1), 30–49.

Herfindahl, O. (1959), *Copper Costs and Prices: 1870–1957*, Baltimore, MD: Johns Hopkins Press.

Hirschman, A. (1964), 'The paternity of an index', *American Economic Review*, **54** (5), 761–72.

Hill, M. (1973), 'Diversity and evenness: a unifying notation and its consequences', *Ecology*, **54** (2), 427–32.

Hubberke, D. (2007), 'Indicators of energy security in industrialised countries', University of Chemnitz, Germany, December.

Hughes, T. (1983), *Networks of Power: Electrification in Western Society 1880–1930*, Baltimore, MD: Johns Hopkins University Press.

International Energy Agency (IEA) (1980), *A Group Strategy for Energy Research, Development and Demonstration*, Paris: IEA.

International Energy Agency (IEA) (1985), *Energy Technology Policy*, Paris: IEA.

International Energy Agency (IEA) (1991), *Energy Policies of IEA Countries: 1990 Review*, Paris: IEA.

International Energy Agency (IEA) (2002), *Electricity Information 2002*, Paris: IEA.

International Energy Agency (IEA) (2007), *Energy Security and Climate Policy-Assessing Interactions*, Paris: IEA.

Jacobsson, S. and A. Johnson (2000), 'The diffusion of renewable energy technology: an analytical framework and key issues for research', *Energy Policy*, **28** (9), 625–40.

Jansen, J., W. Van Arkel, M. Boots (2004), 'Designing indicators of long-term energy supply security', Working paper ECN-C-04-007, Netherlands Energy Research Centre, Petten.

Joint Energy Security of Supply Working Group (JESS) (2004), 'Fourth report', London: DTI, May.

Junge, K. (1994), 'Diversity of ideas about diversity measurement', *Scandinavian Journal of Psychology*, **35**, 16–26.

Kaijser, A., A. Mogren and P. Steen (1991), *Changing Direction: Energy Policy and New Technology*, Stockholm: Statens Energiverk.

Kauffman, S. (1995), *At Home in the Universe: the Search for the Laws of Complexity*, London: Penguin.

Kempton, R. (1979), 'The structure of species abundance and measurement of diversity' *Biometrics*, **35**, 307–21.

King, D. (2006), 'Why we have no alternative to nuclear power', *Independent*, 13 July.

King, D. (2007), Interview on *The Material World*, BBC Radio 4, December.

Kruskal, J. (1964), 'Nonmetric multidimensional scaling of a numerical method', *Psychometrika*, **29**, 115–29.

Landau, R., T. Taylor and G. Wright (1996), *The Mosaic of Economic Growth*, Stanford, CA: Stanford University Press.

Lawson, N. (1992), *The View from Number 11: Memoires of a Tory Radical*, London: Bantam.

Laxton, R. (1978), 'The measure of diversity', *Journal of Theoretical Biology*, **70**, 51–67.

Lovins, A. and L. Lovins (1982), *Brittle Power: Energy Strategy for National Security*, Andover, MA: Brick House.

Lucas, N., T. Price and R. Tompkins (1995), 'Diversity and ignorance in electricity supply investment: a reply to Andrew Stirling', *Energy Policy*, **23** (1), 5–7.

MacArthur, R. (1965), 'Patterns of species diversity', *Biological Reviews*, **40**, 510–33.

Markandya, A., V. Costantini, F. Gracceva and G. Vicini (2005), 'Security of energy supply: comparing scenarios from a European perspective', Fondatione Eni Enrico Matei, Nota Di Lavoro 89.

Matthews, M. and F. McGowan (1990), 'Modes of usage and diffusion of new technologies', European Commission FAST Project, FOP 250, Brussels.

Matthews, M. and F. McGowan (1992), 'Reconciling diversity and scale: some questions of method in the assessment of the costs and benefits of European integration', *Revue d'Economie Industrielle*, **59**, 313–27.

May, R. (1990), 'Taxonomy as destiny', *Nature*, **347**, 129–30.

Mitchell, C. and B. Woodman (2006), 'New nuclear power: implications for a sustainable energy system', Green Alliance, University of Warwick.

Mitchell, C. (2007), *The Political Economy of Sustainable Energy*, London: Palgrave.

National Economic Research Associates (NERA) (1995), 'Diversity and security of supply in the UK electricity market', report to the Department of Trade and Industry, London.

National Economic Research Associates (NERA) (2002), 'Security in gas and electricity markets', report to the Department of Trade and Industry, London, October.

OXERA (2004), 'Results of renewables market modelling', study conducted for UK Department of Trade and Industry.

Parkinson, C. (1989), Secretary of State for Energy, in *Hansard*, 5 April, p. 275.

Patterson, W. (1999), *Transforming Electricity*, London: Earthscan.

Performance and Innovation Unit (PIU) (2001), 'Working paper on generating technologies: potentials and cost reductions to 2020', UK Cabinet Office, London.

Pielou, E. (1969), *An Introduction to Mathematical Ecology*, New York: Wiley.

Pielou, E. (1977), *Mathematical Ecology*, New York: Wiley.

Plummer, J. (ed.) (1983), *Energy Vulnerability*, Cambridge: Ballinger.

Prime Minister's Strategy Unit (PMSU) (2002), 'The energy review', UK Cabinet Office, February.

Prindle, B., M. Eldridge, M. Eckhardt and A. Frederick (2007), 'The twin pillars of sustainable energy: synergies between energy efficiency and renewable energy technology and policy', American Council for an Energy Efficient Economy, ACEEE Report Number E074, Washington, DC.

Rao, C.R. (1982), 'Diversity and dissimilarity coefficients: a unified approach', *Theoretical Population Biology* **21**, 24–43.

Rosenberg, N. (1982), *Inside the Black Box: Technology and Economics*, Cambridge: Cambridge University Press.

Rosenberg, N. (1996), 'Uncertainty and technological change', in R. Landau, T. Taylor and G. Wright (eds), *The Mosaic of Economic Growth*, Stanford, CA: Stanford University Press, pp. 334–53.

Rumsfeld, D. (2002), News Briefing at the Department of Defence, Washington, DC, 12 February.

Runnegar, B. (1987), 'Rates and modes of evolution in the Mollusca', in M. Campbell and R. May (eds), *Rates of Evolution*, London: Allen & Unwin.

Scheepers, M., A. Seebregts, J. de Jong, H. Maters (2007), 'EU standards for energy security of supply: updates on the Crisis Capability Index and the Supply/Demand Index, quantification for EU-27', Netherlands Energy Research Centre, Petten.

Scrase, I. and G. MacKerron (eds), *Climate of Urgency: Empowering Energy Policy*, London: Palgrave.

Sen, P. (2005), 'Gini diversity index, Hamming distance, and curse of dimensionality', *Metron – International Journal of Statistics*, **63** (3), 329–49.

Shannon, C. and W. Weaver (1962), *The Mathematical Theory of Communication*, Urbana, IL: University of Illinois Press.

Simpson, E. (1949), 'Measurement of diversity', *Nature*, **163**, 41–8.

Skea, J. (2007), 'Note on behaviour of diversity indices', personal communication to A. Stirling, October.

Smith, A., A. Stirling and F. Berkhout (2005), 'The governance of sustainable socio-technical transitions', *Research Policy*, **34**, 1491–1510.

Sneath, P. and R. Sokal (1973), *Numerical Taxonomy: The Principles and Practice of Numerical Classification*, San Francisco, CA: Freeman.

Solow, A. and S. Polasky (1994), 'Measuring biological diversity', *Environmental and Ecological Statistics*, **1**, 95–107.

Spicer, M. (1987), Energy Minister, 'Energy 1987', address to the Conference of the British Institute for Energy Economics, Chatham House, London, 7 December.

Stirling, A. (1994a), 'Diversity and ignorance in electricity supply investment: addressing the solution rather than the problem', *Energy Policy*, **22** (3), 195–216.

Stirling, A. (1994b), 'New nuclear investment and electricity portfolio diversity', submission to the UK Government Review of Nuclear Power Policy by Consortium of Local Authorities (COLA), Gwent, September.

Stirling, A. (1995), 'Diversity in electricity supply: a response to the reply of Lucas et al', *Energy Policy*, **23** (1), 8–11.

Stirling, A. (1996), 'Optimising UK electricity portfolio diversity', in G. MacKerron and P. Pearson (eds), *The UK Energy Experience: A Model or a Warning?*, London: Imperial College Press, pp. 135–48.

Stirling, A. (1997), 'Multicriteria mapping: mitigating the problems of environmental valuation?', in J. Foster (ed.), *Valuing Nature: Economics, Ethics and Environment*, London: Routledge, pp. 37–45.

Stirling, A. (1998), 'On the economics and analysis of diversity'. SPRU Electronic Working Paper Number 28, University of Sussex, October, http://www.sussex.ac.uk/Units/spru/publications/imprint/sewps/sewp28/sewp28.pdf.

Stirling, A. (1999), 'On science and precaution in the management of technological risk: Volume I – a synthesis report of case studies', European Commission Institute for Prospective Technological Studies, Seville, EUR 19056 EN, May ftp://ftp.jrc.es/pub/EURdoc/eur19056en.pdf.

Stirling, A. (2003), 'Risk, uncertainty and precaution: some instrumental implications from the social sciences', in I. Scoones, M. Leach and F. Berkhout (eds), *Negotiating Change: Perspectives in Environmental Social Science*, Cheltenham, UK and Northampton, MA, USA: Edward Elgar, pp. 33–76.

Stirling, A. (2006), 'Analysis, participation and power: justification and closure in participatory multi-criteria analysis', *Land Use Policy*, **23** (1), 95–107.

Stirling, A. (2007), 'A general framework for analysing diversity in science, technology and society', *Journal of the Royal Society Interface*, **4** (15), 707–19; also available at http://www.journals.royalsoc.ac.uk/content/a773814672145764/fulltext.pdf.

Stirling, A. (2008a), 'Opening up and closing down: power, participation and pluralism in the social appraisal of technology', *Science Technology and Human Values*, **33** (2), 262–94.

Stirling, A. (2008b), 'Diversity and sustainable energy transitions: multicriteria diversity analysis of electricity portfolios', in M. Bazilian and F. Roques (eds), *Analytical Methods for Energy Diversity and Security*, Oxford: Elsevier, pp. 3–29.

Stirling, A. (2009), 'Foreword', in I. Scrase and G. MacKerron (eds), *Climate of Urgency: Empowering Energy Policy*, London: Palgrave, pp. 251–60.

Suzuki, T. (2001), 'Energy security and the role of nuclear power in Japan', Central Research Institute of the Electric Power Industry.

Ulph, A. (1988), 'Quantification of benefits of diversity from reducing exposure to volatility of fossil fuel prices', evidence to Hinkley Point C Planning Enquiry for Central Electricity Generating Board, 25 October.

Ulph, A. (1989), 'Notes on the use of the CAPM model to value diversity benefits', Document S4165, Hinkley Point C Public Inquiry.

Umbach, F. (2004), 'Global energy supply and geopolitical challenges', in F. Godement, F. Nicolas and T. Yakushiji (eds), *Asia and Europe: Cooperating for Energy Security*, Tokyo: Council for Asia-Europe Cooperation, pp. 58–72.

Verrastro, F. and S. Ladislaw (2007), 'Providing energy security in an interdependent world', *Washington Quarterly*, **30** (4), 95–104.

Vincke, M., M. Gassner and B. Roy (1992), *Multicriteria Decision-Aid*, Chichester, UK; New York, USA; Brisbane, Australia; Toronto, Canada; Singapore: John Wiley & Sons.

Walker, W. (2000), 'Entrapment in large technological systems', *Research Policy*, **29** (7–8), 833–46.

Weitzman, M. (1992), 'On diversity', *Quarterly Journal of Economics*, **107**, 363–405.

Williamson, O. (1993), 'Transaction cost economics and organisation theory', *Industrial Economics and Corporate Change*, **2**, 107–56.

Yoshizawa, G., A. Stirling and T. Suzuki (2008), 'Electricity system diversity in the UK and Japan – a multi-criteria diversity analysis', SEPP Working Paper, University of Tokyo, October.

12 Review of the world and European renewable energy resource potentials

Helena Cabal, Maryse Labriet and Yolanda Lechón

12.1 INTRODUCTION

The availability of resources is crucial in energy planning at local, regional or global level. One of the main advantages of the renewable technologies against the fossil and nuclear ones is the use of non-exhaustible resources. However, the resources vary from one place to another in such a way that one area can be suitable for one renewable technology but not for the others. The estimation of the potentials allows the planning of an energy system in the long term as regards security of supply, sustainability and efficiency.

Some different types of potential can be defined:

- Theoretical potential: total energy content of the resource (wind, water, biomass, sun energy). It is the highest level of potential, the upper limit of what can be produced from a certain energy source from a theoretical point of view (EREC, 2010).
- Geographical potential: resulting from imposing some geographic constraints on the theoretical potential, for instance the exclusion of specific areas (protected areas, urban areas and so on).
- Technical potential: resulting from imposing technical constraints on the geographical potential, for instance accessibility or technical limitations.
- Economic potential: proportion of the technical potential that can be realized economically.

Calculating the potential of any energy resource is not an easy task because many factors can be included or ignored and there are a great number of different tools and methodologies. It is common to find theoretical potential studies in the literature based on geographic information systems (GISs) or on simulation or prediction models. The use of different tools and, above all, different assumptions lead to broad ranges in the data results.

In this chapter, a deep literature review has been carried out to gather the most recent data from the most relevant studies on global and European potentials for wind power, hydropower, biomass, solar power and ocean power.

12.2 WIND POWER

Besides the different potential categories already listed, the European Environmental Agency (EEA, 2009) defines two other potentials: (1) the restricted potential: the part of the total technical potential that could be used when other factors such as biodiversity

---

BOX 12.1   FACTORS AFFECTING WIND ENERGY POTENTIAL CALCULATIONS

Factors to consider:
  Calculation of the conversion potential
  Areas excluded (urban, forest, wetlands, nature reserves, glaciers, sand dunes)
  Excluding of agriculture lands: whether agriculture can coexist with wind farms or can not
  Prevailing wind conditions due to the topography (ravines, basins)
  Slope of the terrain (foundations)
  Minimum distances to settlements
  Local exclusion criteria (smaller nature reserves, infrastructure surfaces, military areas)
Specific to offshore wind power:
  Sea depths and distance to shore
  Average annual ice coverage of the sea
  Regionally varying minimum distance from the coast

---

protection, regulations or social preferences are taken into account; and (2) the economically competitive potential; the part of the technical potential that can be achieved in a cost-effective way considering the projections of future energy costs. Most of the studies start with a top-level theoretical resource that is progressively reduced through consideration of constraints such as those presented in Box 12.1.

The capacity factor and the full-load hours are also two important factors to be considered when calculating wind power potential. The capacity factor is an indication of how efficiently a wind turbine is operating. In this literature review, we have found capacity factors varying from 24 per cent today to 35 per cent in 2050 for onshore wind turbines. Regarding offshore wind turbines, capacity factors are higher because winds are stronger, and at present may reach 30 per cent. Greenpeace and the Global Wind Energy Council (GWEC) (2006) forecast an increase in the average global capacity factor equal to 28 per cent in 2010 and to 30 per cent by 2036.

The full-load hours are the number of hours per year that the wind turbine operates at rated power. This parameter is very site-specific. We have found present data which goes from 2000 hours to 3000 hours, and future data which reaches 4000 hours and even 5000 hours by 2050 (de Vries et al., 2007).

Results may vary a lot depending on all these factors, as may be seen in the next sections.

12.2.1   Global Potential

In October 2008, Greenpeace and the Global Wind Energy Council published the Global Wind Energy Outlook with global perspectives for wind power to 2020 and 2050 (Greenpeace and GWEC, 2008). In this report, three scenarios were set out: a reference scenario based on the 2007 World Energy Outlook projections (IEA, 2007); a moderate scenario including the present and planned renewable energy support policies and assuming that the targets set by many countries for renewables are successfully implemented; and an advanced scenario where all policy options in favour of renewables have been adopted. At the same time, these scenarios are set against two scenarios of projections for the future growth of electricity demand: a reference demand projection scenario and an energy efficiency demand projection scenario. The first is also based on data from 2007 World Energy Outlook while the second assumes the introduction of energy efficiency measures which will lead to an important reduction in the demand. Global technical potential results from this study are:

- for the reference scenario: 864 TWh/y in 2020 and 1783 TWh/y in 2050;
- for the moderate scenario: 1740 TWh/y in 2020 and 4818 TWh/y in 2050; and
- for the advanced scenario: 2651 TWh/y in 2020 and 9088 TWh/y in 2050.

The results for 2050 are more optimistic than those obtained in the previous study of 2006.

Within the framework of the European Fusion Development Agreement (EFDA) and co-funded by the European Commission, CIEMAT (Spain) and the Research Studios (Austria) carried out a review of the global fossil, nuclear and renewable energy resources (Labriet et al., 2009), consisting of a literature review and the use of a geographic information system (GIS). The wind power technical potential was estimated for three different availability factors: low (less than 800 h/a), medium (between 800 and 3000 hs/a), and high (more than 3000 hs/a), taking also into account the distance from the wind farm to the final user. Resulting technical potentials were 56 892 TWh/y for the onshore wind power and 16 474 TWh/y for the offshore, the total amounting to 73 366 TWh/y.

Based on a simulation of global wind fields, Lu et al. (2009) estimated the global wind power theoretical potential in 1.3 million TWh/y, when there is no limit in the capacity factor. Applying a 20 per cent limit of the capacity factor, they estimated a technical potential of 840 000 TWh/y; very high figures when compared with the rest of the literature.

Through a methodology consisting of physical and geographic data gathering and distribution in cells, and the estimation of the availability and suitability of the areas, de Vries et al. (2007) estimated the global wind power technical potential as the product of the power generation density by the available or suitable area. The study focuses on the available land use and its constraints and on the production costs as a function of the availability and the resources exhaustion and the innovation dynamics. Four scenarios are set out, all of them based on Intergovernmental Panel on Climate Change Special Report on Emissions Scenarios (IPCC SRES) scenarios and developed with the IMAGE model tool (Strengers et al., 2004). The technical potential for the four scenarios varies from 62 000 TWh/y to 80 000 TWh/y; and the economic potential, assuming production costs lower than 10c€/kWh, from 23 000 TWh/y to 39 000 TWh/y, in 2050.

The REN21 network, Renewable Energy Policy Network for the 21st Century, estimates a global technical potential of 105 279 TWh/y for the onshore wind power, and 6111 TWh/y for the offshore in 2050 (REN21, 2008; Hoogwijk and Graus, 2008). The

Fourth Report of the IPCC (Sims et al., 2007) gives a global technical potencial of 66 668 TWh/y by 2050. Other previous studies have estimated a technical onshore potential of 96 000 TWh/y (Hoogwijk et al., 2004), 19 000–483 000 TWh/y (WEC, 2004), 53 000 TWh/y (Grubb and Meyer, 1993) and 43 000–86 000 TWh/y (van Wijk and Coelingh, 1993).

As can be seen, there is a great difference among the potentials given by the different studies, making clear the relevance of the hypothesis or assumptions considered. The global technical potential in 2050 can vary between 9000 and 170 000 TWh/y.

12.2.2 European Potential

The most recent study on renewable energy projections in Europe (Beurskens and Hekkenberg, 2010) has been carried out by the Energy Research Centre of the Netherlands (ECN) and financed by the European Environment Agency. It consists of an evaluation of the National Renewable Energy Action Plans (NREAP) of 27 European Union (EU) countries, although this update covers only 19 countries. The report gives aggregate results on the projections for the countries for which NREAP documents are available. Projected total wind power electric capacity by 2020 is 195.5 GW, of which 22 per cent will be offshore wind power. This capacity installed will generate 454 438 GWh.

The European Environmental Agency published in 2009 a report on wind power potential in Europe (EEA, 2009) where the total technical potential in Europe is estimated as 70 000 TWh/y (45 000 TWh/y onshore and 25000 TWh/y offshore) in 2020, and 75 000 TWh/y (45 000 TWh/y onshore and 30 000 TWh/y offshore) in 2030. If some areas are excluded, such as those included in Natura 2000, an EU-wide network of nature protection area (http://ec.europa.eu/environment/nature/natura2000/index_en.htm) or the offshore gas and oil platforms, the resulting restricted potential would be 41 800 TWh/y (39 000 TWh/y onshore and 2800 TWh/y offshore) in 2020 and 42 500 TWh/y (39 000 TWh/y onshore and 3500 TWh/y offshore) in 2030. Finally, the EEA estimates the total economic competitive potential as 12 200 TWh/y (9600 TWh/y onshore and 2600 TWh/y offshore) in 2020, and 30 400 TWh/y (27 000 TWh/y onshore and 3400 TWh/y offshore) in 2030. Areas of Red Natura 2000 have not been excluded in the estimation of this potential.

Ciemat and the Research Studios (Labriet et al., 2009) give a technical potential of 5341 TWh/y for onshore wind power and 3554 TWh/y for offshore, which add up to 8895 TWh/y. The REN21 network estimates a technical onshore potential of 4444.5 TWh/y and an offshore potential of 1388.9 TWh/y in 2050 in the European countries belonging to the Organisation for Economic Co-operation and Development (OECD), and 18 611.3 TWh/y and 11 111.1 TWh/y in the rest of Europe and in former Soviet Union countries, respectively (REN21, 2008; Hoogwijk and Graus, 2008). Finally, Greenpeace published in 2004 a study focused only on offshore wind power (Greenpeace, 2004) where, through a GIS, EU offshore potential is estimated as 82.3 TWh/y by 2010, 259.1 TWh/y by 2015, and 379.6 TWh/y by 2020.

Besides those studies on wind power potential, there are many others on projections of wind power generation in Europe based on policy scenarios such as that recently published by the European Renewable Energy Council (EREC, 2010). In this report, EREC presents a pathway towards a 100 per cent renewable energy system for the EU by 2050. According to EREC, and in this scenario, wind power installed capacity will be 180 GW by 2020, 288.5 GW by 2030, and 462 GW by 2050. Those capacities will produce 477 TWh/y by 2020, 833 TWh/y by 2030, and 1552 TWh/y by 2050.

The European Wind Energy Association (EWEA) has also carried out a study on projections, setting three scenarios for the wind power evolution in Europe up to 2030 (EWEA, 2009; EWEA 2008): low scenario, reference scenario and high scenario. The reference scenario assumes a total installed capacity of 180 GW in 2020 and 300 GW in 2030; the high scenario assumes a capacity of 350 GW (150 GW offshore) in 2030; and the low scenario assumes a capacity of 200 GW (40 GW offshore) in 2030. With those capacities, EWEA estimates a total wind power generation which goes from 176 TWh/y in the low scenario to 179 TWh/y in the high one by 2010; from 361 TWh/y to 556 TWh/y by 2020; and from 571 TWh/y to 1104 TWh/y by 2030. Another relevant conclusion is that the level of development of wind power in Europe mainly depends on the offshore implementation.

Lastly, the 'European energy and transport – trends to 2030' update (EC, 2008) sets a base scenario for each country of the EU-27 including the trends and policies in force at the end of 2006 and, through a modelling exercise, gives a data for wind power generation of 84.82 TWh/y by 2030.

Projections based on different policy scenarios are below the total technical potential data found in the literature.

12.3   HYDROPOWER

Hydropower maximum theoretical potential in each country is commonly defined as the total global energy content of its water resources. It is estimated by taking into account many factors, such as:

- the internally generated surfaced water annual runoff derived from precipitation falling within the nation's boundaries;
- the external flow entering from other nations;
- the external flow leaving to other nations;
- other water uses in competition with electricity generation.

At the same time, the World Energy Council defines the gross theoretical capability as the annual energy potentially available in the country if all natural flows were turbined down to sea level or to the water level of the border of the country (if the water course extends into another country) with 100 per cent efficiency from the machinery and driving waterworks.

Technical potential is the hydropower generated at the geographical potential, defined as the total global amount of land area available for hydro facilities installation. Economic potential is the technical potential that can be realized economically given the cost of the technology. Both potentials are calculated by applying some limitations to the theoretical potential. For instance, floods considerably increase the surface water runoff but cannot be taken as a usable resource. Besides, there are other factors such as the seasonal variability (precipitations, runoff and recharge depend on the season) and the green water, the water that is used to sustain ecosystems and agriculture.

Hydropower can be divided into two categories depending on the size of the plant. Although there are different criteria to distinguish between one and the other, the most popular, at least in Europe, is that which uses the term 'small' for those plants with a capacity up to 10 MW, and 'large' for the rest. However, literature related to hydro potential does not distinguish between large and small hydroelectricity, so that this difference is not included in this review.

Finally, among the literature review, there is an outstanding publication which is frequently referenced by the others: the one from *Hydropower and Dams* (2000).

12.3.1 Global Potential

Hydro resources are widely spread around the world. Most of the economic potential remains to be developed (IHA, 2000), especially in the developing countries. Asia and South America seem to have a big hydropower potential.

According to the International Energy Agency (IEA, 2008), after two decades of stagnation, hydropower could increase production in the coming years. Most of the new capacity will be installed in the developing Asian countries (China, India and Vietnam), while in the developed countries there is almost no possibility of new installations due to the scarcity of good locations and the strict environmental laws. Growth in hydropower generation is foreseen from 3035 TWh/y in 2006, to 3730 TWh/y in 2015, and 4810 TWh/y in 2030.

The REN21 network estimates a global technical potential of 13 889 TWh/y by 2050 (REN21, 2008; Hoogwijk and Graus, 2008), below the one proposed in the Fourth IPCC Report (Sims et al., 2007) that was around 16 667 TWh/y. This document reports a small hydropower potential of 555.6 TWh/y. The energy resources report of the World Energy Council (WEC, 2007) estimates the global theoretical potential as 40 497.5 TWh/y and the technical potential as 16 228 TWh/y. UNESCO publishes a three-yearly report on the water in the world. In the second report (UNESCO, 2006) the global theoretical potential was calculated as 40 293 TWh/y and the technical potential as 15 899 TWh/y. The European Renewable Energy Council (EREC, 2002) gives a global theoretical potential of 14 370 TWh/y based on data from the *World Atlas* of *Hydropower and Dams*.

A consensus on global potential may be reached, being a theoretical potential of around 40 000 TWh/y and a technical potential of around 16 000 TWh/y.

12.3.2 European Potential

The recent study carried out by ECN (Beurskens and Hekkenberg, 2010) gives aggregate results on the projections for the countries for which NREAP documents are available. Projected total hydropower electric capacity by 2020 is 121.97 GW, of which 28 per cent will be pumped storage and 11 per cent will be small hydropower. This capacity installed will generate 335 940 GWh.

Additionally, following the same sources used to estimate the global potential and other European projects and studies, some figures on the theoretical and technical hydropower potential for the EU have also been found. These are between 2597 TWh/y of theoretical potential and 1018 TWh/y of technical potential according to WEC (2007), and 2638 TWh/y of theoretical potential and 1190 TWh/y of technical potential according to UNESCO (2006). EREC EREC (2002) gives a technical potential slightly higher than the former, 1225 TWh/y, and an economic potential of 700 TWh/y. EREC (2008) also estimates that hydropower generation will reach 384 TWh/y by 2020.

The theoretical potential for Europe may be situated at around 2600 TWh/y, and the technical potential at around 1200 TWh/y. This potential is difficult to surpass due to the scarcity of good locations. Nevertheless, the hydropower industry is making big efforts to increase its potential by upgrading of equipment in order to increase the plants' capacity and, in the Eastern European countries, renovating the obsolete plants that were abandoned during the Soviet Union period.

12.4   BIOMASS

In general terms, the energy use of biomass includes energy crops, agricultural residues, forest residues, food processing residues, animal manure, material processing (wood processing) residues, food consumption waste and other waste. This section is focused on the energy crops and agricultural and forest residues, although there is also some estimation on food industry, wood industry and livestock wastes. As regards municipal wastes, their energy role remains debatable. On the one hand, high costs are associated with sorting them, and recycling and packaging policies make very uncertain the future quantity of municipal wastes; on the other hand, waste-to-energy policies are promoted in several countries, for example in Europe. Given these uncertainties, municipal wastes have been kept in the 'other' category. The same logic applies to gas from landfill (directly linked to municipal waste policies) and other industrial wastes. Moreover, the energy potential of these wastes is relatively small (13 EJ in 2050 according to Sims et al., 2007) when compared to energy crops.

Evaluating biomasss resource potentials is a complex task. All the reviewed studies use different scenarios with different hypotheses on, for instance, irrigation, population growth, consumption patterns or technology status. Moreover, data on land areas assigned to energy crop cultivation, yields and geographical distribution differ a lot among the studies. Productivity also has a strong influence on biomass potential: for example, there are studies which show that when productivity increases by 43 per cent, potential increases by 38 per cent (Rogner, 2000).

Several studies assessing biomass potentials available for energy purposes have been consulted in order to gain insights into the future potentials available. More information and a detailed review may be found in Labriet et al., (2010). Some sustainability definitions proposed in the reviewed studies are also presented in this section.

There are three main factors influencing the valuation of energy crop potential:

● Agriculture and animal system production, having an impact on the yields: level of mechanized agriculture systems, optimized varieties with higher harvest indexes or not, irrigation or not, level of utilization of fertilizers and pesticides, landless industrial animal production systems. For instance, according to the UN Food and Agriculture Organization (FAO), organic agriculture reduces production by around 10 per cent to 30 per cent compared with conventional intensive agriculture (Fischer et al., 2007).

● The demand for food and feed that must take priority: the potential will depend on population growth and diet. Turning to a vegetarian diet could release the land currently devoted to livestock (de Wit and Faaij, 2008; Seidenberger et al., 2008). More than one-third of the need for land in the EU is associated to crops for food, while the other two-thirds are devoted to products for livestock, of which half are cultivated with fodder and the other half are pastures (de Wit and Faaij, 2008).

● Type of land considered – arable, grassland, fallow, set-aside, marginal or degraded land – and of biomass and conversion technologies (second-generation or not); these factors have an impact on the yields and the energy content. Many authors think that the crops for the production of second-generation biofuels, such as willow and eucalyptus, will be a better bioenergy option and more sustainable because less intensive management, less fossil energy consumption and less fertilizer are needed and they can be grown an lower-quality land. Moreover, the second-generation energy crop harvest takes place after the leaf fall, allowing nutrients to remain in the soil and facilitating the use of agricultural and forest residues (EEA, 2006; IEA, 2007; Lysen and van Egmond, 2008; Rogner, 2000; Seidenberger et al., 2008).

Box 12.2 summarizes the hypotheses used in the more relevant studies regarding biomass potentials.

There are other studies that do not agree with those hypotheses. Some state that more research and development (R&D) is needed to develop very efficient production and management systems, and also second-generation technologies (Lysen and van Egmond, 2008). At the same time, Doornbosch and Steenblik (2007) consider that there is still enough uncertainty in the set-aside and degraded land potentials to be included in the analysis. They also consider that the potential of unirrigated land to change to irrigated will be limited by water scarcity in many countries. The International Energy Agency (IEA, 2007) also considers that the second-generation technologies will not be commercialized before 2030, and thus they are excluded from its scenarios. However, the IEA also admits that the second-generation biofuels are an environmentally better option than the first-generation ones.

Additionally, in terms of sustainability, there are other factors that have to be taken into account such as deforestation, nature conservation, protection of biodiversity and competition for land between sectors (for example bioenergy production and food production). Whether those parameters are considered or not may change the results a lot. Box 12.3 provides a qualitative description of the sustainability requirements included in some of the studies. Current developments in sustainable biomass certification may also give a detailed framework for the definition of sustainability criteria for the energy use of biomass (GBEP, 2009).

12.4.1   Global and European Potential

Generally speaking, the development of efficient and sustainable agriculture production and management systems is at the heart of the future strategies, and assumptions related to the future production systems (level of yield increasing inputs, irrigation or

---

BOX 12.2   EXAMPLES OF ASSUMPTIONS BEHIND THE ESTIMATION OF BIOENERGY POTENTIALS

Hoogwijk et al. (2003)
Two categories of lands are considered:

- surplus cropland (after the demand for food and fodder is satisfied); a total of 5 Gha agricultural land is considered (3.5 Gha of grassland, 1.5 Gha of cropland, half of the latter being rainfed and the other half irrigated);
- surplus degraded land (deforested or otherwise degraded or marginal land still suitable for reforestation), a total of 430–580 Mha is considered available;

Energy crops are limited to woody short rotation crops like eucalyptus and willow. The yield is 10–20 tonnes/ha/y for surplus agricultural land, 1–10 tonnes/ha/y for degraded land. The HHV energy content is 19 GJ/ton.

Three population projections and three food diet (vegetarian, moderate, affluent). Two production systems: a high external input (HEI) system (best technical means, full fertilizer applications, and so on) and a low external input (LEI) system (minimization of the environmental risks, with no chemical fertilizer application). The HEI system considers high yields (about a factor of two compared to the 2000 yields), while the LEI is close to present global yields.

Seidenberger et al. (2008)
Consider changes in land use, use of fallow (set-aside) areas, area productivity (use of yield increasing agent) and changes in consumption habits. Crops for first-generation biofuels are excluded because they are considered as unsustainable. Scenarios:

- Scenario BAU: Legal and economic conditions apply for the future. Forest clearing, change of grazing and loss of valuable agriculture areas for industry and traffic continue.
- Scenario Basic: No forest clearing. The use of fallow lands is as follow: for countries with political programs for land set-aside, these programs are discontinued and 80 per cent of the fallow areas demarcated for this land will be used for agricultural production in the usual cultivation mix. For the other countries, demarcated fallow areas will be used.
- Scenario Sub 1: (called 'ecological sustainability'). In all the countries, the area productivity increases more slowly than until now. Cultivation methods are more strongly orientated towards sustainability and sparse use of yield-increasing agents. Yield, compared with the base, will be reduced by 10 per cent in 2010, 20 per cent in 2015 and 30 per cent up

to 2020 (with regional variations – for example Russia already has a high percentage of ecological agriculture and low yields, so that an additional high decrease is not appropriate). No change of grassland and pasture takes place.

- Scenario Sub 2: (called 'change in eating habits'). All countries consuming more than 850 grain units per capita per year will reduce it by 30 per cent. Those consuming below 850 grain units per capita per year remain the same. And those consuming between 850 and 1215 grain units per capita per year will reduce it by less than 30 per cent to the level of 850 grain units per capita.
- Scenario Sub 3: Assumes a reduction in area productivity and a change in consumption habits according to the WHO in the entire world.

Smeets et al. (2004)
The computation factors are: population growth, food consumption and composition per capita, food and bioenergy crops yields (from 100 to 700 GJ/ha, depending on the farming practise), efficiency and feed inputs in the animal production system, wood consumption and production, and natural forest growth.

Four scenarios based on: high feed conversion efficiency, mixed (pasture +landless) or landless animal production systems, level of technology (impact of the yield growth) and water supply (irrigation or not).

- Scenario 1: High feed conversion efficiency, mixed animal production system, very high technology, rain-fed.
- Scenario 2: High feed conversion efficiency, mixed animal production system, very high technology, rain-fed + irrigation.
- Scenario 3: High feed conversion efficiency, landless animal production system, very high technology, rain-fed + irrigation.
- Scenario 4: High feed conversion efficiency, landless animal production system, super high technology, rain-fed + irrigation.

not, pasture or industrial animal production system, and so on), translated into yield and energy content of crops, as well as land availability for crops and livestock, are included in (almost) all the proposed scenarios included in the literature review. Values used most often vary between 6.5 and 15 t/ha (with smaller values possible for degraded land, and higher values possible in tropical regions). Moreover, some estimates of the basic annual increases of yields are of 1.2–1.6 per cent, for global average yields up to 2050, and 1.1–2.6 per cent up to 2020 (Hoogwijk et al., 2005).

Most studies recognize that crops appropriated for second-generation biofuel technologies (woody biomass such as willow, eucalyptus, hybrid poplar, miscanthus or switchgrass grasses) would be better suited as feedstock for bioenergy than the conventional (current) ones given their better sustainability performance: they require far less intensive management, low fossil energy inputs and less fertilizer; they can grow on

BOX 12.3   EXAMPLE OF SUSTAINABILITY REQUIREMENTS

Doornbosch and Steenblik (2007)

Although not explicitly included in the estimation, the study presents some possible criteria, like the ones proposed by Netherland's project group for sustainable biomass: GHG emissions, no competition with food and local uses, biodiversity, quality of the soils, water impacts, air impacts and local welfare.

EEA (2007, 2006)

The environmentally compatible primary biomass is the quantity 'that is technically available for energy generation based on the assumptions that no additional pressures on biodiversity, soil and water resources are exerted compared to a development without increased bioenergy production . . . It should be in line with other current and potential future environmental policies and objectives.'

The study includes a list of criteria related to: erosion; soil compaction; nutrient leaching to groundwater and surface water; pesticide pollution of soils and water; water abstraction; fire risk; farmland biodiversity; and diversity of crop types.

Hoogwijk et al. (2005)

Sustainability is not explicitly included in the estimation. However, the categories of land are based on the assumption that the production of energy crops should not affect food and forestry production, nature reserves or biodiversity and animal grazing.

Seidenberger et al. (2008)

In all scenarios: 'Sustainable energy crop production takes place only on the surplus area of arable land and grassland.'

In some scenarios: 'Ecologial goals in land cultivation and landscape utilisation are pursued more strongly . . . Change of pasture is renounced, the use of yield-increasing agricultural chemical aids is successively reduced and ecological cultivation is massively expanded.'

Smeets et al. (2004)

'The production of bioenergy from specialised bioenergy crops must be regarded as unsustainable if one or both of the following criteria is not met:

● The production of bioenergy is only allowed on abandoned or surplus agricultural land (bioenergy production is not allowed to compete with agricultural land use for food production). Surplus agricultural land includes both areas of degraded land no longer suitable for commercial crop production and areas that are taken out of production due to a surplus of productive area;
● Deforestation due to the demand for suitable cropland for bioenergy production is not allowed.'

The authors recognize that many other criteria are thinkable.

De Wit and Faaij (2008)
'At least, no degradation of land or depletion of nutrients occurs and water resources should not deplete or deteriorate in quality.'
There is one scenario with low yield representing a higher share of organic farming.

poor-quality marginal and degraded land (therefore reducing the need to use surplus arable and pasture lands); harvest of some crops that takes place after the nutrient-rich leaves have dropped, which allow most nutrients to remain on the land; and there is the possibility for use of agricultural and forest residues (EEA, 2006; IEA, 2007; Lysen and van Egmond, 2008; Rogner, 2000; Seidenberger et al., 2008).

However, it is also recognized that much more R&D is needed to develop such production and management systems and, especially, the use of feedstocks for the second-generation biofuels (Lysen and van Egmond, 2008). Doornbosch and Steenblik (2007) consider that the estimates of the potential of marginal and degraded land are still too uncertain to be included in their analysis (however, they estimate that the technical potential associated to this land might be in the order of 29–39 EJ). They also consider that the potential for expansion of the irrigated area onto land which is not suited to rain-fed cultivation would be very limited, given the water shortage faced by many countries with arid lands. The IEA (2007) also considers that second-generation biofuel technologies will not be commercially attractive before 2030, and excludes them from their its scenarios. However, it recognizes the second-generation biofuels as very promising, given their environmental advantages compared to conventional biofuels.

As regards the consumption patterns, it is recognized that shifts towards a more vegetarian diet would release substantial amounts of agricultural land for growing non-food crops (De Wit and Faaij, 2008; Seidenberger et al., 2008). For example, about one-third of land area requirements in the EU25 are currently associated with domestic food consumption of crop products, compared to two-thirds associated with livestock products consumption. From the latter, about half of the required area is arable land for the production of feed crops and the other half is pasture (de Wit and Faaij, 2008).

After this literature review, we can conclude that the land available for energy crops and the bioenergy potentials vary widely:

- from 34 (IEA, 2007) to almost 3600 Mha (Smeets et al., 2004), and from to 0 (IEA, 2007) to 1272 EJ (Smeets et al., 2004) for crops at the world level;
- from 2 (Seidenberger et al., 2008) to 100 Mha (Smeets et al., 2004) and from 2.6 (Alakangas et al., 2007) to 17 EJ (Seidenberger et al., 2008) in the EU-25;
- from 10 to 167 EJ for agriculture and food processing residues;
- from 15 to 151 EJ for wood residues.

The high potential for 2050 crops, determined by Smeets et al. (2004), shows potentials under intensive, very highly technologically developed agriculture and animal production systems (mechanized systems, optimized varieties with higher harvest indexes, irrigation, fertilizers, pesticides, landless industrial animal production system). Such assumptions result in both a high availability of land for energy crops, and high yields. More conservative estimations as proposed by Seidenberger et al. (2008), WBGU (2004) or, at the European level, EEA (2006), result from more restrictive assumptions generally motivated by environmental concerns (lower available land, organic farming, and so on).

Regarding the geographical distribution, Smeets et al. (2004) estimate that sub-Saharan Africa, the Caribbean and Latin America are the world regions with the highest potentials for bioenergy production due to the availability of large areas of arable land and pastures, and to the current low productivity and inefficient systems. However, there is a prerequisite for this bioenergy potential revaluation: current inefficient and non-productive systems must be replaced by best practices in technologies and management by 2050. On the other hand, Seidenberger et al. (2008) consider that all the available land in Africa, Central America and Asia is needed to satisfy the food demand, so that there is no land available for energy crops in these regions.

Within Europe, Eastern Europe and the former Soviet Union countries are considered as low-cost biomass sources. The potential growth of production is high, after the drop caused by the fall of communism and the economic restructuring (de Wit and Faaij, 2008; Fischer et al., 2007; Hoogwijk et al., 2005; Smeets et al., 2004).

Tables 12.1–12.3 summarize the bioenergy potentials by type of biomass (crops; agriculture and food-processing residues; and forest and wood-processing residues), at the world level or at a more disaggregated level when available.

Table 12.4 shows, in detail, the potentials for different crops and residues for 30 countries in Europe. Data come from de Wit and Faaij (2008) and have been used in European research projects such as RES2020 (RES2020, 2009).

Finally, the last study carried out on renewable energy projections by ECN (Beurskens and Hekkenberg, 2010) presents data for the projected total biomass electricity capacity by 2020 equal to 36.197 GW, of which 63 per cent comes from solid biomass, 25 per cent from biogas and 12 per cent from bioliquids. This capacity installed will generate 190979 GWh. Regarding biomass thermal energy, total capacity installed by 2020 is estimated as 72917 ktoe.

12.5   SOLAR RESOURCES

Solar energy results from the process of nuclear fusion in the sun and it is available at any location on earth. The amount of solar radiant energy falling on a surface per unit area and per unit time is called irradiance. Peak intensity of solar irradiance at sea level is around 1 kW/m² and the 24 h average is around 0.17 kW/m². This intensity changes from place to place. Some parts of the earth receive up to about 40 per cent more than this annual average. The highest annual mean irradiance of 0.3 kW/m² can be found in the Red Sea area, and typical values are about 0.2 kW/m² in Australia, 0.185 kW/m² in the United States and 0.105 kW/m² in the United Kingdom.

A world map with the annual averaged solar irradiance in W/m² can be found at http://www.soda-is.com/. The total world average power at the earth's surface in the form of solar radiation exceeds the total current energy consumption by 10000 times, but its

Table 12.1   Energy crops potentials

EJ	Smeets et al. (2004)	Berndes et al. (2003)		Hoogwijk et al. (2003)	Hoogwijk et al. (2005)		Ericsson and Nilsson (2006)
Horizon	2050	2050	2100	2050	2050	2100	Long term
World	215–1272	50–240	–	8–1100	311–657	395–1115	–
Part of the world only	–	24–143 (Asia, Africa, Latin America)	107–154 (Asia, Africa, Latin America)	–	–	–	2–15 (EU25)

EJ	Rogner (2000)	WBGU (2004)	De Wit and Faaij (2008) (REFUEL)	Doornbosch and Steenblik (2007)	EEA (2006)	Seidenberger et al. (2008)
Horizon	2050	2100	2030	2050	–	2050
World	276–446	37.4	–	109	–	6–80
Part of the world only	–	–	15.3–18.4 (EU-27 + CH + NO + Ucrane)	10 (Eur + Russia)	6 (EU-25)	4–17 (EU-27)

EJ	IEA (2007)	Wakker et al. (2005) (VIEWLS)	Yamamoto et al. (2001)	Alakangas et al. (2007) (EUBIONET)	Fischer and Schrattenholzer (2001)	Lysen and van Egmond (2008)
Horizon	2050	2020	2050–2100	No date	2050	2050
World	0–851	–	110–22	–	150–200	120–330
Part of the world only	–	6–12 (Central & Eastern Europe)	–	2 (EU-20) 2.6–7.8 (EU-28)	–	–

low density and geographical and time variations pose major challenges to its efficient utilization.

The solar irradiance on the horizontal plane or insolation has two components: direct irradiance and diffuse irradiance. Direct irradiance comes directly from the sun. Diffuse irradiance results from scattering in the atmosphere. A non-horizontal plane also receives reflected irradiance from the ground. The sum of these three components – direct, diffuse and reflected – is the global irradiance.

For clear-sky conditions, solar irradiance at any given time only depends on the latitude, height and ground albedo. Clear skies are not often found in many locations, and dense clouds could reduce the global insolation to a third of the clear sky value, and the direct insolation to zero.

Table 12.2   Agriculture and food-processing residues potentials

EJ	Smeets et al. (2007)			Hoogwijk et al. (2003)		
Horizon	2050			2050		
	Agri residues	Food-process residues	Total	Agri residues	Food-process residues	Total
World	49–69	16	65–85	5–27	5	10–32

EJ	Fischer and Schrattenholzer (2001)	IEA (2007)	Doornbosch and Steenblik (2007)	Kim and Dale (2004)	Lysen and van Egmond (2008)
Horizon	2050 Agri residues only	2050 Agri residues + Dung + Organic waste	2050 Agri residues	2050 Agri residues	2050 Agri + Forestry + Organic wastes
World	35	25–167	34.8	10.4	40–170

Table 12.3   Forest and wood-processing residues potentials

EJ	Berndes et al. (2003)	IEA (2007)	Dessus et al. (1992)	Doornbosch and Steenblik (2007)	Smeets et al. (2007)	
Horizon	2050	2050	2050	2050	2050	
	Forest residues	Forest residues	Forest (residues excluded)	Forest residues	Forest (residues excluded) Theoretical pot.	Forest (residues excluded) Ecological pot.
World	66–113	29–151	65.5	90.6	70.5	7.7

EJ	Sorensen (1999)	Yamamoto et al. (2001)		
Horizon	2050	2050 and 2100		
	Forest (residues excluded)	Forest residues	Wood process residues	Total
World	107.2	2050 4	2050 11	2050 15
		2100 7	2100 17	2100 24

Table 12.4   Biomass potentials in European countries

(PJ/a)	AT	BE	BG	CH	CY	CZ	DE	DK	EE	ES	FI	FR	GR	HU	IE
Rapeseed	24	8	213	6	2	86	247	54	31	174	19	182	29	130	54
Starch crops	30	10	214	6	1	112	318	67	36	215	42	292	35	171	5
Sugar crops	38	21	53	9	4	172	595	115	30	300	15	473	47	107	139
Grassy crops	85	27	538	28	15	325	1014	138	63	438	70	985	65	476	172
Woody crops	84	27	445	30	19	247	928	72	96	235	52	1273	70	405	167
Forestry residues	100	8	44	48	0	38	192	35	12	70	103	380	8	40	6
Wood waste	54	10	4	5	0	10	100	7	36	62	46	113	13	4	2

(PJ/a)	IS	IT	LT	LU	LV	MT	NL	NO	PL	PT	RO	SE	SI	SK	UK
Rapeseed	0	84	125	1	39	0	7	2	329	58	288	47	1	17	76
Starch crops	0	96	8	1	7	0	8	2	452	68	387	68	8	69	86
Sugar crops	0	91	318	1	106	0	17	1	995	156	250	70	3	34	152
Grassy crops	0	197	509	2	198	0	27	6	1285	150	1063	115	12	102	275
Woody crops	0	297	442	2	260	0	21	5	1390	84	860	113	15	130	245
Forestry residues	0	122	30	2	38	0	20	72	90	83	50	138	64	30	40
Wood waste	0	15	4	0	2	0	4	54	66	31	50	111	4	16	33

Source: RES2020 (2009a).

A variety of technical approaches have been developed to harness solar energy:

- Solar energy can be collected for direct thermal end uses such as cooking, heating, water heating and cooling, in different sectors: residential, commercial, industrial and agricultural.
- Solar energy can also be collected to produce electricity. Solar photovoltaic systems (PV) are energy devices that directly convert solar energy into electricity. PV systems make use of the global irradiance (direct and diffuse) to produce electricity.
- Solar thermal power generation systems capture energy from solar radiation, transform it into heat, and generate electricity from the heat. They can only use direct irradiance, so they must be sited in regions having high direct solar irradiance.

In this section, different solar potential categories are analysed:

- The theoretical potential is the yearly solar energy irradiated to the surface of the earth (kWh/y).
- The geographical potential is the yearly irradiance integrated over the terrestrial surface suitable for the installation of solar applications based on geographical constraints (kWh/y).
- The technical potential is the geographical potential after the losses of the conversion from the solar energy to secondary energy carriers such as electricity are taken into account.
- The economic potential is the technical potential up to an estimated production cost of the secondary energy form which is competitive with a specified, locally relevant alternative.

A further category of potentials would be the implementation potential, which is the maximum amount of the economic potential that can be implemented within a time frame, taking into account constraints and incentives.

12.5.1   Global Potential

Geographical solar potential results vary within the range proposed by the Intergovernmental Panel on Climate Change (IPCC), from 437 222 to 13 843 611 TWh/y (IPCC, 2001), taking into account the minimum and maximum values of clear-sky global irradiance and the available land for solar applications – the land category 'other land' of the Food and Agriculture Organization (FAO). The minimum value corresponds to the use of 1 per cent of this unused land; and the maximum, the use of 10 per cent. These values mean from 4 to 117 times the current primary energy consumption. Other important assessments of resources such as the 'World energy assessment' of the UNDP (Goldemberg, 2000) and the 'Survey of energy resources' of the World Energy Council (WEC, 2004) assume the same solar energy potential values proposed by the IPCC. Other estimations of geographic potential (Hoogwijk et al., 2004; Trieb, 2005; Labriet et al., 2009) are within the range of the IPCC proposals.

The technical potential calculation takes into account the efficiency of the technology which transforms the solar energy into the energy vector considered. When talking about power generation, two different technical potentials can be distinguished: one for electricity from photovoltaics, and the other from solar thermal. The technical potential for photovoltaic power found in the literature varies from 369 444 to 900 000 TWh/y (Hofman et al., 2002; Hoowijk et al., 2004; de Vries et al., 2007; Sims et al., 2007; Greenpeace and EREC, 2008; REN21, 2008; Hoowijjk and Graus, 2008). The technical potential for thermosolar power has been estimated between 68 889 and 639 167 TWh/y (Hofman et al., 2002; Trieb, 2005; Greenpeace and EREC, 2008; REN21, 2008; Hoowijk and Graus, 2008). The technical potential for heat generation has been calculated as 34 167 TWh/y (Greenpeace and EREC, 2008; Hoowijk and Graus, 2008).

There are also some studies that calculate the implementation potential of those technologies for different time periods. Figures 12.1–3 show the results of those studies in a comparative way.

Technical potential estimations for both technologies, photovoltaic and thermosolar, are much higher than the implementation potentials found in the literature. The development of those technologies will not therefore be limited by the available resources, but by other restrictive technical or economic constraints. These are not mature technologies that need big efforts in research and development to reach a level of penetration according to their technical potential.

12.5.2   European Potential

Some of the documents analysing the global solar power geographical potentials also give estimations on potentials for Europe. The IPCC (2001) estimates the potential for Europe as between 6944 and 254 167 TWh/y; Hoowijk et al. (2004) as 38 056 TWh/y, and Trieb (2005) as 6944 TWh/y but Trieb referred only to the potential resulting from direct solar irradiance.

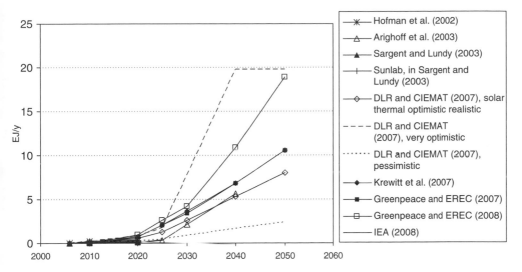

Figure 12.1 Projections of thermosolar power market development

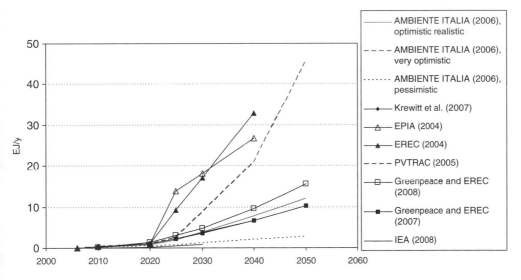

Figure 12.2 Projections of photovoltaic power market development

The recent study carried out by the ECN (Beurskens and Hekkenberg, 2010) presents data for the projected total solar power capacity by 2020 equal to 87.756 GW, of which 92 per cent will be solar photovoltaic. This capacity installed will generate 99 765 GWh. Regarding solar thermal energy, total capacity installed by 2020 is estimated as 5437 ktoe.

Regarding the photovoltaic technical potential, data given by the literature ranges from 1700 TWh/y (Hofman et al., 2002) to 46 000 TWh/y (de Vries et al., 2007), while the thermosolar technical potential ranges from 36 TWh/y (Hofman et al., 2002) to 2222

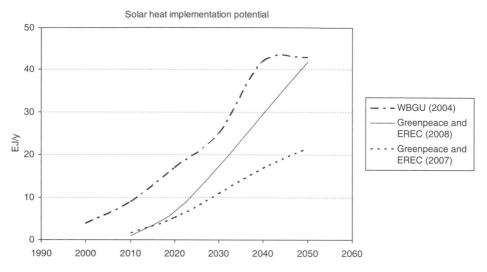

Figure 12.3   Projections of solar heat market development

TWh/y (Trieb, 2005). The solar heat technical potential is estimated as 6389 TWh/y (Greenpeace and EREC, 2008; Hoowijk and Graus, 2008). Finally, the thermosolar and photovoltaic power implementation potential by 2050 estimated by Greenpeace and EREC (2008) is 125 and 410 TWh/y, respectively. EREC has recently updated these data in the 'RE-thinking 2050' report (EREC, 2010) to 385 and 1347 TWh. Solar heat implementation potential given by Greenpeace and EREC (2008) is 1.6 TWh by 2050.

12.6   OCEAN TECHNOLOGIES

Oceans and seas represent around 70 per cent of the planet's surface and store about 1500 x E09 m³ of water. They act as a huge solar collector, being the biggest energy reserve on earth. Most of this energy is stored in the waves, ocean currents and the thermal ocean gradient. These natural phenomena, together with the tides caused by gravity, form a group of renewable energy sources with a very high energy potential. This section is focused on the ocean wave and tide energies because they are the most developed technologies at present.

12.6.1   Ocean Wave Energy

Wave energy can be considered a concentrated form of solar energy. Winds are generated by the differential heating of the earth, and part of their energy is converted into waves (DTI, 2002). The size of the waves depends on the wind speed and duration and the distance over the wind flows (the 'fetch'). In deep water, waves lose energy very slowly and can travel long distances, becoming regular, smooth waves or 'swell'. The coasts exposed to the prevailing wind direction and long fetches tend to have the most energetic wave climates. According to a study by the National Research Council of Canada (Cornett,

2008) this happens at high latitudes of the South Hemisphere (between 40° and 60°), specifically in the Southern Indian Ocean around the Kerquelen Islands and close to the southern coast of Australia, New Zealand, South Africa and Chile; in the Northern Atlantic Ocean to the south of Greenland and Iceland and in the west of the United Kingdom and Ireland; and in the Northern Pacific Ocean in the south of the Aleutian Islands and close to the west coast of Canada and Washington and Oregon states. A world map with the locations of higher swells can be found in Cornett (2008).

As they approach the shoreline, the waves are modified and the wave energy can concentrate in so-called 'hot spots' or, on the contrary, waves can lose energy in bay areas. Shoreline wave energy can be even lower due to additional dissipation mechanisms.

Global potential

The global wave power potential has been estimated as 8000–80 000 TWh/y (1–10 TW) without any consideration of efficiency losses and availability of the resource (IEA-OES, 2003; Pontes and Flacao, 2001). According to Wavenet (2003), this would translate into a conservative technical wave energy resource (total amount of electricity that can be converted from wave energy regardless of economics) of:

- shoreline/near shore devices: 5–20 TWh/y depending on the device; and
- offshore devices: 140–750 TWh/y.

In total, global potential would be between 145–770 TWh/y.

According to the World Energy Council (WEC, 2004, 2007), the economically exploitable resources vary from 140 to 750 TWh/y for current devices and could rise as high as 2000 TWh/y if the potential improvements to existing devices are realized. As cited by Sorensen (2004) a recent study of the UK Department of Trade and Industry (DTI) and the Carbon Trust estimates that in 2050 there could be 200 GW installed, with a total production of 600 TWh/y. According to the Carbon Trust (2006), the practical worldwide wave energy resource would be between 2000 and 4000 TWh/y. The United Nations Development Programme, in its world energy assessment of 2000 (UNDP, 2000), considered a global resource of 18 000 TWh/y. The EREC–Greenpeace study (EREC and Greenpeace, 2007) considered that in 2050, the contribution of wave energy could be 750 TWh/y.

Apart from the data included above, several resource assessments have been carried out in some countries. In the United States, the Electric Power Research Institute (EPRI) performed a feasibility assessment of wave energy (EPRI, 2005). In total, wave power flux was quantified as 2100 TWh/y. Harnessing 20 per cent of the resource at 50 per cent efficiency would result in 210 TWh/y. In Canada, the Ocean Renewable Energy Group (OREG, 2007), estimated in 2007 the Canadian wave resource. Total annual mean power was estimated as 183 GW, producing around 336 TWh/y. In Japan, the wave resource is estimated as 31–36 GW, which could generate 57–66 TWh/y (IEA-OES, 2003).

European potential

The European wave resource has been estimated as 290 GW in the area of the Northeastern Atlantic (including the North Sea), and 30 GW in the coast of the Mediterranean sea. Total wave energy resource in Europe results in 320 GW (EREC and Greenpeace,

2007; EC, 2006). Pontes in 1998 (Pontes, 1998) developed a wave energy atlas for Europe. This has enabled the estimation of the European resource (CEC, 1992) as 120–190 TWh/y offshore and 34–46 TWh/y near shore. In total, the European resource is estimated as 154–236 TWh/y.

There are also national estimations, specifically for the United Kingdom, Ireland, Denmark, Norway and Portugal. In the UK, several assessments of the wave resource have been consulted (Thorpe, 1999; Carbon Trust, 2006), giving data on power generation of 0.2–0.4 TWh/y for shoreline resources, 2.1–7.8 TWh/y for near-shore resources and 50 TWh/y for offshore resources.

In Ireland, Sustainable Energy Ireland (SEI) and the Marine Institute (2005) quantified the practical wave energy resource as 59 TWh/y. Then, the Electricity Supply Board International (ESBI, 2005) quantified the resource in the following terms: theoretical hydrodynamic energy, 11.25–460 TWh/y; technical electrical energy, 2.4–28 TWh/y; practicable electrical energy, 1.2–24 TWh/y; and accessible electrical energy, 1.06–20.76 TWh/y.

In Denmark, the wave resource was estimated as 30 TWh/y (IEA-OES, 2003). According to the same source, the wave energy potential of Norway would be 400 TWh/y. In Portugal, the resource has been estimated as 10 GW. Considering the technical parameters of the wave dragon (www.wavedragon.net), the electricity production could be 18 TWh/y.

12.6.2   Tidal Energy

Tidal energy is the first ocean energy attaining maturity, with 240 MW power plants installed and successfully operating for more than 30 years now. The locations where tidal energy can be developed are few, due to the fact that a strong tidal energy concentration is required for the technology to be competitive.

Tides are generated by rotation of the earth within the gravitational fields of the moon and the sun. There are several tide cycles. In the absence of land, the mean tidal range is 0.5 m. At some locations, this range is increased. Extraction of energy is considered to be practical only where there are large tides (mean tidal ranger larger than 5 m). These sites are not abundant but several possible sites have been identified around the world (Figure 12.4).

Global potential

In 2002, the International Energy Agency-Ocean Energy Systems (IEA-OES) published a document (IEA-OES, 2003) collecting the information available. According to this document, although the total worldwide power of ocean currents is around 5 TW, only a small part of the resource can be extracted. The United Nations Development Programme, in its world energy assessment of 2000 (UNDP, 2000), considered a theoretical potential of 22 000 TWh.

The WEC 2007 Survey of Energy resources (WEC, 2007) reported some additional plans to invest in tidal energy in South Korea and Norway, but without changing the final resource estimation of around 390 TWh/y.

The IEA has signed an Implementing Agreement on ocean energy (IEA-OES) to promote international cooperation in developing wave and tidal energy technologies.

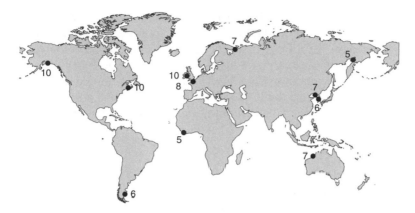

Source: EC (2006).

Figure 12.4 Areas with high tidal resource (tidal range in metres)

According to this body (EA-OES/AEA/SEI, 2006), global tidal energy potential could be higher than 800 TWh/y.

The European Commission recently published a document summarizing the research undergone in the Coordinated Action on Ocean Energy (EC, 2006). According to this document, global tidal resource, in sufficiently shallow waters to be accessible, would be around 1 TW. With a capacity factor of 0.3, this power could supply 2600 TWh/y.

Apart from the data included in the WEC survey and the IEA-OES documents, several resource assessments have been carried out in some countries. In United States, The Electric Power Research Institute (EPRI Ocean Energy Program, 2006) performed a feasibility assessment of the tidal technology in seven states and provinces. In total, extractable power was quantified as 253.5 MW, which could produce 2.2 TWh/y. In Canada, the technical resource is estimated as 110 TWh/y (OREG, 2007). In Russia, Usachev (2000) reported seven sites in the Barents see able to provide 250 TWh/year. The same author states that realization of tidal projects is planned for 139 sites worldwide that would produce 2037 TWh/year. In China, around 7 GW of tidal power would be available (IEA/OES, 2003). With a capacity factor of 0.24, this power could supply 15 TWh/year.

European potential

Tide energy potential in Europe has been estimated between 31 and 48 TWh/y by the European Commission (EC, 1996, 2006). In the UK, the large tidal range along the west coast of England and Wales provides very favourable conditions for the utilization of tidal power, and several documents assessing these resources have been produced. The most recent estimations give a theoretical potential of 110 TWh/y, from which 18 TWh/y would be technically extractable (Black & Veatch, 2005). However, more recent results have shown that the technically achievable maximum energy yield would be 94 TWh/y (ABP Mer Marine Environmental Research, 2007). The study also estimated the exploitable tidal resource in the next five to ten years as 27 TWh/y.

Ireland is another country where detailed assessments of the tidal resource have been performed. In 2005, the SEI and the Marine Institute (2005) quantified the resource as follows: theoretical resource, 230 TWh/y; technical resource (theoretical resource constrained by efficiency), 10.5 TWh/y; practical resource (technical resource constrained by physical incompatibilities), 2.6 TWh/y; accessible resource (practical resource constrained by regulatory constrains), 2.6 TWh/y; and viable resource (accessible resource constrained by economic viability), 0.9 TWh/y. The ECN study (Beurskens and Hekkenberg, 2010) gives added data for both tidal and ocean wave energy. The projected total ocean power capacity by 2020 is 2.118 GW. This capacity installed will generate 5992 GWh.

12.7   CONCLUSIONS

Figure 12.5 shows the renewable energies' technical potentials found in the literature. For biomass, it is assumed that all the potential is for power generation (considering a gasification combined cycle technology with an efficiency of 40 per cent) although part of this potential could be used for heat production and biofuels.

At a global level, the technical potential for power generation will be between 460 and 1900 PWh/y that means 25 to 100 times the global power demand in 2005. The biggest potentials are those for photovoltaic and thermosolar power that together account for between 80 per cent and 90 per cent of the total renewable power potential.

At a European level, however, wind power is the technology with the biggest potential, between 60 per cent and 70 per cent of the total, followed by photovoltaic power. The total renewable power potential in Europe will be between 13 and 130 PWh/y. These figures are 4 to 40 times the power consumption in Europe in 2005.

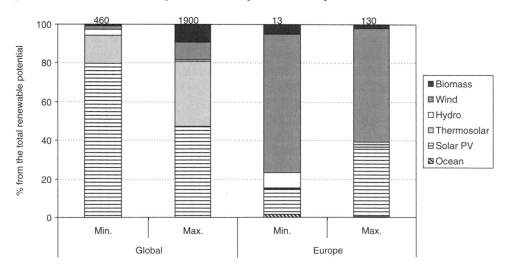

Figure 12.5   Renewable energies technical potential in PWh

REFERENCES

ABP Mer Marine Environmental Research (2007), 'Quantification of exploitable resources in UK waters'.

Alakangas, E., A. Heikkinen, T. Lensu and P. Vesterinen (2007), 'Biomass fuel trade in Europe. Summary Report', EUBIONET2.

AMBIENTE ITALIA (2006), NEEDS (New Energy Externalities Developments for Sustainability) Technical paper no. 11.4 – RS Ia 'Report on technical specification of reference technologies (Photovoltaic systems)'.

Arighoff, R., G. Brakmann, C. Aubrey and S. Teske (2003), 'Solar thermal power. 2020. Exploiting the heat from the sun to combat climate change', Greenpeace and European Solar Thermal Power Industry (ESTIA), http://www.greenpeace.org/raw/content/international/press/reports/solar-thermal-power-2020.pdf.

Berndes, G., M. Hoogwijk and R. van der Broek (2003), 'The contribution of biomass in the future global energy supply: a review of 17 studies', *Biomass and Bioenergy*, **25**, 1–28.

Beurskens, L.W.M. and M. Hekkenberg (2010), 'Renewable energy projections as published in the national renewable energy action plans of the European member status', ECN, Energy Research Centre, the Netherlands.

Black & Veatch (2005), 'Phase II. UK Tidal Stream Energy Resource Assessment'.

Carbon Trust (2006), 'Future marine energy. Results of the marine energy challenge: cost competitiveness and growth of wave and tidal stream energy'.

CEC (1992), 'An assessment of the state of art, technical perspectives and potential market of wave energy', prepared by the Energy Technology Support Unit (ETSU, UK) and Commission of the European Communities DG XVII (CCE).

Cornett, A.M. (2008), 'A global wave energy resource assessment', *Proceedings of the Eighteenth International Offshore and Polar Engineering Conference Vancouver, BC, Canada*, 6–11 July.

Dessus, B., B. Devin and F. Pharabed (1992), 'World potential of renewable energies: actually accessible in the nineties and environmental impact analysis', La Huille Blanche No. 1, Paris.

DLR and CIEMAT (2007), NEEDS (New Energy Externalities Developments for Sustainability), Technical paper no. 12.4 – RS Ia, 'Technology report (including road mapping, technology specification of current and future systems, development of costs) for solar thermal power plant technologies'.

Doornbosch, R. and R. Steenblik (2007), Organisation for Economic Co-operation and Development SG/SD/RT(2007)3, 'Biofuels: is the cure worse than the disease?', Round Table on Sustainable Development, Paris, 11–12 September, SG/SD/RT(2007)3.

DTI (2002), 'DTI Technology Road-map – wave energy'.

EA-OES/AEA/SEI (2006), 'Review and analysis of ocean energy systems development and supporting policies'.

EC (1996), European Commission, 'The exploitation of tidal and marine currents. Wave energy. Project results', Technical report EUR 16683 EN, Commission of the European Communities. Directorate-General for Science, Research and Development.

EC (2006), European Commission, Centre for Renewable Energy Sources Directorate for Renewable Energy Sources 'Ocean energy conversion in Europe – recent advancements and prospects'.

EC (2008), 'European energy and transport – trends to 2030', update 2007.

EEA (2006), 'How much bioenergy can Europe produce without harming the environment?', EEA Report No. 7/2006, European Environment Agency.

EEA (2007), 'Estimating the environmentally compatible bioenergy potential from agriculture', EEA Technical report No. 12/2007, European Environment Agency.

EEA (2009), 'Europe's onshore and offshore wind energy potential: an assessment of environmental and economic constraints', EEA Technical report No 6/2009, European Environment Agency.

EPIA (2004), 'EPIA roadmap', May.

EPIA (2006), 'EPIA, Greenpeace, solar generation report', September.

EPRI (2005), 'Wave energy feasibility study', Final project briefing.

EPRI Ocean Energy Program (2006), http://www.epri.com/oceanenergy/attachments/streamenergy/reports/008_Summary_Tidal_Report_06-10-06.pdf.

EREC (2002), 'Background information to the European renewable energy export strategy', European Renewable Energy Council.

EREC (2004), 'Renewable energy scenario to 2040', May.

EREC (2008), 'Renewable energy technology roadmap 20 per cent by 2020'.

EREC (2010), 'RE-thinking 2050. A 100 per cent renewable energy vision for the European Union'.

Ericsson, K. and L.J. Nilsson (2006), 'Assessment of the potential biomass supply in Europe using a resource-focused approach', *Biomass and Bioenergy*, **30**, 1–15.

ESBI (2005), 'Accessible wave energy resource atlas: Ireland: 2005', Report 4D404A-R2, Ireland.

EWEA (2008), 'Pure power: wind energy scenarios up to 2030', March.

EWEA (2009), *Wind Energy: The Facts. A Guide to the Technology, Economics and Future of Wind Power*, London: Earthscan.

Fischer, G., E. Hizsnyik, S. Prieler and H. van Velthuizen (2007), 'Assessment of biomass potentials for biofuel feedstock production in Europe: methodology and results', Work Package 2 – Biomass potentials for biofuels: sources, magnitudes, land use impacts, REFUEL project, IIASA.

Fischer, G. and L. Schrattenholzer (2001), 'Global bioenergy potentials through 2050', *Biomass and Bioenergy*, **20**, 151–9.

GBEP (2009), '5th meeting of the GBEP Task Force on Sustainability', Paris, 9 July, United Nations Environment Programme (UNEP) Chair Conclusions, http://www.globalbioenergy.org/fileadmin/user_upload/gbep/docs/2009_events/5th_TF_Sustainability_Paris_July_2009/5th_TF_on_Sustainability_-_Conclusions.pdf.

Goldemberg, J. (2000), 'World energy assessment: energy and the challenge of sustainability', UNDP / UN-DESA / World Energy Council.

Greenpeace (2004), 'Sea wind europe'.

Greenpeace and EREC (2007), 'The energy (r)evolution – a sustainable world energy outlook'.

Greenpeace and EREC (2008), 'Energy (r)evolution. A sustainable world energy outlook', update, Amsterdam, http://www.energyblueprint.info/fileadmin/media/documents/energy_revolution2009.pdf.

Greenpeace and GWEC (2006), 'Global wind energy outlook 2006'.

Greenpeace and GWEC (2008), 'Global wind energy outlook 2008'.

Grubb, M.J. and N.I. Meyer (1993), 'Wind energy: resources, systems, and regional strategies', in T.B. Johansson, H. Kelly, A.K.N. Reddy, R. Williams and L. Burnham (eds), T.B. Johansson, Washington, DC: Island Press, pp. 157–212.

Hofman, Y., D. de Jager, E. Molebroek, F. Schiling and M. Voogt (2002), 'The potential of solar electricity to reduce CO_2 emissions', IEA Greenhouse Gas R&D Programme.

Hoogwijk, M., A. Faaij, B. Eickhout, B. de Vries and W. Turkenburg (2005), 'Potential of biomass energy out to 2100, for four IPCC SRES land-use scenarios', *Biomass and Bioenergy*, **29**, 225–57.

Hoogwijk, M., A. Faaij, R. Van den Broek, G. Berndes, D. Gielen and W. Turkenburg (2003), 'Exploration of the ranges of the global potential of biomass for energy', *Biomass and bioenergy*, **25**, 119–33.

Hoogwijk, M. and W. Graus (2008), 'Global potential of renewable energy sources: a literature assessment', Background report.

Hoogwijk, M., B. de Vries and W. Turkenburg (2004), 'Assessment of the global and regional geographical, technical and economic potential of onshore wind energy', *Energy Economics*, **26**, 889–919.

Hydropower and Dams (2000), 'World atlas industry guide', available at: http://www.hydropower-dams.com/world-atlas-industry-guide.php?c_id=159.

IEA (2007), 'World energy outlook 2007, China and India insights', International Energy Agency, http://www.iea.org/textbase/nppdf/free/2007/weo_2007.pdf.

IEA (2008), 'World energy outlook 2008'.

IEA-OES (2003), 'Status and research and development priorities, 2003'.

IHA (International Hydropower Association (UK)), International Commission on Large Dams (F), Implementing Agreement on Hydropower Technologies and Programmes (F) and Canadian Hydropower Association (CA) (2000), 'Hydropower and the world's energy future: the role of hydropower in bringing clean, renewable, energy to the world'.

IPCC (2001), 'IPCC Third Assessment Report. WG III. Mitigation. Technological and economic potential of greenhouse gas emissions reduction'.

Kim, S. and B.E. Dale (2004), 'Global potential bioethanol production from wasted crops and crop residues', *Biomass and bioenergy*, **26**, 361–75.

Krewitt, W., Sonja Simona, Wina Grausb, Sven Teskec, Arthouros Zervosd and Oliver Schaferd. (2007), 'The 2°C scenario – a sustainable world energy perspective', *Energy Policy*, **35**, 4969–80.

Labriet, M., H. Cabal and Y. Lechón (2010), 'Present and future biomass production potentials and routes in EU27+ countries and the rest of the world', Colección Documentos Ciemat.

Labriet, M., H. Cabal, Y. Lechón, M. Biberacher, S. Gluhak, D. Zocher and N. Dorfinger (2009), 'Review of the world energy resource potentials', Colección Documentos Ciemat.

Lu, X., M.B. McElroy and J. Kiviluoma (2009), 'Global potential for wind-generated electricity', *Proceedings of the National Academy of Sciences of the United States of America*, **106** (27), 10933–938.

Lysen, E. and S. van Egmond (2008), 'Climate change scientific assessment and policy analysis. Biomass assessment – assessment of global biomass potentials and their links to food, water, biodiversity, energy demand and economy', Main report. Netherlands Research Programme on Climate Change.

OREG (2007), 'Ocean energy in Canada', Presentation of Project Director J. Jhonson, http://www.all-energy.co.uk/UserFiles/File/2007JessicaJohnson.pdf.

Pontes, M.T. (1998), 'Assessing the European wave energy resource', *Transactions of ASME – Journal of Offshore Mechanics and Arctic Engineering*, **120** (4), 226–31.

Pontes, T. and A. Flacao (2001), 'Ocean energies: resources and utilization', *Proceedings 18th World Energy Congress*, Buenos Aires, Paper 01-06-12, October.

PVTRAC- EC/PV-TRAC (2005), 'A vision for photovoltaic technology', February.

REN21 (2008), 'Renewable energy potentials in large economies – summary report'.

RES2020 (2009a), EU27 Synthesis Report Deliverable D.4.2, 'Monitoring and evaluation of the RES directives implementation in EU27 and policy recommendations for 2020', http://www.res2020.eu/files/fs_inferior01_h_files/pdf/deliver/RES2020_Synthesis-Report.pdf.

Rogner, H.-H. (2000), 'Energy resources', *World Energy Assessment: Energy and the challenge of sustainability*, Washington, DC and New York: UNDP/WEC/UNDESA.

Sargent & Lundy LLC Consulting Group (S&R) (2003), 'Assessment of parabolic trough and power tower solar technology cost and performance forecast', NREL/SR-550-34440.

SEI and Marine Institute (2005), 'Ocean energy in Ireland: an ocean strategy for Ireland', submitted to the Department of Communications, Marine and Natural Resources, Ireland.

Seidenberger, T., D. Thrän, R. Offermann, U. Seyfert, M. Buchhorn and J. Zeddies (2008), 'Global biomass potentials. Investigation and assessment of data. Remote sensing in biomass potential research. Country-specific energy crop potential', German Biomass Research Centre (DBFZ), for Greenpeace International.

Sims, R.E.H., R.N. Schock, A. Adegbululgbe, J. Fenhann, I. Konstantinaviciute, W. Moomaw, H.B. Nimir, B. Schlamadinger, J. Torres-Martínez, C. Turner, Y. Uchiyama, S.J.V. Vuori, N. Wamukonya and X. Zhang (2007), 'Energy supply', in B. Metz, O.R. Davidson, P.R. Bosch, R. Dave, L.A. Meyer (eds), *Climate Change 2007: Mitigation*, Contribution of Working Group III to the Fourth Assessment Report of the Intergovernmental Panel on Climate Change Cambridge University Press, Cambridge, UK and New York, USA: pp. 251–322.

Smeets, E., A. Faaij and I. Lewandowski (2004), 'A quickscan of global bio-energy potentials to 2050: an analysis of the regional availability of biomass resources for export in relation to the underlying factors', Report NWS-E-2004-109.

Smeets, E., A. Faaij, I. Lewandowski and W. Turkenburg (2007), 'A bottom-up assessment and review of global bio-energy potentials to 2050', *Progress in Energy and Combustion Science*, **33**, 56–106.

Sorensen, B. (1999), 'Long-term scenarios for global energy demand and supply: four global greenhouse mitigation scenarios', Roskilde University, Institute 2, Energy and Environment Group, Denmark.

Sorensen, H.C. (2004), 19th World Energy Congress, Sydney, 'World's first offshore wave energy converted wave dragon connected to the grid'.

Strengers, B.J., R. Leemans, B. Eickhout, B. de Vries and A.F. Bouwman (2004), 'The land use projections in the IPCC SRES scenarios as simulated by the IMAGE 2.2 model', *GeoJournal*, **61**, 381–93.

Thorpe, T.W. (1999), 'A brief review of wave energy', a report produced for the UK Department of Trade and Industry, ETSU R 120.

Trieb, F. (2005), 'MED-CSP. Concentrating solar power for the Mediterranean region', BMU/DLR. http://www.dlr.de/tt/med.csp.

UNDP (2000), 'World energy assessment: energy and the challenge of sustainability', http://www.energy-andenvironment.undp.org/undp/index.cfm?module=Library&page=Document&DocumentID=5037.

UNESCO (2006), 'Water, a shared responsibility', United Nations World Water Development Report 2.

de Vries, B.J.M, D.P. van Vuuren and M. Hoogwijk (2007), 'Renewable energy sources: their global potential for the first-half of the 21st century at a global level: an integrated approach', *Energy Policy*, **35**, 2590–2610.

Wakker, A., R. Egging, E. Van Thuijl, X. Van Tilburg, E. Deurwaarder, T. De Lange, G. Berndes and J. Hansson (2005), 'Biofuel and bioenergy implementation scenarios. Final report of VIEWLS WP5, modelling studies', Energy Research Centre of the Netherlands and Chalmers University of Technology, Energy and Environment.

Wavenet (2003), 'Results from the work of the European Thematic Network on Wave Energy', EESD.

WBGU (2004), *World in Transition. Towards Sustainable Energy Systems*, German Advisory Council on Global Change, London: Earthscan.

WEC (2004), World Energy Counci 'Survey of energy resources', http://www.worldenergy.org/documents/ser2004.pdf.

WEC (2007), World Energy Council 'Survey of energy resources'.

van Wijk and Coelingh (1993), 'Wind power potential in the OECD countries', Department of Science, Technology and Society, Utrecht University, Netherlands.

de Wit, M.P. and A.P.C. Faaij (2008), 'Biomass resources potential and related costs. Assessment of the EU-27, Switzerland, Norway and Ukraine', REFUEL Work Package 3 final report, Copernicus Institute, University of Utrecht, The Netherlands.

Yamamoto, H., J. Fujino and K. Yamaji (2001), 'Evaluation of bioenergy potential with a multi-regional global-land-use-and-energy model', *Biomass and Bioenergy*, **21**, 185–203.

13 The cost of renewable energy: past and future
Kirsten Halsnæs and Kenneth Karlsson

13.1 POLICY GOALS AND PERSPECTIVES

The costs of renewable energy and the penetration of options must be seen in the context of global policy settings in terms of energy security, climate change and economic development, where main issues in a global context include energy access and affordability. This is the case because renewable energy in most cases is not the lowest-cost energy supply option, so penetration depends on economic, environmental and political priorities.

There is a lot of variation between estimates of fossil resources across studies, in particular for oil and gas resources, and the literature includes both studies that foresee a very short remaining time-frame of oil resources, and others that expect relatively plentiful coal and gas resources to be available with new discoveries of gas, implying that fossil fuels can still play a large role in global primary energy consumption throughout this century if no special efforts are made to promote the penetration of renewable energy options. Energy security, however, despite uncertainties about fossil fuel resources, for many countries implies a specific preference for using renewable energy options which are considered both to meet environmental policy priorities and to reduce the dependence on scarce imported oil resources.

Climate change stabilization policies have some commonality with energy security in terms of the energy options recommended. Fossil fuel consumption will tend to be decreased in both cases compared with a reference case, and more renewable energy will be used, but climate change stabilization goals can also be pursued by using carbon capture and storage (CCS) on coal, oil and gas power plants, which can compete with the introduction of renewable energy options. CCS options, however, will increase the demand for fossil fuel resources and in this way can be in conflict with energy security concerns if resources are expected to be scarce.

Energy access is a major issue in developing countries, where about 80 percent of the world population lives today. Access is closely related to costs, since affordability is a major issue for low-income families. The costs of renewable energy are therefore a critical factor in penetration rates, particularly in developing countries, where global and local environmental problems are not as yet among the highest policy priorities.

The penetration of renewable energy depends both on the costs of individual options, and on how a portfolio of options can be integrated in energy systems in such a way that energy access, energy security and climate change policy goals are met. In this way, the costs of renewable energy become very policy- and context-specific, and the chapter illustrates how this has been played out in international scenario studies and in a study for Denmark, where the aim is to cover all Danish energy consumption by renewable energy in 2050.

13.2   COST CONCEPTS

The costs of renewable energy in this chapter will be considered from the society's point of view, and social cost estimates are presented. The major focus is on the costs of renewable energy as an integrated energy system element in order to reflect what is required to achieve energy security and climate change stabilization. There are significant differences between what this implies – to include large shares of renewable energy – and the low-scale penetration of renewable options. This particularly is the case for fluctuating renewable energy such as wind or solar power, and this is further illustrated subsequently related to the Danish 100 percent renewable energy supply scenario.

In addition to system interactions between fluctuating renewable energy options and other options, the costs of renewable energy in many cases are also very context-specific. Renewable energy options, when not internationally traded, vary in terms of potential and costs in different geographical locations; specific potentials of wind and solar depend on the site, and the same is the case with hydropower and other options. These differences in many cases create difficulties when including renewable energy options appropriately in global long-term energy–economic models, where potentials and costs are typically represented by a few aggregate parameters that do not reflect the benefits of using renewable options at the most attractive sites in a regional and national setting. This is one of the reasons why there are often very large differences between estimated renewable energy potentials in global and national studies.

Renewable energy includes options that are fully commercialized today, where costs and potentials are relatively well known, and emerging technologies where larger-scale applications still rely on further research, development and demonstration. In long-term climate change scenarios over centuries, this imposes a particular uncertainty on the long-term cost of stabilization scenarios for low temperature targets, where some of the emerging technologies play a significant role. Examples will be given in the following, based on international studies.

Renewable energy is also characterized by other key cost elements including externalities related to local air pollution, domestic employment and other social benefits. These costs are of a short-term nature and are not included in global climate change scenarios, so information about these is not given in this chapter.

Some lessons can be drawn from marginal abatement cost studies of stabilization scenarios. Renewable energy options will be high-cost options and marginal costs therefore reflect the cost of the last technology coming in. The marginal CO_2 reduction costs of international scenario studies are discussed in relation to assumptions about renewable energy potentials and costs.

13.3   RENEWABLE ENERGY IN GLOBAL POLICY SCENARIOS

This section will present the results of international climate change scenario studies that include a broad range of low carbon technologies with renewable energy options, nuclear power and CCS, along with other studies that cover renewable energy options only. The range of studies for the costing review aims at reflecting variety in the literature regarding expected renewable energy potentials and costs. The studies included

Table 13.1   Global electricity production by energy source and scenario

	2007	Baseline 2050	BLUE Map 2050	Blue High Renewable 2050
Nuclear	2719	4825	9608	4358
Oil	1117	311	226	197
Coal	8216	20459	238	330
Coal + CCS	0	0	4746	910
Gas	4126	10622	4263	2983
Gas + CCS	0	0	1815	771
Hydro	3078	5344	5749	6043
Biomass/Waste	259	1249	2149	2488
Biomass + CCS	0	0	311	146
Geothermal	62	297	1005	1411
Wind	173	2149	4916	8193
Ocean	1	25	133	552
Solar	5	905	4958	9274
Total	19756	46186	40137	37656
Share of renewables	18%	22%	48%	75%

Source: ETP 2010, Table 3.1 (IEA, 2010).

are the International Energy Agency (IEA) Energy Technology Perspectives (ETP), 2010 (IEA, 2010), the ReMIND RECIPE study (Luderer et al., 2009), the Energy Revolution study (Greenpeace, 2005) and the MiniCam EMF 22 study (Calvin et al., 2009).

The scenarios generated by the IEA Energy Technology Perspectives (ETP), 2008 covers the period 2010 to 2050. The BLUE Map scenario reduces global CO_2 emissions from the energy sector by 50 percent in 2050 from 2005 level, which corresponds to the low end of the reductions required to meet a 2° stabilization scenario. Very large changes are here seen in the fuel composition compared with the Baseline. Fossil fuel consumption is reduced by 59 percent from the Baseline to the BLUE Map scenario in 2050, and of this, coal consumption is reduced by as much as 36 percent below the 2007 level, despite CCS technologies assumed to be available. Biomass is the largest renewable energy source, amounting to 10 percent of total primary energy consumption today, and biomass consumption also triples in the BLUE Map scenario and is slightly higher in 2050 than the coal use of today. However, it should be recognized that fossil fuels in 2050 will still contribute almost 50 percent of primary energy consumption, while the renewable share will be just below 30 percent in the BLUE map scenario and of this 30 percent, biomass contributes two-thirds. The fossil fuel share is projected to be 84 percent of total primary energy consumption in 2050 in the baseline.

The marginal CO_2 reduction cost in the BLUE Map scenarios is US$175 per tonne of CO_2 in 2050. The ETP 2010 also includes specific BLUE Map scenarios that assume restricted or more optimistic availability for specific power production technologies. The scenarios are: no CCS, high nuclear power, high renewable energy potential, and low discount rate. Table 13.1 shows the fuel composition in the high renewable energy scenario compared with the reference case and the BLUE Map case.

Table 13.2   Renewable energy share in alternative climate stabilization scenarios

	Scenario name	Renewable energy share in 2030 (%)	Renewable energy share in 2050 (%)
Baseline up to 600 ppm CO_2 equivalents	IEA World Energy Outlook 2009	14	15
Between 400 and 600 ppm CO_2 equivalents	ReMIND RECIPE	22	34
Below 400 ppm CO_2 equivalents	Energy Revolution	39	80
	Minicam 450 CO_2 eq.	24	31
	EMF 22		

Table 13.3   The contribution of renewable energy options to power production in 2050 in various scenarios

TWh/year	Genera-tion IEA WEO 2009	Genera-tion ReMIND RECIPE	Genera-tion EMF 22	Genera-tion Energy Revolution 2010	% of global demand WEO 2009	% of global demand ReMIND RECIPE	% of global demand EMF 22	% of global demand Energy Revolution 2010
PV	640	20790	822	6846	1.4	32.8	1.3	15.6
CSP	254	0	1545	9012	0.5	0.0	2.5	20.5
Wind	2516	14290	7848	10841	5.4	22.6	12.5	24.7
Geo-thermal	265	na	1197	2968	0.6	na	1.9	6.8
Bioenergy	994	4217	5847	580	2.1	6.6	9.3	1.3

Source: IPCC, Renewable Energy Report, Table 10.3.3.

The BLUE Map high renewable energy scenario has 75 percent coverage of global electricity production in 2050, with solar and wind each amounting to more than 20 percent of the supply. These two options almost cover the full expansion of renewable energy consumption required when moving from the BLUE Map 2050 to the BLUE Map high renewable case.

The range of estimated renewable energy potentials as part of scenarios for the global energy system is illustrated in Table 13.2, which shows high and low case scenarios. The scenarios included are the WEO 2009 reference case, the ReMIND RECIPE study, the MiniCam EMF 22 study and the Energy Revolution study. The three first mentioned are based on global studies covering the energy sector in the case of WEO 2009 (IEA, 2009), and with a linked macroeconomic and energy sector framework in the case of ReMIND RECIPE and Minicam. The Energy Revolution is based on a detailed bottom-up energy scenario analysis.

In these studies the renewable energy share of global primary energy consumption varies between 15 percent in the WEO 2009 baseline case and 80 percent in the Energy Revolution scenario for stabilization below 400 ppm in 2050.

Table 13.3 shows the contribution of individual renewable energy options to primary energy consumption for 2050, and it highlights that the above-mentioned studies are very different in which renewable they consider to have the major role in primary energy consumption.

The contribution of various renewable energy sources varies a lot in the different scenario studies, as shown in Table 13.3. The ReMIND RECIPE and the Energy Revolution 2010 studies cover a large share of power demand with solar energy; and this potential is followed by wind, with a large potential in the ReMIND and RECIPE studies, and somewhat lower potential in the EMF 22 study. This latter study, relative to the other studies, covers a large share of global electricity demand with bioenergy.

The differences between these studies reflect both varying assumptions about the costs of the individual options, and also how alternative low carbon technologies such as nuclear power and CCS are represented in the studies. The Energy Revolution 2010 study, for example, excludes the use of nuclear power and CCS.

One way to represent the costs of renewable energy across the different scenarios is to consider the carbon price for the scenarios represented in Tables 13.2 and 13.3. There is a large difference between the studies included, which is also to be expected given the different roles of renewable energy sources shown in Table 13.3. The CO_2 price of the WEO baseline with concentrations below 600 ppm is US$54 in 2030; the ReMIND RECIPE study has a CO_2 price between US$100 and US$430 for a 450 ppm scenario in 2050 (Luderer et al., 2009), The EMF 22 study has a CO_2 price of between almost US$100 and US$300 per tonne of CO_2 in 2050 for a scenario corresponding to 450 ppm dependent on assumptions about availability of technologies. The Energy Revolution study does not include CO_2 prices, but concludes that the climate policy scenario has low costs due to gains from energy efficiency improvements and from employment created.

The CO_2 prices are critically dependent on assumptions about the availability of low carbon technologies. The ReMIND RECIPE study includes a sensitivity analysis, where CO_2 prices are calculated for scenarios with the following constraints on technologies:

- 'No CCS'.
- 'Fix Biomass' where biomass use is fixed to baseline level.
- 'Fix Renewable' with no expansion of renewable other than biomass beyond baseline levels.
- 'Fix Nuclear' with no expansion of nuclear energy beyond baseline level.
- 'No CCS' and 'Fix Nuclear' in combination.

The increase in CO_2 prices is dependent on the different constraints. The CO_2 price is not very sensitive to the Fix Nuclear case, because most of this potential is already implemented in the baseline case (price stays at 2005US$50–100/t$CO_2$). No CCS and Fix Biomass increase the CO_2 price significantly, in particular at the end of the period (2005US$200–300/t$CO_2$), where green house gas (GHG) emission reduction goals become more stringent; and turning to the Fix Renewables case, this increases the CO_2 price very dramatically (2005US$250–400/t$CO_2$). The high CO_2 price corresponding to fixed renewable energy potentials reflects that the options are relatively expensive, so the expansion from the reference to the policy case is very large.

13.4   DANISH RENEWABLE ENERGY SYSTEM

The Danish government in 2008 formed a Climate Commission with the aim to assess how and when Denmark could introduce an energy system with no use of fossil fuel resources, including all domestic energy consumption in transportation, business, households and the public sector. The work was supported by technical and economic scenario analysis using a combination of bottom-up sectoral modeling and macroeconomic modeling.[1] The macroeconomic modeling was used for energy demand forecasts and the assessment of gross domestic product (GDP) impacts of increasing energy prices and investments, while the energy sector models were used to assess the costs of energy efficiency improvements and of substituting fossil fuel-based energy supply with renewable energy. Alternative scenarios for 100 percent use of renewable energy were developed for 2050.

The Climate Commission work includes two sets of scenarios, namely:

- Case 1: a reference and policy scenario assuming non-ambitious international climate policy with free import of bioenergy resources.
- Case 2: a reference and policy scenario assuming ambitious international climate policy with bioenergy use in Denmark restricted to match domestic resources.

Case 1 with non-ambitious international climate policy has relatively high fuel prices and low CO_2 prices, while the ambitious Case 2 conversely has low fuel prices and high CO_2 prices.[2] It is important to recognize that the CO_2 prices applied in the scenarios are relatively low compared with the prices reported from the international studies in the previous section. The CO_2 price, for example, is assumed to be DKK 863 per tonne of CO_2 (€116/tCO_2) in the ambitious Case 2, and DKK 380 per tonne of CO_2 (€51/tCO_2) in the non-ambitious Case 1 in 2050, which must be considered as a low price estimate given that international climate policy introduces more stringent targets in this period. Some general results of the energy system composition and related costs are given for the two cases in the following. Figure 13.1 shows final energy consumption in the non-ambitious scenario Case 1.

The final energy consumption is projected to increase by about 13 percent from 2008 to 2050 in the non-ambitious reference case (Ref1) and is reduced by about 11 percent from 2008 to 2050 in the policy scenario (Scen1) as seen in Figure 13.1. This low demand projection reflects that the reference case assumes that previous energy efficiency improvement trends in Denmark are continued, and it is assumed that energy efficiency improvements in the building sector are reducing demand for heating by 34 percent. Final energy consumption in the reference case is supplied by electricity, district heating and gas as the major sources, and without gas in the policy case. Electricity is contributing an increasing share of final energy consumption, reflecting that power supply with wind energy offers a relatively low-cost carbon-free energy supply option in the Danish case. Electricity is produced using 55 percent coal and 30 percent wind in the reference, and using 60 percent wind and 35 percent biomass in the policy case; see Figure 13.2.

The ambitious Case 2 has the same level of final energy demand in the reference case as the ambitious, but uses less biomass and continues using oil (see Figure 13.3). The policy scenario has more electricity use and less district heating compared with the

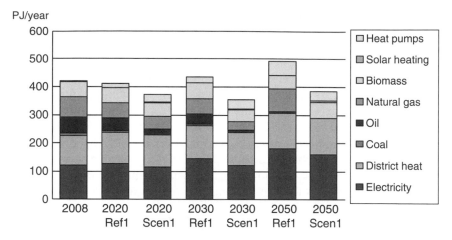

Figure 13.1   Non-ambitious Case 1 final energy consumption without transport

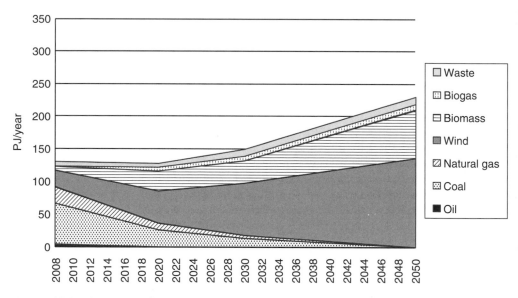

Figure 13.2   Power production in non-ambitious Case 1

non-ambitious case, reflecting that centralized power stations with combined heat and power (CHP) based on biomass play a smaller role. The power system is based on more than 80 percent of wind power in the policy case as shown in Figure 13.4, and very limited biomass resources are used, only the amount that has been assessed to be needed in order to stabilize the power system against the high share of wind.

Inclusion of as much as 80 percent wind energy in the Danish power system in the ambitious Case 2 is a huge challenge and the stability of the system has therefore been studied in detail by the Climate Commission. Detailed hour-by-hour simulations of the power demand and supply have been conducted with the Balmorel model (www.

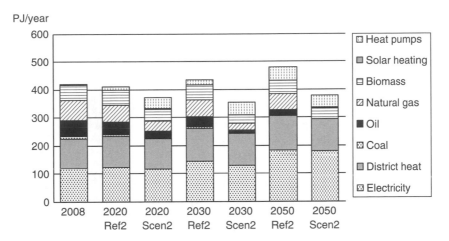

Figure 13.3   Energy without transport: ambitious case

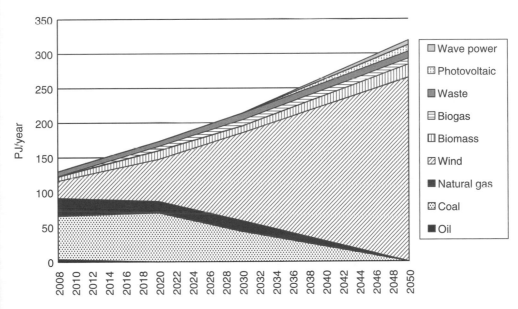

Figure 13.4   Power production in the ambitious case

balmorel.com) which is an optimization model for the centralized heat and power production system in Denmark, Scandinavian countries and Germany. Figure 13.5 shows
power demand and production hour by hour for two weeks. Such a balance is made for
the whole year and illustrates how the power system is balanced so that demand can be
met in every hour of the day.

Danish electricity trade (export and import) is assessed to amount to about 30 percent
of total annual consumption, and it is assessed that 4.7 percent of the production from
wind turbines annually cannot be sold to the market and thereby has to be curtailed.

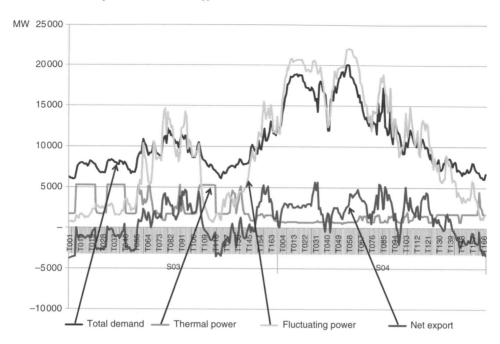

Note: Total electricity demand ('classical' electricity and electricity used for industries, heat pumps and transport), net import of electricity through the connectors to surrounding countries and thermal production in week 3 and 4 2050.

Figure 13.5   Simulated total fluctuating electricity production in Denmark

These interactions with the electricity market are of course very sensitive to assumptions about power supply structure in neighboring countries.

It has been concluded that security of supply can be maintained by implementing the following options:

- Implementation of heat pumps and heat storage facilities in district heating.
- Small dynamic power production units for backstop in periods with low wind speed, for example gas turbines and motors based on biogas.
- Flexible loading of electric cars to balance power demand on an hourly basis during the day.
- Production of biofuels and methanol for transport in periods with high power production, to work as energy storage.
- Increased flexibility in industrial power consumption.
- Investment in transmission lines in order to facilitate more electricity trade with the Scandinavian and German markets. Curtailing power production from wind turbines in part of the year.

An indicator of the costs is the power production price, which is shown for the ambitious case in 2050 in Figure 13.6. Relative to the non-ambitious case there is a small cost difference in the whole period up to 2050. In the non-ambitious case the power price is 4–5 øre cheaper per kWh, which is less than 1 euro cents/kWh. It is an interesting result

(0.01*Dkr/kWh)

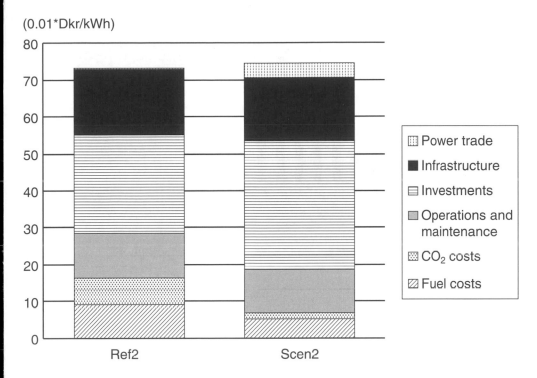

Figure 13.6   Yearly average power production costs in ambitious Case 2

that in both scenarios the power price is around 75 øre/kWh (10 euro cents/kWh) all through the period, in spite of huge differences in fuel mix, fuel prices and CO_2 cost.

The average power production cost as shown in Figure 13.8 is about 75 øre/kWh in 2050 in the ambitious policy case, and this is about the same level as the cost in the reference case. The costs of the policy case with 100 percent renewable energy supply are dominated by large investments, infrastructure costs required for international trade, and also costs of electricity trade reflecting that the international market price can be low in periods with very large wind electricity production.

One of the most difficult sectors to turn into 100 percent renewable energy is transportation. Some details are given in the following on renewable energy options and different transportation modes. The challenge in this context is to find cost-effective solutions that, given scarce bioenergy resources and interrelationships with fluctuating power production, can facilitate the services demanded. The costs of a 100 percent renewable transportation system are very uncertain since some of the key technologies involved such as electric car batteries and fuels cells are still emerging, and major breakthroughs in costs and efficiency could come in the period until 2050. Furthermore, biofuel prices are very uncertain.

The transportation sector in the Danish study has been considered in an integrated assessment, where the sector is considered both in its role as delivering transportation services, and as a key element in balancing power demand in a system with a high share of fluctuating energy. A consequence of this is that some of the options included in the transportation sector assessment have relatively high costs compared with, for example, energy

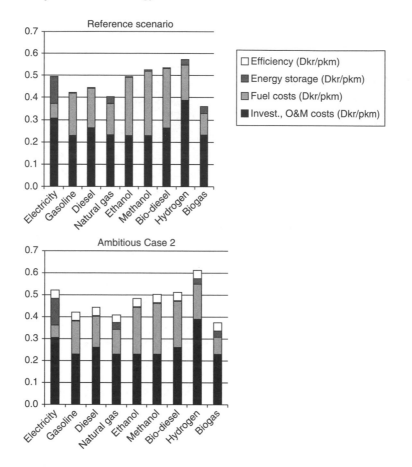

Figure 13.7   Costs of alternative cars in DKK/km in 2050

saving options in the buildings sector. These relatively expensive options (for example electric cars and methanol cars) however play a role as stabilizers in the power system.

Figure 13.7 shows the costs of alternative cars in 2050. It can be seen that gasoline and diesel cars are the most cost-effective in 2050, and electric cars are slightly more expensive. Ethanol and methanol cars are more expensive, but can support a stable power system, and the fuel can be produced with a relatively low power production price. It can be seen from Figure 13.7 that the costs of an electric car in 2050 are slightly higher than for conventional gasoline and diesel cars, and part of the cost of the electric car can be attributed to its role as an energy storage.

The fuel composition of all transportation modes in the ambitious Case 2 are shown in Figure 13.8. It can be seen that 90 percent of all private cars are expected to be electric, and electric vehicles are also introduced for buses, trains and trucks, while domestic aircraft, ferries and ships are supplied with biofuels.

The total costs of the ambitious Case 2 with 100 percent renewable energy in Denmark are shown in Figure 13.9. The costs are measured as extra costs compared to the reference scenario and with a 5 percent rate of interest.

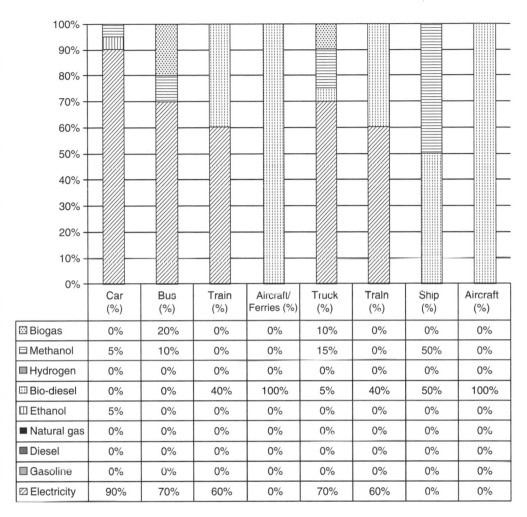

	Car (%)	Bus (%)	Train (%)	Aircraft/ Ferries (%)	Truck (%)	Train (%)	Ship (%)	Aircraft (%)
Biogas	0%	20%	0%	0%	10%	0%	0%	0%
Methanol	5%	10%	0%	0%	15%	0%	50%	0%
Hydrogen	0%	0%	0%	0%	0%	0%	0%	0%
Bio-diesel	0%	0%	40%	100%	5%	40%	50%	100%
Ethanol	5%	0%	0%	0%	0%	0%	0%	0%
Natural gas	0%	0%	0%	0%	0%	0%	0%	0%
Diesel	0%	0%	0%	0%	0%	0%	0%	0%
Gasoline	0%	0%	0%	0%	0%	0%	0%	0%
Electricity	90%	70%	60%	0%	70%	60%	0%	0%

Figure 13.8   Fuels used in ambitious transportation Case 2 for 2050

The total costs of introducing the policy case where all Danish energy consumption is covered with renewable energy sources is around €1.6 billion as seen in Figure 13.9, or 0.5 percent of the expected GDP in 2050. The costs are composed of large fuel cost savings and investments that are a little bit larger than the fuel cost savings. Relative to this case the total costs of the non-ambitious case have been assessed to be about €1.3 billion in 2050 or 0.3 percent of GDP. The lower costs of this case reflect higher investment costs of wind power stations, transmission lines and electric cars in the ambitious case, where less low-cost biomass options are available than in the non-ambitious case.

The non-ambitious and the ambitious case both live up to the goal of ensuring 100 percent coverage with renewable energy sources in 2050, but despite this they are rather different in their timing of corresponding CO_2 reductions, as can be seen from Table 13.4.

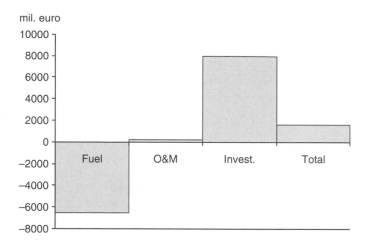

Figure 13.9   Total costs of the energy system in the ambitious Case 2 in 2050

Table 13.4   Danish CO_2 emissions, % of 2008 emissions

Year	Ambitious Case 2		Non-ambitious Case 1	
	Reference (%)	Policy case (%)	Reference (%)	Policy case (%)
2008	100	100	100	100
2020	89	81	90	57
2030	81	56	81	35
2040	64	31	87	21
2050	48	7	92	7

The non-ambitious case results in a 43 percent CO_2 reduction in 2020 compared with 2008, while the ambitious case only results in a 19 percent reduction in 2020, and they both end up with a 93 percent reduction of CO_2 emissions in 2050 compared with 2008. The faster CO_2 reduction in the non-ambitious case reflects that assuming the availability of a large potential of inexpensive biomass that can be imported to the Danish energy system implies that coal-fired power plants in 2020 are already substituted by biomass.

The Danish 100 percent renewable energy scenario is an interesting example of how special requirements for power system stability are managed in the case where very large shares of wind are introduced to the system. Due to its geographical location and size, Denmark has a particularly good potential for large-scale use of wind energy. Regional electricity trade facilitates that as much as 30 percent of total annual electricity consumption can be exchanged with the neighboring countries, and the interactions between for example Norwegian hydropower resources and wind are expected to be beneficial. Despite these good conditions for establishing a 100 percent renewable energy system in Denmark, the scenario is also expected to include some uncertainties that have to be taken into consideration. These include uncertainties about the costs and performance of car batteries, stability of the power system and energy security issues, and also implementation rates for energy efficiency improvements in end-use sectors. These improvements play a key role because the biomass resources that are available for the Danish system are considered to be quite scarce, and a given biomass consumption in the power system and in transportation demand is needed in the scenarios.

The cost assessment concludes that a 100 percent renewable energy system is not going to be very expensive in Denmark. This reflects that fossil fuel prices and CO_2 prices in any case are expected to increase, and that the costs of renewable energy technologies are not very much higher than for fossil fuels. Furthermore, it is also expected that a large share of the energy efficiency improvements included will have very low or even no costs. It should here be recognized that there are uncertainties about the policy instruments required and about implementations costs.

13.5   CONCLUSIONS

The international studies and the Danish scenario confirm that large shares of renewables can be integrated in the energy system, and the actual use will largely depend on policy priorities such as energy security, climate change stabilization and renewable energy targets. The renewable energy included in the studies reviewed, and the associated costs, depend on assumptions about the availability of low-cost options such as nuclear power, CCS and bioenergy that in the climate change stabilization scenarios assessed are expected to be at the low-cost end of the options.

Despite differences in assumptions, the international studies conclude that CO_2 prices required to meet stabilization targets associated with a 2°C temperature change are between US$100 per tonne of CO_2 and US$400 per tonne of CO_2 in 2050, dependent on assumptions about technology availability. This sensitivity is confirmed by specific partial assessments of technologies in the ReMIND RECIPE study showing that the costs of not having plenty of renewable energy resources ready for climate change stabilization are high. The study also concludes that nuclear energy and bioenergy options to a large extent will already be included in the reference case.

The Danish study concludes that 100 percent renewable energy is not very costly given the favorable local conditions for high penetration of wind, and large electricity trade with Scandinavian countries and Germany. This reflects that fossil fuel prices and CO_2 prices in any case are expected to increase, and that the costs of renewable energy technologies are not very much higher than for fossil fuels.

Furthermore, it is also expected that a large share of the energy efficiency improvements included will have very low or even no costs. There are some special challenges in introducing a 100 percent renewable energy system in Denmark in order to stabilize the power system in the case where as much as 80 percent of the electricity is produced with wind energy, and the scenarios include investments in flexibility and storage options such as heat pumps, car batteries, production of biofuels in high-load periods, investment in transmission lines, and intensive electricity trade with Germany and Scandinavian countries.

NOTES

1. The modeling work was undertaken by the DTU Climate Centre jointly with the EA Energy Analysis Consultancy in Denmark.
2. Fuel prices and CO_2 prices follow the assumptions in the WEO 2009 (IEA, 2009).

REFERENCES

Calvin, K., J. Edmonds, B. Bond-Lamberty, L. Clarke, Son H. Kim, P. Kyle, S.J. Smith, A. Thomson and M. Wise (2009), 'Limiting climate change to 450 ppm CO_2 equivalent in the 21st century', *Energy Economics*, **31** (Supplement 2).
Greenpeace (2005), 'Energy revolution: a sustainable pathway to a clean energy future for Europe. A European Energy Scenario for EU 25', Greenpeace International.
IEA (2009), 'World energy outlook 2009', OECD IEA.
IEA (2010), 'Energy technology perspectives 2010. Scenarios and strategies to 2050', OECD IEA.
Luderer, G.V., V. Boseti, J. Stechkel, H. Eaisman, N. Bauer, E. Decian, M. Leimbach, O. Sassi and M. Tavoni (2009), 'The economics of decarbonization – results from the RECIPE model intercomparison', RECIPE Background Paper, available at hht://www.pik-potsdam.de/recipe.

14 Valuing efficiency gains in EU coal-based power generation

Luis María Abadie and José Manuel Chamorro

This chapter deals with the valuation of investments to enhance energy efficiency in a coal-fired station operating in a carbon-constrained environment. It is well known that losses in large-scale energy conversion processes turn out to be substantial. At the same time, conversion facilities typically have useful lives extending over decades. Therefore, the arguments for saving limited resources and avoiding greenhouse gas (GHG) emissions through efficiency gains are compelling. Yet there seem to be a huge number of profitable projects that are not undertaken. A potential explanation has to do with the difficulty in determining the economic value of future savings in fossil fuels consumed and emission allowances surrendered; the upfront cost of the projects, though, is usually much more certain. Our aim is to shed some light on the valuation of uncertain savings. We seize upon the information provided by futures markets and use it in a Real Options framework. We derive numerical estimates of the potential savings that result from enhancing energy conversion efficiency at a coal plant level. Some policy implications for energy regulators are also suggested.

14.1 INTRODUCTION

Energy efficiency (henceforth EE) is becoming an increasingly important component of energy policy. It can help to alleviate supply and demand constraints, reduce energy costs by avoiding wastefulness, and contribute to the mitigation of climate change. Its potential also seems to be large and attractive. Furthermore, since energy flows sequentially through several steps (from primary sources to energy carriers to end-use services) potential EE gains are multiplicative and not additive.

Research by the McKinsey Global Institute and McKinsey's Climate Change Special Initiative shows that we can shift to a clean-energy economy while continuing to grow.[1] The key is dramatically increasing the 'carbon productivity' of the economy. The first step in this revolution is to improve EE. Through a variety of measures we have the potential to cut world energy demand growth by more than 64 million barrels of oil a day, equivalent to one and a half times the current annual US energy consumption. Best of all, improvements in EE more than pay for themselves: EE is the low-hanging fruit of the clean-energy revolution. However, the deep changes required across the global economy will not happen without new incentives and policies at the national and international level.

A recent survey of utility companies across the globe by PricewaterhouseCoopers shows similar concerns.[2] Survey respondents are quite bullish about the pace at which technological innovation may herald improvements in EE and reduced emissions. Yet economic signals, in the form of higher energy prices stimulating end-user savings and financial incentives to invest in EE measures, will play a key role in determining the extent to which greater EE becomes a reality. At the moment, companies are investing both in their own and in end-user EE but nearly 60 percent say governments should take the wider lead.

More recently the European Climate Foundation has launched its report on decarbonizing Europe by 2050.[3] EU leaders have stated their goal to cut GHG emissions by 80 percent by that year. The 'Roadmap 2050' sets out viable, cost-effective scenarios for that goal to be achieved. With regard to EE, specific policy recommendations include a binding European Union (EU)-wide energy saving target, national EE obligations, and incentives beyond the EU Emissions Trading Scheme (ETS) to drive EE improvements.

What are the appropriate government policies toward energy-efficient investment? The proper policy depends on what the problem is. Calls for increased EE are motivated by: (1) a concern for energy security; (2) negative externalities associated with energy consumption or production; and (3) an inability to measure the benefits of conservation, among other things (Metcalf, 1994). We do not address our vulnerability to disruptions in fossil fuel supply from exporting countries. As for environmental externalities, more direct ways to address the problem, such as a tax on the specific externality, can be suitable instruments. Instead, our main focus has to do with the private value of EE and hence with the last issue. In particular, people may be poorly informed about the magnitude of savings that can be obtained from different investments (not least because future energy prices are uncertain).

The term 'EE' in this chapter concerns the technical ratio of the maximum quantity of energy services to the quantity of (primary or final) energy input. Thus, it refers to the adoption of a specific technology that reduces overall energy consumption without changing the relevant behavior.[4] As Gillingham et al. (2009) point out, from the outset one must distinguish between EE and economic efficiency. Maximizing the latter, which is typically operationalized as maximizing net benefits to society, is generally not going to imply maximizing the former, which is a physical concept and comes at a price. An important issue arises, however, regarding whether private economic decisions about the level of EE chosen for processes or products are economically efficient.

From an economic perspective, EE choices fundamentally involve investment decisions that trade off higher initial capital costs and uncertain lower future energy operating costs (relative to an otherwise equivalent but less efficient investment). In other words, the decision whether to make the energy-efficient investment requires weighing the initial extra cost (which is typically substantial) against the expected future savings (that accrue over the lifetime of the deployed measures); Gillingham et al., 2009.

Fossil fuels supplied 80 percent of world primary energy demand in 2004 and their use is expected to grow in absolute terms over the next 20–30 years in the absence of policies to promote low-carbon emission sources (Sims et al., 2007). In 2005, coal accounted for around 25 percent of total world energy consumption; hard coal and lignite fuels were used to generate 40 percent of world electricity production. Electricity is the highest-value energy carrier because it is clean at the point of use and has so many end-use applications to enhance personal and economic productivity. The demand for coal is expected to more than double by 2030 and the International Energy Agency (IEA) has estimated that more than 4500 GW of new power plants (half in developing countries) will be required in this period.

On the other hand, fossil energy use is responsible for about 85 percent of the anthropogenic CO_2 emissions produced annually. The implementation of modern high-efficiency and clean-utilization coal technologies is key to the development of economies if effects on society and environment are to be minimized. The average thermal efficiency for electricity generation plants has improved from 30 percent in 1990 to 36 percent in 2002, thereby reducing GHG emissions. Most installed coal-fired electricity generating plants are of a conventional subcritical pulverized fuel design, with typical efficiencies of about 35 percent for the more modern units. Reductions in CO_2 emissions can be gained by improving the efficiency of existing power generation plants by employing more advanced technologies using the same amount of fuel. For example, a 27 percent reduction in emissions (gCO_2/kWh) is possible by replacing a 35 percent efficient coal-fired steam turbine with a 48 percent efficient plant using advanced steam, pulverized coal technology. Supercritical steam plants are in commercial use in many developed countries and are being installed in greater numbers in developing countries such as China.

Future investment in state-of-the-art technologies in countries without embedded infrastructure may be possible by 'leapfrogging' rather than following a similar historic course of development to that of Organisation for Economic Co-operation and Development (OECD) nations. Yet there is a tendency for some countries, particularly where regulations are lax, to select the cheapest technology option (at times using second-hand plant) regardless of total emissions or environmental impact. These low-carbon energy technologies and systems are unlikely to be widely deployed unless they become cheaper than traditional generation, or if policies to support their uptake (such as carbon pricing or government subsidies and incentives) are adopted.

Now, we focus on EE-enhancing investments that are undertaken by power firms.[5] Investments to enhance EE in power plants affect their performance in two important ways. First, they reduce consumption of (fossil) fuels for a given amount of electricity output. Second, they reduce the emission of GHGs and other pollutants (sulfur dioxide, nitrogen oxides and dust particles, among others). Firms are assumed to have precise information about the magnitudes of physical savings that can be obtained from different investments. However, they face a problem in translating them into monetary units. The difficulty with processing this information is that future energy and emission prices are uncertain and, thus, so is the return on the investment. Fuel prices cannot be predicted with much confidence because of swings in market forces. The same holds for emission allowances, whose prices are also particularly sensitive to changes in regulation. Thus, in principle the anticipation of further restrictions on future emissions enhances the appeal of investments to increase EE.

As long as there are markets for fuels and for emissions it is possible to assess the savings in both bills. Key information in this regard comes from futures market prices.[6] These markets allow the proper assessment of an EE-enhancing project (that is, accounting for the risk premium) and hence inform managers about its appeal. By the same token, despite uncertainty some operations can be hedged in the markets, that is, if there are futures prices available on organized markets (or over-the-counter markets) for the required maturities. Thus, uncertain profit margins can be made more certain.

Of course, EE investments will only afford noticeable profits when enhanced power plants operate during an acceptable number of hours over the year. Otherwise, a lower availability rate will preclude the project cost to be recovered and the investment will be shunned. As a consequence EE-enhancing projects are particularly well suited for baseload power plants (that is, those that meet continuous or minimum demands by producing at an almost constant rate); they would hardly be undertaken in peaking stations (those that meet short-lived peaks in power demand).

In order to illustrate better the issue, this chapter develops a practical case. Specifically, we consider an investment to increase EE in a coal-fired power plant that operates under the EU ETS. At the EU level, GHG emissions are regulated by Directive 2003/87/EC;[7] other pollutants are regulated by the Large Combustion Plant Directive (Directive 2001/80/EC).[8] The remaining useful life of the plant is assumed to be either 5, 10 or 15 years. Two different operation scenarios are analyzed: (1) the plant operates full time, that is, irrespective of whether the profit margin is positive or negative, at 80 percent of its full capacity (this is the base case); (2) the plant operates (at 80 percent) only when the profit margin is positive.

The opportunity to invest in EE is framed as a now-or-never decision, that is, the firm has no option to wait and delay the investment outlay. Three valuation methods are adopted. At one level, we derive analytical solutions; on the other hand, we resort to numerical methods, namely a three-dimensional binomial lattice and Monte Carlo simulation. Parameter values and other data are taken from Abadie et al. (2009), where computations are explained in greater detail. The specific question we address is: how much is a 5 percent increase in EE worth? This figure can then be compared with the investment cost, and the firm may react accordingly.

Regarding the sources of risk, and their complete characterization as stochastic process, we proceed as follows. As a matter of fact, natural gas-fired power plants usually set the price in electricity markets (Federico and Vives, 2008; Sensfuß et al., 2008), or their bid price is very close to the actual marginal price. Therefore, we assume (and compute) a fixed profit margin over production costs of gas plants (the 'clean spark spread') as a long-term average. In addition, we consider three stochastic processes: natural gas price, coal price and carbon allowance price. Gas price and carbon price, along with the fixed spark spread, jointly determine the electricity price and therefore the expected revenues of coal-fired plants; their costs are determined by coal price and carbon price. The difference between revenues and costs determines the 'clean dark spread' of these plants, which can become negative in some instances.

Our results show that the value of an investment to enhance EE is lower when the plant operates in a flexible manner than under a rigid pattern. Intuition suggests that any drop in the number of hours in operation undermines the potential to exploit the enhanced system fully. On the other hand, higher carbon prices (due perhaps to more restrictive abatement policies) affect the less efficient plant relatively more severely. In other words, enhanced facilities may cope better with increasing carbon prices. Last, lower allowance volatility increases the appeal of EE-enhancing technologies. Since coal is really a filthy fuel, those with a say in its use can find some lessons to be learnt in these findings (there are also specific technologies – for example gasification, carbon capture and storage – that may ease some of the shortcomings attached to its use).

The chapter is organized as follows. Section 14.2 introduces some preliminaries about the different stages from primary energy sources to end-use energy services. These processes are fraught with energy losses. The scope for potential savings is thus enormous. Perhaps surprisingly, however, a huge number of profitable projects that enhance EE seem to go unrealized. We briefly discuss some barriers to EE investments that have been put forward alongside market failures and policy measures aimed at them. In Section 14.3 we explain the stochastic processes that are assumed for each of the variables that are relevant for the coal-fired power plant. We also describe our sample data and derive numerical estimates of the relevant parameters. In section 14.4 we address the profit margin of the coal plant in our particular setting. Specifically we look at the margin increase that can result from a given improvement in conversion efficiency. Section 14.5 assesses an investment that raises the efficiency rate from 30 percent to 35 percent. We consider two scenarios, one with (rigid) full-time operation of the plant, and the other with flexible operation. Section 14.6 concludes.

14.2   SOME BACKGROUND ON EE

14.2.1   The Physical Setting

As stated above, the potential of EE seems to be large. According to Cullen and Allwood (2010a), 'conversion devices' (for example power stations, engines and light bulbs) upgrade energy into more useful forms. It is possible to set theoretical efficiency targets for such devices. On the other hand, in 'passive systems' (for example houses, vehicles) useful energy is transformed into low-grade heat to the environment, in exchange for final services (transport, thermal comfort, illumination). In passive systems no meaningful theoretical efficiency target can be calculated.

Potential savings from enhanced EE in conversion devices or systems is calculated as follows:

$$\text{Potential for energy saving} = \text{Scale of energy flow} \times (\text{target efficiency}$$

$$- \text{current efficiency}), \tag{14.1}$$

where the energy terms are measured in joules (J) and the efficiency terms in percentages. Thus, improvement potential is computed using an absolute physical basis, which is independent of drivers in today's market. The distinction between conversion devices and passive systems is shown schematically in Figure 14.1 (adapted from Cullen and Allwood (2010a)). The flow of energy can be traced from energy sources (left) to final services (top right) through three key conversion stages: fuel transformation, electricity generation and end-use conversion. At each stage the energy is upgraded into a more useful form, resulting in significant energy 'losses'.[9]

Cullen and Allwood (2010b) aim to determine the efficiency terms from equation (14.1) for conversion devices. Selecting a suitable 'target efficiency' that is both objective and technically defensible is essential if the full potential for EE gains is to be gauged. They opt for a theoretical potential which is based on thermodynamic efficiency limits. The input and output energy flows for the upstream conversions – fuel transformation

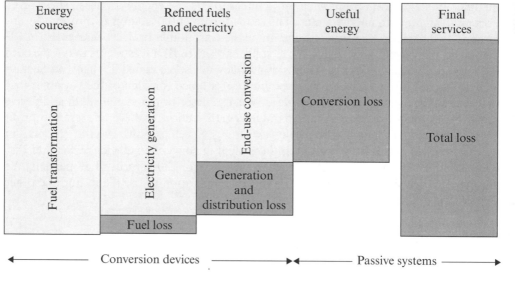

Source: Adapted from Cullen and Allwood (2010a).

Figure 14.1   The flow-path of energy

and electricity generation – are well defined in the energy literature, allowing current conversion efficiencies to be deduced. However, finding representative efficiency values for the global stock of end-use conversion devices is difficult. After a thorough review of the literature and some adjustments they come up with a global map of conversion. As it turns out, only 11 percent of primary energy is converted into useful energy (across all devices); thus the theoretical gains available are substantial.

A sizeable body of literature concerns the so-called 'EE gap' (Jaffe and Stavins, 1994a). In this investment context, the gap takes the form of underinvestment in EE relative to a description of the socially optimal level of EE. This situation can also be portrayed by adoption rates of EE technologies that are 'too slow' (Jaffe and Stavins, 1994b; DeCanio, 1993; Diederen et al., 2003). But the whole issue is the subject of much debate (Gillingham et al., 2009).

A number of explanations that account for part or all of the apparent gap have been posited. Thus, some economists resort to market barriers (that is, any disincentives to the adoption or use of efficient technologies) as a potential explanation of the gap. These include (but are not restricted to) low energy prices,[10] changing energy prices, high technology costs, and systematic biases in consumers' decision-making that drive them away from cost minimization.[11] Meanwhile others are skeptical about the existence, let alone the ubiquity, of unrealized profit opportunities; consequently they deny the very existence of the EE gap by drawing on simple arbitrage arguments.

As a telling example, since the 1970s, refrigerators in the US have swelled in size while at the same time their energy consumption has dropped by 75 percent, saving countless tons of coal from being burned. According to US Department of Energy Secretary Steven Chu: 'Energy efficiency is truly a case where you can have your cake and eat it too. [But] it was driven by standards; it didn't happen on its own' (Biello, 2009).

14.2.2 Barriers to EE

Claims abound that investments that at first glance seem worthwhile usually are not undertaken. For example, around 40 percent of the potential energy savings from the IEA (2009) recommendations, or measures that achieve similar outcomes, remains to be captured. Why? EE continues to face multiple and persistent barriers that are present at the individual opportunity level and the overall system level. McKinsey (2009b) groups them into three broad categories:

- Structural barriers: they prevent an end-user from having the choice to capture what would otherwise be an attractive efficiency option: for example landlord–tenant issues, other principal–agent problems (IEA, 2007), pricing distortions, and so on.
- Behavioral barriers: they include situations where lack of awareness, insufficient information, or end-user inertia block pursuit of an opportunity.[12]
- Availability barriers: they refer to situations when an end user interested in and willing to pursue a measure cannot access it in an acceptable form (for example due to a lack of access to capital).

Several approaches are being used to address these issues. The array of proven, piloted, and emerging solutions fall into four broad categories: (1) information and education; (2) incentives and financing; (3) codes and standards; (4) third-party involvement. For most opportunities, a comprehensive approach will require multiple solutions to address the entire set of barriers faced. Note also that many of the economically attractive opportunities can be achieved at 'negative' marginal costs (that is, investing in these options would generate positive economic returns over their life cycle) and are 'time-perishable' (every year we delay producing energy-efficient commercial buildings, motor vehicles and so forth, the more negative-cost options we lose); see McKinsey (2009a). In addition, an aggressive EE program would reduce demand for fossil fuels and the need for new power plants and other types of infrastructures.[13] The fact that these EE savings are not being fully captured today attests to the fundamental character of many barriers.

It is worth noting, though, that market barriers may or may not be market failures in the traditional Welfare Economics sense (market failure analysis assumes individual rationality); see Jaffe and Stavins (1994b). If they are, then there is an economic rationale for public policies aimed to overcome market barriers. Gillingham et al. (2009) subdivide potential failures into five main categories: (1) energy market failures; (2) information problems; (3) liquidity constraints in capital markets; (4) innovation market failures; and (5) behavioral failures. They discuss these potential concerns and review experience of policies aimed at them.

14.2.3 EE Policy

For policy responses to improve EE they must successfully reduce these failures and the associated benefits must exceed the cost of implementing the policy. In what follows we briefly introduce recent advances along some of these lines with our case study in mind.

Information programs

By providing greater and more reliable information, these programs aim to lessen issues of uncertain future returns and asymmetric information. Take first the consumers' perspective. EE depends on both available technologies and users' choices. Regarding the latter, typical policies to affect these choices focus on (relative) price changes and information disclosure. This focus is consistent with traditional economic models of rational behavior. But a growing body of empirical evidence suggests that they are a bit of a straitjacket. Regarding EE, many studies hint at people's failures to adopt measures that would save them money. Jaffe and Stavins (1994a) assess different explanations for this finding. Yet some barriers seem to be behavioral. Allcott and Mullainatham (2010) report on some recent work by a power company according to which a particular behavioral program compares favorably with estimates of the average cost of other EE programs. If so, behavioral programs should be eligible to receive government support as part of any EE policy. Regulators should also pay attention to how information is conveyed to final consumers and how to 'nudge' them towards reducing energy use. Moreover, as Stern et al. (2010) point out, even larger opportunities can be realized by combining behaviorally sensitive features with financial incentives and information to address a wider range of behaviors.

As for the industry level, Anderson and Newell (2004) examine energy audits. While plants only accept about half of the recommended projects, most of the plants respond to the costs and benefits presented in the energy audits. Provided with the additional information, they adopt investments that meet hurdle rates consistent with standard investment criteria that the audited firms say they use. Muthulingam et al. (2009) investigate the adoption and non-adoption of EE initiatives resulting from recommendations made to US small and medium-sized manufacturing firms. They use a database of over 100 000 recommendations provided to more than 13 000 firms. The recommendations usually have very attractive rates of return and their average payback period is just over a year. As it turns out, however, just providing information on profitable EE opportunities does not seem to be enough for the success of a program: over 50 percent of recommendations are not implemented. Worse still, the percentage of those implemented is decreasing over time. The authors identify several decision biases in the adoption of these recommendations. They further draw implications for enhancing adoption of EE initiatives. Since managers are more influenced by a project's upfront costs than by net benefits when evaluating any initiative, it may be prudent to present lifetime savings of a recommendation rather than the annual savings along with the implementation costs. Another possibility is to provide options to spread the implementation costs over the lifetime of the initiative.

Financial incentives

Sometimes lack of funds is not the main culprit. The EU's regional policy flows to a large extent through the so-called structural and cohesion funds. These funds set aside some money to be spent by the 12 newest member states in renewables and EE measures for 2007–13. To date (March 2010), only around 6 percent of renewable energy funding has been spent, along with slightly more than 16 percent of EE funding.[14] The explanations have to do with limited capacity, complicated application processes and lack of co-financing from national governments and banks.[15]

Starting with households lacking capital, the initial expense of installing solar panels on residential homes may put many people off. A possible response to this situation by municipalities is the following. Local governments issue a particular type of bond (a Property Assessed Clean Energy or PACE bond). Of course the bond's interest rate along with its seniority and fiscal status must fit investors' requirements. The money raised by the municipality is then lent to homeowners so as to install the panels. The borrowers can repay their loans over a period of years through an extra charge on their local property tax bill. One such pilot scheme has just been completed in Berkeley (California). Different versions of the scheme are also possible. For one, homeowners would get the panels for free (from private investors) and then buy the electricity produced by their own rooftop at a rate that is less, per kilowatt-hour, than they would pay for electricity from the grid. This way investors get a safe investment and homeowners get a break on their monthly bills. Alternatively, panel installers could lease the panels to homeowners who would then receive the electricity for free. The total for both may come to less than the old bill.

Some purchasers of equipment may choose a less effective EE product due to lack of access to credit, resulting in underinvestment in EE (and reflected in an implicit discount rate that is above typical market levels). This applies to any capital-intensive investment, not just EE products. In some cases, such as for industrial customers, energy service providers pay the capital cost and receive a share of the resulting savings. When the customer can borrow at a lower interest rate than the energy service provider, the latter recommend EE improvements, guarantee the operating cost savings, and pay the difference if those savings are not realized (Gillingham et al., 2009).

On the other hand, if companies are to be enticed into undertaking more EE investments, the return on these investments must be enhanced. Energy efficiency certificates (EECs) can be a useful tool. Thus, voluntary savings may in principle be measured, audited and verified to certain standards. These savings can later be credited to the company involved with EECs. As a final step, there may be a registry for EECs operating similarly to those in carbon markets.[16] There are several possible routes from here. First, EECs could be used to count towards mandated renewable energy targets (as is the case in several US states). Also, a market for voluntary EECs may develop (in the same way as that for carbon credits). Last, regulatory agencies can seize on them to establish incentives and rebates.

Regulatory instruments
Here there is room not only for public regulators. German Bank WestLB has recently adopted an environmental policy for private sector lending to coal-fired power plants. The policy sets minimum standards whereby new plants (or any expansion or refit) must use best available technology delivering at least 43 percent efficiency. This goes beyond the Carbon Principles adopted by a number of US banks. They encourage clients to invest in cost-effective demand reduction (taking into consideration the value of avoided carbon emissions) caused by either increased EE or other means; but they fall short of setting minimum standards. Note that both approaches aim at the overall goal of assessing project economics and financing parameters related to carbon risks faced by the power industry.

Regarding the EU, Phase III of the ETS, which will start in 2013, is presumed to set tighter caps on carbon emissions and give away fewer emissions allowances. Companies will try to mitigate their exposure to this phase by following different strategies. They can invest in internal projects (or otherwise alter products or processes) to reduce carbon emissions. They can also move to locations with less stringent climate regulations (when this is possible; in the case of electricity production this is hardly an option). They can engage in investments that serve as carbon offsets. And they will try to pass through higher costs to end consumers.

14.3   STOCHASTIC MODELS AND THEIR ESTIMATION

According to Cullen and Allwood (2010b) efforts to improve the EE of coal-fired power stations will deliver the most (energy and carbon) savings in the upstream fuel conversion and electricity generation processes, because coal dominates electricity generation at the world level.[17] We restrict ourselves to process improvements in electricity generation. Even though we develop our analysis at the plant level, it must be noted from the outset that the efficiency gains as such may result from different causes. They can range from a physical upgrade in the processes or products involved (with the corresponding upfront outlay) to an overall transmission and distribution system overhaul whereby the plant better exploits its possibilities.[18]

From the several types of hypotheses that have been advanced to explain the EE gap, we focus upon uncertainty about the future. Two types of uncertainty are covered in the literature (Diederen et al., 2003). First of all, there is uncertainty regarding future technological development: firms may postpone a profitable investment in new technology because they expect an even better technology to come along in the near future (Grenadier and Weiss, 1997; Van Soest and Bulte, 2001). Second, there is uncertainty regarding the profitability of the technology to be adopted itself, for instance due to the stochastic nature of energy prices: firms may postpone an investment that would be profitable, given today's energy prices, because they take into account the possibility of a future drop in energy prices which would render their investment unprofitable (Hasset and Metcalf, 1993).

In our model below we consider three stochastic prices for natural gas G_t, coal C_t, and emission allowance A_t, respectively. We assume a constant profit margin of natural gas-fired power plants (M_e) in the long run.

14.3.1   Profit Margin of a Gas-fired Station

The profit margin M (in /MWh) of a natural gas-fired power plant that operates under a cap-and-trade system is given by:[19]

$$M = S - \frac{G}{E_G} - A \times I_G, \tag{14.2}$$

where S denotes electricity price (€/MWh), G is the price of natural gas (€/MWh), E_G is the net thermal efficiency of a gas plant, and A is the price of a EU emission allowance (€/tCO$_2$). Lastly, I_G stands for the emission intensity of the plant (tCO$_2$/MWh); this in turn depends on the net thermal efficiency of each gas-fired plant. According to the Intergovernmental Panel on Climate Change (IPCC, 2006), a plant burning natural gas has an emissions factor of 56.1 $kgCO_2/GJ$. Since under 100 percent efficiency conditions 3.6 GJ would be consumed per megawatt-hour, we get:

$$I_G = \frac{0.20196}{E_G} \frac{tCO_2}{MWh}. \tag{14.3}$$

Thus the complete formula for the margin M (or clean spark spread, CSS) at time t is

$$M_t = S_t - \frac{1}{E_G}(G_t + 0.20196 \times A_t). \tag{14.4}$$

We adopt an efficiency rate $E_G = 0.55$ in the gas plant.

Regarding behavior over time, we assume that the current margin M_t evolves according to a mean-reverting (Ornstein–Uhlenbeck) stochastic process. This allows for the margin to take on negative values on some occasions. Specifically,

$$dM_t = k_M(M_e - M_t)dt + \sigma_M dW_t^M, \tag{14.5}$$

where M_e is the long-term equilibrium value (that is, the level to which the current margin tends over time), and k_M denotes the reversion speed. σ_M is the instantaneous volatility of the margin, and dW_t^M is the increment to a standard Wiener process. From this behavior in the physical world, the expected value at time t (as considered from time 0) is given by:

$$E(M_t) = M_0 e^{-k_M t} + M_e(1 - e^{-k_M t}).$$

For high speeds of reversion k_M, and also for times t that are far into the future, we get:

$$E(M_t) \cong M_e.$$

Equation (14.5) is the continuous-time version of a first-order autoregressive process AR(1) in discrete time:

$$M_{t+\Delta t} = M_e(1 - e^{-k_M \Delta t}) + M_t e^{-k_M \Delta t} + \varepsilon_{t+\Delta t}^M = a_M + b_M M_t + \varepsilon_{t+\Delta t}^M, \tag{14.6}$$

where $\varepsilon_t^M \cong N(0, \sigma_\varepsilon^M)$, and the following notation holds:

$$a_M \equiv M_e(1 - b_M) \Rightarrow M_e = \frac{a_M}{1 - b_M}, \tag{14.7}$$

$$b_M \equiv e^{-k_M \Delta t} \Rightarrow k_M = -\frac{\ln b_M}{\Delta t}. \tag{14.8}$$

Data

In order to estimate the model in equation (14.6) we collect (weekly) prices from the markets involved. Thus, electricity prices (S_t) are taken from PowerNext (France), natural gas prices (G_t) from Zeebrugge (Belgium), and spot carbon prices (A_t) from BlueNext. Figure 14.2 shows the evolution of the gas plants' profit margin as computed from equation (14.4). The margin of coal-fired stations (the so-called 'clean dark spread', or CDS, for two different efficiency rates assumed) is also displayed. The time horizon spans over 180 weeks from 14 May 2006 to 18 October 2009; each price corresponds to the weekly average.

The estimates of the coefficients in equation (6) are the following (*t*-statistics in brackets):

$$\hat{a}_M = 5.62611; \ (4.9102)$$
$$\hat{b}_M = 0.57720; \ (9.3668).$$

Hence we can compute the long-term profit margin of gas stations (which is assumed constant):

$$\hat{M}_e = \frac{\hat{a}_M}{1 - \hat{b}_M} = 13.30691.$$

14.3.2   Prices of Inputs and Emissions

Natural gas

We analyze the stochastic behavior of natural gas prices through futures prices on the European Energy Exchange (EEX, Leipzig). Note that over the year actual prices of natural gas display a seasonal pattern. This seasonality also shows up in the prices of futures contracts according to their maturity. In our model, seasonality of natural gas would sometimes push upward the profitability of coal plants when (because of seasonality) the electricity price rises.

The risk-neutral behavior of natural gas price is assumed to be governed by the following stochastic process:

$$dG_t = df(t) + [k_G G_m - (k_G + \lambda_G)(G_t - f(t))]dt + \sigma_G(G_t - f(t))dW_t^G.$$

In this setting, G_t denotes the price of natural gas at time t, while G_m stands for the level to which natural gas price tends in the long run. λ_G is the market price of risk related to changes in gas price. $f(t)$ is a deterministic time function. Since we are interested in reflecting the seasonal pattern on the gas price time series throughout the year, we resort to a sinusoidal function like the cosine function: $f(t) = \gamma \cos(2\pi(t + \varphi))$. Here *cos* stands for the cosine function measured in radians, and γ is a constant parameter (Lucia and Schwartz, 2002). The cosine function has annual periodicity, hence the time is measured in years. At time $t = -\varphi$ we have $f(t = -\varphi) = \gamma$ and seasonality is highest. k_G is the speed of reversion towards the 'normal' level of gas price; it can be computed as $k_G = \ln 2/t_{1/2}^G$, where $t_{1/2}^G$ is the expected half-life of (deseasonalized) gas price, that is, the time required for the gap between G_t and G_m to halve. σ_G is the instantaneous volatility of gas price.

Natural gas futures prices on EEX refer to months, quarters, semesters, and years.

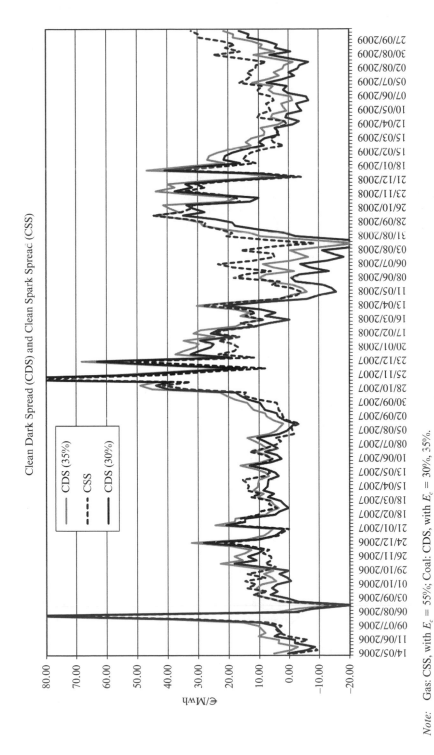

Note: Gas: CSS, with $E_c = 55\%$; Coal: CDS, with $E_c = 30\%$, 35%.

Figure 14.2   Profit margin of natural gas-fired stations and coal-fired stations, May 2006–September 2009

Using all the prices available since 2 January 2009 to 27 November 2009 non-linear least-squares estimation yields the following parameter estimates (95 percent confidence intervals in brackets):

$$\frac{k_G G_m}{k_G + \lambda_G} = 25.04; \; [24.04;26.04]$$

$$k_G + \lambda_G = 0.85; \; [0.65;1.05]$$

$$\varphi(days) = -21.7; \; [-33.20; -10.24]$$

$$\gamma = 3.29; \; [2.64;3.93].$$

Note that $k_G G_m / (k_G + \lambda_G)$ is measured in €/MWh, and φ in days.

We have not derived the value of individual parameters ruling the behavior in the physical world, such as k_G or λ_G. They are not necessary for the computations below; it suffices to know their sum $k_G + \lambda_G$, which we have estimated from futures prices (see Appendix Section A). Two additional parameters will be important in our Monte Carlo simulations, namely the starting (deseasonalized) gas price $G_0 - f(0)$, which amounts to €7.2419/MWh on the last day in our series, and the estimated volatility, $\sigma_G = 0.6356$.

Coal

Regarding coal price we adopt a stochastic process that is similar to that for natural gas but does not display seasonality:

$$dC_t = [k_C(C_m - C_t) - \lambda_C C_t]dt + \sigma_C C_t dW_t^C;$$

the notation runs akin to that for the dynamics in gas price. Using EEX coal futures prices over the same time horizon as before we get the following parameter estimates (95 percent confidence intervals in brackets):

$$\frac{k_C C_m}{k_C + \lambda_C} = 105.27; \; [101.57; 108.96]$$

$$k_C + \lambda_C = 0.69; \; [0.58; 0.79].$$

Note that $k_C C_m / (k_C + \lambda_C)$ is measured in $/t.

In this case the estimated volatility is $\sigma_C = 0.4144$, and the starting value for simulations is $C_0 = 74.7898$ ($/t).

Carbon dioxide

The price of the emission allowance in a risk-neutral world (or under the equivalent martingale measure) is assumed to follow a standard geometric Brownian motion(GBM):

$$dA_t = (\alpha - \lambda_A)A_t dt + \sigma_A A_t dW_t^A.$$

A_t is the carbon price at time t, and α is the instantaneous drift rate (assumed constant). σ_A denotes the instantaneous volatility of carbon price changes (assumed constant), which determines the variance of A_t at t. And λ_A is the market price of carbon price risk.

Futures contracts on emission allowances are traded on different platforms (in addition to over-the-counter markets). Due to its volume of operations and liquidity, the European Climate Exchange (ECX) stands apart. It manages the European Climate Exchange Financial Instruments (ECX CFI), which are traded at the London-based International Petroleum Exchange (later acquired by the InterContinental Exchange).

Futures price data from ICE have been used here with the following results (95 percent confidence interval in brackets):

$$\alpha - \lambda_A = 0.054; [0.048; 0.061].$$

In this case, the estimated volatility is $\sigma_A = 0.4879$, and the starting value for simulations is $A_0 = 13.18$ €/tCO_2.

Correlation coefficients
Correlations between prices are estimated from spot prices of coal, natural gas and carbon allowances. In the particular case of natural gas we use deseasonalized spot prices. Our numerical estimates are:

$$\rho_{CG} = 0.2652; \rho_{AG} = 0.2572; \rho_{CA} = 0.2797.$$

14.4 PROFIT MARGIN OF A COAL-FIRED STATION

The profit margin MC (in €/MWh) of a coal-fired power plant that operates under a cap-and-trade system is given by:[20]

$$MC = S - \frac{C}{E_C} - A \times I_C, \tag{14.19}$$

where C is the price of coal (€/MWh), E_C is the net thermal efficiency of a coal plant, and I_C stands for the emission intensity of the plant (tCO_2/MWh). Following IPCC (2006) a plant burning bituminous coal has an emission factor of 94.6 kgCO_2/GJ under 100 percent efficiency conditions; then:

$$I_C = \frac{0.34056}{E_C} \frac{tCO_2}{MWh}. \tag{14.10}$$

Thus the complete formula for the clean dark spread is:

$$MC_t = S_t - \frac{1}{E_C}(C_t + 0.34056 \times A_t). \tag{14.11}$$

We adopt two efficiency rates $E_C = [0.30; 0.35]$ in the coal plant.

We can solve for S_t in equation (14.4) and then substitute into equation (14.11), thus linking the two spreads. This yields:

$$MC_t = M_t + \frac{1}{E_G}(G_t + 0.20196 \times A_t) - \frac{1}{E_C}(C_t + 0.34056 \times A_t). \quad (14.12)$$

Hence in the long term $(E(M_t) \approx M_e)$ the profit margin is expected to be:

$$MC_t = M_e + \left(\frac{G_t}{E_G} - \frac{C_t}{E_C}\right) + \left(\frac{0.20196A_t}{E_G} - \frac{0.34056A_t}{E_C}\right). \quad (14.13)$$

Thus it is a function of three stochastic variables at time t (namely G_t, C_t, and A_t) along with the efficiency level of the coal plant E_C and that of the marginal gas plant E_G. Regarding numerical application, the original price of ARA (Amsterdam–Rotterdam–Antwerp) coal (in \$/tcoal) must be previously converted to €/tcoal at an exchange rate €1.4934/\$. Then we must transform the units €/tcoal into €/MWh by means of the following equivalences: 1 tcoal = 29.31 GJ, 1 GJ = 0.27777 MWh. Therefore we have $C_t(€/tcoal)/29.31/0.2777 = C_t(€/MWh)$.

The difference between fuel prices for producing 1 MWh is an advantage or positive element for coal plants, which in the above formula is given by the first set of parentheses. This advantage, however, is weaker when the efficiency E_C is lower. On the other hand, the higher consumption of emission allowances by coal plants is a negative element for them. This relative disadvantage appears in the second set of parentheses. Again, a lower efficiency E_C increases the (carbon) bill for the coal plant. If fuel (relative) prices behave according to their long-run expected value $(G_m/E_G - C_m/E_C)$ the evolution of profit margins in coal plants will be determined by the evolution of allowance prices; these could well grow over time and severely damage their margins.

Next we will assess potential investments in a coal plant to enhance its EE rate. We adopt $E_G = 0.55$ for the marginal gas plant, and $E_C = 0.30$ for the base coal plant; should the investment be undertaken, efficiency would jump to $E_C = 0.35$. The long-run cost would then fall from 105.27 (\$/tcoal) / [0.30 × 1.4934 (\$/€) × 29.31 (GJ/tcoal) × 0.2777 (MWh/GJ)] = €28.87/MWh to €24.74/MWh. Regarding its carbon emissions, with the initial efficiency (30 percent) they amount to 1.135 t CO_2/MWh; however, after the enhancement (35 percent) they fall to 0.973 t CO_2/MWh. Thus, in principle the resulting increase in the profit margin would be given by:

$$\Delta MC_t = (C_t + 0.34056A_t)\left(\frac{1}{0.30} - \frac{1}{0.35}\right) = 0.47619(C_t + 0.34056A_t); \quad (14.14)$$

note that this improvement only applies when the coal plant is upgraded from 30 percent efficiency to 35 percent.

14.5   VALUATION OF INVESTMENTS

We assess the decision to invest at an initial time or not to invest. In other words, there is no option to delay the decision until a later time. Whether investing to increase EE at the plant level is a sound decision or not depends on the prices G_t, C_t, and A_t prevailing at the time of the investment. Valuation will proceed along a three-dimensional binomial lattice or alternatively by Monte Carlo simulation. On some occasions we will also use analytical solutions, which will serve as a check for the proper functioning of the numerical methods.

The installed capacity of the coal station is assumed to be 500 MW. For simplicity, the remaining useful life of the plant can only take on three values: 5, 10 and 15 years. Two different scenarios for potential EE enhancement are considered: scenario (a), where the plant operates full time at 80 percent capacity (whatever the margin happens to be); and scenario (b), where it operates at 80 percent capacity only when the profit margin is positive.

14.5.1 Scenario (a): Full-time Operation at 80 Percent Capacity

A unit of time is subdivided into monthly periods; thus, for 15 years to expiration we have 180 time steps, each of duration $\Delta t = 1/12$. Monthly production of electricity by the plant amounts to:

$$P_M = (500\text{MW} \times 0.80 \times 365\text{d/y} \times 24\text{h/d})/12\text{m/y} = 292\,000\text{MWh}. \quad (14.15)$$

Binomial lattice

First consider the plant with $E_C = 35$ percent. At the final nodes (for example with $t = 15$) we assume that the (residual) value of the plant is zero: $W = 0$. At earlier times we get (see Appendix section B):

$$W = P_M \times MC_t + e^{-r\Delta t}(p^*_{uuu}W^{+++} + p^*_{uud}W^{++-} + p^*_{udu}W^{+-+} + p^*_{udd}W^{+--}$$

$$+ p^*_{duu}W^{-++} + p^*_{dud}W^{-+-} + p^*_{ddu}W^{--+} + p^*_{ddd}W^{---}). \quad (14.16)$$

In words, each MWh of monthly production P_M generates a time$-t$ unit profit MC_t. In addition to the resulting current profits during that period, the plant owners receive the present value of continued operation over the next period under the prevailing circumstances (that is, the expected value of the eight possible nodes discounted at the riskless interest rate r over that period). p^*_{udu} denotes the (risk-neutral) probability that (from the current node to the next one) G_t will rise, C_t will fall, and A_t will rise. W^{+-+} stands for the plant value at this particular node of the tree.

We proceed iteratively backwards until the initial time ($t = 0$) is reached, at which time we compute the enhanced plant's value $W_0(E_C = 0.35)$. Following the same procedure under the assumption $E_C = 30$ percent we derive $W_0(E_C = 0.30)$. The difference between them, $W_0(E_C = 0.35) - W_0(E_C = 0.30)$, shows the value of the EE investment under these circumstances.

The same value of the investment results from directly computing (see equation 14.14):

$$\Delta MC_t = 0.47619(C_t + 0.34056A_t)$$

and then developing the lattice as follows:

$$W = P_M \times \Delta MC_t + e^{-r\Delta t}(p^*_{uuu}W^{+++} + p^*_{uud}W^{++-} + p^*_{udu}W^{+-+} + p^*_{udd}W^{+--}$$

$$+ p^*_{duu}W^{-++} + p^*_{dud}W^{-+-} + p^*_{ddu}W^{--+} + p^*_{ddd}W^{---}). \quad (14.17)$$

Table 14.1   Gross value of the EE investment: three-dimensional lattice

Efficiency (%)	5 years	10 years	15 years
35	256.99	513.48	686.68
30	155.29	309.58	383.47
Value (M€)	101.70	203.90	303.21

Table 14.2   Gross value of the EE investment: analytical solution

	5 years	10 years	15 years
Value (M€)	100.66	201.44	299.01

Note that at any time we add the corresponding seasonality in gas prices. In this scenario, though, since the plant operates full time, this addition will have no impact on the net valuation results.

When the only possibility is either to invest in EE initially or not, the investment values in Table 14.1 result. Both $W_0(E_C = 0.35)$ and $W_0(E_C = 0.30)$, in million euros, are computed as a function of the station's remaining life. As could be expected, the value of the enhancement in EE grows consistently with the useful life of the coal plant.

Analytical solution
We can derive an analytical solution for the value of an 'annuity' (Appendix section A). In particular, we refer to the sustained flow of increases in the (unit) profit margin (ΔMC_t) that result from the improvement in EE:

$$12P_M \int_{\tau_1}^{\tau_2} E_0(\Delta MC_t)e^{-rt}dt, \tag{14.18}$$

where E_0 denotes the time -0 expectation. τ_1 and τ_2 stand for the times when the first increase and the last one take place.

With our numerical parameters values we get:

$$\int_{\tau_1}^{\tau_2} E_0(\Delta MC_t)e^{-rt}dt = 0.47619\left(\int_{\tau_1}^{\tau_2} C_t e^{-rt}dt + 0.34056\int_{\tau_1}^{\tau_2} A_t e^{-rt}dt\right). \tag{14.19}$$

Using the formulas in Appendix section A for the valuation of annuities we get Table 14.2.

The differences between the values in Table 14.1 (derived from our complex three-dimensional lattice) and in Table 14.2 turn out to be very small. In principle, we do not need this lattice in this scenario; yet we will need it in the next one (where an analytical solution is no longer available). Anyway the comparison makes us confident that the lattice performs well. With a minor change in the programming we will be able to undertake the valuation when the operation of the plant is flexible (that is, it can be switched on or off).

Table 14.3   Gross value of the EE investment: Monte Carlo simulation

	5 years	10 years	15 years
Value (M€)	100.96	201.73	297.95

Table 14.4   Gross value of the EE investment: binomial lattice

Efficiency	5 years	10 years	15 years
35%	282.00	599.79	875.45
30%	212.69	470.31	697.51
Value (M€)	69.31	129.48	177.94

Table 14.5   Valuation of EE (M€): sensitivity to carbon price

A_0 (€/tCO_2)	5 years	10 years	15 years
13.18	69.31	129.48	177.94
20.00	64.19	122.98	170.24
30.00	53.81	111.44	156.65
40.00	48.14	100.33	143.49

Monte Carlo simulation

We can also develop a Monte Carlo simulation following the scheme in Appendix section C. We consider two sources of risk (coal and carbon prices) and 100000 paths (each of them with monthly time steps). We get the results in Table 14.3. Again, the differences from the above values are pretty small.

14.5.2   Scenario (b): Operation at 80 Percent Capacity Only When the Profit Margin is Positive

Now the plant managers have the option to keep the plant running (at 80 percent capacity) or to close it temporarily; we assume that switching costs are zero. Restricting ourselves to the use of the binomial lattice, when the plant reaches maturity its value falls to zero: $W = 0$. At earlier times:

$$W = \max(P_M \times MC_t, 0) + e^{-r\Delta t}(p^*_{uuu}W^{+++} + p^*_{uud}W^{++-} + p^*_{udu}W^{+-+}$$

$$+ p^*_{udd}W^{+--} + p^*_{duu}W^{-++} + p^*_{dud}W^{-+-} + p^*_{ddu}W^{--+} + p^*_{ddd}W^{---}). \quad (14.20)$$

As before, the difference $W_0(E_C = 0.35) - W_0(E_C = 0.30)$ provides the (gross) value of the EE investment under these circumstances; see Table 14.4.

Comparison with the values in Table 14.5 shows that the value of the flexible system is higher. Thus, with $E_C = 0.35$ and ten years to maturity, the value rises from €513.48

Table 14.6   Gross value of the EE investment: $A_0 = €20/tCO_2$

Efficiency	5 years	10 years	15 years
35%	232.20	514.19	764.97
30%	168.01	391.21	594.73
Value (M€)	64.19	122.98	170.24

million to 599.79. This is no surprise since the utility now can switch the plant off whenever it operates at a loss.

However, the economic impact of the EE improvement turns out to be lower. For the same time horizon, it falls from €203.90 million to €129.48 million. Presumably, the flexibility to operate (or not) involves a lower number of hours in operation than in scenario (a), which in turn undermines the potential to exploit the capabilities of the enhanced system fully. This effect is stronger for longer lifespans of the plant. With five years to expiration the new investment value is 68 percent of the old value (69.31/101.70). However, with 15 years to expiration, this ratio falls to 58 percent (that is, 177.94/303.21).

Next we analyze the sensitivity of these values to changes in allowance price and allowance volatility. As before, the figures below have been computed by means of a three-dimensional binomial lattice.

As could be expected, the improvement in EE is less valuable with higher carbon prices; see Table 14.5. The effect can be observed for any of the maturities considered. Intuitively, as the (initial) allowance price increases, the value of the plant (whatever its efficiency happens to be) will decrease. This can be seen in Table 14.6, for the case of $A_0 = €20/tCO_2$.

Compare these figures with those in Table 14.4, computed under $A_0 = €13.18/tCO_2$. Take for instance five years to expiration of the plant. In absolute terms, the (more) efficient plant is less valuable: €232.20 million – €282.00 million = –€49.80 million; the same holds for the less efficient plant: €168.01 million – €212.69 million = –€44.68 million. So the penalty of a higher allowance price is more severe for the plant which runs over a higher number of hours, namely the more efficient plant. This is why the difference $W_0(E_C = 0.35) - W_0(E_C = 0.30)$ is slightly lower than before (64.19 instead of 69.31). In relative terms, however, the picture is somewhat different. For the efficient plant, the new value amounts to 82 percent of the old one; instead, for the less efficient plant this proportion approaches 79 percent (the same happens with the other useful lives). Therefore, the efficient plant is relatively better fitted to withstand the impact of higher carbon prices. Note that again $W_0(E_C = 0.35)$ exceeds $W_0(E_C = 0.30)$ by a wide margin, so the EE investment adds good value.

Besides, this effect becomes stronger as the useful life of the plant is longer. With five years to expiration the new investment value is 92 percent of the old value (64.19/69.31). However, with 15 years to expiration, this ratio rises to 95 percent (that is, 170.24/177.94). This suggests that younger plants are better candidates for enhancing EE than older ones (similarly, base load plants seem better candidates than peaking plants).

The impact of changes in carbon volatility is explored in Table 14.7. A (permanent) fall in the allowance price volatility raises the value of the EE investment irrespective of

Table 14.7   *Valuation of EE (M€): sensitivity to carbon volatility*

σ_A	5 years	10 years	15 years
0.40	70.99	133.38	183.37
0.4879	69.31	129.48	177.94
0.60	67.23	125.20	172.29

Table 14.8   *Gross value of the EE investment:* $\sigma_A = 0.40$

Efficiency (%)	5 years	10 years	15 years
35	278.91	584.62	842.41
30	207.92	451.24	659.04
Value (M€)	70.99	133.38	183.37

the time to expiration of the plant. Conversely, a (permanent) increase in carbon volatility implies a lower value of EE investments.

This effect can be traced back to the impact of lower volatility on the value of both types of plants. Table 14.8 shows these values for the case of $\sigma_A = 0.40$. Take for instance 15 years to expiration of the plant. In absolute terms, the (more) efficient plant is less valuable: €842.41 – €875.45 million = –€33.04 million; the same holds for the less efficient plant: €659.04 million – €697.51 million = –€38.47 million. So now the lower allowance volatility affects the less efficient plant more severely. Consequently the difference $W_0(E_C = 0.35) - W_0(E_C = 0.30)$ is slightly higher than before (€183.37 million instead of €177.94 million). In relative terms the situation remains the same. For the efficient plant, the new value amounts to 96 percent of the old one, while for the less efficient plant this proportion approaches 94 percent (the same happens with the other useful lives). Therefore, the efficient plant is relatively better fitted to cope with lower allowance volatility. $W_0(E_C = 0.35)$ exceeds $W_0(E_C = 0.30)$ by a wider margin than before, so the EE investment adds even better value. A policy implication, therefore, is that measures to reduce uncertainty in future allowance prices are welcome in that they actually encourage (less-efficient) coal plants' owners to upgrade their facilities.

Again, in percentage terms the rise in the EE investment value is (slightly) higher for longer lifespans. With five years we have 70.99/69.31, or 102 percent. With 15 years, the ratio is 183.37/177.94, or 103 percent. This means that a lower carbon volatility is relatively better news for younger plants. Given the pace of new openings of coal stations, regulators should strive to shed clarity about the future playing field as soon as possible.

14.6   CONCLUDING REMARKS

This chapter assesses investments to enhance energy conversion efficiency in coal-fired power plants that operate under carbon constraints. Typically these plants have a useful life ranging from 30 to over 50 years; as a consequence, there is a slow rate of turnover of around 1–3 percent per year. Thus, decisions taken today that support the deployment of carbon-emitting technologies (especially in countries concerned with security of supply) could have profound effects on GHG emissions for the next several decades. In turn, higher ambient temperatures may affect the efficiency and capacity ratings of fossil fuel-powered combustion turbines. Thus, there is scope for some circular relationships in this issue.

Firms tend to focus on incremental technology improvements to gain profits in the short term. Research and development (R&D) spending by firms in the energy industry is particularly low, with utilities investing only 1 percent of total sales in the US, the UK and the Netherlands compared with the 3 percent R&D-to-sales ratio for manufacturing, and up to 8 percent for pharmaceutical, computer and communication industries (Sims et al., 2007). At the same time, governments provide financial incentives or make direct investments to stimulate the development and deployment of new innovative energy conversion technologies and create markets for them.

Many GHG emission-reduction policies undertaken to date (March 2010) aim to achieve multiple objectives. These include market and subsidy reform, particularly in the energy sector. Governments are also using a variety of approaches to overcome market barriers to EE improvements and other 'win–win' actions. This suggests that there is a role for the public sector in increasing investment directly and in correcting market and regulatory obstacles that inhibit investment in new technology through a variety of fiscal instruments such as tax deduction incentives.

At one level we can mention the public-good nature of knowledge, whereby individual firms are unable to capture fully the benefits from their innovation efforts, which instead accrue partly to other firms and consumers. Thus, the social rate of return to R&D is approximately two to four times higher than the private rate of return (Gillingham et al., 2009). The problem is magnified in the context of EE technologies if energy is under-priced relative to the social optimum.

On the other hand, positive externalities associated with learning-by-using can exist where the adopter of a new energy-efficient product creates knowledge about the product through its use, and others freely benefit from the information generated about the existence, characteristics and performance of the product.

In addition, there are co-benefits of mitigation policies. These include the mitigation of air-pollution impacts, energy supply security (by increased energy diversity), technological innovation, reduced fuel cost, employment and reducing urban migration. Reducing GHG emissions in the energy sector yields a global impact, but the co-benefits are typically experienced on a local or regional level. Benefits of GHG mitigation may only be expected by future generations, but co-benefits are often detectable by the current generation. Co-benefits of mitigation can be important decision criteria in analyses by policy-makers, but are often neglected (Jochem and Madlener, 2003).

Regarding potential explanations of the EE gap at the firm level, we focus on the difficulty in evaluating uncertain future savings derived from enhanced EE. Both fuel prices and carbon price evolve stochastically over time. Therefore, coming up with a numerical value that can be weighted against a certain up-front cost is no easy task. We adopt a Real Options valuation framework. We collect futures prices from the markets involved to estimate the underlying parameters. Then we assess a specific investment that raises the conversion efficiency of the plant by five percentage points. In our first scenario the plant is assumed to operate full time, whereas in the second one operation is flexible (that is, it operates only when the profit margin is positive).

According to our results, the value of an investment to enhance EE is lower in a flexible system than in a rigid system. Presumably, any drop in the number of hours in operation undermines the potential to exploit the enhanced system fully. If so, in decentralized electricity markets, low-efficiency coal plants that do not operate as baseload stations can hardly be expected to engage eagerly in projects to improve their performance. Thus a first suggestion for energy regulators is: when promoting EE in coal plants, first address the most active ones.

On the other hand, more stringent emission limits in the future may push allowance prices upward. As could be expected, higher carbon prices have a negative impact on the value of an EE investment to the adopting firm. However, less efficient plants fare comparatively worse than more efficient ones. And lower allowance volatility (perhaps from a far-reaching, well-established framework) certainly enhances the value of EE technologies, thus promoting their adoption.

ACKNOWLEDGEMENTS

We gratefully acknowledge financial support from the Spanish Ministry of Science and Innovation through the research project ECO2011-25064.

The final version of this chapter was completed while José Manuel Chamorro was a Visiting Scholar at MIT Engineering Systems Division (ESD). He acknowledges financial support from the Basque Government (through the programmes for the upgrade and mobility of researchers, 2011). He also expresses his deepest gratitude and respect to Professor Richard de Neufville and the whole ESD community for their hospitality and assistance. The usual disclaimer applies.

NOTES

1. http://www.mckinsey.com/mgi/mginews/skyhigh.asp.
2. http://www.pwc.com/en_GX/gx/energy-utilities-mining/pdf/2008-utilities.pdf.
3. http://www.europeanclimate.org/index.php?option=com_content&task=view&id=15&Itemid=30.
4. 'Energy conservation', instead, implies merely a change in consumers' behavior. It is strongly influenced by regulation and lifestyle changes. 'End-use energy saving' addresses the reduction of final energy consumption through EE improvement or behavioral change (Oikonomou et al., 2009).
5. Electricity production is but one area where EE improvements are particularly important. For example, in the US the EE resource recoverable from enhanced buildings is equivalent to more than one-third of the coal-fired power production in the country (US DoE, 2009).
6. Gillingham et al. (2006) review the literature on environmental externalities from the production of electricity. They provide a brief synopsis of the relevant programs, along with available existing estimates of energy savings, costs and cost-effectiveness at the US level. They find that past policies to reduce electricity use provided monetized benefits from the reduction in carbon dioxide, nitrous oxides, sulfur dioxide and fine particulate matter that were about 10 percent of the direct value of the electricity savings.
7. Amended by Directive 2004/101/EC, Directive 2008/101/EC, Regulation (EC) No 219/2009, Directive 2009/29/EC.
8. Amended by Directive 2006/105/EC.
9. Sizeable primary energy is lost during transformation to other forms and also in transmission, distribution and transport to end-users. The latter losses are an important energy-saving opportunity but one that is outside the scope of this chapter (even though they could be assessed following similar procedures).
10. Several studies suggest that higher energy prices are associated with significantly greater adoption and innovation of energy-efficient equipment.

11. The perception that consumers apply unreasonably high hurdle rates to energy-saving investments has been dubbed the 'energy paradox'. The actual magnitude of this paradox is much debated. According to Metcalf and Hassett (1999), the case for the energy paradox is weaker than previously thought.
12. Firms may also face some of the same issues (that is, deviations from perfect rationality), although competitive forces serve to moderate the significance of behavioral failures for firms (Gillingham et al., 2009). Managers may cope with uncertainty and complexity in decision-making by adopting simplifying heuristics which may lead to systematic biases. Muthulingam et al. (2009) use actual field data and find that managers are myopic as they miss out on many profitable EE initiatives. On the other hand, because EE investments are frequently generic (improving the lighting efficiency of office space, for example) the go–no-go decision on these investments should be independent of any other characteristics of the firms undertaking them. However, DeCanio and Watkins (1998) find that the characteristics of firms do affect their decision to commit to a voluntary program of investments in lighting efficiency.
13. For instance, all else equal end-use improvements in EE would reduce the need for investments in new infrastructures to transport electricity, or alternatively would render those already existing more secure.
14. http://www.environmental-finance.com/news/view/1042.
15. Organizational and institutional factors as a reason for the paradox of low adoption rates of profitable EE improvements are analyzed by DeCanio (1998).
16. http://www.environmental-finance.com/news/view/1064.
17. However, prioritizing EE measures for end-use conversion devices over fuel transformation and electricity generation delivers more than five times the potential gain.
18. Note, though, that increased EE at the microeconomic level, while leading to a reduction of energy use at this level, does not necessarily lead to a reduction in energy use (and hence reduced CO_2 emissions) at the national or macroeconomic level; indeed it may lead to the opposite, that is, an increase in energy use (Herring, 2006; Huber and Mills, 2005). There is intense dispute over the magnitude of the 'takeback' or 'rebound' effect resulting from the fact that every improvement in EE is reflected in a decrease in the relative unit price of energy services and hence makes their use more affordable. Anyway, EE will save consumers money and promote a more efficient and prosperous economy.
19. It is sometimes referred to as the 'clean spark spread', that is, the spark spread adjusted for the price of the emission allowance.
20. It is sometimes referred to as the 'clean dark spread', that is, the dark spread adjusted for the price of the emission allowance.

REFERENCES

Abadie, L.M., J.M. Chamorro, and M. González-Eguino (2009), 'Optimal investment in energy efficiency under uncertainty', Basque Centre for Climate Change (bc3), Working Paper 2009-11.
Allcott, H. and S. Mullainathan (2010), 'Behavior and energy policy', *Science*, **327**, 5 March, 1204–05.
Anderson, S. and R. Newell (2004), 'Information programs for technology adoption: the case of energy efficiency audits', *Resource and Energy Economics*, **26**, 27–50.
Bastian-Pinto, C., L. Brandão and J.H. Warren (2009), 'Flexibility as source of value in the production of alternative fuels: the ethanol case', *Energy Economics*, **31**, 411–22.
Biello, D. (2009), 'US unveils a $350-million energy efficiency initiative at Copenhagen', *Scientific American*, December.
Boyle, P.P., J. Evnine and S. Gibbs (1989), 'Numerical evaluation of multivariate contingent claims', *Review of Financial Studies*, **2** (2), 241–50.
Cullen, J.M. and J.M. Allwood (2010a), 'The efficient use of energy: tracing the global flow of energy from fuel to service', *Energy Policy*, **38** (1), 75–81.
Cullen, J.M. and J.M. Allwood (2010b), 'Theoretical efficiency limits for energy conservation devices', *Energy*, **35** (5), 2059–69.
DeCanio, S.J. (1993), 'Barriers within firms to energy-efficiency investments', *Energy Policy*, **21** (9), 906–14.
DeCanio, S.J. (1998), 'The efficiency paradox: bureaucratic and organizational barriers to profitable energy-saving investments', *Energy Policy*, **26** (5), 441–54.
DeCanio, S.J. and W.E. Watkins (1998), 'Investment in energy efficiency: do the characteristics of firms matter?', *Review of Economics and Statistics*, **80**, 95–107.
Diederen, P., F. Von Tongeren and H. Van Der Veen (2003), 'Returns on investments in energy-saving technologies under energy price uncertainty in Dutch greenhouse horticulture', *Environmental and Resource Economics*, **24**, 379–94.

Federico, G. and G. Vives (2008), 'Competition and regulation in the Spanish gas and electricity markets', Reports of the Public–Private Sector Research Center, IESE-Orkestra.

Gillingham, K., R.G. Newell and K. Palmer (2006), 'Energy efficiency policies: a retrospective examination', *Annual Review of Environment and Resources*, **31**, 161–92.

Gillingham, K., R.G. Newell and K. Palmer (2009), 'Energy efficiency economics and policy', *Annual Review of Resource Economics*, **1**, 597–620.

Grenadier, S.R. and A.M. Weiss (1997), 'Investment in technological innovations: an option pricing approach', *Journal of Financial Economics*, **44**, 397–416.

Hasset, K.A. and G.E. Metcalf (1993), 'Energy conservation investment: do consumers discount the future correctly?', *Energy Policy*, **21** (6), 710–16.

Herring, H. (2006), 'Energy efficiency – a critical view', *Energy*, **31**, 10–20.

Huber, P.W. and M.P. Mills (2005), *The Bottomless Well*, New York: Basic Books.

IEA (2007), *Mind the Gap: Quantifying Principal–Agent Problems in Energy Efficiency*, Paris: OECD/IEA.

IEA (2009), *Implementing Energy Efficiency Policies: Are IEA Member Countries on Track?*, Paris: OECD/IEA.

IPCC (2006), 'Guidelines for national greenhouse gas inventories'.

Jaffe, B.A. and R.N. Stavins (1994a), 'The energy efficiency gap: What does it mean?', *Energy Policy*, **22** (10), 804–10.

Jaffe, B.A. and R.N. Stavins (1994b), 'The energy paradox and the diffusion of conservation technology', *Resource and Energy Economics*, **16** (2), 91–122.

Jochem, E. and R. Madlener (2003), 'The forgotten benefits of climate change mitigation: innovation, technological leapfrogging, employment, and sustainable development', ENV/EPOC/GSP(2003)16/FINAL, Paris: OECD.

Lucia, J. and E.S. Schwartz (2002), 'Electricity prices and power derivatives: evidence from the Nordic Power Exchange', *Review of Derivatives Research*, **5** (1), 5–50.

McKinsey Climate Change Special Initiative (2009a), 'Reducing US greenhouse gas emissions: how much at what cost?', McKinsey & Company.

McKinsey Global Energy and Materials (2009b), 'Unlocking energy efficiency in the US economy', McKinsey & Company.

Metcalf, G.E. (1994), 'Economics and rational conservation policy', *Energy Policy*, **22** (10), 819–25.

Metcalf, G.E. and K.A. Hassett (1999), 'Measuring the energy savings from home improvement investments: evidence from monthly billing data', *Review of Economics and Statistics*, **81** (3), 516–28.

Muthulingam, S., C.J. Corbett, S. Benartzi and B. Oppenheim (2009), 'Managerial biases and energy savings: an empirical analysis of the adoption of process improvement recommendations', available at http://ssrn.com/abstract=1347150.

Oikonomou, V., F. Becchis, L. Steg and D. Russolillo (2009), 'Energy saving and energy efficiency concepts for policy making', *Energy Policy*, **37** (11), 4787–96.

Sensfuß, F., M. Ragwitz and M. Genoese M. (2008), 'The merit-order effect: a detailed analysis of the price effect on renewable electricity generation on spot market prices in Germany', *Energy Policy*, **36**, 3086–94.

Sims, R.E.H., R.N. Schock, A. Adegbululgbe, J. Fenhann, I. Konstantinaviciute, W. Moomaw, H.B. Nimir, B. Schlamadinger, J. Torres-Martínez, C. Turner, Y. Uchiyama, S.J.V. Vuori, N. Wamukonya and X. Zhang (2007), 'Energy supply', in B. Metz, O.R. Davidson, P.R. Bosch, R. Dave, L.A. Meyer (eds), *Climate Change 2007: Mitigation. Contribution of Working Group III to the Fourth Assessment Report of the Intergovernmental Panel on Climate Change*, Cambridge, UK and New York, USA: Cambridge University Press, pp. 251–322.

Stern, P.C., T. Dietz, G.T. Gardner, J. Gilligan and M.P. Vandenbergh (2010), 'Energy efficiency merits more than a nudge', *Science*, **328**, 16 April, 308–9.

US Department of Energy (DoE) (2009), 'Buildings energy data book'.

Van Soest, D.P. and E.H. Bulte (2001), 'Does the energy-efficiency paradox exist? Technological progress and uncertainty', *Environmental and Resource Economics*, **18**, 101–12.

APPENDIX

A   Expected Values and Valuation of Annuities

The time-t expectations, under the equivalent martingale measure, of fuel prices and carbon prices are given by the following formulas:

$$E_0(G_t - f(t)) = \frac{k_G G_m}{k_G + \lambda_G}(1 - e^{-(k_G + \lambda_G)t}) + (G_0 - f(0))e^{-(k_G + \lambda_G)t},$$

$$E_0(C_t) = \frac{k_C C_m}{k_C + \lambda_C}(1 - e^{-(k_C + \lambda_C)t}) + C_0 e^{-(k_C + \lambda_C)t},$$

$$E_0(A_t) = A_0 e^{(\alpha - \lambda_A)t}.$$

Now the value of an annuity received from time τ_1 and time τ_2 is, respectively:

$$\int_{\tau_1}^{\tau_2} E_0(G_t - f(t))e^{-rt}dt = \frac{k_G G_m}{r(k_G + \lambda_G)}(e^{-r\tau_1} - e^{-r\tau_2})$$

$$+ \frac{(G_0 - f(0)) - \dfrac{k_G G_m}{k_G + \lambda_G}}{r + k_G + \lambda_G}(e^{-(r + k_G + \lambda_G)\tau_1} - e^{-(r + k_G + \lambda_G)\tau_2}),$$

$$\int_{\tau_1}^{\tau_2} E_0(C_t)e^{-rt}dt = \frac{k_C C_m}{r(k_C + \lambda_C)}(e^{-r\tau_1} - e^{-r\tau_2})$$

$$+ \frac{C_0 - \dfrac{k_C C_m}{k_C + \lambda_C}}{r + k_C + \lambda_C}(e^{-(r + k_C + \lambda_C)\tau_1} - e^{-(r + k_C + \lambda_C)\tau_2}),$$

$$\int_{\tau_1}^{\tau_2} E_0(A_t)e^{-rt}dt = \frac{A_0}{\alpha - r - \lambda_A}(e^{(\alpha - r - \lambda_A)\tau_2} - e^{(\alpha - r - \lambda_A)\tau_1}).$$

Regarding natural gas, note that $\int_{\tau_1}^{\tau_2} f(t)e^{-rt}dt \approx 0$ when $\tau_2 - \tau_1$ is an integer number of years.

B   The Three-dimensional Lattice

B.1   Building the lattice

First we take natural logarithms of the prices:

$$x_G \equiv \ln G_t; \ x_C \equiv \ln C_t; \ x_A \equiv \ln A_t.$$

Applying Ito's Lemma, for the dynamics of carbon price we have:

$$dx_A = \left(\alpha_A - \lambda_A - \frac{1}{2}\sigma_A^2\right)dt + \sigma_A dW_t^A \equiv v_A dt + \sigma_A dW_t^A.$$

For the long-run dynamics of natural gas price we have:

$$dx_G = \left[\frac{k_G(G_m - G_t)}{G_t} - \lambda_G - \frac{1}{2}\sigma_G^2\right]dt + \sigma_G dW_t^G \equiv v_G dt + \sigma_G dW_t^G,$$

which can be rewritten as:

$$dx_G = \left[\frac{1}{G_t}\frac{k_G G_m}{k_G + \lambda_G}(k_G + \lambda_G) - (k_G + \lambda_G) - \frac{1}{2}\sigma_G^2\right]dt + \sigma_G dW_t^G \equiv v_G dt + \sigma_G dW_t^G.$$

And for the long-run dynamics of coal price we have:

$$dx_C = \left[\frac{1}{C_t}\frac{k_C C_m}{k_C + \lambda_C}(k_C + \lambda_C) - (k_C + \lambda_C) - \frac{1}{2}\sigma_C^2\right]dt + \sigma_C dW_t^C \equiv v_C dt + \sigma_C dW_t^G.$$

Note that, except for volatilities, all the parameters required for using the above formulas can be estimated in the risk-neutral world from futures prices.

With three dimensions in each node of the lattice, it is possible to move to $2^3 = 8$ different states of nature. Thus there are eight probabilities to be computed, in addition to three incremental values (Δx_A; Δx_G; Δx_C). For this purpose we have ten equations.

The first equation establishes that the probabilities must sum to one. The next three impose the conditions for consistency regarding the second moment; they allow to compute:

$$\Delta x_A = \sigma_A\sqrt{\Delta t};\ \Delta x_G = \sigma_G\sqrt{\Delta t};\ \Delta x_C = \sigma_C\sqrt{\Delta t}.$$

The next three equations require the probabilities to be consistent with observed correlations. We thus have seven equations and eight unknowns. In principle, several solutions are possible. However, we adopt the method suggested by Boyle et al. (1989). This way we get the following probabilities, which satisfy the above equations:

$$p_{uuu} = \frac{1}{8}\left[1 + \rho_{AG} + \rho_{AC} + \rho_{GC} + \sqrt{\Delta t}\left(\frac{v_A}{\sigma_A} + \frac{v_G}{\sigma_G} + \frac{v_C}{\sigma_C}\right)\right],$$

$$p_{uud} = \frac{1}{8}\left[1 + \rho_{AG} - \rho_{AC} - \rho_{GC} + \sqrt{\Delta t}\left(\frac{v_A}{\sigma_A} + \frac{v_G}{\sigma_G} - \frac{v_C}{\sigma_C}\right)\right],$$

$$p_{udu} = \frac{1}{8}\left[1 - \rho_{AG} + \rho_{AC} - \rho_{GC} + \sqrt{\Delta t}\left(\frac{v_A}{\sigma_A} - \frac{v_G}{\sigma_G} + \frac{v_C}{\sigma_C}\right)\right],$$

$$p_{udd} = \frac{1}{8}\left[1 - \rho_{AG} - \rho_{AC} + \rho_{GC} + \sqrt{\Delta t}\left(\frac{v_A}{\sigma_A} - \frac{v_G}{\sigma_G} - \frac{v_C}{\sigma_C}\right)\right],$$

$$p_{duu} = \frac{1}{8}\left[1 - \rho_{AG} - \rho_{AC} + \rho_{GC} + \sqrt{\Delta t}\left(-\frac{v_A}{\sigma_A} + \frac{v_G}{\sigma_G} + \frac{v_C}{\sigma_C}\right)\right],$$

$$p_{dud} = \frac{1}{8}\left[1 - \rho_{AG} + \rho_{AC} - \rho_{GC} + \sqrt{\Delta t}\left(-\frac{v_A}{\sigma_A} + \frac{v_G}{\sigma_G} - \frac{v_C}{\sigma_C}\right)\right],$$

$$p_{ddu} = \frac{1}{8}\left[1 + \rho_{AG} - \rho_{AC} - \rho_{GC} + \sqrt{\Delta t}\left(-\frac{v_A}{\sigma_A} - \frac{v_G}{\sigma_G} + \frac{v_C}{\sigma_C}\right)\right],$$

$$p_{ddd} = \frac{1}{8}\left[1 + \rho_{AG} + \rho_{AC} + \rho_{GC} + \sqrt{\Delta t}\left(-\frac{v_A}{\sigma_A} - \frac{v_G}{\sigma_G} - \frac{v_C}{\sigma_C}\right)\right].$$

These probabilities have the same structure as those derived by Boyle et al. (1989); the terms v_A, v_G, v_C, though, are different. Our development allows for mean-reverting stochastic processes, and is later used to value American-type options (unlike Boyle et al., 1989, who value European-type options).

Negative probabilities cannot be accepted. To avoid this possibility we apply Bayes's Rule which decomposes the former probabilities into a product of conditional and marginal probabilities. We adopt a procedure which is similar to that in Bastian-Pinto et al. (2009). However, we consider three sources of risk (instead of two). We denote these probabilities by means of an asterisk as a superscript, for example p^*_{uuu}.

Next we are going to value the investment which depends on three different stochastic processes by means of a three-dimensional binomial lattice.

B.2   Deploying the lattice

The time T until maturity is subdivided into n steps each of size $\Delta t = T/n$. In our case, after the first step the initial value A_0 moves to one of two possible values, $A_0 u_A$ or $A_0 d_A$, where $u_A = e^{\sigma_A\sqrt{\Delta t}}$ and $d_A = 1/u_A = e^{-\sigma_A\sqrt{\Delta t}}$. Starting from initial values (A_0, G_0, C_0) after the first step we can compute the values $(A_0 e^{\sigma_A\sqrt{\Delta t}}, G_0 e^{\sigma_G\sqrt{\Delta t}}, C_0 e^{\sigma_C\sqrt{\Delta t}})$ with probability p^*_{uuu}. Similarly we derive the remaining nodes that arise in the first step, for example $(A_0 e^{-\sigma_A\sqrt{\Delta t}}, G_0 e^{-\sigma_G\sqrt{\Delta t}}, C_0 e^{-\sigma_C\sqrt{\Delta t}})$ with probability p^*_{ddd}.

After i steps, with j_A, j_G and j_C upside moves, the values $(A_0 e^{\sigma_A\sqrt{\Delta t}(2j_A - i)}, G_0 e^{\sigma_G\sqrt{\Delta t}(2j_G - i)}, C_0 e^{\sigma_C\sqrt{\Delta t}(2j_C - i)})$ will be reached. It can easily be seen that the tree branches recombine; thus, the same value results from a rise followed by a drop or the other way round. At the final time T the possible combinations of values can be represented by means of a cube. At the earlier moment $T - \Delta t$ another less-sized cube describes the set of feasible values. There will be some probabilities of moving from each node to eight possible states of the cube at time T.

This lattice is used to valuate the possibility to invest in enhancing the energy efficiency (thus saving input fuel and emission allowances) of a physical facility already in place (such as an operating coal-fired plant). Therefore, the saving opportunity is linked to the remaining life of the facility to be upgraded.

C   Monte Carlo Simulation

Correlated (deseasonalized) random variables are generated according to the scheme:

$$C_{t+\Delta t} \cong \frac{k_C C_m}{k_C + \lambda_C}(1 - e^{-(k_C + \lambda_C)\Delta t}) + C_t e^{-(k_C + \lambda_C)\Delta t} + \sigma_C C_t\sqrt{\Delta t}\varepsilon_t^1,$$

$$A_{t+\Delta t} \cong A_t e^{(\alpha - \lambda_A)\Delta t} + \sigma_A A_t \sqrt{\Delta t} [\varepsilon_t^1 \rho_{CA} + \varepsilon_t^2 \sqrt{1 - \rho_{CA}^2}],$$

$$G_{t+\Delta t} \cong f(t + \Delta t) + \frac{k_G G_m}{k_G + \lambda_G}(1 - e^{-(k_G + \lambda_G)\Delta t}) + (G_t - f(t))e^{-(k_G + \lambda_G)\Delta t} +$$

$$\sigma_S(G_t - f(t))\sqrt{\Delta t}[\varepsilon_t^1 \rho_{CG} + \varepsilon_t^2 \frac{\rho_{AG} - \rho_{CG}\rho_{CA}}{\sqrt{1 - \rho_{CA}^2}} + \varepsilon_t^3 \sqrt{1 - \rho_{CG}^2 - \frac{(\rho_{AG} - \rho_{CG}\rho_{CA})^2}{1 - \rho_{CA}^2}}].$$

ε_t^1, ε_t^2 and ε_t^3 are standardized Gaussian white noises with zero correlation. The first expression above is derived after replacing σ_ε^S in terms of σ_S. Similarly in the second expression. At the same time, if samples from a standardized bivariate normal distribution are required, an appropriate procedure is the one shown above, where $\rho_{S,C}$, $\rho_{S,A}$ and $\rho_{C,A}$ are the correlation coefficients between the variables in the multivariate distribution.

To get numerical estimates, 100 000 sample paths for G_t, C_t and A_t are generated. Each path comprises monthly time steps; with 15 years to expiration of the plant, this amounts to 180 steps. In the rigid (that is, full-time operation) scenario, the value with 15 years in operation is given by:

$$\frac{1}{100\,000} \sum_{i=1}^{100,000} \sum_{j=1}^{180} P_M \times MC_j \times e^{-\frac{rj}{12}}.$$

With the same useful life but under flexible operation (that is, scenario (b)), the value is:

$$\frac{1}{100\,000} \sum_{i=1}^{100,000} \sum_{j=1}^{180} P_M \times \max(MC_j, 0) \times e^{-\frac{rj}{12}}.$$

15 Energy use in the transport sector: ways to improve efficiency
Kenneth Button

15.1 INTRODUCTION

It takes energy to move anything. Transport is thus inevitably a major consumer of energy in the modern world where higher incomes mean that individuals want to travel more, and specialization in production means an increasing tendency to move goods and components longer distances. In an ideal world, with perfect competition, complete markets and ubiquitous information, the use of energy by transport would be efficient, the benefits from the marginal unit of energy consumed being equated to the full costs of producing that unit. Unfortunately, many of the underlying assumptions of perfectly competitive markets escape the realities of the world, energy consumption in general is not optimal, and there are even wider deviations from optimality in term of types and sources of energy and its use across the various sectors of the economy.

This chapter looks at the various ways in which society may go towards improving the efficiency of fuel use by the various transport industries. It is, by the necessity of limited space, focused on the direct use of energy by the transportation sector, and does not, for example, dwell in any detail on issues involving concerns over the environmental challenges of transporting such things as gasoline, or on the problems of collecting fuels safely, such as relating to deep sea oil drilling or nuclear power generation. A comprehensive analysis would embrace these and other wider considerations.

I also do not dwell for too long on sustainability issues as discussed in the Brundtlund Report (World Commission on Environment and Development, 1987), and less explicitly by Boulding (1966) with his idea of 'Spaceship Earth'. There are clear intellectual merits in treating global resources in a holistic manner, in considering intergenerational effects, and in tying environmental concerns with social and narrower economic considerations. The challenge is to move from this broad perspective to a firm theoretical foundation, policy development and ultimately policy implementation. Indeed, the trend is to ignore the types of energy trade-offs that are implicit in the holistic philosophy underlying Brundtlund (for example, there is nothing inconsistent in using more fuel in transport if there are appropriate reductions in energy use in, say, agriculture or household heating and cooling) and instead still to think of policies in traditional sector-based stovepipes, and to revert to notions such as 'sustainable transport'.

15.2 MARKET AND GOVERNMENT DISTORTIONS

In a first-best world, economics tells us that there would be no need for external assessments of ways to improve the efficiency of energy use; market mechanisms would ensure that use was optimal. The abundance of market and also government regula-tory[1] failures that typify reality mean that inefficiency in energy use in transport, as well as other sectors, is widespread; and also often difficult to disentangle from other imperfections.

The basic problem that results, as with all market and government failures, is the lack of adequate and effective signals to producers and consumers as to the genuine opportunity cost of their actions. Essentially, those using transport are unaware of – or even if they are aware are unwilling to react to – the full costs of what they are doing. Prices, when there are no market distortions (essentially to allocate scarce resources) indicate where more resources should be developed and provide the means by which to increase the supply. Energy markets are singularly badly developed because of their intrinsic nature, or because of policy manipulations, to generate the appropriate prices to ensure these three functions are fulfilled. The result is long- and short-term misuse of energy.

Transport makes use of a diverse range of energy, but in the twenty-first century oil-based fuels dominate in most parts of the world. However the supply of oil is limited to a few countries, and the companies that develop oil-fields and refine their outputs are small in number. There is certainly competition in the market, but it is far from perfect. Governments also manipulate the price of oil, either as the raw product or 'at the pump' where it is sold as gasoline, diesel or kerosene. At the macro level, control of oil is a centerpiece of geopolitical game-playing. In some cases at the micro level, government intent is to collect revenues from sumptuary taxes and in others taxes are a way of limit-ing the external environmental costs of transport. Governments also regulate the nature of the oil-based fuel delivered by controlling additives or stipulating the octane of the fuel for health and safety reasons.

The market is also distorted in terms of the complementary inputs that go into trans-port: the vehicles, ships and planes, and the roads, rail networks and ports. These are often quasi-natural monopolies where economies of scale and scope tend to produce, for example, large, global automobile manufactures and massive hub airports. The types of mobile plant – the cars and planes – and the nature of the infrastructure available influ-ence the costs of travel in terms of the amounts and types of fuel required to provide access to the transport system.

There are alternatives to transport to meet some of the needs of consumers and busi-nesses, and the markets for these alternatives, and the government interventions in them, influence the amount of energy used in transport. There are land-use considerations that are important because substitute, alternative locations can affect the amount of transport consumed, but interference in land markets, and some of their intrinsic features, again hardly means that accurate costs of energy are imposed on transport users. Zoning and land taxes, for example, affect where people live and where businesses locate. Markets are seldom allowed to function, and direction by planning authorities is the norm. But even without that, the economies of scale that are found in many extractive industries, and in some agricultural activities, distort the markets for transportation and the energy consumed.

15.3   THE ENERGY USED IN TRANSPORT AND ITS IMPLICATIONS

Transport is a major consumer of carbon-based fuels, either directly in the movement of vehicles or indirectly in terms of the generation of electricity and the manufacture of vehicles and the infrastructure that they use. The usage of fuel by transport, and its type, can be looked at in several ways: for example, by mode (car, boat or bus), by type of travel (urban, interurban or international), or by trip purpose (work or leisure). Efficiency is in this sense contextual, because it requires consideration of the benefits enjoyed from, say, car travel or from movements in urban areas as well as the nature and type of fuel used. The tendency of much analysis, and *ipso facto* policy-making, is however to focus almost purely on the cost side, rather than to treat efficiency as a full cost–benefit calculation. I follow this trend here, in part for admitted convenience but also because there are probably fewer imperfections on the benefit side.

Looking at a few simple sets of data one can get some idea of the wide variations that exist in the use of fuel in transport. Table 15.1 sets out the total fuel consumption by mode in the United States since 1980. The dominance of highway use by both cars and trucks is clear, as is the long-term upward trend. Table 15.2 offers data on the amount

Table 15.1   Fuel consumption in the United States by the main transport modes

	1980	1990	2000	2003	2004
Highway					
Gasoline, diesel and other fuels (million gallons)	114960	130755	162555	170069	173750
Truck:	19960	24490	35229	32696	33968
Single-unit 2-axle 6-tire or more truck	6923	8357	9563	8880	9263
Combination truck	13037	16133	25666	23815	24705
Truck (% of total)	17.4	18.7	21.7	19.2	19.6
Rail, Class I (freight service)					
Distillate/diesel fuel (million gallons)	3904	3115	3700	3826	4059
Water					
Residual fuel oil (million gallons)	8952	6326	6410	3874	4690
Distillate/diesel fuel oil (million gallons)	1478	2065	2261	2217	2140
Gasoline (million gallons)	1052	1300	1124	1107	1005
Pipeline					
Natural gas (million cubic feet)	634622	659816	642210	591492	571853

Sources: US Department of Transportation, Federal Highway Administration; Association of American Railroads; US Department of Energy, Energy Information Administration, US Department of Energy.

Table 15.2   Average United States light gasoline and diesel vehicle emissions rates (grams per mile)

	1990	1995	2000	2005	2007
Light-duty gasoline vehicles					
Exhaust HC	2.79	1.57	0.97	0.52	0.42
Non-exhaust HC	1.21	1.05	0.91	0.72	0.62
Total HC	0.68	0.77	0.80	1.25	1.04
Exhaust CO	42.89	26.60	18.53	0.58	10.28
Exhaust NO_x	2.70	1.78	1.29	0.92	0.73
Light-duty diesel vehicles					
Exhaust HC	0.68	0.77	0.80	0.60	0.36
Exhaust CO	1.49	1.69	1.78	1.57	1.21
Exhaust NO_x	1.83	1.89	1.81	1.32	0.85

Source: United States Bureau of Transportation Statistics.

of emissions that various types of vehicle in the United States emit per unit of travel; the data for, say, the UK are slightly different for each vehicle category because travel distances are shorter and cities are more compact, leading to more fuel-intensive stopping and starting.

The major global environmental concern since the 1990s has been with the scale of carbon emissions from the combustion and storage of oil-based transportation fuels, and the implications that this release has on climate change. The available physical scientific evidence is that the correlation is not perfect (other natural and man-made actions are also thought to be relevant) but it is significant. Table 15.3 provides same basic information on the share of national carbon dioxide attributed to transportation in various parts of the world. The numbers vary considerably due to factors such as national income levels, patterns of natural and human geography, the age distribution of the population and the modes of transportation used in each country.

15.4   POLICY OPTIONS

I now spend time looking in detail at some economic aspects of transport energy policy; the coverage is selective and looks in most detail at those policies that have attracted the most attention. In many cases the instruments used are from a generic toolkit the elements of which are outlined in Table 15.4. This categorizes instruments according to a two-dimensional matrix.

In simple terms, policy approaches can be divided into two broad types: those that direct the actions of individuals and companies by working to affect the prices levied in the market; and those that set the legal boundaries (command-and-control instruments) as to what vehicles, travel patterns and fuels are permitted. The boundary is slightly artificial in the sense that command-and-control instruments will inevitably affect prices. The broad types of approach are also subdivided according to whether they act directly or indirectly on fuel use. The other dimension of the matrix considers the target of energy

Table 15.3   Share of transport CO_2 emissions

	1971	1990	1998
OECD			
North America	25	29	30
Europe	14	20	23
Pacific	16	20	22
Non-OECD			
Africa	20	18	17
Middle East	14	20	18
Europe	10	9	13
Former USSR	9	9	8
Latin America	31	33	34
Asia (exc. China)	14	16	18
China	4	6	8
World	19	22	24

Source: International Energy Agency.

Table 15.4   Taxonomy of main policy instruments to control the use of fuels in transport

	Market-based incentives		Command-and-control instruments	
	Direct	Indirect	Direct	Indirect
Vehicle	Fuel taxes Tradable permits	Differential taxation by fuel type Tax allowances for fuel efficient vehicles	Fuel consumptions standards	Compulsory inspection and maintenance of fuel systems Mandatory use of fuel efficient vehicles Compulsory scrapping of older vehicles
Fuel		Differential fuel taxation High fuel taxes	Fuel composition Phasing out of high polluting fuels	Fuel economy standards Speed limits
Traffic		Congestion pricing Parking charges Subsidies for more fuel efficiency modes	Physical constraints of traffic Designated areas	Restraints on vehicle use Bus lanes and and other priorities Information systems

policies and categorizes according to whether the policies are directed at the vehicles that consume the fuel, at the fuel itself or at traffic patterns. Again the distinction is often opaque – fuel taxes, for example, affect the types of fuel used as well as directly affecting fuel consumption – but it is a useful one.

From the table, it can be seen that there is theoretically a wide range of policy tools that can be deployed to affect the energy use of transport. Each has its particular characteristics, with their usefulness depending on the background assumptions that are adopted and their costs of implementation. Here I am selective and focus on some of the more important efforts that have been made to influence energy consumption in transport. In particular, longer-term policies involving land use and such policies as 'compact-city' design are explicitly omitted. These are large and multidimensional topics in their own rights and go beyond the boundaries of a chapter such as this. I also do not expend much space on details of individual applications, unless this provides particularly important insights.

Although theoretically there are numerous ways to influence energy use listed in the table, a wide range of practical and political factors determine the policies that have been initiated to influence the level and form of energy consumption in transport (Flynn, 2002). In some cases the costs of introducing, monitoring and policing policies, as with some specific environmental policies, simply make them impractical, or at least in their purest forms. In other cases there may also be trade-offs between improving energy efficiency and meeting other objectives, such as removing pollutants from the atmosphere or ensuring an acceptable level of traffic safety. An example of the former has been the removal of lead from gasoline in many countries to combat child brain damage, and that has reduced the fuel efficiency of internal combustion engines and in some countries has led to alternative health problems when aromatics have been used as a substitute.

The policy tools that are in place, or have been used in the past, to influence the type of fuel used in transport, as well as the aggregate consumption, are nevertheless quite extensive. The following review is not intended to be comprehensive in its coverage, but rather should be viewed as illustrative of the types of measures that have been put in place or seriously considered.[2]

15.4.1 Leaving Things to the Market

One policy option that is often forgotten in energy debates is to leave things to the market. After all, while there are market failures, there are also many government intervention failures that may either result in worsening an original market failure, or cause serious and unexpected distortions elsewhere in the system. The assumption that government somehow knows better than the market, and can therefore interfere in it to improve its performance, suggests that it has better information and that there are no transaction costs involved in the intervention. These are often very strong assumptions.

In practice, the market has been a significant influence on the types and amount of energy used by transport. Historically, for example, changes in prices have demonstrable medium- and long-term impacts on overall energy consumption in transport, most of which have only appreciated in retrospect. Not all these, however, have been directly related to the price of fuel. A simple transmission mechanism illustrates the difficulty of policy-makers trying to foresee energy changes and plan for the development of new technologies.

At the beginning of the twentieth century, automobiles were expensive and coal-powered (either directly or after transformation into electricity) railway systems dominated surface transport. A subsequent 'energy effect' was brought about by the

Table 15.5   Fuel efficiency of United States cars following the 1973 and 1979 'oil crises'

	Miles per US gallon		Real price of gasoline (1967=100) Harmonic mean	
	City	Highway		
1968	12.59	18.42	14.69	97.3
1969	12.60	18.62	14.74	95.4
1970	12.59	19.01	14.85	98.0
1971	12.27	18.18	14.37	87.6
1972	12.15	18.90	14.48	85.9
1973	12.01	18.07	14.15	88.7
1974	12.03	18.23	14.21	108.3
1975	13.68	19.45	15.79	106.0
1976	15.23	21.27	17.46	105.3
1977	15.99	22.26	18.31	103.7
1978	17.24	24.48	19.89	100.5
1979	17.70	24.60	20.25	122.2
1980	20.35	29.02	23.51	149.6
1981	21.75	31.12	25.16	150.8
1982	22.32	32.76	26.06	134.7
1983	22.21	32.90	26.01	126.1
1984	22.67	33.69	26.59	119.2

Source: Crandall et al. (1986).

introduction of mass production of cars, initially by Fiat in Italy, but on a larger scale by Henry Ford in the United States. This occurred to take advantage of the high car prices of the time and brought down the costs of car production and subsequently the price of cars, from $910 for a touring Model T in 1910 to $367 in 1925). In turn, this led to more use of cars and trucks (sales of the touring models were 16890 units in 1910, rising to 691212 in 1925) with a resultant switch in transport away from coal as the primary energy source, to oil. In the East German economy of the 1960s, market forces were largely ignored when policy moves towards greater car ownership at administered prices were initiated. The resultant centrally planning outcomes were the Wartburg and Trabant cars and, by the time the Berlin Wall came down, there was a waiting time of nearly ten years to receive these not very comfortable, reliable or efficient vehicles. The complexity of centrally planning the design and production of cars proved too complex even for the highly skilled planners of East Germany.

Fuel prices themselves are also powerful influences on consumption. Where there have been shortages of some forms of energy, because of either physical factors or institutional difficulties, markets can bring about changes. This has happened when there have been shortages of oil for political reasons, and more recently as the price of oil has become more volatile. While there are short-term adjustment issues, the long-term effect of fuel shortage, and resultant price rises, is that it is used more efficiently.[3]

As an example, Table 15.5 shows the impact on the fuel efficiency of the United States car stock after the oil crises of 1973 and 1979. It is clear that the average energy efficiency of vehicles (and possibly the skill with which they were driven) increased following both crises, albeit with a lag as the adjustment took place. A more recent survey bringing together work on long-term gasoline fuel price elasticities indicates that about 20 percent to 60 percent appear to be due to changes in the vehicle miles driven, with 40 percent to 80 percent being due to changes in fleet composition (Parry et al., 2007). A more general rule of thumb, suggested by Goodwin et al. (2004) after reviewing numerous empirical studies, is that fuel consumption elasticities are greater than relative traffic sensitivities, mostly by factors of 1.5 to 2. People travel about the same amount but use smaller cars.

Energy, because of the relative inelasticity of aggregate demand for its use, has traditionally been the subject of taxation. In many cases this has been for purely sumptuary purposes, but in other cases, as with the federally earmarked gasoline tax in the United States, it has been used as a proxy charge for some related consumption item – in the United States case, to pay for the use of the road. In other cases, there have been environmental motivations, for example the differential taxes applied to gasoline and diesel fuels in many countries.

15.4.2 Fuel Taxation

The Pigouvian solution (Pigou, 1920) to negative market distortions, and especially environmental externalities, is to act by imposing a tax or charge on those responsible for the cost. While an optimum tax of this type technically equates the full costs of fuel use in our case with the benefits derived from it, there are both theoretical and practical problems that can stymie its effect.[4] Not least of these is the need to calculate the level of taxation that is required for fully efficient fuel use. This is not a market-based parameter but has to be estimated exogenously. But even if it can be argued that a reasonable approximation is possible, and a hypothetical Pareto improvement would transpire, there remains the issue that the revenues from a taxation regime would go to the taxing authorities. Conventional microeconomic theory suggests that an authority serving the public interest would use this revenue to enhance the utility of its citizens; that is, to spend it in ways that would maximize the net social welfare of society. Other arguments based upon ideas of regulatory capture and rent-seeking by political decision-makers and bureaucrats suggest, however, that is this not always the situation.

Examples of taxes on the energy used by transport abound, although the exact motivations underlying them are not always clear, claims of being designed to improve fuel efficiency in terms of fuel burn or chemical composition often being entangled with the revenue-raising motive. The United States Energy Tax Act of 1979, for instance, was a law passed as part of the National Energy Act. One element of the Act created the 'gas-guzzler' tax applying to sales of vehicles with official estimated gas mileage below certain levels. In 1980, the tax was $200 for a fuel efficiency of 14 to 15 miles per gallon, and this was increased to $1800 in 1985. In 1980, the tax was $550 for fuel efficiencies of 13 mpg and below, and was changed in 1986 to $3850 for ratings below 12.5 mpg. The gas-guzzler tax only applied to cars under 6000 lbs, which made sports utility vehicles and other large passenger cars exempt.[5]

In terms of using taxation as an instrument for encouraging energy conservation, or changes in the energy source used for environmental reasons, carbon taxes have been adopted in a number of countries. These are not transport-specific but are more holistic in their intent of making optimal use of resources more generally, although their impact on transport is often large. In 1991 for example, Sweden placed a tax of $100 per tonne on the use of oil, coal, natural gas, liquefied petroleum gas, gasoline, and aviation fuel used in domestic travel. Industrial users paid half the rate (between 1993 and 1997, 25 percent of the rate), and certain high-energy industries such as commercial horticulture, mining, manufacturing and the pulp and paper industry were fully exempted from these new taxes. In 1997 the rate was raised to $150 per tonne of CO_2 released. Finland, the Netherlands and Norway also introduced carbon taxes in the 1990s.

In other cases, however, efforts at introducing such policies have failed. In 2005, New Zealand proposed a carbon tax to take effect from April 2007 across most economic sectors, but the policy was abandoned in December 2005. Similarly, in 1993, President Bill Clinton proposed a British thermal unit (BTU) tax that was never adopted.

15.4.3   Cap-and-trade

The cap-and-trade approach to fuel use has its roots in Coasian economics with its focus on the allocation of property rights (Coase, 1962). Basically the idea applied to, say, gasoline is that the fuel is made available for transport use through some mechanism such as an auction or lottery, and that potential users would then be able to buy and sell amongst themselves. How the initial allocation is distributed is irrelevant, save for the distribution of initial windfall gains; the key point being that those who would ultimately gain the most utility from the fuel would end up buying it. How the aggregate amount of fuel is to be determined is seldom discussed by economists and is largely seen as a 'scientific' decision. Ideally, the regime would cover all types of energy and be spread across all potential uses and not just be confined to transport – essentially the aim of the global trading concepts implicit under the Kyoto Protocol.

While theoretically the cap-and-trade approach has intellectual merit, it has practical limitations. The politics of the initial allocation cannot simply be ignored in practice, given the proclivity for policy-makers to be concerned with who gains from their actions as well as the aggregate effects. There is also the issue of policing to ensure that contracts agreed upon are upheld; and there are the transaction costs involved in the buying and selling of the rights themselves. Efficient trade also assumes that markets work perfectly, and there are no monopoly or other distortive powers present.

Despite these challenges, tradeable permits, or at least policies containing their main elements, have been used to deal with specific transport-related fuel issues, in addition to more generic problems such as global warming gas emissions.[6] In particular, they were used in the United States as a component of the policy to remove lead from gasoline (Hahn, 1989). It was decided that lead was an undesirable additive and in 1982 America established a trading program, which in 1985 was modified to allow banking (the holding over of allocations from one period to another). Lead credits were allocated according to the existing use of lead and standards regarding the lead content of gasoline. (For example, if the standard was 1.1 grams per gallon then a firm producing 100 gallons of lead would receive up to 110 grams of lead that it may either use or sell.) The standards and, thus, the amount of lead available, declined to zero by 1987.

In terms of property rights, because current production of energy is correlated to past production, the system implies that the existing distribution is taken as the starting point. Whether the program created net environmental benefits beyond the standards that were

Table 15.6   New automobile emissions standards (grams per mile)

Model year	Hydrocarbons	CO	NO$_x$
Pre-control	10.6	84.0	4.1
1970–71	4.1	34.0	–
1972	3.4	39.0	–
1973–74	3.4	39.0	3.0
1975–76	1.5	15.0	3.1
1977–79	1.5	15.0	2.0
1980	0.41	7.0	2.0
1981–93	0.41	3.4	1.0
1994–2003	0.25	3.4	0.4
2004	0.09	4.2	0.07

also set at the time is difficult to determine, although trading was extensive (about 15 percent of the allocation), indicating that the transition was achieved relatively efficiently in terms of which refineries were used. Those with more efficient capacity had an incentive to buy from those with less efficient capacity.

More recently, the European Union has moved for the extension of its existing carbon cap-and-trade program (that had been initiated in 2005) to air transport (Council of the European Union, 2008), which became effective in 2009 with the system becoming operational in 2012. It will affect virtually all flights departing from or arriving at European Union airports, with domestic flights being subject to the same rules as international air traffic. Aircraft operators will be obliged to hold and surrender allowances for CO_2 emissions, although flights performed under visual flight rules and rescue flights are exempt, as are public service obligation routes and carriers that operate only a few services. In the first year, the allocation of allowances will be equivalent to 97 percent of aviation's annual emissions for 2004 to 2006; 95 percent by 2013. Under the rules of the scheme, airlines will be given free permits to cover 85 percent of their emissions with the remainder auctioned. The total allowance will decline over time.

15.4.4   Vehicle Fuel Standards

Rather than directly regulate on the composition of fuels, or use the pricing mechanism, there have been efforts to influence energy consumption and resultant pollution by legislating on the design of vehicles. The details adopted vary and here I highlight just some of the issues by looking at the recent European and North American experiences.

There has been a diversity of detailed approaches to conventional gasoline car design adopted in different countries. The approach of the Europeans has been to act largely on the environmental emissions themselves by stipulating in agreements with their own industry, and with the manufacturers in Japan and Korea, the maximum levels of pollution that new vehicles may emit (Table 15.6).

In contrast to this, the Corporate Average Fuel Economy (CAFE) regulations, first enacted by the United States Congress in 1975, are federal regulations that have sought to improve fuel economy in the wake of the 1973 Arab oil embargo. In other words, they

Table 15.7   *Canadian Company Average Fuel Consumption (CAFC) goals and the United States corporate average fuel economy (CAFE) standard (litres/100 km)*

Model year	Passenger car		Light-duty trucks	
	CAFC	CAFE	CAFC	CAFE
1978	13.1	13.1		
1980	11.8	11.8		
1982	9.8	9.8		13.4
1984	8.7	8.7		11.8
1986	8.6	9.1		11.8
1988	8.6	9.1		11.7
1990	8.6	8.6	11.8	11.8
1992	8.6	8.6	11.6	11.8
1994	8.6	8.6	11.5	11.7
1996	8.6	8.6	11.4	11.7
1998	8.6	8.6	11.4	11.7
2000	8.6	8.6	11.4	11.7
2002	8.6	8.6	11.4	11.7
2004	8.6	8.6	11.4	11.7
2006	8.6	8.6	11.4	10.9

Source: Adapted from Perl and Dunn (2007).

impact on the industry. The regulations initially applied to the sales-weighted average fuel economy, expressed in miles per gallon, of a manufacturer's fleet of current model year passenger cars or light trucks with a gross vehicle weight rating of 8500lbs or less, manufactured for sale in the United States. Light trucks not exceeding 8500 lbs gross vehicle weight rating do not have to comply with CAFE standards; some half a million vehicles in 1999. From early 2004, the average new car has had to exceed 27.5 mpg and light trucks 20.7 mpg. Trucks under 8500 lbs had to average 22.5 mpg in 2008, 23.1 mpg in 2009 and 23.5 mpg in 2010. After this, new rules set varying targets based on truck size 'footprint'.

Whereas the United States regime is statutory, the Canadians have a voluntary scheme to foster vehicle fuel economy; namely the Company Average Fuel Consumption (CAFC) agreement that was established between government and auto manufactures in 1978. Details of the joint goals set against the CAFE standards are seen in Table 15.7.

The United States National Highway Traffic Safety Administration (NHTSA) regulates CAFE standards and the United States Environmental Protection Agency (EPA) measures vehicle fuel efficiency. Congress specifies that CAFE standards must be set at the 'maximum feasible level' given consideration for technological feasibility, economic practicality, effect of other standards on fuel economy, and need of the nation to conserve energy. If the average fuel economy of a manufacturer's annual fleet of car and/ or truck production falls below the defined standard, the manufacturer pays a financial penalty; thus there is a very crude pricing mechanism involved. Fuel efficiency is negatively highly correlated to vehicle weight, but weight has been considered by many safety experts to be highly positively correlated with safety, intertwining the issues of fuel economy, road traffic safety, air pollution, global warming and greenhouse gases. Hence, historically, the EPA has encouraged consumers to buy more fuel-efficient vehicles, while NHTSA has expressed concerns that this leads to smaller, less safe vehicles. More recent studies tend to discount the importance of vehicle weight to traffic safety, concentrating instead on the quality of engineering design of vehicles.

While there have been changes in the standards over time, the table shows that these have tended to be infrequent and that by, for example, European standards the American fleet is relatively fuel-inefficient. Part of the problem seems to be difficulties in building political alliances strong enough to carry though measures that tighten the prevailing standards.

A further problem within the CAFE standard approach is that it may lead to an increase in other types of highway environmental externalities. Disregarding any effects that it has on the sales of light trucks and, most notably, sports utility vehicles that are much less rigorously controlled, the increased fuel efficiency of individual automobiles will make them cheaper to drive per mile. Additional mileage may then, for example, add to congestion and local environmental damage associated with noise nuisance.

15.4.5 Traffic Demand Management

The routing, timing, mode and amount of traffic can affect fuel burned, and in some cases the type of fuel. The concern with traffic management has not traditionally been focused explicitly on the fuel allocation question, but rather on dealing with increasing traffic volumes and with handling the congestion that accompanies this; this applies as much to congestion at seaports, airports and on some public transit systems as it does to the more discussed matter of road congestion. In the past the focus was largely on increasing capacity through infrastructure expansions rather than by controlling the flow of traffic. More recently, the financial costs of providing more infrastructure, combined with a realization that it did not resolve the congestion problem and, in addition, often created significant environmental damage, has led increasingly to strategies of traffic demand management.

These measures of traffic management range from traffic segregation (bus lanes, one-way streets, bans on daytime deliveries, and so on), through telematics (route guidance systems, and synchronized junction signals, and so on) and traffic prioritization (junction design) to fiscal incentives (congestion pricing and transit subsidies); all, as secondary effects, have implications for fuel efficiency. In many cases, this can result in a positive fuel efficiency effect that complements the primary objective of speeding up traffic.

Focusing on the increasing adoption of road pricing aimed at making motorists cognizant of the costs of congestion they impose on each other, there is clear evidence that this, even when applied in a fairly crude way as in London, can have significant beneficial traffic effects (Button and Vega, 2008).[7] Table 15.8 provides details of the outcomes on the main applications. The exact effects vary, as would be expected with the considerable diversity in the geography and the nature of the transport infrastructure in the urban areas involved, and the variety of charging regimes adopted.

The impacts on fuel consumption, however, are seldom considered in detail and, although they may well be positive effects, they are difficult to measure because of the ways in which road pricing affects traffic. Reduced traffic flows and faster, more

Table 15.8   The effects of major road pricing schemes

City	Traffic effects	Congestion effects	Public transport effects
Singapore, 1975–98[1]	−44%; −31% by 1988	Average speed increased from 19 to 36 km/h	Modal shift, from 33% to 46% trips to work by city bus, 69% in 1983
Trondheim, 1991	−10 %	n.a.	+7% city bus patronage
Singapore, 1998[2]	−10 to −15%	Optimized road usage, 20 to 30 km/h roads, 45 to 65 km/h expressways	Slight shift to city bus
Rome, 2001	−20 %	n.a.	+6%
London, 2003	−18% 2003 vs 2002, 0% 2004 versus 2003	−30%. 1.6 min/km typical delay 2003, 2004 versus 2002 (2.3 min/km)	+18% during peak hours bus patronage 2003, +12% in 2004
London, 2005[3]	Small net reductions −4% 2005/06	−22%. 1.8 min/km typical delay	bus patronage steady
Stockholm, 2006	−30% 2006 versus 2004	−30 to −50% journey time	+6%

Notes:
1. Although called Area Licensing Scheme, the system was a cordon toll rather than an area license.
2. Electronic fee collection introduced.
3. New rate introduced.

consistent traffic speeds in cities in most conditions will result in local fuel savings and a switch, as seen in the table, to more fuel-efficient modes of transport.[8] Offsetting this, the spread of traffic to formerly less congested times of the day and the diversion of through traffic to longer bypass routes will counter this to some extent. One also has to consider how the road agency spends the money that it collects from the road congestion charge; this may well be on activities that, even if remote from transport, will inevitably entail some fuel consumption. Tracing out these wider implications is well outside of any of the impact studies of congestion charges that have been conducted to date.

Physical traffic management measures may actually in some contexts lead to greater fuel inefficiency if introduced on their own. The aim of most of these measures is to increase the traffic flow in a city or corridor. In practice, there is little evidence that they do this in the longer term, although they may result in more traffic using a network. The problem is that improving traffic lights sequencing, modifying junction design, adjusting road space priorities, and so on does little to prevent what traffic engineers often call 'latent demand' for road use being met, with new traffic being attracted by faster road speeds.[9] The trend back towards the same levels of congestion prior to the management initiatives, but with more traffic volume, inevitably means greater energy use.

15.5   CONCLUSIONS

Markets for energy are neither complete nor perfect and this extends well beyond transportation. The policy challenge is thus not simply to allocate energy efficiently within transport, but also to allocate its use across all activities in the economy so that users are more fully aware of the opportunity costs of their consumption, be they potential shippers of goods or users of air conditioning. Much of the policy debate, however, has been stovepiped into second-best considerations and has been largely sector- and industry-based, with such notions as sustainable transport, sustainable agriculture, sustainable manufacturing, and so on being substituted for the holistic notion of sustainable development in the Brundtland sense. But even within the narrow confines of a sustainable transport focus, there has been only limited success in optimizing energy use between modes or across transport activities or users. The result has been a third-best outcome, determined by a diversity of factors including a degree of regulatory capture and the influence of monopoly and monopolistic elements in the market.

NOTES

1. For a general discussion of the difference between government and market failures see Winston (2006), and for a more specific transport–environment overview see Button (1992).
2. In particular, I have not looked in any detail at the use of alternative fuels that are gradually finding niche markets, or at hybrid technologies where a number of different energy sources are combined. These are all being considered and some are gradually being adopted, but space constraints have led to an inevitable selectivity. Equally, most of the discussion is about passenger transport, but there have been a number of major initiatives on the freight side, and in particular with regard to the use of informatics by the American freight rail road companies, and in ship design to enhance fuel efficiency in response to coping with volatile oil prices.
3. This has been particularly pronounced in the logistics field where there have been major advances in 'green logistics' driven by the quest for energy efficiency (see Rodrigue et al., 2001).
4. A particular problem is when countries, or states or provinces, compete with each other to gather fuel tax revenue. Some states, such as Luxembourg, Andorra, and Gibraltar, have strategically reduced fuel tax rates to attract more cross-border fill-ups, which ultimately increase tax revenue. There are ways of minimizing this, however, and gas stations in Argentina near the Brazilian border list two prices for gasoline, one for cars with Argentinean license plates and another for foreign plates, to restrict Brazilian drivers from buying the cheaper fuel.
5. To encourage a switch to cleaner and more efficient fuels, fuel taxes in Germany in 2009 were €0.4704 per litre for ultra-low sulfur diesel and €0.6545 for conventional unleaded gasoline, plus value-added tax on the fuel and the fuel tax making €1.03 per litre for ultra-low sulfur diesel and €1.22 per litre for unleaded gasoline.
6. The CAFE standards discussed below have some elements of a cap-and-trade approach to fuel efficiency but involve trading within a business rather than within a market.
7. The arguments offered here apply in broad terms to other areas where fiscal policies have been advanced and sometimes applied, for example see Button (2008) on airport congestion.
8. There also is some evidence that road pricing was, along with other factors, one reason for a significant uptake of alternative energy vehicles after congestion charging was introduced in Stockholm (City of Stockholm, 2009).
9. In economic terms this just reflects that the aggregate demand for road use in an area is not perfectly inelastic with respect to the generalized costs of using the network,

REFERENCES

Boulding, K. (1966), 'The economics of the coming spaceship Earth', presented at the Sixth Resources for the Future Forum on Environmental Quality in a Growing Economy, Washington, DC.

Button, K.J. (1992), *Market and Government Failures in Environmental Policy: The Case of Transport*, Paris: Organisation for Economic Co-operation and Development.

Button, K.J. (2008), 'Issues in airport runway capacity charging and allocation', *Journal of Transport Economics and Policy*, 42, 563–85.

Button, K.J. and H. Vega (2008), 'Road user charging', in S. Ison and T. Roe (eds), *The Implementation and Effectiveness of Transport Demand Management Measures: An International Perspective*, Aldershot: Ashgate.

City of Stockholm (2009), 'Analysis of traffic in Stockholm with special focus on the effects of the congestion tax, 2005–2008', Stockholm.

Coase, R.H. (1962), 'The problem of social cost', *Journal of Law and Economics*, **2**, 1–40.

Council of the European Union (2008), 'Proposal for a Directive of the European Parliament and of the Council Amending Directive 2003/87/EC so as to include Aviation Activities in the Scheme for Greenhouse Gas Emission Allowance Trading within the Community – Outcome of the Parliament's Second Reading (Strasbourg, 7–9 July 2008)', 11498/08, Brussels.

Crandall, R.W., A.K. Gruenspecht, T.E. Keller and L.B. Lave (1986), *Regulating the Automobile*, Washington, DC: Brookings Institution.

Flynn, P.C. (2002), 'Commercializing an alternate vehicle fuel: lessons learned from natural gas for vehicles', *Energy Policy*, **30**, 613–19.

Goodwin, P., J. Dargay and M. Hanly (2004), 'Elasticities of road traffic and fuel consumption with respect to price and income: a review', *Transport Reviews*, **24**, 275–92.

Hahn, R.W. (1989), 'Economic prescriptions for environmental problems: how the patient followed the doctor's orders', *Journal of Economic Perspectives*, **3**, 95–114.

Parry, I.W.H., M. Walls and C. Harrington (2007), 'Automobile externalities and policies', *Journal of Economic Literature*, **45**, 373–99.

Perl, A. and J.A. Dunn (2007), 'Reframing automobile fuel economy policy in North America: the politics of punctuating a policy equilibrium', *Transport Reviews*, **27**, 1–35.

Pigou, A. (1920), *The Economics of Welfare*, London: Macmillan.

Rodrigue, J-.P., B. Slack and C. Comtois (2001), 'Green logistic', in A.M. Brewer, K.J. Button and D.A. Hensher (eds), *Handbook of Logistics and Supply-chain Management*, Oxford: Pergamon, pp. 339–50.

Winston, C. (2006), *Government Failure versus Market Failure: Microeconomic Policy Research and Government Performance*, Washington, DC: Brookings Institution.

World Commission on Environment and Development (1987), *Our Common Future*, Oxford: Oxford University Press.

PART IV

OTHER ENERGY AND
SUSTAINABILITY ISSUES

16 Nuclear power in the twenty-first century*
Geoffrey P. Hammond

16.1 INTRODUCTION

Energy systems pervade industrial societies and weave a complex web of interactions that affect the daily lives of their citizens. Human development is therefore heated and powered by energy sources of various kinds, but they also put at risk the quality and longer-term viability of the biosphere as a result of unwanted, 'second-order' effects (Hammond, 2000). Many of these adverse consequences of energy production and consumption give rise to resource uncertainties and potential environmental hazards on a local, regional and global scale. Global warming, predominately caused by the enhanced 'greenhouse effect' from combustion-generated pollutants, is viewed by many as the most serious of the planetary-scale environmental impacts. Carbon dioxide (CO_2), the main greenhouse gas (GHG), is thought to have a 'residence time' in the atmosphere of around 100 years. For example, CO_2 accounts for some 80 per cent of the total GHG emissions in the UK, and the energy sector is responsible for around 95 per cent of this (Hammond, 2000; IPCC, 2007; Houghton, 2009). The emphasis of energy strategies around the world has consequently been on so-called 'low or zero carbon' (LZC) energy options: energy efficiency improvements and demand reduction measures, fossil-fuelled power stations with carbon capture and storage (CCS), combined heat and power (CHP) plants, nuclear power and renewable energy systems. The nuclear power industry has been facing contrasting fortunes in different parts of the world. In the United States of America (USA), the investment in new plants has been hindered by public concern over environmental and safety issues for over three decades since the Three Mile Island 2 (TMI-2) nuclear power plant accident (Elliott, 2003; Tester et al., 2005) in 1979. Despite the encouragement of recent US federal administrations to build new nuclear power stations, none have actually started construction over this period. This can be contrasted with the major programme for the construction of nuclear facilities that is under way in the newly industrialized countries, particularly on the Asia-Pacific Rim.

In this chapter, the historical evolution of the civil nuclear power generation industry and future trends are examined in the light of the need to meet conflicting energy supply and environmental pressures. Nuclear power from fission reactors is seen as a near zero carbon resource that is potentially available on a large scale. It might consequently provide an important component of a 'decarbonized' electricity system for the industrialized and emerging economies by around 2030–50. Many argue that this is required in order to mitigate climate change (RCEP, 2000) and stabilize the atmosphere to within an average surface temperature rise of 2°C (IPCC, 2007; Houghton, 2009). The present assessment is made against a background of forecasts of the likely growth of nuclear power in various geopolitical regional groupings out to the mid-twenty-first century. These provide a framework for discussing the global prospects for, and risks of, nuclear power in the twenty-first century. No attempt is made to be judgemental about the merits or otherwise of this energy technology. The aim is simply to identify possible medium-term futures for the global nuclear power industry and the factors that are likely to influence its development. In many ways, it represents an update of the earlier work of Hammond (1996), with the benefit of an additional 15 years of power sector development and experience.

16.2   HISTORICAL BACKGROUND

16.2.1   Gestation (1945–65)

In 2004 the nuclear industry celebrated the sixtieth anniversary of the first self-sustaining controlled nuclear chain reaction at the University of Chicago, under the direction of Enrico Fermi (Collier, 1994; Ramage, 2003; Tester et al., 2005). This led subsequently to the Manhattan project to develop an atomic bomb, which was dropped on both Hiroshima and Nagasaki in 1945. Those involved in the project might contend, with some justification, that this induced its own 'peace dividend' in the sense that the major military alliances – the North Atlantic Treaty Organization (NATO) and the then Warsaw Pact countries – did not engage in full-scale warfare with each other over the intervening period. Whilst nuclear weapons and their delivery systems have undergone great advances since the end of the Second World War, the peaceful uses of nuclear power for electricity generation have also developed apace in those countries with their own independent nuclear 'deterrent': initially the USA, the Former Soviet Union (FSU), the United Kingdom (UK) and France. Internationally, the main concern is now focused on the proliferation of nuclear weapons in those countries that have not hitherto possessed such devices, particularly in the so-called developing countries of the populous or 'majority' South of the planet (The Brandt Commission, ICIDI, 1980; Hammond, 2006). The early Magnox civil nuclear power plants in the UK were developed with a view, in part, to co-producing electricity and plutonium for its atomic weapons programme (Ramage, 2003). They included the Calder Hall power station on the site now known as Sellafield in Cumbria. HM Queen Elizabeth II formally opened this station in 1956. In contrast, the route to civil nuclear power generation in the USA progressed via the development of the pressurized water reactor (PWR) as a propulsion unit for submarines (Ramage, 2003). Governments therefore saw civil nuclear power programmes as being part of a wider strategic development.

16.2.2   Bringing to Maturity (1965–85)

By the end of the 1960s, civil nuclear power programmes had evolved a life of their own (Hammond, 1996). In 1973 the global share of electricity output held by nuclear power was about 3 per cent (see Table 16.1). The peaceful use of nuclear electricity generation has since spread to such an extent that some 30 countries operate nuclear power stations. Nuclear power was initially seen as an important electricity source that was potentially 'clean, cheap and abundant', or 'too cheap to meter' (Elliott, 2003), in comparison to the traditional fossil fuels: coal, oil and natural gas. In this regard, it was subject to the vagaries of the global energy market, which was determined by geopolitical events. The

Table 16.1 *Security and diversity energy sources for electricity generation: a UK–world perspective*

Resources	Fuel extraction		Resource origin and lifetime (R/P ratio, years)			World electricity output shares (%)	
	Geography	Safety and external impact	UK	European Union	World	1973	2007
Depleting fossil fuels							
• Solid fuel (coal, peat, etc.)	UK mines & imports	Opencast & deep mining	9	55	119	38	42
• Oil	UKCS/Norway (NCS)	Offshore drilling	6	8	46	25	6
• Natural Gas	UKCS/Norway (NCS)	Offshore drilling	5	14	63	12	21
Primary electricity							
• Hydropower	N/A	Countryside disruption	N/A	N/A	N/A	21	16
• Nuclear power	UK stockpile recycling & imports	Uranium mining (non-UK)	N/A	N/A	N/A	3	14
• Renewables (e.g. wind power)	N/A	Visual intrusion	N/A	N/A	N/A	1	3

Notes: European Union = EU-27. N/A = not applicable.

Source: Adapted from Hammond (2000); data updated using BP (2010) for the R/P ratios and IEA (2009) for the world electricity output shares.

first of these was the oil embargo by the Organization of Petroleum Exporting Countries (OPEC) in the aftermath of the Middle-East War of 1972, and the oil price hike following the Iran-Iraq war in 1979/80. This induced a great sense of insecurity over oil supplies, particularly in Western Europe and Japan. It motivated the Organisation for Economic Co-operation and Development (OECD) to establish the International Energy Agency (IEA), one of whose tasks was to encourage the holding of buffer oil stocks. The OECD countries also took steps to conserve energy in general, and to develop alternative energy sources to petroleum. Many of them, particularly those with meagre indigenous energy resources, were attracted to the idea of the development or rapid expansion of nuclear energy programmes for electricity generation. However, this 'dash for nuclear' was short-lived (Hammond, 1996).

Several factors acted to discourage the planned rapid construction of nuclear power plants. Firstly, Western efforts to conserve the use of oil were spectacularly successful. World oil demand fell, and the OPEC nations found that they could not control the production and price of crude oil in the 1980s and beyond as they had during the 1970s. Secondly, people in the West became far more aware of the possible disadvantages of nuclear power encapsulated in Amory Lovins's phrase 'the hard energy path' (Lovins, 1977). Two events reinforced this perception: the reactor failure at Three Mile Island 2 (USA) in 1979, which was contained, and that at Chernobyl (FSU, Ukraine) in 1986, which was not (Elliott, 2003; Tester et al., 2005). In addition, many people became concerned about handling radioactive materials, principally radioactive emissions during the fuel cycle, the subsequent disposal of high- and intermediate-level wastes, and the eventual need to decommission civil nuclear power stations. The storage of high-level waste and decommissioning of plants would require facilities that could operate safely over many centuries. Finally, the prospect of cheap nuclear electricity seemed much less certain than had been claimed by its early advocates (Hammond, 1996). The capital costs are high and they have not been offset by low running costs, particularly if waste storage and decommissioning costs are taken into account.

16.2.3   Environmental Imperatives (1985–2010)

The gloomy prospects for nuclear electricity generation in the 1980s were partially transformed in the 1990s and 2000s. This came about owing to the realization that there may be serious global environmental consequences of the continued burning of fossil fuels. Combustion of these fuels produces the principal so-called 'greenhouse' gas: carbon dioxide. A consensus amongst scientific opinion in the climate science community is now that CO_2 contributes significantly to global warming, leading to possible serious climate change (IPCC, 2007; Houghton, 2009). The amount of CO_2 produced by electricity generating plant using different fuels is shown in Table 16.2, where the benefit of nuclear power is clear. This is also reflected in the bar chart shown in Figure 16.1. It is evident from this chart that, in order to mitigate climate change, humanity needs to shift from coal-fired power plants (on the left-hand side of Figure 16.1) towards low carbon technologies: nuclear power or renewables. The reason that the latter technologies are not 'zero carbon' is because of the upstream or embodied energy and carbon (Hammond and Jones, 2008) associated with power plant construction. Similar arguments apply to the fossil fuel exhaust gases that contribute to acid rain: sulphur dioxide and the nitrogen

Table 16.2 *Environmental and safety issues arising from electricity generation*

	Full power cycle emissions and environmental impact (numeric values: tonne/TWh)				Plant decommissioning/waste disposal (long-term engineering problems)
	Global warming potential (CO_2 equivalent)	Resources	Acid rain precursors		
			SO_2	NO_x	
Depleting fossil fuels					
• Solid fuel	800 000–1 200 000	Low with modern plant	925–14 000	495–4545	Disposal of plant/PFA 'waste'
• Oil	680 000–950 000	Low with FGD	1630–16 000	2100	Disposal of offshore drilling rigs and power plants/no back end waste
• Natural gas	430 000	Low	460	860	
Primary electricity					
• Hydropower	15 000	Negligibly small	2	N/A	Disposal of plant/no back end waste
• Nuclear	8000–60 000	Negligibly small	–	55	Long-lived radioactive plant and waste
• Renewables (e.g. wind power)	11 000–75 000	Negligibly small	N/A	N/A	Disposal of plant/no back end waste

Source: Adapted from Hammond (2000).

335

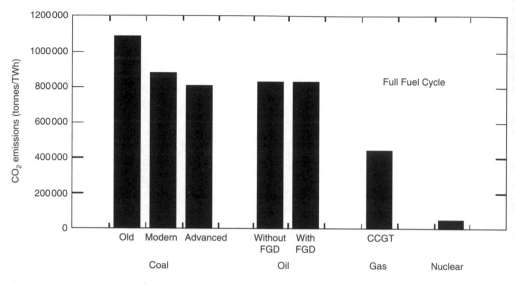

Source: Hammond (1996); adapted from Nuclear Electric plc (1994).

Figure 16.1   Carbon dioxide emissions attributable to UK electrical power plants

oxides. They are also effectively eliminated via the use of the nuclear fuel cycle (see again Table 16.2).

In addition to the problem of pollutant emissions from fossil fuel combustion, there is now a better awareness of the finite nature of oil and natural gas reserves. Thus, by 2007 the global share of electricity output held by nuclear power had risen to about 14 per cent (see Table 16.1); albeit on something of a plateau. The OECD governments are now anxious about the global distribution of fossil fuel resources, which beyond 2030 will be located largely outside their control. Many of them now talk about a 'Nuclear Renaissance', although it is rather slow in taking off. In the European Union two new Evolutionary Power Reactor or European Pressurised Reactor (EPR™) plants – an evolution of the American PWR design – are currently under construction in Finland (the Olkiluoto 3 1600 MW reactor, sited in the municipality of Eurajoki) and France (a 1330 MW reactor at Flamanville in Normandy). By mid-2010, they were both some way behind their construction schedules. The Olkiluoto 3 plant, for example, has been delayed by three years and is now expected to be operational only in about 2012, with a resulting loss of around $2.8 billion. Similarly, the Flamanville 3 plant has been delayed by two years, and now aims to be operational in 2014 (at a construction cost overrun of ~40 per cent).

16.3   TECHNOLOGIES AND TIMESCALES

The present contribution focuses on the medium term: the likely development of nuclear energy up to 2010–30. Over this period, the market will be dominated by fission reactors, principally thermal (or 'burner'), rather than fast (or 'breeder'), types. They require plutonium and/or uranium fuel to operate. Indeed, the prototype fast reactors in France and the UK were designed to take as their first charge of plutonium fuel that produced in the earlier thermal devices (Collier, 1994). In the 1970s there was quite a wide variety of alternative thermal reactor designs, but that of choice (Collier, 1994) has now mainly become the pressurised water reactor (PWR), developed from an American original design (Ramage, 2003; Tester et al., 2005). These give rise to the problem of radioactive emissions and waste disposal highlighted in the previous section, although concentration on one principal reactor type is likely to result in enhanced operational safety and cost competitiveness. Due to the fall in the demand for nuclear-generated electricity compared with the exaggerated projections of the 1970s, concern over the finite nature of uranium supplies has diminished. Consequently the impetus to move towards fast breeder reactors has greatly declined (Elliott, 2003).

There has been much speculation over the possibility of generating electricity using a fusion reaction with heavy water as an abundant source of fuel. Although the USA, the European Union and the Commonwealth of Independent States (CIS – the FSU) all have experimental facilities, the feasibility of a cost-effective reactor appears remote. Perhaps in the long term (say 2050–2100) this might be attainable, but fusion is highly unlikely to have an impact on the energy scene during the period considered here (Elliott, 2003). Even John Collier (1994) – the late chairman of the main British nuclear generator (then known as Nuclear Electric plc) – has argued that commercial exploitation of fusion will be 40 or more years away. However, its great attraction would be the near limitless source of raw material as a supply of fuel. It also appears to be, in principle, much safer and more environmentally benign than the current generation of fission reactors. The amount of radioactive material that might be discharged in a fusion accident would be small, owing to the short residence time of fuel in the reactor, and under normal operating conditions the reaction products in wastes are themselves short-lived (Elliott, 2003; Tester et al., 2005).

16.4   ELECTRICITY AS AN ENERGY SOURCE

The nuclear fuel cycle produces a high-grade energy source, namely electricity. This is a 'capital' resource that is primarily based on the use of either depleting fossil fuels or nuclear fuels. These can be contrasted with the renewable (or 'income') energy sources, such as solar energy and tidal, wave and wind power. In the aftermath of the 1992 Earth Summit in Rio de Janeiro (the UN Conference on Environment and Development) there has been a greater awareness of the need to devise strategies aimed at sustainable development. The World Commission on Environment and Development (WCED) in the influential Brundtland Report (WCED, 1987) defined sustainable development as meeting 'the needs of the present without compromising the ability of future generations to meet their own needs'. Hence, governments have been encouraged to conserve depleting fuel resources, and to make greater use of renewable energy sources. The protagonists on both sides of the nuclear energy debate have argued that the concept of sustainability supports their position. Thus, Collier (n.d.) has suggested that nuclear power is one means of helping to maintain the energy resource balance of the planet at the same time as the environmental quality. He had the issue of greenhouse gas emissions and global warming specifically in mind. On the other hand, Greenpeace (1994) argues that nuclear power will leave future generations with a legacy of nuclear waste. The two positions could, of course, be somewhat reconciled if long-term safe storage of waste could be assured, but this is difficult to prove a priori (Hammond, 1996).

Electricity is a high-grade energy carrier in the sense that it can be used to provide either power or heat. In a thermodynamic sense it has a high 'exergy' (as outlined by, for example, Hammond and Stapleton, 2001). Large energy losses occur during generation unless used in conjunction with combined heat and power (CHP) systems. It is wasteful in thermodynamic terms to convert fuels to electricity only to employ it for heating. If process or space heating were required, then it would be far more efficient to burn fossil fuels (for example) to produce heat directly. Electricity is also difficult to store on a large scale, and is mainly used instantaneously (Hammond and Waldron, 2008). But electricity has other benefits. There is an increasing end-use demand for high-grade and controllable energy carriers (Hammond, 2000). But heat is wasted and energy is 'lost' at each stage of energy conversion and distribution, particularly in the process of electricity generation (Hammond, 2000; Hammond and Stapleton, 2001). There are many feedback loops in which primary energy sources (including fossil fuels, uranium ore and hydro-electric sites) and secondary derivatives (such as combustion and nuclear-generated electricity) themselves provide upstream energy inputs into the 'energy transformation system' (Hammond, 2000; Hammond and Jones, 2008; Hammond and Stapleton, 2001). The latter is that part of the economy where a raw energy resource is converted to useful energy, which can meet downstream 'final' or 'end-use' demand. 'Renewable' energy sources are taken to mean those that are ultimately solar-derived: mainly solar energy itself, biomass resources and wind power.

Another limitation of electricity generation in terms of meeting global energy demand is its requirement for a high-technology infrastructure. It is often argued that energy demands will grow in the future to meet the rising needs and expectations of the developing countries, which typically have rapidly growing populations. However, these countries are unlikely to have the financial resources or expertise to follow a high-technology route, certainly not in Africa (except South Africa) or in much of Latin America (Hammond, 2006). Even in mainland Asia, where the industrial base is in some instances well developed, the proportion of electricity actually produced by nuclear power is very small.

16.5   NUCLEAR POWER ECONOMICS

Nuclear power has not proven to be 'too cheap to meter' with the benefit of hindsight (Elliott, 2003). The avoidable (or variable) costs of gas-fired power stations are currently very low compared with those of their coal-fired or nuclear counterparts (see Table 16.3). This was part of the reason for the so-called 'dash for gas' during the initial phase of UK energy market liberalization 1985–97 (Hammond, 2000). The capital or construction cost of PWR plants is typically higher by a factor of three than modern combined cycle gas turbine (CCGT) plants (Elliott, 2003). However, the 'full cycle costs' taking account of environmental impacts (obtained from the ExternE project: see http://www.externe. info) are more favourable to LZC energy technologies: principally nuclear power or renewables (see again Table 16.3).

Table 16.3 *Power generation efficiency and economics: a UK perspective*

Resources	Electricity generation efficiency (%)		Waste heat loss (%)	CHP potential	Power cycle economics	
	Energy	Exergy			Avoidable costs (p/kWh)	Full cycle costs (EUR-cent/kWh)
Depleting fossil fuels						
● Solid fuel (coal, peat, etc.)	35.1	33.5	64.9	High	1.3–1.6	4–7
● Oil				High	1.4	3–5
● Natural gas	33.0	32.1	67.0	High	1.6–2.5	1–2
Primary electricity						
● Hydropower	92.1	78.0	7.9	Nil	Low	N/A
● Nuclear	33.0	37.0	67.0	Undesirable	1.3–3.6	0.25
● Renewables (e.g., wind power)	N/A	N/A	N/A	Generally Low	Low	0.15

Note: N/A = not applicable.

Source: Adapted from Hammond (2000); updated data on the full cycle costs based on the 2001 ExternE project (http://www.externe.info).

The Stern Review on the *Economics of Climate Change* (Stern, 2005) advocated a realistic carbon price as a vital part of future policy; indeed it argued that failure to take account of environmental externalities (such as climate change) ensures that there will be under provision and slower innovation. However, carbon pricing is still in its infancy (Allen et al., 2008), and even where it is implemented uncertainties remain about the durability of the price signals over the long term. The EU Emissions Trading Scheme (ETS) has shown considerable volatility in terms of the carbon price over recent years and has been generally much lower than is required to encourage the take-up of low carbon energy technologies. The next generation of European nuclear power stations is likely to come on stream during the third trading phase of the ETS. Incentivization of such plants would require an adequate carbon price, which would be critically dependent on there being tighter National Allocation Plans under the ETS (Stern, 2005). Regulation and alternative policy approaches may therefore be required in order to promote the required investment in sustainable technology innovation (Allen et al., 2008).

The private sector has generally proved unwilling to meet nuclear liabilities, including reactor decommissioning and waste storage costs, without government financial support in one form or another (Hammond, 1996). Watson (2005) collated data on the alternative CO_2 abatement options derived from the UK Cabinet Office's Energy Review by the then Performance and Innovation Unit (PIU, 2002). He indicated that, for example, household energy efficiency measures would cost between –£300 and +£50/ tCO_2 abated, whilst comparable onshore wind farm costs would be between –£80 and +£50/ tCO_2, and nuclear power costs would be between +£70 and +£200/ tCO_2 abated (see also Figure 16.2). They can be contrasted with the power plant/CCS estimates for the cost of CO_2 captured with enhanced oil recovery (EOR) storage of between +£7 and +£24/tCO_2 abated (Hammond et al., 2011). However, the costs of electricity generation depend on whether or not countries have access to relatively inexpensive fossil fuel supplies; for example, Western USA coal and southern North Sea natural gas (Hammond, 1996).

The costs of power plants are often compared in terms of their so-called levelized cost of producing a given amount of electricity (LCOE): p/kWh or equivalent (POST, 2003). Recent LCOE estimates by the US Energy Information Administration (EIA, 2010) indicate that advanced coal has a total system levelized cost of $110/MWh, in contrast to advanced coal with CCS at $130/MWh, CCGT at $83/MWh, advanced CCGT with CCS at $113/MWh, modern nuclear at $119/MWh, onshore wind at $149/MWh, and offshore wind at $191/MWh. On this basis, it would appear that nuclear power could compete quite successfully with low carbon competitor technologies. Tester et al. (2005) noted that nuclear power costs from light-water reactors (of which the PWR is the main type) are typically split at about 57 per cent capital, 30 per cent operations and 13 per cent fuel. This contrasts with the recent EIA estimates (EIA, 2010) that suggest 80 per cent capital, 10 per cent operations and 10 per cent fuel. The UK Sustainable Development Commission in its survey of the role of nuclear power in a low carbon economy (SDC, 2006) noted that it is much cheaper to supply nuclear power continuously, than to switch off plant; that is, the avoidable cost (see again Table 16.3) is relatively high. Nuclear power is therefore a 'price-taker', and is unable to determine the competitive price in a liberalized market. The SDC (2006) argue that it is for this reason as well as technical considerations that nuclear power is commonly used for 'baseload' generation in the absence of large-scale electricity storage capacity on the grid.

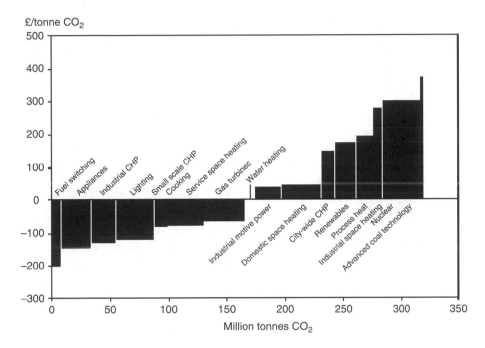

Note: Only some of the main aggregated energy options are labelled.

Source: After that developed by the Stockholm office of McKinsey & Company.

Figure 16.2 An indicative carbon dioxide 'abatement cost curve' for the UK

16.6 NUCLEAR POWER PROJECTIONS, 2010–35

16.6.1 Background

In analysing the likely future demand for energy, it is useful to disaggregate the world into geopolitical or regional groupings. Thus, although the global average consumption of primary commercial energy per capita in 2000 was about 2 kW, the regional variations were dramatic. North America consumed 9 kW (a fall of nearly 1 kW on that in the 1970s). In contrast, other regions have stabilized with Western or OECD Europe at about 4.5 kW and the Former Soviet Union also at about 4.5 kW, whilst the 'rest of the world' only uses 1 kW (Hammond, 1996, 2006). Obviously, the developing countries (which make up a significant proportion of the latter group) use non-commercial energy sources, such as wood and dung, for heating and cooking. Nevertheless, correcting for these fuel sources will alter the figures only slightly. The data clearly reflect the enormous disparities in affluence across the globe. When the per capita energy consumption data are multiplied by the respective populations, they give rise to upward trends in regional energy demands. The proportion of this demand that is supplied by electricity obviously varies from grouping to grouping, as does the fraction met by nuclear power. This nuclear share of electrical power generation within the 30 or so 'nuclear power' countries in 2007 indicates a range from 2 per cent (India) to 76 per cent (France). When these are aggregated into geopolitical groupings, they indicate general upward trends, with the global share increasing. This world total levelled out around 2007 according to recent EIA data (2010); corresponding to a worldwide nuclear share of electricity generation of some 14 per cent.

16.6.2 Nuclear Power Generation

The US Energy Information Administration (EIA, 2010) has produced its own *International Energy Outlook* (hereafter denoted as IEO2010), which projects a range of energy and electricity projections over the period 2007–2035. It employs estimates of population and economic growth in the various regions, together with projections of changes in energy intensities (the energy consumed per unit of gross domestic product at constant prices), to enable forecasts of primary energy requirements and the fuel mix to be made as far as 2035. An EIA database for commercial energy sources only is utilized, and the impact of pending regulatory developments is excluded (including the EU Emissions Trading Scheme, ETS). In the IEO2010 'Reference Case', world electricity output increases by 87 per cent with the non-OECD countries (mainly the developing and emerging economies of the majority South), accounting for 61 per cent of global electricity use by 2035 (EIA, 2010). Here the scenario assumes that high fossil fuel prices and environmental concerns (principally global warming induced by GHG emissions) improve the prospects for alternative energy sources, such as nuclear power and renewables. Electricity generation from nuclear power therefore increases from 2.6 PWh (peta-watt hours) in 2007 to 4.5 PWh in 2035 (EIA, 2010). However, the administration recognizes that there is considerable uncertainty over these projections, notwithstanding the renewed interest in the nuclear option as a means of ensuring diversity of power supplies and the decarbonization of the electricity sector. The projections up to 2020 were in line with the announced plans of national governments and power generating companies. Afterwards, a range of drivers for and barriers to nuclear power were incorporated into the IEO2010 Reference Case: for example, technological innovations, the need for environmental protection, economic issues, likely geopolitical developments, and uranium availability (EIA, 2010).

The IEO2010 Reference Case shows the strongest growth in nuclear new build occurring in non-OECD Asia. This is in line with the development trends previously discussed by Hammond (1996), based on the early projections by the World Energy Council (WEC, 1993). The Reference Case suggests an increase in nuclear power generation in China of 8.4 per cent per annum (pa) and 9.5 per cent pa in India (EIA, 2010). In 2009 China led the way in terms of nuclear new build with some 43 per cent of the worldwide active construction projects. New nuclear installations are thought likely to be brought online in Indonesia, Pakistan and Vietnam by 2020. Outside Asia, the administration believes (EIA, 2010) that the main development will probably be in Central and South America, where a growth rate of around 4.3 per cent pa may be achieved. Worldwide nuclear generation is projected to grow by some 2 per cent pa. By 2035 power supply from nuclear power plant in the USA is assumed to reach 17.1 per cent of electricity generation overall. Over the forecast period (2007–35), 8.4 GW of new capacity is likely to be added, whilst 4.0 GW will be obtained from the life and performance extension of existing plants (EIA, 2010). This will be driven by the likely rise in natural gas prices and environmental pressures. Notwithstanding the increased interest in nuclear power across Europe (with a number of countries reversing policies aimed at retiring plant and inhibiting new build), nuclear capacity is only projected to rise modestly in OECD Europe over the forecast period. In 2035 natural gas and nuclear electricity generation are projected to reach rough parity. The IEO2010 projections take a relatively conservative view of the prospects for nuclear generation in the Russian Federation. Its existing 23 GW capacity is thought likely to be enhanced by just 5 GW in 2015 and a further 20 GW in 2035 (EIA, 2010).

16.6.3   Uranium Supply and Demand

At the time of the earlier study of nuclear futures by the author (Hammond, 1996), the World Energy Council (WEC, 1993) had been arguing that the development of the nuclear power industry up to 2020 would not be constrained by shortages of uranium fuel. However, the Western World's output of newly mined uranium fell significantly in the mid-1990s. The shortfall in supply over demand in the early 1990s was offset by drawing down on inventories, and by the recycling of reprocessed fuel products (Hammond, 1996). Stocks held by Western producers amounted to some 15 000 tU (tonnes uranium), whereas forecast supplies of newly mined uranium depend on the mine capacity utilization rates assumed. In 2009 over half of global uranium mining and production took place in just three countries (NEA and IAEA, 2010): Australia, Canada and Kazakhstan. Australia is home to ~1.7 MtU of recoverable uranium (31 per cent of known world supply); Kazakhstan has 12 per cent of known supplies; Canada and the Russian Federation both have 9 per cent; and Brazil, Namibia and South Africa have about 5 per cent each. The USA has only around 4 per cent of world supplies. But the relationship between uranium supply and demand is complex and speculative, and consequently there can be no definitive figure for 'reserves' (Elliott, 2003); they depend on the prevailing economics, as well as the availability of uranium ore. The latest uranium resources 'Red Book' (NEA and IAEA, 2010) indicates that, at 2008 consumption rates, known uranium resources are likely to last for around 100 years. Even under a high-growth scenario, less than half of the total reserves would be consumed by 2035.

16.7   NUCLEAR ENERGY ISSUES FOR THE OECD COUNTRIES

16.7.1   Strategic Issues

Perhaps the main factor that will determine the extent to which Western Europe will embrace the nuclear option to meet a significant fraction of its electricity needs into the next century is its public acceptability. This is determined by the degree to which the public are convinced that nuclear power stations can operate safely, and that radioactive by-products can be securely stored over long periods. The extent to which people in different European countries appear to accept the nuclear option varies quite widely, presumably due in part to cultural factors. In France, which has very little in the way of indigenous fossil fuel reserves, the nuclear share of electricity generation is already about 76 per cent (EIA, 2010). Certainly, the French have made little resistance to this, in contrast with other countries where there are active and vocal opponents of nuclear power who have a significant influence on the public via their access to the media. The safety record of the nuclear energy industry compares favourably with that of its competitors (Fremlin, 1987), but it is often perceived as being more life threatening; arguably out of proportion to the actual risks. Obviously, it is incumbent on the civil nuclear industry to reassure the general public over its long-term operating safety, a task that is undoubtedly daunting.

The next most significant inhibitor of nuclear power in the world after the collapse of communism, where private ownership has become the dominant mode of economic organization, is the cost issue (Hammond, 1996). The relatively high capital cost of nuclear plant in comparison with the alternative fossil-fuel generators was discussed earlier under 'Nuclear power economics' (Section 16.5). Nevertheless, this is counterbalanced to an extent by the uncertainty over fossil fuel supplies in the medium term (see Table 16.1). The lifetime and global distribution of these vary enormously:

- Oil: OPEC (Middle East) dominated, ~45-year nominal life;
- Natural gas: CIS (Russian) dominated, ~65-year nominal life;
- Coal: widely distributed, ~120-year nominal life.

These figures are rough estimates, assuming current rates of consumption (BP, 2010), and new reserves are quite frequently found. However, they indicate that the sources of fossil fuel supplies for OECD countries, with the exception of coal, are rather insecure. If depletion of oil and natural gas at anything like this rate actually occurred, then the price of these fuels would rise. This would make the financial case for nuclear energy look much brighter. It has often been argued since the oil crises of the 1970s that nuclear power should be adopted as an insurance policy against the insecurity of the oil market. In reality, the two resources are not substitutable (Hammond, 1996), particularly in the transport sector (without a very large expansion in the use of electric vehicles).

In the industrialized regions of the world, it has often been considered important to keep a technical capability in civil nuclear power in order that the OECD countries do not fall behind in this area of high technology. It is also seen as a technology with considerable export potential. Considerations of this type have certainly influenced the industrial strategy of countries such as France (and Japan).

16.7.2   Global Warming

The prospects of global climate change induced by the GHG emissions from fossil-fuel combustion is an issue of considerable interest and concern to those nations sensitive to climate change. The European Union is therefore striving to meet its obligations to reduce CO_2 emissions under the UN Framework Convention on Climate Change agreed at the Rio Earth Summit. The main focus of this activity is concentrated on an examination of economic instruments, such as carbon pricing (for example as implied by the EU ETS), to discourage emissions. These would be favourable to nuclear power, but also to energy efficiency and renewable energy sources. Friends of the Earth and Greenpeace

Note: Each dot represents the equivalent of ten 1-GWe nuclear power plants.

Source: Adapted from Keepin (1990).

Figure 16.3 Nuclear power and climate change mitigation – 1990 actual

(Greenpeace, 1994; Keepin, 1990) have argued that nuclear power is one of the least cost-competitive means of CO_2 abatement. An energy efficiency strategy, for example, displaces between 2.5 and 20 times more carbon dioxide than nuclear power per dollar (euro or pound) invested (Keepin, 1990; Hammond, 1996).

In any case, in order to have a serious effect on global warming, the investment in nuclear power would need to be enormous. One study (Keepin, 1990) has suggested that to replace coal-fired power stations with nuclear ones by 2025 would require over 5000 new 1-GWe nuclear plants to be built worldwide, with one new station being commissioned every two-and-a-half days. Nearly half of these would need to be located in the emerging economies (of the populous or 'majority' South). The actual nuclear power plant inventory and locations in 1990 were depicted by Keepin (1990) as shown in Figure 16.3, where the exact locations have been altered to provide graphical clarity. The number of civil nuclear reactors in the South of the planet is obviously low. However, under an extreme scenario aimed at climate change mitigation by 2025 (see Figure 16.4), over 5000 new nuclear power plants would need to be constructed. This is clearly unrealistic as it would necessitate, for example, a 155-fold increase in installed capacity in the South. A similar study based on earlier IEA data (Keepin, 1990; Kouvaritakis, 1989) concluded that to shift OECD countries to a 70 per cent nuclear share of electricity generation (roughly the proportion in France) by 2010 would require the construction of 800, 1 GWe stations at a rate of one every nine days. These figures are clearly implausible (Hammond, 1996), and serve only to illustrate that a nuclear-only option for tackling global warming is not feasible. It may well, of course, play a part in a programme to reduce CO_2 emissions, together with other measures. However, the choice

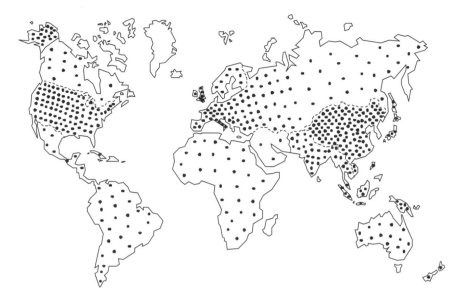

Note: Each dot represents (as in Figure 16.3) the equivalent of ten 1-GWe nuclear power plants.

Source: Adapted from Keepin (1990).

Figure 16.4   Nuclear power and climate change mitigation – 2025 extreme nuclear scenario

of nuclear energy is likely to be dictated by considerations other than global warming, against a background of a European Union (EU-ETS) and similar market frameworks (Hammond, 1996).

16.7.3   The Decommissioning of Nuclear Power Stations

The last nuclear power station to be constructed in the UK was the PWR plant known as Sizewell B in Suffolk, commissioned in 1995. No new nuclear plants have been subsequently constructed, and many of the existing ones are coming to the end of their operating lives. Britain has therefore progressively decommissioned its older nuclear power stations over the last decade or so. The lives of nuclear power stations may be extended from their typical 25-year design life to nearer 40 years (Hammond, 2000). New PWR (or EPR) stations are said to have an operating life of 60 years. The first-generation Magnox plants (named after the magnesium non-oxidizing cladding on natural uranium fuel rods) have now all been decommissioned. There were seven advanced gas-cooled reactors (AGRs) designed and built in the UK, but these are all due to be decommissioned by around 2025. This will leave only the Sizewell B PWR station, with nuclear power holding a considerably reduced share of electricity generation (perhaps as low as 3 per cent by 2020); see Figure 16.5 (Royal Society and Royal Academy Engineering, 1999; also Elliott, 2003). The loss of such a significant proportion of generation capac-

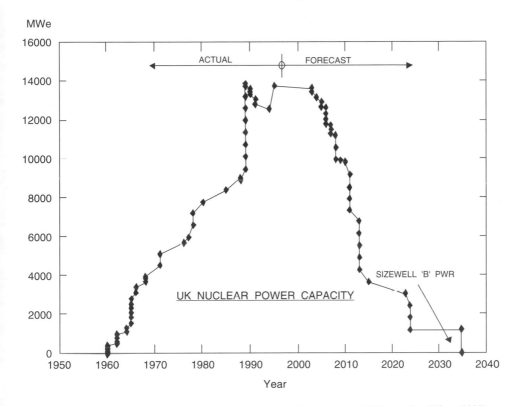

Source: Adapted from Royal Society and Royal Academy of Engineering (1999); see also Elliott (2003).

Figure 16.5   Nuclear power plant capacity in the UK over the period 1950–2040,
illustrating the impact of scheduled reactor decommissioning

ity will require a great deal of investment in alternatives to ensure continued security of supply. Nuclear power is also the main source of near zero carbon electricity in Britain. It is going to be difficult for renewables (principally wind power) to fill the electricity 'gap' (Hammond, 2000). Unless the UK government decides to replace nuclear by nuclear (that is, build a next generation of modular nuclear reactors to replace those being decommissioned), it might need to revert to electricity generation via coal-fired plant coupled with carbon capture and storage technologies (IPCC, 2005) or increase its reliance on imported natural gas for CCGT plants.

Nuclear plant exhibit high life-cycle costs, when 'back-end' (decommissioning and waste disposal and storage) costs are taken into account (Hammond, 2000; Hammond and Waldron, 2008). The private sector has generally proved unwilling to meet nuclear liabilities without government financial support of one kind or another (see section 16.5 on 'Nuclear power economics' above). Future nuclear plants commissioned over the period 2010–20 could produce electricity at a cost of between 2.5 and 4.0 p/kWh depending on the number of plants (SDC, 2006), whilst CCGT power stations can produce electricity at 2.3 p/kWh (at 2005 prices). However, further improvements to gas-fired stations may make it difficult for nuclear to remain competitive. Planning permission for any energy development can be difficult, especially in the case of nuclear. The UK Government confirmed its decision to allow new nuclear power stations to be built in its 2008 *White Paper on Nuclear Power* (BERR, 2008) for the reasons of climate change mitigation and to enhance security and diversity of energy supply. But it continues to state that the nuclear industry will need to invest private funds in new capacity without state aid, although it will streamline the planning process for new large energy projects. Thus, energy companies will need to fund new build in the UK, including 'the full cost of decommissioning and their share of waste management costs' (BERR, 2008). It also intends to help strengthen the EU-ETS in order to ensure that companies have confidence in the long-term future of the carbon market (or price for carbon) when making investment decisions. Action needs to be taken soon, if nuclear power is to contribute to enhanced energy security or carbon abatement out to around 2035.

16.7.4   The Storage of Nuclear Waste

The civil nuclear power industry in the UK feels that radioactive waste management no longer presents any technical problems, but the difficulty lies with the reluctance of the general public to accept an appropriate regime (Elliott, 2003; Ramage, 2003; Hammond and Waldron, 2008). The UK government therefore established in 2003 a Committee on Radioactive Waste Management (CoRWM) to make recommendations for the long-term management of the UK's high- and intermediate-level radioactive waste (CoRWM, 2007). It undertook a two-and-a-half-year process of engagement with the public, stakeholders and the scientific community. They recommended that the long-term disposal of radioactive waste should be deep underground: an option known as 'geological disposal'. In response to the CoWRM final report (2007), the government decided that responsibility for securing this option should fall on a new public authority, the Nuclear Decommissioning Authority (NDA). It therefore brought together all the stages of the nuclear waste management chain under the umbrella of one organization. But it is still unknown whether relieving the industry of these liabilities will change public perception of it and encourage investment. The selection of a suitable site, via community involvement, might take several decades. Storage techniques, such as using concrete blocks and vitrification, could be used to store nuclear waste over centuries.

Waste management liabilities should obviously be added to the full fuel cycle costs of nuclear power generation, and an appreciation of that may be one reason why investors have been reluctant to invest in new nuclear build within a liberalized energy market (see section 16.5 on 'Nuclear power economics' above). CoRWM (2007) argued for the construction of a robust means of interim storage as an insurance against the risk of delay or failure in the long-term repository programme. There is also a significant risk posed by an international terrorist attack (see section 16.7.5 on 'Terrorism' below). Clearly storage of nuclear waste in a deep underground repository would obviously alleviate this risk.

16.7.5   Terrorism

International terrorism is a risk that is considered far more likely since the al-Qaeda attack on the Twin Towers in New York on 11 September 2001; the events of so-called '9/11' (Hammond and Waldron, 2008). Electricity powers much of the infrastructure of

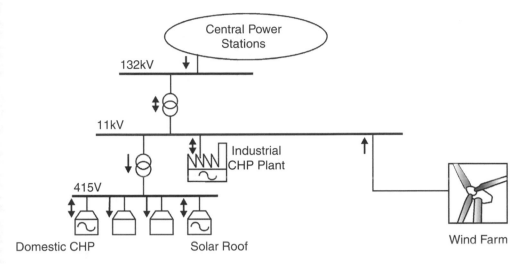

Source: Hammond and Waldron (2008).

Figure 16.6   *A schematic representation of a possible grid-connected 'distributed generation' network*

industrialized nations, from telecommunication systems to waterworks (Douglas, 2005); an attack on the grid therefore provides the possibility of bringing down large areas by a strike on a major generating plant, substations or transmission lines. The urban chaos surrounding such an event could also be used to aid the terrorists, as transport and medical evacuation could become very difficult. Hospitals without power would no longer be able to operate and financial losses in a city would be considerable. However, such a 'ripple attack' may only cause short-term blackouts. The Electricity Supply Industry (ESI) needs to conduct ongoing assessments of the risks posed by terrorism to Britain and elsewhere. These are currently performed to evaluate the safety of nuclear power stations, but not for the risk involved with terrorist attack. Decentralized generation (Allen et al., 2008; see Figure 16.6) might yield increased security, albeit over a rather long-term implementation period. A single attack would obviously then have less impact on the grid as a whole if a significant proportion of distributed generation (~30 per cent) were installed.

It is not surprising that the greatest concern, as well as the most advanced crisis planning, relating to potential terrorist attacks is in countries like Israel and the USA (Douglas, 2005). They recognize that this need not just be a physical strike – one aimed at computer or communications networks might have an equally devastating effect. A successful cyber attack could enable terrorists or hackers to disable electricity grid protective relays or to gain control over parts of the network. Indeed, countries such as China, Russia and North Korea are believed by the US Department of Homeland Security (DHS) to have strategies in place to launch cyber attacks on the Internet (Douglas, 2005). The DHS has been working with the Electric Power Research Institute (EPRI) on an Infrastructure Security Initiative that seeks to protect both physical and communications networks. It has identified high-voltage transformers as a particularly vulnerable target for terrorists. They have a multi-million dollar price tag, and take between one and two years to procure, build and install (Douglas, 2005). The sort of actions that would help to reduce the risk of attack and minimize any damage include (Douglas, 2005): securing the grid from cascading failure; monitoring channels for attack, sealing them off, and compartmentalizing systems potentially subject to attack; installing secure critical controls and communications against penetration by hackers and terrorists; and the adoption of ongoing security assessments to ensure that the industry stays several steps ahead of changing vulnerability to terrorist attack.

In the aftermath of the 9/11 atrocity, and the subsequent 'War on Terror', the UK Parliamentary Office on Science and Technology (POST) reviewed the particular risks of a terrorist attack on nuclear facilities in Britain (POST, 2004). The Office for Civil Nuclear Security (OCNS) has prepared a classified document that identifies the type of terrorist attack that might be launched against UK civil nuclear sites (POST, 2004). Commercial reactors give rise to the bulk of the total radioactive material inventory. This is, in turn, stored at the Sellafield reprocessing plant in Cumbria and at Dounreay in Scotland. Nuclear power stations are dispersed around the UK, both operating plants and those being decommissioned. Although existing nuclear power stations were not specifically designed to withstand a terrorist attack, their containment systems are robust. The POST briefing (2004) suggests that the worst-case scenario might involve an impact from a large aircraft (on the model of 9/11), which might then result in the release of significant quantities of radioactive material. They do not believe that it is possible to estimate accurately the likelihood and severity of a terrorist attack on nuclear facilities. A culture based around existing safety and security regimes clearly needs to be adapted to these new threats.

16.8   RISK ASSESSMENT OF THE ELECTRICITY SECTOR

The major risks associated with a rapidly changing UK power sector were recently identified and quantified with the aid of various stakeholder groups (academic researchers, civil servants, electricity companies, 'green' groups, power system engineers, and various others) via an Internet questionnaire (Hammond and Waldron, 2008). Each stakeholder ranked potential risks according to the perceived 'severity of impact' and 'likelihood of occurrence' – each on a three-point scale. This data were then used to perform a ranking of the risks by multiplying scores for impact and occurrence. There was some variation between the different stakeholder groups, but the same risks were ranked highly by each group.

The greatest risks (see Table 16.4) were identified as being energy security issues (the reliance on imported fossil fuels for electricity generation; the highest score), lack of investment in new infrastructure, decommissioning of nuclear plant leading to reduced network capacity, severe weather events and inadequate spare capacity margins generally. These trial results represent a snapshot of risks to the rapidly evolving UK power sector in the early years of the twenty-first century (Hammond and Waldron, 2008). The stakeholders who responded to the online questionnaire identified decommissioning of the current nuclear plants as the third most serious risk overall. This process is inevitable, so significant generation capacity needs to be built in order to replace these lost,

Table 16.4   Overall ranking of risks to the UK electricity network

Rank	Risk type	Overall score
1	Reliance on primary fuels for electricity generation	6.0
2	Lack of investment in new infrastructure	5.4
3	Decommissioning of nuclear – reducing capacity	5.2
4	Severe weather conditions	4.9
5	Spare capacity margins	4.7
6	Difficulties in the storage of nuclear waste	4.6
7	Market liberalization	4.3
8	Environmental legislation	4.2
9	Decaying infrastructure	4.2
10	Growth in loads causing power quality problems	3.9
11	Cross border transactions	3.9
12	Terrorism	3.6
13	Loss of expertise	3.5
14	Decentralized generation	2.8
15	New generation technologies	2.5

Source: Adapted from Hammond and Waldron (2008).

zero carbon power stations. The options that could be adopted are renewable electricity generators (principally wind farms), fossil fuel-fired power stations with CCS, or a new (replacement) generation of nuclear power stations. The British government appears to favour all three options, which it regards as potential zero carbon power sources. Coal-fired power stations with CCS and nuclear power stations are both available technologies, although the former is unproven at a large scale. It would therefore require a phase of operation with commercial demonstrators.

There were some interesting differences over nuclear power risks in terms of the individual stakeholders groups surveyed by Hammond and Waldron (2008). Electricity companies viewed the closure of old coal and nuclear plants, leading to reduced network capacity, as having the highest risk score (see, for example, Figure 16.7), whereas the 'academic evaluators' ranked it second (along with the category designated as 'engineers and miscellaneous individuals'). The latter group ranked the storage of nuclear waste as the fifth-highest risk to the development of the power sector. In contrast, the 'pro-renewable' (or 'green') group ranked this as the second greatest risk. These differentiated stakeholder findings have potentially important implications for the public acceptability of nuclear power.

Risk assessment of this type is likely to be subject to framing and semantic effects commonly identified in the domain of social science research (Hammond and Waldron, 2008). The term 'framing' means the inevitable process of selective influence over an individual's perception (in this case the various expert respondents) in such a way as to encourage particular (that is, biased) interpretations and to discourage others. Such issues were addressed, in part, by using the pilot study and by enabling respondents to comment on the content of the online questionnaire, as well as identifying additional risks. But the present trial has certainly illustrated the potential of using risk assessment techniques to evaluate developing risks to the power sector. Clearly such an exercise

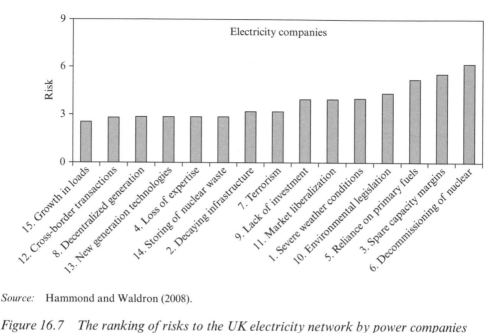

Source: Hammond and Waldron (2008).

Figure 16.7   The ranking of risks to the UK electricity network by power companies

would need to be carried out periodically if it were to maintain its value to the industry, its stakeholders and to policy-makers. Hammond and Waldron (2008) suggest that this might perhaps be undertaken at a frequency of something like every five years.

There is an alternative, and rather more controversial, type of 'energy risk assessment'. This is based on the work of Herbert Inhaber (1982), who was then at the Atomic Energy Control Board of Canada in the late 1970s. He allocated life-cycle risk scores to the production and use of different energy technologies. There are some rather informative diagrams in the original book that place the occupational and public fatality and health risks in a comparative framework for a range of energy technologies (Hammond and Waldron, 2008). They show that, according to Inhaber's life-cycle methodology, natural gas has the lowest health risk, with nuclear next, whereas methanol has the highest incidence of days lost and deaths (followed by coal and some of the renewables). This needs to be taken with a 'health warning', but it is nevertheless instructive. Inhaber's work went through several revisions, and he published the latest in 1982. This book contains (as Appendix Q) details of the correspondence that Inhaber had with his many detractors. The approach leads to some useful, if uncomfortable, insights that could warrant revisiting in a modern (and UK) context. Whatever the merits of Inhaber's method of 'energy risk assessment', it clearly needs to be distinguished from the type of study reported by Hammond and Waldron (2008).

16.9   CONCLUDING REMARKS

The prospects for global nuclear electricity generation have been examined in the light of the need to meet conflicting energy supply and environmental pressures over recent decades. It is argued that fission (thermal) reactors will dominate the market up to the period 2010–35, with fusion being relegated to the latter part of the twenty-first century. A number of issues affecting the adoption of nuclear electricity generation have been considered, including its cost and risks, industrial strategy needs, and the public acceptability of nuclear power. The contribution of nuclear power stations to achieving CO_2 targets aimed at relieving global warming is shown not to be cost-effective in comparison with alternative strategies for sustainable development, such as renewable energy sources and energy efficiency measures (Hammond, 1996). However, nuclear energy can play a useful role in CO_2 abatement if countries adopt it for other reasons; for example, because of lack or insecurity of other fuel supplies.

Trends in the generation of nuclear electricity from fission reactors have been critically analysed in terms of the main geopolitical or regional groupings that make up the world in the early part of the twenty-first century. This has been set in the context of IEO2010 forecasts of the role of nuclear power in the fuel mix over the period 2010–35 (EIA, 2010). The IEO2010 Reference Case indicates that world electricity output may increase by 87 per cent on this timeframe, with the non-OECD countries (mainly the developing and emerging economies of the majority South) accounting for 61 per cent of global electricity use by 2035 (EIA, 2010). As with the earlier study by Hammond (1996), the major expansion in generating capacity is seen to take place in non-OECD Asia. Here the IEO2010 Reference Case is shown to give rise to an increase in nuclear power generation in China of 8.4 per cent pa and 9.5 per cent pa in India (EIA, 2010). In 2009 China led the way in terms of new nuclear build with some 43 per cent of the worldwide active construction projects. New nuclear installations are thought likely to be brought online in Indonesia, Pakistan and Vietnam by 2020 (EIA, 2010). These countries typically have very limited reserves of indigenous fossil fuel supplies (except China). In addition, they are generally attracted by a high-technology energy option, and arguably have rather lower concern for environmental protection than do European and North American OECD nations (Hammond, 1996).

In the 1980s and early 1990s there was widespread concern about handling radioactive materials (Hammond, 1996), principally radioactive emissions during the nuclear fuel cycle, the subsequent disposal of high and intermediate-level waste, and the decommissioning of reactors. These would require disposal and storage facilities which would operate safely over many decades. The risk assessment of the UK electricity sector by Hammond and Waldron (2008) found that still to be the case amongst stakeholder groups that were designated as the 'pro-renewable' (or 'green') and 'engineers and miscellaneous individuals' groups respectively.

Another factor that has discouraged nuclear power development is that the prospect of cheap nuclear energy seemed much less certain than had been claimed by its early advocates (Hammond, 1996). The capital costs are high (and uncertain) and they are typically not offset by low running costs, particularly if waste disposal and storage and decommissioning (or 'back-end') costs are taken into account. The private sector has generally proved unwilling to meet these liabilities without government financial support in one form or another (Hammond, 1996). But nuclear power is an energy technology with near zero CO_2 emissions that is available now on a large scale. It could therefore play a useful role as part of a low carbon energy strategy.

However, energy efficiency measures currently displace between 2.5 and 20 times more carbon dioxide than would new nuclear power plant per dollar (euro or pound) invested (Keepin, 1990). Indeed the UK Sustainable Development Commission (SDC, 2006) has more recently argued that a new nuclear programme would give out the wrong signal to both consumers and businesses. It would not justify public subsidy and, if implemented on a significant scale, might hold out a risk that the taxpayer will be have 'to pick up the tab' (SDC, 2006). They also suggest that such a construction programme could imply that a major 'technological fix' might be all that is needed on the grounds of carbon mitigation, thereby weakening the case for more urgent action on energy efficiency improvements and demand reduction. Likewise, they argue that nuclear would lock-in the UK to 'a central-ised distribution system for the next 50 years' (SDC, 2006), at a time when opportunities for micro-generation and local distribution networks (Allen et al., 2008; see Figure 16.6) are perhaps becoming more attractive. It has often been argued, over the period since the oil crises of the 1970s, that nuclear power should be adopted as an insurance policy against the insecurity of the oil market (Hammond, 2000). In order to meet the special demands of the (currently oil-dominated) transport sector, the electricity generated would need to be used in support of a new generation of electric or hydrogen-fuelled vehicles.

The lives of existing nuclear plant may be extended to nearer 40 years. Nevertheless Britain, as with other nuclear-powered European countries, will progressively need to decommission its older nuclear power stations during the next decade or so (see Figure 16.5). This will leave only the Sizewell B PWR station in the UK, with nuclear power holding a considerably reduced share of electricity generation (perhaps as low as 3 per cent by 2020, from around 20 per cent in 2005; Hammond and Waldron, 2008). A new generation of nuclear power stations may need to be part of an energy strategy aimed at decarbonizing the electricity sector by around 2030–50. In Europe these plants are likely to be variants of the EPR design. Emerging (novel) nuclear reactor designs that may prove to be inherently safer and less costly (POST, 2003), perhaps having a 25 per cent lower generating cost than present systems. However, the research by the UK Sustainable Development Commission (SDC, 2006) suggests that a doubling of Britain's existing nuclear capacity would only yield an 8 per cent cut on CO_2 emissions by 2035 (and nothing before 2010).

Over the longer term, it is likely that the European governments will want to keep a watching brief on advanced nuclear reactor designs, currently being developed in France and Germany, South Africa and the USA (Elliott, 2003). Nevertheless, they will no doubt want to be reassured that such new technologies will be commercially viable in the next 10–15 years. The adoption of either short- or medium-term technologies would obviously be critically dependent on public attitudes to nuclear power (Hammond, 2000; Hammond and Waldron, 2008; PIU, 2002; RCEP, 2000).

ACKNOWLEDGEMENTS

The present contribution builds on a programme of research at the University of Bath on the technology assessment of low carbon energy systems and transition pathways that is supported by a series of UK research grants and contracts awarded by various bodies. The author is jointly leading a large consortium of university partners (with Professor Peter Pearson, now Director of the Low Carbon Research Institute in Wales) funded via the strategic partnership between E.ON UK (the electricity generator) and the Engineering and Physical Sciences Research Council (EPSRC) to study the role of electricity within the context of 'Transition Pathways to a Low Carbon Economy' (under Grant EP/F022832/1). In addition, the author is also a Co-Investigator of the EPSRC SUPERGEN 'Highly Distributed Energy Futures' (HiDEF) Consortium (under Grant EP/G031681/1). This consortia has been coordinated by Professor Graeme Burt and Professor David Infield, both now with the Institute for Energy and Environment at the University of Strathclyde. The author is therefore grateful to these external colleagues for their role in the coordination of large consortia of university and other partners. However, the views expressed in this chapter are those of the author alone, and do not necessarily reflect the views of the collaborators or the policies of the funding bodies. Finally, the author wishes to acknowledge the care with which Gill Green (University of Bath) prepared the figures.

NOTE

* This chapter was written before the Fukushima Daiichi nuclear power plant accident caused by the Japanese earthquake and subsequent tsunami in March 2011. However, it is arguably still too early to determine the full impact of this natural disaster on nuclear power plant developments worldwide. In any case, the current chapter considered previous major reactor accidents, and stressed the importance of maintaining public acceptability for the continued growth of nuclear power.

REFERENCES

Allen S.R., G.P. Hammond and M.C. McManus (2008), 'Prospects for and barriers to domestic micro-generation: a United Kingdom perspective', *Applied Energy*, **85** (6), 528–44.
BP (2010), 'BP statistical review of world energy', London: BP plc, June.
Collier, J.G. (1994), *The Role of Nuclear Power in a Changing World*, Bristol: Bristol University.
Collier, J.G. (n.d.), *Nuclear Power – Clean Energy for the 21st Century*, Barnwood, Gloucester, UK: Nuclear Electric.
Committee on Radioactive Waste Management [CoRWM] (2007), *Managing Our Radioactive Waste Safely: CoRWM's Recommendations to Government*, CoRWM Doc 700, London: CoRWM.
Department for Business Enterprise and Regulatory Reform (BERR) (2008), *Meeting the Energy Challenge: A White Paper on Nuclear Power*, CM 7296, London: TSO.
Douglas, J. (2005), 'Grid security in the 21st Century', *EPRI Journal*, Summer, 26–33.
Elliott, D. (2003), 'The future of nuclear power', in G. Boyle, B. Everett and R. Ramage (eds), *Energy Systems and Sustainability: Power for a Sustainable Future*, Oxford: Oxford University Press, pp. 435–73.
Energy Information Administration (EIA) (2010), *International Energy Outlook 2010* DOE/EIA-0484, Washington, DC: US Department of Energy.
Fremlin, J.H. (1987), *Power Production: What are the Risks?*, Oxford: Oxford University Press.
Greenpeace (1994), *No Case for a Special Case: Nuclear Power and Government Energy Policy*, London: Greenpeace.
Hammond, G.P. (1996), 'Nuclear energy into the twenty-first century', *Applied Energy*, **54** (4), 327–34.
Hammond, G.P. (2000), 'Energy, environment and sustainable development: a UK perspective', *Trans IChemE Part B: Process Safety and Environmental Protection*, **78** (4), 304–23.
Hammond, G.P. (2006), '"People, planet and prosperity": the determinants of humanity's environmental footprint', *Natural Resources Forum*, **30** (1), 27–36.
Hammond, G.P. and C.I. Jones (2008), 'Embodied energy and carbon in construction materials', *Proceedings of the Institution of Civil Engineers: Energy*, **161** (2), 87–98.
Hammond, G.P., S.S. Ondo Akwe and S. Williams (2011), 'Techno-economic appraisal of fossil-fuelled power generation systems with carbon dioxide capture and storage', *Energy*, **36** (2), 975–84.

Hammond, G.P. and A.J. Stapleton (2001), 'Exergy analysis of the United Kingdom energy system', *Proceedings of the Institution of Mechanical Engineers Part A: Journal of Power and Energy*, **215** (2), 141–62.

Hammond, G.P. and R. Waldron (2008), 'Risk assessment of UK electricity supply in a rapidly evolving energy sector', *Proceedings of the Institution of Mechanical Engineers Part A: Journal of Power and Energy*, **222** (7), 623–42.

Houghton, J.T. (2009), *Global Warming: The Complete Briefing*, 4th edn, Cambridge: Cambridge University Press.

Independent Commission on International Development Issues (ICIDI) – The Brandt Commission (1980), *North-South: A Programme for Survival*, London: Pan Books.

Inhaber, H. (1982), *Energy Risk Assessment*, New York: Gordon & Breach.

Intergovernmental Panel on Climate Change (IPCC) (2005), *IPCC Special Report on Carbon Dioxide and Storage*, (prepared by Working Group III of the IPCC), Cambridge: Cambridge University Press.

Intergovernmental Panel on Climate Change (IPCC) (2007), *Climate Change 2007 – The Physical Science Basis*, Cambridge: Cambridge University Press.

Keepin, B. (1990), 'Nuclear power and global warming', in J. Leggett (ed.), *Global Warming: the Greenpeace Report*, Oxford: Oxford University Press, pp. 295–316.

Kouvaritakis, N. (1989), *Exploring the Robustness of Energy Policy Measures Designed to Reduce Long-term Accumulations of Carbon Dioxide: An Approach*, Paris: IEA.

Lovins, A.B. (1977), *Soft Energy Paths*, Harmondsworth, UK: Penguin.

Nuclear Electric plc (1994), *Submission to the UK Government Review of Nuclear Energy, Volume 2: The Environmental and Strategic Benefits of Nuclear Power*, Barnwood, Gloucestershire, UK: Nuclear Electric plc.

Nuclear Energy Agency (NEA) and the International Atomic Energy Agency (IAEA) (2010), *Uranium 2009: Resources, Production and Demand*, Paris: OECD.

Parliamentary Office of Science and Technology (POST) (2003), 'The nuclear energy option in the UK', Postnote No. 208, London: POST.

Parliamentary Office of Science and Technology (POST) (2004), 'Terrorist attacks on nuclear facilities', Postnote No. 222, London: POST.

Performance and Innovation Unit (PIU) (2002), *The Energy Review*, London: Cabinet Office.

Ramage, J. (2003), 'Nuclear power', in G. Boyle, B. Everett and R. Ramage (eds), *Energy Systems and Sustainability: Power for a Sustainable Future*, Oxford: Oxford University Press, pp. 393–434.

Royal Commission on Environmental Pollution (RCEP) (2000), *Twenty-second Report: Energy – The Changing Climate*, London: TSO.

Royal Society and The Royal Academy of Engineering (1999), 'Nuclear energy: the future climate', Document 10/99, London: Royal Society.

Stern, N. (2005), *The Economics of Climate Change: The Stern Review*, Cambridge: Cambridge University Press.

Sustainable Development Commission (SDC) (2006), *The Role of Nuclear Power in a Low Carbon Economy*, London: SDC.

Tester, J.W., E.M. Drake, M.J. Driscoll, M.W. Golay and W.A. Peters (2005), *Sustainable Energy: Choosing Among Options*, Cambridge, MA: MIT Press.

Watson, J. (2005), 'Advanced cleaner coal technologies for power generation: can they deliver?', in *Proceedings of the British Institute of Energy Economics (BIEE) Academic Conference: European Energy – Synergies and Conflicts*, Oxford: St John's College.

World Commission on Environment and Development (WCED) (1987), *Our Common Future (The Brundtland Report)*, Oxford: Oxford University Press.

World Energy Council (WEC) (1993), *Energy for Tomorrow's World*, New York: St Martin's Press.

17 Carbon capture technology: status and future prospects

Edward John Anthony and Paul S. Fennell

17.1 INTRODUCTION

Fossil fuels are currently the primary energy source worldwide, considerably exceeding nuclear and all renewable fuels, providing about 86 per cent of the world's energy needs.[1] The nature of fossil fuel use varies widely from country to country – for example ~18 per cent of electricity generation is provided by coal in Canada, ~50 per cent in the US and ~80 per cent in China, but they are used ubiquitously worldwide. Moreover, there appears to be no prospect, in the near future, of completely replacing coal, natural gas or petroleum-derived fuels with renewables or nuclear. Indeed, this will be exaggerated by the likely growth in energy demand over the coming decades as the world's population edges up from around 6 to 9.5 billion people.

In particular, coal, which emits the most CO_2 per unit of energy but is relatively inexpensive and widely available, will find increasing use, particularly in countries like China and India; unless there are major technological breakthroughs or a general strategy agreed for pricing CO_2 emissions, provoking the adoption of more efficient technologies that will minimize fossil fuel use. Thus, the US Energy Information Administration estimates that coal consumption rose from 5890 to 6565 million tonnes from 2005 to 2009[2] despite the increasing acceptance of the contribution of CO_2 to anthropogenic climate change. To put such figures in perspective it is interesting to note that the physicist Svante Arrhenius, who was the first scientist to suggest that global warming could be caused by anthropogenic CO_2 emissions in a seminal paper published in 1896 (Arrhenius, 1896), gives an estimate of worldwide coal use in 1904 as 900 million tonnes, but noted that it was rapidly increasing (Arrhenius, 1908).

Svante Arrhenius suggested that a doubling of atmospheric CO_2 levels (which is described as carbonic acid according to the fashion of the time) would produce a temperature increase of 4°C, which is reasonably close to current estimates which suggest values in the range of 2°–4.5°C (Edwards, 2010). At the time, Arrhenius looked forward to the benefits of a warming climate produced by increasing anthropogenic CO_2 levels:

> By the influence of the increasing percentage of carbonic acid in the atmosphere, we may hope to enjoy ages with more equable and better climates, especially as regards the colder regions of the earth, ages when the earth will bring forth much more abundant crops than at present, for the benefit of rapidly propagating mankind. (Arrhenius, 1908)

Unfortunately, no similar happy expectations exist among the scientific community of today. Instead, the consensus has steadily grown since the 1960s that man-made global warming or climate change is a major problem (Edwards, 2010), and that if unchecked will cause dramatic and effectively irreversible changes to the biosphere which will result in significant damage to our environment, food producing capabilities, and the safety and well-being of populations – especially in some of the more densely populated regions of the world. Typical of this type of analysis is the Stern Report which suggests that if unchecked the overall costs and risks of abrupt and large-scale climate change will be equivalent to losing 5 per cent of global gross domestic product (GDP), with the possibility of damage rising to the levels of 20 per cent of GDP or more (Stern, 2006). Many of these impacts, such as abrupt and large-scale climate change, are more difficult to quantify. With 5–6°C warming – which is a real possibility for the next century – existing models that include the risk of abrupt and large-scale climate change estimate an average 5–10 per cent loss in global GDP, with poor countries suffering costs in excess of 10 per cent of GDP. To avoid the worst potential for climate change, in terms of irreversibilities (that is, the potential for methane release from Siberian permafrost), it seems important to limit global temperature rise to no more than 2°C (Warren, 2010). It is important to note the implications of this limit for when the global peak in CO_2 emissions can occur and how rapidly emissions can be reduced after this point, since this demonstrates why the rapid development and deployment of carbon capture and storage (CCS) is critical. Recent work conducted in the UK as part of the AVOID program (Warren, 2010) indicates that in order to hit the 2°C target with emissions peaking in 2018, it is likely to be necessary to reduce global CO_2 emissions by 2 per cent per annum thereafter. However, if emissions peak just three years later in 2021, to hit the same target they would then have to reduce at around 5 per cent per year. Thus, bearing in mind the lifetime of the power plants that are being built at the moment (see below), it is critical that we act swiftly to reduce CO_2 emissions from these and future plants, which will require the development of CCS technologies.

A complete analysis of the history of global warming and the methods whereby the scientific community has developed the current consensus on global warming are outside of the scope of this chapter, but the interested reader is referred to the book by Paul Edwards for such a discussion (Edwards, 2010).

17.2   POSSIBLE ENERGY SOLUTIONS IN A CARBON-CONSTRAINED WORLD

While some countries may be able to effectively reduce their use of fossil fuels for thermal power at least, by converting most of their electrical power generation to nuclear, or perhaps replacing coal with less damaging fuels such as natural gas, it is almost certain that such solutions cannot universally be adopted based on cost considerations, regulatory demands and time constraints to move national economies to a fossil fuel free basis. Thus for example Thomas (2010) argues that the costs of nuclear have been rising rapidly, and suggests that energy efficiency and renewables are far more cost-effective than nuclear power and, moreover, that it would be difficult to increase the proportion of energy met by nuclear power much above 10 per cent. Similarly, the use of natural gas as a replacement for coal cannot be a perfect solution, because ultimately natural gas itself is a fossil fuel, albeit the least emitting in terms of CO_2 emissions per unit of energy, and a whole-scale translation of coal-fired plants worldwide to natural gas would almost certainly cause the price of natural gas to rise substantially with increasing demand, and ultimately require the use of more expensive unconventional gas reserves such as gas hydrates and sources such as coal bed methane to be extensively exploited.

Thus, while policies such as the one recently discussed by the Canadian Government:

> Under Ottawa's proposal, power companies would have to close their coal-fired facilities at 45 years of age, or the end of their power purchase arrangements, whichever is later. Companies would be prohibited from making investments to extend the lives of those plants unless emission levels can be reduced to levels equivalent to those of natural gas combined cycle plant.[3]

could make a significant difference in terms of a given national economy, it seems unlikely that such policies can be easily adopted worldwide.

In consequence, this leaves renewables as an alternative to fossil fuels, of which the most important, in terms of current overall contribution to electrical energy, are hydroelectricity, biomass, wind, solar and a variety of other sources including geothermal and wave energy (Letcher, 2008). Unfortunately, hydroelectricity is essentially near its maximum capacity in most Organisation for Economic Co-operation and Development (OECD) countries (Letcher, 2008), and other promising technologies such as solar still have to be regarded as being in the development stage. Probably the most promising sources of renewable primary energy are currently biomass and wind energy, which both have limitations. The issue of availability of sufficient quantities is a major problem for biomass as is the issue of intermittency for wind and solar. Thus, while the authors are strongly supportive of alternate energy development in all forms, the issues of availability, cost and the constraints to restructure the energy supplies of most countries will likely mean that fossil fuel use will remain a major route for meeting energy demands worldwide in the foreseeable future, failing a major breakthrough in energy technologies or energy storage. Furthermore, the difficulties of achieving strong international consensus and action plans also remain a stumbling block to the early elimination of fossil fuels from the world's energy mix and will similarly likely slow the introduction of CCS technology in what could be a lose–lose scenario.

In addition to the purely negative factors that will delay the elimination of fossil fuels from the world's energy mix, a further factor to be considered is that fossil fuels are a resource with significant impacts for numerous national economies worldwide and, depending on how they are exploited, they have benefitted and can continue to benefit humanity. These ideas have recently been elegantly expressed by the economist Paul Collier in terms of the following simple equations (Collier, 2010):

$$Nature + Technology + Regulation = Prosperity \tag{17.1}$$

$$Nature + Regulation - Technology = Hunger \tag{17.2}$$

In this context, the question is whether or not there are solutions which will permit us to use our fossil fuel resources over the coming decades, both to generate wealth and to allow us to develop longer-term energy solutions which do not involve using fossil fuels. The most compelling such solution is carbon capture and storage, namely the idea that fossil fuels can be converted into energy, and that the resulting CO_2 can be stored in deep geological formations, thus preventing the augmentation of current CO_2 levels in the atmosphere (IEA, 2004; IPCC, 2005). In 2008, the International Energy Agency (IEA) and the Carbon Sequestration Leadership Forum (CSLF) recommended to leaders of G8:

> G8 heads of government are urged to recognize the critical role of CCS in tackling global climate change and demonstrate the political leadership necessary to act now to initiate widespread deployment of this technology. CCS can achieve substantial reductions in CO_2 in a world faced with increased demand for fossil fuels. With CCS, fossil fuels will become part of the solution, not part of the problem.[4]

It is this concept and possible methods of achieving adoption of CCS technology which will be discussed in the rest of this chapter. Before discussing specific technologies, one other important point ought to be made: it is generally recognized that something like 70 per cent of the costs of CCS technology are associated with producing a 'pure stream' of CO_2 (Rao and Rubin, 2002). Thus, the focus of this chapter will be on the combustion and conversion of the fossil fuels themselves, rather than the purity requirements of the CO_2 stream either prior to compression, or in transportation to the geological media (Pipitone and Bolland, 2009), or for the implications of the effect of impurities on storage (IEA, 2009a), although all of these are legitimate and important issues in themselves.

Most of the technologies being considered for CCS applications can be divided into three categories: pre-combustion, in which the carbon is removed from the fuel before its final conversion, of which the most important of those technologies is gasification; post-combustion, in which the CO_2 is removed from the flue gas after energy production – this class of technologies is the most likely route for use with the majority of existing power stations; and oxy-fuel combustion, in which the air is first separated (usually cryogenically) to provide nearly pure oxygen in which the fuel is then burned, giving a pure stream of CO_2 at the back-end. An exception to this is a technology known as chemical looping combustion whereby usually a metal or metal oxide is oxidized to a higher oxidation state, and then reduced by a fuel gas to produce heat and either synthesis gas, or CO_2 and H_2O from combustion applications, thus producing a pure stream of CO_2 suitable for sequestration upon condensation of the water. As a final caveat it should be noted that implicit in considering CCS technology is the idea that this technology will be restricted to large-scale applications, since the demands of processing, transporting and storing CO_2 demand economies of scale. Thus the subsequent discussion will be strongly focused on large-scale power plant applications and it will be generally assumed that CCS cannot be practised with small plants, based on reasonable economic cost expectations.

17.3   THE CURRENT ENERGY PICTURE

In order to apply CCS economically, the authors believe that it must be applied to plants that are sufficiently large that significant economies of scale are available, both within the plant operation and for the pipelines leading from the plant. In 2005, worldwide, there were more than 8000 large stationary CO_2 emission sources (defined as emitting greater than 100000 tonnes of CO_2 per year); their cumulative emissions were reported (IPCC,

2005) as being 13 466 megatonnes of CO_2 per year. It is these sources of CO_2 which we believe should be tackled as a priority.

The first important thing to realize about the current thermal power plants available throughout OECD countries is that most of these units are relatively old. According to the IEA, more than half of the operating coal-fired power plants have been in service for more than 25 years and 80 per cent or more are subcritical plants (that is, plants operating with lower steam temperatures, thus achieving efficiencies of electrical energy production in the low 30 per cent range instead of the low 40 per cent range, as would be the case for supercritical power plants). Old power plants are also very much a feature of Canadian and US power installations, with ages typically in the 30–40 year, or more, range. Plants built in the 1960s and 1970s were built with the expectation that they would have a life of around 25 years or so (Ambrosini, 2005). As a rule of thumb, one might now propose that such plant would be replaced by the time they reached 40–50 years of age and it is interesting to note that in a recent analysis of coal-fired boilers operating in the US, boilers built after 1971 were classified as 'new' boilers (Harding et al., 2010).

Unfortunately, the first strategy to deal with older power plants is likely to be life extension (Harding et al., 2010), especially given the current economic difficulties. Therefore, in the absence of strong regulation, the likely method of dealing with such a plant is not to replace it with a higher efficiency supercritical plant, but to try and maintain it so that it can last for a longer period of time. However, this is not a perfect economic solution, despite increasing and substantial experience in the utility sector on the methods and techniques of life extension. This is illustrated by the fact that the costs to the US power industry of power plant failure are now in the order of several billion dollars annually, and this is indicative of the fact that in the next 20 years or so a considerable number of utility boilers will have to be replaced in OECD countries.

The replacement of most boilers in the next 20 or 30 years will have important ramifications. The first is that if this happens in the near-term future the clear preference will be to replace these boilers with well-established technology; in that case the most likely candidate for larger plant will be for PF boilers, presumably of supercritical design, offering overall electrical cycle efficiencies of over 40 per cent. Secondly, if regulation and economic incentives move owners to adopt CCS technology, the most likely solution will be a back-end or retrofit technology, which will permit the new boiler to continue to operate. Since every new major installation can be expected to have an associated cost of the order of several billion dollars, it is difficult to believe that there would be a widespread agreement to abandon such plants. It is therefore necessary that new-build power plant is carbon capture ready (CCR), and while this concept has received a certain amount of criticism in that it can easily be represented as the equivalent of doing nothing, providing the plant is properly designed such that back-end CO_2 control can be readily implemented, it represents a logical solution (Irons et al., 2007), especially in the absence of clear guidelines on CO_2 or clear national and international policies on CO_2 pricing. Importantly, the guidelines for carbon capture readiness are frequently heavily slanted towards the assumption that plant will operate using 'standard' amine solutions such as 30 wt per cent monoethanolamine (MEA). However, 'new' solvents will be developed and to be truly carbon capture ready the plant should have the potential to be efficiently operated with these new solvents when available (Lucquiaud et al., 2009). It is important to note that only changes to the plant which have very good long-term potential pay-offs are worth making to make a plant carbon capture ready (Irons et al., 2007). The modifications required for carbon capture readiness for solvent scrubbing are discussed later.

The next most likely set of solutions represent the use of gasification technology, in the form of integrated gasification combined cycle (IGCC) and oxy-fuel technology, most likely pulverized fuel (PF) but also possibly in the fluidized bed combustion (FBC) variety. What this means is that technologies which are not presently being explored at the demonstration size (say 30 MW_{th} and preferably larger) may well have a very difficult time in achieving significant market penetration during the next wave of boiler replacement. This also means that some very promising technologies such as chemical looping combustion (CLC) may not achieve significant penetration in the utility sector and, in the authors' view, more exotic technology involving, for example, high-pressure oxy-firing or total boiler redesign will not likely be able to take a significant market share in the foreseeable future. It is with this thought in mind that the discussion of potential technologies will be restricted to PF with back-end capture, IGCC and atmospheric pressure oxy-fuel firing.

17.4   SUPERCRITICAL PF BOILERS WITH BACK-END CONTROL

Pulverized fuel-fired boilers are the most well-established coal burning technology, with their origins in the 1950s (Gunn and Horton, 1989), and the vast majority of large utility boilers employ PF technology. At its heart, a PF boiler depends on the production of a high-temperature flame burning small coal particles (<50 μm) at temperatures up to about 1600°C. Steam is raised in a Rankine cycle, and this in turn can be used to produce electrical power. Combustion efficiencies are extremely high and the major pollutants associated with this technology are SO_2 and NO_x (components of acid rain). Pollution control to reduce and remove acid gases is now well established and consists of either selective catalytic reduction (SCR) or selective non-catalytic reduction (SNCR) for NO_x and wet flue gas desulphurization (FGD). While the technology now faces some new challenges, such as Hg (mercury) control, it is robust and universally well established, and in its supercritical mode is probably the most cost-effective coal conversion technology for all but very high-ash or high-sulphur fuels, for which fluidized bed combustion (FBC) or possibly gasification are likely to be cost-competitive.

Before discussing individual technologies for post-combustion CCS, it is important to make three clear points. Firstly, when discussing any CO_2 capture technology likely to be deployed in the near term (within 20 years), there will be some form of penalty in terms of energy efficiency. This means that the power station must burn more fuel (and generate more CO_2) than it would do otherwise, meaning that there is a difference between the CO_2 captured (that actually stored underground – around 90 per cent of the total CO_2 generated by the process) and that avoided, that is, the CO_2 from the existing plant which is not emitted to the atmosphere owing to the application of the CO_2 capture technology. It is invariably the case that more CO_2 is stored than avoided, and this is an important distinction when CO_2 credits are allocated. Generally, costs are discussed on a CO_2 avoided basis, which will be employed below.

Secondly, in order to transport the pure CO_2 produced as either a liquid or a super-

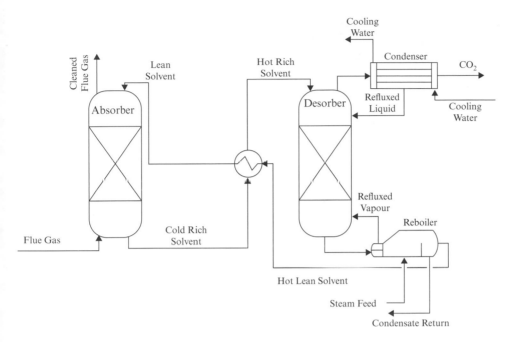

Source: Macdowell et al. (2010).

Figure 17.1 Schematic of CO₂ capture using a basic amine-based chemical adsorption process

critical fluid, and then to inject it into geological strata, it is necessary that the CO$_2$ be pumped up to around 100 bar. This compression process is a significant and often over-looked aspect of CO$_2$ capture, accounting for around 3 per cent (Macdowell et al., 2010) of the total energy penalty imposed upon the plant. This means that: (1) a capture process producing a pure stream of CO$_2$ at 1 bar and imposing an energy penalty of 6 percentage points on the process to generate it will impose an overall penalty of 9 percentage points after compression; and (2) it is important to consider the CO$_2$ compression penalty when optimizing a CO$_2$ capture system – there is frequently a trade-off between the energy required to produce a pure stream of CO$_2$ and that required to produce a pure stream of CO$_2$ at 100 bar.

Finally, the scale of the challenge involved in the development and deployment of CCS should not be underestimated. The IEA technology roadmap (IEA, 2009b) calls for 100 full-scale CCS projects by 2020 and 3400 projects by 2050, spread across a wide range of industries and not just reserved for power generation.

17.4.1 Amine or Solvent Scrubbing

Along with NO$_x$ and SO$_2$ removal, back-end control for CO$_2$ via amine scrubbing is also a possibility, albeit that such technology has yet to be deployed on any PF unit on a commercial scale. Figure 17.1 shows a configuration for amine scrubbing (Macdowell et al., 2010) which could sit at the back-end of a PF boiler. Flue gases pass through a stripping column, where CO_2 is absorbed into the solvent, which is then passed back to a second column where steam, diverted from the turbines (where it would otherwise produce electricity) is condensed, providing the energy required to regenerate the solvent. The loss of the energy in the steam from the steam cycle is the reason why the electricity produced by a CCS-enabled power station with amine scrubbing associated is lower than that for the equivalent unabated station, for the same amount of fuel burned. One major difference between conventional PF, and a unit with CO_2 capture, is that acid gases must be suppressed to very low levels; say for SO_2 down to about 10 ppm, and in this type of system a typical amine scrubber design might remove perhaps 80 per cent or more of the CO_2. (IPCC, 2005). Challenges for this technology include the large energy penalty associated with the final scale-up to full utility scale, and the cost and sensitivity of the amines to degradation from a variety of causes. Also, the work of Bailey and Feron (2005) shows that the make-up of amines is expected to be between 0.35 and 2.0 kg of solvent per tonne of CO_2 captured – potentially a very high cost. Furthermore, fugitive losses from the scrubbing system are a worry, given the toxicity of MEA to marine life, with a recent IEA report indicating that around 3.2 kg of solvent (that is, slightly higher than the estimate of Bailey) will be emitted to the atmosphere per tonne of CO_2 captured (IEA GHG, 2004). There are considerable efforts currently being made to develop new types of solvent that are more stable and less costly, but one new technology which is being developed is chilled ammonia, in which instead of an amine (RNH_2), NH_3 slurry is used. The latter technology looks particularly promising, although it has so far only been tested on a slipstream of a power station. The choice of solvent for CO_2 scrubbing is quite complicated; as discussed above, it is necessary to optimize the entire capture and compression system, rather than optimizing them in isolation (Rochelle, 2009). The greatest limitation to amine or solvent scrubbing, besides the cost and sensitivity of amines to degradation, is that the power requirements are high and losses of 9 per cent efficiencies or more (including compression) are anticipated should the technology be deployed with a PF system.

Significant work has been done on how to make plants CCS-ready for amine scrubbing. The most important findings have been that it is necessary to leave sufficient space for the capture equipment, that the location of this space is important, that provision for steam extraction from the steam cycle is necessary (the details of this extraction have important ramifications for how flexible the plant is for subsequent operation, for example if the solvent used is changed) (Irons et al., 2007). However, there are significant points of disagreement in this area, particularly with regard to the minimum space requirements for the plant (Florin and Fennell, 2010).

17.4.2   Calcium Looping

In addition to amines there are, of course, various alternative back-end methods for scrubbing CO_2 from flue gases and one of the most interesting is calcium looping, which can be considered to be a high-temperature CO_2 scrubbing technology. As a result of the CO_2 removal at much higher temperatures (around 600°C rather than near ambient), it is associated with a much smaller energy penalty than amine scrubbing. This technology affects CO_2 capture via the reaction:

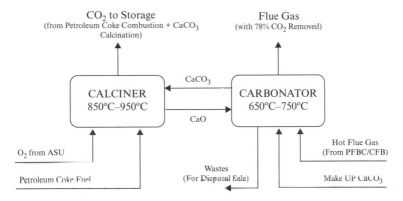

Figure 17.2 Schematic diagram of a Ca looping system for hot CO_2 scrubbing

$$CaO + CO_2 = CaCO_3 \qquad (17.3)$$

Here, CO_2 from flue gas is captured in a vessel called a carbonator, with regeneration of a pure stream in a vessel designated as a calciner, via the reverse of reaction (17.3), and Figure 17.2 (Blamey et al., 2010) shows the essential features of such a system. In addition to high-temperature CO_2 scrubbing, the spent lime produced during the calcination phase has a potential use in the cement industry, helping to decarbonise that industry as well. The biggest problem associated with this technology appears to be the rapid loss of activity or CO_2 carrying capacity of the lime-derived sorbent during multiple carbonation and calcination cycles, from levels of around 80 per cent to around 10 per cent after about 30–50 cycles, which is associated with the sintering and loss of small pores in the Ca-based sorbent. As reaction (17.3) is a reversible reaction, practical CO_2 removal is probably limited to about 90 per cent with this cycle, but as indicated above this is not essentially different for the situation with amine scrubbing (Rao and Rubin, 2002).

This system has been described in detail in a recent review (Blamey et al., 2010). Obvious advantages, besides the lower energy penalty, include the fact that limestone is, after water, arguably the cheapest industrial chemical. Furthermore, it is widely available, non-toxic and a material with which humanity has wide experience. Currently, there is a €6.86 million EU programme known as CaOling[5] under which a 1.7 MW$_{th}$ demonstration plant is being built to demonstrate this technology, prior to commercial-scale demonstrations. If this is successful, then it can reasonably be anticipated that this technology will be developed rapidly, and it is reasonable to anticipate that it might be available to meet the carbon capture requirements of a new generation of OECD fossil fuel-fired boilers.

17.5   IGCC AND GASIFICATION TECHNOLOGY

Gasification can be regarded as fuel-rich combustion, and is a process whereby a solid or liquid fuel is converted to synthetic fuel gas, normally described as syngas, by combusting it with sub-stoichiometric amounts of air or oxygen so that the resulting gas has a heating value (Higman and van der Burgt, 2008). The basis of the technology is old, and in its earlier forms it can be traced back to the 1800s when it was used to make town gas from coal. Indeed the name 'natural gas' arises from the fact that prior to the 1900s the gasification route was the principal way of making large amounts of calorific fuel gas and, in order to distinguish gas obtained from geological formations, the name was coined. Typically, about 30 per cent of the stoichiometric oxygen requirement is used, and the resulting gas can be used both for the production of chemicals, heating and power generation, which is sometimes known as tri-generation or, more usually, polygeneration. In principle at least, gasification can be used for any hydrocarbon source regardless of its original nature, including biomass and all fossil fuels. It can also be used as part of the process to convert, for example, natural gas to a syngas, containing primarily CO and H_2, suitable for Fisher-Tropsch processes; and the unlovely phrase of 'gasification of gas' may occasionally be found. There also exists the possibility of reacting carbon-rich fuels with H_2, in a process which is normally described as hydrogasifcation, but by far the most common applications are ones in which a solid or liquid fuel is reacted with either air or oxygen to produce a syngas.

For coal, gasification is normally carried out with oxygen rather than air, in part because of the lower reactivity of coals compared with (say) biomass, and at high pressures (up to 50 bar) and temperatures (up to 1600–1800°C) in an entrained flow reactor. Heat can therefore be raised not only via raising steam in a Rankine cycle, but also after cleaning by expanding the burning fuel gases through a turbine; hence the name IGCC. Currently, electrical conversion efficiencies of over 40 per cent are possible with this technology, although in the future much higher efficiencies are anticipated. In addition to its ability to produce chemicals, gasification has a large number of additional advantages. Firstly, because of the relatively small volumes of high-pressure nitrogen-free syngases, very deep cleaning of those gases is possible, allowing removal of effectively all pollutants, including any CO_2 produced in the initial gasification step. Secondly, the technology has the potential to be used with almost any hydrocarbon feedstock provided it can be prepared in such a fashion as to allow it to be introduced into the gasifier. Thirdly, the technology is well established worldwide with perhaps more than 140 large gasification plants, and there are no major developments required for its use with coal-fired systems. For an overview of the advantages and potential of gasification, the interested reader is directed to the book by Higman and van der Burgt (2008), or organizations like the Gasification Technology Council.[6]

The major concerns over gasification as a cost-competitive technology with PF technologies are the costs. Gasification is more expensive, and in addition there are concerns over reliability, normally expressed as availability. This is the time the plant actually operates divided by the time it is expected or required to operate. Currently, such availabilities are in the low 80 per cent range. However, to put this in perspective it is worth noting that supercritical PF systems had availabilities in the low 60 per cent range during their first five years of operation and have now achieved availabilities in the 85 per cent plus range (Phillips, 2007). However, as a caveat it must be noted that while the CO_2 being produced in the gasification step could be captured and sequestered relatively readily, a true near-zero-emissions gasification plant would have to have either a shift reactor and/or a solvent removal unit (amine scrubbing) to remove the CO_2 that would otherwise be produced when burning the syngas. Both of these steps mean that the technology is not necessarily more attractive than PF with amine scrubbing. Currently, gasification is making very little progress in the utility sector and it is therefore uncertain as to how much market share it is likely to take over the next 20 years.

17.6 OXY-FIRED TECHOLOGIES

17.6.1 Oxy-fuel PF

Given that the primary goal of CCS technology is to capture an effectively pure stream of CO_2 following the combustion process, direct firing with oxygen is a particularly compelling method of achieving this goal; a schematic for a typical oxy-fuel system is shown in Figure 17.3. Existing boiler designs cannot be permitted to experience the high temperatures that would result from direct firing of a hydrocarbon fuel with pure oxygen; flue gas must be recycled. As a result, the flue gas volumes in an oxy-fired PF system are only about 80 per cent of the air-fired system, and not the 21 per cent that would occur if only pure oxygen were used. Another issue is that using the current best method of cryogenic air separation (ASU), the economics of air separation means that pure oxygen is not used and, instead, oxygen purities of 90–95 per cent are employed. Improvements in this technology or the advent of reliable membrane separation technology for oxygen production could in principle dramatically improve the efficiencies of such boilers and concentrations of CO_2 produced, but for the moment all existing plans are based on the use of ASU. It should also be added that there are side benefits of oxy-fuel combustion, the most obvious being that NO_x levels will be lower since N_2 in the air is largely absent from the oxidant and so thermal NO_x formation will be significantly reduced. Doosan Babcock, conducting tests as part of the Oxycoal-UK phase 1 project, concluded that the overall emissions of NO_x were reduced by approximately 50 per cent using oxy-fuel technology (Seneviratne, 2009; Sturgeon et al., 2009). The effects of oxy-fuel firing on SO_2 are interestingly complex and will not be discussed here in further detail, beyond saying that SO_2 and SO_3 levels in the boiler may be higher, due in part to the reduction in volume of flue gases (Wall et al., 2009). Another important pollutant from coal-fired power generation is mercury, which can be detrimental both to human health and to

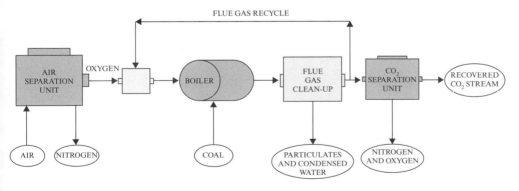

Figure 17.3   A generic oxy-fuel-fired system

Table 17.1   Major pilot plant and pre-commercial oxy-fuel PF demonstrations worldwide

Country	Project
Germany	30 MW_{th} demonstration unit (Strömberg et al., 2010)
USA	Jupiter Oxygen Corporation, 15 MW_{th} burner test facility, National Energy Technology Laboratory (NETL) (Och et al., 2009)
	Demonstration of a 30 MW_{th} oxy-fired unit (the Clean Environmental Development Facility in Alliance, Ohio), Babcock & Wilcox Power Generation Group, Inc. (B&W PGG) and Air Liquide (McDonald et al., 2008)
United Kingdom	40 MW_{th} Doosan Babcock Demonstration unit (Hesselmann et al., 2009)
France	30 MW_{th} Lacq Project, Total, in partnership with Air Liquide (Total, 2010)
Australia	Callide Oxy-fuel Project (30 MW_{th} unit) and various other projects (Cook, 2009)

CO_2 compression systems owing to its rapid depassivation of and subsequent reaction with aluminium. Owing to the previously discussed high concentration of SO_3 under oxy-firing conditions, it may be more difficult to remove mercury from the flue gas. This is because free Cl can react with either SO_3, or Hg to form $HgCl_2$, which is more easily removed from flue gases than unreacted Hg.

One potential problem with oxy-fuel technology is that for safety reasons modern boilers are in general run at slightly below atmospheric pressure. Since the aim is to produce a pure stream of CO_2, any leakage of air into the system could potentially lead to significant and undesirable contamination of the CO_2 by nitrogen. Very small leaks can lead to large drops in overall purity. Potential solutions advocated (Meyer et al., 2009) include running the plant at closer to atmospheric pressure, and potentially sealing key areas with a mantle of CO_2. It remains to be seen how well either solution will work in practice. Table 17.1 lists the major oxy-fuel PF demonstration projects.

Oxy-fired technology would have the enormous advantage of being based on well-established PF technology, and is therefore something which utilities could in principle easily adapt to. More speculative advantages of such technology exist in terms of the idea of retrofits to existing boilers (Farley, 2006), but such advantages are more difficult to quantify since the economics of retrofitting all but fairly new boilers (<20~25 years) are often much less promising on closer inspection. Another issue with this technology is that it is not necessarily more cost-effective than back-end amine scrubbing, and the results of a major Canadian study carried out in 2007 (Xu et al., 2007), performed for the Canadian Clean Power Coalition (CCPC),[7] seem fairly typical of the type of conclusions reached for oxy-fuel PF studies:

- Oxy-fuel was found to have technical and environmental benefits compared to post-combustion capture. It was the most economic option in one case studied, which was based on a greenfield site in Alberta with low-sulphur coal where an FGD was not required for oxy-fuel but was for post-combustion. The assumption was that most SO_2 emissions in this case would be captured in the CO_2 compression phase. In the two other sites studied (Saskatchewan and Nova Scotia) the amine-based post-combustion options were found to be more economic, but the difference was marginal.

● It was also found that parasitic energy losses directly related to CO_2 capture were the largest single cost item, closely followed in most cases by capital charges. These costs were similar for both oxy-fuel and amine-based capture if FGDs were present in both cases. The capital and operating cost of the ASU plant was a major problem for oxy-fuel economics, and improvements in this area will have major benefits to oxy-fuel economics.

● Operations and Maintenance (O&M) costs were found to make up a relatively minor portion of the total charges.

● Oxy-fuel was expected to capture slightly more CO_2 than post-combustion, which tended to help its cost per tonne captured figures.

● Oxy-fuel-based retrofits were found to be significantly more expensive compared to back-end retrofits that left the existing boiler plant intact.

The authors have not been able to find any well-documented availability figures for a proposed oxy-fuel PF plant, and the only figure they have heard discussed, without substantial documentation, was 70 per cent, which would put expectation for the first demonstration plants in line with those for the first supercritical PC units, and the first utility-scale IGCC units. Nonetheless, it seems likely that boiler manufacturers would have considerable confidence in building large-scale oxy-fuel PF units, while the same could probably not be said for utility-scale amine scrubbing or IGCC with shift reactors and/or solvent cleaning at the back-end.

17.6.2   Oxy-fuel Circulating Fluidized Bed Combustion

Fluidized bed combustion is a well-established thermal power technology, which started to achieve commercial importance in the late 1970s. In a fluidized bed a fuel is burned in a bed of particulate solids, suspended in a stream of gas (normally air) at velocities of up to 2.5 m/s, such that they behave like a fluid. It is now an established technology for burning biomass in its low-velocity or bubbling FBC variety, primarily for smaller applications. For larger boilers the preferred version of this technology is the circulating fluidized bed variety (CFBC). The name 'circulating' FBC arises because higher gas velocities of up to 8 m/s are employed, and the hot bed solids are now entrained and must be recycled in a primary reaction loop. This consists of a riser, where the bulk of the combustion reactions occur, and a downcomer, which returns the hot solids to the riser (Grace et al., 2007). There are already several hundred large utility-scale CFBC boilers operating throughout the world, and other large-scale developments in countries like China and Poland (Stamatelopoulos and Darling, 2008; Hotta, 2008).

Typically, with high-quality coals, PF has been preferred by utilities, and CFBC has been reserved for difficult fuels, for example those with high ash and/or high sulphur; and also situations where significant blending of fuels occurs, since properly designed fluidized beds have considerable fuel flexibility. The reason for the preference is owed in part to a number of additional limitations for CFBC technology: first, until recently, CFBC was only available at smaller sizes (300 MW_e or less); and second, the technology was only available in a subcritical mode. This situation has significantly changed with the advent of Foster

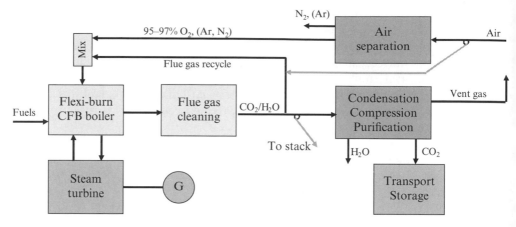

Source: Used with permission of the Foster Wheeler North American Corporation.

Figure 17.4   Schematic of a flexi-burn CFB power plant

Wheeler's 460 MW$_e$ Lagisza CFBC boiler, now successfully operating in Poland, which has a net electrical efficiency of 44 per cent. In addition, the company is now developing its Flexi-burn concept, which is a CFBC boiler which could operate in both air-fired and oxy-fired modes (Hack et al., 2008). The design of such a concept is shown in Figure 17.4.

Oxy-fuel fluidized beds would have considerable advantages in terms of fuel flexibility, and the option of co-firing biomass might even allow a net reduction in anthropogenic CO_2 production, if biomass is regarded as CO_2 neutral. Moreover, this technology could also allow the use of more marginal fuels if, as has been suggested elsewhere, premium fossil fuels become in short supply in the next 30 years or so (Mohr and Evans, 2009). Another advantage is, due to the fact that because fluidized beds circulate hot solids, it would be possible to remove heat from the primary combustion loop via fluidized bed heat exchangers, thus allowing lower flue gas recycle ratios; as a result, there exists the possibility that an oxy-fuel CFBC system might be built 30–40 per cent smaller for any given power output, helping such units achieve larger sizes appropriate to utility applications without substantial redesign.

Recently, one of the authors' groups has recently finished 650 hours of successful oxy-fuel CFBC trials using a 0.8 MW$_{th}$ pilot plant (Kuivalainen et al., 2010) under contract to Foster Wheeler. The general conclusions of these trials appear to be that the technology maintains all of its advantages in terms of emissions when compared with air-fired CFBC systems, and that there are no evident showstoppers for the technology. Foster Wheeler AG has now signed a grant agreement with the European Commission partnering it with ENDESA and the Spanish Foundation, Fundacion Ciudad de la Energia (CIUDEN).[8] The grant agreement stipulates the terms of €180 million ($222.4 million) of EU funding under the EC's European Energy Programme for Recovery (EERP) to support a carbon capture and storage technology development of a 300 MW$_e$ oxy-combustion power plant in Spain.[9] The first steps of the CIUDEN project are the establishment of a 30 MW$_{th}$ oxy-fired CFBC and a similar scale PF unit; if successful it seems likely that there will be no major technical barrier to commercializing oxy-fuel CFBC.

17.7   GLOBAL DEMONSTRATION

A recent briefing paper (Florin and Fennell, 2010) summarizes the progress towards deployment of CCS technologies worldwide. The EU has recently recognized CCS within its nascent emissions trading scheme, and will allow the revenue from the auctioning of 300 million credits to be used to support CCS. In addition, €1.05 billion has been allocated to CCS demonstration (see Table 17.2) within Europe from the European Energy Programme for Recovery. The USA has committed the most money of any country to CCS demonstration, announcing US$3.4 billion in the American Recovery and Reinvestment Act for clean coal and CCS technology development. In Australia, there is AU$2 billion available for flagship projects demonstrating CCS, with additional funding of AU$100 million for a global CCS institute. There is CA$3.3 billion available (CA$1.3 billion federal, CA$2 billion from the province of Alberta) in Canada for research, development and deployment. China has also begun to investigate CCS seriously, via a number of international collaborations. The UK has one of the most advanced CCS demonstration programmes and most significant commitments to reducing CO_2 emissions in the world, with a legally binding target to reduce CO_2 emissions by 80 per cent by 2050. There are plans to fund the additional cost of fitting CCS equipment to up to four power stations. The winner of the UK CCS competition, launched in 2007 with a focus on oxy-fuel and post-combustion capture technologies, will be the first power station to fit CCS at this scale (capturing the CO_2 from the equivalent of 300 MW of electricity production) in the world for a post-combustion system.

It is encouraging that there is public funding available for CCS development and there are a large number of CCS demonstration plants being constructed at the present time, with pilot plants being constructed around the world, including at the MW scale for amine scrubbing, chilled ammonia, oxy-fuel and calcium looping. The interested reader is referred to the recent briefing note of Florin and Fennell (2010) for a detailed overview.

There are a number of areas which hold out the potential to improve CCS technologies. The main improvements which the authors believe are necessary are (in rough order

Table 17.2   Summary of EU funding allocations for CCS demonstration

Proposal	Applicant	Demonstration technology	Funding under EEPR (M euros)
Jaenschwalde, Germany	Vattenfall Europe Generation*	Oxy-combustion/ post-combustion	180
Hatfield, UK	Powerfuel Products Limited*	IGCC with physical sorbent	180
Porto Tolle, Italy	Enel Ingegneria e prod.*	Post-combustion	100
Rotterdam, Netherlands	Maasvlakte CCS Pjt CV	–	180
Belchatow, Poland	PGE Elektrownia Belchatow	Post-combustion	180
Compostilla, Spain	ENDESA Generacion S.A.	Oxy-combustion	180

Note: * Grant agreement signed as of May 2010.

Source: Florin and Fennell (2010).

of priority): to reduce the energy penalty imposed on the plant; to replace the current solvents with less toxic alternatives; to reduce the rate of deactivation of sorbents and solvents in the presence of, for example, O_2 and SO_2. In addition, it is necessary to build rapidly a large number of plants capturing CO_2 from real flue gases to prove the technology at commercial scale.

17.8   SUMMARY

This chapter has attempted to outline the various CCS technologies that might be deployed in the next few decades to meet the requirements of a carbon-constrained world. In particular, the focus has been on technologies which could reasonably be expected to be commercially available in the next ten to 20 years. These include: supercritical PF boilers with some kind of back-end CO_2 solvent scrubbing technology, most likely amine-based but also possibly based on a calcium looping cycle; gasification technology such as IGCC, possibly with shift reactors and possibly again solvent scrubbing technology; and oxy-fuel fired boilers.

The alternative to such CCS technology is, in the authors' view, not a 'leave it in the ground approach', but rather a steady increase in CO_2 levels from the 400 ppm which is now inevitable, to the more dangerous levels of 450 ppm. However, correctly applied CCS technology will buy time for a transition to systems with increased energy efficiency, and the large-scale use of renewable and nuclear power. This view, which the Canadian environmental economist Marc Jaccard has described as 'sustainable fossil fuels' (Jaccard, 2005), is arguably the most promising solution to meet mankind's major energy needs without grossly exacerbating the effects of anthropogenic climate change.

NOTES

1. http://tonto.eia.doe.gov/cfapps/ipdbproject/IEDIndex3.cfm?tid=1&pid=7&aid=6.
2. http://tonto.eia.doe.gov/cfapps/ipdbproject/IEDIndex3.cfm?tid=1&pid=7&aid=6.
3. http://www.transalta.com/newsroom/news-releases/2010-06-23/transalta-responds-federal-government%E2%80%99s-recent-policy-announcement.
4. http://www.iea.org/work/2008/NEARTERMCCS/ResultWorkshop.pdf.
5. http://caoling.eu/.
6. http://www.gasification.org/.
7. http://www.canadiancleanpowercoalition.com/.
8. http://ciuden.es/index.asp?lang=en.
9. http://epoverviews.com/articles/visitor.php?keyword=Foster%20Wheeler.

REFERENCES

Ambrosini, R.M. (2005), 'Life extension of coal-fired power plants', IEA Clean Coal Centre Report.
Arrhenius, S. (1896), 'On the influence of carbonic acid in the air upon the temperature of the ground', *Philosophical Magazine and Journal of Science*, **41**, 237–76, available at www.rsc.org/images/Arrhenius1896_tcm18-173546.pdf.
Arrhenius, S. (1908), *Worlds in the Making: The Evolution of the Universe*, transl. H. Borns, London: Harper & Brothers Publishers.

Bailey, D.W. and P.H.M. Feron (2005), 'Post-combustion decarbonization processes', *Oil and Gas Science and Technology*, **60**, 461–74.

Blamey, J., E.J. Anthony, J. Wang and P. Fennell (2010), 'The calcium looping cycle for large-scale CO$_2$ capture', *Progress in Energy and Combustion Science*, **36**, 260–79.

Collier, P. (2010), *The Plundered Planet: Why We Must – And How We Can – Manage Nature for Global Prosperity*, Oxford: Oxford University Press.

Cook, P.J. (2009), 'Demonstration of carbon dioxide capture and storage in Australia, GHGT-9', *Energy Procedia*, **1**, 3859–66.

Edwards, P.N. (2010), *A Vast Machine: Computer Models, Climate Data, and the Politics of Global Warming*, Cambridge, MA: MIT Press.

Farley, M. (2006), 'Developing oxyfuel capture as a retrofit technology', *Modern Power Systems*, **26** (4), 20–22.

Florin, N. and P.S. Fennell (2010a), 'Assessment of the validity of "Approximate minimum land footprint for some types of CO$_2$ capture plant"', provided as a guide to the Environment Agency assessment of Carbon Capture Readiness in DECC's CCR Guide for Applications under Section 36 of the Energy Act 1998.

Florin, N. and P.S. Fennell (2010b), 'Carbon capture technology: future fossil fuel use and mitigating climate change', Grantham Institute Briefing Paper, Imperial College London.

Grace, J.R., A.A. Avidan and T.M. Knowlton (2007), *Circulating Fluidized Beds*, London: Blackie Academic and Professional.

Gunn, D. and R. Horton (1989), *Industrial Boilers*, London: Longman Scientific & Technical.

Hack, H., Z. Fan, A. Seltzer, T. Eriksson, O. Sippu and A. Hotta (2008), 'Development of integrated flexi-burn dual oxidant CFB power plant', 33rd International Technical Conference on Coal Utilization and Fuel Systems, Clearwater, FL.

Harding, N.S., J.C. Sanchez and D.C. O'Connor (2010), 'Ash deposition impacts in the power industry', conference on Impact of Fuel Quality on Power Production and the Environment, Lapland, Finland, August 29 to September 3.

Hesselmann, G., E.D. Cameron, D.W. Sturgeon, C. McGhie and F.D. Fitzgerald (2009), 'Oxyfuel firing and lessons learned from the demonstration of a full-sized utility scale 40 MW oxycoal combustion system', South African Carbon Capture and Storage Conference, Johannesburg.

Higman, C. and M. van der Burgt (2008), *Gasification*, 2nd edn, Burlington, MA: Gulf Professional Publishers, Elsevier.

Hotta, A., K. Nuortimo, T. Eriksson, J. Palonen and S. Kokki (2008), 'CFB technology provides solutions to combat climate change', *Proceedings of the 9th International Conference on Circulating Fluidized Beds, in Conjunction with the 4th International VGB Workshop on Operating Experience with Fluidized Bed Firing Systems*, Hamburg, Germany.

International Energy Agency (IEA) (2004), *Prospects for CO$_2$ Capture and Storage*, Paris: IEA/OECD.

IEA (2009a), 'Greenhouse gas R&D report, impact of impurities on CO$_2$ capture, transportation and storage', Report No. PH4/32, August.

IEA (2009b), 'Technology roadmap, carbon capture and storage', available at http://www.iea.org/papers/2009/CCS_Roadmap.pdf.

IEA GHG (2004), 'Improvement in power generation with post-combustion capture of CO$_2$', Report PH4/33.

Intergovernmental Panel on Climate Change (IPCC) (2005), 'IPCC special report on carbon dioxide capture and storage', B. Metz, O. Davidson, H.C. de Coninck, M. Loos and L.A. Mayer (eds), Cambridge: Cambridge University Press.

Irons, R., G. Sekkapan, R. Panesar, J. Gibbins and M. Lucquiaud (2007), 'CO$_2$ capture ready plants', IEA Technical Study Report No. 2007/4, May.

Jaccard, M. (2005), *Sustainable Fossil Fuels: The Unusual Suspect in the Quest for Clean and Enduring Energy*, Cambridge: Cambridge University Press.

Kuivalainen, R., T. Eriksson, A. Hotta, A.S.-B. Sacristán, J.M. Jubitero, J.C. Ballesteros, M. Lupion, V. Cortes, B. Anthony, L. Jia, D. McCalden, Y. Tan, I. He, Y. Wu and R. Symonds (2010), 'Development and demonstration of oxy-fuel CFBC technology', 35th International Technical Conference on Clean Coal and Fuel Systems, Clearwater, FL.

Letcher, T. (ed.) (2008), *Future Energy: Improved, Sustainable and Clean Options for Our Planet*, Oxford: Elsevier.

Lucquiaud, M., H. Chalmers and J. Gibbins (2009), 'Steam turbines for operating and future-proof upgrading flexibility', workshop on operating flexibility of power plants, Imperial College London, November.

McDonald, D.K., T.J. Flyn and D.J. DeVault (2008), '30 MWt Clean Environmental Development Oxy-Coal Combustion Test Program', 33rd International Conference on Coal Utilization and Fuel Systems, Clearwater, FL.

Macdowell, N., N. Florin, A. Buchard, J. Hallet, A. Galindo, G. Jackson, C. Adjiman, C.K. Williams, N. Shah and P.S. Fennell (2010), 'An overview of CO$_2$ capture technologies', *Energy and Environmental Science*, **3**, 1645–69.

Meyer, H., M. Radunsky, M. Glausch, T. Witter and D. Hultqvist (2009), 'Vattenfall Oxyfuel Power Plant development – engineering of a coal fired power plant with oxyfuel technology', 1st IEA International Conference on Oxyfuel Combustion.

Mohr, S.H. and G.M. Evans (2009), 'Forecasting coal production until 2100', *Fuel*, **88**, 2059–67.

Och, T., D. Oryshchyn, R. Woodside, C. Summers, B. Patrick, D. Gross, M. Schoenfield, T. Weber and D. O'Brien (2009), 'Results of initial operation of the Jupiter Oxygen Corporation Oxyfuel 15 MWth Burner Test Facility, GHGT-9', *Energy Procedia*, **1**, 511–18.

Pipitone, G. and O. Bolland (2009), 'Power generation with CO_2 capture: technology for CO_2 purification', *International Journal of Greenhouse Gas Control*, **3**, 528–34.

Phillips, J. (2007), 'IGCC 101', Gasification Workshop, Indianapolis, IN, 12 June.

Rao, A. and E.S. Rubin (2002), 'A technical, economic and environmental assessment of amine based CO_2 capture for power plant greenhouse gas control', *Environmental Science and Technology*, **36**, 4467–75.

Rochelle, G.T. (2009), 'Amine selection to reduce energy use for CO_2 capture by aqueous scrubbing', 42nd IUPAC Congress, Glasgow.

Seneviratne, R. (2009), **'**Doosan Babcock oxyfuel R&D activities**'**, 20th Annual Meeting and Meetings of the Combustion and Advanced Power Generation Divisions.

Stamatelopoulos, G.-N. and S. Darling (2008), 'Alstom's CFB technology', *Proceedings of the 9th International Conference on Circulating Fluidized Beds, in Conjunction with the 4th International VGB Workshop on Operating Experience with Fluidized Bed Firing Systems*, Hamburg, Germany.

Stern, N. (2006), *The Economics of Climate Change – The Stern Review*, Cambridge: Cambridge University Press.

Strömberg, L., G. Lindgren, M. Anheden, N. Jentsch, O. Biede and K. Damen (2010), 'Vattenfall's R&D program on CO_2 capture technology in support of scale-up and commercialisation of oxyfuel, postcombustion and precombustion technology', 35th International Technical Conference on Clean Coal and Fuel Systems, Clearwater, FL, 6–10 June.

Sturgeon, D.W., E.D. Cameron and F.D. Fitzgerald (2009), 'Demonstration of an oxyfuel combustion system', *Energy Procedia*, **1**, 471–78.

Thomas, S. (2010), 'The economics of nuclear power: an update', report to the Heinrich Böll Foundation, March, http://www.boell.de/downloads/ecology/Thomas_economics.pdf.

Total (2010), 'Total Lacq Fact Sheet: Carbon dioxide capture and storage project', http://sequestration.mit.edu/tools/projects/total_lacq.html.

Warren, Rachel (2010), 'UEA. The AVOID programme', Joint UK/USA meeting of the AVOID network, Washington, DC, September.

Wall, T., R. Stanger and J. Maier (2009), 'Sulfur impacts in coal fired oxy-fuel combustion with CCS (impacts and control options)', 1st IEA International Conference on Oxyfuel Combustion, Cottbus, Germany.

Xu, B., R.A. Stobbs, V. White, R.A. Wall, J. Gibbins, M. Iijima and A. MacKenzie (2007), 'Future CO_2 capture options for the Canadian market', Report No. Coal R309 BERR/Pub URN 07/12251, March.

18 Environmental, economic and policy aspects of biofuels

Peter B.R. Hazell and Martin Evans

18.1 INTRODUCTION

Total global energy consumption for transportation has grown rapidly in recent decades and is expected to increase by about two-thirds by 2030 (IEA, 2007), with growth being particularly fast in China and India. Nearly all this demand is currently met by oil. Rapid growth in oil demand, finite oil supplies and political instability in many of the major oil-exporting countries are pushing up oil prices and making them more volatile. This trend seems set to continue. As a result, many importing countries are looking to expand and diversify their energy sources and are looking at biofuels as a potentially attractive prospect within their broader energy portfolios.

Biofuels have a number of attractions. They are a sustainable energy source that may help counter rising energy prices, address environmental concerns about greenhouse gas emissions, and offer new income and employment to farmers and rural communities around the world. For many rich countries, the benefits to farmers are also perceived as a good way to reduce the costs and market distortions of their existing farm support policies. Moreover, whereas oil and coal are unevenly distributed among countries, many countries could generate some biofuels from domestically grown biomass of one type or another, thereby helping to reduce their dependence on imported fossil fuels. Some countries with tropical climates may have a comparative advantage in growing energy-rich biomass and could become major exporters.

Biofuels' potential will also increase as second-generation technologies come on line, enabling more efficient conversion of cellulose-rich biomass to transport fuels and electricity. Technology advances will not only help make biofuels more competitive with fossil fuels on price, but will also expand the range of feedstock that can be used, some of which (like fast-growing grasses and trees) can thrive in less fertile and more drought-prone regions and are less competitive with food and livestock feed than current feedstock like sugarcane, maize and oilseeds.

Adding to the interest in biofuels is growing concern about global climate change and the need to reduce greenhouse gas emissions. As the discussions at the United Nations Climate Change Conference 2009 (Conference of the Parties 15 – COP15) demonstrated, many countries seem willing to take steps to cut their emissions, even if this has associated economic costs. Biofuels are attractive because they are a renewable energy source that has the potential to reduce significantly or at least slow growth in carbon emissions without involving much change in the way energy is used (for example, in internal combustion engines and combustion-fueled electric power plants).

Finally, despite the recent food price crisis, farmers in many rich food-surplus countries continue to face low export prices, and diverting some agricultural resources to the production of bioenergy offers an attractive way of helping such farmers. For example, the diversion of part of the maize crop to ethanol production in the United States helps to maintain the maize price, reducing the need for price compensation and export subsidies.

All this seems very promising, but just how realistic are these hopes and expectations? And what are their implications for the poor and the environment? Bioenergy uses resources (land, water and labor) that compete with food and feed production. This could lead to higher food prices in many poorer countries, but also around the globe if major food-exporting countries like the United States, the European Union or Brazil were significantly to divert additional agricultural resources to bioenergy production. Higher food prices would hurt poor people, who are net buyers of food, while benefiting farmers who produce net surpluses. In those countries that grow more biomass, the rural poor might also gain from greater employment and income in the bioenergy sector. For example, small farmers might grow feedstock for bioenergy, and rural workers might be employed in its transportation and processing, especially if the processing can be conducted at small scales and in rural areas. But how would all these pros and cons balance out, and what would be the net impact on the poor?

While international trade could in principle create opportunities for some countries to develop new exports and for importing countries to diversify their energy supplies, trade in biofuels still faces important barriers. Unless changed, these barriers will retard development of the bioenergy sector in countries with a comparative advantage (often developing countries with tropical climates) and encourage the development of protected and more costly bioenergy production in many rich countries with temperate climates. Removing these barriers now, during the early stages of bioenergy development, should be much easier than trying to remove them once powerful national interests have become entrenched.

Although biofuels are in principle a carbon-neutral source of energy that could help reduce carbon emissions, they also require some fossil fuels for their production and distribution. Depending on the type of feedstock, and on where and how it is grown, processed and used, the net carbon balance can vary widely. Net carbon and fossil fuel savings are not at all assured. Some current first-generation feedstock and technologies have carbon balances not much better than oil, although some (like ethanol from sugarcane and biodiesel from oil palm) can be much better, but not if they are associated with tropical deforestation or the draining of wet peatland. Second-generation feedstocks and technologies promise to bring large improvements. For example, many fast-growing trees and grasses are perennials and require little cultivation once established, while sequestering much more carbon than alternative land uses. Part of this carbon will be retained in the soil on a long-term basis. Beyond issues related to carbon balances, bioenergy crops and plantations present their own local environmental challenges for soil, water and biodiversity management.

In sum, despite the exciting prospects for biofuels, many important questions remain unresolved about their implications for the poor, food security, the environment and international trade. Moreover, because most of the environmental and social benefits and costs of biofuels are not priced in the market, leaving biofuels development entirely to the private sector and the market will lead to biofuel production and processes that fail to achieve the best environmental and social outcomes. To ensure better outcomes, the public sector has important roles to play. But what are these roles, and what policies, technologies and investments are needed to ensure that biofuels are developed in ways that are economically efficient as well as compatible with reducing poverty and global warming?

This chapter attempts to answer these questions. We first briefly describe the state of biofuels today and then structure the discussion under five heads: the economics of biofuels; environmental costs and benefits; potential trade-offs against food supply; implications for developing countries; and appropriate policies for growing the industry. This is followed by our conclusions.

18.2   BIOFUELS TODAY

In total, biofuels account for almost 2 per cent of total transport fuels. About 90 per cent of biofuels are bioethanol and 10 per cent are biodiesel. The dominant producers are the US, European Union (EU) and Brazil. As a pioneer in biofuels production, Brazil obtains over 40 per cent of its transport fuel from biofuels, whereas the US and EU obtain 4 7 per cent. Table 18.1 summarizes the main users of biofuels and the types of feedstock that they use.

The principal biofuel feedstocks in commercial production today are maize and sugarcane for bioethanol; and palm oil, rapeseed oil, soya oil and animal fats for biodiesel. Other crops used are cassava, wheat, rye, sorghum, barley and sugar beet for bioethanol; and coconut, sunflower, castor bean and other oilseed trees for biodiesel. Among the last, the most important is *Jatropha curcas*, which has been heavily promoted in recent years because of its relatively high content of non-edible oil (indeed, it is toxic) and, in particular, its apparent ability to grow (but not necessarily grow well) in dry areas on marginal soils under a low-input regime. Large areas of jatropha are under development in a number of African countries and India.

18.3   CAN BIOFUELS BE ECONOMICALLY VIABLE?

18.3.1   Viability Criteria

Since biofuels are intended to substitute, partly or wholly, for gasoline and diesel, then any assessment of biofuel viability must be based on a comparison of the prices of these fossil fuels with the supply prices of biofuels (most of which incorporate fossil fuel costs in their own value chains). Early analyses of the potential competitiveness of

Table 18.1   Main types and users of biofuels for transport today

Type	Use	Replaces	Raw material	Main users
Bioethanol	Transport	Petrol	Sugar, Maize	Brazil and US
Biodiesel	Transport	Diesel	Oilseeds	EU, especially Germany and France

different biofuels, for example by the Food and Agriculture Organization (FAO, 2008), the International Energy Agency (IEA, 2005, 2006), Johnston and Holloway (2007) and Schmidhuber (2006), often based on experimental, pilot project or small sample cost data for some biofuels, have generally subsequently held true as more operational experience has accumulated. In these assessments the use of the 'world oil price' or international gasoline or diesel prices as the comparator abstracts from the pricing complexities of fossil fuels in most national markets, where domestic prices include (often very substantial) taxes and subsidies on domestic ex-refinery costs or the landed costs of imported petroleum. Any government wishing to create a level playing field for biofuels in fuel markets will have to apply the same measures (on an energy equivalence basis) to biofuels as to fossil fuels.

The only truly integrated biofuel market is in Brazil, which in 1974–75 pioneered the large-scale use of sugarcane bioethanol in the automobile energy market through a combination of subsidies, directives and the promotion of biofuel-using auto engines. Since the 1990s Brazil has emerged as the world's largest exporter of bioethanol, and domestic automotive fuel markets are effectively liberalized. Ethanol production in Brazil is now more or less unsubsidized (OECD, 2007; FAO, 2008) and its cost is the lowest in the world (as indeed is the cost of Brazil's sugar, much of which is co-produced with ethanol). Sugarcane land in Brazil, most of which is rainfed, is reasonably productive by best international standards in terms of yields of sugar or ethanol per hectare, but the main reasons for the sugar and ethanol industry's low production cost are that it operates on a very large scale (like much of Brazilian agribusiness), and is serviced by relatively sophisticated input and output markets.

Worldwide, biofuels have generally not been competitive with fossil fuels in domestic markets without government intervention to support biofuels, except in the case of bioethanol in Brazil in recent years, as already noted. However, this could change if there were another substantial and sustained increase in crude oil prices. Sugarcane bioethanol in Brazil has been estimated to be competitive at oil prices of $0.3–0.4/litre ($48–64/barrel) gasoline equivalent (ge), but in other countries the threshold may be more than $0.4–0.5/l ($64–79/brl) ge. The cheapest source of biodiesel is animal fat resulting in a biodiesel cost in the range $0.4–0.5/lde, while vegetable oil biodiesel costs about $0.6–0.8/lde (IEA, 2007). Malaysia, Indonesia and the Philippines are among the lowest-cost producers of biodiesel from vegetable oil. Among biodiesel feedstocks, palm oil yields by far the most biodiesel per hectare and is also among the cheapest to produce: compare (at the plantation or farm level) Malaysian palm oil at around US$380/ton with Argentinean and Brazilian soya oil at US$400/ton and Western European rapeseed oil at US$1000–1200/ton (Agribenchmark, 2010). Malaysia, Argentina and Brazil all produce vegetable oil on a large scale, as does Indonesia, which is also a low-cost producer.[1]

18.3.2   Economics of Biofuel Production

Feedstock costs form a high proportion of total biofuel production costs (60–80 per cent) and high-feedstock-yielding vegetable oil and sugar industries (beet as well as cane) tend to have lower costs of production than do low-feedstock-yielding industries. Where this is not the case, it is usually due to exceptional economies of scale at the feedstock processing stage or exceptionally low distribution costs for the final product. Yield also has a considerable influence on the cost of feedstock harvesting and transport, since the higher the yield, the lower the labor or machine input required to harvest per ton and the shorter the haulage distance to the factory[2] (important for some feedstocks such as sugarcane or beet and palm oil). Thus, a key parameter in assessing the feasibility of a new investment in tree crop oil or sugar production is the expected farm yield of recoverable feedstock. The other critical determinant of cost is the scale of operation. Economies of scale are very important in biofuel production, particularly at the feedstock processing and biofuel manufacturing stage; less so for feedstock production itself. The value of by-products, particularly various forms of animal feed, are also important in reducing the net cost of biofuel production from grains and oilseeds, as glycerol can be in the case of the latter.

18.3.3 Requirements for a Viable Biofuel Industry

To have the best chance of being commercially viable at oil prices of US$60–70/barrel (which some oil industry experts consider to be a reasonable long-term trend price band going forward), a biofuel industry will generally need the following:

1. Access to sufficient suitable land within a reasonably compact area (with water availability), and to a large enough market that will allow production enterprises, particularly their processing and manufacturing components, to operate on a large scale.
2. A low-cost (which invariably means means high-yielding) feedstock.
3. The option of vertical integration, particularly in situations where the feedstock supply segment of the value chain is relatively unorganized and operates on a small scale.

The third is not necessarily essential but, in the absence of strong organization of small-scale feedstock producers, is likely to become increasingly desirable as a business model to help ensure raw material supply security and lower total cost in the value chain.

There is an exception to the minimum market requirement. The existence in many developing countries of remote rural communities that are costly to reach by road, are expensive to link to national power grids, and for which micro-hydro power is not feasible, may offer another route to viability for biofuel production. Vegetable oil grown by local farmers and converted to biodiesel in small-scale plants can be competitive with 'imported' fossil fuels in some such circumstances. Indeed, for many rural energy applications, particularly the use of static engines such as generators, mill engines and pump sets, straight vegetable oil (SVO) will suffice. Bioethanol is not an option here because it is technically more difficult to manufacture than biodiesel and has more severe diseconomies of scale. A similar situation can exist with 'in-house' SVO or biodiesel production, where a transportation company or municipality provides a sufficiently large internal market to justify dedicated biofuel production. Again, this is decentralizing away from the national fuel market.

18.3.4 Some Country Examples

Countries' differing economic circumstances mean different prospects for the potential viability of domestic biofuel production, but the three almost necessary conditions (with the local market proviso for vegetable oil-based biofuel) referred to above will apply to most. India, Mozambique and Senegal, all of which have biofuel development programs in various stages of advancement, between them illustrate the main opportunities and constraints for developing counties, while Germany and the USA can be used to typify rich-country situations.

India

India is pinning its hopes on tree-borne oilseeds, particularly jatropha (Raju et al., 2009; Gopinathan and Sudhakaran, 2009; USDA, 2009). India already produces bioethanol from sugarcane molasses, but fundamental policy reforms of the sugar sector will be required before this becomes a big transport fuel contributor. The government's biodiesel strategy focuses on the 13 Mha of waste or marginal land estimated to be available for jatropha planting without encroaching on foodcrop land, the use of which for biofuels is strongly discouraged. With government support, many states are planting jatropha under a variety of institutional arrangements between the public, private and civil society sectors. India certainly has a large enough market, although land tenure arrangements will preclude large-scale jatropha farms. However, this is less critical than large-scale biodiesel plants for supplying the national market. Condition (1) is therefore effectively met.

The problem is meeting condition (2), due to the feedstock used (see below). It is becoming apparent that the yields originally expected of jatropha grown on waste or marginal land were unrealistically high. Like most crops, jatropha yields best under a reasonably intensive input regime. Low yields will mean high-cost harvesting and transport costs, and as a consequence of this emerging truth the planting programme has faltered and is now far behind what is needed to reach the blending targets set out in the government's strategy. It is probably not a coincidence that it is difficult to find an agribusiness of any commercial significance anywhere in the world that is based on low-yielding feedstock.

Mozambique

Mozambique has abundant land and water resources, much of it available for large-scale development, and so it partly fulfils condition (1). The potential productivity of oilseed and sugarcane-based biofuel production is high and the cost of producing these feedstocks can become internationally competitive in time (Government of Mozambique, 2008, 2009). With appropriate investment, nearly all of which is expected to come from the private sector, condition (2) can therefore be met (even probably for jatropha – see below). Mozambique's domestic market is far too small to support an unsubsidized biofuel industry, but the likely free on board (f.o.b.) cost of bioethanol and SVO (and maybe biodiesel too) will open up international markets in which Mozambique should be very competitive. Thus condition (1) will be fully met. There will also be scope for vertical integration within the biofuel sector. A well-planned strategy for biofuel development in Mozambique is in the early stages of implementation with private sector investment leading the way.

Senegal

Senegal has neither a large domestic market, nor much prospect of producing exportable biofuel at an internationally competitive cost. Its government-led jatropha planting campaign came to a halt when farmers realized there was no assured market for the crop. However, private entrepreneurs are continuing to plant, largely on a speculative basis, and many of them with the intention of making and selling SVO or biodiesel into their immediately local market.

Germany

Germany is the world leader in production and consumption of biodiesel,[3] most of which is produced from domestically produced rapeseed oil, but is only a relatively small player in the global bioethanol economy. In 2009, producing biodiesel in Germany cost around $0.67/l compared with $0.29/l for mineral diesel, but taxes on the latter took its market cost to $0.84/l.[4] Up to 2006, biofuels were exempted from any tax, which resulted in a rapid increase in biofuels' share of the total fuel market. This was also helped by the auto manufacturers issuing warranties for biodiesel use. However, to compensate for the expected loss of fuel revenue, government began to tax biofuels in 2006, with step-wise increases for biodiesel and SVO planned to 2015, when the taxes on biodiesel and mineral diesel would roughly equate. A similar magnitude tax on blends was levied in a single step. Bioethanol was tax-exempt. At the same time, mandatory tradable quotas (minimum blending requirements) were introduced, fixed for biodiesel at 4.4 per cent by 2015, stepwise increasing for bioethanol to 3.6 per cent, and stepwise increasing for bio-fuels in total to 8 per cent (which exceeded the EU's directive of 5.75 per cent).

Since 2008, the biodiesel market in Germany has been shrinking as the tax increases and large imports of heavily subsidized US blended fuel (subsequently choked off by EU anti-dumping duties) undermined its competitiveness. Growing concern over the consequences of EU biofuel policy for the global environment resulted in the introduc-tion of sustainability standards into the market. In October 2008, Germany proposed removing soy and palm oil from the list of qualified biofuels effective 2009 (although this draft legislation was subsequently challenged) and the biofuels quota was reduced. From July 2010, companies must prove the sustainability of their biofuels to qualify for tax deductions or to have their products countable towards mandatory renewable energy targets. Despite recent scaling back of tax increases, pure biodiesel (B100) is reportedly now barely able to compete with mineral diesel (Rosillo-Calle et al., 2009) and biodiesel production in Germany is running at considerably less than 60 per cent of total national capacity of 5 million tons. Given the various twists and turns in national biofuel policy, it is not surprising that many biofuel producers and consumers are uncertain about the outlook for their businesses.

USA

The USA is the world's largest producer of bioethanol (with Brazil by far the second largest), nearly all of it made from maize, and the second-largest global producer of biodiesel, most of it made from soya or recycled cooking oil. Bioethanol accounts for around 90 per cent of all biofuels consumed in the US. Although US maize bioethanol is, internationally, one of the cheaper biofuels to produce, it cannot compete on price with Brazilian sugarcane ethanol (which comes closest of any national biofuel industry to real commercial viability), which is why there is a substantial tariff applied to imports of the latter into the US market. There are other substantial subsidies and supports of various kinds for biofuel production in the US, including blenders' tax credits, small-producer income tax credits, fuel excise tax exemptions and farm payments. The total cost of the major tax subsidies in 2007, for example, was \$3.5 billion for bioethanol and \$0.7 billion for biodiesel (OECD, 2007).[5]

After initial very rapid growth in bioethanol production capacity, expansion levelled off in early 2009, with several ethanol producers going out of business or suspending operations (O'Brien, 2009). This followed the maize price spike of 2007 and 2008 and the approach of national production levels towards the limit set by the US government's renewable fuels standard mandate. The general view is that US bioethanol production will continue to grow, but the outlook is perhaps less clear for biodiesel. In the absence of a significant and sustained rise in the oil price, neither bioethanol or biodiesel production in the US look likely to become commercially viable in the foreseeable future, and the industries will require continuing government intervention in the national biofuels market.

18.3.5   The Special Case of Jatropha

Nearly all jatropha currently planted for the production of biofuel is the progeny of undomesticated (wild) plants that have at best been subject to selection on the basis of parental characteristics. Systematic varietal improvement programs by genetic methods are in their infancy, although agronomic trials have been under way for some time to determine the best methods of cultivation and harvest for the cultivars already in use. At present, there is very little commercial-scale experience with jatropha to provide any certainty about the economics of producing it. What is becoming apparent is that the early claims that it will yield well under marginal conditions for minimum input cost are almost certainly unfounded. However, the lessons of the jojoba experience, a case of another 'new' crop that was heavily promoted on the basis of little research, probably apply here: systematic improvement programs subsequently resulted in large increases in commercial-scale yields. The same will undoubtedly be true for jatropha, and those companies and governments that can afford to invest in long-term research and development (R&D) will eventually benefit substantially.

18.4   ARE BIOFUELS GOOD FOR THE ENVIRONMENT?

One of the supposed attractions of biofuels is that they are a renewable energy source that might significantly reduce greenhouse gas (GHG) emissions from the transport sector. This is because the CO_2 emitted when biofuels are combusted in engines is simply a recycling to the atmosphere of the CO_2 absorbed during production of the feedstock, so biofuels should be carbon-neutral when viewed over their entire production cycle (from field to tank). However, this overlooks the fact that some fossil fuels are used in the production and distribution of biofuels. Agricultural machines and nitrogen fertilizers both require substantial amounts of fossil fuel, and additional fossil fuel is used in transporting the feedstock and biofuels and, in some cases, processing the feedstock.[6] Hence there is need to look at energy ratios, defined as the ratio of available energy delivered per liter of biofuel to the total fossil fuel energy used in its production, calculated over the full production cycle. Ratios of less than 1 imply that more fossil fuel energy is used than

Table 18.2   A United States Department of Agriculture (USDA) estimate of the energy ratio for bioethanol from maize in the US

Production phase	BTU/gallon
Maize production	21 598
Maize transport	2 263
Ethanol conversion	51 779
Ethanol distribution	1 588
Total energy used	77 228
Ethanol energy content	83 960
Co-product energy content	14 372
Energy ratio w/o co-products	1.08
Energy ratio with co-products	1.27

Source: Shapouri et al. (2002).

Table 18.3   Pimental's estimate of the net energy ratio of bioethanol from maize in the US

	BTU/gallon
Farm production (machinery, fertilizers, electricity, transport, etc.)	40 221
Ethanol production	58 898
Total energy used (not including final distribution to petrol stations)	99 119
Ethanol energy content	77 000
Energy ratio	0.78

Source: Pimental (2003).

is contained in the biofuels produced. Since additional fossil fuels are also required to refine and distribute gasoline or diesel, they also have energy ratios less than 1. Gasoline, for example, has an energy ratio of 0.8, so this rather than 1 is a relevant benchmark for comparisons with biofuels.

Although the concept of energy ratios or balances seems straightforward, controversies have arisen over the way they are measured. Should, for example, the energy used in making agricultural machines or feeding farm workers be included, or just the energy content of direct inputs like diesel and fertilizer used in the cultivation and harvesting of biofuel crops? Also, what energy credit should be given to co-products that can be used for cattle feed or burnt to generate electricity? Different assumptions can lead to very different results. For example, Shapouri et al. (2002) estimated the energy ratio for bioethanol from maize in the US to be 1.27, but if co-products were excluded, the ratio fell to 1.08 (Table 18.2). In a comparable estimate that excluded co-products but included many indirect energy uses, Pimental (2003) obtained a very low energy ratio of 0.78 (Table 18.3), implying that bioethanol requires more non-renewable energy to produce than there is energy in the final product.

Using consistent assumptions across different types of feedstock and fuels that include co-products but exclude most indirect energy use, the Worldwatch Institute (2007)

Table 18.4   Energy balances by fuel type

Fuel (feedstock)	Fossil energy balance
Cellulosic ethanol	2 to 36
Biodiesel (palm oil)	≈ 9
Bioethanol (sugar cane)	≈ 8
Biodiesel (soybeans)	≈ 3
Biodiesel (rapeseed)	≈ 2.5
Bioethanol (wheat, sugar beets)	≈ 2
Bioethanol (maize)	≈ 1.5
Gasoline and diesel	0.75–0.8

Source: Worldwatch Institute (2007).

compiled the energy ratios shown in Table 18.4. These show that of the first-generation technologies, biodiesel from palm oil and bioethanol from sugar cane are far superior to the main feedstocks grown in the US and the EU.

Energy ratios are improving over time with advances in the technologies for processing feedstock and the development of crop varieties that have better energy attributes. They could also be improved by using more biomass for generating the power for processing feedstock (as with sugar cane and oil palm), using biofuels in farm machines, and by reducing the use of nitrogen fertilizer in feedstock production by using nitrogen-fixing feedstock or rotating feedstock with nitrogen-fixing crops. Indications are that second-generation biofuels based on cellulose-rich biomass will have more favorable energy ratios, but most of these are still too costly to process with available processing technologies.

A more direct measure of the contribution of biofuels to carbon reduction compared to fossil fuels is their net carbon savings. When blended with gasoline or diesel, most biofuels from grains can reduce carbon emissions by 20–30 per cent per mile traveled, and the savings are greater the higher the fuel blend (Worldwatch Institute, 2007). However, biodiesel from soybeans can save 40 per cent and bioethanol from sugar cane can save 90 per cent.

These energy and carbon savings do not allow for any changes in land use. They assume the same crop would have been grown anyway. The results would be much worse if, for example, forest is cleared to grow biofuels, as has happened with oil palm in Malaysia, or with sugarcane pushing more soybeans into the Brazilian Amazon. Searchinger et al. (2008) calculated that the US corn ethanol program leads to huge emissions of GHGs relative to use of gasoline when induced land-use changes around the world are taken into account. More recent work using global agricultural models to simulate market-mediated responses in the US and overseas through trade and price changes suggests less dramatic indirect impacts, but still large enough to offset any initial reduction in GHGs (Hertel et al., 2010). While part of the problem arises from clearing forest and grasslands for biofuels production and releasing huge amounts of previously sequestered carbon, additional emissions also arise when additional land is converted to food production in many poor countries in response to higher food prices. If these adverse environmental impacts are to be avoided, then either more existing agricultural land must be diverted to biofuels – leading to a possible trade-off against food security (see later) – or the average yields of food and fuel crops must be increased beyond current trends. Second-generation technologies may also have more favorable carbon balances if they enable cellulose-rich crops to be established on already degraded lands, as is happening in India, or if perennial feedstocks that sequester large amounts of carbon in the soil were to replace annual crops.

As with all crops, bioenergy crops need to be grown and managed responsibly to avoid creating local environmental problems of their own. For example, removing all the biomass can exacerbate shortages of organic matter for returning to the soil, leading to nutrient mining and land degradation. Inappropriate cultivation of bioenergy crops can mine water resources, expose land to greater erosion, pose problems with the intensive use of pesticides and fertilizers, and threaten local biodiversity. On the other hand, grown under the right conditions, bioenergy crops can contribute to better environmental management. For example, dedicated energy plantations grown on degraded lands may actually help restore the soil and biodiversity.

18.5   BIOENERGY AND FOOD SECURITY

The rapid expansion in biofuels production in recent years has contributed to an increase in world food prices, though it was only one of several drivers that led to the unusual price spike of 2007–08 (Piesse and Thirtle, 2009). The world has significant potential to expand total agricultural output, even over the course of just a couple of years. For example, the *Wall Street Journal* (Monday, 21 June 2010, p. A2) reports that globally, an extra 82 million acres of grains and oilseeds had been brought into production by 2010 in response to the 2007–08 food price spike, equivalent to creating another US corn belt. Higher prices also attract additional long-term investment in agricultural research, irrigation and land improvements that can add to total production over the longer term. However, the world does not have an infinite capacity to expand total agricultural production, and there is growing evidence that water and land constraints, higher oil prices and climate change are slowing agricultural growth rates. Given these supply-side challenges, and growing populations and demand for livestock products in the developing world, the FAO is predicting that in the future, world food prices are likely to trend upwards in contrast to past decades of decline, and they will become more volatile. Further expansion of biofuels production would add to these price pressures, and while good for many farmers, it would have adverse implications for the food security of many poor people around the world.

Consumers in rich countries like the US and EU would not be greatly affected. They mostly live in food-exporting countries so do not have to worry about food supplies. Moreover, even if food prices rise, they will still be able to afford it because food is just a tiny share of the average household budget, and the price of raw materials is also a tiny share of the supermarket price of food. Most poor people in rich countries are also protected by public safety net programs. The real problem lies in developing countries. Many of the poorest developing countries are net importers of food and cannot easily afford higher import prices, and poor people in developing countries are particularly vulnerable to price increases since they spend large shares of their meager incomes on food.

According to the World Bank, a 1 per cent increase in the price of food directly correlates with a 0.5 per cent decrease in calorie consumption by the poor. So even modest food price increases can lead to a substantial reduction in the calorie consumption of the poor.

The primary food security problem in most developing countries will not arise from their own diversion of crop land to biofuels, but from international trade. All the major regions of the developing world are net cereal importers, and in fact about half of global cereal trade is between the North and South. Hence trade plays a key role in shoring up the food security of much of the developing world. This North–South trade pattern is not new and has built up over several decades, driven by country differences in natural resource endowments, extensive agricultural subsidies in the North, and escalating demands for food and livestock feed in the South because of growing populations and changing diets. Land conversion away from food production in the US and EU will lead to a decline in cereal exports, which will have a large impact on the availability and price of food in the developing world. The same will happen if other major agricultural export countries like Brazil follow suit.

How large will this problem be and what can be done to avoid it? In 2004, about 14 million hectares worldwide were devoted to the production of biofuels, equivalent to about 1 per cent of global cropland (Chakravorty et al., 2009). From this relatively small base, the answer depends on several key factors: how quickly the rich countries expand their biofuels demand, how much of this demand they want to meet from domestic production rather than imports, how much more land can be brought into agricultural production around the world, and how fast yields can be increased.

The recent food crisis led to some moderating of biofuel mandate targets in the EU, but both the US and the EU still plan rapid increases in biofuels consumption by 2020. These countries also seem determined to produce much of the feedstock at home, and there are substantial trade barriers against imports. One significant trade barrier erected by the EU is its sustainability criteria for imported biofuel feedstock. The EU seems rather belatedly to have realized that its mandate targets for biodiesel have stimulated great interest in palm oil expansion in countries where sustainability is often questionable (such as Indonesia). Given the relatively low energy yield per hectare of biofuels from temperate, first-generation technologies, this will aggravate the diversion of cropland away from food to biofuels. An obvious strategy is to slow down the mandates until more second-generation technologies are profitable, and to allow greater international trade in biofuels, thereby expanding the available land base that can be tapped.

How much new cropland can be brought into production around the world? Estimates vary widely (Berndes et al., 2003), from no effective land constraint to little additional land that can be brought into crop production outside a few countries like Brazil, Malaysia, DR Congo and Indonesia that have large amounts of remaining tropical forest or grassland. There is greater scope for expanding the crop area for cellulose-rich biomass as second-generation technologies develop. About 60 per cent of the global agricultural domain can be classified as less-favored lands (Hazell and Wood, 2008). But much of this land is sloping, has low and uncertain rainfall, poor and degraded soils, poor infrastructure and market access, and supports many poor people who cannot easily be displaced. Attempts to develop these lands for biofuels production will face many of the same challenges that have bedeviled past attempts to intensify them for food production and poverty alleviation.

Yield growth offers a more viable option for biofuels expansion. The world has managed to feed itself over the past 50 years despite a virtual doubling of the human population by increasing crop yields. In fact, the cropped area only increased by 12.2 per cent between 1963 and 2002 (Hazell and Wood, 2008). Underlying this success has been a dramatic technological revolution brought about by significant and mostly public investments in agricultural R&D and rural infrastructure that led to the increased use of improved crop varieties, fertilizers, pesticides and irrigation. There are concerns today that the high yields attained in many intensively farmed areas have peaked and cannot be sustained, but there is still scope for further yield increases through agricultural R&D and better management of water and soil (Hazell and Wood, 2008).

There are several ongoing modeling efforts to explore the food fuel trade off and to evaluate appropriate policy options. Chakravorty et al. (2009) review 13 such efforts. In one early study, the International Food Policy Research Institute (IFPRI) (Rosegrant et al., 2006) estimated that if there were a global attempt to replace 10 per cent or more of transport fuels with biofuels, then the world prices of some major food crops could increase by as much as 50 per cent by 2020. Similar exercises using other models – for example the Organisation for Economic Co-operation and Development (OECD)'s Aglink and sugar models, FAO's COSIMO model, EC's ESIM model, the extended and modified Global Trade Analysis Project (GTAP) model used by the Agricultural Economics Research Institute, Wageningen Research Centre, Netherlands, and US Department of Agriculture (USDA)'s models – have generally indicated less dramatic food price changes, even including price falls for protein feeds as a consequence of rapid oilseed production growth. The results are also sensitive to assumptions about yield growth, and hence implicitly to underlying trends in the levels of public and private investment in agriculture around the world. With increased agricultural investment and the emergence of second-generation technologies, the trade-off between biofuels and food can be reduced considerably (for example Rosegrant et al., 2006).

There are several ways to reduce the potential trade-off between bioenergy and food production:

- Develop biomass crops that yield much higher amounts of energy per hectare or unit of water, thereby reducing the resource needs of bioenergy crops.
- Focus on food crops that generate by-products that can be used for bioenergy, and breed varieties that generate larger amounts of by-products (for example sweet sorghum).
- Develop and grow biomass in less-favored areas rather than in prime agricultural lands. Second-generation technologies that enable cost-effective conversion of cellulose-rich biomass, like fast-growing trees, shrubs and grasses that can grow in less fertile and low-rainfall areas, will greatly expand this option within the next 10–15 years.
- Invest in increasing the productivity of the food crops themselves, since this would free up additional land and water for the production of bioenergy crops.
- Remove barriers to international trade in biofuels. With the right investments, the world probably has enough capacity to grow all the food that is needed as well as large amounts of biomass for energy use, but not in all countries and regions. Trade is a powerful way of spreading the benefits of this global capacity while

enabling countries to focus on growing the kinds of food, feed or energy crops for which they are most competitive. Trade would also allow bioenergy production patterns to change in the most cost-effective ways as new second-generation technologies come on line. An important challenge facing a more open trade regime in biofuels and related feedstocks is the need to ensure that exporting countries meet internationally agreed sustainability standards so that primary forests and peatlands are not being destroyed.

18.6   IMPLICATIONS FOR DEVELOPING COUNTRIES

Developing countries should be most concerned about a scenario in which the US, the EU and rapidly industrializing countries like Brazil, China and India continue to pursue aggressive biofuel strategies without an accompanying global effort to increase the levels of investment in agriculture to significantly raise crop yields. Faced with this scenario, a high priority for many developing countries must be to ensure that they have adequate safety nets in place to protect the poorest and most vulnerable.

A second priority is to invest in their agricultural growth to improve their food security in the medium to long term. For example, many African countries have considerable potential to grow more of their own food, but past performance has been poor, constrained by lack of public investment in agricultural development. Now is the time for governments and donors to reverse that neglect. There are encouraging signs that this is beginning to happen – for example, the African heads of state have committed to doubling agriculture's share in total government spending to 10 per cent by 2015; the Rockefeller Foundation and the Bill and Melinda Gates Foundation have financed the establishment of the African-led Alliance for a Green Revolution in Africa (AGRA); and the New Economic Partnership for African Development (NEPAD) is leading an Africa-wide initiative for agricultural development called the Comprehensive African Agricultural Development Programme (CAADP). But so far none of these initiatives have led to the levels of investment needed to make a significant change to Africa's agricultural productivity. Renewed agricultural growth would help many small farmers increase their incomes and help lower food prices for all.

A third priority is to invest in new bioenergy opportunities that could benefit small farmers and poor people. Most rural people in developing countries already get most of their household energy from biomass, but their main sources are crop by-products and animal manures – which are also badly needed to maintain soil organic matter and fertility – and charcoal, whose production underlies a lot of deforestation and woodland degradation. Therefore, development of suitable energy crops for use in local energy production could provide a source of new income for many small farmers, improve management of soils and woodlands, and provide an affordable and secure source of energy for local people and communities. Carbon payment schemes may yet grow beyond the constraints of the current Clean Development Mechanism (CDM) of the Kyoto Treaty to include possibilities for payments for carbon sequestration on small farms. If so, planting of perennial crops for bioenergy might be able to attract an additional source of income for poor farmers.

Furthermore, in developing countries where viable commercial opportunities exist for bioenergy production, whether in the form of biofuels for transport or biomass for large-scale electricity generation, it may be possible to link groups of small farms into the market chain. Where successful, such a linkage could play an important role in raising incomes and employment.

A fourth priority is to strengthen environmental regulation and management in developing countries to prevent encroachment of agriculture, whether for food or energy, into environmentally valued areas like the Brazilian rainforest. This was already a growing problem before biofuels appeared on the scene and they are just another driver of land conversion. The underlying problem is more fundamental; it arises from a divergence of interest between individual countries and the larger international community. The local people have a need to make a living and feed their families, even if that involves chopping down forests and clearing land to grow crops, whereas the international community wants the forest to stay intact. Biofuels merely reinforce the need for more effective international collaboration over environmental externalities, and this requires positive incentives for countries and local people, such as can be offered through carbon and other environmental payments or green labeling arrangements, not just negative attempts to enforce trade embargoes.

Rich countries could help in several ways:

- Provide more financial support for agricultural development in poor countries, especially Africa.
- Support the development and international transfer of improved agricultural technologies to enhance productivity and sustainability.
- Slow down on biofuels mandates until: (1) policies and investments are in place to permit the required expansion in food production in developing countries; and (2) second-generation cellulose-rich feedstocks become more commercially viable.
- Contribute financially to the cost of protecting forest and peatland and for carbon sequestration in developing countries, either directly through grants or through market-mediated approaches like carbon markets.
- Remove trade barriers and tariffs on agricultural commodities and biofuels to allow more efficient use of global agricultural resources.

18.7 GROWING THE INDUSTRY

Launching and developing a new industry like biofuels poses difficult challenges for the private sector. The substantial investments that must be made up front can yield little return until sufficient scales of production and demand have been achieved to slash unit costs. But achieving those scales depends on complementary investments throughout the market chain, and these investments may not be forthcoming until bioenergy costs have fallen to a level competitive with alternative energy sources. A viable biofuel industry requires large and coordinated investments not only by farmers and processors, but also by car manufacturers, consumers, fuel distributors and garages. Until these investments are in place, biofuel sales are destined to be low, and economies of scale in production and distribution cannot be exploited. Given higher costs, biofuels may remain uncompetitive with oil.

The solution to this problem is for governments to provide initial incentives to help launch the industry. The public sector can help achieve critical market size by offering tax rebates on biofuels (but not on oil-based gasoline and diesel); by mandating fuel blending requirements; by offering investment incentives such as tax exemptions or holidays on bioenergy investments by industry and subsidies to consumers (to buy flex-fuel cars, for instance); and by investing directly in research and development and relevant infrastructures. Brazil began using these kinds of interventions in the mid-1970s and has now built up a viable biofuels industry that not only contributes a significant share of the country's energy requirements for transportation, but also exports to other countries. The European Union and the United States began later and are in the process of building up their own domestic industries. Many other countries seem likely to follow.

The optimal design of domestic policies for promoting biofuels can be complex, with unanticipated consequences (Gorter and Just, 2010). For example, on its own, subsidizing biofuels (for example through a tax credit) reduces the average price of fuel and can lead to a net increase in total fuel consumption. This in turn can increase total GHG emissions even if biofuels emit less GHG per liter than gasoline or diesel. Worse, a biofuels mandate in combination with a subsidy or tax credit can actually increase the consumption of gasoline as well as total fuel. A better policy is to combine a mandate with a fuel tax that keeps the total use of fuel unchanged.

Not all countries can grow and process biofuels at costs that are competitive with oil prices, and the domestic biofuel industries that are being so carefully nurtured in some of these countries may not be able to compete in the future without sustained trade protection and subsidies. Is this worthwhile? Infant-industry subsidies and trade protection can sometimes be justified if they are phased out after a successful transition period, but if they are to be sustained for the long haul then they need to return other social benefits. Politically, a drive for greater energy security has been important in the US and the EU, and it is hard to put an economic value on this benefit. Other benefits that have been claimed include reducing GHG emissions and substituting for alternative farm support policies. Given the relatively low carbon balance obtained with many first-generation biofuels, the cost of achieving net reductions in carbon emissions from biofuels can be high, and there may be more cost-effective alternatives for achieving the same gain. For example, it might be more cost-effective for some countries to continue to use fossil fuels and buy carbon offsets, or to import biofuels from countries that can grow them with more favorable carbon balances. Savings in the cost of farm support programs are only likely to be significant if the current types of support are tied to commodity prices. Since there has been a shift in recent years to more indirect forms of farm support, such as environmental payments, the potential savings from biofuel support programs may only be modest, and may anyway not be as cost-effective as some other kinds of income support policies. They have the added disadvantage of requiring additional government support for a whole new biofuels processing industry. A key question for policy-makers in many countries that are not cost competitive is how much they are willing to pay to achieve other perceived benefits of bioenergy. These costs might decline over the next 10–15 years as second-generation technologies come on line, but for many countries, especially in temperate climates, there is a danger that some expensive white elephants have recently been created.

18.8 CONCLUSIONS

The promise of bioenergy is that it may help cope with rising energy prices, address environmental concerns about greenhouse gas emissions, and offer new income and employment to farmers and rural areas. But the rapid development of bioenergy also poses risks and has the potential to result in difficult trade-offs for the poor and the environment. There is already a food–fuel trade-off and biofuels are impacting on world food prices with severe consequences for many poor people. Africa will be the biggest loser from the biofuel–food trade-off unless significant new investments are made in its agricultural growth. The environmental benefits of many commonly used biofuels are also marginal in terms of their energy and carbon balances, and are disastrous where biofuels production leads to agricultural encroachment into remaining primary forest and peatlands. Technological advances in growing and processing first-generation feedstock and in second-generation technologies based on cellulose-rich feedstock may transform the situation, but major advances at scale seem unlikely within the next 10–15 years.

In this situation, countries should be encouraged to slow down on their biofuels mandates, allowing time for: (1) new investment in agricultural development in poorer countries to reduce the trade offs with food; (2) second-generation technologies to come online that could greatly reduce competition with food crops, enhance environmental benefits and lower the costs of biofuels compared to oil; and (3) development of binding international agreements and compensation schemes to protect remaining primary forest and peatlands from conversion to agriculture.

NOTES

1. In palm oil production, large scale is compatible with the incorporation of smallholders within the value chain. Much of the oil palm fruit in both Indonesia and Malaysia is produced by smallholders, but these are usually organized in large, consolidated blocks of land and supply large central mills.
2. Higher yields can support a higher density of factories, and hence are associated with shorter average haulage distances.
3. Europe as a whole is expected to produce a little under half of the world's biodiesel in 2010 (*Farmers Weekly*, 4 May 2010).
4. 'Taxation takes its toll on the German biodiesel industry', www.power-technology.com/features/features58850, 20 July 2009.
5. The EU subsidizes its own biofuel industry with a similar amount: total excise tax exemptions in 2006 were $1.2 billion for bioethanol and $3.0 billion for biodiesel (IEA, 2007b).
6. In some cases, biomass residues are used to provide the energy for processing, making them self sufficient in energy. This is the case with sugar cane and oil palm and may also apply to jatropha.

REFERENCES

Agribenchmark (2010), 'Cash crop report 2009', Institute of Farm Economics, Johann Heinrich von Thünen-Institut (vTI), Germany.
Berndes, Göran, Monique Hoogwijk and Richard van den Brock (2003), 'The contribution of biomass in the future global energy supply: a review of 17 studies', *Biomass and Bioenergy*, **25**, 1–28.
Chakravorty, Ujjayant, Marie-Hélène Hubert and Linda Nøstbakken (2009), 'Fuel versus food', *Annual Review Resource Economics*, **1**, 645–63.
FAO (2008), *The State of Food and Agriculture 2008. Biofuels: prospects, risks and opportunities*, Rome: FAO.

Gopinathan, M.C. and R. Sudhakaran (2009), 'Biofuels: opportunities and challenges in India', *In Vitro Cell Development Biology – Plant*, **45**, 350–71.

Gorter, H. de and D.R. Just (2010), 'The social costs and benefits of biofuels: the intersection of environmental, energy and agricultural policy', *Applied Economic Perspectives and Policy*, **32** (1), 4–32.

Government of Mozambique (2008), 'Mozambique biofuels assessment', Final Report, Econergy, 1 May, Ministries of Agriculture and Energy.

Government of Mozambique (2009), *National Biofuel Policy and Strategy*, (drafts in English).

Hazell, P. and S. Wood (2008), 'Drivers of change in global agriculture', *Philosophical Transactions of the Royal Society B*, **363** (1491), 495–515.

Hertel, Thomas W., Alla A. Golub, Andrew D. Jones, Michael O'Hare, Richard J. Plevin and Daniel M. Kammen (2010), 'Effects of US maize ethanol on global land use and greenhouse gas emissions: estimating market-mediated responses', *BioScience*, **60** (3), 223–31.

IEA (2005), *Biofuels for Transport – An International Perspective*, Paris: OECD.

IEA (2006), 'The outlook for biofuels', *World Energy Outlook*, Paris: OECD.

IEA (2007a), *World Outlook 2007*, Paris: International Energy Agency.

IEA (2007b), 'IEA energy technology essentials, biofuel production', Paris: International Energy Agency, January.

Johnston, M. and T. Holloway (2006), 'A global comparison of national biodiesel production potentials', *Environmental Science and Technology*, **41** (23), 7967–73.

O'Brien, D. (2009), 'Trends in fuel ethanol production capacity 2005–2009', *AgMRC Renewable Energy Newsletter*, October.

OECD (2007), 'Biofuels: linking support to performance', Round Table 7–8 June, Paris: OECD/International Transport Forum Joint Transport Research Centre.

Piesse, Jenifer and Colin Thirtle (2009), 'Three bubbles and a panic: an explanatory review of recent food commodity price events', *Food Policy*, **34** (2), 119–29.

Pimental, David (2003), 'Ethanol fuels: energy balance, economics, and environmental impacts are negative', *Natural Resources Research*, **12** (2), 127–34.

Raju, S.S., P. Shinoj and P.K. Joshi (2009), 'Sustainable development of biofuels: prospects and challenges', *Economic and Political Weekly*, **44** (52), 65–72.

Rosegrant, Mark W., Siwa Msangi, Timothy Sulser and Rowena Valmonte-Santos (2006), 'Biofuels and the global food balance', in P. Hazell and R.K. Pachauri (eds), *Bioenergy and Agriculture: Promises and Challenges*, 2020 Focus 14, Washington, DC: International Food Policy Research Institute.

Rosillo-Calle, F., L. Pelkmans and A. Walter (2009), 'A global overview of vegetable oils, with reference to biodiesel', a Report for the IEA Bioenergy Task 40, June.

Schmidhuber, Josef (2006), 'Impact of an increased biomass use on agricultural markets, prices and food security: a longer-term perspective', International Symposium of Notre Europe, Paris, 27–29 November.

Searchinger, Timothy D., Ralph E. Heimlich, Richard A. Houghton, Fengxia Dong, Amani Elobeid, Jacinto Fabiosa, Simla Tokgoz, Dermot J. Hayes and Tun-Hsiang Yu (2008), 'Use of US croplands for biofuels increases greenhouse gases through emissions from land use change', *Science*, **319** (5867), 1238–40.

Shapouri, Hosein, James A. Duffield and Michael Wang (2002), 'The energy balance of corn ethanol: an update', Agricultural Economics Report Number 814, United States Department of Agriculture, Washington, DC.

USDA (2009), 'India', *Biofuels Annual 2009*, Global Agricultural Information Network, USDA FAS.

Worldwatch Institute (2007), *Biofuels for Transport: Global Potential and Implications for Sustainable Energy and Agriculture*, London: Earthscan.

PART V

ENERGY AND CLIMATE POLICY

19 The European carbon market (2005–07): banking, pricing and risk-hedging strategies
Julien Chevallier

19.1 INTRODUCTION

The European Union Emissions Trading Scheme (EU ETS) was created on 1 January 2005 to reduce by 8 per cent CO_2 emissions in the European Union by 2012, relative to 1990 emissions levels. This aggregated emissions reduction target in the EU has been achieved following differentiated agreements, sharing efforts between member states based on their potential of decarbonization of their economy. The introduction of a tradable permits market has been decided upon to help member states in achieving their targets under the Kyoto Protocol, and entered into force on February 2005 following the ratification of Iceland. It aims at reducing the emissions of six greenhouse gases (GHGs) considered as the main cause of climate change. Among the members of Annex B, these agreements include CO_2 emissions reductions for 38 industrialized countries, with a global reduction of CO_2 emissions by 5.2 per cent. These agreements have been fostered by the United Nations Framework Convention on Climate Change (UNFCCC, 2000) which recognizes three principles: the precautionary principle,[1] the principle of common but differentiated responsibilities,[2] and the principle of the right to development.[3] A total of 174 countries, Australia being the latest on 3 December 2007, have ratified the Protocol, with the notorious exception of the United States. The first commitment period of the Kyoto Protocol goes from 1 January 2008 to 31 December 2012.

This political will has been reaffirmed at the international level during the UN Conference that took place in Bali in December 2007, where a roadmap of negotiations that should lead to a post-Kyoto agreement was adopted. The United States is expected to cooperate, given the initiatives of emissions reduction introduced at the regional level.[4] The next round of negotiations will take place in Durban in December 2011. As the Clean Development Mechanism (CDM)[5] has revealed the strong potential for CO_2 emissions abatement in countries such as Brazil, China and India, the main issue of these negotiations is linked to achieving the highest possible level of cooperation, in order to avoid the well-known free-rider behaviour, and to preserve the global public good that constitutes the climate. On this matter, the European Union has clearly adopted a leadership position, which contrasts with its early reluctance during the first steps of the negotiation of the Kyoto Protocol.

In January 2008, the European Commission extended the scope of its action against global warming by 2020 with the 'energy and climate change' package. This package aims at reducing GHG emissions by 20 per cent, at increasing the use of renewable energy in energy consumption to 20 per cent, and at saving 20 per cent of energy by increasing energy efficiency. The European carbon market, which has currently entered its Phase II (2008–12), has been confirmed until 2020 also. Its scope has been extended to major sectors in terms of CO_2 emissions growth, such as aviation and petro-chemical industries during 2013–20. The creation of the EU ETS as well as the adoption of the EU Energy – Climate Package aim at correcting the negative externality attached to the release of uncontrolled GHG emissions in the atmosphere and thus, according to the well-known principle in economics, at internalizing the social cost of carbon. At the same time, these initiatives reveal the difficulty of creating a scarcity condition regarding CO_2 emissions. These emissions indeed were not limited in the pre-existing institutional environment, and thus could not be considered as a scarce resource.

The European Union being at the forefront of environmental regulation dedicated to climate policies, this chapter reviews the market rules of the European carbon market during Phase I. It investigates the role played by the regulator, among the various choices at stake when creating a tradable permits market, on the behaviour of firms. Thus, this chapter contributes to the literature on the 'birth' of the European carbon market (Convery et al., 2008; Convery, 2009; Ellerman and Buchner, 2008), by focusing attention on the study of several key provisions in a moving institutional context, and by identifying learning effects.

The remainder of the chapter is organized as follows. Section 19.2 presents the key design issues of the EU ETS. Section 19.3 examines the effects of banking restrictions between 2007 and 2008. Section 19.4 discusses the price fundamentals of CO_2 allowances. Section 19.5 details market participants' risk behaviour. Section 19.6 concludes.

19.2   KEY DESIGN ISSUES OF THE EU EMISSIONS TRADING SCHEME

This section reviews the scope, allocation methodologies, calendar, transactions levels and penalties associated to non-compliance of the EU ETS.

19.2.1   Scope

Directive 2003/87/CE defines the scope of the EU ETS.[6] This scheme concerns around 10 600 installations in Europe, mainly in the production sectors of combustion, iron and steel, pulp and paper, refineries and cement. Installations in these sectors are eligible to emissions trading when their energy consumption is superior to the threshold of 20 MWTh. This threshold has been decided by the European Commission so as to target the most energy-intensive industries during the first phases of the programme. This choice was justified initially by the will of the European Commission to minimize political resistance, and to enforce a quick implementation of the scheme in 2005. To increase the environmental performance of the scheme, the debate is now centred on the progressive extension of its scope. The EU ETS Review[7] has revealed that other sectors will soon be included, such as aviation as of 2013.

19.2.2   Allocation

The CO_2 emissions reduction target of each member state has been converted into National Allocation Plans (NAPs). Each government is in charge of deciding the amount

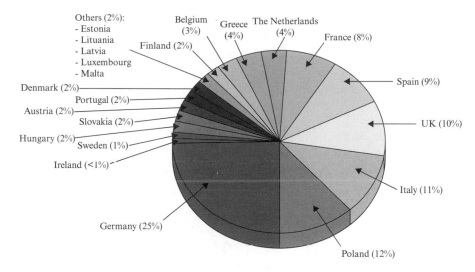

Others (2%):
- Estonia
- Lituania
- Latvia
- Luxembourg
- Malta

Belgium (3%)
Greece (4%)
The Netherlands (4%)
France (8%)
Finland (2%)
Denmark (2%)
Portugal (2%)
Austria (2%)
Slovakia (2%)
Hungary (2%)
Sweden (1%)
Ireland (<1%)
Spain (9%)
UK (10%)
Italy (11%)
Germany (25%)
Poland (12%)

Source: CITL (2007) and CDC (2006).

Figure 19.1   EU ETS National Allocation Plans – Phase 1 (2005–07)

of quotas available for trading, after negotiating with industrials, and after the validation by the European Commission. The role of the Environment DG is central in this scheme in order to harmonize NAPs among member states, and to recommend stricter NAPs validation criteria. The NAPs submissions may be rejected by the European Commission, and sent back to member states for revision before the final decision. The sum of NAPs determines the number of quotas distributed to installations in the EU ETS. During 2005–07 2.2 billion quotas per year were distributed, and 2.08 billion quotas per year will be distributed during 2008–12, which corresponds to a more restrictive allocation, given some changes in the scope of the market with the inclusion of new member states. Figures 19.1 and 19.2 represent, respectively, the repartition of quotas (in million tonnes of CO_2) between Member States during the commitment periods 2005–07 and 2008–12.[8] Germany, Poland, Italy, the UK and Spain total around two-thirds of allowances distributed.

The allocation methodology consists in a free distribution of quotas in proportion to recent emissions, also known as grandfathering. With a value of around €20 per quota, the launch of the EU ETS corresponds to a net creation of wealth of around €40 billion. The environmental constraint during 2005–07 has not been considered sufficiently binding for most market observers, and the allocation methodology has been criticized for distributing rents to pre-existing market players, as some of them may make a net profit simply by selling their unused allowances.

During 2005–07, allowances distributed more than covered verified emissions, with a net cumulated surplus of 156 million tonnes. This surplus however decreased, going from 83 million tonnes in 2005 to 37 million tonnes in 2006, and finally 36 million tons in 2007. Emissions increased by 0.4 per cent in 2007 compared to 2006, and reached 2043 million tonnes with respect to 2080 million allowances distributed.

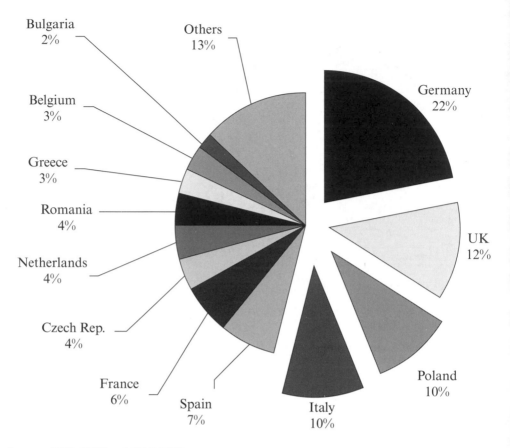

Source: CITL (2008) and CDC (2008).

Figure 19.2   EU ETS National Allocation Plans – Phase II (2008–12)

19.2.3   Calendar

The EU ETS Phase I may be considered as a warm-up phase, during 2005–07. Phase II corresponds to the commitment period of the Kyoto Protocol, 2008–12. Phase III is supposed to correspond to a post-Kyoto agreement, that is, 2013–20. During each of these phases, the delivery of allowances is made on a yearly basis, and follows a precise calendar:

- On February 28 of year N, European operators receive their allocation for the commitment year N.
- 31 March of year N is the deadline for the submission of the verified emissions report during year $N-1$, from each installation to the European Commission.
- 30 April of year N is the deadline for the restitution of quotas utilized by operators during year $N-1$.

- 15 May of year N corresponds to the deadline of the official publication by the European Commission of verified emissions for all installations covered by the EU ETS during year $N - 1$.

The annual frequency of verified emissions, imposed by the European Commission, thus corresponds to a central event, structuring the diffusion of reliable information at the aggregated level on the European carbon market.

19.2.4   Transactions

One allowance exchanged on the EU ETS corresponds to 1 tonne of CO_2 released in the atmosphere, and is called a European Union Allowance (EUA). Allowance trading is recorded electronically by national registries. The information contained in these registries is centralized by the European Commission in the European registry, called the Community Independent Transaction Log (CITL).[9] The CITL contains exhaustive information on CO_2 emissions for all installations covered by the EU ETS, and is used to record the compliance position of each firm. The information contained in the CITL is available at the installation level. As a first step, data compilation appears necessary to reconstruct the ownership structures between subsidiaries and parent companies, which yield a more precise analysis for the evaluation of the scheme (McGuinness and Trotignon, 2007).

To comply with their emissions target, installations may exchange quotas either over-the-counter, or through brokers and marketplaces. Bluenext, formerly Powernext Carbon, is the marketplace dedicated to CO_2 allowance trading based in Paris. The European Climate Exchange (ECX) is the marketplace based in London, which is the leader for derivatives products. NordPool represents the marketplace common to Denmark, Finland, Sweden and Norway, and is based in Oslo. The prices of products exchanged in these marketplaces are strongly correlated; that is, they conform to other marketplaces like stock markets. Moreover, the European carbon market is characterized by an increasing sophistication of financial instruments using a quota of CO_2 as the underlying asset, and the development of option prices or swaps.[10]

Figure 19.3 indicates the total volume of allowances exchanged in the EU ETS during Phase I. This graph reveals that the number of transactions multiplied by a factor of four between 2005 and 2006, going from 262 to 809 million tonnes. This increasing liquidity of the market was confirmed in 2007, where the volume of transactions recorded equals 1.5 billion tonnes. This peak of transactions may be explained by the increase in the number of contracts valid during Phase II, with delivery dates going from December 2008 to December 2012, which amount for 4 per cent of total exchanges in 2005, and 85 per cent in 2007. These transactions reached €5.97 billion in 2005, €15.2 billion in 2006, and €24.1 billion in 2007, thereby confirming the fact that the EU ETS represents the largest emissions trading scheme to date in terms of transactions.

19.2.5   Penalties

During 2005–07, if an installation does not meet its emissions target during the compliance year under consideration, the penalty was equal to €40/tonne in excess, plus the

Total volume

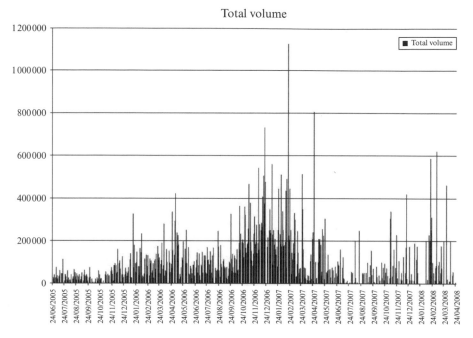

Source: BlueNext.

Figure 19.3   Volume exchanged for the spot price valid during 2005–07 from 24 June 2005 to 25 April 2008 in tons of CO₂

restitution of one allowance during the next compliance period. During 2008–12, this amount corresponds to €100/tonne, following the same principle.

Following this review of the institutional context on the European carbon market, the next section details the allowance price development and associated banking strategies.

19.3   BANKING PROVISIONS

This section details the banking borrowing provisions adopted in the EU ETS. Intertemporal emissions trading allows firms to smooth their emissions over time, and offers a greater flexibility in order to meet the emissions target. Therefore, banking and borrowing allow firms to achieve their depolluting objectives at least cost, if these provisions are adequately configured by the regulators and their effects have been sufficiently discussed, evaluated and understood. Let us first examine the allowance price development in the EU ETS during 2005–07.

19.3.1   Price Developments

In Figure 19.4, we may observe that from January to July 2005, CO_2 prices increased due to the perceived scarcity of allowances: demand comes mainly from power producers,

Source: BlueNext and ECX.

Figure 19.4 EUA spot and futures prices from July 2005 to May 2007

while most other market participants did not take advantage of buying or selling carbon allowances. From August 2005 to March 2006, the volume of transactions increased, driving the equilibrium allowance price up to €25/tonne of CO_2. Demand continued to come primarily from power operators, and increased during the winter due to the rise in energy prices – especially gas prices. During April to May 2006, the allowance market encountered a sharp drop in prices of all maturities, due to the first compliance report by the European Commission revealing that the market was oversupplied by approximately 4 per cent. The allowance price was divided by a factor of two within a window of only four days. Following this reversal of expectations from market operators, allowance prices stabilized at around €15/tonne of CO_2 during June to September 2006.

From October 2006 until the end of 2007, we finally observe a divorce between spot and futures prices of validity during Phases I (2005–07) and II (2008–12): while spot prices fell to €0.5/tonne of CO_2, futures prices remained in the range of €15 to €20/tonne of CO_2. The motives for such a disconnection between allowance prices of different maturities are explained in the next section. Since March 2007, allowance prices valid during 2008–12 have stabilized over €20/tonne of CO_2, following the decision by the European Council to maintain the EU ETS at least until 2020, and the decision to enforce stricter validation criteria for NAPs II (Convery and Redmond, 2007).

19.3.2 Banking Restrictions

In the EU ETS, allowances are valid during a specific compliance year. However, an installation may have banked allowances during year N to cover its emissions during year $N+1$, if years N and $N+1$ correspond to the same phase. The same mechanism applies for allowances borrowed from year $N+1$ in order to comply with the emissions target of the installation during year N. Thus, allowances banked or borrowed are fungible within the same phase. However, allowances distributed during Phase I are not valid during Phase II. Allowances distributed during Phases II and III are fungible between the different phases.

Phase I is characterized by a full intertemporal flexibility, like Phases II and III. Yet, given the simultaneity of the commitment periods between the Kyoto Protocol and Phase II, the intertemporal transfer of allowances has been strictly limited between Phases I and II, in effect banning the transfer of allowances between 31 December 2007 and 1 January 2008.

Alberola and Chevallier (2009) develop a statistical analysis showing that the disconnection between Phase I prices, decreasing towards zero, and Phase II prices, stabilized around 20€/tonne, may be explained by the restriction on the interperiod transfer of allowances enforced during Phase I. Indeed, the cost-of-carry relationship between EUA spot and futures prices for delivery during Phase II does not hold after the enforcement of the interperiod banking restrictions around October 2006.

The inefficiency of the EUA price signal to reflect correctly the social value of carbon until the end of Phase I may be explained by the restrictions enforced by member states concerning the transfer of quotas, banked or borrowed, from Phase I to Phase II. This sacrifice of the intertemporal flexibility mechanism may be interpreted by the will of the European Commission to limit the transfer of inefficiencies from the creation of the allowance market to Phase II, which simultaneously corresponds the Kyoto Protocol commitment period. Between Phases II and III of the EU ETS, the transfer of allowances has been authorized. Therefore, it appears possible to identify institutional learning effects between Phases I and II, as the early inefficiencies due to the youth of the European carbon market during 2005–07 do not seem to have been transferred to the subsequent periods. Preliminary analyses of the 2005–07 data concerning the extent of the use of banking in the EU ETS may be found in Ellerman and Trotignon (2008) and Chevallier et al. (2008).

To further develop this analysis of price developments in the EU ETS, I conduct in the next section a review of the main price fundamentals of EUAs during 2005–07.

19.4 CO_2 PRICE FUNDAMENTALS

This section focuses on the price fundamentals of CO_2 allowances. These fundamentals are mainly linked to regulatory decisions, energy prices and extreme temperatures events (Christiansen et al., 2005).

19.4.1   Institutional Decisions

First, it is worth noting that political and institutional decisions on the overall cap stringency have an impact on the carbon price setting through initial allocation. Also, any decision or announcement from regulators may induce changes in market players' behaviour. From this perspective, official communications by the European Commission are essential in order to reach a better information flow on installations' net short–long positions[11] (Ellerman and Buchner, 2008).

Whereas on energy markets the question of price formation is closely related to commodity storage, on the EU ETS the essential issue is the expected 'emission shortfall' during each compliance year. The emission shortfall, defined as the difference between verified emissions during the compliance year and allocated allowances, depends on the actual amount of emissions abatements required (which are unknown, but estimable based on reliable recent data) by the stringency of the cap (which is known). This information is publicly disclosed each year by the European Commission by mid-May, as detailed in section 19.2.3. It has a strong market effect on allowance price changes of all maturities, as it provides market participants with reliable information to update their expectations about future market developments.

Alberola et al. (2008) develop an original method to identify structural breaks in the CO_2 price series. They provide statistical evidence that two institutional events – in April 2006, following the disclosure of 2005 verified emissions, and in October 2006, following the European Commission announcement of the stricter Phase II – occurred during 2005–07. Those events had a sharp effect on market participants' expectations changes, and further allow the isolating of distinct energy and weather influences on carbon prices as discussed below.

19.4.2 Energy Prices

Second, energy prices are the most important price drivers in the short term of the EUA demand due to the ability of power generators to switch between their fuel inputs. This fuel-switching behaviour at the installations level applies especially in the power sector, which was endowed with more than 50 per cent of EUAs during 2005–07. The EU ETS price formation is indeed largely influenced by the electricity power market, since its participants are the main traders on the carbon market.

Figure 19.5 shows the price development of natural gas and coal prices from July 2006 to June 2007. The natural gas price (in €/MWh) is the daily futures *Month Ahead* natural gas price negotiated on Zeebrugge Hub. The price of coal (coal in €/tonne) is the daily coal futures *Month Ahead* price CIF ARA.[12] During 2005–07, natural gas prices exhibited strong volatility compared to coal prices. During the months of November–December 2005, natural gas prices soared to €50/MWh and steadily declined afterwards to €20/MWh during 2006, and to €10/MWh during the first quarter of 2007. The competitiveness of natural gas compared to coal therefore improved during 2006 and the first quarter of 2007 compared to the end of 2005.

Figure 19.6 shows the price development of the electricity price, as well as the 'clean dark' and 'clean spark' spreads from July 2006 to June 2007. The price of electricity from, Powernext (in €/MWh) is the contract of *Month Ahead* futures base. To take into account abatement options for energy industrials and relative fuel prices, it appears also important to introduce two specific spreads.[13]

The clean dark spread (in €/MWh) represents the difference between the price of electricity at peak hours and the price of coal used to generate that electricity, corrected for the energy output of the coal plant and the costs of CO_2:

$$clean\ dark\ spread = elec - \left(coal* \frac{1}{\rho_{coal}} + p_t * EF_{coal} \right) \tag{19.1}$$

with ρ_{coal} the net thermal efficiency of a conventional coal-fired plant,[14] and EF_{coal} the CO_2 emissions factor of a conventional coal-fired power plant.[15]

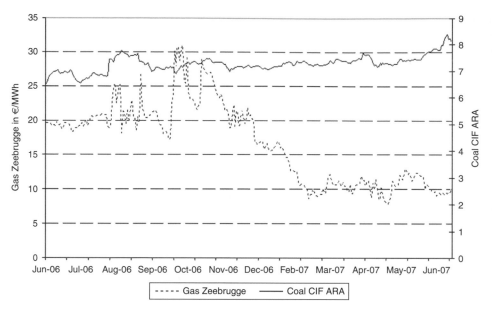

Source: Thomson Financial Datastream.

Figure 19.5   Gas Zeebrugge and coal CIF ARA prices from July 2006 to June 2007

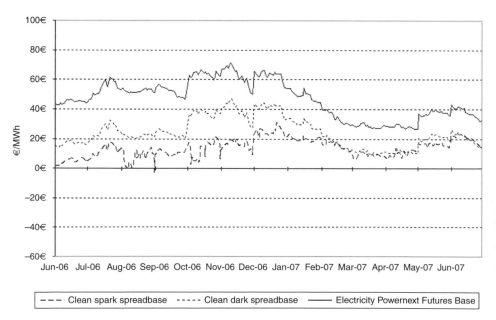

Source: Thomson Financial Datastream.

Figure 19.6   Clean spark spread, clean dark spread and Powernext electricity prices from July 2006 to June 2007

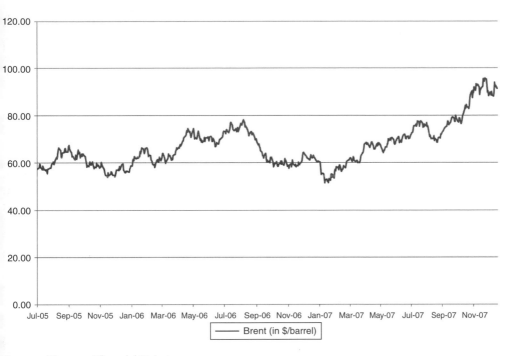

Source: Thomson Financial Datastream

Figure 19.7   Brent ICE prices from July 2005 to November 2007

The clean spark spread (in €/MWh) represents the difference between the price of electricity at peak hours and the price of natural gas used to generate that electricity, corrected for the energy output of the gas-fired plant and the costs of CO_2:

$$clean\ spark\ spread = elec - \left(ngas * \frac{1}{\rho_{ngas}} + p_t * EF_{ngas} \right) \qquad (19.2)$$

with ρ_{ngas} the net thermal efficiency of a conventional gas-fired plant,[16] and EF_{ngas} the CO_2 emissions factor of a conventional gas-fired power plant.[17]

During 2005–06, the use of coal appeared more profitable than gas. Since the beginning of 2007, the difference between the clean dark and clean spark spreads has been narrowing. This situation encourages consequently electric companies to decrease the use of coal to the profit of natural gas.

Figure 19.7 shows the price development of Brent prices from July 2005 to November 2007. The oil price (in $/barrel) is the daily Brent crude futures *Month Ahead* price negotiated on the Intercontinental Futures Exchange.

Alberola et al. (2008) show that energy prices forecast errors have basically driven the CO_2 price over 2005–07, but their influence changed over the period depending on regulatory changes. High levels of natural gas led power operators to realize a switch in fuel utilization from gas to coal. The natural gas price got higher from October 2005 to April 2006 and thereby positively influenced the EUA price. As the most CO_2-intensive variable, coal plays a negative role in carbon price changes: when confronted with a rise in the price of coal relative to other energy markets, firms have an incentive to adapt their energy mix towards less CO_2-intensive energy sources. Brent prices have a positive effect on EUA price changes, which channels through the natural gas price (Kanen, 2006). These energy influences on carbon prices are also in line with Mansanet-Bataller et al. (2007).[18]

The next section discusses weather influences on the EUA price formation.

19.4.3 Extreme Weather Events

Weather conditions have an impact on EUA price changes by influencing energy demand. Previous literature focuses on the most important dimension of weather: extremely hot and cold degree-days (Roll, 1984). In addition, this section discusses the non-linearity of the relationship between temperatures and carbon price changes.

Weather influences may be captured by using the daily data of Powernext Weather indices (expressed in °C) for four countries: Spain, France, Germany and the United Kingdom. These indices are computed as the temperature average at the representative regional weather station weighted by regional population:

$$\Theta = \frac{\sum_{i=1}^{N} pop_i * \Theta_i}{\sum_{i=1}^{N} pop_i} \tag{19.3}$$

with N the number of regions in the country under consideration, pop_i the population of region i, and Θ_i the average temperature of region i during the month under consideration in °C.

The European temperature index published by Tendances Carbone[19] may also be used. It is equal to the average of national temperatures indices provided by Powernext weighted by the share of each NAP in the previous four countries:

$$T = \frac{\sum_{j=1}^{4} Q_j * \Theta_j}{\sum_{j=1}^{4} Q_j} \tag{19.4}$$

with Q_j the number of allowances allocated by the NAP in country j, and Θ_j the national temperature index of country j. The national shares of allocation during Phase I in total allocation of EUAs are equal to 14.55 per cent for France, 46.40 per cent for Germany, 22.82 per cent for the UK, and 16.23 per cent for Spain, according to the European Commission.

Figure 19.8 represents the European Temperatures Index from July 2005 to April 2007. To take into account extreme weather conditions, Alberola et al. (2008) compute the deviation of the temperatures value from their seasonal average expressed by absolute value. They depart from previous literature by showing that unanticipated temperatures changes have a statistically significant effect on EUA prices only during some specific

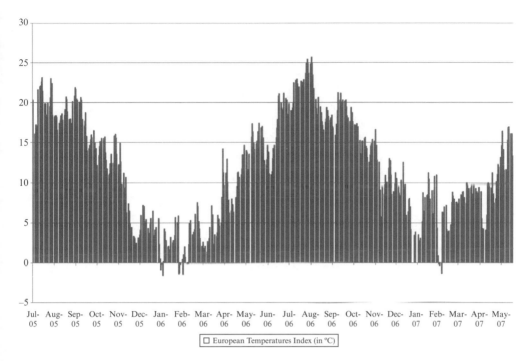

Source: Mission Climat – Caisse des Dépôts.

Figure 19.8   European Temperatures Index from July 2005 to May 2007

extreme weather events. During the winter of 2006, colder temperatures than decennial averages have had a positive impact on EUA price changes. Similarly, during the summer of 2006 and the winter of 2007, warmer temperatures negatively affected carbon price changes. The economic rationale behind this analysis is that when extremely cold events are colder (hotter) than expected, power generators have to produce more (less) than they forecasted, which conducts to an increase (decrease) of allowances demand and finally to an increase (decrease) of CO_2 price changes.

In this section, I have identified energy prices, weather events and institutional decisions as CO_2 price drivers during 2005–07. Linked to the influence of political, energy, climatic and economic uncertainties on CO_2 price changes, in the next section I discuss adequate risk-hedging strategies on the EU ETS.

19.5   RISK-HEDGING STRATEGIES

This section deals with the risk-hedging strategies used by firms. Investors naturally attempt at hedging against a variation of the risk attached to allowance trading, especially given the institutional amendments to the functioning of the scheme. I discuss first the introduction of option prices in the EU ETS, and then the consequences on market participants' hedging strategies.

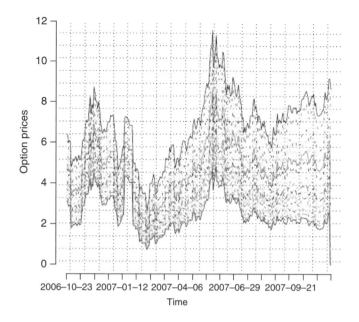

Source: ECX.

Figure 19.9   *Option prices available along with several strikes from October 2006 to October 2007*

19.5.1   Carbon-based Derivatives Products

Investors need to manage the risk of holding CO_2 allowances in the European carbon market among a portfolio of diversified investments. As in financial markets, the uses of derivatives products allow reduction of the risk of a position in emissions markets. Indeed, the European Climate Exchange launched in October 2006 derivatives products trading carbon allowances as the underlying asset.

Figure 19.9 displays option prices available along with several strikes from October 2006 to October 2007 on ECX. As detailed in section 19.2.4, ECX is the most liquid trading platform with approximately 86.5 per cent of the total exchange-based trades of allowances. The underlying assets of the contracts are first- and second-period spot prices. The maturity of the contracts typically ranges from December 2008 to 2013, Phase II contracts (2008–12) being more actively traded than post-Kyoto contracts. Option prices on the carbon market lead to pricing errors that are usual for commodity or equity markets. On such a commodity market, the easiest way to hedge against the risk of allowance price changes is by selling calls: call prices are more actively traded than puts.

Since option prices transfer the risk of financial exposure between market agents, in the next section I further detail agents' behaviour with respect to risk on this newly created commodity derivatives market.

19.5.2 Investors' Risk Aversion

Chevallier et al. (2009) estimate changes in investors' risk aversion on the European carbon market around the 2006 compliance event.[20] They recover investors' risk aversion by using the existing relationship with the risk-neutral and historic probabilities. This methodology has proved to be robust for stock markets. First, the risk-neutral distribution is recovered from ECX option prices. Second, the historical distribution is approximated by the historical return distribution of futures allowance prices. Third, the risk aversion is obtained as a by-product (Leland, 1980).

Figures 19.10 and 19.11 represent changes in the risk-neutral distribution for the futures contracts of maturity for December 2008 and December 2009, respectively. For both figures, the left panel denotes the risk neutral density before the 2006 compliance event, while the right panel denotes the risk neutral density after this institutional event. The left panel has a steeper slope than the right panel, which induces more volatility. These results are consistent with the role of information in lowering volatility on financial markets.

Figures 19.12 and 19.13 represent changes in the implied volatility for the futures contracts of maturity, respectively, December 2008 and December 2009. These two figures illustrate the dramatic changes in investor's risk aversion around the 2006 compliance event, as the implied volatilities exhibit dramatically different slopes depending on the sample considered. By extracting the information contained in option and futures prices, these results uncover a dramatic shift in investors' anticipation around the 2006 compliance event.

Overall, this study provides an efficient tool to quantify the effects of risk aversion on the European carbon market which, during the period under consideration, has been higher than on the stock market. This situation underlines the necessity for investors to manage adequately the risk attached to holding CO_2 allowances. With the start of Phase II on a sound institutional framework, risk aversion on the European carbon market is likely to tend progressively towards the values found on stock markets.

19.6 CONCLUDING REMARKS

This chapter reviews the market rules of the European carbon market during 2005–07. The synthesis of theoretical and empirical approaches developed here has been fruitful for the analysis of banking, pricing and risk-hedging strategies. These results teach us that institutional learning has indeed occurred within Phase I, from the viewpoint of both market agents and the regulator.

The banking restrictions enforced between December 2007 and January 2008 in the EU ETS led to the disconnection between spot prices valid during Phase I, which plummeted to zero, and futures prices valid during Phase II, which remained stable around €20 throughout the period. This particular episode of the EU ETS highlights the necessity to understand the underlying mechanisms of CO_2 price changes.

Like other commodity markets, the amount of allowances available for trading and thus the EUA price are driven by the balance between supply and demand (energy prices, weather variables, and so on), and other factors related to market structure and

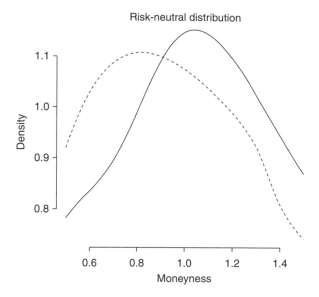

Source: Chevallier et al. (2009)

Figure 19.10   *Changes in the risk-neutral distribution for the December 2008 futures contract*

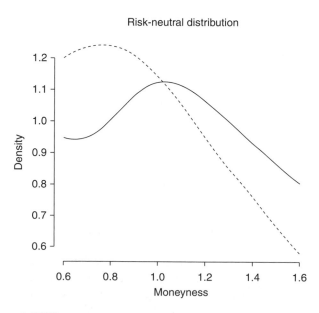

Source: Chevallier et al. (2009).

Figure 19.11   *Changes in the risk-neutral distribution for the December 2009 futures contract*

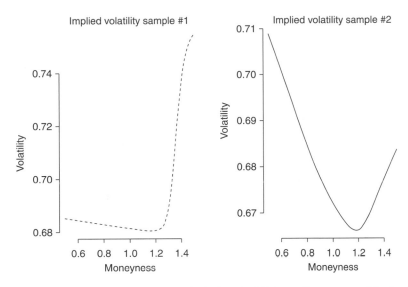

Note: The left panel denotes the risk-neutral density before the 2006 compliance event, while the right panel denotes the risk neutral density after this institutional event.

Source: Chevallier et al. (2009).

Figure 19.12 *Changes in the implied volatility for the December 2008 futures contract*

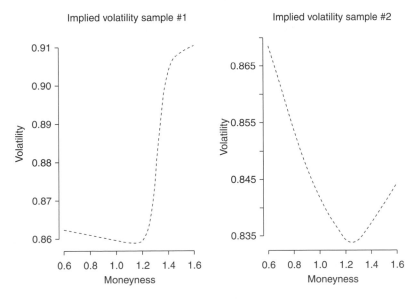

Note: The left panel denotes the risk-neutral density before the 2006 compliance event, while the right panel denotes the risk neutral density after this institutional event.

Source: Chevallier et al. (2009).

Figure 19.13 *Changes in the implied volatility for the December 2009 futures contract*

institutional policies. Decision changes within the regulatory environment have a sharp market effect on allowance prices of all maturities. These structural breaks may be statistically identified within the time series of carbon prices, and are linked to yearly compliance events and to official communications by the European Commission. It is also possible to isolate the influence of carbon price fundamentals linked to energy prices and extreme weather events, which vary before and after institutional events.

Carbon allowances therefore form another asset in commodities against which industrials and brokers need to hedge. Studies based on methods used on stock markets prove to be robust in quantifying changes in investors' anticipations.

Overall, this chapter highlights the inefficiencies following the creation of the European carbon market that prevented the emergence of a price signal leading to effective emissions reductions by industrials. The early design inefficiencies of the European carbon market, linked to initial allocation or the interperiod transfer of allowances, seem to have been corrected for the period 2008–12, thereby limiting the transfer of inefficiencies to Phase II.

ACKNOWLEDGEMENTS

The author is thankful to Emilie Alberola, Benoit Cheze and Florian Ielpo for fruitful collaboration. I wish to thank also Simon Buckle, Derek Bunn, Thierry Brechet, Denny Ellerman, Pierre-Andre Jouvet, Valerie Mignon, Christian de Perthuis, and Gilles Rotillon for discussions on this topic. Helpful comments were received from seminar audiences at Imperial College, LSE, MIT-CEEPR, Oxford, Toulouse; and conferences at CORE, EAERE, EEA-ESEM, Envecon, IAEE. Finally, I wish to acknowledge financial support from the French Ministry of Research and Superior Education. The usual disclaimer applies.

NOTES

1. Scientific uncertainty concerning the precise impacts of climate change does not justify delaying immediate action.
2. Each signatory country recognizes the impact of its GHG emissions on climate change. The most industrialized countries carry a heavier historical responsibility, given their prior GHG-intensive development, which translates into tighter targets.
3. Action will be taken in accordance with the economic development of each country.
4. These include the Regional Greenhouse Gas Initiative (RGGI), which contains several GHG reduction objectives in nine north-eastern states, and the Assembly Bill 32 in California which aims at reducing CO_2 emissions by 25 per cent by 2020 relative to 1990 emissions levels, and by 80 per cent by 2050. At the federal level, the Climate Stewardship Act introduced by Senator Lieberman-McCain did not find sufficient political support to become legally binding.
5. According to Article 12 of the Kyoto Protocol, CDM projects consist in achieving GHG emissions reduction in non-Annex B countries. After validation, the UNFCCC delivers credits that may be used by Annex B countries for use towards their compliance position.
6. See the European Commission Environment DG website at http://ec.europa.eu/environment/ets/.
7. Available at http://ec.europa.eu/environment/climat/emission/reviewen.htm.
8. The data come from the Mission Climat Caisse des Depots, available at http://www.caissedesdepots.fr.
9. Available at http://ec.europa.eu/environment/ets.
10. Note that there exists also financial instrument with a Clean Development Mechanism (CDM) credit

on the secondary market as the underlying asset, stemming from the Kyoto Protocol and fungible with quotas traded in the EU ETS with a maximum limit of around 13.4 per cent.

11. Note that an installation is defined as short (long) when it records a deficit (surplus) of allowances allocated with respect to actual emissions.
12. CIF ARA (cost, insurance and freight Amsterdam–Rotterdam–Antwerp) denotes the price of coal inclusive of freight and insurance delivered to the large North Western European ports, for example Amsterdam, Rotterdam or Antwerp.
13. As calculated by the Mission Climat of the Caisse des Depots for Tendances Carbone. The methodology is available at http://www.caissedesdepots.fr/IMG/pdf/_Document_Methodologie_Tendances_Carbone_EN_V4-2.pdf.
14. 40 per cent according to the NEA/IEA (2005) report, *The Projected Costs of Generating Electricity*.
15. 0.86 tCO$_2$/MWh (NEA/IEA, 2005).
16. 55 per cent (NEA/IEA, 2005).
17. 0.36 tCO$_2$/MWh (NEA/IEA, 2005).
18. Note that their study covers a shorter time period, going from 1 January 2005 to 30 November 2005.
19. As calculated by the Mission Climat of the Caisse des Depots for Tendances Carbone. The methodology is available at http://www.caissedesdepots.fr/IMG/pdf/_Document_Methodologie_Tendances_Carbone_EN_V4-2.pdf.
20. Given the central role played by the 2005 compliance event highlighted in section 19.3.1, I focus on the 2006 compliance event, which is the only event empirically observable following the introduction of options trading on ECX.

REFERENCES

Alberola, E. and J. Chevallier (2009), 'European carbon prices and banking restrictions: evidence from Phase I (2005–2007)', *The Energy Journal*, **30** (3), 107–36.
Alberola, E., J. Chevallier and B. Cheze (2008), 'Price drivers and structural breaks in European carbon prices 2005–2007', *Energy Policy*, **36** (2), 787–97.
Chevallier, J., J. Etner and P.A. Jouvet (2008), 'Bankable pollution permits under uncertainty and optimal risk-management rules: theory and empirical evidence', Working Paper EconomiX-CNRS 2008-25.
Chevallier, J., F. Ielpo and L. Mercier (2009), 'Risk aversion and institutional information disclosure on the European carbon market: a case-study of the 2006 compliance event', *Energy Policy*, **37** (1), 15–28.
Christiansen, A., A. Arvanitakis, K. Tangen and H. Hasselknippe (2005), 'Price determinants in the EU emissions trading scheme', *Climate Policy*, **5**, 15–30.
CDC (2006), 'Research Bulletin Number 8', Mission Climat of Caisse des Dépôts, Paris, available at http://www.caissedesdepots.fr.
CDC (2008), 'Research Bulletin Number 20', Mission Climat of Caisse des Dépôts, Paris, available at http://www.caissedesdepots.fr.
CITL (2007), 'Community independent transaction log', European Commission, available at http://ec.europa.eu/environment/ets.
CITL (2008), 'Community independent transaction log'. European Commission, available at http://ec.europa.eu/environment/ets.
Convery, F.J. (2009), 'Reflections – the emerging literature on emissions trading in Europe', *Review of Environmental Economics and Policy*, **3** (1), 121–37.
Convery, F.J., D. Ellerman and C. de Perthuis (2008), 'The European carbon market in action: lessons from the first trading period', Interim Report, MIT-CEEPR, Mission Climat Caisse des Dépôts and University College Dublin.
Convery, F.J. and L. Redmond (2007), 'Market and price developments in the European Union Emissions Trading Scheme', *Review of Environmental Economics and Policy*, **1** (1), 88–111.
Ellerman, D. and B. Buchner (2008), 'Over-allocation or abatement? A preliminary analysis of the EU ETS based on 2005 emissions data', *Environmental and Resource Economics*, **41**, 267–87.
Ellerman, D. and R. Trotignon (2008), 'Compliance behavior in the EU ETS: cross border trading, banking and borrowing', MIT-CEEPR Working Paper 2008-12.
Kanen, J.L.M. (2006), *Carbon Trading and Pricing*, London: Environmental Finance Publications.
Leland, H.E. (1980), 'Who should buy portfolio insurance?', *Journal of Finance*, **35** (2), 581–94.
Mansanet-Bataller, M., A. Pardo and E. Valor (2007), 'CO$_2$ prices, energy and weather', *Energy Journal*, **28** (3), 67–86.

McGuinness, M. and R. Trotignon (2007), 'Technical memorandum on analysis of the EU ETS using the Community Independent Transaction Log', MIT-CEEPR Working Paper 2007-12.

NEA/IEA (2005), *The Projected Costs of Generating Electricity: 2005 update*, Nuclear Energy Agency, International Energy Agency, Paris: OECD.

Roll, R. (1984), 'Orange juice and weather', *American Economic Review*, **74** (5), 861–80.

UNFCCC, (2000), 'Procedures and mechanisms relating to compliance under the Kyoto Protocol: note by the co-chairmen of the Joint Working Group on Compliance', United Nations Framework Convention on Climate Change, Bonn, Report.

20 The Clean Development Mechanism: a stepping stone towards world carbon markets?

Julien Chevallier

20.1 INTRODUCTION

Industrial operators may reduce their CO_2 emissions by using credits issued from the Kyoto Protocol Clean Development Mechanism (CDM), called Certified Emissions Reductions (CERs).[1] These CERs correspond to one tonne of avoided CO_2 emissions in the atmosphere, and may be obtained through projects development in non-Annex B countries of the Kyoto Protocol that allow to reduce emissions compared to a baseline scenario. Once credits have been issued by the CDM Executive Board of the United Nations, they may be sold by project developers on the market, and thus become secondary CERs (sCERs). The European Union Emissions Trading Scheme (EU ETS) is the EU's flagship climate policy forcing industrial polluters to reduce their CO_2 emissions in order to help the European Union member states to achieve their Kyoto Protocol target.[2]

The compliance of industrial operators requires a balance between verified emissions and allocated allowances. Both European Union Allowance (EUA) and sCER prices may be used towards compliance within the EU ETS due to the partial fungibility between these two carbon assets. Indeed, to provide more flexibility to carbon-constrained installations, the European Commission has allowed industries covered by the EU ETS to use both assets for compliance. However, it has established a limit on the use of CERs (primary or secondary) up to 13.4 per cent of their allocation from 2008 to 2012 on average. To comply with their emissions cap, industrial emitters may thus adopt various strategies: (1) surrender EUAs (allocated either to the plant or to other plants of the same company); (2) reduce real emissions (either at the installation level or abroad, using the Kyoto Protocol's flexibility mechanisms); (3) buy EUAs or/and sCERs; (4) borrow EUAs from future allocation; (5) surrender banked EUAs from past allocation. Trotignon and Leguet (2009) document that, in 2008, 96 per cent of the surrendered allowances were EUAs, and only 3.9 per cent were sCERs.[3] The trade-offs between using EUAs or sCERs towards compliance in the EU ETS depend on their respective price trends, and the price difference between them.

In theory, as the sCERs are free of project delivery risks, the prices of EUAs and sCERs should be equal since they represent the same amount of CO_2 emissions reduction (1 tonne). However, due to the limit of 13.4 per cent on average of the credits surrendered, the sCER 'exchange rate' is smaller than that for EUAs, and therefore sCERs are discounted with respect to EUAs. This premium represents the opportunity cost of using EUAs for compliance instead of sCERs. Beyond prices, regulatory issues may also explain the price between variation between these two carbon assets in the long run. First, with the European Energy Climate package, the EU ETS is confirmed until 2020. However, the details concerning the import of CDM credits within Phase III (2013–20) are not known with certainty. Indeed, the European Union establishes particular conditions of the emissions trading scheme in Phase III that are dependent on the achievement of a post-Kyoto international agreement at the December 2011 United Nations Framework Convention on Climate Change (UNFCCC) Durban Summit. Thus, there exists a wide range of uncertainties arising around the status and recognition of CERs (both primary and secondary) in a revised EU ETS beyond 2012. Second, carbon assets form another class of commodities against which traders need to define specific hedging strategies (Chevallier, 2009; Chevallier et al., 2009).

The central goals of this chapter are twofold: (1) to study the price drivers of sCERs; and (2) to explain the links with EUAs. Compared to previous literature, I provide the first empirical analysis of sCERs drivers. Indeed, Mansanet-Bataller et al. (2007), Alberola et al. (2008) and Alberola and Chevalier (2009) have already analysed the price fundamentals of EUAs, but not the drivers of sCERs. In addition, I review the main characteristics of EUAs and sCERs through cointegration analysis, vector error correction (VEC) and vector autoregression (VAR) modelling (as detailed previously by Chevallier, 2010).

My central results show that EUAs and sCERs share the same price drivers, that is, these emissions markets prices are mainly determined by institutional events, energy prices, weather events, and macroeconomic variables. Moreover, EUAs are found to determine significantly the price path of sCERs, by accounting for a large share of the explanatory power of sCERs. This result emphasizes that EUAs remain the main 'money' in the field of emissions market, which is exchanged broadly as the most liquid asset for carbon trading. The trading of sCERs, while growing exponentially, is still mostly determined by the fact that the EU ETS remains the largest emissions trading scheme to date in the world.[4] This result also explains why sCERs are traded at a discounted price from EUAs: with the project risk which is characteristic of primary CERs, sCERs are still limited by the import limit set within the EU ETS.

Regarding the links between EUAs and sCERs, it is found that both price series are cointegrated. The VEC model indicates that EUAs are leading the price discovery in a dynamic system with sCERs prices, while the VAR model reveals interdependencies between the two markets (as both variables are found to influence each other statistically in a static framework). Taken together, the results indicate that while the fungibility between emissions markets worldwide is quickly developing, there remain significant opportunities for price arbitrage.

The remainder of the chapter is organized as follows. Section 20.2 details compliance strategies in carbon markets. Section 20.3 develops a cointegration analysis between EUAs and sCERs prices. Section 20.4 covers the specific sCERs price drivers. Section 20.5 concludes.

20.2   COMPLIANCE STRATEGIES IN CARBON MARKETS

This section briefly reviews background information on the EU ETS, which was launched in 2005 according to Directive 2003/87/EC to facilitate EU compliance with its Kyoto commitments. Phase I was introduced as a training period during 2005–07. Phase II coincides with the commitment period of the Kyoto Protocol (2008–12). Phase III will cover the period 2013–20. Around 11 000 energy-intensive installations are covered by the scheme, which accounts for nearly 50 per cent of European CO_2 emissions (Alberola et al., 2009a, 2009b). Emissions caps are determined at the installation level in National Allocation Plans (NAPs). In what follows, I examine EUAs and CERs contracts more closely, as well as their respective price developments.

20.2.1 EUAs and CERs Contracts

On the one hand, EUAs are the default carbon asset in the EU emissions trading system. They are distributed by European member states throughout NAPs, and allow industrial owners to emit one tonne of CO_2 into the atmosphere. The supply of EUAs is fixed in NAPs, which are known in advance by market participants (2.08 billion per year during 2008–12).[5]

On the other hand, CERs, which also compensate for tonne of CO_2 emitted by their owners, are much more heterogeneous than EUAs. Primary CERs represent greenhouse gases emissions reductions achieved in non-Annex B countries of the Kyoto Protocol. These certificates are issued by the United Nations Clean Development Mechanism Executive Board (CDM EB). CDM projects may associate various partners (ETS compliance buyers, Kyoto-bound countries, project brokers, profit-driven carbon funds, international organizations such as the World Bank, and so on). CDM projects partnerships are governed by emissions reduction purchase agreements (ERPAs).[6] The price of primary CERs will depend on the risk of each project, and on its capacity to issue primary CERs effectively. This price will be the cost of the project divided by the number of primary CERs actually issued. Thus, primary CERs from different projects will have different prices.

Once issued by the CDM EB, primary CERs may either be used by industrial firms for their own compliance, or sold to other participants in the market. In the latter case, it becomes a secondary CER (sCER). Note that as the sCERs are CERs that have been already issued by the CDM EB, their project delivery risk is null. As stated in the introduction, the main difference between the use of EUAs and CERs (both primary and secondary) for compliance in the EU ETS lies in the 13.4 per cent (on average) import limit set by the European Commission on CERs, while EUAs may be used without any limit. The CERs import limit for compliance is equal to 1.4 billion tonne of offsets being allowed into the EU ETS from 2008 to 2012.[7]

In this chapter I focus on the price relationships between EUAs and sCERs. Next, I describe the EUAs and sCERs price developments.

20.2.2 Price Development

In this section, I examine Phase II EUA and sCER prices, which reflect the price of reducing emissions during the commitment period of the Kyoto Protocol (2008–12).[8] The sCER price series used for this study is the longest historical price series existing for sCERs: the sCER Price Index developed by Reuters. It has been built by rolling over two sCERs contracts with different maturity dates (December 2008 and December 2009). Similarly, I have rolled over EUA futures contracts traded at the European Climate Exchange (ECX) of the corresponding maturity dates (December 2008 and December 2009) to match them with the sCER price series.[9] The sample period considered starts

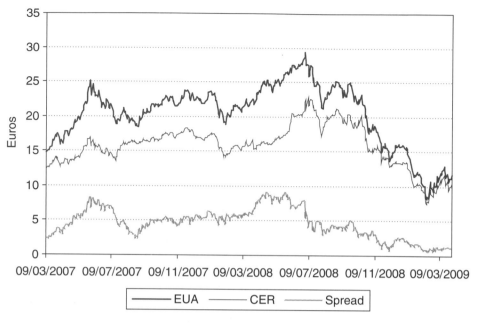

Source: Reuters.

Figure 20.1   Time-series of ECX EUA Phase II futures, Reuters CER Price Index and CER–EUA spread from 9 March 2007 to 31 March 2009

with the beginning of the sCER Price Index (9 March 2007) and ends on 31 March 2009. As shown in Figure 20.1, the EUA and the sCER price series follow a similar price path.

EUAs were traded at €15 in March 2007, then stayed in the range of €19–€25 until July 2008, and decreased steadily afterwards to achieve €8 in February 2009. sCERs started at €12.5 in March 2007, evolved in the range of €12–€22 to July 2008, and continued to track EUA prices until €7 in February 2009. Thus, sCERs have always remained below EUAs and consequently the spread has been positive during all of the sample period. Descriptive statistics for EUAs, sCERs and the spread may be found in Table 20.1.

Given the price paths observed in historical data, it appears interesting to investigate the presence of one cointegrating relationship between EUAs and sCERs in the next section.

20.3   COINTEGRATION ANALYSIS

Following the methodology used in Manzoni (2002) and Ramchander et al. (2005), who studied the relationship between bond spreads, I proceed in a first step by identifying the possible cointegration relationship between the two types of assets considered (EUAs and sCERs). I will then analyse the EUA–sCER spread drivers.

Table 20.1   Summary statistics for all dependent variables

Variable	Mean	Median	Max.	Min.	Std. Dev.	Skew.	Kurt
Raw prices series							
EUA_t	20.40389	21.52000	29.33000	8.20000	4.459218	–	3.031938
$sCER_t$	15.85798	16.6875	22.8500	7.484615	2.986495	–	3.135252
$Spread_t$	4.545912	4.620000	9.043571	0.647857	2.108445	0.047792	2.292397
Natural logarithms							
EUA_t	2.986643	3.068983	3.378611	2.104134	0.255164	–	4.275898
$sCER_t$	2.743941	2.776476	3.128951	2.012850	0.505511	–	4.182189
Log returns							
EUA_t	–	0.0001	0.113659	–	0.026833	–	4.868026
$sCER_t$	–	0.0001	0.112545	–	0.024441	–	5.961950
VAR(4) residuals							
EUA_t	0.00001	0.001242	0.108251	–	0.05903	–	4.522629
$sCER_t$	0.00001	0.000390	0.111584	0.097672	0.023742	–	5.520998
First-differences							
Δ	–	–	1.070000	–	0.295605	–	6.262861

Note: EUA_t refers to ECX EUA Futures, $sCER_t$ to Reuters sCER Price Index, and $Spread_t = EUA_t\text{-}sCER_t$ spread. Std.Dev. stands for Standard Deviation, Skew. for Skewness, and Kurt. for Kurtosis. The number of observations is 529. The VAR(4) specification is detailed in section 20.2.3.

20.3.1   Unit Roots and Structural Break

A necessary condition for studying cointegration involves that both time-series are integrated of the same order. I thus examine the order of integration, noted d, of the time-series under consideration based on Zivot and Andrews's (1992) unit root test. This test allows examining the unit root properties of the time-series, while simultaneously detecting endogenous structural breaks for each variable. Figure 20.2 presents the Zivot–Andrews unit root test statistics for the two EUA and sCER variables transformed to log-returns.

The model estimated is a combination of a one-time shift in levels, and a change in the rate of growth of the series. The null of unit root is clearly rejected in favour of the break-stationary alternative hypothesis. One estimated break point is identified for each of the time-series: 13 February 2009 for the EUA variable, and 20 February 2009 for the sCER variable. These breakpoints may be due to a delayed effect of the 'credit crunch' crisis on the carbon market (see Chevallier, 2009 for a discussion). Both time-series are integrated of order 1 ($I(1)$). The existence of a structural break in the time-series considered, while remaining stationary, means that cointegration tests need to be developed that explicitly include potential breaks, as they have been developed by Lutkepohl et al. (2004).

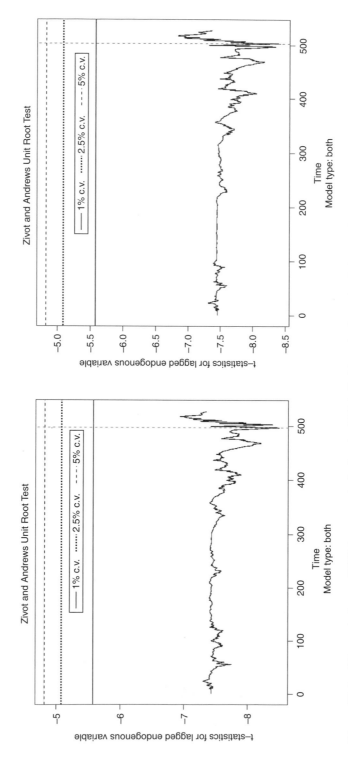

Figure 20.2   Zivot–Andrews (1992) test statistic for the EUA (left) and sCER (right) variables

Table 20.2   Johansen cointegration rank trace statistic, cointegration vector, model weights and VECM with structural break for the EUA and the CER variables

Panel A: Johansen cointegration rank trace statistic

Hypothesis	Statistic	10%	5%	1%
$r \leq 1$	5.26	5.42	6.79	10.04
$r = 1$	16.95	13.78	15.83	19.85

Panel B: Cointegration vector

Variable	EUA (1)	sCER (1)
EUA (1)	1.0000	1.0000
sCER (1)	−0.4955009	−1.519945

Panel C: Model weights

Variable	EUA (1).	sCER (1)
ΔEUA	−0.06163548	0.00734759
ΔsCER	−0.04490726	0.0182197

Panel D: VECM with structural break (r = 1)

Variable	ΔEUA	ΔsCER
Error Correction Term (ect)	−0.0197908	−0.0282009
Deterministic constant	0.0106349	0.0154190
Lagged differences		
ΔEUA (1)	−0.0641515	−0.0504123
ΔsCER (1)	0.2307197	0.1423340

Notes: EUA refers to ECX EUA Phase II Futures, sCER to Reuters sCER Price Index, transformed to natural logarithms.
Critical values are reported in Lutkepohl et al. (2004).
Lag order in parenthesis.
The number of observations is 529.

20.3.2   VECM and Structural Break

After having validated the necessary condition for studying cointegration (which requires that both time-series should be integrated of the same order), I now investigate the existence of a long-term relationship across these two carbon prices by employing a cointegration analysis with the maximum-likelihood test procedure established by Johansen and Juselius (1990) and Johansen (1991). Results for the cointegration test with one structural shift at unknown time (Lutkepohl et al., (2004) are shown in Panel A of Table 20.2. The trace statistic result indicates a cointegration space of $r = 1$, given a

5 per cent significance level. We may conclude that there exists one long-term cointegrating vector between the EUA and sCER variables taken in natural logarithm form.

Next, I proceed to the estimation of the vector error correction Model (VECM), which is useful in making causal inferences among the variables of the system.[10] As shown in Panel D of Table 20.2, the coefficients of the error correction terms for the EUA and sCER variables are negative, and thus I validate the error correction specification. In terms of short-run dynamics, the error correction terms emerge as important channels of influence in mediating the relationship between the different EUAs and sCERs prices. Notice in Panel D of Table 20.2 that the error correction term appears stronger for sCERs than for EUAs. This implies that the sCER variable has a stronger behaviour to adjust to past disequilibria by moving towards the trend values of the EUA variable. This specification confirms that EUAs constitute a leading factor in the price formation of sCERs. It can also be seen that changes in the respective prices of EUAs and sCERs have a significant causal influence (in the Granger sense) on each other.[11]

20.3.3   VAR(*p*) Modelling

In light of the previous results, and in order to proceed with the suitable identification of the price drivers for each variable, I use a VAR(*p*) in differences with an intervention dummy for February 2009 to model the data-generating process of the EUA and sCER log-series. The VAR(*p*) model is specified as follows:

$$\Delta y_t = A_0 + A_1 \Delta y_{t-1} + A_2 \Delta y_{t-2} + \ldots + A_p \Delta y_{t-p} + \varepsilon$$

where

$$\Delta y_t = \begin{bmatrix} \Delta EUA_t \\ \Delta sCER_t \end{bmatrix}$$

is a vector of EUA and sCER log-returns,

$$A_0 = \begin{bmatrix} b_{10} \\ b_{20} \end{bmatrix}$$

is a vector of constants, and

$$A_1 = \begin{bmatrix} \gamma_{11} & \gamma_{12} \\ \gamma_{21} & \gamma_{22} \end{bmatrix},$$

and so on are the coefficient matrices.

To determine the appropriate lag structure, I computed the following information criteria: Akaike ($AIC(n) = 4$), Schwarz ($SC(n) = 1$), Hannan–Quinn ($HQ(n) = 1$), and Final Prediction Error ($FPE(n) = 4$). Since the Ljung–Box–Pierce Portmanteau test on the residuals of the VAR(1) model indicated the presence of autocorrelation, I choose to retain a lag of order $p = 4$. As shown in Table 20.3, residuals are not auto-correlated for the VAR(4) model.

The ARCH effect is very strong, which indicates the necessity to use a GARCH model for further analysis. Figure 20.3 plots the log-returns and the VAR(4) residuals of the ECX EUA Phase II Futures and sCER Price Index time price series.

Table 20.3   Diagnostic test of VAR(4) model

Test	Statistic	DF	*p*-value
Portmanteau	57.4878	48	0.16
ARCH VAR	97.1946	9	0.01
JB VAR	147.6817	4	0.01
Kurtosis	143.5005	2	0.01
Skewness	4.1811	2	0.12

Note: Portmanteau is the asymptotic Portmanteau test with a maximum lag of 16, ARCH VAR is the multivariate ARCH test with a maximum lag of order 5, JB is the Jarque Bera Normality test for multivariate series applied to the residuals of the VAR(4). Kurtosis and Skewness stand for separate tests for multivariate kurtosis and skewness. DF stands for degree of freedom of the test statistic.

Figure 20.4 shows the OLS-based CUSUM tests for the VAR (4) residuals. Despite some structural instability around the February 2009 breakpoints, the residuals stay within the interval confidence levels.

Additional impulse response analysis reveals the traditional 'hump' shape between EUAs and sCERs, as shocks pass on both variables and fluctuations dampen at the horizon of ten lags.[12] The variance decomposition indicates that the variance of the forecast error for the EUA price is due to its own innovations up to 90 per cent. For the sCER price, the variance of the forecast error is due to EUAs up to 70 per cent, and only 30 per cent to its own innovations. These results confirm the findings in section 20.2.2.

In the next step of this empirical analysis, I proceed by fitting a suitable GARCH model to the residuals of the VAR(4) model for the EUA and sCER variables.

20.4   SCERs PRICE DRIVERS

It is important to distinguish between demand and supply factors affecting sCERs. Contrary to the allocation of EUA, the supply of sCERs is unknown. The main sources of uncertainty are due to the facts that: (1) the supply of primary CERs is unknown and difficult to estimate (as it depends on several risks related to the issuance of primary CERs); and (2) the amount of primary CERs that will be converted into sCERs is also difficult to assess (see Trotignon and Leguet, 2009). On the demand side, whereas on the EU ETS the demand comes from private financial or industrial operators, for sCERs the demand comes from a larger number of participants (investors, industrials and Annex B countries). Most of the CERs demand to date comes from European industrials, which are limited to 13.4 per cent (on average) of surrendered allowances for compliance during Phase II of the EU ETS. Annex B countries of the Kyoto Protocol may also use CERs for compliance. Countries with a potential deficit of Assigned Amount Units (AAUs valid under the Kyoto Protocol) in 2012 – such as Japan – are involved in sCERs purchasing.

Among other factors that may impact upon sCERs prices, the same factors as those affecting EUAs prices can be identified, since both assets may be used for compliance in the EU ETS. Announcements relative to the strictness of NAPs have been shown

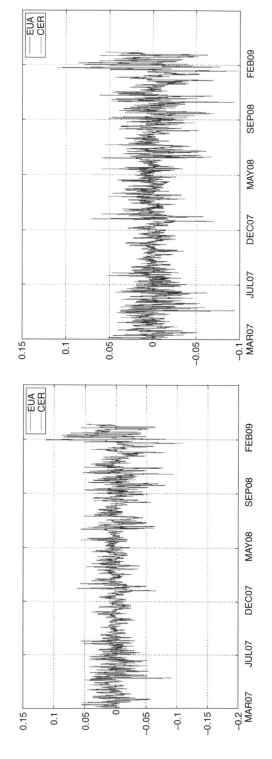

Figure 20.3   Log-returns (left) and VAR(4) residuals (right) of ECX EUA Phase II futures and Reuters sCER price index, 9 March 2007 to 31 March 2009

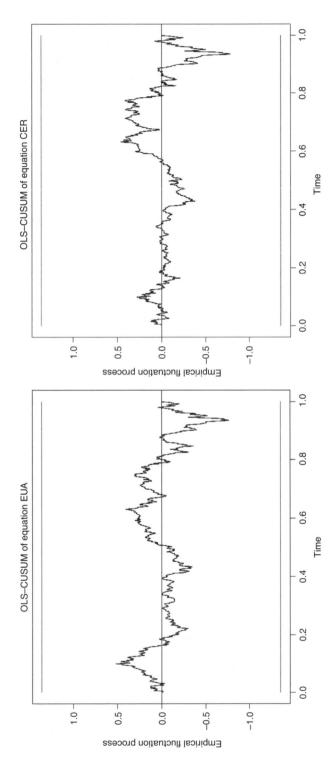

Figure 20.4   OLS-CUSUM test for the EUA (left) and sCER (right) variables of the VAR(4) model

to have a strong influence on EUA prices (Alberola et al., 2008; Chevallier et al., 2009; Mansanet-Bataller and Pardo, 2009). Concerning demand factors, previous literature identifies energy prices, weather events and the level of industrial production as being the main drivers of EUAs during Phase I (Mansanet-Bataller et al., 2007; Alberola et al., 2009a, 2009b).

20.4.1   Database

I include as CERs price drivers the most representative energy prices in Europe. That is, the daily Brent and natural gas futures prices traded at the International Petroleum Exchange (IPL) and coal prices CIF ARA.[13] The time-series have been built by rolling over the nearest month ahead contract. As the futures contract on Brent is quoted in USD per barrel, the futures contract on natural gas is quoted in GBP per therm, and the coal contract is quoted in USD per metric ton, I have converted all price series to euros by using the daily exchange rate data available from the European Central Bank.[14] Figure 20.5 shows these energy prices.

I use the CO_2 switch price between coal and gas in €/tonne, as computed in the *Tendances Carbone* database.[15] This variable represents the fictional daily price that establishes the equilibrium between the 'clean dark spread' and the 'clean spark spread'.[16] It therefore represents the price of CO_2 above which it becomes profitable in the short term for an electric power producer to switch from coal to natural gas. The economic logic behind the use of these spreads lies in the central role played by power producers in the determination of the EUA price, since they receive around half of the allowances distributed in the EU emissions trading system (Delarue et al., 2008; Ellerman and Feilhauer, 2008). The CO_2 switch price, clean dark and clean spark spreads are displayed in Figure 20.6.

To take into account weather influences, I use the *Tendances Carbone* European temperatures index, which is an average of national temperatures indices of four European countries (France, Germany, Spain and the United Kingdom), weighted by the share of each National Allocation Plan. From this index, I have created three new variables: *tempec* represents the difference between the value of the temperatures index and the decennial average; *temphot* is a dummy variable for extremely hot temperatures (equal to 1 if the value of the temperatures index is higher than the third quartile of the series, and 0 otherwise); and *tempcold* is a dummy variable for extremely cold temperatures (equal to 1 if the value of the temperatures index is lower than the first quartile of the series; and 0 otherwise). The temperatures index and its deviation from decennial average are shown in Figure 20.7.

I have also introduced exogenous variables impacting upon CO_2 emissions levels. First, we consider the *Tendances Carbone* European Industrial Production index indicator, which uses Eurostat production indices and is a backward-looking indicator tracking past economic trends. Second, I use the Economic Sentiment Index published by Eurostat, which reflects overall perceptions and expectations at the individual sector level in a single aggregate index. This index is a forward-looking indicator used to mirror economic sectors' sentiment. Finally, the 'credit crunch' crisis may also have an impact on CO_2 emissions levels. To detect this potential influence, I have created the variable *crisis* as a dummy variable equal to 1 from 17 August 2007 onwards and 0 otherwise.

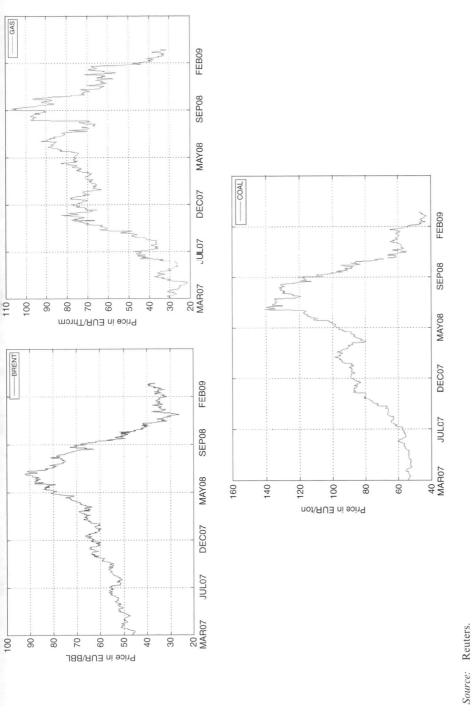

Source: Reuters.

Figure 20.5   IPE crude oil Brent, IPE natural gas and coal CIF ARA prices, 9 March 2007 to 31 March 2009

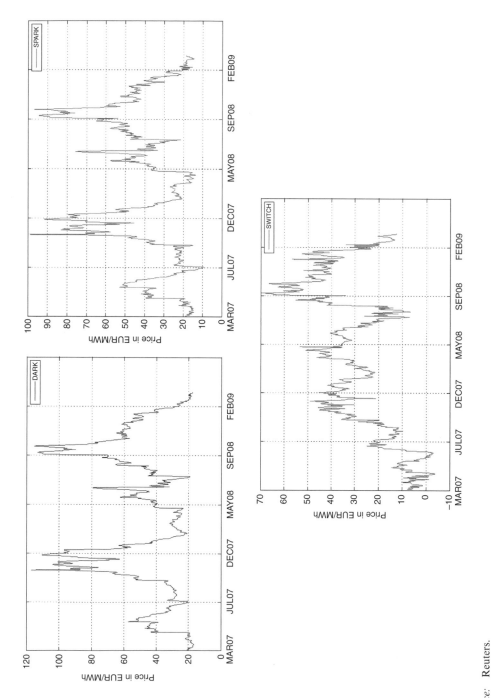

Source: Reuters.

Figure 20.6 Clean dark spread, clean spark spread and switch price, 9 March 2007 to 31 March 2009

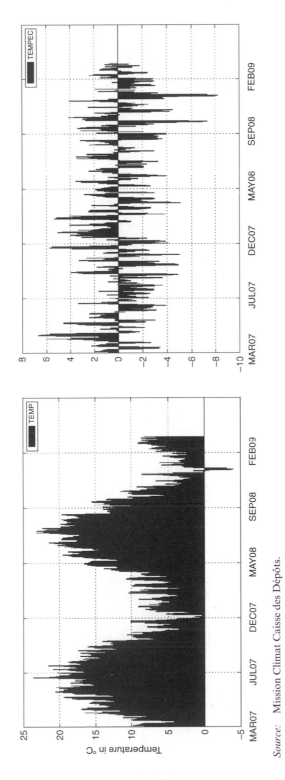

Source: Mission Climat Caisse des Dépôts.

Figure 20.7 European Temperatures Index and deviation from decennial average, 9 March 2007 to 31 March 2009

This date corresponds to the first cut in interests rates by the US Federal Reserve, and may be considered as the beginning of the financial crisis (Chevallier, 2009). Figure 20.8 shows the European Industrial Production Index and the European Sentiment Index variables.

Additionally, three other variables relevant to market trends are considered. First, to take into account the slope of the euro area yield curve, I have used the *yield* variable, which is available from the European Central Bank.[17] This series is built as the spread between the five- and the two-year interest rates. A positive (negative) value of the variable *yield* is expected to indicate an upward-sloping (downward-sloping) interest rate term structure, and hence a trend to cool down (stimulate) the economy (Collin-Dufresne et al., 2001). Second, I have computed the $momentum_{EUA}$ variable. This variable represents the difference between ECX EUA Phase II Futures prices at time t and at time t-5, thereby indicating bullish or bearish carbon market trends. Finally, *VIX* is the volatility index published by the Chicago Board Options Exchange (CBOE), which is widely recognized as an indicator of aggregate market volatility among financial practitioners (Collin-Dufresne et al., 2001). Figure 20.9 presents the evolution of the three variables.[18]

Regarding news variables that may impact upon the supply of EUAs, three types of events are considered. First the arrival of new information concerning Phase II NAPs is taken into account. Second, news related to the extended development of the EU ETS during Phase III is considered. These two dummy variables have been constructed by filtering the most reliable and significant announcements on EU ETS developments from the European Commission website.[19] Third, I also take into account the likely impact on EUA prices provoked by the connection between the Kyoto Protocol's International Transaction Log (ITL) and the EU ETS Community Independent Transaction Log (CITL) on 10 October 2008 thanks to the *ITL-CITL* dummy variable. This variable takes the value of 1 when news concerning the connection occurred and 0 otherwise. Note that I have computed a new specific variable, called $momentum_{sCER}$, for the indication of bullish and bearish periods. Similarly to the case of the $momentum_{EUA}$ variable, the $momentum_{sCER}$ variable is obtained as the difference between the sCER variable at time t and at time t-5.

I also add three variables that take into account the specificities of sCERs (mostly related to the supply side): *CDM EB meeting*, *linking* and *CDMpipeline*. The dummy variable *CDM EB meeting* is equal to 1 on the publication date of CDM EB's reports, and 0 otherwise. This variable indicates the arrival of new information when CDM projects are validated for the delivery of CERs credits. The dummy variable *linking* is equal to 1 when there is an announcement date related to the linking of emissions trading schemes worldwide, and 0 otherwise.[20]

Finally, the *CDMpipeline* variable is the forecast error concerning the number of primary CERs actually delivered by the CDM EB. Each month, the United Nations Environment Programme (UNEP) Risoe announces how many primary CERs are expected to be delivered in the CDM pipeline.[21] This variable is computed following the approach developed by Kilian and Vega (2011):

$$CDMpipeline_t = \frac{Realised_t - Expected_t}{\hat{\sigma}}$$

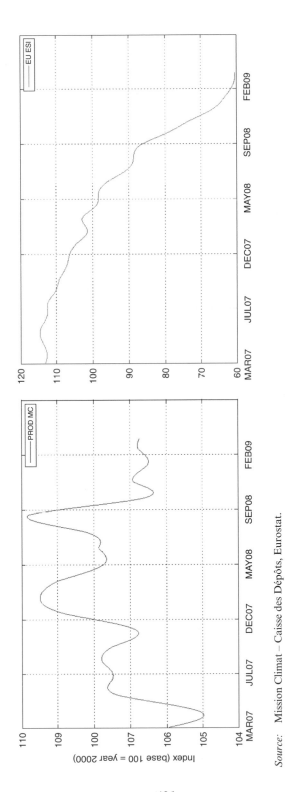

Source: Mission Climat – Caisse des Dépôts, Eurostat.

Figure 20.8   Mission Climat Industrial Production Index (weighted by the share of NAPs) and EU Economic Sentiment Index

431

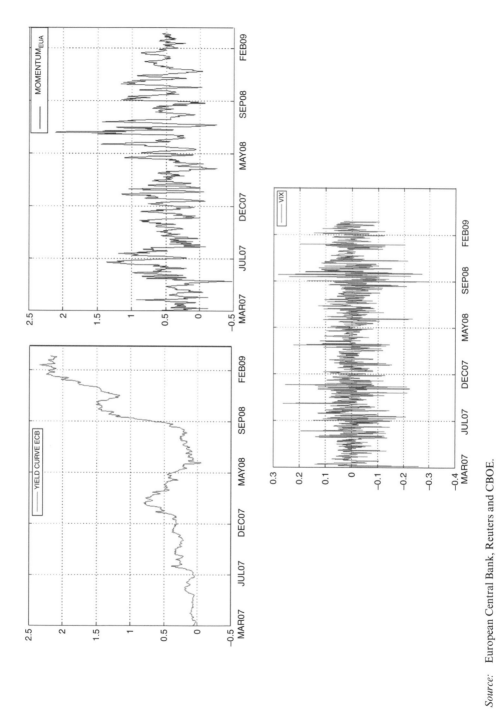

Source: European Central Bank, Reuters and CBOE.

Figure 20.9 Slope of yield curve, market momentum and VIX index, 9 March 2007 to 31 March 2009

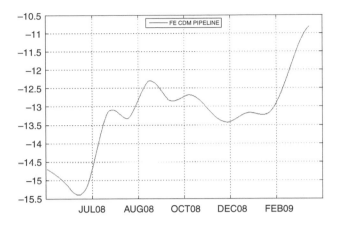

Source: UNEP Risoe and Mission Climat Caisse des Dépôts.

Figure 20.10   Forecast errors for the number of CERs available in the CDM pipeline, May 2008 to March 2009

With *Realised,* the announced value of the amount of primary CERs delivered by the UNEP Risoe, *Expected,* the market's expectation of the amount of primary CERs to be delivered prior to the announcement, calculated by Trotignon and Leguet (2009), and $\hat{\sigma}$ the sample standard deviation of the 'surprise' component. Figure 20.10 shows the forecast errors for the number of primary CERs available in the CDM pipeline.

After transforming, when necessary, the exogenous variables of our database into stationary variables, I detail in the next section the GARCH modelling for the EUA variable.

20.4.2   GARCH Modelling

I focus in this section on the modeling of the sCER variable defined as the residuals of the VAR(4) model for the sCERs.[22] To the best of my knowledge, this constitutes the first empirical analysis of sCER price drivers.

I model the EUA variable by using the asymmetric TGARCH (p,q) model by Zakoian (1994) with a Student's t innovation distribution, estimated by quasi maximum likelihood with the BHHH algorithm:

$$sCER_t = \alpha + \rho brent_t + \chi coal_t + \delta gas_t + \xi switch_t + \phi temphot_t + \gamma tempcold_t$$

$$+ \eta MCprod_t + \omega EUESI_t + \iota yield_t + \kappa momentum_{sCERt} + \lambda crisis_t + \vartheta VIX_t$$

$$+ \mu EUETSphaseIII_t + \nu ITL\_CITL_t + \pi CDMpipeline_t + \varsigma CDMEBmeeting_t$$

$$+ \upsilon linking_t + \varepsilon_t$$

$$\sigma_t = \alpha_0 + \alpha^+(L)\varepsilon_{t-1}^+ - \alpha^{-(L)}\varepsilon_{t-1}^- + \beta(L)\sigma_{t-1}$$

where $sCER_t$ are the residuals of the VAR(4) model related to the sCERs at time t $momentum_{sCERt}$, $CDMpipeline_t$, $CDMEBmeeting_t$, and $linking_t$ exogenous variables spe cific to sCERs defined as above. α is the constant, $brent_t$, $coal_t$ and gas_t are the returns o the brent, coal and gas series, $switch_t$ the switch variable, $tempec_t$, $temphot_t$ and $tempcold$ the temperatures variables, $MCprod_t$ the industrial production index from *Tendance. Carbone*, $EUESI_t$ the EU Economic Sentiment Index, $yield_t$ the slope of the Euro Area yield curve, $momentum_{EUAt}$ the momentum variable concerning the EUA market, $crisis$ the dummy variable accounting for the 'credit crunch', VIX_t the CBOE volatility indica tor, $EUETSphaseIII_t$ the dummy variable for Phase III news, $NAPphaseII_t$ the dummy variable for Phase II news, ITL_CITL_t the dummy variable for the ITL-CITL connec tion, ε_t the error term, σ_t the conditional volatility, the subscript index t refers to date t $(L)\varepsilon^*_{t-1}$ and $(L)\varepsilon^-_{t-1}$ are the positive and negative errors of the mean equation lagged one period respectively, and $(L)\sigma_{t-1}$ is the conditional volatility lagged one period. Note tha in this model $(L)\varepsilon^*_{t-1}$ and $(L)\varepsilon^-_{t-1}$ capture asymmetric effects.

20.4.3 Estimation Results

By estimating the TGARCH model presented in section 20.4.2 and removing one by one non-significant exogenous variables, it is possible to identify three different sets o regression results. The quality of the regressions is verified following several diagnostic tests: the Adjusted R^2, the log-likelihood ratio, the ARCH Lagrange Multiplier (LM test, the Ljung–Box Q-test statistic with a maximum number of lags of 20 (Q(20) sta tistic), the Akaike Information Criterion (AIC) and the Schwartz Criterion (SC). For both models, the Ljung–Box–Pierce test indicates that residuals are not autocorrelated and the Engle ARCH test indicates that heteroskedasticity is adequately captured by the structure of the TGARCH model. I have investigated the presence of multicolinear ity by computing the matrix of partial cross-correlations and the inflation of variance between explanatory variables.[23] These calculations did not reveal serious problematic multicolinearities.

In regression (1) in Table 20.4, we observe that energy prices (*brent*, *coal* lagged one period, and *gas*) have a statistically significant impact on sCER prices. Energy variables have an impact on the *sCER* variable at statistically significant levels, which conforms to previous literature on EUAs (Mansanet-Bataller et al., 2007; Alberola et al., 2008).[24] *Brent* and *gas* have a positive impact on EUA price changes: increases in fuel prices are directly transmitted to the CO_2 allowance market. As the most CO_2-intensive fuel, $coal$ has a negative impact on CO_2 prices. This implies that when the coal price increases, industrials have an incentive to use less CO_2-intensive fuels, which decreases the demand and the price of CO_2 allowances. This first result confirms that EUAs and sCERs share basically the same price fundamentals with respect to the interaction with energy markets.

In regression (2) in Table 20.4, $momentum_{sCERt}$ and *linking* are statistically significant at the 1 per cent and 10 per cent levels, respectively. The positive sign of $momentum_{sCERt}$ may be explained by the fact that CER price changes responded positively to carbon market trends during our study period. The positive sign of *linking* suggests that news about the future connection between the European and international credits carbon markets tend to increase sCERs prices. Note that sCERs are fungible across regional and domestic

Table 20.4 TGARCH (1,1) regression results for the sCER price drivers

Variable	sCER$_t$		
	(1)	(2)	(3)
Constant	0.0008 (0.007)	0.0007 (0.0007)	0.0007 (0.0013)
brent$_t$	0.0009*** (0.0002)		0.0005* (0.0003)
coal$_{t-1}$	0.0008** (0.0001)		−0.0017*** (0.0003)
gas$_t$	0.0002*** (0.0001)		0.0002* (0.0001)
momentum$_{CERt}$	0.0093** (0.0009)	0.0098*** (0.0009)	
Linking$_t$		0.0194* (0.0111)	
CDM pipeline$_t$			0.0005 (0.0013)
Adjusted R^2	0.1582	0.1427	0.0469
Log-Likelihood	1344.581	1339.208	660.743
ARCH LM Test	0.9195	0.9730	0.7560
Q(20) Statistic	25.137	24.396	20.724
AIC	−5.1074	−4.8026	−4.5827
SC	−5.0341	−4.7783	−4.4542
N	529	529	529

Notes: sCER$_t$ refers to the residuals of the VAR(4) model related to sCERs (Reuters sCER Price Index).
***, (**), (*) denotes 1%, (5%), (10%) significance levels.
The quality of regressions is verified through the following diagnostic tests: the adjusted R-squared (Adjusted-R^2), the Log-Likelihood, the ARCH Lagrange Multiplier (ARCH LM Test), the Ljung–Box Q-test statistic with a maximum number of lags of 20 (Q(20) statistic), the Akaike Information Criterion (AIC), and the Schwarz Criterion (SC).
The 1% (5%) critical value for the Ljung–Box portmanteau test for serial correlation in the squared residuals with 20 lags is 37.57 (31.41).
N is the number of observations.

markets. Thus, this positive sign is coherent with what we would expect: as the global demand of sCERs increases, the price of sCERs also increases.

In regression (3) in Table 20.4, we note that *CDMpipeline* is not significant in explaining sCERs price changes. This result conforms to the view that sCERs have distinct fundamentals from the delivery of primary CERs, since they are free of project delivery risk.

20.5 CONCLUSION

This chapter provides the first complete empirical analysis of sCERs price drivers. To the best of my knowledge, no previous empirical study has focused on the determination of sCERs drivers. The findings are summarized in Table 20.4.

My analysis of sCERs (that is, CERs already issued by the CDM Executive Board of the United Nations) has confirmed that EUAs determine significantly the sCERs price path. I show that there exists one long-term cointegrating vector between EUAs and sCERs taken in natural logarithm transformation. The sCER variable has a stronger tendency to adjust to past disequilibria by moving towards the trend values of the EUA variable, which confirms that EUAs are the leading factor in the price formation of sCERs. This result emphasizes that EUAs remain the most widely recognized 'money' on the emissions market. EUAs are exchanged widely as the most liquid asset for carbon trading, which may be explained by the fact that Europe remains to date the major source of demand for that kind of credits. It was also found that energy prices, variables referring to the linking of international carbon markets, and *momentum*$_{sCER}$ variables have an impact on sCERs prices. I conclude that sCERs pricing differs from EUAs since it embodies a greater level of uncertainty. Market participants are lacking the exact information concerning either the supply of CERs, or the total expected demand by 2012. Indeed, the future of credit offset mechanisms beyond 2012 looks rather bleak, while the use of CERs in Europe is confirmed only until 2020.

NOTES

1. Emission Reduction Units (ERUs) generated through the Joint Implementation mechanism (JI) of the Kyoto Protocol fall beyond the scope of this chapter, and are left for future research.
2. That is, to reduce greenhouse gas emissions on average by 8 per cent with respect to 1990 levels.
3. Note that 0.01 per cent were ERUs. No CERs were used towards compliance before that period, due to the lack of connection between the Kyoto Protocol's International Transaction Log (ITL) and the EU ETS Community Independent Transaction Log (CITL).
4. Note that this situation could change with the future developments from the US federal cap-and-trade scheme and other regional initiatives.
5. However on 23 September 2009, the European Court of First Instance (CFI) overruled the decision of the European Commission concerning NAPs for the second period submitted by Estonia and Poland. The Commission will explore two options: (1) issue a new decision based on 'proper' criteria before 23 December 2009; and (2) appeal against the CFI ruling, on a point of law, before 23 November 2009. Six other Eastern European countries may contest NAPs as well. In total, this represents a potential additional 162 million allowances.
6. The ERPA basically sets forward the duties and rights of the partners. Among the rights of the partners is the right to receive a pro rata quantity of the primary CERs.
7. In the absence of a satisfactory international agreement, installations subject to allowances during Phase III will only be able to use the credits left over from Phase II (2008–12), or a maximum amount corresponding to 11 per cent of the Phase II allocation. These measures are equivalent to capping the potential demand for Kyoto credits to 1510 Mt between 2008 and 2020. If a post-Kyoto international agreement is achieved, the ceiling on the use of credits from project mechanisms towards the compliance of EU ETS installations will be raised to 50 per cent of the additional reduction efforts. Beyond this issue, the introduction of a new international agreement on climate change would introduce 'high quality' as a condition for project credits coming from countries which have signed the international agreement. This would translate into a reduced supply of credits originated from project mechanisms to EU ETS compliance buyers.
8. Note that banking and borrowing of allowances are allowed within Phases II and III of the EU ETS, contrary to Phases I and II (Alberola and Chevallier, 2009).
9. Carchano and Pardo (2009) analyse the relevance of the choice of the rolling-over date using several methodologies with stock index future contracts. They conclude that regardless of the criterion applied, there are not significant differences between the series obtained.
10. The VECM is specified as follows:

$$\Delta y_t = A_0 + A_1 Ecm_{t-1} + A_2 \Delta y_{t-1} + \varepsilon$$

where

$$\Delta y_t = \begin{bmatrix} \Delta EUA_t \\ \Delta sCER_t \end{bmatrix}$$

is a vector of first differences of EUA and sCER prices,

$$A_0 = \begin{bmatrix} b_{10} \\ b_{20} \end{bmatrix}$$

is a vector of constants,

$$A_1 = \begin{bmatrix} b_{11} \\ b_{21} \end{bmatrix}$$

is a vector measuring the speed of the adjustment to the long-run relationship, and

$$A_2 = \begin{bmatrix} \gamma_{11} & \gamma_{12} \\ \gamma_{21} & \gamma_{22} \end{bmatrix}$$

is a coefficient matrix.

11. These results are not reproduced in the chapter to conserve space, and may be obtained upon request.
12. These results are not reproduced here to conserve space, and may be obtained upon request. See Chevallier (2010) for more details.
13. CIF ARA (cost, insurance and freight Amsterdam–Rotterdam–Antwerp) defines the price of coal inclusive of freight and insurance delivered to the large North West European ports, for example Amsterdam, Rotterdam or Antwerp.
14. Data available at http://www.ecb.int/stats/exchange/eurofxref/html/index.en.html
15. *Tendances Carbone* is a monthly newsletter on the EU ETS, produced by the Caisse des Dépôts, finance carbon research department. It can be found at http://www.caissedesdepots.fr/missionclimat
16. Note that the clean dark spread represents the difference between the price of electricity at peak hours and the price of coal used to generate that electricity, corrected for the energy output of the coal plant. The clean spark spread represents the difference between the price of electricity at peak hours and the price of natural gas used to generate that electricity, corrected for the energy output of the gas-fired plant. Both spreads are expressed in €/MWh.
17. Data can be found at http://sdw.ecb.europa.eu.
18. Note that I leave for further research the investigation of other potential explanatory variables, such as EUA forward curves and the return on investment for EUAs growing at the Euribor rate.
19. See the European Commission website, http://ec.europa.eu/environment.
20. Please see Appendix Table 20A.1 for detailed information on both variables.
21. Available at http://cdmpipeline.org.
22. Note that as the drivers of primary and secondary CER are not the same, it is important to remember here that we are considering secondary CER prices.
23. This table is not reproduced here to conserve space, and may be obtained upon request.
24. Note that the energy variables are considered here as contemporaneous variables. Including lags did not fundamentally change the results obtained.

REFERENCES

Alberola, E. and J. Chevallier (2009), 'European carbon prices and banking restrictions: evidence from Phase I (2005–2007)', *Energy Journal*, **30** (3), 51–80.
Alberola, E., J. Chevallier and B. Chèze (2008), 'Price drivers and structural breaks in European carbon prices 2005–07', *Energy Policy*, **36** (2), 787–97.
Alberola, E., J. Chevallier and B. Chèze (2009a), 'Emissions compliances and carbon prices under the EU ETS: a country specific analysis of industrial sectors', *Journal of Policy Modeling*, **31** (3), 446–62.
Alberola, E., J. Chevallier and B. Chèze (2009b), 'The EU Emissions Trading Scheme: the effects of industrial production and CO_2 emissions on European carbon prices', *International Economics*, **116**, 95–128.
Carchano, O. and A. Pardo (2009), 'Rolling over stock index futures contracts', *Journal of Futures Markets*, **29**, 684–94.
Chevallier, J. (2009), 'Carbon futures and macroeconomic risk factors: a view from the EU ETS', *Energy Economics*, **31** (4), 614–25.
Chevallier, J. (2010), 'EUAs and CERs: vector autoregression, impulse response function and cointegration analysis', *Economics Bulletin*, **30** (1), 558–76.
Chevallier, J., F. Ielpo and L. Mercier (2009), 'Risk aversion and institutional information disclosure on the European carbon market: a case-study of the 2006 compliance event', *Energy Policy*, **37** (1), 15–28.
Collin-Dufresne, P., R.S. Goldstein and J. Spencer Martin (2001), 'The determinants of credit spread changes', *Journal of Finance* **56** (6), 2177–208.

Delarue, E.D., A.D. Ellerman and W. D'haeseleer (2008), 'Short-term CO_2 abatement in the European power sector', *MIT CEEPR Working Paper* #2008-008.

Ellerman, A.D. and S. Feilhauer (2008), 'A top-down and bottom-up look at emissions abatement in Germany in response to the EU ETS', MIT CEEPR Working Paper #2008-017.

Johansen, S. (1991), 'Estimation and hypothesis testing of cointegrating vectors in Gaussian vector autoregressive models', *Econometrica*, **59**, 1551–80.

Johansen, S. and K. Juselius (1990), 'Maximum likelihood estimation and inference on cointegration with applications to the demand for money', *Oxford Bulletin of Economics and Statistics* **52**, 169–210.

Kilian, L., and C. Vega (2011), 'Do energy prices respond to US macroeconomic news? A test of the hypothesis of predetermined energy prices', *Review of Economics and Statistics*, **93** (2), 660–71.

Lutkepohl, H., P. Saikkonen and C. Trenkler (2004), 'Testing for the cointegrating rank of a VAR with level shift at unknown time', *Econometrica*, **72** (2), 647–62.

Mansanet-Bataller, M. and A. Pardo (2009), 'Impacts of regulatory announcements on CO_2 prices', *The Journal of Energy Markets*, **2** (2), 1–33.

Mansanet-Bataller, M., A. Pardo and E. Valor (2007), 'CO_2 prices, energy and weather', *Energy Journal*, **28** (3), 67–86.

Manzoni, K. (2002), 'Modeling credit spreads: an application to the sterling Eurobond market', *International Review of Financial Analysis*, **11**, 183–218.

Ramchander, S., M.W. Simpson and M.K. Chaudhry (2005), 'The influence of macroeconomic news on term and quality spreads', *Quarterly Review of Economics and Finance*, **45**, 84–102.

Trotignon, R. and B. Leguet (2009), 'How many CERs by 2013', Mission Climat Working Paper 2009-5.

Zakoian, J.M. (1994), 'Threshold heteroskedastic models', *Journal of Economic Dynamics and Control*, **18** (5), 931–44.

Zivot, E. and D.W.K. Andrews (1992), 'Further evidence on the Great Crash, the oil-price shock, and the unit-root hypothesis', *Journal of Business and Economic Statistics*, **10** (3), 251–70.

APPENDIX

Table 20A.1   Dummy variables for news announcements

Date	NAPs Phase II	EU ETS Phase III	CDM EB meeting	CDM mktdvlpt
25/03/2007			1	
26/03/2007	1			
26/03/2007	1			
02/04/2007	1			
16/04/2007	1			
18/04/2007			1	
04/05/2007			1	
04/05/2007	1			
15/05/2007	1			
30/05/2007			1	
01/06/2007				1
04/06/2007	1			
05/06/2007				1
06/06/2007			1	
11/06/2007			1	
11/06/2007				1
22/06/2007			1	
11/07/2007			1	
13/07/2007	1			
18/07/2007			1	
18/07/2007	1			
27/07/2007			1	
13/08/2007				1
29/08/2007			1	
31/08/2007	1			
01/10/2007			1	
05/10/2007			1	
19/10/2007			1	
22/10/2007	1			
26/10/2007	1			
26/10/2007	1			
12/11/2007			1	
14/11/2007			1	
20/11/2007			1	
30/11/2007			1	
07/12/2007	1			
16/01/2008			1	
23/01/2008		1		
01/02/2008			1	
06/02/2008				1
21/02/2008				1
26/02/2008				1
27/02/2008			1	
28/02/2008		1		

Date	EU ETS Phase III	CDM EB meeting	CDM mktdvlpt	Linking	ITL-CITL
14/03/2008			1		
23/04/2008			1		
30/04/2008		1			
16/05/2008		1			
19/05/2008			1		
20/05/2008				1	
23/05/2008			1		
29/05/2008			1		
30/05/2008		1			
05/06/2008	1				
26/06/2008	1				
06/06/2008				1	
09/06/2008			1		
11/06/2008	1				
17/06/2008		1			
04/07/2008					1
09/07/2008	1				
16/07/2008		1			
02/08/2008		1			
04/08/2008			1		
06/08/2008					1
12/08/2008			1		
12/08/2008			1		
10/09/2008		1			
26/09/2008		1			
07/10/2008	1				
08/10/2008		1			
08/10/2008	1				
15/10/2008			1		
20/10/2008	1				
24/10/2008			1		
25/10/2008		1			
28/10/2008					1
28/11/2008		1			
28/11/2008		1			
04/12/2008	1				
08/12/2008			1		
17/12/2008	1				
17/12/2008	1				
17/12/2008	1				
28/01/2009		1			
02/02/2009			1		
03/02/2009			1		
10/02/2009		1			

Table 20A.1   (continued)

Date	NAPs Phase II	EU ETS Phase III	CDM EB meeting	CDM mktdvlpt	Date	EU ETS Phase III	CDM EB meeting	CDM mktdvlpt	Linking	ITL-CITL
03/03/2008		1			13/02/2009		1			
14/03/2008			1		16/02/2009			1		

Note: The dummy variables refer to new information disclosure concerning NAPs Phase II (NAPs Phase II), the development of the EU ETS during Phase III (EU ETS Phase III), the day of publication of the CDM Executive Board report (CDM EB meeting), the CER market development (CDM mktdvlpt), the linking of emission trading schemes worldwide (linking) and the ITL-CITL connection (ITL-CITL).

Sources: UNFCCC, European Commission, European Council, European Parliament, European Economic and Social Committee, Committee of the Regions, Nordpool, ECX, EEX, Bluenext, ICE, Point Carbon, CNN.

21 Second-best instruments for energy and climate policy
Xavier Labandeira and Pedro Linares

21.1 INTRODUCTION

The debate seems to be well settled among economists that the best policy instrument to reduce carbon emissions is a carbon tax (for example Newell and Pizer, 2008). Indeed, following from Pigou's ideas about the correction of externalities (Pigou, 1932), and adding Weitzman's about the choice between price and quantity instruments (Weitzman, 1974), carbon taxes seem to be the best policy instrument to induce a reduction in carbon emissions. Another issue of course is the right value for this tax (for a discussion of the social value of carbon see, for example, Tol, 2010).

However, and surprisingly, most of these discussions have taken place in a first-best setting. Surprisingly because, first, there is a widespread recognition that first-best situations are mythical at best. Baumol and Bradford had already stated in 1970 that: 'generally, prices which deviate in a systematic manner from marginal costs will be required for an optimal allocation of resources, even in the absence of externalities'. In other words, any level of tax revenue to be collected by a government will ultimately produce some price distortion, and will therefore cause the economy to deviate from the first-best. To this we may add a significant list of additional real-life distortions: previously existing, and not necessarily efficient, subsidies or other regulations for energy and economic activities; vested interests; additional market failures; and so on.

The second reason is that there is already a well-developed literature on the choice of policy instruments in second-best settings such as those in which modern economies must make their decisions, since the seminal work of Lipscy and Lancaster (1956).[1] Although forgotten for some time, these ideas about instrument choice in second-best settings are returning to the debate – as we will show in the chapter – particularly for energy and climate policies, which are usually subject to many of the problems that are considered deviations from first-best situations.

Traditionally, the debate has generally focused on efficiency concerns. However, efficiency is not the only criterion for choosing instruments for environmental policies: other criteria such as distributional impacts, effectiveness and ease of implementation must also be considered, particularly if political and social acceptability is to be achieved. Dietz and Atkinson (2010), for example, found that people give the same weight to equity as to efficiency concerns when designing environmental policies. The evaluation of environmental policy instruments under these different criteria has already been addressed by other authors (for example Goulder and Parry, 2008).

In the end, a second-best analysis may result in having to modify the value of the first-best carbon tax, but it may also conclude in the need to use alternative and, in some cases, multiple instruments. There is a risk, however, of taking this as an excuse to step aside and dismiss traditional economic analysis, and by doing this to justify any type of policy and any level of political intervention which, in fact, is a common argument in many discussions about energy and climate policies.

We think instead that second-best settings should be incorporated into the analysis with at least the same level of rigour. Therefore, a careful assessment under second-best conditions of all the available instruments, and their combination, seems well deserved. Here we build on the existing literature and try to provide an integrated, (although preliminary) approach, merging efficiency with other concerns, with the final goal of giving indications on which instrument (or instruments) seem better suited for climate policy.

But before starting, we would like to set the boundaries for our analysis. First, we will only address the reduction in carbon emissions, and not other possible policy objectives (such as general issues of energy security, or economic development, which would also interact with climate policies). Second, we will not deal here with international second-best issues, such as the political economy of climate negotiations (see for instance Bosetti and Victor, 2011), carbon offsets (De Cian and Tavoni, 2010) or trade imperfections (for example Ulph, 1996). Finally, we will not address the problems related to uncertainty about costs and benefits of environmental instruments, which have been already well covered in the literature (for example Weitzman, 1974) or about uncertainty in the future costs of abatement measures (for instance, Liski and Murto, 2010).

The chapter is organized as follows. First, we identify and classify the reasons for deviating from a first-best analysis both under efficiency concerns and also under other criteria; we then look at the issue of multiple instruments and coordination. Finally, we provide some recommendations on instruments for climate policy under second-best.

21.2   SECOND-BEST SITUATIONS

Lipsey and Lancaster wrote in 1956 that: 'if there is introduced into a general equilibrium system a constraint which prevents the attainment of one of the Paretian conditions, the other Paretian conditions, although still attainable, are, in general, no longer desirable'. This is the formal definition of a second-best setting. Although it was conceived basically for efficiency concerns, it may be easily extended to other constraints (distributional requirements, administrative ease of use, and so on). Therefore, we will start by analysing the different constraints that may arise in climate policies which prevent attaining the first-best solution.

We may distinguish two types of second-best situations: the 'pure efficiency' one, in which the design of environmental instruments tries to maximize their economic efficiency under distorting prices, taxes, and so on; and a more comprehensive one, in which equity, political acceptability, behavioural issues or other non-efficiency aspects come into play. In this chapter we will take this 'comprehensive' approach for the analysis of second-best instruments.

21.2.1   Distortionary Taxes

Optimal taxation

As indicated before, there has been a gradual departure from the initial interpretations which considered carbon taxes only to solve the externality problem towards more comprehensive approaches that related these taxes to the general fiscal structure. This is linked to the fact that a distortionary tax system is in place and, therefore, the first-best Pigouvian prescription (a carbon tax rate equal to the marginal damage caused by emissions, whose revenues are returned lump sum) does not hold any more.

Thus two separate issues are related to the presence of existing distortionary taxation. The first has to do with the effects on the carbon rate structure, and the second with the use of tax revenues. Revenue recycling will be also considered in the next sections, and the debate here is mainly related to the gains (a second dividend) from reducing those existing distortionary taxes through the carbon tax revenues. In general, the literature favours the use of environmental tax revenues to reduce other distortionary taxes, although without precluding a net welfare gain with respect to the situation without the tax (only a welfare gain with respect to the non-recycling alternative: a 'weak' double dividend). This will depend of course on many factors, such as the pre-existing relative prices and taxes (for example Babiker et al., 2003), and the stringency of the climate policy (Anger et al., 2010).

With respect to the effects of distortionary taxes on the carbon rate structure, the literature was originally favourable to tax rates that would guarantee larger revenues (to exploit distortionary tax reductions). Lee and Misiolek (1986), for instance, argued that this could lead to higher or lower environmental tax rates, depending on the tax elasticity of emissions at the environmental optimum achieved through the Pigouvian tax. However, an important shift in the understanding of this issue took place in the early 1990s. Bovenberg and de Mooij (1994), through the use of a static general equilibrium model, showed that in most cases environmental taxes exacerbate the distortions brought about by conventional taxes. As a consequence of the extra costs, it was argued that environmental tax rates should be below the Pigouvian rates.

Other instruments

In addition to the impact of pre-existing distortionary taxes on the determination of the value of the carbon tax, this second-best situation may even recommend the choice of a different instrument. Although Quirion (2004) concludes that, under pre-existing distortions, the case for a tax compared to a quota is even stronger, there is a more comprehensive discussion on this issue, which is based on two elements: the revenue-generating capacity (Parry and Williams, 1999) and the creation of rents (Fullerton and Metcalf, 2001).

Parry and Williams (1999) argue that the revenues generated by the different instruments can be used to reduce pre-existing distortionary taxes, and therefore to improve the efficiency of the instrument. Indeed, Goulder et al. (1997) argue that the discussion should not be centred on taxes versus other instruments, but rather on the presence or absence of revenue recycling. If, for example, taxes are returned lump-sum, then their efficiency may be lower than that of an auctioned cap-and-trade system. In another study, Goulder et al. (1999) found that standards may be more efficient than emission permits if the permits are not auctioned. Parry and Williams (1999) in turn show that, with no revenue recycling, performance standards are more cost-effective than taxes or cap-and-trade. This would also lead us to the conclusion that subsidies, which need additional revenue, would be the worst instrument on efficiency grounds.

However, Fullerton and Metcalf (2001) argue that raising revenue is neither necessary nor sufficient as an attribute for an optimal policy under a second-best setting. Instead, they point to the creation of privately owned scarcity rents as the reason for the differences between policies. Revenue recycling will only be relevant when scarcity rents are created. Therefore, non-revenue recycling policies such as standards, or even subsidies, might be superior when compared to policies which create scarcity rents but which are not captured by the government and therefore recycled. That is, the advantages of economic instruments regarding setting prices right (correcting externalities, dynamic efficiency) may be negated if instead of reducing the pre-existing distortions in the economy, they exacerbate them when their rents are not recycled.

Therefore, the conclusion of this analysis is that, under pre-existing distortionary taxes, the most efficient instrument will be one in which the scarcity rents created – which are good for other purposes – are captured by the government (taxes, auctioned cap-and-trade). If other issues preclude this recycling, then other instruments should be contemplated.

21.2.2 Knowledge Spillovers

Jaffe and Stavins (1995) argued that policies with large economic impacts – climate policies are a clear example – should be designed to foster rather than inhibit technological change. The reason, suggested by their evidence, is that environmental policies alone are not strong enough to overcome technology market failures.

In fact, technological change is expected to play a key role in mitigation and adaptation policies, and should therefore be promoted strongly. However, although a price for carbon should in principle promote the use of carbon-free technologies, this does not take into account the market failures present in this sector, namely knowledge spillovers, and also credibility problems, learning-by-doing effects (Fischer and Newell, 2008), or risk of decreasing costs in the future.

Therefore, specific technology-promotion instruments will be required. Fischer (2008) has shown that the social return from technology policies depends on the degree of spillover, and also on the share of priced marginal social costs. If there is no social cost pricing (carbon pricing, in this case) there will be no incentive to innovate. On the other hand, if we assume a minimum rate of carbon pricing, the higher is the knowledge spillover, the higher will be the social return from public investments in technology.

However, Fischer also shows that technology policy cannot substitute completely for mitigation policy: waiting for costs to decrease requires huge investments and forgoes cost-effective emissions reductions. Therefore, we may need both carbon prices and technology policies. Indeed, Fischer and Newell (2008) conclude that multiple instruments will be required to correct both externalities, and will in fact be cheaper than a single instrument.

The challenge of course is to determine the right combination of instruments, accounting for the interactions between them. For example, Weber and Neuhoff (2010) argue that when innovation is included in the model, innovation effectiveness may change the optimal carbon price, and also may make quantity instruments more attractive than taxes. More research is clearly needed on this critical issue.

21.2.3 Other Market Failures

Carbon policies respond to a market failure, the non-internalization of the damages of carbon emissions. Innovation market failures have also been addressed. However, there are other market failures which also create a second-best setting for the definition of carbon policies.

The first one is the asymmetry of information, which gives rise to option values which differ from 'optimal' ones (Metcalf, 1994) or to the principal–agent problem, which in turn explain part of the energy efficiency paradox (Linares and Labandeira, 2010). Another of those is the coordination problem (Rodrik, 1996), that is, the lack of complete information transfers between the different parties affected by carbon policies, which is further complicated by the time lags and uncertainties involved. In the presence of this problem, complementary measures may be required, targeted at the different agents who may play a role in the reduction of carbon emissions (Hanemann, 2010).

Network externalities (for example related to the supporting infrastructure required by carbon-free technologies, such as smart grids, fuel distribution, and so on) may also be in play, creating path-dependence or technology lock-in, and therefore specific measures may be required to change path. Scale economies may also be an issue which requires specific support. Both should be discontinued once the problem is removed. Finally, other authors have also identified other market barriers (not necessarily market failures) or pre-existing advantages for fossil fuels which may prevent the efficient deployment of carbon-free technologies (Sovacool, 2009).

21.2.4 Behavioural Issues

Behavioural issues may be another reason for modifying our first-best choice for climate policy. We term as such the seemingly irrational behaviour of consumers when making decisions, which is also known as bounded rationality (Simon, 1955): the lack of capacity of decision-makers to incorporate all the information and criteria available when making their decisions. This fact has been shown in several studies regarding energy efficiency (Linares and Labandeira, 2010), which is a similar framework to the one being discussed here.

Jaffe and Stavins (1995), for example, found that even after correcting for different market failures, the effect on technology diffusion of up-front technology costs was much greater than equivalent longer-term energy prices. Therefore, technology adoption subsidies had much larger effects than equivalent Pigouvian taxes.

In general terms, bounded rationality reduces the response to economic instruments. Therefore, when emissions reductions are required, we may require product-specific instruments, such as building codes or standards for energy efficiency, which move the load of the proof away from consumers. However, these standards should be as flexible as possible, to allow cost-effective alternatives and ongoing incentives for improvement.

21.2.5   Political Acceptability

Although not very frequently addressed in the economic literature, the political accept-ability of environmental policy instruments plays a large role. In general terms, carbon prices are not well accepted by voters, and therefore are usually politically difficult to swallow. Even considering the same attributes, the word 'tax' provokes an instantane-ous rejection by people (Brännlund and Persson, 2010). To the contrary, subsidies and standards are much more popular, although they will typically be more inefficient, as mentioned in previous sections (Metcalf, 2009a).[2]

If the price (the implicit tax) results from a cap-and-trade system, the government will be strongly tempted to intervene – generally to keep prices at a reasonable level[3] – thus destroying the credibility of the market; if it results from a carbon tax, then the tax itself may not pass through the legislative body (there are plentiful examples of this, such as the recent French fiasco with the proposed carbon tax).

Thus, governments may be interested in lowering *ex ante* the perceived cost of the policy to an acceptable level, and this can be done with complementary measures, which basically make more elastic supply and demand curves for emission reductions (see for example Linares et al., 2008). This can be achieved with public investments or support policies for carbon-free technologies, as has been done in many countries with support policies for renewable energy.[4]

Another option for the government is to hedge against uncertain costs of the policy, but without losing credibility. One way of doing this is to use hybrid instruments, com-binations of price and quantity instruments. Roberts and Spence (1976) showed that a price and quantity instrument, such as the safety valve proposed in the US for climate policy, may be superior to a single price or quantity instrument. Metcalf (2009b) presents a more sophisticated version of this instrument. Webster et al. (2010) show that indexed caps may be superior to safety valves when there is a high correlation between the cost uncertainty and the index uncertainty. Quirion (2004), in turn, argues that contingent instruments such as indexed caps are even less appropriate under second-best (pre-existing taxes) conditions.

A second element besides the perceived cost that determines acceptability is the distri-bution of the cost of the policy among the different segments of society. As mentioned before, equity concerns may drive policy as much as efficiency ones, as shown by Dietz and Atkinson (2010) and Brännlund and Persson (2010): a climate policy will be more acceptable when it is not regressive, and when it is shared by others. This is one of the reasons for the opposition to carbon taxes: they fail to distribute explicitly the rents created, and also make polluters pay for the whole of emissions, not only for the cost of abatement.[5] Instead, a cap-and-trade system, by separating the efficiency and equity concerns, gives more room for adjusting the latter (Stern, 2009). For example, Goulder (2000) found that a cap-and-trade system with only a small degree of grandfathering can create enough rent to eliminate the opposition from the hardest-hit sectors, while being only slightly more costly.

Therefore, more equitable policies (either directly or through redistribution of the revenue) will probably be more acceptable. Also, since public research and development (R&D) costs are usually distributed more broadly than carbon prices, falling on a more specific set of actors, their political acceptability may also be higher.

21.2.6 Government Failures

Finally, we should not forget that the correction of market failures should be balanced against the possibility of government failures. Pigou (1932) noted that regulation will be inevitably imperfect when proposing the use of environmental taxes: 'governments may not have the necessary expertise, may be subject to pressures, and prone to corruption'. Government failures may arise for a number of reasons. With a different wording than Pigou's, we may argue that the most common are the disalignment of incentives, and the lack of complete information by the regulator on the regulated activities.

Regarding the former, Anthoff and Hahn (2010) argue that, in energy and environmental policy, evidence shows that governments are not driven by efficiency or even distributional concerns. Instead, they tend to prefer instruments which are easier to understand, which hide the costs of the policy while emphasizing the benefits, and which offer a greater degree of control over the distributional impacts. As a result, governments tend to prefer standards rather than economic instruments, which are moreover imposed only on new sources. When choosing between taxes and quotas, they also choose quotas. These choices are backed by industry, which demands regulation to restrict entry, to support prices, to provide subsidies or to capture scarcity rents (Keohane et al., 1998).

Another type of government failure is the impossibility of governments to commit in the long term. This will make carbon prices, or carbon quotas, scantly credible to investors. This is presented by Stern (2010) as an argument to promote technology policies directly.

As for the lack of information, when the capacity of regulators to observe output measures is limited, voluntary approaches such as management-based regulation may be more effective (Bennear, 2007). Two-part instruments – generalizations of deposit–refund systems, as proposed by Fullerton and Wolverton (2000) – or combinations of taxes on various inputs and outputs may also increase welfare in these situations.

21.3 MULTIPLE INSTRUMENTS AND COORDINATION

The previous analysis has assumed the choice of a single instrument to address the single environmental problem of reducing carbon emissions (as advised by Tinbergen). However, it is true that in this case other market failures may exist. Of course, multiple market failures do not necessarily require multiple instruments: sometimes one instrument can be designed to address multiple problems (Bennear and Stavins, 2007). For example, a carbon tax could be used to raise revenue, reduce emissions, induce innovation, and so on. However, the same authors argue that in a second-best setting the use of multiple policy instruments may be optimal.

In a previously mentioned paper, Fischer and Newell (2008) found that the combination of carbon prices and technology policies will be cheaper for reducing emissions than each single instrument. Following the same line, Acemoglu et al. (2009) also showed that the optimal climate policy should combine a carbon tax with research subsidies or profit taxes to direct research towards carbon-free technologies. However, in this case, the authors argue that a carbon tax alone would lead to excessive distortions in the economy, so the optimal policy should rely more on direct encouragement to the development of clean technologies, to counteract the market size effect of innovation.

Again, these results were obtained under an efficiency-maximization paradigm. If we add to this the already mentioned existence of other market failures and barriers, of bounded rationality, or of government failures, we may need more instruments to address these issues. Therefore, a major conclusion is that coordination is critical, because of the frequent unintended and negative consequences of instrument interaction (for example Metcalf, 2009a).

21.4   CONCLUSIONS

Climate policy is an appropriate environment in which to paraphrase Simon (1955): in the presence of incomplete information, limited resources, multiplicity of goals, and so on, the decision-maker is not able to optimize anything, and it may be advisable instead to try to attain non-optimal but reasonable solutions.

In this chapter we have described the many reasons that justify this second-best approach to climate policy: pre-existing distortionary taxes, knowledge spillovers, information asymmetry, network externalities, bounded rationality, political acceptability, equity concerns, government failures. Indeed, it seems that a traditional first-best setting (and therefore the optimal instrument under it) is absolutely unrealistic. That is, a carbon price will not reach the level required to compensate the externality, and even in that case it may not be sufficient or appropriate due to other distortions. Therefore, a carbon tax alone may not be the best instrument to deal with climate policy. It seems rather that, as we have said at the beginning, complex problems really require complex solutions. Climate policy requires a combination of instruments to address the multiple market failures and other second-best situations that arise in the real world.

We will still need carbon prices, as a necessary companion to other policies. For example, a carbon price is required along with technology policies, in order to provide the right incentives for these policies to work. How are we to generate these prices? Here carbon taxes may be more attractive theoretically. However, auctioned cap-and-trade systems, while retaining the rent-capturing feature of taxes, also allow for redistributing a part of the cost more explicitly and more easily than taxes, and may therefore be more politically acceptable. Their acceptability would be even higher if they are combined as hybrid instruments, such as safety valves, to hedge against unexpected high costs.

These more efficient instruments should probably be coupled in some sectors – those closer to the final customer – with technology standards to account for bounded rationality and also to improve acceptability; with technology policies (both market-pull and market-push, depending on their situation in the learning curve) to counteract knowledge spillovers; with education and training policies to reduce bounded rationality and to decrease perceived costs; and with voluntary approaches when performance is not easily observable. As mentioned before, given the different objectives addressed by these complementary measures, they may be required only temporarily, and might eventually be phased out, once the transition to a carbon-free technology has been achieved.

The ideal approach should therefore combine political pragmatism, economic efficiency, distributional concerns, environmental effectiveness and behavioral aspects. And of course, the multiplicity of instruments will require a strong coordination, to look for synergies and to avoid unexpected effects.

Along this line, and based on our conclusions, it would be highly recommended to check how revenues from taxes or allowance auctioning could be used to finance R&D policies or other policies, instead of using them to reduce labour taxes. Given that the evidence for the double dividend of green tax reform is sometimes not that strong, recycling revenues through these complementary instruments might also prove to be recommendable. This is a specific area where research is much needed, particularly on the political economy of coupling these instruments automatically.

To conclude, the challenge now is to determine, using sound economic analysis, which is the right combination of these multiple instruments, accounting for these interactions. Some research is already being produced on this issue, and the ideas presented here only reinforce this need. In this sense, we would like to insist that a second-best setting does not preclude the use of rigorous economic analysis, but rather reinforces it. Using the existence of multiple market or government failures as an argument for whatever may fit political or social demands is not acceptable. For example, requesting drastic changes in the way we generate our energy by assuming an infinite benefit from doing this does not make much sense. Therefore, we still need economic analysis to determine the goals for these second-best instruments. Of course the complexity of the problem will make it less tractable, and less precise. But rigour is, and should still be, required.

ACKNOWLEDGEMENT

The authors would like to acknowledge financial support from the Spanish Ministry of Science and Innovation (ECO2009-14586-C02-01).

NOTES

1. Indeed, this discussion started from Pigou himself, who was perfectly aware of the difficulty of attaining first-best conditions in real-life economies (Pigou, 1932).
2. Although it should be remembered that a more accepted instrument may in practice become more efficient if acceptability increases its effectiveness.
3. It may be argued, even based on official documents, that the EU Emissions Trading System (ETS) has a determined allowance allocation so as not to result in higher than acceptable carbon prices.
4. Although these policies are also justified by other policy objectives: energy security or industrial development.
5. This can be fixed through a careful design of the tax – for example partial exemptions – but complex tax structures are not well received either.

REFERENCES

Acemoglu, D., P. Aghion, L. Burztyn and D. Hemons (2009), 'The environment and directed technical change', NBER Working Paper No. 15451.

Anger, N., C. Böhringer and A. Löschel (2010), 'Paying the piper and calling the tune? A meta-regression analysis of the double-dividend hypothesis', *Ecological Economics*, **69** (7), 1495–1502.

Anthoff, D. and R. Hahn (2010), 'Government failure and market failure: on the inefficiency of environmental and energy policy', *Oxford Review of Economic Policy*, **26**, 197–224.

Babiker, M., G. Metcalf and J. Reilly (2003), 'Tax distortions and global climate policy', *Journal of Environmental Economics and Management*, **46** (2), 269–87.

Baumol, W. and D. Bradford (1970), 'Optimal departures from marginal cost pricing', *American Economic Review*, **60** (3), 265–83.

Bennear, L. (2007), 'Are management-based regulations effective? Evidence from state pollution prevention programs', *Journal of Policy Analysis and Management*, **26** (2), 327–48.

Bennear, L.S. and R.N. Stavins (2007), 'Second-best theory and the use of multiple policy instruments', *Environmental and Resource Economics*, **37** (1), 111–29.

Bosetti, V. and D.G. Victor (2011), 'Politics and economics of second-best regulation of greenhouse gases: the importance of regulatory credibility', *Energy Journal*, **32** (1), 1–24.

Bovenberg, A.L. and R. de Mooij (1994), 'Environmental levies and distortionary taxation', *American Economic Review*, **94**, 1085–9.

Brännlund, R. and L. Persson (2010), 'Tax or no tax? Preferences for climate policy attributes', CERE Working Paper 2010–4.

De Cian, E. and M. Tavoni (2010), 'The role of international carbon offsets', FEEM Nota di Lavoro 2010-033.

Dietz, S. and G. Atkinson (2010), 'The equity–efficiency trade-off in environmental policy: evidence from stated preferences', *Land Economics*, **86**, 423–43.

Fischer, C. and R. Newell (2008), 'Environmental and technology policies for climate mitigation', *Journal of Environmental Economics and Management*, **55** (2), 142–62.

Fischer, C. (2008), 'Emissions pricing, spillovers, and public investment in environmentally friendly technologies', *Energy Economics*, **30** (2), 487–502.

Fullerton, D. and G. Metcalf (2001), 'Environmental controls, scarcity rents, and pre-existing distortions', *Journal of Public Economics*, **80** (2), 249–67.

Fullerton, D. and A. Wolverton (2000), 'Two generalizations of a deposit refund system', *American Economic Review*, **90**, 238–42.

Goulder, L.H. and I.W. Parry (2008), 'Instrument choice in environmental policy', *Review of Environmental Economics and Policy*, **2** (2), 152–74.

Goulder, L.H., I.W. Parry and D. Burtraw (1997), 'Revenue-raising versus other approaches to environmental protection: the critical significance of preexisting tax distortions', *RAND Journal of Economics*, **28** (4), 708.

Goulder, L., I. Parry, R. Williams III and D. Burtraw (1999), 'The cost-effectiveness of alternative instruments for environmental protection in a second-best setting', *Journal of Public Economics*, **72** (3), 329–60.

Goulder, L.H. (2000), 'Confronting the adverse industry impacts of CO_2 abatement policies: what does it cost?', Climate Issues Brief 23, Resources for the Future, Washington, DC.

Hanemann, M. (2010), 'The compact of climate change: an economic perspective', in E. Cerdá and X. Labandeira (eds), *Climate Change Policies: Global Challenges and Future Prospects*, Cheltenham, UK and Northampton, MA, USA: Edward Elgar, pp. 9–24.

Jaffe, A. and R. Stavins (1995), 'Dynamic incentives of environmental regulations: the effects of alternative policy instruments on technology diffusion', *Journal of Environmental Economics and Management*, **29** (3), 43–63.

Keohane, N., R. Revesz and R. Stavins (1998), 'Choice of regulatory instruments in environmental policy', *Harvard Environmental Law Review*, **22** (2), 313–67.

Lee, D. and W. Misiolek (1986), 'Substituting pollution taxation for general taxation: some implications for efficiency in pollution taxation', *Journal of Environmental Economics and Management*, **13**, 338–47.

Linares, P. and X. Labandeira (2010), 'Energy efficiency: economics and policy', *Journal of Economic Surveys*, **24**, 573–92.

Linares, P., F.J. Santos and M. Ventosa (2008), 'Coordination of carbon reduction and renewable energy support policies', *Climate Policy*, **8**, 377–94.

Lipsey, R.G. and K. Lancaster (1956), 'The general theory of second best', *Review of Economic Studies*, **24** (1), 11–32.

Liski, M. and P. Murto (2010), 'Uncertainty and energy saving investments', MIT-CEEPR Working Paper 2010-005.

Metcalf, G. (2009a), 'Tax policies for low-carbon technologies', NBER Working Paper No. 15054.

Metcalf, G. (2009b), 'Cost containment in climate change policy: alternative approaches to mitigating price volatility', *Virginia Tax Review*, **29** (2), 381–405.

Metcalf, G.E. (1994), 'Economics and rational conservation policy', *Energy Policy*, **22**, 819–25.

Newell, R.G., W.A. Pizer (2008), 'Indexed regulation', *Journal of Environmental Economics and Management*, **56**, 221–33.

Parry, I. and R. Williams (1999), 'A second-best evaluation of eight policy instruments to reduce carbon emissions', *Resource and Energy Economics*, **21** (3), 347–73.

Pigou, A.C. (1932), *The Economics of Welfare*, London: Macmillan.

Quirion, P. (2004), 'Prices versus quantities in a second-best setting', *Environmental and Resource Economics*, **29** (3), 337–60.

Roberts, M.J. and M. Spence (1976), 'Effluent charges and licenses under uncertainty', *Journal of Public Economics*, **5**, 193–208.

Rodrik, D. (1996), 'Coordination failures and government policy: a model with applications to East Asia and Eastern Europe', *Journal of International Economics*, **40**, 1–22.

Sovacool, B.K. (2009), 'The importance of comprehensiveness in renewable electricity and energy-efficiency policy', *Energy Policy*, **37** (4), 1529–41.

Simon, H.A. (1955), 'A behavioral model of rational choice', *Quarterly Journal of Economics*, **69**, 99–118.

Stern, N. (2010), 'Presidential address: Imperfections in the economics of public policy, imperfections in markets, and climate change', *Journal of the European Economic Association*, **8** (2/3), 253–88.

Tol, R.S.J. (2010), 'International inequity aversion and the social cost of carbon', *Climate Change Economics*, **1** (1), 21–32.

Ulph, A. (1996), 'Environmental policy instruments and imperfectly competitive international trade', *Environmental and Resource Economics*, **7** (4), 333–55.

Webster, M., I. Sue Wing and L. Jakobovits (2010), 'Second-best instruments for near-term climate policy: intensity targets vs. the safety valve', *Journal of Environmental Economics and Management*, **59** (3), 250–59.

Weber, T.A. and K. Neuhoff (2010), 'Carbon markets and technological innovation', *Journal of Environmental Economics and Management*, **60**, 115–32.

Weitzman, M.L. (1974), 'Prices vs. quantities', *Review of Economic Studies*, **41**, 477–91.

22 Addressing fields of rationality: a policy for reducing household energy consumption?
Hege Westskog, Tanja Winther and Einar Strumse

The reason stateways fail to modify folkways is that policymakers often get it wrong – the experiences the state manipulates are not the experiences that produce the habits that produce the visible patterns they seek to change. (Stephen Turner, 1994: 104)

22.1 INTRODUCTION

In this chapter we analyse effective strategies for changing household energy behaviour. We synthesize insights from psychology, anthropology and economics and develop an interdisciplinary model for understanding energy behaviour and change. More specifically we develop a model which includes a concept of 'fields of rationality', understood as the modus operandi that individuals act within. We suggest that this approach, which attempts to be holistic by taking factors at different levels of analysis into account, might be important for understanding people's energy use behaviour. Also, we discuss how insights into fields of rationality might give guidance for the design of effective policy instruments.

Different disciplines have different perspectives on what influences energy behaviour. In economic theory it is assumed that a consumer chooses what they prefer, given what is feasible with the given prices of goods and their income. The preferences of the consumer are considered as given, and the main instruments for change are considered to be the price level of the products consumed and the income level of the consumers.[1] Schiffer (1979) for instance relies on these perspectives when he underlines that: 'The demand for energy depends critically on the elasticity of substitution, as well as upon the income and price elasticities of various energy-intensive amenities'.[2] Based on these perspectives, there is a strong argument in economic literature that the use of economic instruments is superior to command-and-control instruments when it comes to their cost-effectiveness (see for instance Baumol and Oates, 1988). The recommendations for cost-effective policies for reduced energy consumption would thus be for policies that focus on 'getting the price' of electricity to the right level to meet the goals set by policy-makers.

Like economists, psychologists study people's preferences, but from a slightly different angle. Whereas the economist's approach is normative, the psychological approach is descriptive, defining preference as whatever likes or dislikes the individual may have in a certain domain, with the explanation of the observed preference being a matter of empirical investigation. Also, psychologists question the assumption that the preferences of the individual are a function of maximizing gain and avoiding loss. Environmental psychologists (Kaplan and Kaplan, 1989) have instead suggested what they label the Reasonable Person Model as a conceptual framework both for understanding individual preferences and for environmental decision-making in general. In this model, not being

452

rational in terms of maximizing one's gain on one single value does not mean that people are irrational. The Reasonable Person Model builds on three principles: (1) people can be reasonable, depending upon the circumstances that surround them, implying that reasonableness is the outcome of an interaction between person and situation; (2) people actively seek to understand their world, but often possess extremely limited information; (3) people's needs are many and varied, and thus not reducible to any single unitary value: we are not maximizers but 'satisficers' (Simon, 1972). From this it follows that people, when constructing their preferences, relate to several sets of values at the same time and that they are likely to draw on, refer to and evaluate such values contextually dependent on the situation. The existence of multiple values is also commonly acknowledged in sociology and anthropology.

Another important assumption shared by parts of psychology and other social sciences concerns the unit of analysis. Rather than seeing individuals as autonomous consumers who act independently of each other, relational aspects are highlighted. From a psychological perspective, a holistic and relational perspective on individual behaviour is found in the so-called transactional approach to environmental psychology, aiming to study how, mutually, 'individuals change the environment and their behaviour and experiences are changed by the environment' (Gifford, 2007). Here, persons-in-environments are treated as the basic unit of analysis without any further dividing into smaller entities. Persons, processes and environments are seen as parts of a whole, not as independent components that are combined in an additive fashion to make the whole. Thus, the transactional approach focuses on shifting processes in person – environment configurations, studying acting, doing, talking, thinking instead of studying personal states, structures and static entities, and aims at explaining specific psychological phenomena on the basis of the most appropriate theoretical principles at any given time (Altman and Rogoff, 1987; Stokols and Altman, 1987; Pepper, 1942; Werner et al., 2002).

Similar to the way the transactional approach in psychology centres not on individuals but on persons-in-context (for example social environment), sociology and anthropology are disciplines concerned with relationships between people. This focus on relationships complements the economic perspective by enhancing our understanding of some of the central factors that condition people's preferences and thus their energy behaviour and consumption at large. To provide some examples: an investment in a heat pump; a family's yearly consumption of kilowatt hours; or a teenager watching television in her room with her friends – these are different types of energy behaviours that tend to be highly socially conditioned. This is so both with respect to the intra-household relationships and dynamics played out in the place where consumption takes place (people's homes), and also through the wider social networks that household members feel a belonging to and seek approval from (see for example Douglas and Isherwood, 1996[1979]; Miller, 1998; Carrier and Miller, 1999; Henning, 2005). Moreover, cultural values come into play and are connected to the social norms (or 'rules') that guide energy behaviour. Grasping such values may be particularly illuminating for understanding energy practices, which tend to vary according to the cultural context (Wilhite et al., 1996; Henning, 2005; Wilhite, 2008a; Winther, 2008).

In addition, material factors condition any kind of practice, and particularly so in the area of energy, with its heavy, costly and enduring technologies of production and supply (Holden, 2002; Shove, 2003). Recent contributions in sociology and anthropology combine a people-centred focus on notions such as comfort, cleanliness and convenience, with an analytic emphasis on the systems of provision which significantly contribute to shaping consumption (Shove, 2003; Wilhite, 2008a, 2008b; Winther, 2008). Here, the sociocultural dynamics, together with the push from markets, suppliers and providers of infrastructure, produce new patterns and often increasing levels of consumption. One implication of such works on how systems change over time is that material and social conditions are not treated as contextual factors but actively brought into the analysis.

There have been many attempts to synthesize insights from different disciplines for understanding pro-environmental behaviour on a general basis and energy behaviour specifically. Faiers et al. (2007) draw together key issues from consumer behaviour theories relevant for energy use to aid policy-making, and develop a discussion ground around integrated theories of consumer behaviour. Their theory overview indicates that the issues of learning and awareness, coupled with accessibility to simple technologies, are central factors for formulating effective policy for energy use. Also, Wilson and Dowlatabadi (2007) review different perspectives on drivers for individual behaviour and apply these insights to decisions affecting residential energy use. They conclude by pointing to the necessity for collaboration between disciplinary approaches to energy efficiency. Kallbekken et al. (2011) combine social-psychological and economic theory in a model of environmentally significant consumption. Their perspectives open up the way for a wider hypothesis on policy instruments through showing the potential for the use of a diversified policy (information measures and incentives).

As Wilson and Dowlatabadi (2007) point out, there is a need for syntheses that specifically combine the economic perspectives with social and behavioural determinants of energy use. In our study we synthesize insights from economic, anthropological and psychological theories to combine the foci on individual factors for explaining behaviour within the sociocultural and material context in which the individual operates. This allows us to open up the under investigated field in economic theory on what forms preferences.

In the following we present our model for explaining energy behaviour and define the concept of 'fields of rationality' within this model. We then illustrate the model by pointing to two contrasting fields of rationality relevant for the present purposes, that is, the consumer – citizen dichotomy. By addressing this dichotomy we discuss strategies for policy for reducing people's energy consumption. The chapter ends by pointing to the need for knowledge of how the context and different groups respond to different policy instruments and their design. This knowledge is often crucial to get the desired outcomes of the interventions

22.2   RELEVANT FACTORS FOR EXPLAINING BEHAVIOUR: THREE LEVELS

In economic theory, consumer preferences are a central element for understanding energy consumption. Commonly, preferences are understood as the set of assumptions relating to a real or imagined choice between alternatives. Based on the degree of utility they are assumed to provide, the consumer will rank the alternatives. Energy behaviour is thus a result of the individual consumer's optimizing process between the utility provided by the various alternatives available and the individual's budget restriction. The preferences behind such behaviour are here treated as an exogenous variable in the model, thus as a given. In our model, however, we seek to highlight and develop factors that may explain energy behaviour and thus the preferences behind them.

22.2.1 The Basis: Structures Surrounding Us

As a first step towards explaining energy behaviour we base our model on practice theory as outlined by Bourdieu (1977) and later elaborated by Sewell (1992), Warde (2005) and Ortner (2006).[3] This will result in the identification of four central factors which in sum characterize the overall social structures that condition people's energy behaviour. We will later show that these four factors also have relevance for capturing the dynamics at the group or household level, which is no less important for understanding individual behaviour.

The cluster of approaches referred to as practice theory has shown to have merit in the field of energy. This is so due to the everyday, repetitive characteristic of household energy use and the significant role played by material and sociocultural factors in shaping what people do with energy (Wilhite, 2008a, 2008b; Winther, 2008). We are interested in the interplay between individual consumers on the one hand and, on the other, the sociocultural and material conditions that affect their behaviour.

Within practice theory structures are considered as 'the principles' that form social practices (Sewell, 1992: 8). Structures come in two forms. First of all they constitute the ideas, values, norms, conventions and codes for human interaction that exist in any social group. We may denote this part of structures as the 'frameworks of meaning' (Gullestad, 1992; Ortner, 2006; also referred to as 'cultural schemas' by Sewell, 1992). Such frameworks exist tacitly among and within people. They are the culturally informed tools with which we think, act and feel (Sewell, 1992). For our purposes we follow the common separation between social norms (how things should be) and cultural values (what is valued).

Secondly, and in contrast to the 'virtual' characteristic of frameworks of meaning, structures also contain elements which Sewell refers to as 'real'. Here we find natural resources, material objects and assets of various kinds, and also regulations and formalized procedures. Sewell refers to this category as 'non-human resources' (Sewell, 1992: 13). Also, real structures contain human resources such as bodily strength and knowledge. Resources are unevenly distributed and controlled within a given society, but Sewell asserts, following Giddens (1979), that all members of a society possess resources of both the human and the non-human kind. 'Indeed, part of what it means to conceive of human beings as *agents* is to conceive of them as *empowered* by access to resources of one kind or another' (Sewell, 1992: 10). Again we see the significance of knowledge and resources as a premise for action, as noted above (Simon, 1972; Faiers et al., 2007).

With this model of what surrounds an individual and their given energy behaviour we have a conceptual tool for beginning to understand why people do as they do. The notion of 'practice' entails this interrelationship between individual action (behaviour) and structures. Following this, the concept of 'habitus' highlights the way in which such structures become internalized in individuals through the routines and practices of everyday life (Bourdieu, 1977). This does not mean that individuals are only subjected to the power of established social norms, values and a given set of resources. As we discuss below, individuals have emotions, ambitions and skills. They also have personal norms, values and identities that may or may not contradict those of the larger group. We may rather consider an individual's potential to influence a given behaviour personally along a continuum, stretching from pure routine (non-reflexive behaviour) at one end. At the other end, where behaviour is marked by a high degree of reflexivity and intentionality, we have 'agency' (Ortner, 2006).

In a given society a multitude of structures and practices exist in parallel (Sewell, 1992: 16–17). The conventions and values (and indeed material factors) guiding the practice of waste management, for example, may be different from those shaping people's shopping habits. One of Sewell's central points for explaining how structures may change over time is that people may transfer the conventions (schemas) attached to one given practice over to the next. Thus if 'go green' characterizes the conventions, values and norms related to waste recycling practices, there is a potential that people would start applying a similar set of values and conventions for modifying their shopping practices. For such a shift to become a new established practice, however, one also has to consider how such (new) conventions interplay with other (old) parts of the total structure; that is, their material and knowledge-related aspects.

In sum, four structural factors condition and partly make up a given practice: cultural values, social norms, human resources and non-human resources. The other significant contributors to the practice are those who perform them: acting people. Positioned between the overall structures and the individual level are the households, groups and social networks to which individuals belong or otherwise relate.

22.2.2   The Group or Household Level

Social relationships significantly influence people's energy behaviour, as we have argued in the introduction. To grasp these and other factors relevant to the study of energy use at the group and household level, the four-factor model described for the larger structures also yields significance. Cohabiting families tend to share certain norms and values (Aune, 2007). Members of a family also have significant groups which they relate to outside the household, whose norms may be in line with or contradict those of the household. Such norms may nonetheless come into play when individuals behave at home, for instance when taking a shower. Negotiations may occur when various members' values, norms and preferences vary (Henning, 2005; Winther, 2008). We nevertheless presume that individuals to a considerable extent share norms and values with the relevant groups in question, which vary according to the practice in question. This results in particular family or group conventions (or debate) as to how a certain type of behaviour, such as showering, should be carried out. Furthermore, such conventions are also formed by the set of resources possessed by the group; both human (knowledge, capacity, competences) and material (for example available shops in the neighborhood, household income, type of dwelling). Finally, we recall that the larger structures work on groups and individuals whose engagement and behaviour within various practices is often characterized by a high degree of routine. In sum, households and groups are largely influenced by the outside structure and the wider social network, but they also possess a capacity to negotiate and form their own practices according to the internal power relations and the group's common aspirations, possibilities and constraints.

22.2.3   The Individual Level

Both psychology and economics are preoccupied with explaining behaviour (consumption in traditional economic theory) through a focus on the individual. However, the two disciplines provide different explanatory factors for behaviour. Economics puts effort into understanding how external factors, income and prices, influence choices (termed 'material conditions' in our model), while in psychology the internal factors are the main focal points. An internalist approach sees behaviour: 'mainly as a function of processes and characteristics which are conceived as being internal to the individual: attitudes, values, habits and personal norms' (Jackson, 2005). Building on this internalist approach, we will also draw on psychological approaches discussed in the introduction which put an analytic focus on the individual–environment relationship, acknowledge the existence of individuals' multiple values and preferences, and give attention to the various and often limited types of knowledge that people possess. In our model we therefore put weight on internal individual factors such as skills and knowledge; attitudes and personal norms; and beliefs, values and identities. We also include material conditions or constraints as perceived by the individual for explaining energy behaviour. Together with the habitual aspects embedded in what people do with energy, these are important cognitive, affective and material factors partly accounting for the motivations pushing or pulling the individual to perform various measurable behaviours, and could as a first step be seen as the underlying factors for understanding preferences in economic theory. The main factors influencing behaviour on the individual level are illustrated in Figure 22.1.

It is important to keep in mind that the influences of psychological factors on environmentally significant behaviour such as energy use are more varied than commonly

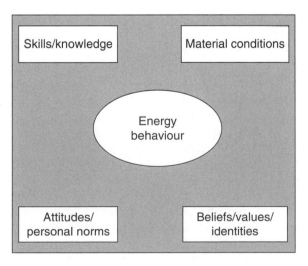

Figure 22.1   Factors influencing behaviour on the individual level

assumed, and that the strongest influences are often determined by structural factors. The more a behaviour is shaped by technology, infrastructure, regulation, financial cost, convenience and other contextual factors, the weaker the effect of personal variables (Stern, 1999). However, even when energy use in the household is to a large extent determined by physical or structural features, there is considerable room for the influence of psychological factors. Some examples of such factors are personal commitment, perceived personal costs and benefits of actions, and behaviour-specific beliefs and norms. Energy behaviour may also be influenced by non-environmental attitudes and beliefs (Stern, 2008). Although the effects of many psychological variables on specific behaviours are highly indirect, they can potentially influence a wide variety of behaviours. Psychological interventions aiming at behavioural change in the home should target the 'niches' between powerful structural variables, when behaviour is not strongly constrained by regulation, habit, matters of economic cost, convenience and the like. Niches thus refer to the moments in which the people–environment configuration is particularly open to change. Examples of such situations when people's values, attitudes, personal norms or knowledge bases could be receptive to psychological interventions are: when a regulatory regime is transformed; when the purpose for spending money is not clearly identified; when there are contradictions embedded in a system as to how people 'should behave'; or when significant material changes are to be made, such as when planning for the construction of a new home (Stern, 2008).

22.3   OUR MODEL

In the discussion above we have elaborated the factors that we find important for explaining people's energy behaviour. We have introduced four main factors on three levels, where the factors are related to human resources, non-human resources, and norms and values, and the levels are the society, the group and the individual, respectively. The model is illustrated in Figure 22.2.

As can be seen, each level (square) has been given a specific shading. This is to illustrate that the individual is positioned within a particular type of group structure and dynamics, which in turn is located in a larger societal structure. A field of rationality in this model comprises a particular configuration of factors (and their content) on all three levels. For example, if you are living in Norway, you relate to a given set of knowledge areas, energy regulations and technologies, social norms and cultural values which are different from those found in France. Thus if you move from Norway to France you would have to relate to the structures of the French society. This is represented in our model by the shading of the background of the societal level. Hence, the background shading might vary; it depends on the type of society the individual relates to.

A change in cultural context may thus produce changed behaviour. To take the example of people's uses of light sources: in Norway the habit of keeping many lights on might be valued for the cosiness and heat this produces, and also be morally accepted. In the French context, the cultural and social conditions are likely to be different. An individual who shifts context may thus experience a discrepancy between their own behaviour and that promoted by and embedded in the new cultural context. After a while, for example when visitors have given hints of dislike of the individual 'wasting' electricity,

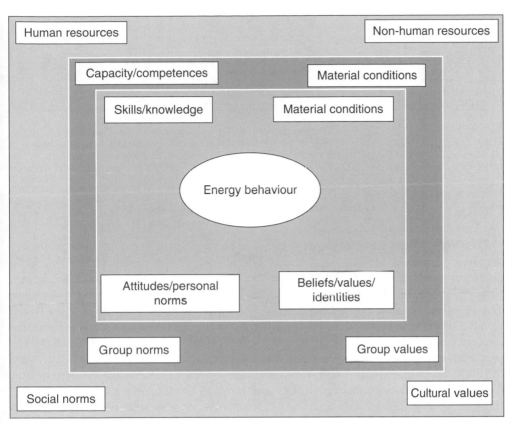

Figure 22.2   Field of rationality: factors influencing behaviour

this discrepancy may be adjusted by the individual who modifies their behaviour to a kind that fits with the new context (for example, turning off the lights when leaving a room). If we add other possible differences between Norway and France to the picture, such as electricity prices (being higher in France) and the supply of movement detectors in local shops (more available in France), the likeliness of the Norwegian individual switching to the French field of rationality for light behaviour becomes higher (Figure 22.3).

With the same kind of reasoning the shading of the group or household level varies depending on which group you relate to in your energy behaviour. The groups you relate to (significant others) may also differ from one type of behaviour to the next (for example showering before meeting friends or baking pizza for a family's joint home evening). Hence, the background shading of the group level signifies what group you relate to when performing a specific energy behaviour.

Following from the above, the shading of the inner square, the individual level, illustrates the position taken up by the individual who is acting or performing a given behaviour. More specifically we let this shading represent the logic from the individual's perspective (realised by them or not), which in turn results in a given behaviour. A person who has moved to France may also insist on keeping the lights on despite the potential social and economical costs such behaviour may imply.

Notes: The box to the left illustrates a Norwegian individual positioned in the Norwegian cultural/
energy context who behaves in compliance with this. The box in the middle indicates the situation where
the Norwegian individual has moved to the French cultural/energy context. In the box to the right, the
Norwegian individual has adjusted to the French context and changed his/her behaviour accordingly. In this
example, the *white* color represents the group level and is left out of the discussion.

Figure 22.3   Shift of cultural context as a cause for changed behaviour

The field of rationality is thus made up of factors in the inner, middle and outer squares. We have underlined the close interrelationship that exists between an individual and their social, cultural and material environment. Within a given type of behaviour, such as lighting, the field of rationality is what in sum provides a rationale for behaviour. The field may be more or less congruently constituted (same shading on all three levels) or have separate logics/shading on the different levels. The individual level is nevertheless privileged in the model, in that it is the individual and their logic that ultimately produces a given behaviour. We argue that the concept of a field of rationality is useful for understanding individual behaviour, and the varying ways in which this behaviour is partly conditioned by contextual factors and partly by the socially, culturally and materially positioned individual. In section 22.5 we will treat strategies for reduced energy consumption through a focus on fields of rationality.

The model is relevant both for plain habitual behaviour and more cognitively informed behaviour. Pertinent here is Ortner's (2006) axis for behaviour which ranges from plain routine informed by established conventions (reproducing habitus) at one end, to agency with a high degree of reflexive thought and intentionality at the other. At each extreme, people have a given rationale for doing what they do, and energy is, as mentioned, a realm where habitual aspects are strong. However, we will in the following concentrate on energy behaviour that involves reflection. We are concerned with situations in which people either think of what they do or consider changing their behaviour with the purpose of fulfilling certain goals. We will use two examples of fields of rationality to illustrate our points concerning possible strategies for policy: the consumer field and the citizen field. More generally, we hold that an important strategy for changing energy behaviour towards sustainability is to focus on the field of rationality of the individual. This strategy is discussed further in section 22.5.

22.4   CONSUMER AND CITIZEN: EXAMPLES OF FIELDS OF RATIONALITY RELEVANT TO POLICY DESIGN

There is a body of literature which shows that behaviour is linked to the perspective or rationale that forms the basis for it. A well-known example is the study conducted by Gneezy and Rustchini (2000) on parents picking up their children after closing time in the nursery. In a field experiment they introduced a monetary fine for latecoming parents. As a result, the number of latecoming parents increased significantly, contrary to what had been expected when the fine was introduced. This is referred to as the 'crowding-out effect' of taxes and might be interpreted as a shift from one 'decision area' to another or, to use the jargon developed in the present work, as a shift in the field of rationality.[4] Before the fine was introduced, parents might have considered the act of picking up children on time as a moral duty, but when they were fined for being late they appeared to consider latecoming as part of a market transaction where latecoming was morally acceptable. The field of rationality here shifted from moral obligation towards a market-based thinking. Hence, parents' likeliness for picking up children after closing time depended on which field of rationality they thought this decision belonged to.

A similar dichotomy between different fields of rationality is that between 'the consumer' and 'the citizen'. Sagoff (1988) defines the citizen role as the one you take when you are concerned with and behave according to the public interest. Thus as a citizen you are concerned with the community rather than your own well-being. On the other hand, as a consumer you are primarily concerned with personal or self-regarding wants and interests, and put aside the community-regarding values which you take seriously as a citizen (Sagoff, 1988: 8).[5] Nyborg's (2000) *Homo economicus–Homo politicus* distinction has much in common with Sagoffs's consumer–citizen dichotomy. *Homo economicus* is a person who maximizes their own well-being; that is, a consumer field of rationality, whereas a person acting in the *Homo politicus* mode puts themself in the role as a ethical observer, and tries to consider what is best for the society (Nyborg, 2000: 309–10), that is, a citizen field of rationality.

An example of a product that addresses people as citizens is renewable electricity certificates.[6] Electricity suppliers who offer such guarantees for an extra fee expect customers to be positioned within a citizen field of rationality, which in turn will trigger their decision (behaviour) to buy the guarantee. We return below to a study on such certificates and the fields of rationality found in Norway. We note, however, that the purchase of a certificate does not affect the electrons the customer gets in their outlet; the electricity product they receive is exactly the same as before. The reason why some people still pay extra for such a product is because they want to contribute to renewable electricity development, and not as a result of their consumer rationality (Winther and Ericson, forthcoming). The field of rationality in which they operate is the citizen field.

A citizen field of rationality is not always of equal relevance or significance. The literature indicates that the citizen field is more dominant for behaviour that is civic-oriented than purely consumer-oriented (see for instance Berglund and Matti, 2006 or Westskog, 1997 for a discussion of this). One example of civic-oriented behaviour is pro-environmental behaviour (that is, recycling or use of public transport instead of a private car). From social psychology, we know that pro-environmental behaviour is influenced by more citizen-oriented values like altruistic and biospheric values (for example Stern's value–belief–norm model – Stern et al., 1999; Schwartz's norm activation theory – Schwartz, 1977).

In the above discussion, we have been focusing on the field of rationality that triggers an act. However, it is also important to note that individuals tend to have values, norms and skills that are not act-provoking in a given situation. Latent rationalities exist and

Sagoff (1988), Sen (1985) and Nyborg (2000) all include the notion of different rationalities in their discussions. The dichotomy between consumer and citizen, *Homo economicus* and *Homo politicus*, points to the possibility of individuals operating within different rationalities depending on the context. Even if you act within the citizen field of rationality, you might at the same time have your *Homo economicus* rationality intact. In this regard, changing the field of rationality that provokes the act might change the outcome or behaviour, as the Gneezy and Rustchini (2000) study on nursery fines indicates.

The existence of latent rationalities brings a new perspective to policy. The way a policy-maker designs the policy instrument, or that a decision problem is framed for the decision-maker, could influence the outcome of the behaviour. A specific design or context could realize the desired outcome.[7] In our model this is referred to as changing the field of rationality through design of the policy instrument or the decision problem. In the next section we will look at findings from an ongoing study that will illustrate the point further.

22.5   FIELDS OF RATIONALITY: A POTENTIAL FOR CHANGE

In the following we will use examples from qualitative interviews conducted in Kirkenes and Vadsø (towns in Northern Norway) and a focus group study in Kirkenes and Oslo to illustrate how our field of rationality thinking might be useful in policy formation. The study focused on people's electricity behaviour and the potential for change. More specifically, we asked what may make people purchase renewable electricity certificates and save electricity, and what role information could play in this respect.[8]

Specific behaviours and specific products might be more related to one field of rationality than other, for example related to the citizen field rather than the consumer field. People's decision to purchase a renewable electricity product will tend to be based on a citizen field of rationality, and by using this logic in the promotion of the product in specific ways, one may increase the number of people wanting to buy this kind of product, as illustrated below. The supplier of electricity in Kirkenes offers a renewable electricity certificate for $5 per month. The renewable certificate does not imply that you receive a benefit as a consumer, as the quality of the electricity (as noted earlier) remains the same. Among customers who have bought the certificate, they said that their motivation for doing so was related to a concern for the environment. Some explained in more detail how they want to make use of their customer power and send a signal to the market, so as to stimulate development of renewable production at the cost of fossil and nuclear production. As 'Sonja' concluded: 'It is like a statement. It will not have effect if only a few people buy it. But it is a start.'[9] This is clearly a citizen-motivated act.

The content and design of information intended to trigger people's motivation to purchase the certificates appear to affect their likeliness for doing so, and thus also the possibility that this will lead to increased production of renewable electricity. In our focus group study in Kirkenes we tested how different designs of information regarding the renewable electricity product were understood, and how the participants expected that this might influence their behaviour. One of the designs tested provided the customer with a short text on renewable certificates and the Norwegian 'standard mix' on their invoice, and also a link to NVE[10] for further information. In another treatment the customer was given information about how much CO_2-emissions were generated from the consumption of electricity without a renewable certificate. These emissions were compared to the emissions resulting from driving a car. A third treatment gave a narrative where a story was told, focusing on why a named person had chosen to buy the certificate. The narrative highlighted the relevance of a citizen field of rationality for the interviewee's motivation for buying the product: 'if more people required this guarantee it would mean less CO_2 emissions in the long term. What matters is to use our power as customers to influence the development.'

The first treatment produced considerable confusion: 'I do not quite understand what renewable energy is, where it comes from. Is it smart or is it stupid? . . . Do they mean a heat pump, that one should install one of those?'[11] Most participants did not have an idea of the certificate before arriving for the discussion, and this information (which they were exposed to first) did not seem comprehensible. Instead, this information appeared to make people focus on their lack of understanding, thus producing a kind of ignorance. In all four focus groups participants expressed disinterest in the message and even distrust in the product, apparently preventing them from wanting to seek further information or to choose the certificate. 'To be honest, I do not think I would have even read it.'[12] In this treatment you provide people with information that is not made specifically relevant for the citizen field, and as such you do not trigger people's citizen motivation to act. On the other hand the treatments focusing on environmental effects were more positively evaluated. These two other treatments facilitated the translation of what the purchasing of certificates may mean in terms of environmental consequences. Hence, the information received about the product was made consistent with the citizen field of rationality in a transparent way which seemed to open up the potential for obtaining a change – through the purchase of certificates for those that are citizen motivated.

For electricity-saving behaviour, people referred to a double set of motivating factors: to save money (economical factors) and environmental concerns. Thus we found two fields of rationality (consumer and citizen, respectively) to exist in parallel. The same individual would highlight both aspects: 'To me both the price and environmental concerns would have been important'. Other participants stressed economic factors as most important, highlighting the need for 'getting the expenses down'. Another person put weight on 'environmental concerns' only. Hence, this might imply that when promoting energy-saving behaviour, effects might be increased both by using cost-saving arguments and by pointing to environmental effects. There might also be gender differences related to which field of rationality more easily comes into play. As Henning (2005) found for Swedish homes in a study of heat pumps, men tended to be more concerned with economic arguments (which have hegemony in the Swedish context, according to Henning). Women were more attuned towards demonstrating environmental values when deciding on such investments, but since heat pumps in Sweden belonged to men's domain, women had a less direct say in such cases.

22.5.1 Facilitate Shifts in the Field of Rationality

Based on the Kirkenes study we emphasized the importance of providing clarity and consistency in the way citizen-oriented information and products are presented (or translated) in order to appeal to people as citizens and in turn trigger the desired behaviour.

The information that got the most positive evaluation in the focus groups was that which was consistent with a citizen field of rationality. Shifts in the fields of rationality themselves may also potentially produce more sustainable energy behaviour. We recapture some points that have been mentioned. Above we discussed how a person's change of cultural context (Norway–France) may inform their changed behaviour. We explained this shift by pointing to the new field of rationality that appeared in which the individual's (changed) behaviour took place. Policies may also induce changes in the field of rationality. In the general discussion of the consumer – citizen dichotomy we drew on cases in the literature to show how the introduction of new policies (for example money for blood donations, or fines for latecoming parents) negatively affected people's behaviour in relation to what had been intended. The crucial point in this process was that the new policies induced a shift in which decision area people felt that the actual behaviour belonged to.

In our general model (see Figure 22.2), a change in any one of the shadings of the three squares (collective, group and individual level) indicates a change in the field of rationality. This could potentially (but not necessarily) affect the logic (shading) of the inner square and thus the behaviour of the individual. For example, a policy could make an individual modify their attitudes; or a significant group to which the individual belongs could be gaining new knowledge and cause the individual to change behaviour; or the arrival of new regulations or taxes on the societal level could affect the overall field of rationality and also the area of decision of the individual.

That being said, societal structures may also contain considerable fixity and therefore be resistant to change. Socio-material structures and cultural values have been noted as relatively enduring. For example, in a cross-cultural comparison of people's energy use in the 1990s, Japanese and Norwegian households referred to distinct sets of cultural values in their ways of using energy (Wilhite et al., 1996). The Japanese put cultural emphasis on bathing practices (with a link to notions of purity), but their lighting practices appeared less culturally significant. In contrast, Norwegians stressed the cultural value of providing a 'cosy home' by the use of an average of 14 light points in the living room, preferably by the use of incandescent light. The authors concluded that behaviours that are deeply culturally anchored are less likely to be influenced by policies. Correspondingly, they advised that policy makers should focus on the 'lighter' practices that are not so culturally embedded (Wilhite et al., 1996). Following this, and interpreted within our model of fields of rationality, in Japan it would make more sense to try to change the field of rationality for lighting behaviour than bathing practices.

But cultural values may also change. The field study in Norway referred to above was conducted in 2009, just before an EU Directive putting a ban on incandescent light was introduced. Norwegians had for some time started to readjust to the lack of available incandescent bulbs in the shops, and the situation had also received attention in the media. What we found in Kirkenes was that the 11 households interviewed seemed less negative towards energy-saving bulbs (compared to incandescent light) than Wilhite et al. had found for Oslo in the 1990s.[13] If this discrepancy is not only caused by local variation between towns in Norway, this could indicate that Norwegians, possibly due to the ban on incandescent bulbs and a generally increased environmental awareness, may currently be modifying their perceptions of what the value of 'cosiness' entails.

This indication of a shift in the degree of fixity in Norwegian lighting practices points to the realization that cultural practices are also open to change. The concept of fields of rationality may help us further understand the dynamics embedded in such processes of change. In the case of light sources in Norwegian homes, the individuals who express a concern for the environment when justifying their choice of light source reflect that they consider this decision to be located in a citizen field of rationality associated with environmental concerns. This reference hence replaces the previous 'cultural' one, associated with the value of cosiness.[14] If more people adapt to the new field of rationality, new, shared conventions guiding practices and behaviours may be invented (see section 22.2.1 above, 'The basis: structures surrounding us'). We also notice how a ban or regulation may have an effect on people's norms and values. The introduction of the ban on smoking in public places is a striking example of how regulations may induce shifts in norms. Nyborg and Rege (2003) show how an introduction of a smoking regulation may move a society from a situation where most smokers are non-considerate to a situation where they are considerate towards non-smokers even in the unregulated zone. They also refer to empirical evidence that social norms on considerate smoking behaviour have changed dramatically in Norway. In 1999, only 10 per cent of non-smokers reported that smoking guests would usually smoke indoors in their home without asking for permission. As much as 74 per cent reported that this was the most common behaviour among smoking guests 10–15 years earlier.

Importantly for the presented model, the distinct factors on all levels together make up the field of rationality and are mutually influencing each other. Changes in one factor potentially affect the others. The embedded fundamental assumption is that for understanding the individual, we need to see the world as they do. This is what we try to capture with the concept of fields of rationality.[15]

22.6 CONCLUSION AND REMARKS

In this chapter we have focused on how energy behaviour can be understood and explained by integrating perspectives from different disciplines. We have introduced a model which includes the concept 'field of rationality' signifying the configuration of factors on the individual, group and societal levels which are relevant for energy behaviour. We argue that knowledge of the fields of rationality that guide energy behaviour for groups and individuals might be an important tool for designing policies for change. Although the model has general appliance, and may be used to assess the relevance of tools ranging from economical incentives, regulations, to enhancing people's access to new technologies, we have primarily discussed information measures which presupposes that people through conscious or reflexive processes change behaviour based on the information provided. Specifically, we have focused on the dichotomy between the citizen and the consumer fields of rationality and showed how these fields might form behaviour and we have indicated how insights into such dynamics maybe used in shaping information that will have effect. The material from Norway on people's electricity use and information interventions that address either 'citizens' and/or 'consumers' shows that it might matter how messages are designed. We argue that for information to have the desired effect, the message must be translated so as to match the relevant field of rationality on the part of individuals. Thus promotion of green electricity certificates (citizen field) for instance was evaluated more positively if activating the citizen field of rationality in individuals. Thus we highlight consistency between the logic in message and the field of rationality of the individual.

An effective policy for sustainable energy consumption needs to be based on knowledge of how energy behaviour most likely might be influenced, as illustrated above. To follow Stephen Turner (1994), whose quote opened this chapter, in order to form policies for changed behaviour that work, one needs to address the experiences that actually produce the habits one seeks to change. A central problem in the realm of energy has been policy-makers' enduring trust in energy efficiency as a means to reduce consumption. In Wilhite's words: 'What the record shows is that efficient technologies may well increase the efficiency of energy throughout but that promised reductions in energy demand seldom pan out' (Wilhite, 2008b: 121). Technologies alone do not change behaviour in a sustainable direction. Instead, as embedded in our model, one should regard them in connection with the behaviours that produce the patterns one seeks to change. Referring to a similar point, that energy efficiency depends not only on technology but also on the choices of the users, Allcott and Mullainathan (2010) observe the considerable investments that go into research and development (R&D) for providing energy-efficient technologies. They argue that a corresponding R&D effort should be made in the field of behavioural science in order to develop policy innovations and even large-scale businesses that promote energy efficiency in practice, and thus energy conservation (Allcott and Mullainathan, 2010). They also point to the cost-effectiveness of such programmes.

Within an R&D programme based in behavioural and social sciences one should develop various policy instruments and focus on what kinds of designs would produce the desired effects on people's behaviour. Policy instruments like taxes and subsidies are often believed to be of a more general nature with no need for specific testing. However, as illustrated above, effects of policy instruments are dependent on both the sociocultural and material context, and the individual whom these instruments are used towards. When focusing on the effects of the instruments, knowledge of how 'the context' and various groups and individuals would respond to different instruments and their design is often crucial to achieving the desired outcome. Hence, effective interventions often need to be tailor-made.

Acquiring the necessary knowledge and tailoring interventions might of course be very costly, and as such not a strategy that can be followed fully. However, in the same way as R&D is used to transform 'hard science' into useful technological solutions, so systematic effort needs to be made on developing pilot projects informed by behavioural and social sciences and, successively, innovative policy tools for sustainable energy consumption. According to the premises in our model, successful innovation at large would most likely be achieved through efforts which include both hard and soft sciences. As a start, testing policy instruments before implementing them is one way to understand how behaviour is influenced by instrument use, and might be worth the money invested in field studies – just as pilot projects are worth it for hard sciences to become useful technologies.

ACKNOWLEDGEMENTS

We thank Sylvie Douzou, Torgeir Ericsson, Catherine Grandclement, Isabelle Moussaoui and Håkon Sælen for helpful comments and suggestions. Funding from the Norwegian Research Council is gratefully acknowledged.

NOTES

1. Presentations of this theory can be found in most standard textbooks on microeconomics, like Gravelle and Rees (1992).
2. The elasticity of substitution measures the ease or difficulty of substituting between commodities, while price elasticity tells how demand for a good changes with the price of that good or alternatively with the price of another good (cross-prices elasticity).
3. Sewell theorizes about how structures and practices change, Warde elaborates on practice theory's relevance in the realm of consumption, and Ortner provides a model on the relationship between people's routine habits on the one hand and intentional, reflective behaviour or 'agency' on the other.
4. The effects observed are coupled to Frey's (1997) crowding theory. Frey argues that an external incentive might also affect the internal incentives for behaviour to crowd out moral motivation.
5. Sen (1985) discusses people's preferences and finds a dichotomy between well-being and agency which has resemblances to Sagoff's notions of consumer and citizen, respectively (Sagoff, 1988). Well-being is connected to an individual's individualistic preferences or self-interests (Westskog, 1997), whereas the agency aspect of individual choice refers to the opinions and beliefs that an individual has, and this is often connected to the individual's participation in a society.
6. When a customer buys a renewable electricity certificate, the supplier guarantees that the customer's total consumption will be purchased from production that is 100 per cent based on renewable sources. A system of 'guarantees of origin' has been established in the European Union (EU) area, including Norway. A disclosure regulation puts an obligation on suppliers of electricity to state on the customer's bills which fuel mix is behind their supply (percentage of fossil, nuclear, renewable and so on).
7. There is a lot of literature from the field of behavioural economics and social psychology on the effects of design of policy instruments or framing of decision problems for behavioural outcomes. Much of this literature is concerned with how a decision problem is understood by the decision-maker, and is termed 'framing' (for example Tversky and Kahneman, 1981).
8. The project is titled 'Do customer information programs influence household electricity consumption?' and is financed by the Norwegian Research Council (2009–11). We conducted 11 in-depth interviews in Kirkenes and Vadsø in September and October 2009. The focus group discussions were held in Oslo (November 2009) and Kirkenes (December 2009) and had eight sessions with eight persons in each group.
9. From interview with 'Sonja' in Kirkenes, 6 October 2009. Her name has been changed.
10. NVE: Norsk vassdrags- og energidirektorat (Norwegian Water Resources and Electricity Directorate).
11. Young, female focus group participant no 8, Kirkenes 7 December 2009, Transcript 88322 Cicero Gr.6-Kirkenes, page #6.
12. Middle-aged female focus group participant no 3, Kirkenes 7 December 2009, Transcript 88322 Cicero Gr.5-Kirkenes, page #7.
13. Our informants referred to the cost and environmental benefit from energy-saving bulbs; or they said that incandescent was more beautiful; or they said that the quality of the light from the two types of sources was the same. This shows more variation in people's attitudes towards energy-saving lights than was documented in Oslo in the 1990s.
14. To an anthropologist, this distinction between environmental and cultural values may appear a bit strange, since environmental values can be considered as cultural as any other value. The distinction is made to clarify the argument which seeks to highlight how policies may affect fields of rationality in ways that produce sustainable consumption.
15. It follows that we do not adapt the notion of rationality as premised on 'the rationally optimizing man', as opposed to that informed solely by norms, which embedded much social science discourse some decades ago. See for example Elster's discussion of actions being influenced 'both by rationality and by social norms' (1989: 102). The present work does not operate with such distinctions, but regards fields of rationality as encapsulating norms and other factors relevant to people's ways of behaving.

REFERENCES

Allcott, H. and S. Mullainathan (2010), 'Behavior and energy policy', *Science*, **327**, 1204–5.
Altman, I. and B. Rogoff (1987), 'World views in psychology: trait, interactional, organismic, and transactional perspectives', in I. Altman and D. Stokols (eds), *Handbook of Environmental Psychology*, New York: Wiley, pp. 7–40.
Aune, M. (2007), 'Energy comes home', *Energy Policy*, **35**, 5457–65.
Baumol, W.J. and W.E. Oates (1988), *The Theory of Environmental Policy*, Cambridge: Cambridge University Press.
Berglund, C. and S. Matti (2006), 'Citizen and consumer: the dual role of individuals in environmental policy', Working Paper 6, Sharp Research Program, Luleå University of Technology.
Bourdieu, P. (1977), *Outline of a Theory of Practice*, Cambridge: Cambridge University Press.
Carrier, J.G. and D. Miller (1999), 'From private virtue to public vice', in H. Moore (ed.), *Anthropological Theory Today*, Cambridge: Polity Press, pp. 24–47.
Douglas, M. and B. Isherwood (1996 [1979]), *The World of Goods: Towards an Anthropology of Consumption*, London: Routledge.
Elster, J. (1989), 'Social norms and economic theory', *Journal of Economic Perspectives*, **3** (4), 99–117.
Faiers, A., M. Cook and C. Neame (2007), 'Towards a contemporary approach for understanding consumer behaviour in the context of domestic energy use', *Energy Policy*, **35**, 4381–90.
Frey, B.S. (1997), *Not Just for the Money: An Economic Theory of Personal Motivation*, Cheltenham, UK and Northampton, MA, USA: Edward Elgar.
Giddens, A. (1979), *Central Problems in Social Theory. Action, Structure and Contradiction in Social Analysis*, Berkeley and Los Angeles, CA: University of California Press.
Gifford, R. (2007), *Environmental Psychology: Principles and Practice*, 4th edn, Colville, WA: Optimal Books.
Gneezy, U. and A. Rustichini (2000), 'A fine is a price', *Journal of Legal Studies*, **29**, 1–17.
Gravelle, H. and R. Rees (1992), *Microeconomics*, London: Longman, 2nd edn.
Gullestad, M. (1992), *The Art of Social Relations. Essays on Culture, Social Action and Everyday Life in Modern Norway*, Oslo: Scandinavian University Press (Universitetsforlaget AS).
Henning, A. (2005), 'Equal couples in equal houses: cultural perspectives on Swedish solar and bio-pellet heating design', in S. Guy and S.A. Moore (eds), *Sustainable Architectures: Cultures and Natures in Europe and North America*, New York, USA and London, UK: Spon Press, pp. 89–104.
Holden, E. (2002), 'Boligen som grunnlag for bærekraftig forbruk', Doctoral dissertation, Trondheim, Norwegian University of Science and Technology (NTNU).
Jackson, T. (2005), 'Motivating Sustainable Consumption – a review of evidence on consumer behaviour and behavioural change', a report to the Sustainable Development Research Network.
Kallbekken, S., J. Rise and H. Westskog (2011), 'Combining insights from economics and social psychology to explain environmentally significant consumption', in K.D. John and D.T.G. Rübbelke, *Sustainable Energy*, Abingdon: Routledge, pp. 109–28.
Kaplan, R. and S. Kaplan (1989), *The Experience of Nature: A Psychological Perspective*, Cambridge: Cambridge University Press.
Miller, D. (1998), *A Theory of Shopping*, New York: Cornell University Press.
Nyborg, K. (2000), 'Homo Economicus and Homo Politicus: interpretation and aggregation of environmental values', *Journal of Economic Behavior and Organization*, **42**, 305–22.
Nyborg, K. and M. Rege (2003), 'On social norms: the evolution of considerate smoking behavior', *Journal of Economic Behaviour and Organization*, **52**, 323–40.
Ortner, S. (2006), *Anthropology and Social Theory: Culture, Power, and the Acting Subject*, Durham, NO: Duke University Press.
Pepper, S.C. (1942), *World hypotheses: A Study in Evidence*, Berkeley, CA: University of California Press.
Sagoff, M. (1988), *The Economy of the Earth*, Cambridge: Cambridge University Press.
Schiffer, L. (1979), 'Another look at energy conservation', *American Economic Review*, **69** (2), 362–8.
Schwartz, S. (1977), 'Normative influences on altruism', *Advances in Experimental Social Psychology*, **10**, 222–79.
Sen, A. (1985), 'Well-being, agency and freedom: the Dewey Lectures 1984', *Journal of Philosophy*, **82** (4), 169–221.
Sewell, W.H. (1992), 'A theory of structure: duality, agency, and transformation', *American Journal of Sociology*, **98** (1), 1–29.
Shove, E. (2003), *Comfort, Cleanliness and Convenience: The Social Organization of Normality*, Oxford: Berg.
Simon, H. (1972), 'Theories of bounded rationality', in C.B. McGuire and R. Radner (eds), *Decision and Organization: A Volume in Honor of Jacob Marscham*, Amsterdam North-Holland, pp. 161–76.

Stern, P. (1999), 'Information, incentives and pro-environmental behavior, *Journal of Consumer Policy*, **22**, 461–78.

Stern, P.C. (2008), 'Environmentally significant behaviour in the home', in A. Lewis (ed.), *Psychology and Economic Behaviour*, Cambridge: Cambridge University Press, pp. 363–82.

Stern, P., T. Dietz, T. Abel, G. Guagnano and L. Kalof (1999), 'A value–belief–norm theory of support for social movements: the case of environmental concern, *Human Ecology Review*, **6**, 81–97.

Stokols, D. and I. Altman (eds) (1987), *Handbook of Environmental Psychology*, New York: Wiley.

Turner, Stephen (1994), *The Social Theory of Practices. Tradition, Tacit Knowledge, and Presuppositions*, Chicago, IL: University of Chicago Press.

Tversky, A. and D. Kahneman (1981), 'The framing of decisions and the psychology of choice', *Science*, **211** (4481), 453–8.

Warde, A. (2005), 'Consumption and theories of practice', *Journal of Consumer Culture*, **5** (2), 131–53.

Werner, C.M., B.B. Brown and I. Altman (2002), 'Transactionally oriented research: examples and strategies', in R.B. Bechtel and A. Churchman (eds), *Handbook of Environmental Psychology*, New York: John Wiley & Sons, pp. 203–21.

Westskog, H. (1997), 'The use of cost–benefit analysis to decide environmental policy', *Journal of Interdisciplinary Economics*, **8**, 185–208.

Wilhite, H. (2008a), *Consumption and the Transformation of Everyday Life: A View from South India*, New York, USA; London, UK; and Delhi, India: Palgrave Macmillan.

Wilhite, H. (2008b), 'New thinking on the agentive relationship between end-use technologies and energy-using practices', *Energy Efficiency*, **1**, 121–30.

Wilhite, H., H. Nakagame, T. Masuda, Y. Yamaga and H. Haneda (1996), 'A cross-cultural analysis of household energy-use behaviour in Japan and Norway', *Energy Policy*, **24** (9), 795–803.

Wilson, C. and H. Downlatabi (2007), 'Models of decision making and residential energy use', *Annual Review of Environment and Resources*, **32**, 169–203.

Winther, T. (2008), *The Impact of Electricity. Development, Desires and Dilemmas*, Oxford: Berghahn Books.

Winther, T. and T. Ericson (forthcoming), 'Norwegian customers' responses to the introduction of an electricity disclosure regulative'.

23 The role of R&D+i in the energy sector
Alessandro Lanza and Elena Verdolini

23.1 INTRODUCTION

Secure, reliable and affordable energy supplies are fundamental to economic stability and development. The current energy prospects are however unsustainable: concerns about energy security, the threat of disruptive climate changes and the growing energy needs of the developing world pose major challenges to energy decision-makers.

In recent years, fossil fuel prices have risen considerably and, at the same time, oil and gas resources remain concentrated in a small number of countries. This situation raises concerns about energy security and the prospect that sustained high energy prices may harm economic growth. Reducing fossil fuel dependency is a key policy target in many countries.

Energy security concerns are linked to the need to mitigate greenhouse gas emissions. From 1990 to 2000, the average annual increase in carbon dioxide emissions was 1.1 percent per year while, between 2000 and 2005, this rate rose to 2.9 percent. High economic growth, particularly in coal-based economies such as China and India, and higher oil and gas prices (which have led to an increase in coal-fired power generation) are the main reasons for the increase.

About 69 percent of all CO_2 emissions (more than 20 Gt per year) are energy-related. The International Energy Agency (IEA hereafter) forecasts that, unless current policy changes, global energy-related emissions will continue to grow and fossil fuels will remain the dominant source to feed the world's incremental energy needs. Higher emission levels will imply more significant impact from climate change: the United Nations Intergovernmental Panel on Climate Change (IPCC) has concluded that only scenarios resulting in a 50 percent to 80 percent reduction of global CO_2 emissions by 2050, compared to today's level, can limit the global mean temperature rise between 2 and 2.4°C

At the 2009 meeting in L'Aquila (Italy), the G8 leaders agreed to join forces in a global response to achieve a 50 percent reduction in global emissions by 2050. To avert the worst consequences of global warming, leaders of the Group of Eight wealthiest nations agreed in principle to limit global warming to 2°C and cut their greenhouse gas emissions by 80 percent by 2050. The 2°C goal was accepted for the first time by Canada, Japan, Russia and the United States. It had already been adopted by the European Union and its G8 members Britain, France, Germany and Italy.

The options proposed to mitigate climate change underline the importance of early action and a long-term vision. Capital stock currently installed may be still in use by 2050: this means that the transition towards cleaner energy technologies should start as soon as possible.

The aim of this chapter is twofold. On the one hand, it seeks to provide an analysis of the status and future prospects of key energy technologies, which is developed in section 23.2.[1] On the other hand, it will give a picture of the magnitude of past and current energy R&D activity (section 23.3). To this end, we make extensive use of publicly available data sources and of recent reports and descriptive analyses of energy-related research and development (R&D). Review of past and current spending in energy-related R&D investment will help to put into perspective the effort necessary to 'walk the talk' of addressing climate change and energy issues.

If the analysis of future prospects of energy technology is confined to a simple description, the reasons why a specific technology will be eventually introduced are not easy to understand. On the other hand, if R&D efforts are simply described as investments and patents, the comparison with the analysis of future prospects is not straightforward. Although the data available allow us only to paint a partial picture with respect to research and development plus investments (R&D+i), both public and private energy-related investment and patenting seem to be responding to the call for higher innovation in less carbon-intensive (or carbon-free) technologies.

23.2 STATUS AND FUTURE PROSPECTS OF KEY ENERGY TECHNOLOGIES

In this first part of the chapter we will refer extensively to the Energy Technology Perspectives (ETP) of the International Energy Agency published in 2008 (IEA, 2008). The IEA-EPT focuses on three different scenarios with respect to GHGs emissions: the Baseline scenario, the ACT scenario and the BLUE scenario. In the Baseline scenario, the transition from 2005 to 2050 is presented considering what would happen if no climate policies were enacted. In this case, CO_2 emissions are forecast to increase by 130 percent with respect to 2005. In the ACT scenario the transition is presented considering what should be done to bring emissions in 2050 down to 2005 levels. The ACT scenario is based on the assumption of new technologies incentives of US$50 per tonne of CO_2 saved. The BLUE scenario is the more challenging scenario and presents the changes necessary to achieve a halving of emissions by 2050 with respect to 2005 levels. This last scenario is compatible with incentives ranging from at least US$200 to US$500 per tonne of CO_2 saved.

Achieving either the ACT or the BLUE scenario proposed by the IEA requires substantial changes to the world's energy system, both with respect to supply-side technologies, namely electricity sources, and with respect to demand-side technologies, namely all processes and appliances using energy and electricity. While the ETP (IEA, 2008) provides a broad and detailed overview of all the technologies that could help reduce greenhouse gas (GHG) emissions, in this chapter we focus on supply-side technologies, which include fossil fuel power plants, nuclear, biomass and bioenergy, wind power, solar photovoltaic (PV), concentrated solar power (CSP), hydro, geothermal and ocean energy.

This choice is determined by two main considerations: first of all, electricity generation is the sector with the highest GHG reduction potential (Table 23.1). Secondly, data availability for the power sector is relatively better than for other sectors. This allows us to provide a more detailed picture of private and public investment in R&D in the second part of this chapter. The stringent emission targets envisioned by the ETP (2008) require the development of negative emission technologies, such as Carbon Capture and Storage

Table 23.1   Emission reductions by sector and technology option in the ACT map and BLUE map scenarios in 2050

	CO$_2$ reduction ACT Map (Gt CO$_2$/yr)	CO$_2$ reduction BLUE Map (Gt CO$_2$/yr)
Power Generation	14.1	18.3
CCS	2.9	4.8
Fuel Switching Coal to Gas	3.8	1.8
Nuclear	2.0	2.8
Wind	1.3	2.1
Solar –PV	0.7	1.3
Solar – CSP	0.6	1.2
BIGCC and Biomass Co-combustion	0.2	1.5
IGCC	0.7	0.7
Ultra/Supercritical Coal	0.7	0.7
Gas Efficiency	0.8	0.4
Hydro	0.3	0.4
Geothermal	0.1	0.6
Buildings	7.0	8.2
Transport	8.2	12.5
Industry	5.7	9.2

Note: BIGCC is biomass integrated gasification combined cycle; IGCC is integrated gasification combined cycle.

Source: IEA (2008).

(CCS) and a substantial restructuring of the electricity systems. For this reason we also briefly highlight these special technologies.

23.2.1   The Supply Side of Energy Technologies

Fossil-fueled power plants
The current mix of natural gas and coal in electricity generation varies by country and region depending on resource availability. Total electricity production from coal and gas worldwide is around 11 TWh per year: overall 40 percent of the world's electricity production comes from coal and 20 percent from gas.

The IEA Baseline scenario indicates that, without proper CO$_2$ reduction incentives, emissions from the power sector alone (where coal and gas will account for 75 percent of total power generation) will increase to 27 Gt in 2050. This would equal the total CO$_2$ emissions in 2005. With respect to fossil-fuel energy production, the way to achieve lower CO$_2$ emissions is through the increase of energy efficiency. Currently, many of the existing plants are obsolete and highly polluting, and the potential for economically appealing improvements is high. Table 23.2 lists the current technologies used for fossil fuel electricity production.

To date, the major fossil fuel input for electricity production is coal. Pulverized coal

Table 23.2 New thermal power plant technologies

Plant type Fuel		PCC Hard coal	PCC Hard coal	PCC Hard coal	PCC Hard coal	NGCC Natural Gas	IGCC Hard coal
Steam cycle		Subcritical	Typical super-critical	Ultra-super-critical (best available)	Ultra-super-critical (AD700)	Triple pressure reheat	Triple pressure reheat
Steam conditions		180 bar 540°C 540°C	250 bar 560°C 560°C	300 bar 600°C 620°C	350 bar 700°C 700°C	124 bar 566°C 566°C	124 bar 563°C 563°C
Gross output	MW	500	500	500	500	500	500
Auxiliary power	MW	42	42	44	43	11	67
Net output	MW	458	458	456	457	489	433
Gross efficiency	%	43.9	45.9	47.6	49.9	59.3	50.9
Net efficiency	%	40.2	42	43.4	45.6	58.1	44.1
CO_2 emitted	t/h	381	364	352	335	170	321
Specific CO_2 emitted	t/MWh net	0.83	0.8	0.77	0.73	0.35	0.74

Source: Loyd (2007).

combustion (PCC) is the most diffused technology for power generation from coal. It is divided into subcritical, supercritical and ultra-supercritical, depending on the pressures and the temperatures of the steam cycle and the consequent efficiencies achieved (higher in ultra-supercritical plants). Integrated gasification combined cycle (IGCC), on the other hand, uses the fuel gas generated from the combustion of coal to run a turbine generator. The residual heat contained in the turbine is then used to produce electricity in a steam generator. The IGCC is a recently developed technology in the power sector. Clean coal technology was considered in Obama's presidential memorandum of February 2010 as the technology that would allow the US to take the lead 'in the global clean energy race'.

Capital costs for coal plants are higher: a typical coal plant requires US$900 to US$2800[2] per kW installed, while a typical gas plant requires US$520 to US$1800 per kW. The efficiency of most coal-fired power plants is well below the levels that are already possible. Efficiency gains can be realized not only by building new plants but also by improving the existing ones: moving from 30 percent efficiency (of older plants) to 40 percent would significantly reduce harmful emissions.

Gas is gaining an increasingly important role in the production of electricity. The most widespread technology with respect to gas is natural gas combined cycle (NGCC) with best-available efficiencies of around 60 percent. Today combined cycle power generation using natural gas is the cleanest source of power available using fossil fuels. The adoption of already available fossil fuel technology can make a significant contribution to containing the growth of CO_2 emissions from fossil fuel power generation. Power generation using natural gas is competitive with coal at today's prices: total generation costs[3] are in fact between US$0.067 and US$0.142 per kWh for coal and between US$0.076 and US$0.120 per kWh for gas.[4] Fuel costs account for around 67 percent of total expenditures in natural gas plants, compared to 20 percent in coal plants. However, rising gas prices, together with increasing concerns about its supply security, have resulted in a switch from gas to coal generation in recent years. If the development of new natural gas plants strains gas production and transmission systems, this would result in further natural gas price increases.

With respect to fossil fuel power generation, further technological developments are needed to achieve the objectives of more sustainable scenarios. Together with research aimed at increasing the efficiency of the existing technologies (aerodynamic turbines, control equipments, achievement of higher temperatures), new fossil fuel plant conceptions are currently in the development phase.

For example, combined heat and power (CHP) is the simultaneous utilization of heat and power from a single fuel source. This new technology could achieve an overall conversion efficiency of 90 percent. CHP is well suited for all fossil fuels (and biomasses) and provides district heating along with electricity generation. Installation costs for CHP range from US$700 to US$9000 per kW, while generation costs are similar for coal and natural gas: US$0.04–0.120 per kWh.

Another very promising technology is fuel cells, which generate electricity and heat using hydrogen-rich fuels together with oxygen from the air. Their incorporation in gasification-based power plants can greatly improve overall efficiency. The cost of these integrated plants is still very high (US$12000 to US$15000 per kW) and a lot of R&D is still needed to diminish these costs.

One of the main challenges for cleaner fossil fuel power plants is the existence of abatement incentives, strict regulation frameworks and international cooperation to avoid carbon leakage activities. Maybe the most important abatement activity is the global diffusion of plants with integrated carbon sequestration devices (carbon capture and storage – see section 23.2.2).

Biomass and bioenergy

Biomass – that is, organic material for energy use – is a source of renewable hydrocarbons that can be converted to provide energy carriers (heat, electricity and fuel) as well as materials and chemicals. The total annual demand for biomass has increased steadily over recent years, particularly in members of the Organisation for Economic Co-operation and Development (OECD), and currently it represents 10 percent of global primary energy consumption. Only one-third of this amount is consumed for industrial purposes; the rest is used in developing countries for domestic cooking and heating.

Bioenergy is the largest renewable energy contributor to global primary energy today and has the highest technical potential of all renewable sources under both the ACT and BLUE scenarios. Biomass use could increase fourfold by 2050, accounting for around 23 percent of total world primary energy. Different technologies are currently available for the exploitation of biomasses (their global capacity is around 400 GW) and their costs are summarized in Table 23.3

While the potential of biomass for electricity is very high, strong efforts are needed to fully exploit this potential. Whether biomass supply will increase depends on many different factors, such as the availability of land for the diffusion of energy crops, the improvement of the average biomass yield thanks to better managements, fertilizers, plant hybrid varieties, irrigation. In addition, the ability to convert waste, agricultural residues and wood to energy will play a very important role. Finally, since biomass tends to rapidly deteriorate over time and have a low energy density (with respect to fossil fuels), improvements in the logistic and in the pre-treatment phases are needed.

Current bioelectricity generation costs range from US$0.06 per kWh for traditional grate boilers to US$0.40 per kWh for CHP plants. The IEA expectations are in line with a 30 percent reduction of these generation costs: this would be possible with technology learning and economies of scale. Supply costs of crops cultivation, wood chipping and biomass transportation and storage are expected to decline over time.

Biomass R&D activity aims at making the existing technologies competitive with respect to fossil fuel power plants. Together with the usual efficiency-seeking R&D activities (new turbine devices, control systems and biomass treatments to achieve higher energy density), many efforts are currently focusing on energy crops: new plant breeding systems and genetically modified organisms (GMOs) are being studied.

Another important branch of biomass R&D is represented by recycling activities, such as increasing the quality, the availability and the suitability of urban waste, industrial waste, agriculture waste, wood chips, ashes and pulp and paper residuals. The continuous supply of biomass is very important for energy production: R&D aims at improving the whole logistic chain.

All these activities will be at their best only with technological transfers, from countries where biomass is already used successfully, to all the rest of the world; cooperation is fundamental here. It should also be clear that the reduction of emissions remains a

Table 23.3 Biomass technologies

Technology	Description	Typical Capacity	Net Efficiency	Costs
Grate boilers	Grate firing is the oldest combustion principle and was the most common design of small-size boilers until the beginning of the 1980s. Capacity of such plants rarely goes beyond 50 MW.	10–100 MW	20–40%	US$2000–3500 per kW
Fluidised bed combustion	This technology resembles grate firing but offers better temperature control and is more suitable for non-homogeneous biomass.			
Co-firing	Firing biomass with coal-fired boilers for electricity production can make a contribution to CO_2 emissions reductions, while achieving higher efficiency than that of dedicated biomass facilities.	5–100 MW existing > 100 MW new plants	30–40% 40%	US$100–1500 per kW Plus power station costs
Gasification	Biomass is oxidised at high temperatures (using air, oxygen or steam) and the gas obtained is used in Biomass Integrated Gasification Combined Cycle (BIGCC)	50–500 kW_{th}	80–90%	US$~1000 per kW_{th}
Combined heat and power	These generators are very efficient (90% efficiency compared to 30% efficiency of grate boilers) and simultaneously exploit heat and power from a single fuel source. They can be used to produce both electricity and district heating.	0.1–1 MW 1–50 MW	60–90% 80–100%	US$3500–5000 per kW US$~4000 per kW
BIGCC for power	Biomass can be used in industrial CHP units to produce a range of high value products alongside heat and power. Stand-alone BIGCCs for power generation are unlikely to gain a significant market position unless their costs can be driven down with the aid of supporting policies and R&D investment over the next decade	5–10 MW demos 30–200 MW future	40% 50%	US$4500–7000 per kW US$1500–3000 per kW

Source: IEA (2008).

priority: biomasses must not become more pollutant than the fuel they substitute (for example in terms of NO_x or methane emissions). One last point to remark upon is that biomass use must be sustainable: it cannot dramatically increase deforestation or decrease the availability of food from agriculture.

Wind power

Wind power has grown rapidly since the 1990s. Global installed capacity in 2007 was 94 GW, just less than 1 percent of global electricity supply. European and USA markets continue to dominate, while India and China are experiencing impressive growth. The current standard technology is a three-bladed horizontal axis, upwind and grid-connected wind turbine. The largest wind turbines are 6 MW units with a rotor diameter of up to 125 meters. Turbines have doubled in size nearly every five years, but a slow-down in this rate is expected in the near term, as transport and installation constraints become binding.

Most of the world's wind power capacity is land-based. Larger turbines can usually deliver electricity at a lower average cost than smaller ones; they also use wind resources more efficiently. Therefore, the repowering of many early wind farms with larger units has yielded higher outputs. The efficiency of electricity production, measured as annual energy production per unit of swept rotor area (kWh per m^2), has improved over time (2 percent annually). The costs of onshore wind energy projects are dominated by the price of the wind turbine. They are typically between US$1000 and US$3700 per kW installed. Generation costs range from US$0.100–0.234 per kWh at sites with low average wind speed, to US$0.070–0.090 per kWh at sites with high average wind speed. Of these generation costs, about 16 percent is due to operation and maintenance activities.

Offshore wind power technology is less mature and currently 50 percent more expensive than onshore installations. Yet, offshore installations produce up to 50 percent more output than onshore machines due to better wind conditions. New approaches in foundation technologies, larger turbines and more reliable components have increased the attractiveness of offshore wind energy. However, there are also factors that hinder offshore developments, the more significant of which are difficulties in site approvals, availability of installation vessels and constraints in the manufacturing supply chain. Offshore costs are largely dependent on foundations activities and grid connection; wind speeds, weather and wave conditions, water depth and distance to the coast matter as well. Offshore costs range between US$2540 and US$5550 per kW installed. Overall generation costs go from US$0.146 to US$0.261 per kWh produced. Operation and maintenance costs are higher in offshore wind energy generation with respect to onshore wind energy generation.

The cost of wind power has decreased steadily in recent years but a lot of R&D remains to be done to exploit its full potential. Further reductions in wind costs are expected if the high learning rate and technological improvements of the past years is to continue. As a result, in the ACT and BLUE Map scenarios, wind accounts for 9 percent and 12 percent respectively of global electricity supply in 2050 (between 1400 GW and 2000 GW).

R&D in wind technologies is focused on exploiting new resource opportunities and bringing down costs. Thus, R&D is focused mostly on design and construction of components, increased production volumes and innovations in materials, improving site assessments, aerodynamics, providing lighter structures and more resistant components, more efficient generators converters and grids, blades with larger diameters and electricity storage turbines. All these development could improve the ability to provide constant supply even with low wind activity.[5] Other branches of R&D concentrate specifically on offshore wind power: they try to improve the reliability of structures and to optimize the logistics necessary for the platforms to function well. In addition to traditional wind technologies, hybrid power plants are also being deployed: as offshore oil and gas decline, their production facilities can be transformed to hybrid sites by adding wind (as well as solar or wave) devices.

Even though wind power generates no CO_2 emissions, it has environmental impacts; ways to minimize the noise from the blades, to improve its suitability as regards wildlife and to protect the marine environment are studied by researchers. It is worth saying that nowadays wind power deployments have been granted support from governments; to achieve a sustainable future, this support should continue. The increase in the demand for wind power must be accompanied by an increase in the whole supply system. If not, prices will tend to increase and this would bring about a loss in attractiveness of this technology.

Nuclear
As of March 2010, 437 nuclear power plants worldwide had a total capacity of 371 GW. Fifty-five new reactors were under construction for an additional 50 GW of power (ENS, 2010). Currently, nuclear power supplies 16 percent of the world's electricity (the largest share is in OECD countries, in particular the USA, France and Japan).

Nuclear reactors that operate by fission are classified by neutron energy (thermal or fast), by coolant fluid (water, gas or liquid metal), by moderator type (light water, heavy water or graphite) and by reaction generation. Generation I reactors were developed in the 1950s and 1960s. Very few of them are still operational. Generation II reactors were built in the 1970s as large commercial power plants and many of them are still operating today. Generation III reactors were developed in the 1990s with evolutionary design and advances in safety and efficiency. Generation III+ and Generation IV reactors are the modern challenge of research: they should be more efficient, safer and they are expected to minimize waste production.

Projected costs of generating electricity show that in many circumstances nuclear energy can be competitive against coal and gas generation. As a result, a number of countries are reconsidering the role of nuclear energy, particularly in view of its advantages in reducing CO_2 emissions. Three factors contribute to high direct costs of nuclear power: construction, operating and back-end costs.

Construction costs depend on the length and complexity of the pre-construction period (the time to secure permits and planning approvals), on the long construction times (they are decreasing nowadays) and on the cost of capital. Operating costs relate to the safe running of a power station (including the costs of inspections, safeguard and insurance costs), while fuel-cycle costs are related to the costs of energy production and waste disposal. The estimation of back-end costs is on the other hand rather controversial, as waste management costs are not incurred until the end of the reaction's life, allowing the operation to accumulate funds from revenues. They do not therefore impact much on the levelized costs

Available estimates of levelized costs range from US$30 to US$75/MWh, with an average of US$40. Broadly, investment costs represent around 70 percent of the total levelized cost, while operation and maintenance and fuel contribute respectively for 20 percent and 10 percent. Electricity generating costs projections from nuclear plants vary from US$0.042 to US$0.137 per kWh.[6] Reduction cost opportunities in nuclear technologies are mainly related to an increase in the level of output due to efficiency improvements: 90 percent efficiency is currently achieved by OECD countries.

Nuclear electricity generation depends on the availability of uranium for fuel. Such fuel is projected to last for at least 85 years. In addition, nuclear technology improvements (new reactors can extract 50 times more energy per kilogramme of uranium) and the discovery of new sites would increase the projection to 270 years of availability. Thorium fuel is the new frontier of nuclear plants: it is more abundant than uranium and therefore some countries (India in particular) are carrying out a significant amount of research on this issue. However, one major issue of concern is that uranium fuel comes from politically unstable countries.

In addition to the direct costs of nuclear operations, it is also important to assess external costs and benefits linked with nuclear technologies, that is, those costs and benefits that are not internalized in the market prices and paid by consumers, but are paid by society as a whole. While the production of nuclear energy is almost CO_2 free, one of the main costs linked with nuclear electricity production is connected to the disposal of radioactive waste. Proper siting (often deep geological disposals) and transmutation technologies reducing the radioactivity of wastes should always follow any nuclear development program. In addition, the nuclear industry is often concerned with safety issues. However, empirical data show that nuclear plants represent the technology with the smallest rate of accidents. The recent Japanese earthquake, the tsunami and the accident in the Fukushima nuclear power station have recently increased the debate on the safety issues. Both in the US and Europe new and more stringent controls have been advocated.

According to the ACT and BLUE scenarios, the switch to nuclear power can contribute with 6 percent of CO_2 savings with respect to the Baseline scenario. However, safety, weapons proliferation and waste remain as a constraint to the wide spread use of this technology. Most current reactor designs have large power outputs. These large reactors are unsuited to many developing countries, where there may be limited electric grid capacity. Consequently, small and medium-sized reactor (SMR) designs are being considered globally: their reduced size and complexity translates to lower capital costs, shorter construction times and advantages for countries with limited nuclear experience.

R&D focuses on the new SMR designs together with the possibility of alternative uses for nuclear energy, which include district heating and hydrogen electrolytic creation. Other R&D efforts are carried out for efficiency improvements (Generation IV reactors) and for partitioning and transmutation technologies which would reduce the radioactive life of dangerous waste reaction isotopes. These programs are likely only with substantial international cooperation and knowledge diffusion. Intense research activity is also carried out to obtain energy from fusion of light elements (for example hydrogen) instead of fission of heavy elements (for example uranium). While this is possible in theory, a great deal of effort and collaboration are needed to succeed in this field.

One challenge to nuclear deployment is related to the proliferation of nuclear weapons: there is a global need for the reinforcement of the non-proliferation treaties.

Finally, there are concerns about nuclear education (which has decreased over time in many countries): a new generation of engineers must be trained for nuclear programs.

Solar

Solar energy is the most abundant energy resource on earth. Its low energy density and intermittency, however, make large-scale exploitation difficult and expensive. Solar energy currently (2010) provides less than 1 percent of the world's commercial energy. In the IEA sustainable scenarios solar power provides between 6 percent and 11 percent of global electricity. This is equivalent to a thousand-fold increase with respect to today's levels. Photovoltaics (PV) and concentrated solar power (CSP) will each account for about 50 percent of solar electricity.

Photovoltaics   The basic building block of photovoltaic systems is the PV cell, namely a semiconductor device that converts solar energy into direct-current electricity. PV systems are modular: they can be linked together to provide power in a range of several watts (typically 150 peak watts/m²). The cells can be grid-connected or stand alone, they can be ground-mounted or integrated into buildings. The majority of them currently belong to the latter category.

Since 2000, PV capacity in IEA countries increased by a factor of eight, reaching 5.7 GW in 2006. Investments are expected to grow again, due to government incentives and to the end of a shortage in the supply of purified silicon (the main component of the cells). However, the most important obstacle to PV system deployment is their high costs. Total PV costs are currently around US$3067–7381 per kW. Between 2006 and 2010 the increase in demand has raised PV prices (although, as mentioned above, they are expected to diminish as soon as new silicon production plants become operative).

The penetration of thin films modules in the market will help to drive down total PV systems costs: they are produced at US$2000 per kW, but costs show a decreasing tendency. PV plants' expected generation costs are estimated to range from US$0.333 to US$0.600 per kWh. If electricity produced in building-integrated systems is fed into the distribution grid it can compete with electricity retail prices. PV systems require very little operation and maintenance activities (5 percent of generation costs).

As said, the modules are mainly based on silicon which is a well-established and reliable technology. The cost-efficiency of this technology can however be improved further: switching to single-crystalline silicon from multi-crystalline silicon yields better efficiency but also higher costs. Thin films are the new frontier of PV: they require less raw material, they are less difficult to manufacture and less sensitive to overheating. The drawbacks are lower efficiency and limited experience of lifetime performance. Recent developments try to overcome these problems. The thin films market is expected to grow significantly. New generations of PV are emerging: organic solar cells, third-generation PV cells and dye-sensitized nano-crystalline solar cells are the latest outputs of research. Their contribution to the energy industry remains uncertain however. Further techno-logical development is needed to achieve higher efficiencies and much larger production volumes. The target cost to reach should be lower than US$1250 per kW installed.

Currently, R&D for PV is mainly focused on materials, equipment and device con-cepts. Substituting costly materials, improving the quality of silicon and generating recycling options are some important R&D goals. There is concern for achieving fully automated processing, less energy-intensive production units, a longer lifetime of PV systems and low waste processes. With respect to device concepts, research is focusing on new modules designed for easy assembly, increasing the modules' lifespan and achieving higher cell efficiency (from the current 15–20 percent to 30 percent). For the new PV generations, the research focus is mainly on improving efficiency. One of the major challenges for PV development is the presence of incentive schemes, which remain fundamental to support the research, cost reduction and deployment of these technologies.

Concentrated solar power   Concentrated solar power (CSP) uses direct sunlight, concentrating it several times to reach higher energy densities and higher temperatures. The heat is used to operate a conventional power cycle, namely a steam turbine which drives a generator. This technology is clearly best suited for areas with intense direct solar radiation. Because CSP uses a thermal energy intermediate phase, it has the potential to deliver power on demand (heat storage used to run steam turbines). CSP is much cheaper than PV in those areas characterized by good-quality sunlight, although it is not yet competitive with fossil fuel or even with wind power.

CSP installations are typically large, and the technology has evolved in three different types: troughs, towers and dishes. Troughs are parabolic trough-shaped mirror reflectors which linearly concentrate sunlight into receiver tubes, heating a thermal transfer fluid. They represent the more mature technology with a solar-to-thermal efficiency of 60 percent and solar-to-electric efficiency of 12 percent. Hybrid plants use both fossil fuel and solar energy to have continuous generation. Costs for trough plants are now in the range of US$4000 to US$9000 per kW, depending on local production costs, on the desired yearly electrical output and on local solar conditions.

Towers are made up of numerous heliostats that concentrate sunlight onto a central receiver on the top of a tower. Towers are characterized by many different designs and new plants are being considered worldwide from USA to South Africa. Costs for towers currently start at US$9000 per kW. Parabolic dish-shaped reflectors, on the other hand, concentrate sunlight in two dimensions and run a small engine or turbine at the focal point. They are usually associated with a Stirling engine: a 300 MW plant would cover approximately 3 square miles. Costs for dishes are above US$10 000 per kW.

CSP plants currently under construction are expected to generate electricity at a cost between US$0.135 kWh and US$0.243 per kWh, mostly depending on the location. Further cost reductions can be achieved only through massive R&D and deployment investments. The USA is trying to improve the technologies and volume of production of such plants and to make them competitive with fossil power generation systems by 2020, achieving an average cost of US$0.06 per kWh.

The main challenges to CSP are: the high cost it entails (thus the strong incentives it needs to create mass production, economies of scale and learning-by-doing), its limited deployment options to certain areas, and the great surface areas needed to produce a reasonable amount of energy. R&D for CSP is carried out to improve the conversion rate (sun–heat–electricity) even with a smaller mirror surface. Thinner mirrors are also being studied to prevent dust deposition. For trough plants, the replacement of expensive heat carriers, such as mineral oil with water, would reduce investments and operating costs: the challenge here is that superheated steam may create unacceptable material stresses. For towers, projects are trying to increase the temperature of solar heat further (even burning fossil fuels) and then run a gas turbine. This would achieve higher power conversion efficiencies. Finally, R&D efforts are also concentrated on trying to use CSP to produce hydrogen from water solar thermolysis; this requires improvements in materials capable of withstanding the very high temperature required for water thermolysis.

Hydropower

Hydropower is an extremely flexible technology which is already well established. The IEA estimates that the hydro potential worldwide is around 6000 TWh per year. Of this potential, only 5 percent has yet been exploited. Nowadays capacity is estimated around 200 GW with 800 GW under construction. OECD countries produce roughly half of worldwide hydroelectricity. The share from non-OECD countries is, however, likely to increase (in China in particular). Currently, hydro reservoirs provide built-in energy storage, and the fast response time of hydropower means that it can be used to optimize electricity production across power grids, meeting sudden fluctuations in demand or helping to compensate the loss of power from other sources. Hydropower can also be generated from pumped storage systems. Pumped hydro consists in two or more reservoirs at different heights: energy is stored when the water is pumped from the low to the high reservoir and released when the opposite occurs. The typical efficiency of these systems is about 80 percent. The IEA sustainable scenarios suggest that by 2050 hydropower production will double (reaching 1700 GW of capacity).

Existing hydropower is one of the cheapest ways to produce electricity. For new plants in OECD countries, installation costs depend on the type of plant and range between US$700 (small plants) and US$19000 (large plants and pumped storage systems) per kW. Total generating costs are around US$0.03 (China and Brazil) to US$0.15 per kWh (EU and USA). The cost of pumped storage systems depends on their configuration and use: they may be up to twice as expensive compared to an equivalent unpumped system.

Hydropower generation produces no CO_2 emissions other than those emitted during its construction. Large-scale hydropower projects can however be controversial because they affect water availability downstream, inundate valuable ecosystems and may require the relocation of populations. Moreover hydropower usually depends on rainfall in upstream catchment areas: reserve capacity may be needed to cover periods of low rainfall and this increases costs. Small-scale hydropower installations are normally designed to run in rivers. Thus, they are an environmentally friendly option: they do not interfere significantly with river flows. In general, concerns over undesirable environmental and social affects have been the main barriers to hydro schemes worldwide. Protection of fish passage, reproduction and migration is also an issue. Proper siting and design can mitigate many of these problems. However, to date, there is no universally accepted method of establishing an agreed minimum flow rate that satisfies both developers and regulators. An emerging issue is the possible impact of climate change on hydropower production: rainfall run-offs may decrease and, as a consequence, power generation diminishes. As plants have a long lifetime their proper allocation (in areas less affected by lower rainfall) is an important issue.

Like other energy technologies, hydropower technologies need to improve efficiencies and reduce costs. For new plants, advanced technologies should be developed in order to minimize environmental impacts, improve control systems, optimize generation and investigate different materials for which less maintenance would be required.

Geothermal

Geothermal plants grew at a constant rate of about 200 MW/yr from 1980 to 2005. Total capacity reached 10 GW in 2007, generating 56 TWh/yr of electricity. Geothermal power plants can provide extremely reliable capacity. Heat is produced from the decay of radioactive material inside the earth and is moved to the surface through conduction and convention. High-temperature geothermal resources can be used in electricity genera-tion, while lower-temperature resources can be tapped for a range of direct uses such as district heating and industrial processing. In what follows, we concentrate on electricity generation.

There are three types of commercial geothermal power plants: dry steam, flash steam and binary cycle. They all use steam to drive a turbine that generates power. Dry steam sites use direct steam resources at 250°C, but only five fields of this nature have been discovered to date (2008). Underground reservoirs that contain hot, pressurized water at 180°C are more common (flash steam plants). Finally, binary-cycle plants use geother-mal resources with 85°C temperatures: the heat is transferred to a fluid that vaporizes at a lower temperature and the vapor drives the power turbine.

All these types of plants are practically emission free. Exploration, drilling and con-struction make a large share of the overall cost of geothermal electricity. Set-up costs vary from US$1700 per kW installed capacity for large high-quality resources (with cheap drilling activities) to US$12800 per kW for small, low-quality resources (with expensive drilling activities). Generation costs depend on a number of factors but par-ticularly on the temperature of the geothermal fluid: they may range from US$0.05 to US$0.27 per kWh.

Costs fell by almost 50 percent from the 1980s to 2000 and new developments are helping geothermal plants to become more and more competitive. We are witnessing an accelerating development in many countries (the USA, Mexico, Indonesia and New Zealand). However, large geothermal power development is limited to tectonically active regions. The IEA prediction is that power production can increase twenty-fold (from 10 GW to 200 GW).

Current R&D is focused on ways to enhance the productivity of geothermal reservoirs and to use new areas with a lot of potential but that are difficult to access (new drilling devices represent the bulk of R&D investment). A new technique called hot dry rock is under development: water is injected into hot underground rocks and, when it returns to the surface, the heat it has accumulated is used to generate electricity in a binary plant. Another line of R&D focuses on the development of deeper wells in volcanically active areas: the deeper the well, the higher is the flow of power to the surface. Reducing the costs of drilling activity is therefore a major goal. Collaboration between governments and industries has been carried out for these purposes. Another line of R&D is trying to improve the exploitation of low-temperature resources and to develop more efficient cooling systems.

Challenges to expanding geothermal energy utilization include long project develop-ment times, the risk and the costs of exploratory drilling and potential undesirable envi-ronmental effects (flows of polluting fluids into groundwater and small earthquakes). Moreover, geothermal energy carries a high commercial risk because of the uncertain-ties involved in identifying and developing reservoirs that can sustain long-term heat flow.

Ocean energy
Ocean energy technologies for electricity generation are at a relatively early stage of development: wave energy and tidal energy are the main areas of interest. The technology required to convert water energy into electricity is very similar to that used in hydro-electric power plants. Capacity worldwide is around 300 MW, with new projects being built or under consideration.

The major challenges to the deployment of such technologies are their costs and risks. Only in recent years has adequate government funding supported pilot projects. Infrastructure typically represents more than half of the total cost for the installation. Because the technologies are still at R&D and demonstration stages, data are often not available or not very informative. A first rough estimation suggests a cost of more than US$6700 per kW. More precise information will be available only in the future. The other main challenge is the environmental impact of underwater barrages compromising water wildlife.

The current focus of R&D is on moorings, structure and hull design, power take-off systems, wave behaviour and hydrodynamics of wave absorption. Research on turbines focuses on cost-efficiency: reliability and ease of maintenance of components that can resist hostile marine environments. Also, plants which exploit both wind and water are being studied.

23.2.2   Key Technologies

Carbon capture and storage
Carbon dioxide emissions from fossil fuel-fired power plants can be strongly reduced by using the CO_2 capture and storage technologies (CCS). New power plants should be designed to be suitable for CCS retrofitting, and located in places where they can be connected to storage sites. With proper development of legal and regulatory frameworks, CO_2 reduction incentives, support for R&D and public outreach CCS may become a mature technology by 2020. Under the IEA scenarios it has the potential to reduce the emissions from the power sector by 20 percent.

CCS involves three main steps: (1) CO_2 capture from the source; (2) transportation to an injection sink; and (3) underground geological injection in suitable strata. Capturing CO_2 includes gas processing, fuel transformation and compression. With post-combustion processes, CO_2 is captured at a low pressure from flue gas. CO_2 can also be captured pre-combustion: reacting the fuel with air or oxygen enables the capture of high CO_2 concentration. Transportation is mostly carried out through pipelines, ships and trucks, with pipelines generally being the most cost-effective. Geological injection is focused on the research of suitable strata such as saline formations or depleted oil and gas reservoirs. Other geological options include basalts, caverns and mines. But these techniques are limited by low storage volumes and possible chemical interactions.

CCS has not been widely deployed yet but it is indeed the technology with the largest potential for CO_2 emissions reduction. CCS is best suited for modern and efficient plants; capturing CO_2 from low-efficiency plants is not economically viable: the higher the efficiency of generation, the lower the cost per kWh of electricity. In 2007, four fully integrated large-scale CCS projects were in commercial operation and injected about 0.5 Mt of CO_2 per year. Three (Sleipner, Norway; In Salah, Algeria; and Snøhvit, Barents Sea) inject CO_2 which is separated from natural gas in a gas production facility. The fourth project captures CO_2 from the Great Plains Synfuels Plants (North Dakota, USA) and transports it by pipeline for about 200 miles to the Weyburn–Midale project (Canada).

Total CCS costs are between US$44 and US$100 per tonne of CO_2 avoided. Using CCS with natural gas-and coal-fired power plants would increase electricity production costs by US$0.02 to US$0.04 per kWh. The bulk costs of CCS projects are associated with capture: on average between US$25 and US$50 per tonne of CO_2 avoided, depending on the fuel and the technology used. The cost of transportation depends on the method chosen and ranges between US$1 and US$16 per tonne of CO_2 carried. Storage costs are estimated from US$1 to US$44 per tonne: storage in saline formations is cheaper than storage in depleted oil and gas fields. For example, the installation costs of a CCS-equipped coal plant is US$3223–6268 per kW installed, while a traditional coal plant costs US$900–2800 per kW. Transport and storage activities may add US$0.010–0.015 per kWh to total generation costs. Finally, operation and maintenance costs double in CCS plants. In the IEA scenarios an incentive of US$50 per tonne of CO_2 saved will result in a reduction of 5.2 Gt of CO_2 per year in 2050 (one-fifth of global CO_2 emissions in 2005).

CCS deployment on a large scale depends on today's efforts. R&D aims at improving the current status of capture technologies; reducing costs and increasing efficiency of the systems are the main issues. New methods are being tested: they include innovative capturing membranes, solid absorbers and thermal processes. Regional pipelines are being developed for CO_2 transportation. CCS success does however need more efforts, including: the development of legal and regulatory frameworks at national and international level; the incorporation of CCS into emission trading schemes and post-Kyoto instruments; and inclusion of CCS in the design of new power plants. In addition, it is necessary to promote public awareness and education on CCS, as well as technological transfers and private and public financial support.

Electricity systems
The characteristics of the electricity system can significantly affect the cost of emission mitigation options. Much more electricity is produced than is ever used. Transmission and distribution (T&D) losses account on average for 14.3 percent of the electricity produced worldwide. Unlike other energy carriers, electricity supply and demand must always be balanced in real time because electricity can only in rare circumstances be stored in large quantities, and always in other energy forms. The current devices to store electricity are batteries (chemical energy), pumped-storage (potential energy) and compressed air. Batteries are an efficient storage system but they remain a costly option applicable only on a relatively small scale: lithium batteries cost US$550 per kWh, but to be cost-competitive with other alternatives they should cost US$160 per kWh at most. Other systems, such as CAES (compressed air energy storage systems) and pumped storage, are more cost-competitive, but there is always room for efficiency improvements.

Transportation of electricity therefore entails significant losses and is costly, with costs varying by country and market segment. The electricity supply for low-voltage users costs (in OECD countries) around US$0.006 to US$0.009 per kWh. Reducing losses is a way to diminish these costs. The improvement of the efficiency of such devices can reduce costs, losses and emissions. Investments in transmission and distribution systems are as important as investments in production plants.

Providing more flexible technologies is a way to improve both electricity transmission and distribution, and to reduce global losses by 10–20 percent. Most grid management systems aim at transporting electricity over as short distances as possible. High voltage direct current (HVDC) systems transmit power at longer distances and with fewer losses than traditional Alternating current (AC) systems. Transformers are then necessary to step voltage down for industry or domestic use. In this case, losses can be reduced by using new core materials, such as amorphous iron.

R&D focuses on the optimization of the grid system (improving the synchronized matching of demand and supply), on the development of superconductors and new materials, and on new storage systems such as hydrogen produced from electrolysis. In addition to R&D aimed at technological improvements, promoting consumers' energy-saving behaviour is another important component of a strategy focused on reducing electricity losses. In particular the use of AC/DC transformers for electronic equipment has increased rapidly in recent years. These transformers are often switched on permanently (stand-by) while the equipment is used only intermittently. Losses connected to this situation may amount up to 10 percent of total electricity use. Spreading information about energy-saving behaviour could help reduce these costs.

23.3   ANALYSIS OF ENERGY R&D ACTIVITY

The previous section highlighted the technological pathways that are necessary in order to address successfully both energy security concerns and climate change issues. These changes to the energy sector call for an unprecedented increase in the amount of R&D investment focused towards energy-related technologies. Transforming markets and reducing barriers to the commercialization and diffusion of nascent low carbon energy technologies requires an innovation-based energy strategy. In this process, both public and private investment will play a major role. This section provides an historical perspective on energy-related and climate-friendly R&D investments.

The main goal of this section is to paint a picture of the magnitude of past and current energy R&D activity and to put into perspective the effort necessary to 'walk the talk' of addressing climate change and energy issues. To this end, we rely on publicly available data sources and on recent reports and descriptive analyses of energy-related R&D. Most of the available information refers to IEA and EU member countries. Where data availability permits it, we will also highlight the role played by other (non-IEA or EU member) countries.

The analysis is organized as follows: section 23.3.1 describes the publicly available data sources. Section 23.3.2 focuses on public energy and environmentally focused R&D. Section 23.3.3 provides a snapshot of the private investment in selected alternative technologies. Section 23.3.4 turns instead to patent data: we summarize recent results from studies looking at the national and international patenting strategies.

23.3.1   Data Challenges

A major problem in providing an overall picture of past investment in R&D activities for energy-related and environmentally friendly technologies is the lack of data. In particular, it is possible to retrieve information on R&D investments for a number of countries, but only a few sources specifically distinguish between energy- and non-energy-related innovation activities. Those sources focus mainly on government-sponsored energy R&D. No database exists which contains both private and public investment focused only on energy-efficiency or climate-friendly technologies.

The IEA Energy RD&D Statistics offer information on governmental RD&D (research, development and deployment) expenditures. The data are very detailed and are broken down into several specific technologies: energy efficiency, fossil fuels, renewable energy sources, nuclear fission and fusion, hydrogen and fuel cells, and other power and storage technologies. In many cases, these general investment areas are also subdivided into more specific technologies, such as wind, solar PV and CSP. One major concern regarding this database is that it includes only government expenditures and does not account for private investments.

The OECD GERD (gross domestic expenditures on R&D) and ANBERD (Analytical Business Enterprise Research and Development) databases include industry-specific R&D investment (based on NACE codes) since 1987.[7] While expenditures are broken down by sources of funds (namely business enterprise sector, government sector, higher education sector, private non-profit sector), data relate to overall investments.

The European Commission also provides information on energy-related RD&D activities and investments. The 2009 Report 'R&D Investment in the priority technologies of the European Strategic Energy Technology Plan' (EC, 2009a) was prepared by the Institute for Prospective Technological Studies (IPTS) of the European Commission's Joint Research Centre (JRC) and offers a benchmark of the environmentally friendly R&D spending in 2007. The report estimates R&D investments in selected low carbon energy technologies in the EU-27 funded by the member states, through the 6th EU Research and Euratom Framework Programmes and by companies with headquarters registered in the EU. The report focuses on the priority technologies of the European Strategic Energy Technology Plan (SET-Plan), which include wind energy, photovoltaics (PV) and concentrated solar power (CSP), carbon dioxide capture and storage (CCS), biofuels, hydrogen and fuel cells, smart grids, nuclear fission (with a focus on Generation IV reactors), and nuclear fusion.

The 'EU Industrial R&D Investment Scoreboard' is part of the Industrial Research Monitoring Activity carried out jointly by the JRC and the Research Directorate-General (DG RTD) of the European Commission. These annual reports collect information on the top 1000 EU companies and the top 1000 non-EU companies ranked by their investments in R&D.[8] The data described in the Scoreboard are drawn from the latest available company accounts. Companies are listed by sector of operation; thus it is possible to distinguish which companies are active in the energy sector. Each company's total R&D investment is attributed to the country in which the company has its registered office. The Scoreboard provides a much-needed and detailed picture of the top energy and non-energy investors. Nonetheless, it is limited to 2000 companies and does not allow distinguishing energy-related and environmental investment for those companies not operating in the energy sector.

Another source of information that has been used to gain understanding on private investment in eco-innovation is patent data. Patents measure the output on the innovation process, namely the outcome of the R&D activity that is carried out within firms.

The shortcomings of patent data are linked with the nature of innovation activity: on the one hand, patents are not the only way to protect an innovation. Patents as a proxy for innovation do not account for other kinds of innovations, such as production practices, organizational changes, or inventions that are kept secret. In addition, simple patent counts do not take into account the quality of the patented innovation. It is well known that patents have a skewed value distribution, with a few having high commercial and innovative value and many having little.[9] Currently two main sources of patent data can be accessed. The National Bureau of Economic Research (NBER) Patent Statistics Database (Hall et al., 2001) includes all patents granted to both USA and foreign innovators by the United States Patent and Trademark Office (USPTO) since the mid-1960s. The European Patent Office (EPO) PATSTAT Database, on the other hand, includes patent applications to more than 90 patenting authorities worldwide.

In addition to the above-mentioned data collection efforts, in many cases it is possible to get country-or region-specific data. For example, data availability for the USA allows a comparison of energy versus non-energy investment. While we will present some of this analysis, we note that since the collection procedure is not extended to other countries, international comparison is impossible.

An additional point of concern relates to the fact that the available databases are limited to developed countries (with the exception of PATSTAT), namely IEA and OECD member states. This does not necessarily pose a major problem with respect to the analysis of past data, since the majority of R&D investment at the global level so far has been carried out in developed countries. In recent years, however, developing and fast-developing nations have started to play a more active role both with respect to general R&D and with respect to energy-related investment. In order to keep track of these dynamics, it is necessary to ensure that data collection efforts be enlarged to include at least the BRICs countries (Brazil, Russia, India, China), and even better all developing countries.

23.3.2   Overall Public Energy and Environmental R&D Spending

National governments have been active in financing energy-related R&D, even if the trend of spending has decreased since the late 1970s. Figure 23.1 provides an overall trend in public energy-related R&D investment for IEA member countries from 1978 to 2007.[10] Energy R&D spending has been categorized under five research areas: nuclear power, fossil fuels and energy efficiency, renewable sources, hydrogen and fuel cells, and other technologies. Public energy-related R&D is lower now than it was at the end of the 1970s. Public spending (in real terms – 2000 USD) decreased until around 2000 and only recently began to rise. Much of the decline in public energy spending can be attributed to lower nuclear investment. On the right axis of Figure 23.1, we report the percentage of nuclear spending over total energy R&D: nuclear accounted for less than 40 percent of investment in 2007, down from 60 percent at the end of the 1970s. Investment in renewable energy technologies was higher at the end of the 1980s than it was in 2000, but has shown an increase in the decade to 2007.

A recent EC report (2009a) provides an overview of the spending for SET Plan technologies in 2007 in EU member states. Figure 23.2 shows both the distribution of investment among member states as well as a comparison with US and Japanese spend-

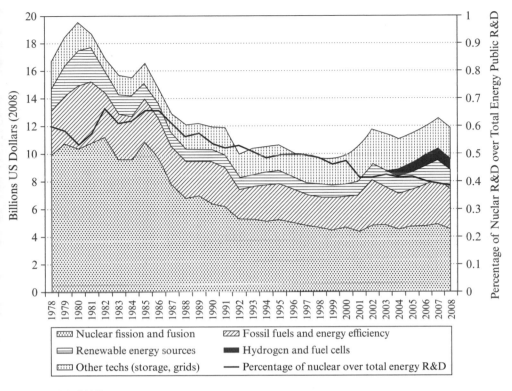

Source: IEA (2009).

Figure 23.1 *Trends in the aggregated public energy R&D funding of IEA member states (1978–2007) including EC funds*

ing in the same research areas. The three top investors in Europe are France, Italy and Germany, accounting for two-thirds of investments. Each of these countries has slightly different funding priorities. Investments related to transportation (namely biofuels, hydrogen and fuel cells) dominate the French funding. Italy's budget is more or less equally divided between transportation-related technologies, solar technologies and smart grids. Germany invests in solar technologies and wind as well as hydrogen.

While the overall spending of EU members is much higher than that of Japan and the USA, a mere comparison between different world regions might be misleading. In fact, there are important differences with respect to the way in which energy R&D is financed in these different regions. In the USA and in Japan energy research has a strong focus and coordination, provided respectively by the Department of Energy (DoE) and the Ministry for Economy, Trade and Industry (METI).

At the European level, on the other hand, R&D activities within member states are often fragmented, due to the complexity created by the involvement of several ministries and agencies in the management of different parts of national programs. In addition, until recently there was no unified European program to foster low carbon technologies (with the exception of nuclear fusion-related research). Cooperation among European

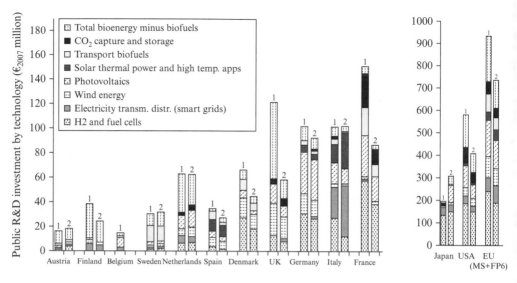

Source: EC (2009).

Figure 23.2   Public R&D spending in non-nuclear SET Plan technologies for EU member countries. 1 – 2007 (gap filled), 2 – average 2002–07

countries is often limited and the possible synergies between member states with respect to the development of new energy technologies have so far not been fully exploited. More recently, initiatives such as the SET-Plan have started to address this problem (EC, 2009a).

23.3.3   Private R&D Investment in Energy-related Technologies

The main challenge in providing a detailed picture of private energy-related R&D investment is the lack of a comprehensive overview on corporate and public R&D investments in selected energy-related technologies. Three main sources provide information on private spending in energy-related innovation: the recent EC (2009a) report assessing spending on SET Plan technologies in EU member states, the 2008 and 2009 Scoreboards providing information on the top 2000 European and non-European firms with respect to R&D spending, and an article by Nemet and Kammen (2007) describing private and public R&D investments in the USA.

Table 23.4 summarizes the investments for public, private and EC funding for selected energy-related technologies for 2007 (EC 2009a).[11] The private sector invested an estimated total of €1656 million in renewable energy sources and €205 million for nuclear technologies. Private investment accounts for 26–81 percent of investment in non-nuclear SET Plan technologies. Investment by the EU accounts for 7 percent of the investment in renewable technologies and 22 percent of nuclear energy investment (mostly focused on nuclear fusion). The share of corporate R&D investments is elevated for the more mature technologies such as wind energy and biofuels. In comparison, the share of corporate R&D investments is lower for PV, hydrogen and fuel cells and CSP,

Table 23.4   Private and public energy-related investments in the European member states, summary of findings from JRC Report to the European Commission

	Amount Total R&D investment	Percentages		
		Corporate R&D investment	Public R&D investment from member states	Public EU
Non-nuclear SET Plan technologies				
Total	2358	0.69	0.24	0.07
Wind	383	0.76	0.21	0.03
PV	384	0.58	0.35	0.07
CSP	86	0.56	0.38	0.06
Biofuels	347	0.77	0.19	0.04
CCS	269	0.81	0.13	0.06
Hydrogen and fuel cells	616	0.61	0.28	0.11
Smart grids	273	0.78	0.17	0.05
Nuclear SET Plan technologies				
Total	940	0.22	0.56	0.22
Nuclear reactor	458	0.45	0.54	0.01
Nuclear fusion	482	0	0.58	0.42

Source: EC (2009a).

as well as for Generation IV nuclear reactors and nuclear fusion. These technologies may be considered as less mature, in particular if we assume that research in PV concentrates on new technologies instead on the more mature crystalline silicon cells. The results describing the distribution of R&D spending by investor must be interpreted with care due to the differences in the nature between corporate and public R&D spending (EC, 2009a).

Private firms are not equally active in all European Union member countries: Figure 23.3 shows the geographical distribution of corporate investment. Companies located in Germany, France, the UK, Denmark, Spain and Sweden account for almost 95 percent of the total corporate estimated R&D investments. Those countries giving strong public support to research into a certain technology simultaneously account for the largest R&D investment of industry in the same technologies. This is a possible indication of a positive correlation between public research support and industrial R&D investment. The dynamics between private and public (as well as energy and non-energy) investments are however not yet fully understood. In particular, the economic literature does not provide clear empirical evidence regarding the fact that increases in public spending can crowd out or crowd in private investments.

Since the EC (2009a) does not report any private investment in fossil fuel energy sources, we complement the previous information on SET Plan technologies with some insights from the EU Scoreboard (EC, 2009b). The 2000 companies listed in the 2008 Scoreboard accounted for about 80 percent of worldwide business enterprise expenditure on R&D (BERD). A special focus was given to energy-related innovation: firms investing in the production and distribution of oil, gas and electricity displayed high

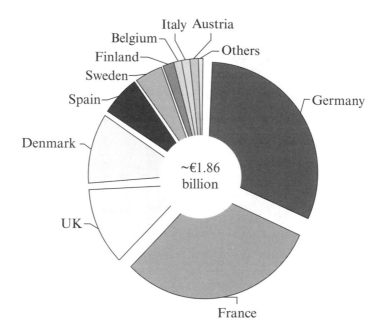

Source: EC (2009a).

Figure 23.3   Regional distribution of private investment in SET Plan technologies

R&D growth rates in the last several years. Figure 23.4 shows the ranking of the top 20 R&D companies in the energy sector (oil and gas producers, oil equipment, services and distribution and electricity) by their total R&D investment.

There are seven EU companies in this list of the top 20, some of them showing high annual R&D growth rates in the period 2005–08, namely Royal Dutch Shell (29.5 percent), AREVA (20.3 percent), Vattenfall (11.6 percent) and BP (8.8 percent). Non-EU firms showing high growth rates of R&D investment were Petroleo Brasiliero (52.6 percent), Gazprom (38.4 percent), Chevron (32.4 percent) and China Petroleum & Chemical (31.1 percent). Very high R&D trends are also shown by other energy-related firms, namely those active in alternative energies. Some examples of companies operating in this area are Vestas Wind Systems and Nordex, with R&D growth rates over 2005–08 of 29.6 percent and 44.4 percent, respectively. In the solar photovoltaic field, Q-Cells showed annual R&D growth rate of 148.6 percent; while in biofuels, Abengoa's R&D investment grew at an annual rate of 32.8 percent. Figure 23.5 shows the growth rate in R&D investments for selected energy companies.

Nemet and Kammen (2007) provide a comparison of public and private energy-related investment for the USA derived from federal budgets and from surveys of companies conducted by the National Science Foundation, which we report in Figure 23.6. Public energy spending reached a peak around the year 1980, when the energy-related budget was around US$ 8 billion (2000 USD). Investment subsequently decreased: by 1995 the USA government was investing the same amount in energy as in 1975. Subsequently, energy-related spending decreased further. Private investment followed a different path:

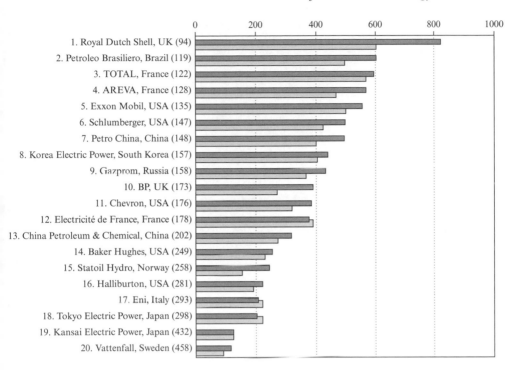

Source: EC (2009b).

Figure 23.4 Top 20 R&D companies in the energy sector

private energy spending was growing, but significantly below the level of public spending, until 1985. From 1985 to 1990 yearly private investment in energy R&D was higher than public investment, but subsequently decreased to reach around US$1 billion in 2004, roughly corresponding to a quarter of the total energy R&D budget.

The decline in energy investment occurred while overall US R&D was growing by 6 percent per year; the comparison between the energy budget and the health and defence budgets is striking. The former was declining while the latter were growing at rates of 10 and 15 percent per year, respectively (Nemet and Kammen, 2007). It is interesting to compare the dynamics of both energy and non-energy research investment (Figure 23.7): energy research accounted for 10 percent of all R&D in the 1980s, while it covered only 2 percent of the research budget in 2004.

23.3.4 Patent Data and Innovation in Energy-related Technologies

In recent years, the dynamics of energy-related innovation in the private sector has been extensively studied using patent data as a proxy both for innovation activity and 'knowledge transfer' between countries. As already pointed out, patent data is a useful measure of the output of the R&D process. Exploring patenting dynamics both in the national and in the international market can help to paint a picture of the innovation effort that is taking place worldwide, and complements data on energy-related R&D investments.

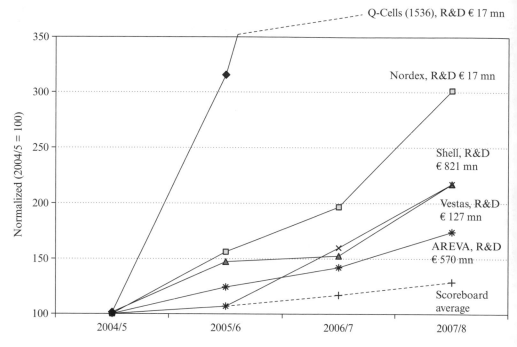

Source: EC (2009b).

Figure 23.5   Growth rate in R&D spending for top EU companies in the energy sector

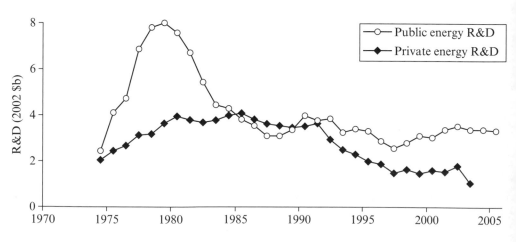

Source: Nemet and Kammen (2007).

Figure 23.6   Public and private energy R&D spending in the USA

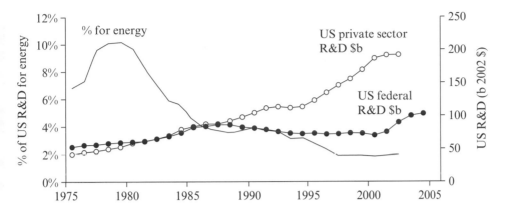

Source: Nemet and Kammen (2007).

Figure 23.7   Total USA R&D and percentage devoted to energy

In addition, patenting strategies can also inform on the flow of knowledge across countries. An innovator who has been granted a patent in his domestic market can also request patent protection in other countries. The original patent, labelled 'mother patent', is thus 'duplicated'. The transfer of technology through the patent system is referred to as 'intended technology transfer' as opposed to (unintended) knowledge spillovers. Since this process is costly, duplicating a patent shows the willingness to market the technology abroad.

Recently, a number of contributions have looked at the trends in and determinants of R&D using patent data as a proxy for innovation activity (Popp, 2002, 2003, 2006; Nemet and Kammen, 2007; Verdolini and Galeotti, 2011; Johnstone et al., 2010 among others). Nemet and Kammen (2007) provide an analysis of renewable energy patents granted by the USPTO, shown in Figure 23.8. In the case of innovators in the USA market, patent applications seem to anticipate public R&D spending in wind, PV and fuel cells technologies, while they seem to follow public R&D investments in the case of nuclear technologies. In addition, for PV, wind and nuclear fusion the rate of patenting was at its highest around 1980, roughly corresponding to the peak of public R&D investment. Unfortunately, the source of patent data for the USPTO includes patents granted up to 2002, therefore it does not allow speculation on the rate of innovation in energy technologies in recent years.

Johnstone et al. (2010) provide similar data for the European Patent Office (EPO). They focus on selected renewable energy (wind, solar, geothermal, wave and tied, biomass and waste) and examine the pattern of patent applications at the EPO by 25 countries over 26 years. Also in the case of the EPO, patent applications show an increase around 1980 (Figure 23.9), but of much lower in magnitude than that analyzed for the USA. Moreover, since the mid-1990s, patent applications in renewable energy technologies have increased significantly.

It is to be kept in mind however that a straightforward comparison between the data available for the USA and for the EPO is not informative. The EPO published information on patents granted, which does not include applications that were dropped or

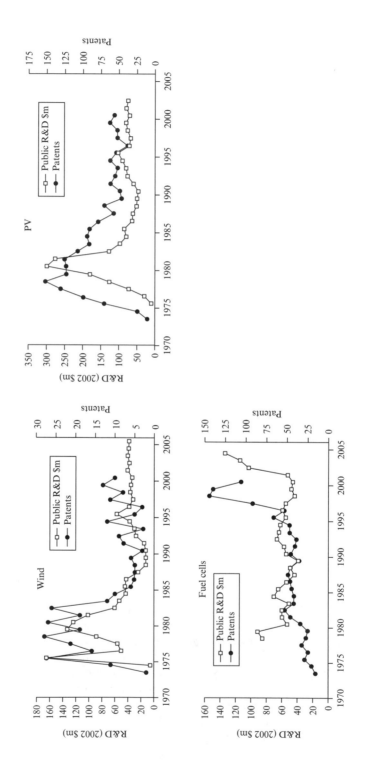

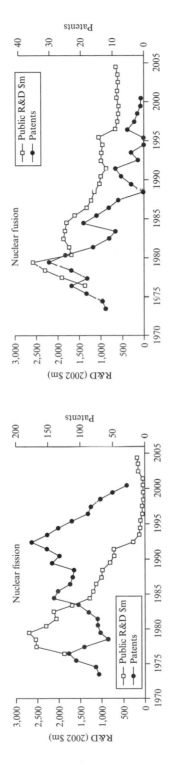

Source: Nemet and Kammen (2007).

Figure 23.8 Patents and public R&D expenditures in selected energy-related technologies

497

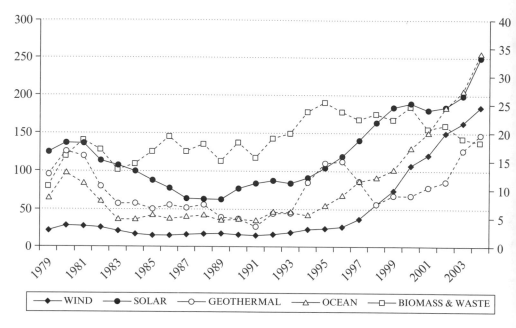

Source: Johnstone et al. (2010).

Figure 23.9   Patenting in selected environmentally friendly technologies, by technology

rejected by the patent office. The EPO data, on the other hand, refer to patent applications but contain no information on how many of these applications were finally granted patent status. Johnstone et al. (2010) also provide information on the patent applications to the EPO by country, shown in Figure 23.10. All countries show increases in their rate of applications to the EPO in recent years. The graph also complements the information on the USA patenting trend, showing that US patent applications at the EPO have increased in recent years, just like those of other European countries.

Dechezleprêtre et al. (2009) explore innovation in 12 energy-efficient and environmentally friendly technologies worldwide.[12] Patent data show that innovation in these areas is concentrated in three countries, namely Japan, Germany and the USA. The export rate of the patents within these technologies is around 25 percent, meaning that one-quarter of innovation is protected in more than one country. Table 23.5 summarizes the rates and direction of 'intended technology transfer'. Most of the transfer happens between developed countries (75 percent). Exports from developed to developing nations are lower (18 percent), but are growing rapidly.

One of the advantages of patent data is that they are also available for developing countries, unlike R&D investment data. Table 23.6 shows the top innovators among developing nations for the technologies analyzed in Dechezleprêtre et al. (2009).

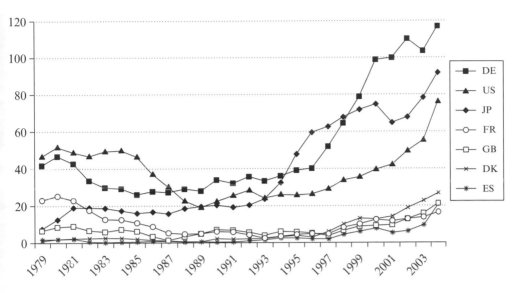

Source: Johnstone et al. (2010).

Figure 23.10   Patenting in selected environmentally friendly technologies, by country

Table 23.5   Patent duplicates and intended technology transfer

Origin \ Destination	Developed countries	Emerging and transition economics
Developed	5812 (75.9 %)	1360 (17.8 %)
Emerging and transition economics	377 (4.9 %)	112 (1.5 %)

Source: Dechezleprêtre et al. (2009).

23.4   CONCLUSION

A restructuring of the energy system implies the development and deployment of a number of energy-efficient and carbon-free technologies that will allow diversification of the portfolio of energy sources. This will necessarily require a significant amount of investment for innovation, adoption, diffusion and transfer of technologies worldwide.

This chapter provides a full analysis of the status and future prospect of energy technology on the one hand, and the analysis of past R&D efforts on the other. The future prospect of energy technology is basically a complex taxonomy, a list of equipment, machinery and tools whose adoption and diffusion is necessary in order to address climate issues and energy security concerns that are facing the world economy. On the other hand, the analysis of past R&D investment trends has followed a more quantitative approach. It is not an easy task to consider these two issues together and reconcile their results and suggestions.

Table 23.6   Top innovators among developing countries

	World rank	Average % of world inventions	Most important technology classes (decreasing order)
China	4	5.8 %	Cement, geothermal solar, hydro, methane
South Korea	5	4.6 %	Lighting, ocean, hydro, biomass, cement
Russia	6	4.2 %	Geothermal, cement, hydro, CCS, ocean
Brazil	10	1.1 %	Ocean, building
Taiwan	18	0.6 %	Ocean, lighting
India	30	0.2 %	Cement
Mexico	34	0.1 %	Ocean
South Africa	53	0.03 %	

Source: Dechezleprêtre et al. (2009).

If the analysis of future prospects of energy technology is confined to a simple description, the reasons why a specific technology will be eventually introduced are not easily understood. On the other hand, if R&D efforts are simply described as investments and patents, the comparison with the analysis of future prospects is not straightforward.

The funding of alternative energy sources was high around 1980, as a consequence of the oil crisis which brought about concerns for energy security. Subsequently, both innovation and investment in R&D stagnated. More recently, in light of the widespread concern relating to environmental problems, both R&D investment and patenting activity in key energy-related fields have increased. Although the data available allows us only to paint a partial picture with respect to R&D+i, both public and private energy-related investment and patenting seem to be responding to the call for higher innovation in less carbon-intensive (or carbon-free) technologies.

The roadmaps highlighted in the previous section of this chapter will require that this trend in energy-related R&D and deployment continues and is sustained by both governments and the private sector. The effort required to redesign the energy systems worldwide is extremely high, and this will create concerns about trade-offs and crowding-out effects of energy with respect to non-energy technologies.

While assessment of the outcome of present and past policy is paramount, lack of data often impairs such efforts. The picture on past R&D trends presented in this chapter suffers from the lack of a comprehensive dataset and this chapter is not immune from this shortage of good information. As far as data on public R&D are concerned, information is more precise and detailed than that on private R&D. Our analysis shows that private energy R&D investments are probably largely underestimated. This is basically due to the lack of a common approach in collecting R&D data from the different companies' balance sheets.

ACKNOWLEDGMENTS

The authors would like to thank Walter Galvalisi for the excellent research assistance. Elena Verdolini acknowledges funding from the European Research Council under the European Community's Seventh Framework Programme (FP7/2007-2013)/ERC grant agreement No. 240895 – project ICARUS 'Innovation for Climate Change Mitigation: A Study of Energy R&D, its Uncertain Effectiveness and Spillovers'.

NOTES

1. It should be clear that the review concentrates its attention only on CO_2, which is the main greenhouse gas (GHG) responsible for climate change. Other GHGs such as methane are important as well, but here the attention is not focused on them.
2. The costs and benefits will all be expressed in 2008 US dollar terms in section 23.2 of this chapter.
3. All the generation costs are levelized at a 10 percent discount rate.
4. Operation and maintenance costs account for 8 percent in coal plants and 5 percent in gas plants.
5. The compressed air energy storage (CAES) is very interesting: electricity is used to compress air when demand is low; the air is then stored and released in a natural-gas-fired turbine only when demand rises. Efficiency increases up to 60 percent.
6. Decommissioning costs are also included in this calculation.
7. The Czech Republic and Poland provide data since 1992 and 1994, respectively.
8. The term 'EU company' refers to companies whose ultimate parent has its registered office in a member state of the EU. The term 'non-EU company' is applied when the ultimate parent company is located outside the EU.
9. Please refer to Griliches (1990) for an overall discussion of the use of patents as indicators on innovative activity
10. Countries included are: Australia, Austria, Belgium, Canada, Czech Republic, Denmark, Finland, France, Germany, Greece, Hungary, Ireland, Italy, Japan, Korea, Luxembourg, Netherlands, New Zealand, Norway, Poland, Portugal, Slovak Republic, Spain, Sweden, Switzerland, Turkey, United Kingdom, United States. Data are not gap-filled.
11. The results for public R&D spending, though taken from the same source used for Figure 23.1, are not directly comparable as the JRC report included gap-filling of the data in its methodology.
12. The technologies considered in this study are: biomass, building, cement, fuel injection, geothermal, hydro, lighting, methane, ocean, solar, waste and wind.

REFERENCES

Dechezleprêtre, A., M. Glachant, N. Johnstone, I. Haščič and Y. Ménière (2009), 'Invention and transfer of climate change mitigation technologies on a global scale: a study drawing on patent data', FEEM Working Paper 2009.082.
European Commission (EC) (2009a), 'R&D investment in the priority technologies of the European Strategic Energy Technology Plan', prepared by the Institute for Prospective Technological Studies (IPTS) of the European Commission's Joint Research Centre (JRC), SEC (2009) 1296.
European Commission (EC) (2009b), 'Monitoring industrial research: the 2008 EU Industrial Investment Scoreboard', prepared by the European Commission's Joint Research Centre (JRC).
European Nuclear Society (ENS) (2010), 'Nuclear Power plants, Worldwide', http://www.euronuclear.org/info/encyclopedia/n/nuclear-power-plant-world-wide.htm, accessed on 20 May 2010
Griliches, Z. (1990), 'Patent statistics as economic indicator: a survey', *Journal of Economic Literature*, **28** (4), 1661–1707.
Hall, B.H., A.B. Jaffe and M. Trajtenberg (2001), 'The NBER patent citation data file: lessons, insights and methodological tools', NBER Working Paper 8498.
International Energy Agency (IEA) (2008), 'Energy Technology Perspectives', Paris.
International Energy Agency (IEA) (2009), 'Energy Technology R&D Database', version 2009, accessed April 2010.

Johnstone, N., I. Haščič and D. Popp (2010), 'Renewable energy policies and technological innovation: evidence based on patent counts', *Environmental Resource Economics*, **45** (1), 133–55.
Loyd, S. (2007), 'New thermal power plant technologies: technology choices for new projects – CCGT, IGCC, supercritical coal, oxyfuel combustion, CFB, etc.', paper presented at the Coaltrans Conference, 2nd Clean Coal and Carbon Capture: Securing the Future, London, September.
Nemet, G.F. and D.M. Kammen (2007), 'US energy research and development: declining investment, increasing need, and the feasibility of expansion', *Energy Policy*, **35**, 746–55.
Popp, D. (2002), 'Induced innovation and energy prices', *American Economic Review*, **92** (1), 160–80.
Popp, D. (2003), 'Pollution control innovations and the Clean Air Act of 1990', *Journal of Policy Analysis and Management*, **22** (4), 641–60.
Popp, D. (2006), 'International innovation and diffusion of air pollution control technologies: the effects of NOx and SO$_2$ regulation in the US, Japan, and Germany', *Journal of Environmental Economics and Management*, **51** (1), 46–71.
Verdolini, E. and M. Galeotti (2011) 'At home and abroad: an empirical analysis of innovation and diffusion in energy-efficient technologies', *Journal of Environmental Economics and Management*, **61** (2), 19–134.

PART VI

OTHER DIMENSIONS OF ENERGY

24 Energy and poverty: the perspective of poor countries
Rob Bailis

24.1 INTRODUCTION

Understanding the role of energy in poverty alleviation requires an understanding of the complex role that energy plays in facilitating individual and collective well-being. It is simple to state that particular forms of energy are required for economic activity, and that such activity contributes to wealth. However, this only partially reflects the interrelationship between energy and well-being. Well-being is not determined purely by wealth or economic activity. Other factors such as freedom from avoidable disease and mortality, economic self-determination, political voice and freedom of cultural expression contribute to, or perhaps define, well-being more accurately than wealth or income.

The idea that well-being is synonymous with freedom, both freedom from deprivation and freedom of self-determination, is based largely on Amartya Sen's notion of 'development as freedom' (1999). Sen conceives of poverty as a deprivation of capabilities: a state in which individuals are unable to 'lead the kind of lives they value – and have reason to value' (Sen, 1999: 18). Sen stresses that a fixation on income-poverty ignores the more complex determinants of human well-being. Thus, without denying that there are frequently strong correlations between low income and 'capability deprivation', Sen encourages us to avoid thinking that 'taking note of the former would somehow tell us enough about the latter' (1999: 20). In this chapter, I wish to use Sen's notion of poverty as capability deprivation in order to analyze the role that energy may play in enhancing human well-being and, conversely, the ways in which lack of access to energy contributes to the deprivation of individual and societal capabilities.

Others have grappled with this question. For example, the conceptual link between energy and poverty dates back at least to the energy crises of the 1970s and 1980s if not before. At that time, energy rose to prominence as a topic of study among the social sciences. In the context of developing countries, energy analyses started to look explicitly at non-commercial sources of energy like woodfuel (for example fuelwood, charcoal, sawdust), which were (and still are) essential to household production. However, early analyses were primarily focused on perceived supply–demand imbalances, which initially led to overly simplistic and largely mistaken conclusions about pending fuelwood crises (Eckholm, 1975; de Montalembert and Clement, 1983; FAO, 1983). These early assessments were largely discredited (Leach and Mearns, 1988; Arnold et al., 2003), although they periodically resurface both in policy discussions and scientific analyses (Arnold et al., 2003).

Like the assessments that supported the idea of a woodfuel crisis in developing countries, early analyses of energy–poverty linkages tended to focus on energy supply.[1] However, more recently, socio-technical studies of energy have shifted focus to the concept of 'energy services'. As Rogner and Popescu (2000: 31–2) write: 'energy is not an end in itself. The energy system is designed to meet demands for a variety of services such as cooking, illumination, comfortable indoor climate, refrigerated storage, transportation, information, and consumer goods. People are interested not in energy, but in energy services.'

Thinking in terms of 'energy services' helps to highlight the material applications to which energy is put, and reminds analysts that they should focus on more than energy supply in order to grasp the interrelationship between energy and well-being. A second, related concept that has emerged alongside the focus on energy services is 'energy poverty'. The use of the term dates to the early 1980s (*Oil and Gas Journal*, 1981), but it has gained prominence since the late 1990s. Like energy services, the concept encourages broad thinking in making the link between energy and well-being. In defining energy poverty, Amulya Reddy makes an explicit link to choice: 'The energy dimension of poverty – energy poverty – may be defined as the absence of sufficient choice in accessing adequate, affordable, reliable, high-quality, safe, and environmentally benign energy services to support economic and human development' (Reddy, 2000: 44).

By explicitly noting the lack of choice as an integral element of energy poverty, Reddy makes an association between energy and well-being that resonates with Sen's notion of development. That is to say, the 'energy poor' are not best identified as those whose energy consumption falls below a certain threshold. Rather, they are best defined as those who lack a range of choices that the 'energy-rich' enjoy. In lacking access to choices, the energy-poor, by extension, cannot access many of the freedoms that are facilitated by having 'adequate, affordable, reliable, high-quality, safe, and environmentally benign energy services'. These freedoms include: freedom from preventable morbidity and mortality caused by dependence on solid biomass fuels used in highly polluting stoves; freedom of mobility and connectivity, which are critical for both economic self-determination and political expression; and freedom from drudgery and low returns to labor associated with lack of access to shaft power provided by reciprocating engines or electric motors. However, overconsumption of energy, like many forms of excess, can also be extremely problematic. It is also worth noting that Reddy's vision of energy poverty includes a lack of access to 'environmentally benign energy services'. By this standard, most of the world could be considered energy-poor, including the high per capita consumers in the Global North. However, the choice not to use environmentally benign energy services in the North is a very different matter than the energy poverty faced by populations with limited access to all forms of energy largely in the Global South. A full examination of these issues, though needed, is beyond the scope of this chapter, which is focused instead on energy poverty in the South.

In the remainder of this chapter, I explore the theme of energy poverty in more detail. Section 24.2 examines the range of energy choices available and draws out some of the contrasts between the energy-poor and energy-rich. Section 24.3 explores the links between energy and economic growth at the national scale. Section 24.4 considers energy poverty at the household level, drawing examples from cooking technologies, agriculture and communications technology. Section 24.5 provides some closing remarks.

24.2   ENERGY CHOICES AND ENERGY POVERTY

This section examines the types of energy that are utilized by people and nation states at different levels of wealth and describes the services that can be derived from them. It is useful to categorize energy into primary sources, energy carriers and energy services (Figure 24.1). Primary energy sources are the natural resources that are extracted, collected or otherwise harvested. These resources include both exhaustible fossil fuels and renewable wind, solar and tidal power. Fossil fuels are finite and occur only in limited geographic locations, but they are storable and easily transported. Biomass occurs much more broadly, and is also storable and transportable, but it has a much lower energy density than most fossil fuels, which limits the extent to which it can be transported economically. Other forms of renewable resources, like wind and solar power, are intermittent and cannot be stored or transported without conversion into other types of energy carriers.

Energy carriers are derived from primary resources, such as coal oil and natural gas, and represent an intermediate step to facilitate transport and storage between extraction or harvest and final use. Others, like intermittent renewable resources, require immediate conversion. They are most frequently converted to electricity, which is the most versatile of energy carriers; however, it is difficult and costly to store.

Energy services represent the collection of end uses that energy carriers provide. Energy carriers vary in quality, or in the ease and efficiency with which they can be converted into useful services.[2] Electricity, liquid and gaseous fuels are typically considered higher-quality energy carriers than solid fuels for their ease and efficiency of use. Liquid fuels most commonly provide shaft power for transportation, although they also provide heat and electric power in some circumstances. Solid and gaseous fuels also provide heat and electric power. Electricity provides a diverse range of services includ-

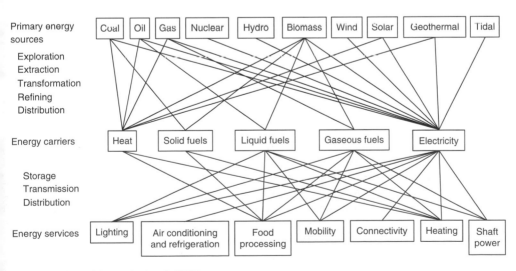

Source: Adapted from Metz et al. (2007).

Figure 24.1   Energy sources, carriers and services

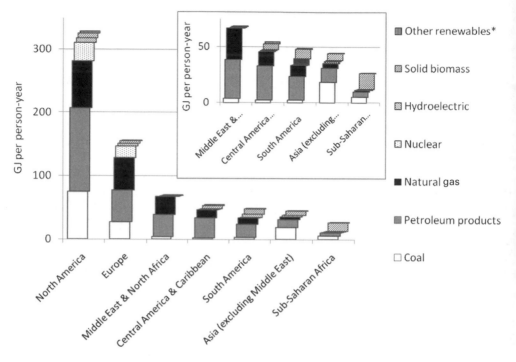

Note: * Does not include hydro or solid biomass.

Source: Data are from World Resources Institute (2010) with the exception of sub-Saharan Africa. That region is based on data from World Bank (2010) and BP (2009).

Figure 24.2   Fuel-specific per capita energy consumption by world region in 2005

ing heat and shaft power, illumination, refrigeration and some transportation. It is also critically important in providing communication services or 'connectivity' (Jacobson, 2007), which has exploded in recent years, as I explore in more detail below.

Figure 24.1 sheds light on the multiple pathways through which energy services can be derived, yet provides no insight on the differential levels of access that currently characterize global energy use. Looking at different world regions, we can identify substantial variation in the composition and quantity of energy resources utilized. Figure 24.2 shows global energy supply composition and per capita use disaggregated by income groups. As the plot indicates, per capita energy use differs dramatically between regions. The average per capita consumption in North America is 6–12 times that of Latin America, Asia and sub-Saharan Africa. The composition of energy supply also varies, with lower-income countries, particularly sub-Saharan Africa, relying far more on solid biomass (wood and wood products, agricultural residues and dung) than medium- and high-income groups.

Thus, there is a close association between biomass dependence and national wealth. This association is mirrored at the household level. Figure 24.3 shows primary cooking fuels used by different wealth quintiles of urban households in six African countries.[3] The data illustrate a common pattern in which people at lower income levels are constrained in their energy options. Despite variation from country to country, there is a systematic trend in which wealthier households more frequently choose higher-quality energy carriers like kerosene, liquefied petroleum gas (LPG) or electricity over solid fuels like wood and charcoal.

Patterns like those shown in Figure 24.3, in which wealthier households opt for higher-quality sources of energy, have been used to develop a simple model of fuel progression called the 'energy ladder' in which households are envisioned to shift from low- to high-quality fuels as income increases (Hosier and Dowd, 1987; Leach, 1992). Its simplicity makes the energy ladder an attractive conceptual tool; however, it has been criticized as too deterministic. Further, empirical evidence has shown that households may expand their energy options as they get wealthier, but they do not necessarily abandon the fuels that outsiders have labeled 'lower-quality' fuels (Saatkamp et al., 1999; Masera et al., 2000). As with the notion of poverty itself, a focus on income leaves analyses blind to the contingencies, cultural preferences and policies that influence fuel choice.

It is also possible to observe variation in sectoral energy use by level of wealth. For example, countries that are heavily dependent on biomass have a limited ability to produce high-quality energy carriers needed for industrial production. With limited industry, the bulk of energy consumption occurs within the residential sector. Sub-Saharan Africa and parts of Asia, where the residential sector accounts for over half of all energy consumption, exemplify this trend. A sectoral breakdown of energy consumption in different world regions is shown in Figure 24.4.

Energy-poor populations rely heavily on biomass resources for their energy needs because it is the most accessible resource in terms of both cost and availability. It is technically possible to convert biomass into advanced energy carriers (Sagar and Kartha, 2007). Indeed, there has been a great deal of discussion about the potential for biomass to replace fossil fuels in transport as well as other applications (Hoogwijk et al., 2005). Tropical regions, where the bulk of energy-poor nations are located, have a relative advantage in producing energy derived from plant matter because of higher productivity, warmer weather and longer growing seasons (Field et al., 2008). Biomass can be used to produce electricity; however, with a few exceptions among energy-poor countries, solid biomass is limited to supplying heat for cooking and some basic industrial processes.[4]

24.2.1 Energy and Well-being at the National Level

Energy enables myriad processes that improve the human condition: agricultural production, manufacturing and transportation. It also indirectly facilitates the provision of essential services such as communications, health and education. The Human Development Index (HDI) is a composite indicator consisting of health, wealth and educational data (UNDP, 2009), which was introduced in 1990 as a direct result of Sen's influential work on capabilities (Haq, 1995; Fukuda-Parr, 2003). A plot of the HDI as a function of per capita energy consumption demonstrates the close association between energy use and human well-being (Figure 24.5).

While the association between energy and HDI is clear, a closer look reveals that countries at a given level of per capita energy consumption perform quite differently

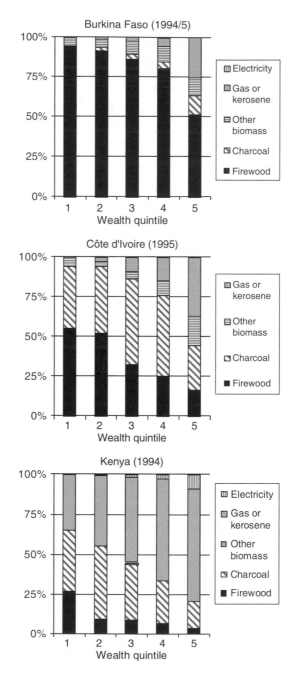

Notes: These data are derived from nationally representative household welfare surveys conducted in the 1990s (see World Bank, 2000 for the original sources).

Source: World Bank (2000).

Figure 24.3   Primary household cooking fuel by wealth quintile in 6 African countries

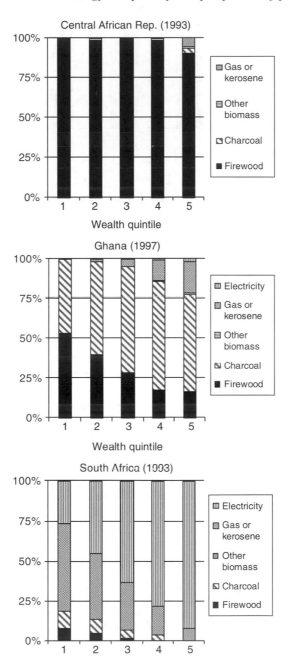

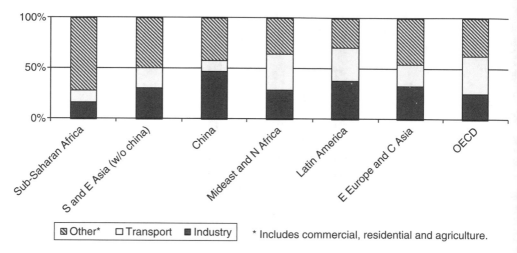

Source: IEA (2007).

Figure 24.4   Shares of energy consumption by sector in world regions

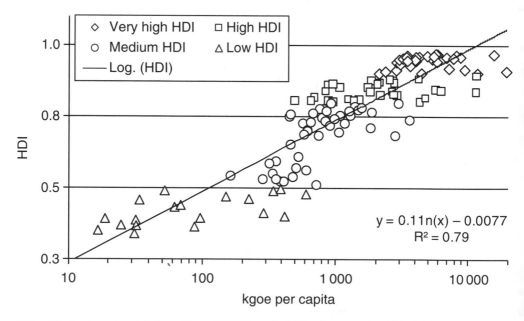

Note: The horizontal scale is logarithmic. HDI data and divisions between Very high; High; Medium; and Low are based on UNDP (2009). Per capita energy consumption data is from World Bank (2010).

Figure 24.5   HDI as a function of per capita energy consumption for 160 countries (based on 2007 data)

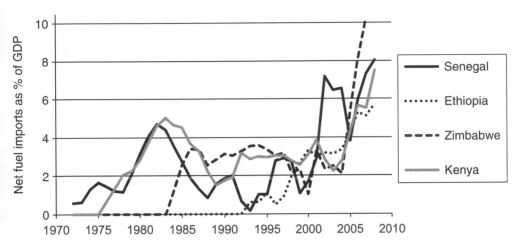

Source: World Bank (2010).

Figure 24.6   Net fuel imports as percentage of GDP from 1970 to 2008 (three-year moving average)

with respect to HDI. For example, the lowest quartile of annual energy consumption, which extends from 20 kgoe to 510 kgoe per capita, includes countries ranging from low to high HDI. The highest quartile of annual energy consumption, which ranges from 3300 kgoe to over 19 000 kgoe per capita, includes countries ranging from medium to very high HDI.

24.2.2   Energy Trade and Energy Poverty

While biomass constitutes a large fraction of primary energy supply in energy-poor countries, it is clear from Figure 24.2 that other energy resources are also utilized. However, in contrast to biomass, which is a locally available resource, most energy-poor countries must import primary energy carriers like coal, oil or natural gas. This can affect the country's balance of trade and weaken the economy, particularly during price shocks. In sub-Saharan Africa, 35 out of 40 countries for which recent data are available are net fuel importers. The recent increase in petroleum prices have hit the region hard. As the price of oil increased, the fraction of national accounts spent on energy rose to unprecedented levels, with many countries spending 6 to 10 percent of gross domestic product (GDP) on net fuel imports (Figure 24.6). For comparison, both the US and China spend roughly 3 percent of GDP on net energy imports (World Bank, 2010).

However, not all energy-poor countries are net energy importers. Indeed, some countries are endowed with large reserves of fossil fuels. For example, in sub-Saharan Africa, Cameroon, Congo, Gabon and Nigeria have been producing and exporting oil for decades. Others, like Sudan and Chad, have started exporting more recently. In each of these countries, oil revenues constitute the majority of export earnings and a considerable fraction of GDP. In the case of Chad, for instance, oil production came on line in 2003. Revenues rose rapidly, and oil sales now constitute 90 percent of export earnings and nearly 40 percent of GDP (IMF, 2009). Between 2002 and 2008, per capita GDP increased 50 percent in real terms. However, 96 percent of the population lacks access to electricity (ICF Macro, 2010). Even in Nigeria, where oil production began decades ago, only 50 percent of the population had access to electricity by 2008 and nearly all of the energy used in the residential sector is still derived from biomass (IEA, 2009). The pattern is similar in each of the region's oil producers as is shown in Table 24.1. Only Gabon has succeeded in providing electricity to over half of its population, and even there the bulk of energy in the residential sector continues to be derived from solid biomass (IEA, 2009a).

Among Africa's oil exporters, the persistent reliance on biomass for residential energy and the lack of access to electricity among the majority of the population highlights the limited degree to which resource wealth can contribute to the well-being of citizens. This is particularly true in polities that lacked democratic institutions at the time that resources were discovered. The failure to share resource wealth with the general population is linked to a broader phenomenon in which resource wealth, particularly petroleum, has been implicated in maintaining and reinforcing anti-democratic tendencies, political repression and violent conflict.[5]

24.2.3   Electricity and Energy Poverty

As Figure 24.1 indicates, electricity is an extremely versatile energy carrier. It can be produced from all solid, liquid and gaseous fossil fuels as well as all common renewable resources. Similarly, electricity is capable of providing every conceivable energy service. However, in countries that rely primarily on biomass, electricity production tends to be relatively low, and access is limited. These limits are expressed in terms of both the quantity that is consumed and who is actually consuming power. Figure 24.7 shows per capita electricity consumption among world regions. The vertical scale is logarithmic in order to illustrate the full spread among the lower-consuming groups. The difference in per capita electricity consumption between the Organisation for Economic Co-operation and Development (OECD) countries and South Asia and sub-Saharan Africa is roughly a factor of 20.

In addition to total consumption, electricity should also be considered from a distributional perspective. Whereas regions like the OECD are characterized by near universal access to electricity, energy-poor countries are characterized by very low levels of access, particularly in rural areas. Figure 24.8 shows the fraction of urban and rural households with access to electricity in sub-Saharan Africa, Latin America and Asia. Entries are disaggregated by region and ordered by level of access in rural areas. Rural households have systematically lower access than urban areas. This should not be surprising: many countries have rural electrification policies, but few have been successful. Grid extension is costly; urban areas, with higher population densities, are much cheaper to serve (Abdullah and Markandya, 2009). Further, urban households tend to have higher incomes, which make them more attractive customers to power providers: they are more likely to own appliances that consume electricity and more likely pay their bills. In contrast, dispersed rural populations are costlier to reach and, once reached, they are likely to consume less power, making the investment in infrastructure more difficult to justify from a business perspective.

Table 24.1 *Oil production and residential energy use in oil-exporting countries of sub-Saharan Africa*

	Oil output in 2008 (10⁶ ton)	Revenue from oil 2006–08 (% of GDP)	Sources of residential energy (%)[a]			Electrification (% with access)[b]			Comments
			Biomass	Petroleum products[a]	Elec	Urban	Rural	Total	
Angola	92.2	80%	91%	6%	3%	66	9	38	Oil production dates to the 1970s, but was hampered by conflict. War ended in 2006, allowing increased production, but with charges of corruption and minimal development benefits (Gettleman, 2009).
Chad	6.7	39%	NA	NA	NA	16	< 1	4	No data for residential energy use. Oil production began in 2003 with completion of controversial pipeline through Cameroon (Polgreen, 2005).
Equatorial Guinea	17.9	75%	NA	NA	NA	NA	NA	NA	Oil production began in the 1990s and contributed to a 16-fold increase in real per capita GDP between 1990 and 2008 (**BP**, 2009; World Bank, 2010). However, no domestic energy data is available and little can be said about energy poverty.
Rep. of Congo	12.9	70%	92%	5%	4%	51	15	34	Oil production began in the 1970s and forms the bulk of the country's economy.
Gabon	11.8	50%	89%	5%	7%	90	30	74	Oil production dates to the 1960s. With Eq. Guinea, Gabon has one of the highest per capita GDPs in the region.
Nigeria	105.3	38%	97%	2%	1%	85	31	50	Nigeria is infamous for the corruption and violent conflict associated with oil wealth (see, for example Watts, 2009).
Sudan	23.7	20%	93%	4%	3%	NA	NA	NA	Sudan began oil production in the late 1990s. Output increased substantially after a peace settlement ended a long civil war that was fueled in part by struggles over oil (Le Billon and Cervantes, 2009).

Notes:
[a] Includes kerosene, LPG and other refined products, but excludes gasoline and diesel used for transportation.
[b] Electrification data comes from ICF Macro (2010) based on surveys conducted in Angola in 2006–07; Chad in 2004; Rep. of Congo in 2005; Gabon in 2000; and Nigeria in 2008.

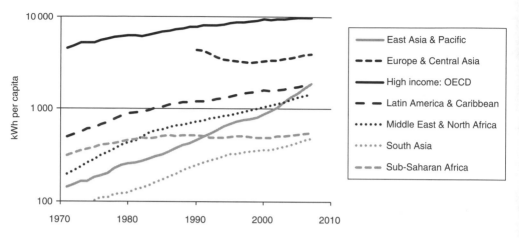

Figure 24.7   Per capita electricity consumption among world regions (1970–2007)

However, some have come to view the provision of electricity as a basic need, even a human right (Thompson and Laufman, 1996; Graham, 2006; Tully, 2006). Once viewed in this way, national governments may implement policies mandating that electricity providers offer service to all, including rural consumers who may be difficult and costly to reach.[6] Thus, despite the possibility of financial losses, some nations have implemented widespread rural electrification programs. For example in the 1990s, both China and South Africa embarked on campaigns to provide universal access to electricity to rural populations with the specific intent of rural poverty alleviation (Jiahua et al., 2006; Niez, 2010). In each case, rural electrification proceeded without requiring full cost recovery. In China, where electrification reached 99 percent of rural households by 2004, the program also supports agriculture and rural industry. In fact, rural industry consumes the majority of electricity provided to rural areas (Jiahua et al., 2006). In South Africa, there is little rural industry and most of the power consumed in rural areas goes to households. Interestingly, reports from South Africa note limited economic opportunities associated with rural electrification. One IEA report states:

> the South African example confirms what some documentation has revealed: that rural electrification in itself does not lead to economic growth or business development. Rural electrification does not generate local jobs – except those jobs created for the implementation of the electrification schemes . . . Rural economic development needs more than just household electrification. (Niez, 2010: 97)

Despite this realization, the post-apartheid state made a commitment to provide service to all, implicitly acknowledging that there are non-economic benefits to be gained. Indeed, the International Energy Agency (IEA) report goes on to state that: 'The long-term effect on the population of smoke-free kitchens, and the benefits of being able to read in the evening, watch television, or charge a cell-phone will be interesting to follow' (Niez, 2010: 97).

The notion of providing electricity as a basic need runs counter to the paradigm that dominated policy among the international financial institutions through the 1990s. With the Washington Consensus of the late 1980s (Krugman, 1995), states fell under increas-

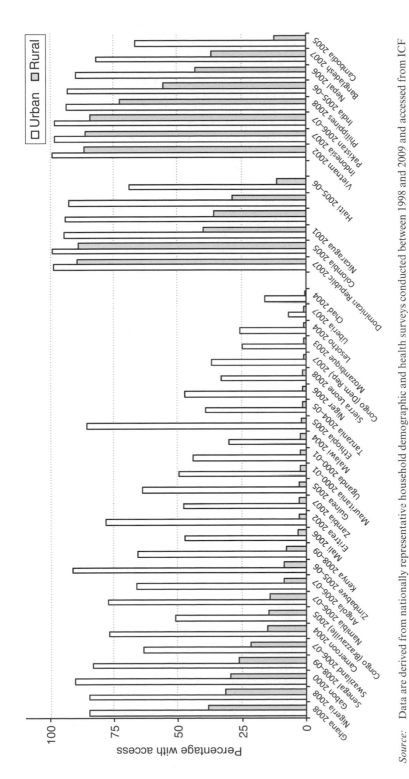

Source: Data are derived from nationally representative household demographic and health surveys conducted between 1998 and 2009 and accessed from ICF Macro (2010). For countries with multiple surveys, the most recent data were used.

Figure 24.8   Percentage of households with access to electricity in sub-Saharan Africa, Latin America and Asia

ing pressure from international financial institutions like the World Bank to liberalize the provision of public services like electricity (Williams and Ghanadan, 2006). In the early 1990s, the World Bank made power sector finance in developing and transition economies contingent on power sector reform, a process with included a 'standard menu' of features including major restructuring: unbundling what was usually a vertically integrated sector; introducing legislation allowing private sector participation in the power sector; divesting state-owned generation facilities; introducing new pricing structures that eliminate most or all of the subsidies previously offered to consumers; enforcing collections and ensuring cost recovery; and eventually moving toward fully competitive markets (Williams and Ghanadan, 2006). It was thought that this set of policies would encourage private sector investment in the power sector of developing and transition economies, and that this would increase efficiency in power provision, leading to higher access and lower prices.

However, these results were rarely seen. There was a flood of private investment in the mid-1990s as a result of liberalization, focused largely in Latin America and East Asia, but this was relatively short-lived. The Asian financial crisis of 1997 shocked many investors and investment fell off in all regions. Soon after, the US state of California experienced serious problems with market manipulation in its power sector after it had deregulated, which nearly brought down the state's power grid along with its economy (Reddy, 2001; World Bank, 2001). California's experience, and similar though less extreme examples of market abuses in other northern economies, led many to question the viability of unfettered power sector deregulation (Woo et al., 2003). Thus, by 2003, foreign investment in the power sectors of developing and transition economies had declined considerably (Williams and Ghanadan, 2006). In those markets that were affected, the promises of increased access and lower prices were not met. As Williams and Ghanadan observe (2006: 840):

> Reforms in many countries currently risk failing the test of social legitimacy on one or more counts. In some, tariff increases and payment enforcement outpace consumer service and public benefits. In others, idealized reform goals and models are outweighed by the corruption and non-transparent dealings that occur beneath the radar screen of formal policy. In still others . . . electricity reform may succeed in policy terms but fail politically if the public views it as part of a larger package of broken economic development promises.

Since that time, private investment in electrification in developing and emerging economies has rebounded. By 2008, investment had reached the level at which it peaked just before the 1997 Asian crisis. However, investment patterns have shifted such that the 'transition' economies in Eastern Europe now capture a much higher share of the investment: Russia alone received 25 percent of global investment in the power sector in 2008 (World Bank and PPIAF, 2010). In contrast, sub-Saharan Africa captured just 1 percent of private investment in the same year (World Bank and PPIAF, 2010).

Thus, while some countries have undertaken concerted efforts to provide universal access to electric power as a basic need, others, which have followed the recommended prescriptions of international financial institutions, have struggled with inconsistent investment patterns and disappointing performance. Of course, there are less capital-intensive pathways to providing electricity service to energy-poor populations, which can be critical in providing the first few units of electricity to households that lack access. These pathways include decentralized energy technologies that rely on renewable resources like solar photovoltaic (PV) panels, micro-hydroelectric systems or biomass-based generation systems, as well as fossil fuel-based systems like diesel-based generators or gen-sets. These can all be sized appropriately for individual households or rural communities and avoid the need for costly grid extension (Barnes and Floor, 1996). However, these technologies require institutional arrangements, financial structures and technical know-how that differ sharply from those that have developed in support of conventional grid-based electricity provision.

24.2.4   Energy and Industrialization

High-quality energy carriers like coal, oil and, more recently, gas and electricity, are essential factors of production and drivers of economic growth (Kümmel, 1982; Ayres and Warr, 2005). Indeed, some historians view the industrial revolution as an 'energy revolution' (Wrigley, 1962; MacLeod, 2004). However, industrial development requires more than access to energy. Scholarly discussions of the role that energy played in early industrialization reveal a complex interplay between newfound sources of energy, resource extraction and processing, industrial production and technological change. For example, the steam engine, a key innovation in early industrialization, was initially developed to improve upon pumps used to drain water from coal mines (as well as mines for copper and tin ores). Improved pumps made it possible to extract previously inaccessible resources (Freese, 2003; Nuvolari, 2004). As British historian Edward Wrigley notes, steam-driven pumps: 'were at once essential to the continued expansion in coal production, and virtually unusable without a supply of coal' (Wrigley, 1962: 11). Only after their widespread deployment as stationary pumps, accompanied by a series of innovations, were steam engines deployed in locomotives, which led to vast increases in haulage capacity, further driving industrialization. Thus the shift to a coal-based energy system simultaneously created demand for technological innovation and facilitated its widespread dissemination.

By the mid-twentieth century, petroleum had substantially amplified this process, allowing for higher efficiency and greater power densities and creating a demand for additional innovation (Yergin, 1991). Thus, historically, access to energy has proven to be a necessary precondition for industrialization and associated economic development. However, the energy or industrial revolution has already occurred. The countries at the forefront of that revolution two centuries ago enjoyed a degree of political and economic dominance at that time, which they maintain to this day. In fact, many enjoyed a century or more of economic prosperity leading up to their respective industrial transitions (Smith, 1988). As leaders in manufacturing and industrial output, the countries at the vanguard of industrialization already had access to regional and global markets for the primary resources and finished goods that they consumed and produced. Indeed, economic historians estimate that per capita income in Western Europe prior to industrialization was several times higher than it was in many developing countries in the latter half of the twentieth century (Kuznets, 1963; cited in Smith, 1988: 16).

While early industrializers still maintain a degree of dominance with respect to both global political economy and industrial output, they now share their position with a selection of countries that, until recently, also ranked among the energy-poor. Take Brazil, China, India and South Korea as examples. These four countries are among the world's most important emerging economies. Collectively, they represent 40 percent of the world's population (primarily in India and China). They account for 25 percent of the world's electricity production and 21 percent of global energy supply. Utilizing these resources have made these countries major suppliers of the world's primary industrial inputs: they collectively produce 60 percent of the world's cement and 50 percent of the world's steel. However, in 1980, China used less primary energy and consumed less electricity per capita than most countries of sub-Saharan Africa. Its GDP per capita, adjusted for purchasing power parity (PPP), was barely one-tenth the global average, and the incidence of poverty approached 100 percent. India was in a similar position and on par with the group of 'low-income' countries.[7] Notably, both India and China were slightly worse off than sub-Saharan Africa in all indicators. At that time, Brazil and South Korea had indicators that were slightly better, but they were below the global average in all cases except Brazil's per capita GDP (see Figure 24.9).

By 2007, the most recent year for which data is available, the situation had changed. Per capita energy consumption in China and India now exceeds the average of low-income countries. Further, although both remain below the global average, they enjoy growth rates that are far above global rates of growth. South Korea has grown at an astonishing rate, going from slightly below the global average in all indicators to levels of energy use and GDP equivalent to the OECD, which it joined in 1996. Brazil, which was close to the global average in all indicators in 1980, changed at a pace similar to the global average in most indicators.

In contrast to the industrializing economies of Brazil, India, China and South Korea, sub-Saharan Africa has seen little change in energy and GDP. Similarly, the region has made limited progress with respect to poverty alleviation. In contrast, rapid industrializers made substantial progress in reducing the incidence of poverty (also shown in Figure 24.9). For example, between 1980 and 2007, the incidence of poverty in China and Brazil decreased by roughly 60 percent. In India, the incidence of poverty decreased by just 15 percent. However, given the large population, that represents over 60 million inhabitants who left the ranks of the poor. In the same period, the incidence of poverty in sub-Saharan Africa remained decreased by just 1 percent, but due to population growth, this represents an expansion of poverty in real terms by nearly 190 million people (World Bank, 2010).

Thus, we see a pattern in which some countries or regions have simultaneously industrialized and managed to achieve some degree of economic growth and poverty alleviation, and industrialization in those places was facilitated by large increases in energy consumption. However, other countries have been able to make progress toward poverty alleviation and, truer to Sen's vision, an increase in capabilities, without necessarily building up their industrial base. For example, both Nepal and Bangladesh have improved their Human Development Indices at a rate of roughly 2 percent per year since 1980 (UNDP, 2009) while their respective industrial sectors have maintained relatively constant shares of GDP (in each case, the service sector grew considerably) (World Bank, 2010).

24.3   ENERGY AND ECONOMIC GROWTH

There are many examples of correlation between growth in energy, growth in GDP and reductions in poverty. However, the causal association is not always clear. While it is true that energy facilitates industrialization and economic growth, the inverse is also true: as economies grow, people consume more. They purchase more durable goods, opt for more individualized modes of transportation, and live in larger, more individualized dwelling spaces, all of which require more energy. Thus, there are logical arguments supporting both the assertions that increasing energy contributes to economic growth and that economic growth drives demand for energy.

Of course, circumstances differ from country to country. A closer analysis reveals a somewhat murky view, which makes it difficult to generalize about the relationship between energy and economic growth. By analyzing time series of national energy supply and economic growth, economists have shed some light on the interactions between energy and economic development. An econometric technique called 'Granger analysis' (Granger, 1988; Stern, 2004) tests whether one variable in a time series (for example GDP or another measure of economic growth) can be causally linked to another time series variable (for example energy or electricity consumption).[8] Results of 74 country studies examining the question of energy–GDP causality are shown in Figure 24.10. The data represent all world regions and include time series that are generally in excess of 30 years.

As shown in Figure 24.10, this technique reveals that the number of cases in which energy use has driven economic growth (*n* = 18) barely exceeds the number of cases in which the opposite is true (*n* = 17). It also reveals a number of cases of 'bidirectional causality'; that is, causality acted in different directions at different periods during the time series under analysis (*n* = 20). The technique also reveals that in nearly one-quarter of the cases analyzed (16 out 74), the causal relationship is indeterminate.

This analysis challenges the conventional wisdom that increased energy consumption is a necessary precondition to economic growth and, by extension, poverty alleviation. While large amounts of energy are necessary to undertake some types of industrial activities such as steel, cement, aluminum or fertilizer production, these energy-intensive industrial activities are not the only path to economic growth and poverty alleviation. While it is true that China has experienced unprecedented economic growth since 1990 largely based on energy-intensive industries, South Korea also built up a large industrial base and substantially expanded energy consumption, but apparently did so with economic growth driving energy consumption (Soytas and Sari, 2003; Oh and Lee, 2004). In addition, other countries included in the analysis, like Tunisia and Indonesia, have made substantial improvements in well-being (UNDP, 2009), but show no causal relationship between energy consumption and economic growth.[9]

24.4   ENERGY POVERTY AT THE HOUSEHOLD LEVEL

The evidence discussed thus far indicates a complex and contingent relationship between energy and poverty at the national scale. Countries considered poor by conventional measures such as low per capita GDP and high poverty headcounts are also energy-poor

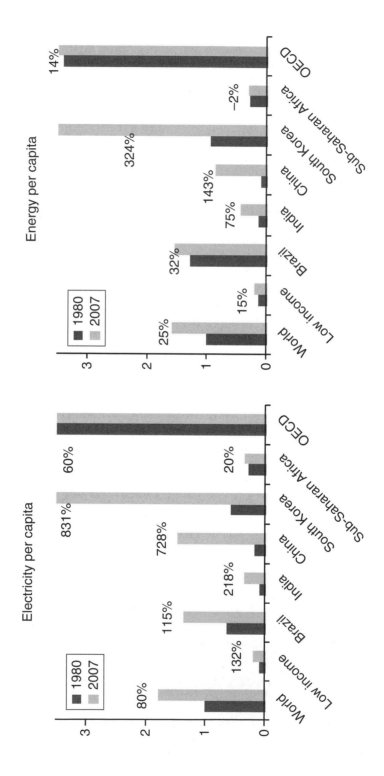

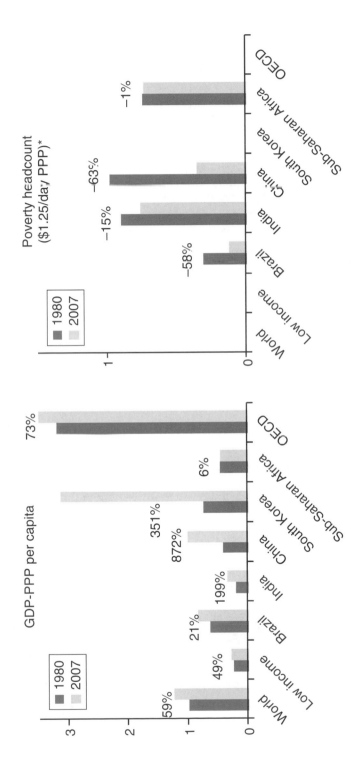

Notes: Data are indexed to 1980 = 1 for all plots except Poverty Headcount. Percentage changes between 1980 and 2007 are shown for each country.
* $1.25/day is the World Bank's benchmark for absolute poverty based on 2005 prices adjusted for purchasing power parity (PPP) (for more details, see World Bank, 2010). Poverty data are not reported for aggregate low-income countries, OECD, S. Korea, or globally. Further, data is not available for every year; thus, 1980 data come from a range between 1978 and 1981 and 2007 data come from 2005–7.

Figure 24.9   Per capita electricity, energy and GDP plus poverty incidence in Brazil, China, South Korea and sub-Saharan Africa, with low-income countries, OECD and global data for reference (1980–2007)

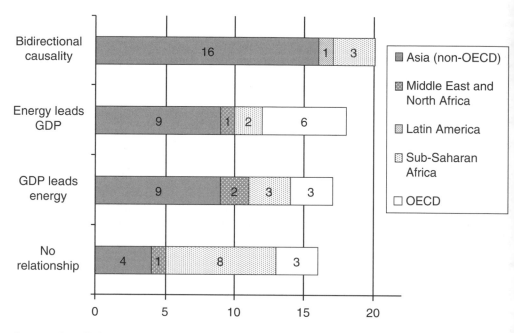

Source: Compiled by the author based on data in Soytas and Sari (2003); Jumbe (2004); Oh and Lee (2004); Paul and Bhattacharya (2004); Shiu and Lam (2004); Wolde-Rufael (2005); Lee and Chang (2008) and Oh and Lee (2004).

Figure 24.10   Analysis of time-series GDP and energy consumption data from 74 studies in 43 countries

in the sense that they lack sufficient choice to access certain energy services. However, as Figure 24.5 indicates, at any given level of energy use a wide range of well-being is attainable. In this sense, energy is analogous to income: it is not the sole determinant of well-being. Nevertheless, access to energy can facilitate gains in welfare and lack of access can make achieving those gains more difficult. This section examines some welfare gains that can be achieved by reducing energy poverty. These are best understood by looking at the local level, where the impact of poverty is most clearly observed. I address this by examining three topics: health, agricultural production and connectivity.

24.4.1   Health

For Sen, human health, particularly longevity, is a critical determinant of well-being and the inability to avoid preventable mortality is a gross deprivation of capabilities (Sen, 1998). In this section, I examine the relationship between energy services and health, with a focus on the issues most relevant to energy-poor populations in the global south.[10] With the exception of electricity supplied from nuclear or hydro resources, energy services are most commonly derived from combustion of some fuel (see Figure 24.1), which results in emissions of some health-damaging pollution. However, human exposure and health impacts resulting from combustion-related emissions vary a great deal depending on spatial and temporal relationships, pollution transport, and so on (Bennett et al., 2002).

By far the largest exposures to combustion-related pollutants occur in rural areas of the developing world, where solid biomass fuels are often utilized indoors, in open hearths or simple stoves characterized by low combustion and heat transfer efficiency (Smith, 1993; Westhoff and Germann, 1995). Such stoves cannot fully combust solid fuel; as a result, they emit large quantities of pollutants including carbon monoxide, particulate matter and aromatic hydrocarbons (Smith, 1993). In these conditions, indoor concentrations of pollutants reach levels far in excess of safety standards established by the World Health Organization (WHO) (Smith et al., 2007). The health impacts of this reliance are profound. Exposure to wood smoke has been causally linked to numerous illnesses including acute respiratory infections, which are particularly harmful for young children, as well as chronic lung disease and lung cancer, which tend to afflict people later in life (Smith et al., 2004).[11] The WHO estimates that exposure to pollution from indoor solid biofuel combustion is the ninth leading cause of illness and death worldwide. However, it ranks among the top five risk factors in developing regions and accounts for as much as 8.5 percent of illness and death in sub-Saharan Africa (WHO, 2008). Importantly, this burden of disease is preventable through simple technological and/or behavioral changes (Bailis et al., 2005).

There have been many efforts to address the health problems associated with the indoor use of solid fuels. The basic three-stone fire is very common, but it can be modified in countless ways: for example, sinking the combustion zone below ground level or constructing a barrier around it to shield the fire from drafts and reduce thermodynamic losses. Additional adjustments involve fully enclosing the combustion zone, which allows for some degree of air-control and directs the hot gases produced by combustion more toward the cooking pots and less to the surrounding environment. Some stoves include a flue or chimney which directs the exhaust out of the room, dramatically reducing indoor air pollution. A well-functioning chimney also creates a natural draft in the stove, which can improve combustion, further reducing pollution. Further modifications include insulating parts of the stove to reduce heat losses and adding dampers or secondary air inlets to improve airflow and combustion characteristics (Bailis, 2004).

Other interventions have attempted to switch fuels entirely. For example, many countries subsidize kerosene and/or LPG to promote its use among poor populations (Nhete, 2007; Fall et al., 2008).[12] However, these interventions typically favor urban households, for which such fuels are more accessible. They also tend to be captured by wealthier consumers or traders despite the stated goals of increasing access for the (energy) poor (Nhete, 2007). As was mentioned above, electricity is also subsidized in many cases. However, unless the subsidies are very high, electricity is still a costly cooking option and it is common for newly electrified households to cook with other sources of energy. In addition, other stove–fuel combinations, including renewable technologies, have been disseminated. For example, biogas-based systems are common in parts of South and East Asia (Chen et al., 2009; Gautam et al., 2009; Ravindranath and Balachandra, 2009). Solar cookers have also been disseminated in numerous places, though with less success than biogas (Suharta et al., 1999; Ahmad, 2001; Wentzel and Pouris, 2007; Büscher, 2009).

There are many non-technical factors that constrain stove design and adoption. Technical considerations weigh equally with many other factors in determining whether a stove is suitable in a particular sociocultural context. With the benefit of three decades

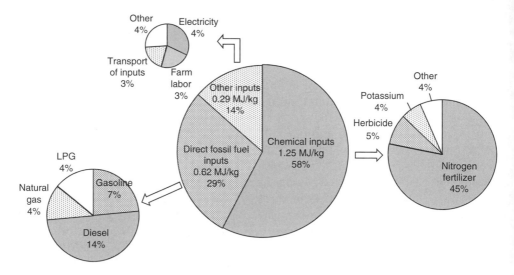

Source: Based on data in Farrell et al. (2006).

Figure 24.11   *Breakdown of energy inputs into current maize production in the US (MJ per kg maize)*

of hindsight, much has become apparent to people working in cookstove design and dissemination. For example, successful stove dissemination is contingent on many factors, including consumer education about the need to avoid exposure to combustion emissions, user feedback on design to ensure that technical features meant to reduce pollution do not interfere with usability, and extensive follow-up to ensure users are satisfied and to enable design and/or programmatic changes if they are not. Further, most successful examples of large-scale stove dissemination began as highly subsidized programs and shifted only gradually to self-sustaining commercial delivery models. Long-term external support from government or donor agencies appears critical, and early withdrawal of support can cause efforts to stall (Bailis et al., 2009).

24.4.2   Agricultural Production

A second arena in which energy is critically important for enhancing capabilities is agricultural production and processing. The majority of the world's poor are directly engaged in agricultural production (World Bank, 2007). Subsistence agriculture common in energy-poor regions utilizes very little energy beyond animal or human power. In contrast, agriculture practiced in energy-rich countries utilizes multiple forms of direct and indirect energy as essential factors of production. The bulk of it is consumed as 'embodied energy' in fertilizer. Energy is also consumed directly by farm machinery, which does the physical work that is performed by animal or human power in most energy-poor agricultural systems: preparing fields, planting and harvesting. Figure 24.11 shows the breakdown of energy inputs in typical maize production in the energy-rich global north.

One kilo of maize produced in US conditions requires roughly 2.16 MJ of energy; 58 percent is attributable to chemical inputs. The bulk of this energy is embodied in nitro-

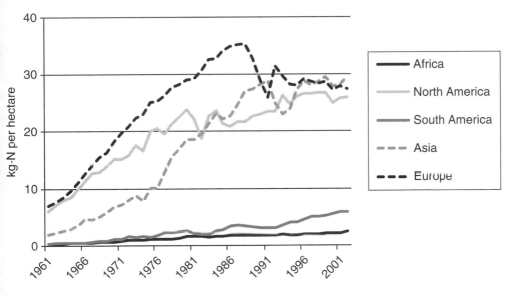

Source: FAOSTAT (2009).

Figure 24.12 *Nitrogen fertilizer use per unit of cultivated land (all crops) by region (1961–2002)*

gen fertilizers (45 percent of the total) and the remainder divided among herbicides, P and K fertilizers, insecticides and agricultural lime.[13] An additional 29 percent of energy is represented by fossil fuels used in farm machinery. The remainder is attributable to many materials and processes ranging from the manufacture of farm machinery to seed production and transportation of inputs. While similar caches of detailed data do not exist for production in the south, the Food and Agriculture Organization (FAO) estimates that traditional maize production utilizes less than 10 percent of the energy that is consumed per ton in the US (FAO, 2000; cited in World Bank, 2007).[14] However, the intensification enabled by fertilizers and other energy-intensive inputs results in tremendously higher yields.

The contrast between conventional and traditional maize exemplifies that the contrast between agricultural production in energy-poor and energy-rich settings is stark, with respect to both inputs and outputs. As was mentioned for maize in the US, the largest use of energy in conventional agriculture is the form of embodied energy in nitrogen fertilizers. Since the introduction of 'green revolution' technologies in the early 1960s (Tilman et al., 2002), the consumption of nitrogen fertilizers has increased by a factor of seven worldwide. However, as with other energy trends, there is substantial variation between world regions. For example, application rates of nitrogen fertilizers per unit of cultivated land vary by a factor of 20, from just over 2 kg per ha of cultivated land in Africa to nearly 50 kg in South and Southeast Asia (Figure 24.12).[15]

The gains from exploiting increasing quantities of energy in agriculture have been substantial. Globally, agricultural yields of staple grains have doubled or tripled, with the largest increases coming in those regions that have increased fertilizer applications.

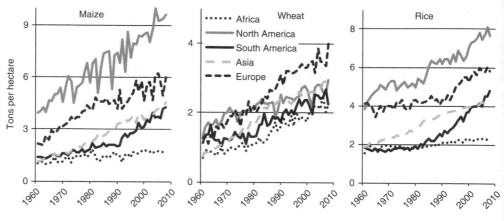

Source: FAOSTAT (2009).

Figure 24.13   Maize, wheat and rice yields by world region from 1961 to 2008 (tons/ha)

As a result, yields vary between energy-rich and energy-poor regions. For example, currently between the highest- and the lowest-yielding regions, maize productivity differs by a factor of five, wheat differs by a factor of two, and rice by a factor of four (Figure 24.13). Of course, yield improvements are not solely the result of increased fertilizer use. Improved seed varieties and, in some cases irrigation, have also contributed to improved productivity (Evenson and Gollin, 2003). However, improved varieties typically depend on fertilizer inputs in order to achieve high yields. Some are also require irrigation.

Regional disparities in yield have led to calls for a second 'green revolution' in order to raise yields in regions of low productivity, particularly sub-Saharan Africa (Rockefeller Foundation, 2006; Sanchez et al., 2009), which suffers a widening 'yield gap' that is apparent in Figure 24.13 (World Bank, 2007). However, these calls have also raised concerns about the sustainability and socio-economic risk associated with the introduction of intensive agriculture with high energy costs and irrigation requirements (Holt-Giménez et al., 2006; Rockström et al., 2007).

Others have tried intervening in 'energy-poor' agriculture in different ways. For example, there have been numerous attempts to introduce small-scale renewable energy and other novel technologies at the farm level (Biswas et al., 2001; Cabraal et al., 2005). However some critics have claimed that a focus on certain renewable technologies such as photovoltaic (PV) systems, which have received a significant amount of support since the mid-1990s, is not the best way to enhance the capabilities of the (energy) poor (Karekezi and Kithyoma, 2002). Indeed, research has shown that the potential for poverty alleviation from PV is limited. For example, Kenya has a thriving PV market, but no policies supporting inclusivity among the poor. There, most household PV systems are owned by middle- and upper-income rural households, which use them for consumptive activities like household lighting and connective applications (see section 24.4.3 below) rather than productive income-generating activities (Bailis et al., 2006; Jacobson, 2007).

One niche in which enhanced access to energy has proven useful is in irrigation. Moving both surface and ground water is extremely energy-intensive. Some regions have made significant investments in irrigation. India, for example, dug over 20 million tubewells and built over 4000 large dams during the twentieth century (World Commission on Dams, 2000; Holt-Giménez et al., 2006). Other regions rely almost entirely on rain-fed agriculture. This contrast is most notable in sub-Saharan Africa, where only 4 percent of cropland is under irrigation, compared to 39 percent in South Asia and 29 percent in East Asia (World Bank, 2007). Of course, irrigation is associated with numerous negative impacts. In addition to high energy costs associated with pumping water, increasing withdrawals have led to water scarcity in many regions (Rockström et al., 2007). This is exacerbated by ill-conceived energy policy. For example, in parts of India, where groundwater withdrawals exceed recharge rates by over 50 percent, irrigation is supported by subsidized or free electricity, which encourages inefficient use of both energy and water (World Bank, 2007). Further, irrigation can increase the incidence of vector-borne diseases like malaria by creating additional habitat (Keiser et al., 2005).

Nevertheless, similar to energy in general, irrigation can be an indispensible tool, particularly in areas suffering from acute scarcity. In some locations, human-powered pumps, which enhance the efficiency of animate power, have been introduced as alternatives to conventional energy-intensive irrigation. These and other capability-enhancing devices are often overlooked in favor of more sophisticated, but potentially less helpful technologies (Biswas et al., 2001; Fisher, 2006).

With irrigation comes the ability to grow higher-value crops for non-local markets, which can bring higher farm income. In many parts of the developing world, farmers now participate in globally extended supply chains for fresh produce. These chains extend from rural farms to the supermarkets of London, Amsterdam and New York where the demand for fresh seasonal produce can now be met year-round (Hughes, 2006). For example, 25 percent of all food imports to the UK are from Africa (Lucas et al., 2006). Kenya, one of the UK's main sources of fresh produce, has seen revenues from horticultural products overtake traditional high-value cash crops like coffee and tea (KNBS, 2010). However, when rural farmers do manage to link to high-value markets by exporting horticultural products to countries in the global North, they face several energy-related barriers. First, there is a direct transportation cost related to shipping fresh produce over vast distances. Unlike traditional cash crops (for example sugar, coffee and cotton), fresh horticultural crops are highly perishable and must travel by air freight. By one estimate, transportation constitutes roughly 7 percent of the cost of a typical high-value basket of produce in the UK (Lucas et al., 2006; cited in World Bank, 2007). A second, closely related barrier arises that is linked to 'food miles', in which commodities like fresh produce and cut flowers from the Global South, shipped to the US and Europe by air, have come under criticism for the carbon emissions and energy use associated with their supply chains (Weber and Matthews, 2008). This raises an interesting situation in which poverty alleviation in energy-poor regions appears to be at odds with environmental sustainability.[16] However, such apparent conflicts can encourage a useful dialog about the social and economic contextualization of different of forms of energy consumption and associated environmental impacts, including greenhouse gas (GHG) emissions.

Enhancing agricultural productivity and associated household capabilities among the poor is linked to energy access in numerous ways: through embodied energy in conventional agricultural inputs and irrigation; through access to transportation, which allows high-value produce to reach non-local markets; and even through technologies that can

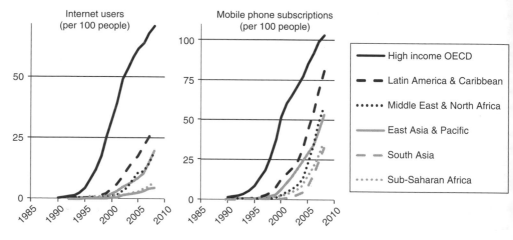

Source: World Bank (2010).

Figure 24.14   Indicators of connectivity by world region (1990–2008)

harness human power in more efficient ways. However, such energy services still remain relatively inaccessible for poor households.

24.4.3   Connectivity

A final area in which energy can directly contribute to the enhancement of capabilities and well-being is with respect to connectivity. Connectivity refers to the multiple ways in which individuals access information, maintain social relationships, and communicate across space.[17] This communication may be unidirectional, as with radio or television, or it may be bi- or multidirectional, as with mobile phones and the Internet. The use of connective technologies in the world's energy-poor regions has exploded in recent years. Figure 24.14 shows penetration rates of Internet and mobile phones between 1990 and 2008. As the graphs show, between 2000 and 2008, Internet use in developing regions grew from less than 1 percent of the population to 5–6 percent in South Asia and sub-Saharan Africa; 20 percent in East Asia, the Middle East and North Africa; and nearly 30 percent in Latin America. Rates of mobile phone subscriptions have grown even more rapidly. In 1997, there was fewer than one subscription per 100 persons in all developing regions; now, there are over 30 in South Asia and sub-Saharan Africa; over 50 in East Asia, the Middle East and North Africa; and over 80 in Latin America.

However, mobile phone data may be somewhat misleading. Whereas data on Internet use is a direct measure of individual access, the rate of mobile phone subscriptions likely includes multiple subscriptions per household. Therefore, it is not necessarily a reliable indicator of access (that is, 30 subscriptions per 100 persons does not mean 30 of every 100 people have access). Nevertheless, the growth rates are indicative of a significant increase in access to connective technologies, which is evidenced by other studies (Silva and Zainudeen, 2007).

Increasing connectivity requires that small but essential quantities of electricity be accessible to the end user. However, electricity does not necessarily need to reach the user's home; for example, the Internet can be accessed from kiosks or shops dedicated to the provision of that service. Such shops are now common in small towns across the developing world (James, 2010). Similarly, mobile phones can be charged at centralized charging stations, an activity that has evolved into a viable business in many rural areas where mobile phone service exists, but access to electricity is limited (Greenberg, 2005). With Internet kiosks and mobile phone charging stations, power may be provided by centralized generation facilities distributed through the conventional grid, or it can be provided by decentralized off-grid technologies.

However, as rates of access to connective technologies climb, it is natural to question whether there is a measurable impact on well-being among energy-poor populations. Proponents of communications technology claim, somewhat uncritically, that access to connective technologies is essential to enhance well-being in an increasingly interconnected world (Greenberg, 2005; World Bank, n.d.), while others maintain that the technologies carry a risk of reinforcing existing inequalities by benefiting populations that already possess high levels of connectivity. Rather than helping the poor, these technologies simply entrench those who are already politically and economically advantaged (Niles and Hanson, 2003; cited in Alampay, 2006). One survey of mobile phone users in Tanzania indicates that users feel the technology contributes little to poverty alleviation and may actually strain household resources in terms of both cost and time (Mpogole et al., 2008). However, there is limited evidence supporting this position. The evidence is equally scant in support of the opposite contention: that connective technologies unambiguously reduce poverty. In an approach that draws heavily on Sen's notion of capabilities as critical determinant of well-being, Alampay describes an intermediate perspective that may be a more accurate description of the contingent role that connective technologies can have on welfare:

> ICTs [information and communication technologies] can play a role in a country's development if applied appropriately . . . There is anecdotal evidence that shows access to the telephone, for instance, can have a dramatic effect on the quality of life of the rural poor. Historically, however, telecommunications roll-out has generally increased inequality, benefited mostly the wealthy, and had little impact on quality of life. (Forestier et al., 2002; citing Soeftestad and Sein, 2003 and Alampay, 2006)

He continues by noting the need for explicit policies that stress social inclusion in order to ensure that connective technologies increase capabilities for people living under acute deprivation. In this way, connective technologies are similar to the energy technologies on which they depend. They have the potential to create capability-enhancing conditions. However, this is not an inherent quality of the technology of the technology; in fact, no technology can deliver preordained socio-economic outcomes (Warschauer, 2004). Rather, the outcomes related to connective technologies are dependent on the sociopolitical conditions into which the technology is introduced.

24.5 CONCLUDING THOUGHTS

This chapter has examined energy poverty and the ways in which energy can contribute to social welfare in developing countries at both national and local scales. By situating energy within Sen's capability framework, energy is conceived broadly as potentially contributing to an expanded range of services, which are, in reality, choices that individuals and communities can draw on in order to enhance well-being. These choices range from expanded industrial production and electric power provision to alternative technologies for household food preparation, mobility, connectivity and agricultural production. However, as with other types of resources, poverty reduction does not automatically result when additional energy supply is made available at a nationally aggregate scale. Distributional issues are critically important. In the absence of policies that promote inclusive access to energy services and associated technologies, additional supply may simply reinforce poverty and inequality across scales.

NOTES

1. An early and notable exception to this is Goldemberg et al.'s (1985) paper on 'Basic needs and much more with one kilowatt per capita'.
2. The notion of 'quality' is often subjective. However, in the context of energy analyses, the quality of an energy carrier can be determined by the quantity of useful work that it enables. This is defined as exergy (Ayres et al., 2007) and it may be determined quantitatively for different materials and systems (Lior and Zhang, 2007).
3. At the time that these data were collected, in the early to mid-1990s, fuels other than wood were largely unavailable to rural households in sub-Saharan Africa. Access was limited because supplies of fuels other than wood did not exist and the overwhelming majority of households used fuelwood regardless of income level. In contrast, most urban areas had a variety of fuels available, albeit at different prices. Thus, access was more a function of affordability rather than availability.
4. The IEA estimates that a little more than 1 percent of global electricity is derived from biomass (IEA, 2009b), although it notes that this is likely an underestimate. Biomass feedstocks are primarily derived from agricultural and timber residues, particularly sugarcane bagasse. Brazil, the world's largest producer of sugarcane, generates roughly 4 percent of its electricity from bagasse (IEA, 2009b). Several smaller countries in Latin America such as Nicaragua and Guatemala, which produce less sugar than Brazil but also have far less electricity demand, generate over 10 percent of their electricity in this way (IEA, 2009b). Similarly, the small Indian Ocean island nation of Mauritius generates roughly 20 percent of its electricity from bagasse (Ramjeawon, 2008).
5. These phenomena, known collectively as the 'resource curse', have been studied in depth by economists, political scientists, geographers, and other scholars (see, for example, Ross, 2001, 2008; Collier and Hoeffler, 2005; Le Billon and Cervantes, 2009; Watts, 2009).
6. The notion of providing service to the poor exists in many developing countries. For example, the concept of a lifeline tariff, which offers a limited number of kilowatt hours each month (typically 50–100 kWh) at subsidized rates, is a common policy among service providers. However, these subsidies only reach those who have obtained household connections.
7. Under the World Bank's definition, the group of 'low-income countries' consists of countries with per capita gross national income (GNI) less than $975 in 2008 dollars.
8. The technique works by doing a regression of the change in one time series (ΔY) on earlier values of the same series. If a lag interval for that time series is significant through a standard statistical test, then subsequent regressions for lagged levels of the other time series ($\Delta X_1, \Delta X_2, \ldots \Delta X_N$) are performed. These lagged variables are added to the regression if they are significant in and of themselves, and if they add explanatory power to the model. This can be repeated for multiple ΔX's (with each ΔX being tested independently of other ΔX's, but in conjunction with the proven lag level of ΔY) (Stern, 2004).
9. Tunisia and Indonesia are among the ten countries showing the most improvement in HDI between 1985 and 2007 (UNDP, 2009). Indonesia is an interesting case in Granger causality. Five published studies, covering overlapping time periods between 1960 and 2002, report conflicting results: one finds energy leads economic growth (Fatai et al., 2004), two find the opposite (Masih and Masih, 1996; Yoo, 2006), one finds bidirectional causality (Asafu-Adjaye, 2000), and one finds no discernable relationship (Soytas and Sari, 2003). Incidentally, South Korea and China are also among the top ten largest improvers in HDI.
10. The overall interplay between energy and health is a complex topic that cannot be treated fully here. For a review see (Ezzati et al., 2004).

11. The link to lung cancer has only been shown inconclusively for exposure to smoke from coal combustion, but there is limited evidence that exposure to wood smoke causes lung cancer as well (WHO, 2006). Associations with other diseases such as asthma, cataracts and tuberculosis are also suspected, but the epidemiological evidence is not sufficient to derive robust estimates of morbidity and mortality (Smith et al., 2004).
12. These interventions were commonly introduced to reduce demand for woodfuels in order to conserve forest resources rather than to reduce indoor air pollution. However, more recently they have been promoted as health measures as well (Smith, 2002).
13. Note that herbicides and insecticides collectively represent 6 percent of the energy used to produce US maize, but additional impacts arise as a result of their use such as eco-toxicity (Tilman et al., 2002).
14. The World Bank (2007) reports much higher energy consumption for US maize than is cited in this text, but the Bank's value is taken from a study by the FAO published in 2000 using data from 1990 (FAO, 2000). In that report, the ratio of energy use in 'modern' US maize to 'traditional' Mexican maize was approximately 33 to 1. Comparing values from Farrell et al. (2006) to the FAO's estimate for energy use in traditional Mexican maize production yields a ratio of 12 to 1.
15. This pattern is similar for phosphates and potash (FAOSTAT, 2009).
16. See, for example, Benito Müller's (2007) provocative commentary about the 'moral obligation' of people in the UK to eat African strawberries in the winter.
17. I borrow this term from Jacobson (2007) who used 'connectivity' to describe one of the major applications of household PV panels among middle-class rural Kenyans.

REFERENCES

Abdullah, S. and A. Markandya (2009), 'Rural electrification programmes in Kenya: policy conclusion from a valuation study', Economics Working Paper, University of Bath 43, at http://www.bath.ac.uk/economics/research/workingpapers.html.

Ahmad, B. (2001), 'Users and disusers of box solar cookers in urban India: implications for solar cooking projects', *Solar Energy*, **69** (Supplement 6), 209–15.

Alampay, E.A. (2006), 'Beyond access to ICTs: measuring capabilities in the information society', *International Journal of Education and Development using ICT [Online]*, **2** (3), 4–22, available at http://ijedict.dec-uwi.edu/viewarticle-php?id=196.

Arnold, M., G. Kohlin, R. Persson and G. Shepherd (2003), 'Fuelwood revisited: what has changed in the last decade?', CIFOR: 37, Bogor, Indonesia.

Asafu-Adjaye, J. (2000), 'The relationship between energy consumption, energy prices and economic growth: time series evidence from Asian developing countries', *Energy Economics*, **22** (6), 615–25.

Ayres, R.U., H. Turton and T. Casten (2007), 'Energy efficiency, sustainability and economic growth', *Energy*, **32** (5), 634–48.

Ayres, R.U. and B. Warr (2005), 'Accounting for growth: the role of physical work', *Structural Change and Economic Dynamics*, **16** (2), 181–209.

Bailis, R. (2004), 'Wood in household energy use', in C.J. Cleveland (eds), *Encyclopedia of Energy*, New York: Elsevier Science, pp. 509–26.

Bailis, R., A. Cowan, V. Berraeta and O. Masera (2009), 'Arresting the killer in the kitchen: the promises and pitfalls of commercializing improved cookstoves', *World Development*, **37** (10), 1694–1705.

Bailis, R., M. Ezzati and D.M. Kammen (2005), 'The role of technology management in the dynamics of greenhouse gas emissions from household energy use in sub-Saharan Africa', *Journal of Environment and Development*, **14** (1), 149–74.

Bailis, R., C. Kirubi and A. Jacobson (2006), 'Searching for sustainability: Kenya's energy past and future', African Centre for Technology Studies (ACTS) Nairobi.

Barnes, D.F. and W.M. Floor (1996), 'Rural energy in developing countries: a challenge for economic development', *Annual Review of Energy and the Environment*, **21** (1), 497–530.

Bennett, D.H., T.E. McKone, J.S. Evans, W.W. Nazaroff, M.D. Margni, O. Jolliet and K.,R. Smith (2002), 'Peer reviewed: defining intake fraction', *Environmental Science and Technology*, **36** (9), 206A–211A.

Biswas, W.K., P. Bryce and M. Diesendorf (2001), 'Model for empowering rural poor through renewable energy technologies in Bangladesh', *Environmental Science and Policy*, **4** (6), 333–44.

BP (2009), 'Statistical review of world energy 2009: historical data', *Statistical Review of World Energy 2009*, at http://www.bp.com/productlanding.do?categoryId=6929&contentId=7044622, accessed 4 June 2009.

Büscher, B. (2009), 'Connecting political economies of energy in South Africa', *Energy Policy*, **37** (10), 3951–58.

Cabraal, R.A., D.F. Barnes and S.G. Agarwal (2005), 'Productive uses of energy for rural development', *Annual Review of Environment and Resources*, **30** (1), 117–44.

Chen, Y., G. Yang, S. Sweeney and Y. Feng (2009), 'Household biogas use in rural China: a study of opportunities and constraints', *Renewable and Sustainable Energy Reviews*, **14** (1), 545–9.

Collier, P. and A. Hoeffler (2005), 'Resource rents, governance, and conflict', *Journal of Conflict Resolution*, **49** (4), 625–33.

Eckholm, E. (1975), 'The other energy crisis: firewood', Worldwatch, Washington, DC.

Evenson, R.E. and D. Gollin (2003), 'Assessing the impact of the green revolution, 1960 to 2000', *Science*, **300** (5620), 758–62.

Ezzati, M., R. Bailis, D.M. Bailis, D.M. Kammen, T. Holloway, L. Price, L.A. Cifuentes, B. Barnes, A. Chaurey and K.N. Dhanapala (2004), 'Energy management and global health', *Annual Review of Environment and Resources*, **29**, 383–419.

Fall, A., S. Sarr, T. Dafrallah and A. Ndour (2008), 'Modern energy access in peri-urban areas of West Africa: the case of Dakar, Senegal', *Energy for Sustainable Development*, **12** (4), 22–37.

FAO (1983), 'Wood fuel surveys', UN Food and Agriculture Organization, Rome, at http://www.fao.org/docrep/X5555E/x5555e00.htm#Contents.

FAO (2000), 'The energy and agriculture nexus', Environment and Natural Resources Working Paper No. 4, UN Food and Agriculture Organization, Rome, at http://www.fao.org/docrep/003/x8054e/x8054e00.htm.

FAOSTAT (2009), 'FAOSTAT agricultural production data', at http://faostat.fao.org/site/339/default.aspx, accessed 7 October 2009.

Farrell, A.E., R.J. Plevin, et al. (2006), 'EBAMM Release 1.1', UC Berkeley – Energy and Resources Group, Berkeley, CA, at http://rael.berkeley.edu/ebamm/EBAMM_1_1.xls.

Fatai, K., L. Oxley and F.G. Scrimgeour (2004), 'Modelling the causal relationship between energy consumption and GDP in New Zealand, Australia, India, Indonesia, the Philippines and Thailand', *Mathematics and Computers in Simulation*, **64** (3–4), 431–45.

Field, C.B., J.E. Campbell and D.B. Lobell (2008), 'Biomass energy: the scale of the potential resource', *Trends in Ecology and Evolution*, **23** (2), 65–72.

Fisher, M. (2006), 'Income is development: KickStart's pumps help Kenyan farmers transition to a cash economy', *Innovations: Technology, Governance, Globalization*, **1** (1), 9–30.

Forestier, E., J. Grace and C. Kenny (2002), 'Can information and communication technologies be pro-poor?', *Telecommunications Policy*, **26** (11), 623–46.

Freese, B. (2003), *Coal: A Human History*, New York: Penguin Books.

Fukuda-Parr, S. (2003), 'The human development paradigm: operationalizing Sen's ideas on capabilities', *Feminist Economics*, **9** (2–3), 301–17.

Gautam, R., S. Baral and S. Herat (2009), 'Biogas as a sustainable energy source in Nepal: present status and future challenges', *Renewable and Sustainable Energy Reviews*, **13** (1), 248–52.

Gettleman, J. (2009), 'Clinton praises Angola, buturges more reform', *New York Times*, p. 6.

Goldemberg, J., T.B. Johansson, A.K.N. Reddy and R.H. Williams (1985), 'Basic needs and much more with one kilowatt per capita', *Ambio*, **14** (4–5), 190–200.

Graham, C. (2006), 'The politics of necessity: electricity and water in Great Britain', *Journal of Consumer Policy*, **29** (4), 435–48.

Granger, C.W.J. (1988), 'Some recent development in a concept of causality', *Journal of Econometrics*, **39** (1–2), 199–211.

Greenberg, A. (2005), 'ICTs for poverty alleviation: basic tool and enabling sector', I.F.D. Secretariat, Swedish International Development Cooperation Agency, at www.sida.se.

Haq, M.U. (1995), *Reflections on Human Development*, New York: Oxford University Press.

Holt-Giménez, E., M.A. Altieri and P. Rosset (2006), 'Ten reasons why the Rockefeller and the Bill and Melinda Gates Foundations' alliance for another Green Revolution will not solve the problems of poverty and hunger in sub-Saharan Africa', Policy Brief No 12, Food First, Oakland, CA, at http://www.foodfirst.org/files/pdf/PB12%2010%20Reasons%20Gates%20Rockefeller%20-%20English.pdf.

Hoogwijk, M., A. Faaij, B. Eikhout, B. de Uries and W. Turkenburg (2005), 'Potential of biomass energy out to 2100, for four IPCC SRES land-use scenarios', *Biomass and Bioenergy*, **29** (4), 225–57.

Hosier, R. and J. Dowd (1987), 'Household fuel choice in Zimbabwe: an empirical test of the energy ladder hypothesis', *Resources and Energy*, **9**, 347–61.

Hughes, A. (2006), 'Learning to trade ethically: knowledgeable capitalism, retailers and contested commodity chains', *Geoforum*, **37** (6), 1008–20.

ICF Macro (2010), 'Measure DHS STATcompiler', *Building Tables with DHS Data*, at http://www.measuredhs.com, accessed 12 July 2010.

IEA (2009a), *Energy Balances of non-OECD Countries 2009*, Paris: IEA.

IEA (2009b), *Energy Statistics of Non-OECD Countries 2009*, Paris: IEA.

IMF (2009), 'Chad: Selected Issues', *IMF Country Reports*, Washington, DC, International Monetary Fund, at http://www.imf.org/external/pubs/ft/scr/2009/cr0967.pdf.

International Energy Agency (IEA) (2007), *Energy Balance of Non-OECD Countries 2004–2005*, Paris: Organisation for Economic Co-operation and Development (OECD).

Jacobson, A. (2007), 'Connective power: solar electrification and social change in Kenya', *World Development*, **35** (1), 144–62.

James, J. (2010), 'Mechanisms of access to the Internet in rural areas of developing countries', *Telematics and Informatics*, **27** (4), 370–76.

Jiahua, P., P. Wuyuan, L. Meng, W. Xiangyang, W. Lishuang, H. Zerriffi, B. Elias, C. Zhang and D. Victor (2006), 'Rural electrification in China 1950–2004: historical processes and key driving forces', Working Paper Series, Program on Energy and Sustainable Development – Stanford University, Palo Alto, CA.

Jumbe, C.B.L. (2004), 'Cointegration and causality between electricity consumption and GDP: empirical evidence from Malawi', *Energy Economics*, **26** (1), 61–8.

Karekezi, S. and W. Kithyoma (2002), 'Renewable energy strategies for rural Africa: is a PV-led renewable energy strategy the right approach for providing modern energy to the rural poor of sub-Saharan Africa?', *Energy Policy*, **30** (11–12), 1071–86.

Keiser, J., M.C. De Castro, M.F. Maltese, R. Bos, M. Tanner, B.H. Singer and J. Utzinger (2005), 'Effect of irrigation and large dams on the burden of malaria on a global and regional scale', *American Journal of Tropical Medicine and Hygeine*, **72** (4), 392–406.

KNBS (2010), 'Leading economic indicators', Monthly Report, at http://www.knbs.or.ke/sectoral/energy/archive_lei.html, accessed 2 August 2010.

Krugman, P. (1995), 'Cycles of conventional wisdom on economic development', *International Affairs*, **71** (4), 717–32.

Kümmel, R. (1982), 'The impact of energy on industrial growth', *Energy*, **7** (2), 189–203.

Kuznets, S. (1963), 'Underdeveloped countries and the preindustrial phase in the advanced countries', in A.N.A. and S.P. Singh (eds), *The Economics of Underdevelopment*, Oxford: Oxford University Press, pp. 135–53.

Le Billon, P. and A. Cervantes (2009), 'Oil prices, scarcity, and geographies of war', *Annals of the Association of American Geographers*, **99** (5), 836–44.

Leach, G. (1992), 'The energy transition', *Energy Policy*, **20** (2), 116–23.

Leach, G. and R. Mearns (eds) (1988), *Beyond the Woodfuel Crisis: People, Land, and Trees in Africa*, London: Earthscan.

Lee, C.-C. and C.-P. Chang (2008), 'Energy consumption and economic growth in Asian economies: a more comprehensive analysis using panel data', *Resource and Energy Economics*, **30** (1), 50–65.

Lior, N. and N. Zhang (2007), 'Energy, exergy, and Second Law performance criteria', *Energy*, **32** (4), 281–96.

Lucas, C., A. Jones and C. Hines (2006), 'Fueling a food crisis: the impact of Peak Oil on food security', The Greens, European Free Alliance in the European Parliament, Brussels, at http://www.carolinclucasmep.org.uk/wp-content/uploads/file/Fuelling%20a%20food%20crisis.pdf.

MacLeod, C. (2004), 'The European origins of British technological predominance', in L.P. de la Escosura and P.K. O'Brien (eds), *Exceptionalism and Industrialisation: Britain and its European Rivals, 1688–1815*, Cambridge: Cambridge University Press, pp. 111–26.

Masera, O., B. Saatkamp and D. Kammen (2000), 'From linear fuel switching to multiple cooking strategies: a critique and alternative to the energy ladder model', *World Development*, **28** (12), 2083–103.

Masih, A.M.M. and R. Masih (1996), 'Energy consumption, real income and temporal causality: results from a multi-country study based on cointegration and error-correction modelling techniques', *Energy Economics*, **18** (3), 165–83.

Metz, B., O.R. Davidson, P.R. Bosch, R. Dave and L.A. Meyer (eds) (2007), *Climate Change 2007: Mitigation Contribution of Working Group III to the Fourth Assessment Report of the Intergovernmental Panel on Climate Change*, Cambridge, UK and New York, USA: Cambridge University Press.

de Montalembert, M.R. and J. Clement (1983), 'Fuelwood supplies in the developing countries', FAO Forestry Paper, UN Food and Agriculture Organization, Rome.

Mpogole, H., H. Usanga and M. Tedre (2008), 'Mobile phones and poverty alleviation: a survey study in rural Tanzania', in J.S. Pettersson (eds), *1st International Conference on 'M4D': Mobile communication technology For Development*, Karlstad University, Sweden, Centre for HumanIT, Karlstad.

Müller, B. (2007), 'Food miles or poverty eradication? The moral duty to eat African strawberries at Christmas', *Oxford Energy and Environment Comment*, Oxford: Oxford Institute for Energy Studies, at http://www.oxfordenergy.org/pdfs/comment_1007-1.pdf.

Nhete, T.D. (2007), 'Electricity sector reform in Mozambique: a projection into the poverty and social impacts', *Journal of Cleaner Production*, **15** (2), 190–202.

Niez, A. (2010), 'Comparative study on rural electrification policies in emerging economies: keys to successful policies', Information Paper, International Energy Agency, Paris, at http://www.iea.org/papers/2010/rural_elect.pdf.

Niles, S. and S. Hanson (2003), 'A new era of accessibility?', *URISA Journal*, **15**, 35–41.

Nuvolari, A. (2004), 'Collective invention during the British Industrial Revolution: the case of the Cornish pumping engine', *Cambridge Journal of Economics*, **28** (3), 347–63.

Oh, W. and K. Lee (2004), 'Energy consumption and economic growth in Korea: testing the causality relation', *Journal of Policy Modeling*, **26** (8–9), 973–81.

Oil and Gas Journal (1981), 'Many LDCS ignore proven remedy for energy poverty', **79**, 107.

Paul, S. and R.N. Bhattacharya (2004), 'Causality between energy consumption and economic growth in India: a note on conflicting results', *Energy Economics*, **26** (6), 977–83.

Polgreen, L. (2005), 'Chad backs out of pledge to use oil wealth to reduce poverty', *New York Times*.

Ramjeawon, T. (2008), 'Life cycle assessment of electricity generation from bagasse in Mauritius', *Journal of Cleaner Production*, **16** (16), 1727–34.

Ravindranath, N.H. and P. Balachandra (2009), 'Sustainable bioenergy for India: technical, economic and policy analysis', *Energy*, **34** (8), 1003–13.

Reddy, A. (2000), 'Energy and social issues', in J. Goldemberg (eds), *World Energy Assessment: Energy and the Challenge of Sustainability*, New York: United Nations Development Programme, pp. 39–60.

Reddy, A.K.N. (2001), 'California energy crisis and its lessons for power sector reform in India', *Economic and Political Weekly*, **36** (18), 1533–40.

Rockefeller Foundation (2006), 'Africa's turn: a new Green Revolution for the 21st century', Rockefeller Foundation, New York, at http://www.rockefellerfoundation.org/uploads/files/dc8aefda-bc49-4246-9e92-9026bc0eed04-africas_turn.pdf.

Rockström, J., M. Lannerstad and M. Falkenmark (2007), 'Assessing the water challenge of a new green revolution in developing countries', *Proceedings of the National Academy of Sciences*, **104** (15), 6253–60.

Rogner, H.-H. and A. Popescu (2000), 'An introduction to energy', in J. Goldemberg (eds), *World Energy Assessment: Energy and the Challenge of Sustainability*, New York: United Nations Development Programme, pp. 30–37.

Ross, M.L. (2001), 'Does oil hinder democracy?', *World Politics*, **53** (3), 325–61.

Ross, M.L. (2008), 'Blood barrels: why oil wealth fuels conflict', *Foreign Affairs*, **87** (3), 2–9.

Saatkamp, B.D., O.R. Masera and D.M. Kammen (1999), 'Energy and health transitions in development: fuel use, stove technology and morbidity in Juaracuaro, Mexico', *Energy for Sustainable Development*, **4** (1), 7–16.

Sagar, A.D. and S. Kartha (2007), 'Bioenergy and sustainable development?', *Annual Review of Environment and Resources*, **32** (1), 131–67.

Sanchez, P., G. Denning and G. Nziguheba (2009), 'The African Green Revolution moves forward', *Food Security*, **1** (1), 37–44.

Sen, A. (1998), 'Mortality as an indicator of economic success and failure', *The Economic Journal*, **108** (446), 1–25.

Sen, A.K. (1999), *Development as Freedom*, Oxford: Oxford University Press.

Shiu, A. and P.-L. Lam (2004), 'Electricity consumption and economic growth in China', *Energy Policy*, **32** (1), 47–54.

Silva, H.D. and A. Zainudeen (2007), 'Teleuse on a shoestring: poverty reduction through telecom access at the "bottom of the pyramid"', Centre for Poverty Analysis Annual Symposium on Poverty Research in Sri Lanka, Colombo, Sri Lanka.

Smith, K. (1993), 'Fuel combustion, air pollution, and health: the situation in developing countries', *Annual Review of Energy and the Environment*, **18**, 529–66.

Smith, K.R. (2002), 'In praise of petroleum?', *Science*, **298**, 1847.

Smith, K.R., R. Edwards, K.N. Shields, R. Bailis, K. Dutta, C. Chengappa, O. Masera and V. Berrueta (2007), 'Monitoring and evaluation of improved biomass cookstove programs for indoor air quality and stove performance: conclusions from the household Energy and Health Project', *Energy for Sustainable Development*, **11** (2), 5–18.

Smith, K., S. Mehta and M. Maeusezahl-Feuz (2004), 'Indoor air pollution from household use of solid fuels', in M. Ezzati, A. Lopez, A. Rodgers and C. Murray (eds), *Comparative Quantification of Health Risks: Global and Regional Burden of Disease Attributable to Selected Major Risk Factors,* Geneva: World Health Organization, pp. 1435–93.

Smith, T.C. (1988), *Native Sources of Japanese Industrialization, 1750–1920*, Berkeley, CA: University of California Press.

Soeftestad, L.T. and M.K. Sein (2003), 'ICT and development: East is East and West is West and the twain may yet meet', in S.M.S. Krishna (eds), *The Digital Challenge: Information Technology in the Development Context*, Burlington, VT: Ashgate, pp. 63–82.

Soytas, U. and R. Sari (2003), 'Energy consumption and GDP: causality relationship in G-7 countries and emerging markets', *Energy Economics*, **25** (1), 33–7.

Stern, D.I. (2004), 'Economic growth and energy', in J.C. Cutler (eds), *Encyclopedia of Energy*, New York: Elsevier, pp. 35–51.

Suharta, H., P.D. Sena, A.M. Sayigh and A. Komarudin (1999), 'The social acceptibility of solar cooking in Indonesia', *Renewable Energy*, **16** (1–4), 1151–54.

Thompson, G. and J. Laufman (1996), 'Civility and village power: renewable energy and playground politics', *Energy for Sustainable Development*, **3** (2), 29–33.

Tilman, D., K.G. Cassman, P. Matson, R. Naylor and S. Polasky (2002), 'Agricultural sustainability and intensive production practices', *Nature*, **418** (6898), 671–7.

Tully, S. (2006), 'The human right to access electricity', *Electricity Journal*, **19** (3), 30–9.

UNDP (2009), 'HDR 2009 statistical tables', Human Development Reports, 2009 edition, at http://hdr.undp.org/en/media/HDR_2009_Tables_rev.xls.

Warschauer, M. (2004), *Technology and Social Inclusion: Rethinking the Digital Divide*, Cambridge, MA: MIT Press.

Watts, M. (2009), 'Oil, development, and the politics of the bottom billion', *Macalester International*, **24** (1), 79–130.

Weber, C.L. and H.S. Matthews (2008), 'Food-miles and the relative climate impacts of food choices in the United States', *Environmental Science and Technology*, **42** (10), 3508–13.

Wentzel, M. and A. Pouris (2007), 'The development impact of solar cookers: a review of solar cooking impact research in South Africa', *Energy Policy*, **35** (3), 1909–19.

Westhoff, B. and D. Germann (1995), *Stove Images*, Frankfurt: Apsel Verlag.

WHO (2006), *Fuel for Life: Household Energy and Health*, Geneva: World Health Organization.

WHO (2008), 'Risk factors estimates for 2004', *The Global Burden of Disease*, at http://www.who.int/healthinfo/global_burden_disease/risk_factors/en/index.html, accessed 26 July 2010.

Williams, J.H. and R. Ghanadan (2006), 'Electricity reform in developing and transition countries: a reappraisal', *Energy*, **31** (6–7), 815–44.

Wolde-Rufael, Y. (2005), 'Energy demand and economic growth: the African experience', *Journal of Policy Modeling*, **27** (8), 891–903.

Woo, C., D. Lloyd and A. Tishler (2003), 'Electricity market reform failures: UK, Norway, Alberta and California', *Energy Policy*, **31** (11), 1103–15.

World Bank (2000), *Africa Development Indicators: 2000*, Washington, DC: World Bank.

World Bank (2001), 'The california experience with power sector reform: lessons for developing countries', Washington, DC, World Bank Energy and Mining Sector Board.

World Bank (2007), *World Development Report 2008: Agriculture for Development*. Washington, DC: World Bank, at http://econ.worldbank.org/WBSITE/EXTERNAL/EXTDEC/EXTRESEARCH/EXTWDRS/EXTWDR2008/0,,contentMDK:21410054~menuPK:3149676~pagePK:64167689~piPK:64167673~theSitePK:2795143,00.html.

World Bank (2010), 'WDI Online – World Development Indicators', at http://ddp-ext.worldbank.org/ext/DDPQQ/member.do?method=getMembers&userid=1&queryId=6, accessed 3 February 2010.

World Bank (n.d.), 'ICT and poverty alleviation', at http://siteresources.worldbank.org/INTSDNETWORK/Resources/1.pdf, accessed 29 July 2010.

World Bank and PPIAF (2010), 'PPI Project Database', *Private Participation in Infrastructure (PPI) Project Database* 2009, retrieved 20 July 2010 at http://ppi.worldbank.org./index.aspx.

World Commission on Dams (2000), *Dams and Development: A New Framework for Decision-making: The Report of the World Commission on Dams*, London: Earthscan.

World Resources Institute (2010), 'EarthTrends: the environmental information portal', at http://earthtrends.wri.org/index.php, accessed 18 May 2010.

Wrigley, E.A. (1962), 'The supply of raw materials in the Industrial Revolution', *Economic History Review*, **15** (1), 1–16.

Yergin, D. (1991), *The Prize: The Epic Quest for Oil, Money, and Power*, New York: Simon & Schuster.

Yoo, S.H. (2006), 'The causal relationship between electricity consumption and economic growth in the ASEAN countries', *Energy Policy*, **34** (18), 3573–82.

25 The role of regions in the energy sector: past and future
Thomas Reisz

25.1   INTRODUCTION

The starting point of this chapter – the economic and ecological necessity of increasing energy efficiency and of abandoning the use of fossil fuels in favour of inexhaustible sources of energy – receives almost daily treatment in the media. If this is not tackled, we face the threat of a global energy emergency – whether it is caused by dependencies (artificially created energy shortages) or simply by resources running out (natural shortages) – that would inevitably lead to a global political and social crisis. Renewables promise energy that can be produced and used economically (prosperity), that is almost infinitely available (security of supply) and that does not harm the environment (inviolability), thus fulfilling the requirements of an energy policy based on the principle of sustainability (NRW, 2008: 24).

The acceleration of information exchange and flows of goods that we call globalization has accorded an increasingly political role to subnational territorial frames of reference; in other words, regions. The regions' new confidence in their ability to solve current energy supply challenges is expressed in the findings of a survey conducted by the Assembly of European Regions (AER), which also shows them asking for the financial wherewithal to utilize this competence (AER press release, 24 November 2009). The present chapter considers the role of regions as actors with considerable problem-solving competence in the field of energy. Taking the German federal state of North Rhine-Westphalia (NRW) as an example, it focuses on specific developments and microstructures in this region and the way they influence one another in order to identify some general patterns. The ecological modernization of North Rhine-Westphalia is following a course that vascillates between economic constraints and ecological necessities. The so-called 'third way' (Hüttenhölscher and Reisz, 2005: 264) emphasizes that while politicians have the power to chart this course they should bear in mind that without economic prosperity there will be neither the psychological willingness nor the technological capability to make a sustainable energy economy a reality.

25.2   WHAT IS A REGION?

In general terms there are two things that characterize a region. Firstly, the term 'region' describes a socially constructed, homogenous area characterized by a concentration of movements of people and goods and of communications and trade (Osterhammel, 2009: 156). At the same time, a region can define its homogeneity on the basis of a variety of criteria such as geography, culture, ethnicity, economic activity or history; and this homogeneity also 'lends unity to a population engaged in the pursuit of common goals and interests' (Standing Conference of Local and Regional Authorities of the Council of Europe, quoted from Hrbek and Weyand, 1994: 17). Secondly, a region is a territory with defined boundaries that set it apart from a superordinate (super-regional or national) spatial entity.

A region is subject to a dynamic of structural changes, which can be perceived retrospectively as an economic, political or cultural rise or decline. The current trend towards regionalization and regional identity is a consequence of globalization whereby national states are losing their ability to control and influence processes (Altemeyer-Bartscher, 2009: 36). The resulting dismantling of barriers, dissolution of boundaries and acceleration of economic and social life (Virilio, 1993: 15ff.) are perceived as reducing the significance of national states as life-world categories. As barriers to the exchange of goods, information and communications fall, so does the significance of national states. The ability of national governments to determine the previously undetermined nature of operations (Luhmann, 2000: 19) – in other words, to exercise power – declines. National states seeking to unify the regions into a single national entity then find themselves at a disadvantage vis-à-vis decentralized structures when it comes to tackling ever more complex tasks, for they lack the flexibility and regional knowledge required to make decisions (Altemeyer-Bartscher, 2009: 27).

Regions are experiencing a boom. The pragmatic approach taken by actors right down to the local level has given rise to a growing number of unstructured, unregulated individual 'regions' organized as territorial units, albeit to different degrees. These range from regions within German federal states such as the Emscher-Lippe region in the northern Ruhr area – with its energy network and research project DynaKlim – to the Ruhr itself; the Rhine-Main region and the 'energy region' Lausitz, which straddle several German federal states; and 'regions' that span several national states like the 'blue banana' or SaarLoLux. As they increase in number, these regions have come to symbolize the transformation of statehood and the growing significance of multilevel governance approaches for policy analysis. At the same time the growing significance of the regions is also an expression of an increasing functional differentiation in the quest for the optimal 'operational size' for a political organizational entity.

For the remainder of this chapter I use the European Union (EU) definition of regions as territorial units located directly below the level of central government in the territorial hierarchy but above that of local authorities. Their assumption of responsibility for making decisions not directly under the jurisdiction of the central state makes them important elements in the territorial organization of the administration (Hrbek and Weyand, 1994: 18f.). The assignment of these responsibilities is based on an assessment of their capacity to act, since regions require a minimum level of infrastructure in order to be economically dynamic enough to hold their own in the global competition to which they are exposed. Local authorities and city councils do not fulfil these criteria (Kaltenbrunner, 2009: 57). Hence the term 'regionalization' refers not to the emergence or creation of new regions but rather to the delegation of competences by the national state to subnational territories. The task of the region (acting on behalf of the community) is to use these competences to protect companies investing in the development and production of new energy technologies (which are assumed to be acting for the good of the community) from the risks (market failure, natural disasters) inherent in such investment. (By contrast, certain theories of classical economics teach us that company profit is the reward to be gained for willingness to take risks and engage in innovation; cf. Schumpeter, 2006 [1912].) By assuming responsibility for research, technology and education and by satisfying demand, in particular for infrastructure (McKinsey, 2009: 57) – originally fields under the purview of the central state – the regions, as substitutes for national spatial entities, are at the same time becoming increasingly important protagonists.

Following the Second World War a conscious decision was taken to enhance the political status of the regions, giving rise to Germany's federal structure. The founding of the state of North Rhine-Westphalia as a subnational territory with considerable leeway for shaping policy was part of this development. In response to the unwieldy and non-transparent administrative structures of the centralized state of the Nazi era, greater value was now attached to life in smaller communities (Hrbek and Weyand, 1994: 21). Later on the regions also acquired greater significance as independent economic and political actors within the framework of the EU, and hence as important elements in the European system of multilevel governance.

In the age when energy became the key to prosperity and energy requirements were naturally determined by economic growth, energy was the leitmotif of development. Energy gained a place in political economics, and the image of the 'human engine', whereby the value of labour was quantified as the difference between energy input and energy output (Osterhammel, 2009: 929f.), became established. The connection between regions and the energy factor and the idea that energy production and consumption could either inhibit or promote the process of regionalization, or indeed even be a constitutive criterion for it, can be traced back to 'the cultural history of energy' (Weber 1951 [1909] citing Ostwald: 407) and to the belief in the early twentieth century that every cultural shift was caused by energy developments. The interdependency of technology and society mean that technological changes – in this case in energy production or consumption – produce cultural change. This socio-technological interaction (Keppler et al., 2009: 11) has a long-term impact that continues to be felt even today and finds expression, for instance, in the choice of sustainable energy production as a theme of RUHR.2010 (the Ruhr region as 2010's European Capital of Culture) or in calls to think beyond coal to other fuel alternatives (Pleitgen, 2009).

25.3   NRW PAST AND PRESENT

If our task were to write an intellectual history of energy efficiency, in other words to go back to the beginning, we would find ourselves in the Age of Enlightenment and its discovery that all technology is based on adherence to the laws of nature (in other words, it is nothing more than applied science or theory put into practice). According to Christian tradition, the reward for obeying the laws of nature or living (as *animal laborans*) in harmony with nature (paradise) is happiness and fulfilment. Over recent decades bending or even breaking the laws of nature (and by implication the divine will) – in other words, consuming energy — has come to be associated with a sense of guilt for which ritual atonement must be made (see Niklas Maak in *Frankfurter Allgemeine Zeitung*, 16 September 2009), for we imagine that by consuming energy we destroy it (*Homo faber*, see Arendt, 1981 [1958]: 202ff.). The common perception is that through the act of being consumed, energy is irretrievably lost. It then follows that if we can manage to tap renewable, inexhaustible sources of energy we can repair the damage; in other words, we can replace what we have used with 'new energy'. Nothing symbolizes this image of an inexhaustible source of energy better than a wind turbine. It transcends its actual function to become a kind of energy *perpetuum mobile*, promising eternal energy as the source of life, and hence symbolizing eternal life on earth.

Against this backdrop we might, rather like Nietzsche, sing the praises of physics. If we put our faith in science, energy consumption soon loses its mystique, for seen in terms of physics energy is not actually consumed at all but merely, and rather profanely, converted – from electricity into heat, for instance. Technical terms like 'conversion losses' are sometimes misleading for those outside the scientific community, because they suggest that energy is lost when it is converted. In fact, though, the term 'conversion losses' simply refers to the number of kilowatt hours (kWh) not converted into the form of energy we desire. So even after energy has supposedly been 'consumed' it is really still there, and in some cases even still available for use. All we need to do (and this is the technically challenging aspect) is to convert it back into a form we can use. Formulated in general terms, this means that just as technical inventions are predetermined by the labour process (Marx, 1957 [1867]: 240), so energy efficiency is predetermined by the way energy is converted.

Viewed retrospectively the Enlightenment is a good place to start in tracing the history of energy culture, as it was the starting point for the Industrial Revolution viewed in terms of a process, and hence for industrialization. It was the latter, in turn, that marked the rise in the consumption of fossil fuels and the emergence of a geography of centres and peripheries, of dynamic and stagnating regions (Osterhammel, 2009: 909). After 1850 the region known today as North Rhine-Westphalia became the centre of Germany's rather late industrialization. Whereas in England, an industrial pioneer, textiles had played a key role, in Germany it was mining and railways that served as engines of development. Right from the beginning industrialization went hand in hand with regionalization, for industries tended to become concentrated in places where raw materials, labour and capital were available. As the economy grew, sources of energy (together with rapid and reliable means of transport) became all important. While some areas like the Ruhr, the Rhineland, Cologne-Bonn, Aachen and Bielefeld underwent a rapid process of industrialization, other proto-industrial areas like the Sauerland and Siegerland, which were traditional centres of iron production, failed to keep pace with the times and instead experienced a process of deindustrialization. These processes were the starting point for an ever greater industrial and economic asymmetry in North Rhine-Westphalia.

The region of North Rhine-Westphalia, Germany's most populous federal state, was established in the wake of the territorial reorganization undertaken in Germany after the Second World War with the democratically motivated intention of decentralizing the formation of the political will and decision-making processes and hence endowing them with greater legitimacy. Today NRW's roughly 18 million inhabitants represent around a quarter of the German population. NRW calls itself Germany's 'number one energy state', for it is both the country's largest energy consumer and its largest energy producer. Thirty per cent of Germany's electricity is produced and 28 per cent of it consumed here (Roels, 2005: 269). In 2005, 22.9 per cent of primary energy consumed came from coal, 32.1 per cent from oil, 22.2 per cent from gas, 20.7 from lignite, and around 2.1 per cent from renewables (source: Ministry for Economic Affairs and Energy of the State of North Rhine-Westphalia). On account of its structure, NRW's economy has been particularly affected by the worldwide rise in energy requirements and by fluctuations in energy prices.

Alongside industrially structured areas, sizeable parts of NRW – in the Sauerland, Westphalia, Münsterland and on the Lower Rhine – are largely rural. Their position as industrially underdeveloped areas in the nineteenth and twentieth centuries is now turning out to be an advantage. Having not had to cope with the structural changes that have affected the rest of the region's economy, they have led the way as suppliers of decentralized energy from locally available resources. A prototype is the biogas plant in Steinfurt. Built in 2005 and consisting of two cogeneration units with output power of 347 kW and 536 kW, it is the result of an informal cooperation project between the University of Applied Sciences Steinfurt, EnergieAgentur.NRW, farmers and local government representatives brought together in a locally based consortium (Arbeitsgemeinschaft Biogas). An operating company formed by a group of farmers financed the plant. The electricity and heat that it generates is sold to the municipality to light and heat its schools and administrative buildings.

If we take the number of new companies founded as an indicator of economic prosperity, innovative power and the pace at which modern technology reaches various economic sectors, then North Rhine-Westphalia ranked in the upper mid-field in 2008. However, the NUI (new company initiatives) Index issued by the Institute for Small Business Research in Bonn recorded across NRW as a whole an average of 152.3 in 2008. This was both below that for the previous year (157.3) and below the average for Germany as a whole (154.6). (The disadvantage of the NUI Index, however, is that it has no breakdown according to sector and would therefore fail to show a possible lead in new company foundings in the energy sector.) Düsseldorf remains the city in NRW with the largest number of new companies (213.3). By comparison, Offenbach am Main, with an NUI index of 379.5, is the leading city nationally for new companies; while the leading region is the city state of Hamburg with an NUI index of 195.1. Until well into the 1990s the Ruhr region was regarded as a prototypical example of the lock-in effect, whereby a strategy of avoiding costs on account of mono-industrial structures (the mining industry) meant that companies were insufficiently able to adapt to economic change. And even today the trend towards forming new companies is weaker in the Ruhr (NUI Index 144.5) than in other regions of NRW. Nevertheless, a number of cities in the Ruhr region, notably Dortmund (176.0), Essen (173.6) and Duisburg (159.3) are well above the NRW average when it comes to founding new companies (May-Strobel, 2009).

The renewables sector has established itself as an engine of growth in NRW in recent years (IWR, 2008: 1). In 2007 the region produced about 10 billion kWh of electricity from renewables, an increase of 16 per cent over the previous year. The heat generation sector was also able to boast double-digit growth figures for energy from renewables (5.6 billion kWh, representing a rise of 10.3 per cent). NRW's share of the total volume of electricity generated from renewables in Germany was around 10 per cent (IWR, 2008: 18). Turnover for companies engaged in building regenerative plants and systems rose in 2007 by 16 per cent to €5.5 billion. Empirical evidence would appear to show that particularly in structurally weak regions the renewables sector, dominated by medium-sized firms, is an important source of employment (Keppler et al., 2009: 13). In 2008 the government of North Rhine-Westphalia presented an energy and climate protection strategy which envisaged a reduction in CO_2 emissions by 2020 (over a baseline of 2005) of 81 million tonnes (NRW, 2008: 26) and was designed to support the positive trend in the renewables sector.

25.3.1   Regional Strategies and Structures

The first oil crisis ushered in a new approach to energy policy. A demand-oriented energy policy that attempted to secure energy supplies by building up surplus capacity increasingly gave way to a supply-oriented energy policy aimed at controlling demand; for instance by systematically creating energy-saving incentives (see NRW, 2008: 9). The collapse of the Bretton Woods regime of fixed currency exchange rates, the consequences of the 'stagflation' brought about by the first oil price shock, the challenge of breaking the vicious circle of demand-deficit unemployment and cost-push inflation led 30 years ago to a change of direction at all levels of governance. The oil crisis coincided with the beginning of monetarism, as a consequence of which price stability increasingly took precedence over full employment as the goal of macroeconomic policy. As stable energy prices became a prerequisite for keeping product prices constant, 'energy' became a higher-profile policy field. Ever since then stable energy prices have been regarded as a condition for macroeconomic stability. At the same time the dividing line between economic and energy policy has become increasingly blurred. Even if renewables would appear to qualify as a policy field in their own right (Hirschl, 2008: 36), the issues of fuel prices and the quest to find the apparently cheapest fuel at any given time also make them a subdiscipline of economic policy (Mez, 2003). For this reason, responsibility for energy in North Rhine-Westphalia is assigned to a department in the regional Ministry of Economics.

   While high growth rates and employment potential are being predicted for the energy sector, it is not immune to changes in the general economic climate. It especially needs to take account of exit and voice options of participants in the energy sector, although their relocation potential obviously varies. Whereas producers of wind energy components, for example, are free to choose their production location (location arbitrage), thus forcing regions to compete for mobile production factors, local authorities or municipalities are by their very nature obviously bound to a particular location.

   At the same time the increasingly complex structures of the knowledge and information society induced by globalization have changed actor structures as well. Whereas 30 years ago competition was still between companies, nowadays it is between organizational or regional networks (Fritz, 2009: 189). In modern regional policy, networks and network formation are considered the key to success and a guarantee of continuing prosperity on account of their reciprocity (Semlinger, 2006: 48; Fritz, 2009: 190f.; Arndt and Kaiser, 2009, 237ff.); rigid competition strategies, on the other hand, prevent economic actors from adapting to changing conditions. What are required therefore are dynamic competition factors. Against the background of regions as policy arenas in the framework of multilevel governance (Hirschl, 2008: 37), regional networks and clusters constitute a way of coordinating and optimizing cooperation between formally independent yet functionally interdependent actors within a region. The German federal state of Baden-Württemberg and the Italian province of Emilia Romagna are widely considered to be two examples of successful regional networks. The NRW government has responded to the changing competition situation by realizing a cluster structure. Its aim is to promote and provoke synergy effects within the region in accordance with the concept of 'social embeddedness'. These in turn promise competition advantages in the form of so-called 'specialized factors' (Porter, 1991). The clusters imply a broadening of the technological basis, increases in turnover, synergies and access to complementary competences and resources; thus strengthening regional competitiveness.

In implementing this policy North Rhine-Westphalia draws a hierarchical distinction between clusters and networks, whereby clusters are the superordinate and networks the subordinate structures. Clusters are particularly dense networks. The clusters and networks in North Rhine-Westphalia are territorially defined by the state borders. The aim of NRW's cluster policy is to promote innovation in future-oriented and climate-friendly energy sectors, so that it becomes a trendsetter in the field of energy both nationally and internationally (press release, Ministry of Economic Affairs and Energy of the State of North Rhine-Westphalia, 13 August 2009). One of the clusters established by NRW is EnergieRegion.NRW (Energy Region NRW). It consists of eight networks working in the following fields:

- power station technology;
- fuel cells and hydrogen;
- biomass;
- energy-efficient and solar construction;
- fuels and engines of the future;
- photovoltaics;
- geothermal energy;
- wind power.

EnergieRegion.NRW brings together 3300 companies and institutions in the energy sector – three-quarters of them small and medium-sized businesses – as well as 64 universities, 107 institutes and 94 professional associations. A total of 5200 skilled staff are involved in working groups and networks in this cluster.

A further energy cluster is Cluster Energie-Forschung.NRW (Energy Research Cluster NRW). Located in the Ministry for Innovation this cluster is devoted chiefly to research, and focuses on the following topics:

- centralized energy production;
- decentralized energy production;
- biological fuel production;
- energy networks;
- the energy economy.

Membership of a cluster gives companies access to the competencies of other members of the cluster while allowing them to specialize in their own key competencies. This yields productivity and cooperation advantages arising from a reduction in business risks and lower control costs. The degree of interlinkage offered by clusters and networks speeds up communication and means that information is disseminated more rapidly. A collective pool of knowledge is created which forms the basis for organizational learning (Fritz, 2009: 193). The latter, in turn, is a prerequisite for an innovative milieu (Ivanisin, 2006: 59). One example of a project being conducted through cluster or network cooperation is the NRW-GeoTechnikum, which emerged from Netzwerk Geothermie (Geothermal Energy Network) as a cooperation project between the Ruhr University Bochum, the RWTH Aachen University, the University of Applied Sciences Gelsenkirchen and the University of Applied Sciences Ostwestfalen-Lippe. NRW-GeoTechnikum is concerned mainly with research and development in the fields of drilling technology, reservoir technology, geophysical measuring technology and deep drilling down to a depth of 5000 metres using coiled tubing drilling rigs.

Integrating NRW in federal or national strategies
Evaluations of the cooperation between the national government and the region differ, depending on who is doing the evaluating. The coordination of energy policy between the national state and the regions is determined by competing interests (foreign policy, economic and security interests, to name but a few) and is therefore not free of conflict. The Assembly of European Regions (AER) came to the conclusion at its annual general meeting in Franche-Comté, France in November 2009 that regional use of renewables is being hampered by national states (AER press release, 24 November 2009). By contrast, it sometimes appears to the German government as if the conservative liberal coalition government of NRW were obstructing federal government efforts to expand renewables (Hirschl, 2008: 182), yet regional actors have welcomed what they see as the generally positive attitude of the Christian Democratic Union–Free Democratic Party (CDU–FDP), state government of NRW to renewables (Keppler, 2009: 49). Hirschl has made the general observation that while regional energy policy tends to be shaped by local energy policy traditions, in most cases a majority favours the expansion of the renewables sector and does not present serious obstacles to national policy efforts in this field (Hirschl, 2007: 152). The German Institute of Economic Research (DIW) ranks NRW in first place when it comes to the promotion of renewables. What is more, a DIW study also finds that the region of NRW offers the best information about using renewables (DIW, 2009: 23–27). There seems to be a general consensus about the need to increase use of renewables, irrespective of territorial organizational entity.

The Integrated Energy and Climate Programme (IEKP, also known as the Meseberg Program) adopted by the German federal government in August 2007 put together a package of measures to define and achieve goals for implementing renewables and energy efficiency. The Meseberg Program commits NRW to reducing energy-related CO_2 emissions by 29 million tonnes a year via efficiency increases and energy savings and by 7 million tonnes a year by increasing the share of renewables (in both cases by 2020). In addition the German federal states (including NRW) are responsible for taking up federal objectives and putting them into practice while taking account of state-specific conditions. A working group entitled Energy Efficiency and Climate Protection has been founded as part of NRW's Economics and Environment Dialogue to address questions related to the implementation of the federal government's energy and climate programme as well as the European Climate Protection Pact.

Within Germany's federal system NRW basically has two levers with which to influence policy processes. First of all, a state can influence the legislative process via its vote in the Bundesrat (the second chamber of the German parliament in which each state or Bundesland has a vote) and can either approve or reject laws that require the approval of the Bundesrat. Secondly, it can also support the implementation of laws via flanking measures, for example an active subsidies policy. In North Rhine-Westphalia all funding and promotional measures come under the programme progres.nrw, which aims to accelerate the launch to a broad market of the diverse technologies designed to utilize inexhaustible energy sources and to facilitate the efficient use of energy. The progres. nrw scheme replaced the REN Program, which expired at the end of 2006 and which had approved more than 51 000 projects since 1989. Funding to the tune of €640 million provided the impetus for investments of more than €3.2 billion.

25.3.2   Instruments and Actors

EnergieAgentur.NRW engages in a broad spectrum of activities that form a central element in the operative implementation of regional energy policy (NRW, 2008: 30; Keppler, 2009: 40). EnergieAgentur.NRW is funded out of the NRW state budget, and NRW is reimbursed by the European Regional Development Fund (ERDF). It advises companies, local authorities and end users on technical aspects of increasing energy efficiency, using renewables and on obtaining funding; it also offers professional training programmes (for example, in the field of heat insulation of buildings) and promotes technology transfer between academia and industry by serving as a coordinator between NRW's energy clusters and networks. An important requirement for this cluster and network management role is EnergieAgentur.NRW's product-neutrality. Whereas in markets where price is the main steering mechanism, or in hierarchies where authority is the central factor, little attention is paid to trust as social capital, in clusters and networks it is a key steering element (Fritz, 2009: 191).

The EnergieAgentur.NRW's projects aim to fulfil the political objective of using know-how transfer to increase energy efficiency and to help bring the technology required for the use of renewables into general use. Its projects are directed specifically at:

- companies (for example, the programmes JIM.NRW, 'Mein Haus spart');
- local authorities (for example Contracting consultation, the European Energy Award, JIM.NRW, 'Mein Haus spart'); and
- end users (for example, 'Mein Haus spart', Gebäude-Check Energie).

JIM.NRW

With its pilot project JIM.NRW (Joint Implementation Model Project NRW) NRW is treading new ground. The aim of the project is to create economic incentives for companies to invest in increasing efficiency and in climate protection. Companies and local authorities which refurbish heating or steam boilers of up to 20 megawatts (MW) can earn tens of thousands of euros worth of revenues by trading their emissions reductions in the form of emissions certificates. For the companies participating in the scheme JIM. NRW has the same effect as a funding programme but draws its financial resources not from the public purse but from the revenues earned by trading CO_2 certificates.

To make it easier to participate in the market, JIM.NRW pools its efficiency projects. North Rhine-Westphalia assumes responsibility for:

- monitoring individual projects;
- pooling the projects;
- verification of emissions reductions by an independent testing facility;
- registration (converting emissions into certificates) with the German Emissions Trading Office;
- marketing (sale of certificates) of the emissions 'saved' through plant modernization at market prices within the framework of the EU emissions trading market.

Local authorities and contracting

Local authorities are among Germany's largest energy consumers. As the operators of buildings such as schools, town halls and swimming pools, they use above-average amounts of electricity, heat and water. For this reason energy costs are a major item of expenditure in their budgets, and have now become a fundamental problem. There are two reasons why large cities are increasingly coming to resemble 'historic centres' in energy terms: (1) outsourcing has led to a loss of specialist knowledge about technological advancements; and (2) local authority personnel are reluctant to make use of variable financing and funding options. It is estimated that 50 per cent of German local authorities still use 1960s-standard street lighting technology. Only 3 per cent of these energy-guzzling 'old-timers' are replaced annually. The savings potential for Germany as a whole is estimated at 2.7 billion kWh, or around €400 million (MWMW press release, 28 November 2009), for viable alternatives do indeed exist. Here EnergieAgentur.NRW acts as a partner to local authorities, providing them with advice and helping their staff to acquire further skills at conferences and in-house training sessions.

EnergieAgentur.NRW participated in a lighting contracting programme of this kind in the town of Mechernich. Spurred on by the economically attractive prospect of obtaining external financing from a contractor, the Mechernich local authority went ahead and replaced all the lighting systems in the town's schools with energy-saving models and presence detectors. The modernization is now saving the local authority around €49 000 annually in electricity bills. The local authority was first made aware of contracting as a form of financing by EnergieAgentur.NRW, which provided advice during the planning and implementation phase. Mechernich has since received an award for lighting modernization from the European Commission.

European Energy Award

EnergieAgentur.NRW has been tasked by the NRW Ministry of Economic Affairs to organize the European Energy Award (EEA) energy management procedure for NRW. The European Energy Award is a European certification system that evaluates the sustainability activities of local authorities, looking at such things as use of solar energy, operation of biomass plants, and so on. Of the region's 396 local authorities, 100 are now participating in the EEA; one citizen in five in NRW lives in a local authority area undergoing certification.

'Mein Haus spart' ('energy saving begins at home')
More than 70 per cent of NRW's residential buildings have little or no heat insulation (NRW, 2008: 29). Around 75 per cent of all the buildings in NRW were built before the first thermal insulation regulations came into force in 1981. In 2008 only 1 per cent of the entire building stock of NRW had been renovated to improve insulation, thus falling well short of the 3 per cent per year required to meet climate protection goals. Therefore the NRW Ministry of Economic Affairs launched a household energy-saving programme that brings together important actors from the housing sector, end users and planners (both architects and engineers). 'Mein Haus spart' awards 'NRW energy saver' (Energiesparer.NRW) plaques to residential buildings which have undergone exemplary energy-saving modernization programmes or which use renewables. The awards are designed to motivate other property owners to carry out similar measures. Since the energy-saving awards scheme was launched in 2004, more than 1400 plaques have been awarded (as of December 2009).

Gebäude-Check Energie
Gebäude-Check Energie (energy check-up for buildings) is an instrument to modernize old buildings with a view to saving energy, as well as to promoting the skilled trades sector. Tradespeople, specially trained by EnergieAgentur.NRW, are sent to assess the actual state of the building and to present possibilities for improvement. Since 1997 more than 2000 building check-ups have been carried out annually, and since 2004 engineers and architects have given 300 extensive consultations in buildings with a maximum of six residential units. Up to now Gebäude-Check Energie has shown that the average energy consumption of existing buildings (built in or before 1980) can be reduced from around 220 kWh/m²a to around 100 kWh/m²a, in other words a savings potential of around 50 per cent. On average an energy check-up of this kind prompts an investment of €7500, so that the approximately 3200 check-ups carried out in 2009 alone led to investments of around €25 million in the energy-saving modernization of buildings. Since the introduction of this scheme more than 28 000 check-ups (as of December 2009) have been carried out, prompting investments of more than €210 million.

25.4   FROM AN INDUSTRIAL DISTRICT TO AN INNOVATIVE MILIEU

Given the correlation between climate change and energy use, defining climate policy goals has had concrete repercussions for energy policy. The cluster policy represents a comparatively new approach: the energy sector cluster EnergieRegion.NRW only came into being in the summer of 2009. NRW has thus crossed the threshold from being an industrial district to becoming an innovative milieu. The latter is characterized by interaction between actors on the basis of learning processes, which in turn generate innovation-specific externality. In an innovative milieu the juxtaposition of shared goals and competition between actors provides a positive atmosphere for innovation. Cooperation between research, education, the private sector and politics reduces innovation costs as a risk inherent in the process of innovation (Ivanisin, 2006: 59). The following factors have been shown to be conducive to the development of an innovative milieu:

- the presence of first-class universities (for example RWTH Aachen, Westfälische Wilhelms-Universität Münster) representing an accumulation of research competence in the region;
- the presence of research-intensive sectors (for example the solar industry, fuel cells and heat pump technology) in the region;
- the networking of research and development (R&D) and industry (for example at the Centre for Fuel Cell Technology at the University of Duisburg, or in RWE's automated algae production plant in Niederaußem);
- extensive international networks and super-regional contacts (for example the cooperation agreement concluded between North Rhine-Westphalia and the Chinese province of Shanxi in 2009 to use mine gas for energy purposes).

The innovative milieu is an expression of the growing conviction of the necessity for structured collective action. Ultimately it is the result of the learning process, which generates opportunities for interaction and the ability to cooperate, in other words to maintain relationships based on mutual dependence. However, the structures thus developed are not static, but actually highly flexible and able to adapt quickly to the changing requirements with which a network or cluster is confronted.

25.5   NRW BETWEEN COOPERATION AND CONFRONTATION

There are some age-old lessons that never change over the centuries. Indeed, in our age faith in the power of the Greek goddess Eris (the goddess of strife and envy), to drive even an inept man to work is stronger than ever. As soon as he, the poor man, sees the rich man, so the theory goes, he will hurry off to sow and to reap and put his house in order just like the rich man (Hesiod, *Works and Days*). Neighbours compete with one another in bettering themselves. Nowadays, we call this by its modern economic equivalent: competition. In North Rhine-Westphalia competition has now been introduced to almost all policy arenas, and plays a role in the following programmes:

- Energieforschung.NRW (research institutions, companies);
- Energie.NRW (research institutions, companies, local authorities);
- Klimaschutz und Klassenkasse (schools);
- Energiesparer.NRW (private households);
- ElektroMobil.NRW (research institutions, companies);
- European Energy Award (local authorities).

Energy is a crucial element in the mechanism of the competitive economy, and competition likewise has an important role to play in the energy sector. It has a positive effect on pricing (for instance, PEPP, 2008: 4). The classic neoliberal market formula is: growth and prosperity through competition. Until 1998 eliminating competition was the preferred instrument for regulating the market; this changed, however, following the liberalization of energy markets in the EU.

Within a cluster, however, there is a danger that too much competition is a bad thing. There is a consensus that the repertoire of human activity extends beyond envy and strife. Psychological studies have shown that people are prepared to accept disadvantages suffered as a result of their own actions if, for example, this means that others are punished at the same time – a phenomenon which occurs when our sense of justice is violated (Matthies, 2009). In this context we might say that the outcome of the December 2009 Copenhagen Conference may have been a disappointment for some, while for others it held out the prospect that after decades of inequality and injustice, the destruction of the world through climate change would finally consign all to the same fate. So *Homo economicus* must be more than a being conditioned by competition. And indeed the cluster concept does seem to take account of needs other than the will to compete. The behaviour of cluster members is determined not by market criteria (concerns with money or profit) but by psychosocial needs (trust). In the cluster, cooperation takes the place of competition. People are motivated to cooperate by their striving for perfection, by the urge to carry out the perfect act, to achieve the perfect result that is inherent in any human action; Plato formulated this as *arete*, the standard of excellence implicit in any act; see Sennet, 2008: 37). How people behave in a cluster is thus only partly subject to the logic of the market. Is a cluster therefore a heterotopia, a differently ordered economy?

The parallel existence of cluster policy (based mainly on cooperation), and the many schemes that involve an element of competition whereby saving energy takes the form of a game or contest, does indeed make it tempting to construct a connection. A working hypothesis might be the following: the implementation of cluster policy creates quasi competition-free zones, making the cluster a kind of protected 'internal area' where the power of the market is restricted. Clusters therefore stand in opposition to the easy market access demanded by those who say that market forces must be given free rein. According to classical liberal competition theory, imposing constraints on competition is undesirable. The cluster moderator or coordinator therefore compensates for the lack of competition based on market logic by creating an 'artificial' competition situation that is like a game or contest. His intention here might be to have the best of both worlds; in other words to use the advantages offered by competition as a motor of innovation without having to give up the benefits offered by the cluster (optimal exchange of information). If the cluster were a heterotopia, then it would be a place where prevailing economic rules are reflected on and called into question – and not another place where they are applied.

NRW's energy policy and the organization of relevant actors in clusters are designed to bind regional actors within the framework of competition, or to recruit new actors for the cluster and to persuade them to establish themselves in the region. The establishment of clusters would thus seem to represent a market response to globalized (regulative) competition that entails suspending market-based competition at the regional level. In practice, though, cooperation can be just as detrimental to the success of clusters as competition can. It is therefore quite realistic to expect cooperation based on trust in a network to give way to confrontation through intense competition (Fritz, 2009: 198; Sennett, 2009: 48f.). Too much competition can have a counterproductive effect on the result; for example if important information that network actors need in order to decide how to act is not exchanged, but kept secret instead (Sennett, 2008: 50). In a national context competition can also lead to discontinuity and wealth differentials, as the example of northern and southern Italy demonstrates. Ultimately, competition strengthens regionalism. Subsidiarity is also an expression of regionalism in the sense that the image of an organism that has evolved historically or naturally is the prevailing one and can in extreme cases mutate into ethno-nationalism, see Gerdes, 1989: 853. In line with the dromological law (Virilio, 1993: 15) one could say that regions with the ability to keep pace with a speeded-up, globalized world or even to set the pace have the edge over 'slower' regions and eventually displace them. Competition is at one and the same time a stimulating impulse and a controversial motivation. Excessive harmony between the cluster actors leads, via an exaggerated desire for consensus or opportunism, to overembeddedness (Altemeyer-Bartscher, 2009: 39).

To what extent the experience of NRW can be applied elsewhere depends largely on how decentralized a given national state is. Hence it seems to be the case that the degree of decentralization – understood as the sum of competences at all subnational levels – has a positive effect on the innovative power of an economy. Or put more succinctly: the more decentralized a state is, the greater the achievements of research and development (AER, 2009: 12). There are some regions, for example in Hungary and France, that have insufficient competences, which diminishes the chances of a cluster or network policy being successfully implemented in practice. After all, the task of politicians with respect to clusters and networks is not limited exclusively to initiating and promoting communication between actors from the energy sector; rather, success depends mainly on how much scope there is to shape and organize clusters and networks. For organization does not just mean communication and securing the flow of information; it also expressly includes the assignment of jurisdiction and responsibilities, the creation of structures, moderation, control and sanctioning (Barkowsky and Huber, 2009: 123). Only in this way can actors be encouraged to act collectively and given the opportunity to participate in political decision-making and implementation processes.

As a result of globalization (regulative competition) the federal state of North Rhine-Westphalia is having to come to terms with an undermining of its ability to tackle problems at a regional level, just as Germany is at a national level. Regional energy policy is therefore clearly characterized by attempts to regain macroeconomic influence. In this respect the European Energy Award – that is, the certification of soft location factors at local authority level – can be interpreted as regulatory intervention designed to correct market failure caused by the 'market for lemons' effect. A basic assumption is that quality of the environment and quality of life in an area are important factors that influence whether companies decide to move there (Salmen, 2009: 179). Regulatory intervention of this kind is justified by the fact that information about differences in the quality of locations is difficult to access. For this reason superior locations may fail to attract companies, despite being more attractive. This kind of market failure can, however, be corrected via certification; in other words, by having a location's quality vouched for by a 'trustworthy authority' (Scharpf, 1999: 88). At the same time certification prevents regulatory competition becoming a 'race to the bottom' (the Delaware effect); in other words the radical lowering of market barriers such as local environmental protection regulations, social standards, collective working conditions, and so on. This is the effect of giving buildings that have been modernized to meet today's energy standards 'energy-saver NRW' plaques. Making the energy status of a building visible gives it a competitive advantage on the housing market.

What will the future bring? Since clusters are not, in theory at least, permanent institutions, there are three possibilities for the future fate of clusters in North Rhine-Westphalia:

- they dissolve into markets;
- they fail or disappear;
- they become institutionalized (Ivanisin, 2006: 150).

In addition the trend towards decentralization of energy supplies would seem to promise a return of politics to economics. At first glance the decentralization of energy production would appear to be an exit option from the economic constraints of regulatory competition and a means to regain macro-economic influence, as thermal and electrical energy will in future be produced in the place where it is used. In accordance with the principle of subsidiarity, regionalization as a consequence of globalization would result in the decentralization of energy supplies. Yet other developments call into question the influence of the regions in the sphere of energy:

- If the energy sector is unable to secure any further (comparative or absolute) cost advantages, the region's significance in economic policy will decline.
- As a result of the financial crisis and the ensuing constraints (massive public debt) the European Commission's policy of transferring competences to private investors under the auspices of public–private partnerships has been accelerated (see Rügemer, 2010: 77). Statehood threatens to become minimalized and there is even a danger that the number of 'failed states' will increase (Risse, 2005: 6ff.).
- The importance of energy as a regional policy field will decline as soon as the national state starts to use it to promote or safeguard its own security or foreign policy interests. Conceivable scenarios range from the destabilization of democratically legitimate systems (Leggewie and Welzer, 2009: 137ff.) to the 'militarization' of energy policy or even a 'climate war' (Welzer, 2008).

To conclude: if the industrialization (as an expression of a desire for hegemony) that took place between 1910 and 1920 created an energy gulf (Osterhammel, 2009: 936f.) between what we might simplistically call the First World and other worlds, then this gulf will become deeper following the energy revolution (referred to variously as the fourth or fifth industrial revolution). The world will become divided into those regions that manage to find substitutes for fossil fuels, and the others – *tertium non datur*.

REFERENCES

AER (2009), 'Durch Subsidiarität zum Erfolg: Der Einfluss von Dezentralisierung auf wirtschaftliches Wachstum', study commissioned by the Assembly of European Regions (AER), Strasbourg and Brussels.

Altemeyer-Bartscher, Daniel (2009), 'Region als Vision', in Marissa Hey and Kornelia Engert (eds), *Komplexe Regionen – Regionenkomplexe, Multiperspektivische Ansätze zur Beschreibung regionaler und urbaner Dynamiken*, Wiesbaden: Verlag für Sozialwissenschaften, pp. 27–52.

Arendt, Hannah (1981 [1958]), *Vita activa oder Vom tätigen Leben*, 7th edn, Munich: Piper.

Arndt, Olaf and P. Kaiser (2009), 'Die Regionen als Laboratorien der Zukunft – Entwicklungsperspektiven deutscher Regionen', in Hans J. Barth and Christian Böllhoff (eds), *Der Zukunft auf der Spur – Analysen und Prognosen für Wirtschaft und Gesellschaft*, Stuttgart: Schäfer-Poeschel, pp. 237–45.

Barkowsky, Kai and A. Huber (2009), 'Politikstil und Interaktionsfähigkeit in der Region als Einflussfaktoren wissenschaftlicher Prosperität', in Marissa Hey and Kornelia Engert (eds), *Komplexe Regionen – Regionenkomplexe, Multiperspektivische Ansätze zur Beschreibung regionaler und urbaner Dynamiken*, Wiesbaden: Verlag für Sozialwissenschaften, pp. 119–35.

DIW (2009), 'Vergleich der Bundesländer: Best Practice für den Ausbau Erneuerbarer Energien', Berlin.

Fritz, Miriam (2009), 'Sozialkapital als weicher Standortfaktor – Das Potential dynamischer sozialer Netzwerke als Wettbewerbsfaktor für Regionen', in Marissa Hey and Kornelia Engert (eds.), *Komplexe Regionen – Regionenkomplexe, Multiperspektivische Ansätze zur Beschreibung regionaler und urbaner Dynamiken*, Wiesbaden: Verlag für Sozialwissenschaften, pp. 189–205.

Gerdes, Dirk (1989), 'Regionalismus', in D. Nohlen (ed.), *Pipers Wörterbuch zur Politik*, Munich: Piper-Verlag, pp. 852–5.

Hirschl, Bernd (2007), 'David gegen Goliath? Die deutsche Erneuerbare Energien-Politik im Mehrebenensystem', in Achim Brunngräber and Heike Walk (eds), *Multi-Level-Governance – Klima, Umwelt- und Sozialpolitik in einer interdependenten Welt*, Baden-Baden: Nomos, pp. 129–60.

Hirschl, Bernd (2008), *Erneuerbare Energie-Politik, Eine Multilevel Policy-Analyse mit Fokus auf den deutschen Strommarkt*, Wiesbaden: Verlag für Sozialwissenschaften.

Hrbek, R. and S. Weyand (1994), *Betrifft. Das Europa der Regionen: Fakten, Probleme, Perspektiven*, Munich: Verlag C.H. Beck.

Hüttenhölscher, Norbert and T. Reisz (2005), 'Energiewende in NRW – Perspektiven der ökologischen Modernisierung', in Heribert Meffert and Peer Steinbrück (eds), *Trendbuch NRW – Perspektiven einer Metropolregion*, Gütersloh: Verlag Bertelsmann Stiftung, pp. 259–67.

IWR (2008), 'Internationales Wirtschaftsforum Regenerative Energien, Zur Lage der Regenerativen Energiewirtschaft in Nordrhein-Westfalen 2007', Münster.

Ivanisin, Marko (2006), *Regionalentwicklung im Spannungsfeld von Nachhaltigkeit und Identität*, Wiesbaden: Deutscher Universitäts-Verlag.

Kaltenbrunner, Robert (2009), 'Das Überörtliche als Strukturprinzip', in Marissa Hey and Kornelia Engert (eds), *Komplexe Regionen – Regionenkomplexe, Multiperspektivische Ansätze zur Beschreibung regionaler und urbaner Dynamiken*, Wiesbaden: Verlag für Sozialwissenschaften, pp. 53–78.

Keppler, Dorothee (2009), 'Fördernde und hemmende Faktoren des Ausbaus erneuerbarer Energien in der Niederlausitz und im Ruhrgebiet', in Dorothee Keppler, H. Walk and H.L. Walk (eds), *Erneuerbare Energien ausbauen – Erfahrungen und Perspektiven regionaler Akteure in Ost und West*, Munich: Oekom-Verlag, pp. 21–71.

Keppler, Dorothee, H. Walk and H.L. Dienel (2009), *Erneuerbare Energien ausbauen – Erfahrungen und Perspektiven regionaler Akteure in Ost und West*, Munich: Oekom-Verlag.

Leggewie, Claus and H. Welzer (2009), *Das Ende der Welt, wie wir sie kannten – Klima, Zukunft und die Chancen der Demokratie*, Frankfurt am Main: S. Fischer Verlag.

Luhmann, Niklas (2000), *Die Politik der Gesellschaft*, Frankfurt am Main: Suhrkamp-Verlag.

Marx, Karl (1957 [1867]), *Das Kapital*, Stuttgart: Alfred Körner Verlag.

Matthies, Ellen (2009), 'Nutzerverhalten und Energieeffizienz sind träge Angelegenheiten', *Innovation&Energie*, **2**, 6.

May-Strobel, Eva (2009) 'Regionales Gründungsgeschehen – das Regionenranking auf Basis des NUI-Indikators', in Wolfgang George and Martin Bonow (eds), *Regionales Zukunftsmanagement Band 3: Regionales Bildungs- und Wissensmanagement*, Lengerich: Pabst Science Publisher, pp. 124–36.

McKinsey (2009), 'Wettbewerbsfaktor Energie – Neue Chancen für die deutsche Wirtschaft', Frankfurt.

Mez, Lutz (2003), 'Energiepolitik', in Uwe Andersen and Wichard Woyke (eds), *Handwörterbuch des Politischen Systems der Bundesrepublik Deutschland*, Opladen: Leske & Budrick, pp. 162–7.

NRW (2008), 'Mit Energie in die Zukunft – Klimaschutz als Chance, Energie- und Klimaschutzstrategie Nordrhein-Westfalen', Düsseldorf.

Osterhammel, Jürgen (2009), *Die Verwandlung der Welt – Eine Geschichte des 19. Jahrhunderts*, 3rd edn, Munich: Verlag C.H. Beck.

PEPP (2008), 'Projektgruppe Energiepolitisches Programm, Effizienz, Transparenz, Wettbewerb – Sichere und bezahlbare Energie für Deutschland', Berlin.

Pleitgen, Fritz (2009), 'Wasser ist die Kohle der Zukunft', in *Innovation&Energie*, **3**, 3.

Porter, Michael E. (1991), 'Towards a dynamic theory of strategy', *Strategic Management Journal*, **12**, 95–117.

Risse, Thomas (2005), 'Governance in Räumen begrenzter Staatlichkeit', *Internationale Politik*, **55** (2), 6–12.

Roels, Harry (2005), 'Strom für Deutschland – Der Energiestandort Nordrhein-Westfalen', in Heribert Meffert and Peer Steinbrück (eds), *Trendbuch NRW – Perspektiven einer Metropolregion*, Gütersloh: Verlag Bertelsmann Stiftung, pp. 269–76.

Rügemer, Werner (2010), 'Public private partnership: Die Plünderung des Staates', *Blätter für deutsche und internationale Politik*, **55**, 75–84.

Salmen, Thomas (2009), 'Kultur – Standortfaktor für die Kulturwirtschaft', in Marissa Hey and Kornelia

Engert (eds), *Komplexe Regionen – Regionenkomplexe, Multiperspektivische Ansätze zur Beschreibung regionaler und urbaner Dynamiken*, Wiesbaden: Verlag für Sozialwissenschaften, pp. 173–88.

Scharpf, Fritz W. (1999), *Regieren in Europa – effektiv und demokratisch?*, Frankfurt am Main: Campus-Verlag.

Schumpeter, Joseph (2006 [1912]), *Theorie der wirtschaftlichen Entwicklung*, Berlin: Dunkler und Humblot.

Semlinger, Klaus (2006*)*, 'Missverständnisse und Fallstricke regionaler Kooperationen – Konkurrenz und Kooperation in Netzwerken', in Burckhardt Kaddatz and Gabriele Nitsch (eds), *Netzwerkwelt 2006. Forschungsthemen, Schwerpunktbranchen, politisches Know-how*, Bielefeld: Kleine Verlag, pp. 48–58.

Sennett, Richard (2008), *The Craftsman*, New Haven, CT: Yale University Press.

Virilio, Paul (1993), *Revolutionen der Geschwindigkeit*, Berlin: Merve-Verlag.

Weber, Max (1951 [1909]), 'Energetische Kulturtheorien', in *Gesammelte Aufsätze zur Wissenschaftsgeschichte*, 2nd revised and expanded edition, Tübingen: J.C.B. Mohr.

Welzer, Harald (2008), *Klimakriege – Wofür im 21. Jahrhundert getötet wird*, Frankfurt am Main: S. Fischer Verlag.

26 California's energy-related greenhouse gas emissions reduction policies

David R. Heres and C.-Y. Cynthia Lin

26.1 INTRODUCTION

Global climate change is expected to greatly disrupt physical and biological systems during the second half of the twenty-first century (Parry et al., 2007). Anthropogenic greenhouse gas (GHG) emissions have long been identified as the drivers of such change and, unless societies prove capable of curbing the continual rise in the concentration of those gases in the atmosphere, the costs of adaptation and of the increased risk of catastrophic events occurring will be large.[1]

In the absence of federal policies to abate GHG emissions, states and regions in the United States of America (US) started to develop GHG emissions reduction targets. One of the first North American regions to establish a reduction goal was that composed by 11 provinces and states from eastern Canada and the US in 2001.[2] Regionally, the goal was established at reducing GHG emissions to a level 10 percent below 1990 emissions by 2020 (NGE-ECP, 2001). Although several states had since set reduction goals, it was not until 2006 that California became the first subnational US entity to establish a statewide enforceable target on total GHG emissions.[3]

Signed into law in September 2006, Assembly Bill 32 (2006), the California Global Warming Solutions Act of 2006,[4] required the California Air Resources Board (CARB)[5] to define strategies to achieve statewide GHG emissions at or below the 1990 levels by 2020 and 80 percent below 1990 levels by 2050. The bill, commonly referred to as AB32, was preceded a year before by a state Executive Order mandating those reduction targets and directing the California Environmental Protection Agency to coordinate the efforts of several agencies and secretaries. A Climate Action Team (CAT), composed of representatives from 17 state agencies, worked on the proposal of GHG emissions reduction strategies. The CAT's reports (CAT, 2006, 2007) describe such measures, several of which are reflected in the final Scoping Plan (SP) adopted by the CARB (CARB, 2008). Aside from the CARB and the 12 subgroups of the CAT, the general public and stakeholders actively participated in the development of the SP through public meetings, workshops and responding to solicitation for ideas.[6]

Concerns in California regarding climate change are however not that recent and have been reflected in law since 1988, when by Assembly Bill 4420 (1988) the California Energy Commission (CEC) was directed to study the impacts of climate change on the state as well as to develop the first GHG inventory and provide policy recommendations.[7] After the establishment of a voluntary registry scheme which started operations in 2002, one of the most important milestones in California climate policy came in 2002 when the passage of Assembly Bill 1493 (2002) triggered the opposition of automakers and the subsequent involvement of the US Environmental Protection Agency (EPA).

This bill required the CARB to develop and adopt regulations to reduce GHG emissions from passenger vehicles, light-duty trucks and other non-commercial vehicles sold in California. A year after, the states of California, Oregon and Washington created the West Coast Global Warming Initiative to promote collaborative work in programs addressing climate change. In the summer of 2010, these and other state members of the Western Climate Initiative released their joint emissions-reduction strategy centred on the implementation of a regional cap-and-trade system (CTS) for GHG emissions of which the first compliance period is planned for 2012–14.[8]

Although either a statewide or a regional CTS will cover sources responsible for about 86 percent of the emissions reduction target, the economic and technology advisory committees to the CARB have recommended the implementation of complementary measures to aid in the technological and behavioral transition towards a lower-carbon economy in the state (MAC, 2007; ETAAC, 2008). This chapter describes and discusses the complementary policies pertaining to the energy sector (that is, commercial and residential natural gas use, electricity and transportation) with the highest projected GHG emissions reductions. Together, the six strategies presented in section 26.3 of this chapter (vehicle GHG standards, low carbon fuel standards, regional transportation targets, energy efficiency, renewable electricity standard, and increasing combined heat and power generation) are expected to contribute to almost 60 percent of California's 2020 reduction target.

While the description of strategies in section 26.3 of this chapter is largely based on CARB's SP, the rapid process of design and implementation of strategies that the state is experiencing resulted in changes to some of the measures considered there. Therefore, projected reductions from strategies contained in the original plan were updated whenever possible, considering the most recent modifications and estimates from CARB and other agencies involved. In section 26.4 we provide discussion regarding current debates related to specific policies, as well as their potential unintended impacts.

Among the worldwide set of subnational policies with global implications, climate policy in California stands out not only for the size of its economy (twelfth in the world in 2008) but also for its contribution to global GHG emissions (seventeenth in the world in 2000).[9] In the past, this state, which accommodates about 12 percent of the US population, has successfully experimented with environmental policies that have ultimately been adopted in the US. Furthermore, climate change impacts, some of which have been already manifesting, will spread across all the regions in the state. Some of the major threats derived from even moderate increases in temperature, precipitation and sea level rise include a higher frequency of wildfires and extreme events, water supply shifts from earlier snowpack melting, damage to infrastructure and entire coastal communities, as well as a range of impacts on the state's agriculture, public health and biodiversity (CNRA, 2009).

The remainder of the chapter is organized as follows: section 26.2 characterizes California's GHG inventory, highlighting the contribution of energy-related sectors, Section 26.3 presents the most important complementary measures to reduce emissions from commercial and residential natural gas use, electricity and transportation. Section 26.4 discusses some of the implications of these policies and section 26.5 concludes.

26.2   THE ROLE OF ENERGY IN THE GENERATION OF GHG EMISSIONS IN CALIFORNIA

The energy used in business establishments and houses, transportation activities, and electricity generation is globally and for most countries the largest source of GHG emissions. As shown in Figure 26.1, these activities together contributed 70 percent of the 477.74 megatonnes of carbon dioxide equivalent ($MtCO_2e$) generated within the state of California.[10] Although there exists a high dependence on private means of transportation in the US that is not only particular to California,[11] the contribution from the transportation sector in the state is larger than that for the US (37 and 27 percent, respectively). This is because California's large share of hydropower and renewable resources generating electricity, together with a long tradition of energy efficiency measures, reduce the responsibility of the electricity sector in the state's GHG inventory compared to that for the nation (24 and 35 percent, respectively).

Under a business-as-usual (BAU) scenario, GHG emissions by 2020 in the state would be about 596 $MtCO_2e$. Interestingly, the two sectors with the largest shares, transportation and electricity, are also among those that would experience the largest increases in the absence of mitigation policies: 30 and 20 percent, respectively. High global warming potential (HGWP) gases, mostly used in refrigeration, are expected to increase by 300 percent under BAU. The emissions reductions of these gases are however not directly related to energy use since they result from leakage in, and disposal of, refrigeration and cooling systems.

As part of CARB's obligations under AB32, the board determined the 1990 GHG statewide emissions at 427 $MtCO_2e$, therefore requiring a reduction of 169 $MtCO_2e$ (or 28 percent) with respect to the BAU scenario by 2020. The following section describes energy-related policies that will together contribute about 60 percent of the reductions required to reach the 2020 target.

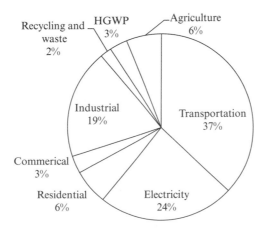

Source: Data from CARB (2010a).

Figure 26.1   California GHG emissions by source category, 2008

26.3   THE OPPORTUNITY FOR ENERGY-RELATED MEASURES TO REDUCE GHG EMISSIONS IN CALIFORNIA

A CTS covering most of the GHG sources is already expected to be in operation in California by 2012. Our focus in this chapter is not on the characteristics of a potential permit market but on the complementary measures that, beyond the price signals from the carbon market, will drive technological and behavioral changes. In particular, transportation emissions are not expected to be largely abated in the absence of sectoral policies. At the current carbon content of the fuel mix used in passenger vehicles, even a high price of US$50 (USD) per tonne of carbon dioxide equivalent (CO_2e) would only translate into gasoline price increases of 18 percent over the average price of gasoline between the summers of 2005 and 2010.[12] Holding everything else constant (for example carbon content of fuels, fuel economy of cars, driving habits and population) and considering a long-run gasoline price elasticity of –0.74, transportation emissions would be only reduced by 12 percent.[13] Without any other policies, a higher price of carbon would be required to achieve considerable GHG emissions reductions in the transportation sector in a wealthier and more populous 2020 California.

The introduction of mandatory standards for vehicle fuel efficiency and fuel carbon content together with the incorporation of GHG emissions projections during the regional planning process are expected to fill the gap left by the limited response of gasoline demand to small gasoline price variations. It should be noted, however, that gasoline taxes in some European countries can be up to eight times as large as those currently in place in California.[14] Although comparable levels of taxation would imply an increase in the retail price of gasoline in California that would translate into a more visible fall in the demand for gasoline demand, the reality is that such a level of taxation on gasoline is very unlikely to be imposed anywhere in the US in the near future.[15]

The most important measures in terms of GHG reductions involving energy use are briefly described below. The mitigation strategies, intended to promote both technological changes on the supply side and behavioural changes on the demand side, are here broadly classified into transportation, and natural gas and electricity. Table 26.1, which presents the mitigation potential of each measure (with their SP identifier in parentheses) as well as its expected cost, shows that, with the exception of the Renewable Portfolio Standard, each of these measures results in net savings. Both the CARB and the CEC had stressed that the aim of increasing the share of renewable sources for electricity generation is not only to reduce GHG emissions but also to diversify energy sources; these benefits are not accounted for in the SP net cost calculation (CARB, 2008).

26.3.1   Transportation

Vehicle greenhouse gas standards (T-1)
After a four-year process characterized by automakers' opposition and the expectations about the EPA's authorization to allow California to set its own vehicle GHG standards, a waiver was finally granted in April 2009 by the EPA.[16] Under Assembly Bill 1493 (2002), the CARB was required to adopt lower vehicle GHG standards for passenger vehicles and light trucks. The standards established for new vehicles from 2009 to 2016

Table 26.1 GHG emissions mitigation strategies from California's Scoping Plan 2008

SP identifier	Strategy	Mitigation potential as a percentage of 2020 Target[+]	Mitigation cost in 2020 (USD per tonne of CO_2e)[++]
T-1	Vehicle GHG Standards	18.8	−349
T-2	Low Carbon Fuel Standards	9.5	0
T-3	Regional Transportation Targets	2.0*	−311
E-1	Energy Efficiency – Electricity	9.0	−109
CR-1	Energy Efficiency – Natural Gas	2.5	−109
E-3	Renewable Electricity Standard	7.1	197**
F-2	Increase Combined Heat and Power Generation	4.0	−196

Notes: Data from CARB (2008).
[+] Shown as a percentage of the 169 $MtCO_2e$ reductions required.
[++] Mitigation costs were obtained from the net annualized costs and emissions reductions calculated by CARB (2008). Co-benefits and adverse impacts described in Appendixes H and J of the SP were not considered in CARB's calculations. Further details regarding the assumptions and formulae used for these estimations can be found in Appendix I of the report.
We updated mitigation potential figures to the most current estimate whenever possible (*CARB, 2010b; **CARB, 2010d).

implied an average fleet reduction of 36 percent in grams of CO_2e per mile for all new vehicles sold in California by 2016 compared to those built in 2009. These standards, also known as Pavley I, will be followed by those resulting from the final amendments to the Low-Emission Vehicles (LEV) regulations (known as Advanced Clean Cars, Pavley II, or LEV III).

The adopted regulation allows compliance flexibility by means of averaging model year fleet emissions within manufacturer, and banking and trading of credits which are in units of grams of CO_2e per mile. Borrowing from anticipated credit generation in future periods is not allowed. However, credits generated for model years 2000–2008 based on 2012 standards can be used at full value to offset shortfalls up to 2012. For future compliance periods, such credits would be only worth a fraction of their original value, ultimately expiring by 2015.

A number of states are also adopting California's standards, including neighbouring ones with the exception of Nevada. With the recently approved federal standards for model years 2012–16, equal to those from Pavley I, automakers delivering for sale in California (and in those states that adopted California standards) will in fact be subject to the federal regulation in that period but to the state regulation before and after.[17]

Overall, compared to a BAU scenario this measure is expected to remove 31.7 $MtCO_2e$ from the atmosphere by 2020, 4 $MtCO_2e$ of which are expected from Pavley II, representing about 19 percent of the total reduction target. According to the SP, these reductions will be achieved at a negative cost (that is, savings) of US$349 per tonne of CO_2e by 2020.

Low carbon fuel standard (T-2)
The California Low Carbon Fuel Standard (LCFS) was adopted in April 2009 and took effect in January 2010. The standard is imposed on all fuel providers requiring them to

attain a decreasing level of GHG emissions per unit of fuel energy sold in California.[18] The reduction requirement increases annually, starting with 0.25 percent in 2011, rising to 5 percent in 2017 and reaching 10 percent in 2020 for both gasoline and diesel fuel substitutes.[19] Since the standard is imposed on a per-unit of fuel energy basis, specifically grams of CO_2e per megajoule (MJ), instead of in terms of an overall GHG emissions target, the overall level of emissions could be higher.[20] Importantly, life-cycle GHG emissions from extraction (cultivation in the case of biofuels) to combustion are considered, including land use conversion, as well as emissions resulting from processing and distribution of all fuels. Given other possible environmental and social impacts resulting from the production of some of the alternative fuels such as electricity and biofuels, the CARB is working with interested stakeholders to include sustainability provisions into the regulation by December 2011.

As with the case of vehicle GHG standards, the regulation allows for trading and banking of credits in order to achieve reductions at the minimum cost and provide compliance flexibility. The calculation of credits in units of $MtCO_2e$ depends on the applicable standard for either gasoline or diesel as well as on the energy and carbon intensities of each alternative fuel. Adjustment factors are applied to energy generated from electricity and hydrogen because of the higher average mileage that each unit of energy from these sources delivers compared to gasoline and diesel.[21] Although credits generated within the LCFS program can be exported to other GHG emissions trading systems, buying credits generated in other programs is not allowed. This provision was made in order to attain the projected emissions reductions within the program. Borrowing from anticipated reductions in subsequent periods is not allowed; however, under certain conditions (for example no deficit reported in the previous period) fuel providers can carry over a deficit to the next compliance period without a penalty.

Gasoline and diesel consumption represented more than 99 percent of the gasoline gallon-equivalent sales for all fuels in the state in 2008. This percentage would be only slightly modified by 2020 even though the use of electricity, ethanol and natural gas to power vehicles experiences large increases under each of the scenarios considered in CEC (2010). The main obstacle for the widespread use of the different types of ethanol as primary fuel (gasoline is currently composed of 6 to 10 percent ethanol) will not be their production costs but the thousands of service stations for high-ethanol-content fuels that would need to be made available throughout the state. Despite these figures, this measure is expected to result in 16 $MtCO_2e$ reductions by 2020 compared to a BAU scenario. The reductions will be mainly brought about by cellulosic and advanced renewable ethanol substituting for gasoline, and advanced renewable biodiesel substituting for diesel (CARB, 2009). According to the analysis carried out along with the SP, the costs of this measure are negligible because the costs of producing alternative fuels are projected to be competitive to those of gasoline and diesel. In section 26.4 of this chapter we explore the arguments from a study that contends that the costs of achieving these reductions are in fact large.

Regional transportation targets (T-3)

California's Senate Bill 375 (2008) required the CARB to set regional passenger vehicle GHG reduction goals for each of the 18 Metropolitan Planning Organizations (MPO) in the state.[22] After recommendations from the Regional Targets Advisory Committee

Table 26.2   Population, CO_2e emissions per capita and regional reduction targets in California MPOs

Region	Population in 2005 (thousands)	Passenger vehicles' CO_2e per capita in 2005 (kilograms a year)	Per capita GHG emissions target by 2020 (% change)
Southern California Association of Governments	17763	3337	−8
Bay Area Metropolitan Transportation Commission	7095	3269	−7
San Diego Association of Governments	3034	4085	−7
Sacramento Area Council of Governments	2057	3547	−7
Eight MPOs in San Joaquin Valley[+]	3751	2585	−5
Six smallest MPOs[+]	1851	2431	7

Notes:
Data from 'Proposed SB 375 Greenhouse Gas Targets: documentation of the resulting emission reductions based on MPO data', available at www.arb.ca.gov/cc/sb375/mpo.co2.reduction.calc.pdf (accessed August 2010).
[+] The eight MPOs in San Joaquin Valley will be complying with the targets as one entity, however, the six smaller MPOs will report individually.

in September 2009, the CARB in conjunction with the MPOs agreed on establishing the targets in terms of percentage reductions on per capita GHG from passenger vehicles, instead of total GHG reductions from passenger vehicles in the region. Practical considerations were taken into account for the adoption of this metric; however it was determinant that this choice would also compensate for the different rates in population growth across the state's regions.

This strategy started out with a large reduction potential of 20 $MtCO_2e$ in the earlier drafts (CAT, 2006, 2007), however further research found that estimate to be very optimistic and it was consequently adjusted.[23] The SP of 2008 projected 5 $MtCO_2e$ reductions and US$1554 million in savings from this strategy by 2020. However, based on current regional targets proposed, the most recent estimate without considering the impact of LCFS and Pavley is of the order of 3.4 $MtCO_2e$ (CARB, 2010b).

In August of 2010, the CARB (2010b) made public the proposed targets for the MPOs in the state. These were set considering the MPOs' own proposals and were adopted in September of the same year. Table 26.2 summarizes baseline emission levels and 2020 targets for the MPOs. Regions generating 95 percent of the state's GHG from passenger vehicles – that is, the four largest MPOs and those in the San Joaquin Valley – would face targets of between 5 and 8 percent reductions. Some small MPOs are allowed to increase their emissions per capita, and although some of the six smallest MPOs have reduction or zero-change targets, in conjunction their emissions per capita will be larger by 2020.

From the individual plans included in CARB (2010c), the array of policies being considered by the MPOs mainly include increasing the number of high-occupancy lanes, improving the extension and service of transit, promoting new compact and mixed land use housing development in areas close to transit services, incentivizing telecommuting, extending bicycle networks, and intensifying campaigns promoting the use of alternative modes of transportation and reduced travel.

The potential reductions from this strategy are not expected to be large in the near term because a large portion of the land use patterns by 2020 will be determined by development already carried out. It is however expected that by 2035 the reduction from this strategy will be of the order of 15 MtCO$_2$e: about a fivefold increase from the 2020 projections.

26.3.2   Electricity and Natural Gas

Emissions of GHG generated in the process of production of electricity for its consumption in California account for 25 percent of the total, and are second to transportation in importance. California, however, is already the state with the lowest consumption of electricity per capita in the US.[24] While consumption per capita has increased by almost 50 percent in the US in the last 30 years, California's has practically remained steady in the same period, resulting in a per capita consumption of almost half of that in the US (CEC, 2007). Commercial and residential emissions (almost entirely the result of natural gas consumption for space and water heating and cooking) are responsible for 9 percent of the total in the state.[25]

Although GHG emissions resulting from electricity generation and natural gas consumption will be covered under a CTS system as of 2012, the state has designed a set of policies that will aim directly at fostering behavioral and technological changes.[26] The main vehicle to achieve the latter will be through energy efficiency measures that require more stringent standards for appliances, and by mandating a larger share of renewable sources for electricity suppliers. Also important will be the support provided to increase combined heat and power installations. As for policies impacting the behavior of energy consumers beyond the price signals derived from the carbon market, education and real-time information about energy consumption will be central to promote conservation.

Energy efficiency: electricity and natural gas (E-1 and CR-1)

The California Long Term Energy Efficiency Strategic Plan (CPUC, 2008) provides the basis for the energy-efficiency reduction measures. The set of strategies includes building and appliance standards, as well as utility efficiency programs and provision of information technologies to help in optimizing energy use and conserving energy.[27]

The Energy Action Plan of 2003 (CEC-CPUC, 2003) has already established energy efficiency together with energy demand reductions as the primary strategy aimed to meet California's energy requirements for 2020. The California Public Utilities Commission (CPUC) and the CEC have estimated reductions of 15.2 and 4.3 MtCO$_2$e from electricity and natural gas respectively. The CPUC and the CEC are the agencies in charge of energy policies and regulation in the state. The former regulates private utilities and providers, while the latter is authorized to regulate publicly owned utilities, as well as to adopt and update buildings and appliance standards.

The GHG emissions reductions from this strategy are based on CEC-CPUC targets of 32 000 gigawatt hours (E-1) and 800 million therms (CR-1). The reductions will be achieved through a series of codes and standards for buildings and appliances leading to

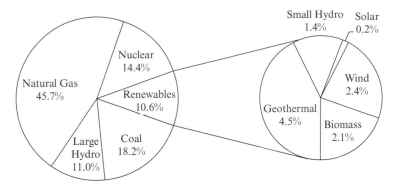

Source: Data from CEC's Energy Almanac at energyalmanac.ca.gov.

Figure 26.2 Sources of electricity generated for retail in California in 2008

'zero net energy' new residential buildings by 2020.[28] Commercial buildings should have improved their efficiency by 2020, however their goals are to be met by 2030.

To be catalogued as a zero net energy building, it must generate enough energy on-site to offset completely the energy consumed within the building in a year. This transition will be supported by intermediate standards targets, adoption of zero energy heating and cooling technologies (for example geothermal heat pumps, solar thermal water heating), integrated and passive solar designs and, importantly, by enabling the supply of energy surplus into the grid (CARB, 2008).

Although future development will have to comply with a set of standards, efforts will be strongly directed towards the energy-efficiency improvement of existing homes and establishments which comprise most of the buildings stock. CPUC's (2008) plan established a 40 percent reduction goal in energy (electricity and natural gas) consumption in existing homes by 2020. This would be achieved through a set of measures including mechanisms aimed at encouraging retrofits and providing education that will promote conservation and efficient use of energy.[29] Mandatory improvements might be imposed at the time of sale of an existing building (residential or commercial), while financing mechanisms to help cover the upfront costs of on-site renewable systems are currently being explored.

The CEC will be continuously adopting and updating standards for new types of appliances, importantly those that require plug-loads which represent a growing percentage of residential energy use. Water use efficiency, and improved compliance and enforcement are among the measures supporting the overall goal which will be achieved at net savings of over US$100 per tonne of CO_2e according to the calculations in the SP.

Renewable energy (E-3)

California's statewide GHG emissions per kilowatt hour consumed are among the lowest in the nation and more than 30 percent below the national ratio due to the high share of renewable sources and natural gas in the generation of electricity. From Figure 26.2, 10.6 percent of the electricity consumed in the state comes from renewable sources (more than one-fifth when large hydroelectric facilities are included), and the largest source of electricity is natural gas (46 percent). Even though per capita consumption of electricity is only 60 percent of that in the US, the generation capacity within the state is exceeded by the demand, and therefore California imports nearly 30 percent of its electricity needs. Because 93 percent of these imports are produced by facilities burning coal which has a carbon intensity almost twice that of natural gas, they generate more than 50 percent of the emissions assigned to this sector.

In 2006, Senate Bill 107 moved the deadline from 2017 to 2010 for private utilities to reach a 20 percent proportion of their electricity provided from renewable sources. Known as the Renewable Portfolio Standard (RPS), this legislation defines as renewable sources those electricity generation facilities that: 'use biomass, solar thermal, photovoltaic, wind, geothermal, fuel cells using renewable fuels, small hydroelectric generation of 30 megawatts or less, digester gas, municipal solid waste conversion, landfill gas, ocean wave, ocean thermal, or tidal current' (Senate Bill 107 2006). Electricity generated by large hydroelectric facilities is not considered as being produced by a renewable source due to other environmental impacts that commonly accompany them.[30] Small hydroelectric facilities and those converting biomass into electricity are required to demonstrate that other environmental impacts are negligible.

Following a recommendation included in the Energy Action Plan II (CEC-CPUC, 2005), Executive Order S-21-09 from September 2009 directed the CARB to adopt regulation that will increase the share of electricity generated from renewable sources to 33 percent by 2020. This is the most ambitious goal among the US states that have set an RPS.[31] In spite of the wide range of eligible renewable alternatives considered, wind, geothermal and solar are projected to provide 85 percent of the renewable electricity by 2020 (CARB, 2010d). This regulation, known as the Renewable Electricity Standard (RES) builds upon the Renewable Portfolio Standard and will apply to both investor-owned and publicly owned utilities.[32] In 2008, the latter type provided about 24 percent of the electricity consumed in the state, and the five (three investor-owned and two publicly owned) largest utilities provided more than 80 percent of the total electricity generated for consumption in California.

According to CARB (2010d) the RES target will translate into CO_2e reductions of 12 Mt by 2020, at a cost of US$200 per tonne.[33] The total costs of the RES are translated into monthly utility bill increases in 2020 of 3–6 percent for both residential users and small businesses. According to the economic analysis of the proposal, the cost estimate is conservative because it assumes that the cost of renewable sources remains constant. The RES will be gradually implemented in the following manner: 20 percent by 2012–14, 24 percent by 2015–17, 28 percent by 2018–19, and 33 percent by 2020 and after. Bankable and tradable renewable energy credits will provide compliance flexibility to regulated parties.

Considering the 7.9 $MtCO_2e$ reductions that will result from the current 20 percent renewable standard, strategies enforcing renewable standards will together contribute 12 percent of the 2020 GHG emissions reduction target.

Increase combined heat and power generation (E-2)
In a CEC-commissioned study, the Electric Power Research Institute (EPRI) estimated that under a 'moderate market access' scenario for 2020, the installed capacity of combined heat and power (CHP) applications could increase by 4400 megawatts (MW).[34]

CARB's slightly smaller targeted recommendation in its SP of 2008 is based on this estimate. According to the CARB, 4000 MW of additional capacity would displace 32000 gigawatt hours (GWh) from the grid.[35] The key change under EPRI's moderate scenario compared to current incentives is to allow energy surplus from CHP applications to be sold to the grid. However, in order to achieve the targeted increase in CHP capacity, current incentives such as payments for self-generation, access to lower gas rates and surcharge exemptions for systems meeting efficiency and environmental standards are assumed to be maintained.[36] In fact, under a scenario in which energy surplus cannot be exported but current policies are maintained, there would still be an increase in CHP capacity of about 2000 MW by 2020.

Generating electricity for its consumption, on-site CHP systems are catalogued as distributed generation systems, as opposed to centralized power generation from utilities. Along with electricity, CHP systems simultaneously produce thermal energy. The latter can be used to heat space and water, as well as to generate more electricity or to run a cooling device, therefore avoiding the consumption of fuel otherwise needed to meet those needs. Non-combined systems of power generation simply dispose of the unused energy in the form of waste heat into the atmosphere. CHP systems are best suited for applications in which the demand for electricity and heating are continuous, such as hospitals, colleges, prisons, hotels and large stores.

According to EPRI's study, 776 sites with an accumulated capacity of 9130 MW were in operation in 2005, 90 percent of which had capacities above 20 MW. It is important to note that most of the current and projected incentives are directed towards installations smaller than 20 MW in which the expansion potential is larger.

Under the Waste Heat and Carbon Emissions Reduction Act (Assembly Bill 1613 2007), the CPUC is authorized to require investor-owned utilities to purchase excess electricity from customers with CHP systems.[37] The Act also made provisions for the adoption of feed-in tariffs and to establish a pilot program in which utilities would finance upfront costs of customers installing CHP systems.[38]

Publicly owned utilities are also required by this Act to establish a program that allows their customers to install CHP systems and creates a market for the excess electricity generated. Such electricity would be purchased at a rate determined by the governing boards of the publicly owned utility instead of the rate determined by the CPUC for investor-owned utilities.

In December 2009, the CPUC ordered electric corporations to adopt contracts for electricity purchases from small (up to 5 MW) and medium (up to 20 MW) eligible customers with CHP systems. The decision also established that the costs of a combined cycle gas turbine will serve as the basis of the rate at which the electricity will be paid.[39]

Although CHP applications can be fueled by renewable sources, 84 percent of installed CHP systems in 2005 depended on natural gas to produce energy. The CARB assumes that natural gas is used to run all of the added CHP capacity in its GHG emissions reductions calculations. The combustion of natural gas to produce electricity generates on-site emissions of pollutants other than carbon dioxide. Emissions of the latter are considered in the calculations of the net GHG change with and without this strategy, while installations must comply with existing regulations for other pollutants and environmental standards. The added installation target would deliver reductions of 6.7 MtCO$_2$e at almost US\$200 savings per tonne.

26.4 DISCUSSION

Even though the six mitigation strategies overviewed in this chapter will reduce California's GHG emissions, not all of them can guarantee that those emissions will not be generated elsewhere as a response to the state's policies. For instance, leakage of GHG emissions can occur under the Low Carbon Fuel Standards, the Renewable Portfolio Standards and the Vehicle GHG Standards programs. In particular, under the LCFS, fuel exporters outside California might divert their high-carbon-content fuels to other regions and the low-carbon ones to California, with a negligible change in global emissions from this regulation. Automakers and electricity producers exporting to California might react in comparable ways. Responses from different industries can take another dimension by relocating in states with laxer regulations. This type of response would not only leave global GHG emissions unchanged, but could inflict further negative impacts on the state's economy. Leakage is minimized and ultimately resolved when more states and countries adopt California's or comparable regulations.

Another concern that is relevant to all of the measures described in this chapter is how environmental and economic impacts will be spatially and socially distributed among Californians. Importantly, most of these programs already have provisions addressing environmental justice,[40] while others are working on its incorporation.[41] These issues become particularly important in the case of siting renewable energy generation facilities such as landfill gas, and wind and solar farms which can involve considerable environmental and aesthetic impacts. As mentioned earlier, large hydroelectric facilities are not considered under the CEC's eligible renewable sources due to the wide range of environmental damage they can entail.

While concerns about leakage of GHG emissions and environmental justice permeate through most of regional mitigation policies, other unintended impacts are specifically related to some of the measures. The higher fuel economy under the vehicle GHG standards program could derive into a more intensive use of cars due to reduced total costs per mile. The magnitudes of this rebound effect calculated by Van Dender and Small (2005) were however taken into account in the estimation of the GHG emissions reductions. On the other hand, this policy has been criticized for its lack of stringency in light of required GHG emissions reductions to stabilize the climate (Johnson, 2007). However, this critique and that regarding the low cost-effectiveness and inefficiency of the LCFS in Holland et al. (2009), are less sustained when these policies are viewed as part of a policy package that includes a CTS covering transportation emissions, and when their individual goals other than GHG reductions are taken into account.

A contrasting finding in Holland et al. (2009) is that the LCFS program will result in costs ranging from 263 to 903 per tonne of CO_2e compared to the negligible cost that the CARB placed on this strategy. It is important to stress that both the LCFS and the RES programs, which could deliver reductions at a cost, have objectives beyond GHG emissions reductions such as diversifying energy sources for security reasons. An investigation on the benefits obtained as a result of meeting these other objectives would be necessary in a complete assessment of the economic efficiency of the policies.

In fact, assuming that achieving the GHG reduction target is the only objective, complementary policies are redundant in the presence of a CTS that covers most of the emissions. For instance, since the global warming impact of 1 tonne of CO_2e is the same regardless of its source, policies aimed at reducing GHG emissions within the transportation sector will not be economically efficient unless other ancillary benefits or policy objectives are relevant, or the overall GHG emissions cap is set at a level higher than that yielding the social optimum.

Regardless of their economic efficiency and cost-effectiveness it is undeniable that the complementary policies will facilitate the transition towards a lower carbon economy in California. Based on their own and other studies' elasticity estimates, Heres and Niemeier (2011) argue that separate increases in the range of 20–25 percent in the price of gasoline or in residential density could deliver the same GHG emissions reductions in the state. Modifying the type of housing development might seem technically more difficult than imposing a tax on gasoline; however the latter option is commonly perceived to involve a large political burden. Bundling the two policies will result in larger reductions; however efficiency gains would depend on the stringency of each policy alone and the full set of objectives. The potential synergies that could develop between an RPS and energy efficiency measures also justify the coordination of policies. In Mahone et al. (2009) the level of investments on energy efficiency that are cost-effective increases in the presence of a higher RPS because the latter increases electricity prices.

Some strategies that would deliver emissions reductions to individuals and businesses at apparently negative costs have not been widely adopted due to other factors beyond cost and savings calculations at private discount rates. As Jaffe et al. (2004) point out, although the diffusion of new technologies is always gradual, the rate of adoption might still be suboptimal. For instance, a potential energy efficiency gap – that is, a slower than optimal substitution of energy-inefficient appliances and systems for high-energy-efficiency ones – can also be hindered by market and non-market failures. Examples of the former are the public-good attributes of information, and principal–agent situations that can arise when the owner of a building chooses the investments in energy efficiency (Jaffe et al., 2004). An important non-market failure in this context would be the uncertainties regarding future benefits from an investment in energy efficiency occurring today. The presence of some of these barriers to adoption calls for policy interventions, such as the implementation of financing mechanisms and measures to provide better access to information, aimed at remedying the efficiency gap. A study prepared for the CEC (EPRI, 2005) found some of these barriers to be highly relevant in the context of the expansion of CHP systems, especially short payback periods demanded by users, which could be the result of uncertainties regarding future energy prices and systems costs.

A crucial hurdle to be cleaned in order to increase the amount of electricity consumed on-site from renewable sources and CHP systems is to allow their electricity surpluses to be exported. In several countries this has been accompanied not only by electricity purchases quotas that utilities have to meet from specific sources (that is, RPS) but importantly with a feed-in-tariff (FIT) by which a price per kilowatt hour of electricity is guaranteed to the seller, therefore resolving part of the uncertainties. The larger costs of electricity generation from most renewable sources have been covered in the past through direct subsidies to the utilities purchasing the electricity, or more commonly by authorizing increases in the price per kilowatt hour delivered to end users. The purpose of the FIT in California, less ambitious than some of the European FITs which were designed as substitutes for enforced RPSs, is to facilitate sales of surplus electricity generated by small renewable energy projects (less than 3 MW). Because the California's FIT is based on the cost per kilowatt hour of a combined cycle natural gas turbine power plant and an overall cap on utilities' required purchases has been set, this particular program does not transfer public funds to any of the parties and should not imply large changes to the retail price of electricity. This program alone might fall short of providing incentives to spread the exploitation of renewable sources to produce electricity that some European countries have experienced. However, California's hybrid approach combining an RPS and an FIT will ensure that the target of 33 percent electricity from renewable sources is met.

26.5 CONCLUDING REMARKS

By passing AB32 in 2006, California became the first subnational US entity to establish a state-wide enforceable target on total GHG emissions. AB32 required the CARB to define strategies to achieve statewide GHG emissions at or below the 1990 levels by 2020, and 80 percent below 1990 levels by 2050 (Assembly Bill 32 2006). Among the worldwide set of subnational policies with global implications, climate policy in California stands out for the size of its economy (twelfth in the world in 2008) and its contribution to global GHG emissions (seventeenth in the world in 2000).

Although either a state-wide or a regional CTS will cover sources responsible for about 86 percent of the AB32 emissions reduction target, the economic and technology advisory committees to the CARB have recommended the implementation of complementary measures to aid in the technological and behavioral transition towards a lower carbon economy in the state (MAC, 2007; ETAAC, 2008). The six energy-related complementary policies presented in this chapter (vehicle GHG standards, low carbon fuel standards, regional transportation targets, energy efficiency, renewable electricity standard, and increasing combined heat and power generation) are expected to contribute to almost 60 percent of California's 2020 reduction target. With the exception of the Renewable Electricity Standard, each of these measures results in net savings.

There are some concerns related to the six mitigation strategies reviewed in this chapter, however. First, until other states and countries adopt similar policies, leakage of GHG emissions can occur. Second, there are potential distributional impacts and environmental justice concerns. Third, the vehicle GHG standards program may lead to a rebound effect. Fourth, the LCFS may not be cost-effective or efficient. Fifth, complementary policies are redundant in the presence of a CTS that covers most of the emissions. It should be noted that these concerns have been, or are expected to be, addressed by the authorities in charge of developing the final regulations. Despite their potential drawbacks, the six mitigation strategies reviewed in this chapter will reduce California's GHG emissions and will facilitate the transition towards a lower carbon economy in California.

NOTES

1. Although their present value calculation is subject to controversy, future economic costs of inaction could be undeniably large. The core of the debate among economists is not about whether or not we should impose restrictions on the emissions of GHG but rather about the timing and magnitudes of such restraints. Heal (2009) provides a review of the main positions in the field.
2. The Canadian provinces of Newfoundland and Labrador, Nova Scotia, Prince Edward Island, New

Brunswick and Quebec, and the states of Connecticut, Maine, Massachusetts, New Hampshire, Rhode Island and Vermont.

3. The Pew Center on Global Climate Change maintains an updated description of regional efforts and rulemaking in the US (www.pewclimate.org/states-regions, accessed August 2010).

4. Hanemann (2008) provides a compelling recount of the events leading to the passage of this legislation in California. He explores the political and legal circumstances reigning during the few years prior to the law enactment but the narration is also enriched by tracing back the seeds to the ahead-of-federal regulations on air pollution in the middle of the twentieth century and the creation of a unique state Energy Commission in 1974. The interested reader will find further interesting details about the different segments and characters along the road to the passage of AB32 in Hanemann (2007).

5. Part of the California Environmental Protection Agency, CARB's mission is: 'to promote and protect public health, welfare and ecological resources through the effective and efficient reduction of air pollutants while recognizing and considering the effects on the economy of the state' (www.arb.ca.gov/html/mission.htm accessed August 2010). Under AB32, aside from developing the scoping plan with reduction strategies to become operative by January 2012, the board is also responsible for setting the 1990 statewide GHG emissions reference level, adopting regulation requiring the mandatory reporting of GHG emissions sources, ensuring that early actions receive appropriate credit, convening an environmental justice advisory committee, and appointing an economic and technology advancement advisory committee (CARB, 2008).

6. CAT's subgroups are the economic sector-specific subgroups for agriculture, cement, energy, forest, green buildings, land use, recycling and waste management, state fleet, and water energy; the other three are multisector subgroups for economics, research and state operations.

7. The first California GHG inventory was published in 1990 and reported only carbon dioxide emissions from 1988. It was not until 1997 that emissions of methane and nitrous dioxide were included, and in 2002 emissions of gases with a high global warming potential were also incorporated. The CEC was in charge of developing the inventory until 2007 when this responsibility was transferred to the CARB.

8. Other partners are the states of Arizona, Montana, New Mexico and Utah, and the Canadian provinces of British Columbia, Manitoba, Ontario and Quebec. Five other US states, three Canadian provinces and six Mexican bordering states participate as observers.

9. Gross domestic product by state and country obtained from the US Department of Commerce Bureau of Economic Analysis at www.bea.gov/regional and from the World Bank database at data.worldbank.org. GHG emissions for California are from CARB (2010a). The earliest year for data on all GHG emissions by country from the World Resources Institute (cait.wri.org) is 2000. Websites were accessed in August 2010.

10. Tonnes (t) and metric tons are unit measures representing 1000 kilograms. Therefore a megatonne (Mt) is equal to both 1 million tonnes and to 1 million metric tons (MMT). The latter is the terminology used in the US inventories, while tonnes is the standard elsewhere, particularly in the reports published by the Intergovernmental Panel on Climate Change (IPCC). We follow IPCC's terminology throughout this chapter.

11. According to data from the Federal Highway Administration of the US Department of Transportation, vehicle miles travelled almost tripled between 1970 and 2007 in both California and the rest of the US ('Highway statistics summary to 1975' and 'Selected highway statistics and charts 2007' available at www.fhwa.dot.gov, accessed August 2010). However, since population growth in California has been larger compared to that in the other 50 states together during the same period ('Population estimates' from the US Census Bureau at www.census.gov, accessed August 2010), the increase in vehicle miles travelled per capita has been more rapid in the rest of the country than in California (87 and 55 percent, respectively). The high reliance on private means of transportation is also reflected in the small share of all person-trips that were made by public transportation – about 4 percent – in both the nation and California (calculated from preliminary data from the '2009 National Household Travel Survey' available at nhts.ornl.gov, accessed August 2010).

12. This example borrows from Sperling and Yeh (2009). Considering that 1 tonne is equal to 2204.6 pounds and that each gallon of gasoline produces 19.4 pounds of CO_2e (from current EPA emission factors), a CO_2e price of US$50 per tonne translates into 44 cents per gallon. The average of weekly prices for regular gasoline from August 2005 to August 2010 in California is US$2.40 per gallon (based on data from the CEC's 'Energy almanac' at energyalmanac.ca.gov).

13. This is the average of the mean values for long-run elasticities from the literature review in Lin and Prince (2009). It should be noted that this is probably a conservative assumption since these elasticities were estimated with long time series and there is evidence of a downward shift in the magnitudes of at least short-run elasticities (Hughes et al., 2008).

14. From data for August 2010 from the International Energy Agency, the gasoline taxes are between 113 and 179 per cent of the pre-tax price for France, Germany, Italy, Spain and the United Kingdom (www.

iea.org). This percentage is only 17 per cent in the US and 21 per cent in California (the latter is calculated from August 2010 prices and taxation levels in energyalmanac.ca.gov). Importantly, since the pre-tax prices are very similar among these regions, the retail prices in the above European countries are about twice as high as those in the US, California inclusive.

15. See Parry and Small (2005) for some conjectures regarding the political factors behind a presumably lower than optimal gasoline tax in the US (higher than optimal in the United Kingdom).

16. California had set standards for motor vehicles prior to the passage of the Federal Clean Air Act of 1970. On this basis, a provision was included in the Act allowing California to set its own stricter than federal emission standards for motor vehicles as long as it meets a set of conditions followed by a waiver granted by the federal government. Once granted, other states are free to abide by Californian or the federal standards. AB32 also considered the development of a 'feebate' program that would achieve the same reductions as this measure in the event of a final rejection of the waiver request. A feebate program would have provided rebates for high fuel-efficiency vehicles and imposed fees on low fuel-efficiency vehicles.

17. Light-Duty Vehicle Greenhouse Gas Emission Standards and Corporate Average Fuel Economy Standards, 75 Fed. Reg. 25324 (2010) (to be codified at 49 CFR Parts 531, 533, 536, 537 and 538).

18. The point of regulation is where finished gasoline is first manufactured or imported. This could be refineries, blenders or importers, but sometimes also wholesalers further downstream. In the latter case the wholesaler would have to report the GHG associated with the ethanol used for blending.

19. Final Regulation Order, Low Carbon Fuel Standard (2010), Title 17, California Code of Regulations, sections 95480–95490.

20. For instance, a provider could be selling fuels in 2020 representing a total carbon content larger than that in 2010 but that is lower in a per-unit of fuel energy basis.

21. For example, the carbon intensity of electricity considering an electricity mix of natural gas and renewable energy sources is larger than that for gasoline (104.71 and 95.86 grams of CO_2e per MJ, respectively). However, once adjusting and according to the average carbon intensity targets, carbon intensity of fuels and the formulae for credit generation in the Final Regulation Order, only 62 MJ from electricity would be necessary to offset 1000 MJ from gasoline by 2020.

22. MPOs are composed of representatives from local government and federal and state transportation authorities. They receive funding from the federal government and are in charge of the design of long-term planning policies.

23. Actions expected from this strategy appeared originally in CAT (2006) under the strategies 'Measures to Improve Transportation Energy Efficiency' and 'Smart Land Use and Intelligent Transportation', with combined reductions of 27 $MtCO_2e$ by 2020. CAT (2007) modified the title of the former to 'Transportation Efficiency' and adjusted the estimate of the latter, resulting in combined reductions of 19 $MtCO_2e$ by 2020.

24. In 2008 California consumed 7.1 thousand kilowatt hours per capita. This number was 12.3 for the US and 31.3 in Wyoming, the state with the largest consumption of electricity per capita in 2008. Electricity sales in kilowatt hours by state are available at the Energy Information Agency (www.eia.gov, accessed August 2010). Population estimates by state for 2008 are available at US Census (www.census.gov, accessed August 2010).

25. Emissions from electricity consumed in homes and businesses are included in 'electricity' emissions. When assigned to commercial and residential, the combined share of these sectors rises to 22 percent. Note that California only produces a small fraction of the natural gas consumed within the state (13 percent), however all of the emissions resulting from consumption are considered as produced within the state.

26. Natural gas GHG emissions will not be incorporated in the first compliance period of the CTS (2012–14) but are expected to be covered in subsequent phases.

27. The groups of strategies can be consulted in detail in CPUC (2008). The Scoping Plan (CARB, 2008) distinguishes 12 strategies that will maximize energy efficiency.

28. By law such standards must be cost-effective. That is, the efficiency savings must be larger than the installation, maintenance and operation costs.

29. Advanced metering infrastructure is currently being deployed in California by some of the largest electricity utilities as part of the energy conservation measures promoted in the state. Also known as smart metering, the purpose of this system is to provide real-time information about electricity consumption and prices through displays installed in homes and businesses.

30. Among these impacts are changes in stream flows and reservoir surface area, groundwater recharge, water temperature, turbidity and oxygen content. Biological impacts, damage to historic sites, changes in visual quality, loss of scenic resources and increased erosion are also important ('Hydroelectric power in California', www.energy.ca.gov, accessed August 2010).

31. Summary map of 'RPS policies from the Database of State Incentives for Renewables and Energy Efficiency' at www.dsireusa.org.

32. Investor-owned and publicly owned utilities provide together almost 95 percent of the electricity con-

sumed in the state. Other type of electricity providers such as joint utility agencies, rural electric cooperatives and self-generators will also be subject to the standard. Any type of provider whose retail sales fall under 200 000 megawatt hours will be exempted from the requirement. The latter group delivers less than 1 percent of total retail sales in the state.

33. This estimate does not include the benefits from energy sources diversification. Environmental impacts from construction and operation of transmission lines and localized air impacts are also not included. The latter will nevertheless be subject to existing legislation.

34. 'Assessment of California CHP market and policy options for increased penetration', EPRI, Palo Alto, CA, California Energy Commission, Sacramento, CA, April 2005.

35. CARB's calculations assume a utilization factor of 85 per cent; that is, the electricity generator is operating $0.85 \times 365 \times 24 = 7446$ hours a year. This number of hours multiplied by the generation capacity of 4000 MW results in 29 784 000 MW hours or about 30 000 GWh. Further assuming a 7 percent loss along the transmission lines from centralized power generation, these additional on-site CHP applications would displace 32 000 GWh from other sources.

36. 'Departing load cost responsibility surcharges' apply to customers of electric utilities that discontinue or reduce their purchases because part of their electricity needs are generated on-site. The purpose of these surcharges is to retain contributions towards the funding of social and energy efficiency programs and previous investments without shifting costs to other customers. Small CHP applications meeting energy efficiency and environmental standards can currently apply to be exempted from these surcharges.

37. An eligible customer must comply with certain criteria such as time of installation and capacity of the CHP system. The latter is also required to meet interconnection specifications and to comply with efficiency and GHG standards.

38. The eligibility criteria under the so-called 'pay-as-you-save' pilot program was amended by Assembly Bill 2791 (2008). It was previously directed solely to non-profit organizations but it now includes government facilities.

39. In January 2010 the three largest investor-owned utilities filed a joint motion to stay (denied in June 2010) the 'Decision adopting policies and procedures for purchase of excess electricity under Assembly Bill 1613' (D.09-12-042). The decision and rest of rule-making catalogued under 08-06-024 are available at http://docs.cpuc.ca.gov/Published/proceedings/R0806024.htm. This document provides an interesting example of the obstacles encountered in the implementation process of a state-wide policy that would differently affect the parties involved.

40. Environmental justice refers to: 'the fair treatment and meaningful involvement of all people regardless of race, color, national origin, or income with respect to the development, implementation, and enforcement of environmental laws, regulations, and policies' (US Environmental Protection Agency at www.epa.gov/environmentaljustice/, accessed August 2010).

41. Appendix J in CARB (2008) identifies potential adverse environmental impacts and provides general recommendations that would support the environmental justice requirements of the strategies in the SP. The regulatory design for strategies such as LCFS and RES was especially recommended to address a large number of potential impacts.

REFERENCES

Assembly Bill 32 (2006), Chapter 488, California Statutes of 2006 (codified at Cal. Health and Safety Code §§ 38500-38599).

Assembly Bill 1493 (2002), Chapter 200, California Statutes of 2002 (codified at Cal. Health and Safety Code §§ 42823, 43018.5).

Assembly Bill 1613 (2007), Chapter 713, California Statutes of 2007 (codified at Cal. Pub. Util. Code §§ 2840-2845).

Assembly Bill 2791 (2008), Chapter 253, California Statutes of 2008 (codified at Cal. Pub. Util. Code § 2842.4).

Assembly Bill 4420 (1988), Chapter 1506, California Statutes of 1988.

CARB (2008), 'Climate change scoping plan: a framework for change', California Air Resources Board.

CARB (2009), 'Proposed regulation to implement the Low Carbon Fuel Standard. Volume I. Staff Report: initial statement of reasons', California Air Resources Board.

CARB (2010a), 'California greenhouse gas inventory for 2000-2008', California Air Resources Board.

CARB (2010b), 'Proposed regional greenhouse gas emission reduction targets for automobiles and light trucks pursuant to Senate Bill 375', California Air Resources Board.

CARB (2010c), 'Draft regional greenhouse gas emission reduction targets for automobiles and light trucks pursuant to Senate Bill 375', California Air Resources Board.

CARB (2010d), 'Proposed regulation for a California renewable electricity standard', California Air Resources Board.

CAT (2006), 'Climate Action Team report to Governor Schwarzenegger and the legislature', Climate Action Team.

CAT (2007), 'Climate Action Team proposed early actions to mitigate climate change in California', Climate Action Team.

CEC (2007), '2007 Integrated Energy Policy report: final report', California Energy Commission.

CEC (2010), 'Transportation energy forecasts and analyses for the 2009 Integrated Energy Policy report: final report', California Energy Commission.

CEC-CPUC (2003), 'Energy Action Plan I', California Energy Commission and California Public Utilities Commission.

CEC-CPUC (2005), 'Energy Action Plan II', California Energy Commission and California Public Utilities Commission.

CNRA (2009), '2009 California climate adaptation strategy', California Natural Resources Agency.

CPUC (2008), 'California long-term energy efficiency strategic plan: achieving maximum energy savings in California for 2009 and beyond', California Public Utilities Commission.

EPRI (2005), 'Assessment of California CHP market and policy options for increased penetration', Consultant Report Prepared for the California Energy Commission, Electric Power Research Institute.

ETAAC (2008), 'Recommendations of the Economic and Technology Advancement Advisory Committee. Final report. A report to the California Air Resources Board', Economic and Technology Advancement Advisory Committee.

Hanemann, M. (2007), 'How California came to pass AB 32, the Global Warming Solutions Act of 2006', Working Paper, Department of Agricultural and Resource Economics, University of California, Berkeley, CA.

Hanemann, M. (2008), 'California's new greenhouse gas laws', *Review of Environmental Economics and Policy*, **2** (1), 114–29.

Heal, G. (2009), 'Climate economics: a meta-review and some suggestions for future research', *Review of Environmental Economics and Policy*, **3** (1), 4–21.

Heres, D.R. and D. Niemeier (2011), 'CO$_2$ emissions: are land-use changes enough for California to reduce VMT? Specification of a two-part model with instrumental variables', *Transportation Research Part B*, **45**, 150–61.

Hughes, J., C. Knittel and D. Sperling (2008), 'Evidence of a shift in the short-run price elasticity of gasoline demand', *Energy Journal*, **29** (1), 93–114.

Holland, P., J. Hughes and C. Knittel (2009), 'Greenhouse gas reductions under low carbon fuel standards?', *American Economic Journal: Economic Policy*, **1**, 106–46.

Jaffe, A., R. Newell and R. Stavins (2004), 'Economics of energy efficiency', in C. Cleveland (ed.), *Encyclopedia of Energy Vol. 2*, Oxford: Elsevier, pp. 79–90.

Johnson, K.C. (2007), 'California's greenhouse gas law, Assembly Bill 1493: deficiencies, alternatives, and implications for regulatory climate policy', *Energy Policy*, **35**, 362–72.

Lin, C.-Y.C. and L. Prince (2009), 'The optimal gas tax for California', *Energy Policy*, **37**, 5173–83.

MAC (2007), 'Recommendations for designing a greenhouse gas cap-and-trade system for California: recommendations of the Market Advisory Committee to the California Air Resources Board', Market Advisory Committee.

Mahone, A., C.K. Woo, J. Williams and I. Horowitz (2009), 'Renewable portfolio standards and cost-effective energy-efficiency investment', *Energy Policy*, **37**, 744–77.

NEG-ECP (2001), 'Climate Change Action Plan 2001', New England Governors – Eastern Canadian Premiers.

Parry, I.W.H. and K. Small (2005), 'Does Britain or the United States have the right gasoline tax?', *American Economic Review*, **95** (4), 1276–89.

Parry, M.L., O.F. Canziani, J.P. Palutikof, P.J. van der Linden and C.E. Hanson, (eds) (2007), *Climate Change 2007: Impacts, Adaptation and Vulnerability. Contribution of Working Group II to the Fourth Assessment Report of the Intergovernmental Panel on Climate Change*, Cambridge: Cambridge University Press.

Senate Bill 107 (2006), Chapter 464, California Statutes of 2006 (codified in scattered sections of Cal. Pub. Res. Code and Cal. Pub. Util. Code).

Senate Bill 375 (2008), Chapter 728, California Statutes of 2008 (codified in scattered sections of Cal. Gov. Code and at Cal. Pub. Res. Code §§ 21155, 21159.28).

Sperling, D. and S. Yeh (2009), 'Low carbon fuel standards', *Issues in Science and Technology*, Winter, 57–66.

Van Dender, K. and K. Small (2005), 'A study to evaluate the effect of reduced greenhouse gas emissions on vehicle miles traveled', Final Report ARB Contract Number 02-336.

27 Regional experiences: the past, present and future of the energy policy in the Basque region

Jose Ignacio Hormaeche, Ibon Galarraga and
Jose Luis Sáenz de Ormijana

27.1 INTRODUCTION

Starting from the fact that 'all life depends on energy for survival and growth' (Fouquet, 2008), human development has been determined by great changes in the way that energy has been produced and consumed. The use of energy has played a crucial role in the development process, from the improvement in survival and expansion of population with the discovery of fire, to the use of the steam engine and of electricity as the foundations of the industrial revolution and any ensuing development (Fouquet, 2008).

There is nowadays clear scientific and political consensus regarding the fact that the climate is changing and that human activity is responsible for this (IPCC, 2007). In fact, many (including the UN Secretary General Ban Ki-moon) have defined climate change as the greatest challenge facing humankind, as it is affecting (or will affect) nearly all human and natural environments (IPCC, 2007). The use of energy is again at the very core of the problem in this challenge, and furthermore, it is central to any potential solution to deal with it.

This book has addressed many of the most important issues related to the sustainable use of energy, including some ideas of what can be achieved at regional and local level (defined as any subnational level of governance). Some interesting examples of other regions have been presented. In this chapter, the idea that the relationship between national and subnational governments has changed dramatically over recent decades is stressed (OECD, 2007); in fact most countries, particularly in Europe, are currently involved in some form of decentralization (Oates, 1999). Energy policy is no exeption to this.

The case of the Basque Country is rather interesting as it illustrates the case of a small region in Europe (a surface area of around 7250 km²) with a high degree of decentralization that combines a significant weight of energy-intensive industrial activity with a very successful economic growth model. The weight of the industrial sector accounts for 29.8 per cent of gross domestic product (GDP), followed by 60.3 per cent for services, 8.9 per cent for construction and 1 per cent for the primary sector. With a population of 2.1 million, it enjoys a high income per capita, 40 per cent above the EU-27 average (Eustat, 2008). In terms of energy intensity of the economy – that is, energy consumed per unit of GDP – the Basque Autonomous Community (BAC) has an index of 183.2, well below the EU-27 average and that of countries such as France, the UK, Sweden, the Netherlands and Belgium (all below the EU-27 average) (Eustat, 2008). In fact, the BAC energy intensity in 2007 was 22 per cent lower than in 1990, mainly due to energy saving and energy efficiency policies (EVE, 2009).

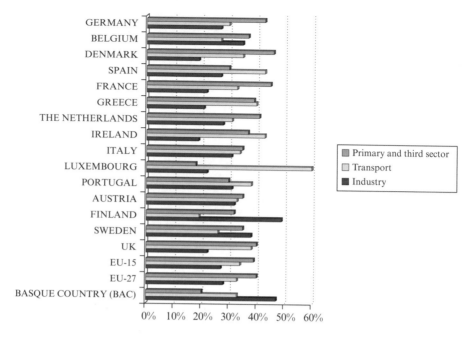

Source: EVE (2009)

Figure 27.1   Final energy consumption in Europe

The great degree of decentralization and the significant room for manoeuvre of the regional government – that is, the Basque Government – together with the proactive attitude of the industrial sectors are the main reason for the improvements. The government set up a Basque energy agency, called the Basque Energy Board (Ente Vasco de la Energía, EVE) in the early 1980s.

In terms of energy consumption, industry accounts nowadays for 48 per cent of the region's consumption, while consumption in buildings (housing, shops and services) is under 20 per cent. Transport is responsible for 33 per cent of final consumption (see Figure 27.1). Similarly to the rest of the world, the highest growth over recent years has been in the energy supply for transport, which now stands at 33 per cent of final energy consumption (EVE, 2009).

Gross domestic energy consumption in 2008 was 7872 ktep while final consumption was 5757 ktep. Renewable energies accounted for 5.4 per cent of total energy demand, while renewable electricity accounted 5.6 per cent of electricity demand (EVE, 2009). In per capita terms, Figure 27.2 shows that energy consumption was higher than the EU-27 average while energy production was almost as low as in Luxembourg. This generates a situation of 95 per cent energy dependency towards external sources; much greater than any other state in Europe and, of course, much greater than that of Spain.

In terms of energy per sources, the low rate of renewable energy generated in the BAC can be easily confirmed, as set out in Figure 27.3.

Regarding CO_2 emissions in 2008, the total emissions (TE) due to the socio-economic activity accounted for 25.2 million tonnes of CO_2e (CO_2 equivalent) in 2008, 18 per cent

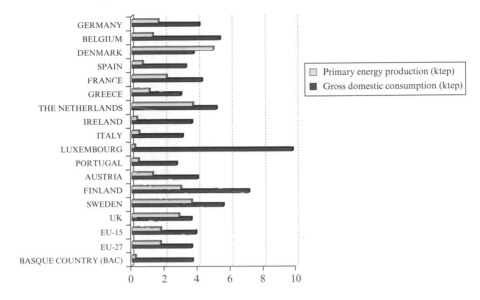

Source: EVE (2009).

Figure 27.2   Per capita primary energy production and domestic gross consumption

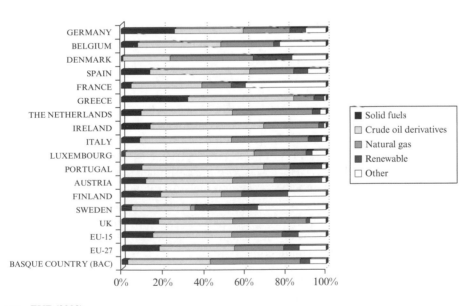

Source: EVE (2009).

Figure 27.3   Gross domestic consumption per energy source

Emissions per sector

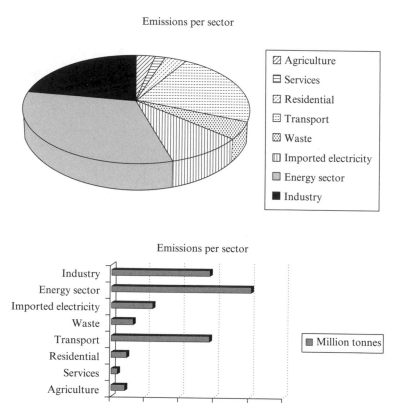

Source: IHOBE (2009).

Figure 27.4 CO_2 emissions per sector in 2008 (million tonnes)

higher than in 1990 (see Figure 27.4).[1] If the direct emissions (DE) are considered – that is, not accounting for the emissions of imported electricity – the figure was close to 22.8 million tonnes of CO_2. In this case, it was 39 per cent higher than 1990. The distinction between TE and DE in this region becomes extremely useful, as in the early 1990s only 10 per cent of the final electricity consumption was generated within the BAC and the other 90 per cent was imported from outside the region. This situation had changed dramatically by 2008, when 62 per cent of the electricity was generated internally. In terms of sectoral distribution, the three main emitters were industry, which accounted for 22 per cent (5.7 Mt) of emissions; the energy sector, with 32 per cent (8.1 Mt); and transport, with 23 per cent (5.7 per cent). These were followed by imported electricity (2.4 Mt, 10 per cent), waste (1.3 Mt, 5 per cent), agriculture (0.8 Mt, 3 per cent), services (0.4 Mt, 2 per cent) and the residential sector (0.9 Mt, 3 per cent) (IHOBE, 2009).

When these emissions are compared with those of EU member states, the BAC had a ratio of 11.71 in per capita terms, above the average for the EU-15 (10.37), higher than for Austria, the UK, the EU-27, Spain, Italy and Sweden. Now, if the ratio of emissions per GDP unit corrected for purchasing power parity (PPP) is considered, the picture changes dramatically. The BAC has the lowest ratio in Europe after Sweden and France, well below the EU-27 and EU-15 averages. States such as Finland, Germany, UK, Italy and even Denmark have a greater ratio.

In terms of policy instruments, the BAC region has also set up the necessary instruments to design an effective climate policy. IHOBE, the publicly owned company, manages environmental issues under the mandate of the Basque government and a climate change office was recently set up to coordinate all efforts in the field.

The case of the BAC is rather interesting due to many factors that the reader might already have worked out:

1. Its high degree of decentralization justifies in many cases its being benchmarked against other member states in EU in terms of the policies that can be implemented.
2. It is a heavily industrialized region where energy intensity has been reduced dramatically thanks to the regional energy policy.
3. Above all, the BAC has a high degree of energy dependency on external sources, up to 95 per cent, although domestic electricity generation has increased up to 65 per cent as a consequence of the energy policy as well.
4. The BAC offers a higher ratio of CO_2 emissions per unit of GDP (PPP corrected), while there is great room for improvement with regard to the ratio in per capita terms.
5. Fighting climate change has become a very serious political priority since 2000, in line with a deep sense of responsibility towards future generations and towards other regions and countries.

27.2   QUARTER OF A CENTURY OF ENERGY POLICY: EFFICIENCY, DIVERSIFICATION AND RENEWABLE

27.2.1   The Energy History in a Few Lines

Although the energy policy in the BAC is closely coordinated with European and Spanish policy due to their ensuing legislation and regulations, Basque policy has been designed to ensure that it can set up its own domestic energy policies in a sustainable way. In a context of an energy market which is global (gas is imported from Trinidad and Tobago, Algeria, Qatar and other countries, and coal from Eastern Europe), an electricity sector that is conceptualized at the level of the Iberian Peninsula, and shortage of local energy resources, the BAC has managed to direct its energy system towards a vision of lowering external dependency (particularly with regard to electricity generation). This has been achieved by creating energy generation market opportunities and developing the infrastructure, providers and distributors that could guarantee a certain degree of autonomy. This is based on three pillars: energy efficiency, diversification of energy sources, and encouragement and support of the development of renewable energies.

In fact, one of the main energy priorities has been to reduce consumption through energy saving and improvements in energy efficiency, primarily in the industrial sector. As a result of the many programmes and actions implemented, Basque energy intensity in 2007 was 22 percentage points lower than in 1990: in other words, the energy consumed per unit of GDP generated in the region fell by 22 per cent over the period. Industry has played a particularly important role in this trend: by incorporating new equipment and innovative energy technologies such as combined heat and power, the sector has managed to improve energy efficiency dramatically.

In the field of supply diversification, the situation in the 1980s was based on fuel oil and coal, with gas networks scarcely reaching the area. The government decided to encourage the use of natural gas, seeing it as a clean energy with good prospects for the future. At the time, however, the Spanish state gas company had other priorities and was not interested in developing the gas network in the region. The Basque government therefore decided to invest in gas distribution infrastructures and created a regional gas distributor, Gas de Euskadi, and a number of local distributors, in which public capital (the government and municipal authorities) had a majority holding. Thus, over 90 per cent of the population nowadays has access to the natural gas distribution networks. These distribution companies subsequently merged, and once they were felt to have fulfilled their original aims they were mostly privatized, with the Basque government retaining a small holding.

The Basque Country is a heavy user of electricity for industrial purposes (two-thirds of all electricity consumption is industrial). Until recently, production at local power stations was only capable of meeting 20 per cent of consumption and the rest had to be imported from the Spanish market. In order to balance power supply and demand, the Basque government provided the means to allow private investment in installing combined-cycle plants in the region. Initially, direct participation from the government through the energy agency was needed to drive development of the first plant (the 800 MW Bahía Bizkaia Electricidad station) and encourage private enterprise to invest. In 2010, 2000 MW have been installed, with projects under way for the erection of two more power stations in the short to medium term. Therefore, in addition to encouraging local economic development, the strategy has meant that electricity imports have fallen from accounting for 80 per cent of the power to supply just 30 per cent, thus achieving a better balance in the electrical system.

As a result of this policy of promoting infrastructures, natural gas has gone from being a practically unused energy source in 1980 to meeting 42 per cent of final energy consumption in the Basque Country in 2007. In order to guarantee supply and ensure that businesses and citizens had access to this primary energy source, the last action in this stage, completed in 2003, consisted of building and commissioning the Bahía Bizkaia Gas (BBG) liquefied natural gas (LNG) storage and regasification plant at Bilbao Port. The government kept a 25 per cent stake through EVE. BBG has a regasification capacity of 800 000 Nm^3 per hour and two 150 000 cubic metre storage tanks, and it will be enlarged with a new tank of the same size and a new regasification line which will be in operation in 2012. The jetty allows the docking of 250 000 cubic metre LNG tankers.

The policy has also been directed at getting wind power off the ground in the region, with the first wind farms being developed by way of a jointly owned 50–50 public–private company. In order to ensure that there was no competition with private enterprises in mature industries, the public stake was sold once an installed capacity of 150 MW was reached.

In other renewable sources, EVE works with municipal authorities to restore small hydroelectric power stations (again through its affiliate companies). A total of 175 photovoltaic facilities have been installed in schools, and more than 100 in municipal buildings. Biomass has also been widely harnessed as a source of heat in industrial processes (paper and cement, for example) and power is being generated from the incineration of municipal solid waste in a combined-cycle plant that also integrates natural gas.

27.2.2   Future Energy Challenges

It is a reality that 95 per cent of all energy consumed in the Basque Country still comes from outside the region. BAC only manages to produce, using its own resources, around 5 per cent of the energy consumed. The great challenge today is very similar to the one faced 26 years ago: what needs to be done to ensure that enough energy is available when users require it, at a competitive price and in conditions of environmental sustainability?

The straightforward answer has remained unchanged over all these years, and it is to save energy, consuming less and more efficiently. In this area, the BAC plans to be just as insistent as it has been in the past. The world's reserves of fossil fuels are reducing rapidly, CO_2 emissions have to be halved in the following years and actual energy consumption has become totally unsustainable. That is, it is not the power plant that produces the electricity that has become unsustainable, but rather the way of life, the energy demanded and consumed every day cannot be sustainable for much longer. The energy mix has to be optimized to comply with many objectives: that is, to cater to consumption, to economic and social benefits, and to match changes on the market and the opportunities created by the development of new energy technologies.

One has to be aware that fossil fuels, oil and gas are most likely to continue to account for the lion's share of BAC consumption over the coming years. Therefore, it is reasonable to ensure that infrastructures are in place to guarantee supply in a diversified and competitive manner. Meanwhile, the search for reserves of natural gas in the Basque Country will continue through SHESA, a publicly owned company, and working with international partners. Although there are well-founded hopes that viable reserves may be found, whatever is found will only help to ameliorate temporarily the existing energy deficit.

The BAC is aware that the future lies in overcoming the technological limitations still facing renewable energies, making them increasingly competitive and giving them a more important role in power generation. In view of these opportunities for development, major commitment to research and development in this field has been made. Building on a strong business sector and a network of technology centres with experience in technological development, the Basque commitment to research and development into energy has now been channelled into the setting-up of a new research centre, the so called EnergiGUNE Cooperative Research Centre, which will develop long-term basic research targeting four priority areas: power storage, high-temperature solar thermal energy and its storage, and wave energy.

27.2.3   The Energy Strategies: Cornerstone of Energy Policy

There have been many energy strategies in the BAC over the last 30 years. The starting point for a structured energy policy was the 1982–90 energy plan which set up the EVE group, made up of a series of companies and public entities in the sector. The framework of the plan for the late 1980s was the post-1973 and 1979 energy crisis era in which energy scarcity and high prices were the main conditioning factors. CO_2, SO_2 and other pollutants had already been reported in a systematic form and an emissions reduction target was included in the strategy.

In the early 1990s the Parliament mandated the EVE to prepare the next strategy. As a result of this, the 3E-2000 strategy was approved in 1992 for the 1992–2000 period with the following goals (EVE, 1992):

- Controlling energy consumption without reducing life quality.
- Reducing the environmental impact of energy use.
- Continuing with energy supply diversification as a means to reduce dependency.
- Maximizing the energy supply guarantee.

The 3E-2000 strategy was reviewed in 1996 to assess the degree of success up to 1995. The review concluded that there had been a significant overachievement of energy goals for renewable, natural gas and electricity generation, while there had been slightly lower achievements for efficiency.

These results paved the way for the following strategy (3-2005), which defined the framework for 1996–2005 period. This one looked closely at the European energy framework, the role of the Spanish state and some interesting examples at EU level such as the Netherlands and Denmark. The main acting principles proposed were (EVE, 1997):

- Increase energy efficiency endeavours with the aim of reducing energy intensity (defined as consumption per unit of service).
- Prioritize solutions with lower global environmental impact.
- Increase the efforts to use domestic renewable resources from the point of view of economic competitiveness and environmental protection.
- Promote resource exploration efforts.
- Improve the Basque energy system by the development of storage, generation, transport and distribution infrastructures.
- Promote a better supply and demand balance by making energy-providing centres closer to demand sources to minimize energy losses.
- Support agreements and collaboration with energy operators in the Basque system.

The main quantitative goals are summarized in Table 27.1.

The so-called 3E-2010 energy strategy was scheduled to be totally implemented by the end of 2010; its development starting point was the year 2000 benchmark. The summary of the goal achievement indicator is depicted in Table 27.2.

Based on the year end of the 2008 energy balance, on 2009 indicator trends and the degree of progress in the current projects and measures, the following general conclusions can be put forward regarding the foreseeable results in 2010 with respect to the targets established in the aforementioned 3E-2010.

The high degree of implementation of savings and efficiency measures will lead to annual savings levels within reach of the set target (around 90 per cent). By sectors, it is noteworthy that there will be greater than forecast savings in the industrial sector.

Table 27.1   3E-2005 synthetic goals

Indicator	1995 situation	Goal 2005	Ratio 2005/1995 (%)
Global energetic indicators			
Energy demand (tep)	5 590 000	6 440 000	15
Final consumption (tep)	4 180 000	4 070 000	−3
Energy intensity (tep/Mpts PIB)	1.05	0.78	−25
Sectoral energy indicators			
Hydrocarbons annual production (tep)	282 000	670 000	138
Renewable resources (tep)	214 000	404 000	89
Energy efficiency: consumption reduction (tep)		532 000	
Natural gas supply (Mte)	9 640	18 760	95
Electricity self supply rate	20	82	
Environmental indicators			
Emissions: global impact index (%)	100	50	−50
Index per consumption unit (%)	100	43	−57
Economic indicators			
Investments (Mpts)		442 000	

Source: EVE (1997).

However, savings in the tertiary sector will be around the target, and will fall far short when it comes to transport.

As far as renewable generation is concerned, the share of the total consumption of the BAC in 2010 will probably be around 6.5 per cent, far below the target of 12 per cent. This difference is mainly due to the existing low levels of wind power compared to the envisaged 624 MW, and to the lower contribution of biomass.

With regard to the energy infrastructures, they have been greatly developed since the late 1990s. Special mention should be made of the Bahía Bizkaia Gas (BBG) regasification plant, of three natural gas combined-cycle plants (BBE, Santurtzi and Boroa), of the Bergara–Irun gas pipeline, including the Euskadour line connecting with France, and work starting on the Petronor project for a fuel oil reduction unit to light products and coke. However, there are other strategic natural gas infrastructures that are still in the initial stages, pending construction getting under way. These include the Transcantábrico (Bilbao–Treto) gas pipeline, the third tank of the BBG regasification plant and the expansion of the Gaviota underground storage facilities

With respect to the target to boost research and development (R&D), the number of R&D projects in the field of energy has increased significantly, thanks to the support programmes run by the Basque Public Administration, the European Commission R&D Framework Programmes, state programmes and private funding. Special mention should be made of the significant involvement of Basque companies in European Research, Development and Innovation (R&D+i) energy programmes, basically through the action lines of the VI Framework Programme (2002–06). Their measure of success is established among other factors by the rate of return, which for energy projects

Table 27.2   3E-2010 goals

Indicator	Situation in 2000	Goal 2010
Energy efficiency		
Energy saving wrt 2000 (tep/year)		975 000
Energy saving index wrt 2000 (%)		15
Energy intensity improvement wrt 2000 (%)		16
Energy supply with cogeneration (%)	10	14
Use of renewable energies		
Use of renewable resources (tep/year)	264 000	978 000
Share of renewable in energy (%)	4	12
Share of renewable in electricity (%)	2	15
Use of cleaner conventional energies		
Consumption of natural gas (bcm)	1.5	4.7
Share of natural gas in demand (%)	21	52
Energy generation capacity		
Electricity self generation index (%)	27	114
Thermal generation (MW)	1 132	2 880
Cogeneration and renewable (MW)	525	1 460
Environmental impact		
Green House Gas index wrt 1990	24	11
Economic impact		4 900
Investment in energy efficiency and renewable (M€)		1 710
Investment in energy infrastructure (M€)		3 190

Source: EVE (2003).

submitted by Basque companies was €12.7 million, mainly through the 48 stakes in companies and technological centres. The setting up of the CIC-energiGUNE Research Centre (Cooperation Research Centre) is also noteworthy. Its mission is to generate basic research focused on energy technologies, with medium- and long-term criteria of excellence. The CIC energiGUNE has defined its priority action lines in the field of electrical and thermal energy storage.

27.3   DEVELOPMENT OF CLIMATE POLICY

The BAC has been recognized by EU institutions[2] and the United Nations[3] as being one of the most active regions in climate change policy over recent years. It can be stated that that climate change policy started with the approval of the Environmental Sustainable Development Strategy 2002–20, where fighting climate change was one of the main five priorities. The detailed setting of policy goals are regularly being updated every four years through the Environmental Framework Programmes (EFP). The first of these was for 2002–06 and the second for the 2007–10 period. The latter gave a decisive push to climate policy by encouraging the development of the Basque Plan to Combat Climate Change 2008–12 (BPCCC, 2009).[4] During 2010–11, work was ongoing to prepare the next EFP 2011–15.

The BPCCC defines the aim of the Basque Country: 'to take irreversible steps towards a socio-economic model non-dependent on carbon by 2020, so that the Basque Autonomous Community is less vulnerable to climate change'. The plan sets two priorities to achieve this: (1) to act against climate change and to be prepared for its consequences; and (2) to promote innovation and research to move towards a low carbon sustainable economy. There are four strategic objectives defined to achieve this. These are:

- Limiting GHG emission growth to 14 per cent with respect to 1990 levels.
- Increasing carbon sinks by 1 per cent with respect to 1990, especially managing forests, agricultural soil and pastures.
- Minimizing the risks on natural resources.
- Minimizing the risks on human health, on the quality of urban habitats and on socio-economic systems.

The plan oversees 120 policy actions in the fields of energy, transport, housing, education, industry, environment, land use planning, water, education and so on, that have been organized under four policy programmes:

- Lower Carbon Programme to deal with mitigation policy actions in the areas of energy efficiency and saving, renewable and carbon sequestration.
- Adaptation Programme to enhance research and monitoring, vulnerability studies, adaptative planning and preparation of infrastructures.
- Knowledge Programme to promote basic and applied high level research.
- Citizens and Administration Programme to manage issues related to green procurement, awareness raising, education and training.

The BPCCC aims to reduce emissions from 22 per cent in 2006 and 18 per cent in 2009 with respect to 1990 emission levels.[5] The situation is better than in Spain (around 50 per cent above 1990 levels) but still is far behind the average EU index of minus 2 per cent[6] (see BPCCC, 2009).

The Basque Country has set itself a very ambitious target: a reduction of 4.30 million tonnes of CO_2e by 2012. Table 27.3 shows how these reductions will be achieved. As can be imagined, the main reductions will be though energy efficiency measures and renewable energies.

The plan also envisages the possibility of reducing an additional 0.5 million tonnes through domestic emission offsetting projects and certified reductions. Carbon sequestration is also included in the plan, aiming at reducing CO_2e by a further 223 163 tonnes by 2010 from afforestation and reforestation, forest management, cultivated land management and grazing land management.

This plan coordinates efforts with other sector-specific plans and targets such as the Energy Strategy 3-E2010, The Environmental Action Programme 2007–2010, the Sustainable Transport Plan and the Science, Innovation and Society Plan. It wishes to mobilize €630.3 million, €79.5 million of which refer to additional resources not included in any other existing plan or strategy. Hoyos et al. (2009) show that the aggregate willingness-to-pay (WTP) to implement the Basque Plan to Combat Climate

Table 27.3   Contribution of each line of action to the emission reduction objective

Lines of action	Reduction target for 2010 (average for 2009–2012 Mt CO_2e)	Measures to 2012
Energy efficiency & savings		
More efficient use of fossil fuels	1.01	All thermal electricity generated by combined cycle natural gas plants
Savings & efficiency in industry	0.57	Improvements in energy efficiency for a saving of 583 Ktoe on 2001 figures by 2010
Savings & efficiency in means & use of transport	0.33	21% improvement in efficiency in transport in terms of CO_2 emissions
Savings & efficiency in residential & service sectors	0.09	Improvements in energy efficiency for a saving of 58 Ktoe on 2001 figures by 2010
Encouragement for CHP	0.08	600 MW of installed capacity from CHP
Encouragement of renewables		
Encouragement of renewables	1.06	Production from renewables to meet 15% of electricity demand
Encouragement of renewables (bio-fuels) in transport	0.53	177 ktoe of consumption requirements met by renewables
Encouragement of renewables in the residential & service sectors	0.02	152,000 m^2 of solar power used for heating
Reduction of non-energy emissions		
Reduction of non-energy GHG emissions from industry	0.31	89% drop in fluorinated gas emissions on 1995 levels by 2012
Reduction non-energy GHG emissions in the waste sector	0.17	Less than 40% of MSW landfilled
Reduction of non-energy GHG emissions from agriculture & forestry	0.12	Construction of 3 livestock waste treatment plants
TOTAL	4.30	

Source: BPCCC (2009).

Change (BPC) is estimated to be €400.6 million, well above the additional investment needed.

It is important to highlight the set of indicators and feedback system (FS) that have been envisaged to guarantee the effective implementation of the plan. Both the indicators and the FS should allow further fine-tuning of the climate policy.

The Basque Climate Change Office (BCCO) was set up in order to coordinate all the efforts to implement the plan effectively. It gathers representatives from the main departments involved in climate policy and is organized at two levels: the decision-making board and the advisory technical committee. The latter has the mandate to analyse and prepare the reports to be approved by the decision-making board. The main four tasks of the office are defined in the BPCCC and are:

- Taking active part in the implementation of the BPCCC.
- Promoting knowledge and research.
- Coordinating climate planning.
- Communicating and awareness-raising.

The effectiveness of the climate plan has not been tested so far. An interesting *ex ante* study of the economic impact of CO_2 mitigation policy in the Basque Country can be found in González and Dellink (2006), which shows that the costs for achieving the Kyoto targets may remain limited if the appropriate combination of changes in fuel-mix and restructuring of the economy is induced. A reduction in emissions of 15 per cent will induce a decrease in GDP of approximately 1 per cent.

The evolution of the GHG emissions is presented together with other environmental indicators annually in the 'Environmental Indicators Report'. As seen in the introduction of this chapter, the emission reduction trend is downwards, with the reduction from a 22 per cent increase with respect to 1990 to 21 per cent in 2007, 18 per cent in 2008 and 6 per cent in 2009. In any case, it might be too early to evaluate this policy properly, and the impact of the global economic crisis in industrial and other polluting activities should also be clearly acknowledged.

In 2010, work was also ongoing for the preparation of a climate change Act and a new BPCCC up to 2020 in line with the new energy strategy, as will be explained below.

27.4   ENERGY AND CLIMATE STRATEGY 2020: THE CHALLENGE AHEAD

The fundamental *raison d'être* of the future 3E-2020 Strategy should be to guarantee that the BAC (with high energy consumption and very scarce natural resources) has sufficient energy in quantity, quality and in time, at a competitive cost and in an environmentally sustainable manner. Having said that, it must be noted, as stated earlier in this chapter, that the Basque system is part of the Spanish gas and electricity systems. Therefore, although the planning and implementing capacity of the Basque authorities is outstanding, it is also true that they have limited regulatory and legislative capacity. This fact strongly suggests that other roles be undertaken and approaches be developed, to be more directly involved in projects and actions with the sector.

Yet, apart from this purely energy aspect, one of the key features of this 2020 strategy should be its integration and constructive overlapping with, on the one hand, the competitiveness policies of the Department of Industry, Innovation, Trade and Tourism, particularly those relating to industrial development and to R&D+i; and on the other hand, the work driven from the Basque Climate Change Office to provide the CO_2 reduction scenarios and adaptation strategy for 2020. This is all in line with the hard work being carried out by Spanish and European public institutions. In practical terms, this means that the programmes and investments linked to energy strategy should also be critical to drive and energize the energy-related business sectors in aspects such as technological development, inter-business cooperation or the creation of new business opportunities. The energy sector is a provider of a central resource for economic activity and particularly for day-to-day living, but it also represents a fairly large niche of business, innovation and research opportunities. Fostering greater interaction between the Department of Industry, the Ente Vasco de la Energía (Basque Energy Board – EVE) and all the stakeholders, companies and entities in the sector through the Energy Cluster seems to be an ideal framework for the integration of industrial and energy policies. Indeed, as the use of energy is the cornerstone of the climate policy, reinforcing the role of the EVE in the Basque Climate Change Office should also be strongly encouraged.

27.4.1   The Context

In order to define the priorities on which the BAC must focus over the coming decade (2010–20) to tackle the new energy challenges, the different contexts in which policies are set will have to be taken into account. These will condition and influence development of the policies to a great extent.

The first context to be considered is, of course, the BAC. Here, the initial focus has to be on analysing the situation in 2010, taking into account demand, availability of infrastructures, the current generation capacity and future potential capacity.

The second level is the context defined by the policies and the regulation established by the Spanish government. The BAC energy system is part of the Iberian Peninsula gas and electricity systems, regulated respectively by the Hydrocarbon and Electricity Market Acts, resulting from the transposition of the relevant EU Directives. All the Spanish legislative developments in the field of energy decisively affect the Basque generation mix. Over the coming years, the new National Renewable Energies Action Plan 2011–20 (PANER) is going to have a particularly significant impact. The Ministry of Industry in Spain has recently sent its PANER proposal to Brussels, which envisages a target that exceeds the mandatory 20 per cent of renewable participation and sets it at 22.7 per cent of the final energy and 38.2 per cent of the electricity generated. The decrees and legislation that develop these targets and define the future compensation for the Special System installations (renewable and cogeneration)[7] will be fundamental when planning the power and production assigned to each technology. The future Sustainable Economy Act to be discussed in Parliament may also play an important role as it includes interesting contributions in the chapter entitled 'Sustainable energy model' (Articles 96 to 107). The successive reviews of the gas and electricity documents of the State Infrastructure Planning will have also a part to play.

And, last but not least, the European energy policies will have to be a clear benchmark. In compliance with the March 2006 European Council mandate, in January 2007, the Commission submitted a Strategic Communication on Energy Policy for Europe, adopted by the Council together with an Energy Priority Action Plan. After different draft legislative 'packages' and 11 months of negotiations, the European Parliament passed the legislative package in December 2008 that will help the EU achieve its targets for 2020: 20 per cent reduction of greenhouse gas emissions, 20 per cent improvement in energy efficiency, and renewable energies having a 20 per cent stake in energy consumption. A new Directive was therefore passed which has introduced national mandatory targets for the member states in terms of producing renewable sources to ensure that 20 per cent is achieved for the EU overall. For a detailed description of the policy (the so-called 20-20-20 energy and climate package), see Gallastegui and Galarraga (2010).

Mandatory targets for generating energy using renewable sources have not been set in the Basque Country as the Directive affects only the Spanish state. The same happens with CO_2 emission targets that are only set for Spanish state within the EU climate policy targets. The CO_2 emission limits established in the Basque Country are unilaterally established by the Basque authorities, clearly in close coordination with the Spanish government. After a thorough assessment of the existing potential, the Basque government will establish coherently ambitious but also realistic targets, and the contribution to the EU target.

27.4.2   Policies in Relation to Demand

In a country with high per capita energy consumption and scarce resources (either fossil or renewable), the main priority has to be ensuring that the region is energy efficient. This is interpreted as increasing the GDP of the productive sectors and improving the quality of life of the citizens by consuming less energy. That is, reducing the energy intensity ratio. When considering the policies and action plans related to energy demand in the BAC, three major consumption sectors should be distinguished as priorities: industry, building and transport.

Industry
Industry and the business sector have placed significant emphasis since the mid-1980s on improving their energy efficiency, by means of introducing measures such as improving their production processes, renewing facilities, introducing cogeneration, using residual heat, and so on. In fact, the industry sector, even though it still represents a great share of the aggregate CO_2 emissions in the BAC (see Figure 27.4), 22 per cent, has managed to reduce its emissions by up to 26 per cent with respect to 1990. Well above any other sector except agriculture.

Nonetheless, this line of work must continue and should be consolidated. Therefore, Basque institutions will continue to back energy audits and the implementation of savings and efficiency measures in industry.

Building
The reduction of energy consumption in buildings (public administration offices, private houses and shops) will be a clear policy priority over the coming years. Therefore, many working areas will be strengthened: implementing and developing legislation and norms (compulsory ones for certification purposes but also other more innovative ones); supporting restoration or renewal of existing building components that contribute to energy consumption (for instance windows, insulation, installations); encouragement of energy service markets to increase investments, monitoring and consumption management; supporting infrastructure renewal and implementation of smart grids in order to reduce energy losses and contribute to effective management of demand.

Transport

Over recent years, the transport sector has become the second-largest consumer of energy in the BAC, behind the industrial sector. It is a sector where practically all the energy demand comes from oil, particularly in the case of road transport. In terms of emissions, it represented 23 per cent of aggregate emissions for 2008, while these had increased by 110 per cent with respect to 1990 emissions (IHOBE, 2009). In order to improve transport energy efficiency over the coming decade (2010–20), great emphasis will be placed on boosting the introduction of the electric vehicle (EV). The EV will offer huge advantages over the conventional vehicle as its energy efficiency is much greater, it is silent, and it does not emit particles or local pollutants through the exhaust[8] when being used (which will help to reduce pollution in the cities). It also reduces greenhouse gas emissions and, if its charging periods are optimally managed, it can be beneficial for the electricity system by absorbing the night electricity surplus, which thus ensures greater use of renewable generation.

The main technological drawback that the electric vehicle is facing is the battery, which is still limited in terms of its duration, range and cost. Yet there are also other non-technological barriers that the electric vehicle must overcome to achieve the degree of penetration sought. The most important of these barriers are: the non-existence of relevant legislation or regulation; the lack of social habits regarding sustainable mobility; the lack of knowledge, or mistrust, regarding the electric vehicle; and, in particular, the need for a new infrastructure with charging points, that currently does not exist. The policies and action plans of the Basque government in this area will be mainly aimed at overcoming and finding solutions for these barriers.

27.4.3 Supplying and Generating Energy

Over the coming decade (2010–20), natural gas is going to continue to be the main fuel of the BAC's energy mix and will be the basic energy source during the transition to a situation when greater use will be made of renewable energy sources. The commitment to fostering the development of renewable energies still requires a complementary supply from the natural gas combined-cycle power stations to guarantee that the transition can be done effectively. Unfortunately, renewables still suffer from problems of intermittence and being less predictable. Therefore, renewables and natural gas is a binomial that must grow and be developed in parallel over the coming years as pillars of a more sustainable energy strategy.

Natural gas has been an important driver of energy change at local level, by contributing to improving the supply for companies and households. Moreover, over the coming years it will be, along with renewable energies, a central pillar of energy strategy, economic development and environmental sustainability in a region such as this one where the industrial sector still plays an important role.

Aware of the aforementioned fact that the generation mix in the BAC is absolutely conditioned by the operating and the legislative framework of the state gas and electricity system, policy actions are likely to be directed to supporting the whole system. They can be classified into two major concepts.

The first concept is that of supporting the energy operators and companies to develop infrastructures that guarantee the supply. The companies must design and undertake the investments in infrastructure that ensure a guaranteed and quality supply of gas and electricity. The government will use all its capacities and competences to support projects in this area, where special mention should be expressly made of those involving natural gas: the Transcantábrico gas pipeline, the French part of the Irun Euskadour branch line, the third and fourth tank of the Bahía de Bizkaia Gas (BBG) regasification plant plus the expansion of the regasification capacity, the expansion of the Gaviota underground storage, and the exploration and possible use of non-conventional natural gas reserves in the Alava basin. Likewise new electricity infrastructures will be supported, such as high-voltage transport networks or the deployment of intelligent distribution grids.

The second concept is that of development of the renewable generation facilities. Even though spanish government decrees as well as EU Directives define the real incentives to promote the renewable facilities, the Basque government will develop an active policy to contribute developing favourable conditions to use renewable resources. Some examples of how this will be done are:

- Wind energy: based on the agreement on wind energy resources reached with the provincial councils and with the association of municipalities (EUDEL), wind farms will be promoted while ensuring the greatest respect for the environment and biodiversity.
- Wave energy: this is one of the core commitments over the coming decade (2010–20). Even though technological development is still very much in its initial stage and the output from these facilities is still in the distant future, new research and demonstration facilities have been developed, such as a plant in Mutriku port or the Armintza BIMEP (Biscay Marine Energy Platform). The final objective of these infrastructures is to start up this type of generation and to boost its development.
- Biomass: There is a need to increase renewable production based on biomass and the BAC has the resources to do so, mainly from agriculture and forestry waste. With this in mind the efforts will be directed in two ways: creating the conditions to ensure the profitability of the plants (tariffs system, distribution system, and so on), and directly supporting plant construction projects that can be proved to be feasible.
- Geothermal gradient: over the coming years, there is going to be a huge deployment in the number of heat and cold generation facilities for buildings based on the geothermal gradient. In fact, this technology has a proven track record of excellent performance in the countries of Northern Europe, and has already begun to spread to other regions of the world. The users that opt for this technology, both in terms of new facilities and when refurbishing existing ones, will receive significant incentives from regional support programmes.

Programmes to support other renewable technologies such as solar photovoltaic, thermal solar installations in existing buildings, mini-hydraulic sources or the distribution and use of biofuel will also continue to be a central part of the forthcoming energy policy for 2020.

27.4.4   Technological and Industrial Development Priorities

As a result of the analysis of the preceding sections and the development of the industrial sector in energy-producing activity represented by the Energy Cluster[9] the following sector and key areas will be strongly encouraged and supported:

- Wind energy: development of state-of-the-art wind turbines and the critical components of their value chain (turbine, multiplier, control systems, alloy structures, blades, and so on); development of wind turbines and components for offshore applications; internationalization of the supplier structures.
- Thermoelectric solar energy: engineering and design of solar concentration systems for electricity generation; mechanical components (structures, tracking systems); thermal energy storage systems (molten salts).
- Wave energy: underwater connection elements, mooring and anchoring systems; energy converters.
- Biomass and geothermal gradient: new technological developments, pilot plants and to demonstrate new applications.
- Smart grids: electronic metering, smart substation and transformation centres, demand management system.
- Electricity storage technologies, mass storage application for both home and automotive generation: electrochemical batteries (ion-lithium, sodium, new composites), ultra-condensers, nano-technologies.
- Equipment and systems to developing charging networks for electric vehicles; new business models in relation to the sale of energy to electric vehicles.
- Energy service companies (ESCOs) to manage, operate and maintain energy facilities.

27.4.5   Interaction with Climate Policy: The Future

That energy use is a vital part of the climate policy is a well known idea. In fact, the aforementioned EU 20-20-20 energy and climate package can be interpreted as the effective materialization of the real integration of these two policies. Other chapters in this book have also shown this evident connection. The efforts being made to integrate the climate and energy policy in the BAC. These explain the prominent role that the Energy Department plays (together with some others) as a member of the Basque Climate Change Office. The office is in charge of drafting the two main policy instruments in progress: the Climate Change Act and the BPCCC 2020. Both will include the basic guidelines to define the energy scenario for 2020 summarized in this chapter. The work was ongoing in 2010 and is expected to be concluded by the end of 2011. The EU 20-20-20 Directive is a guiding principle and the fact of being a subnational or regional government sets up interesting new challenges in this framework.

27.5 FINAL THOUGHTS

This chapter has sought to show the experience of a specific region in Europe in energy and climate policy. It will, perhaps, serve to complement the other contributions gathered in this book on the regional dimension: the case of North Rhine-Westphalia (Europe) and the case of California (USA).

The regional (understood as subnational) level of governance offers highly interesting policy opportunities in many fields, particularly in the areas of climate and energy issues. While both undoubtedly have a global and international dimension, there is still room available for regions to contribute in the endeavours towards more sustainable and low carbon energy use. Many agree on this, but what this chapter aimed to illustrate is that with appropriate coordination and guidance, the role of regions should not be under-estimated. The 20-20-20 energy and climate package in the European case is a general framework within which member states (and their internal regions) have to act. At the same time, the minimum dimension of the energy supply system (electricity and gas) that is rational is the entire Iberian Peninsula or other central parts of Europe. An even greater scale may be advisable. Yet the regional dimension acquires a great importance in a situation where energy efficiency measures are to be applied, energy saving has to be encouraged, habits must be redirected and renewable sources fostered.

Energy supply is important to guarantee economic activity and living standards, but also offers an opportunity to develop new energy-related business activities and create employment. This has to be acknowledged. Whether a region is able to increase compe-tiveness of its economy, generate new activity and promote new research and develop-ment and innovation opportunities, and does so by directing it towards a low carbon economy, depends largely on the ability to design and implement realistic and effective policies. Some ideas of how this can be done have been offered here through the example of the Basque case, and some insights into the ongoing policy preparation discussed.

NOTES

1. According to 2009 data (IHOBE, 2010), the figure had decreased to 22.6 million (+6% with respect to 1990). This significant reduction from 2008 to 2009 is mainly explained by the impact of the worldwide economic crisis.
2. The report 'Regions 2020' of the European Commission. Available at http://ec.europa.eu/regional_policy/sources/docoffic/working/regions2020/index_en.htm.
3. The 3rd UN World Water Development Report, 'Water in a Changing World', 2009. Available at http://www.unesco.org/water/wwap/wwdr/wwdr3/.
4. Available at http://www.ingurumena.ejgv.euskadi.net/r49-6172/en/contenidos/plan_programa_proyecto/plan_cambio_climatico/en_cc/indice.html.
5. Note that according to 2010 data this goal has already been exceeded.
6. Note that as a heavily industrialized region, one expects GHG emissions per capita to be poorer compared with other less industrialized regions, while GHG emissions per unit of GDP show a different reality.
7. This refers to the Feed in Tariff (FiT) system currently in place in Spain. More information can be found in Cazorla Gónzalez-Serrano (2010).
8. Note that a significant share of emissions from cars are not only associated with the combustion process but also with other issues such as oil waste, evaporation of oil, wearing out of tyres, brakes and other com-ponents. More information can be found in Inche (2001).
9. The Energy Cluster brings together approximately 80 companies and institutions (see http://www.clus-terenergia.com).

REFERENCES

BPCCC (2009), 'Basque plan to combat climate change 2008–2012', Basque Government, Vitoria, Spain.

Cazorla Gónzalez-Serrano, L. (2010), 'El Régimen tarifario de las energías renovables', in F. Becker, L.M. Cazorla and J. Martínez-Simancas (eds), *Tratado de Energías Renovables*, Cizur Meno, Spain: Iberdrola–Thomson Reuters-Aranzadi, pp. 119–46.

Eustat (2008), 'Cuentas Económicas', www.eustat.es

EVE (1992), '3E-2000. Estrategia Energética de Euskadi 2000', BI-1.590/92.

EVE (1997), 'Política Energética. Plan 3E-2005. Estrategia Energética de Euskadi 2005',

EVE (2003), 'Estrategia Energética Euskadi 2010: Hacia Un Desarrollo Energético Sostenible (3E-2010)', BI-2025-03.

EVE (2009), 'Energía: Datos Energéticos País Vasco', BI-2287-09.

Fouquet, R. (2008), *Heat, Power and Light: Revolutions in Energy Services*, Cheltenham, UK and Northampton, MA, USA: Edward Elgar.

Gallastegui, M.C. and I. Galarraga (2010),'La Unión Europea frente al cambio climático: el paquete de medidas sobre cambio climático y energía 20-20-20', in F. Becker, L.M. Cazorla and J. Martínez-Simancas (eds), *Tratado de Energías Renovables*, Iberdrola-Thomson Reuters-Aranzadi.

González, M. and R. Dellink (2006), 'Impact of climate policy on the Basque economy', *Economía Agraria y de los Recursos Naturales*, 12, 187–213.

Hoyos, D., A. Longo and A. Markandya (2009), 'Concienciación pública y aceptabilidad de medidas para la reducción de emisiones de gases de efecto invernadero: el caso del País Vasco', *Papeles de Economía Española*, 121, 68–78.

IHOBE (2009), 'Inventario de Gases Efecto Invernadero en la Comunidad Autónoma del País Vasco 1990–2008', available at www.ihobe.net).

IHOBE (2010), 'Inventario de Gases Efecto Invernadero de la Comunidad Autónoma del Pais Vasco 2009', available at http://www.ihobe.net.

Inche, J. (2001), 'Estimación de Emisiones En Vehículos En Circulación', *Industrial Data*, 4 (1), 11–16.

IPCC (2007), *IPCC Fourth Assessment Report (AR4): Climate Change 2007*, Geneva: IPCC.

Oates, Wallace E. (1999), 'An essay on fiscal federalism', *Journal of Economic Literature*, 37 (3), 1120–49.

OECD (2007), *Linking Regions and Central Governments: Contracts for Regional Development*, Paris: OECD.

Epilogue
Anil Markandya

This book has covered a lot of ground, summarizing which is difficult. Perhaps, however, allowing myself not to follow directly the formal structure of the book, it is worth drawing out some key trends in this rich collection of contributions. I would begin by noting that history is important and we can learn a lot about what a sustainable energy future will look like by studying how energy use has evolved. The chapter by Fouquet reminds us that successful long-run economic growth has depended on sound management of demand, and where such management was lacking, development was also curtailed. Clearly there is a message there for the future. The historical perspective is also to be found in the chapter by Gallastegui et al., who examine whether we have indeed been able to decouple greenhouse gas (GHG) emissions from economic growth. The past is not so encouraging in this regard about the future, but there are some positive signs that such a decoupling may be possible.

The historical view is also present in a number of other chapters. Bigano et al. look to see whether the goals of energy security and carbon and energy efficiency - all important to a sustainable energy future – can be achieved by the same instruments. In general they find that we need a combination of instruments to meet these objectives and one cannot do the job with just the few that focus on improved energy efficiency. Hammond and Jones look at indicators of sustainable development and sustainability and how the application of the precautionary principle has been controversial in some recent agreements involving the use of energy. The more recent past has been examined in the chapters by Chevallier, who looks at how the European carbon market has worked over the period 2005–07; and how the secondary market in certified emissions reductions has supported the carbon reductions.

The second main theme of the book has been how to bring about energy and carbon efficiency. Clearly this is central to the goal of a sustainable energy future and there are several areas where actions can be taken. The chapters look at energy markets (Bonacina et al.), electricity transmission (Pérez-Arriaga et al.), renewables (Cabal et al.), wind (Halsnæs and Karlsson), nuclear power (Hammond), carbon capture and storage (Anthony and Fennell) and biofuels (Hazell and Evans). In terms of end use, the chapter by Button looks at the transportation sector. Each chapter notes the importance of its particular sector in contributing to a low carbon future, without saying that other sources of energy efficiency are not important. The sustainable energy future is likely to be composed of advances in a combination of technologies (Yoshizawa et al.), and sources, and we simply do not know now which ones are likely to be the most important.

An important aspect of the move to a low carbon future is the impact it has on the poor. Indeed the whole issue of energy poverty and how to eliminate it is key to achieving the goals of sustainable development more generally and of sustainable energy in particular. In this regard the chapter by Bailis is very important; it looks at energy poverty in a global context and at the reforms needed to eliminate energy poverty, while also respecting the goals of improving energy and carbon efficiency.

That takes us to the third theme of sustainable energy and that is research and development (R&D). The chapter by Lanza and Verdolini reviews the level of R&D activity both past and present and notes the surge in activity in response to the growing interest in a low carbon future, expressed both through international treaties such as the Kyoto Protocol and through specific measures that reward low carbon technologies. Incentives for innovation are also discussed in the chapter by Foxon, who examines the drivers and barriers to low carbon innovation.

The fourth theme is changing behavior. A sustainable energy future will be much easier to achieve if people can be more careful in the way they use energy. These changes in behavior can be achieved through fiscal incentives, such as higher prices, but not only though these instruments. The options available are discussed in the chapters by Madlener and Harmsen-van Hout; Labandeira and Linares; and Abadie and Chamorro (who look at incentives to generators of coal-fired power); and Westskog et al., who focus on individual energy consumption and stress the importance of insights from social psychology and anthropology as well as economics.

The fifth theme is the role of modeling in understanding the options for a sustainable future. Of necessity we need to look into the future and we have limited tools to be able to predict what developments will take place in the energy sphere. So we have to rely on models that help us to estimate the costs of alternative targets towards a sustainable energy future. These models have been reviewed and discussed in the chapter by Rodrigues et al., and (partly) in the chapter by Foxon. A related modeling approach is less focused on estimating the costs of targets but looks more at the broader issues of growth and how sustainable it can be in the context of carbon and pollution constraints. Using models in the endogenous growth framework, Pittel and Rübbelke address these questions.

The last theme explored in this book is the role of regions in helping us to move towards a sustainable energy future. Regions have an advantage over national governments in that they can be more innovative than the latter and can act as 'leaders' in the formation of public opinion in this field. The book looks at how three regions have achieved this: North Rhine-Westphalia (Reisz), California (Heres and Lin) and the Basque Country (Hormaeche et al.).

The reader who has already dipped into (or even read) most of the book will I am sure agree that it offers a lot to researchers in the field, laying out what is known on many dimensions of sustainable energy and in identifying the gaps. The reader who starts the book by looking at the conclusion is encouraged to go and dip into the chapters, perhaps following the themes laid out here.

Index

Leabharlanna Poiblí Chathair Bhaile Átha Cliath
Dublin City Public Libraries